AF697177
WITHDRAWN

Naturalized mammals of the world

Previous works by the author

The naturalized animals of the British Isles (Hutchinson, 1977; Paladin Granada, 1979)

Chapter in Mason, I. L. (ed.) *Evolution of domesticated animals* (Longman, 1984)

Naturalized mammals of the world

Christopher Lever

Foreword by
H. R. H. The Duke of Edinburgh, K. G., K. T.

Illustrations by
Norma Hansell

LONGMAN
London and New York

QL
86
.L48
1985

Longman Group Limited
Longman House, Burnt Mill, Harlow
Essex CM20 2JE, England
Associated companies throughout the world

Published in the United States of America
by Longman Inc., New York

© Christopher Lever 1985

All rights reserved. No part of this publication may be reproduced, stored in a retrieval system, or transmitted in any form or by any means, electronic, mechanical, photocopying, recording, or otherwise, without the prior written permission of the Publishers.

First published 1985

British Library Cataloguing in Publication Data
Lever, Christopher
Naturalized mammals of the world.
1. Mammals
I. Title
599 QL703
ISBN 0-582-46056-5

Library of Congress Cataloging in Publication Data
Lever, Christopher, 1932–
Naturalized mammals of the world.

Bibliography: p.
Includes index.
1. Animal introduction. 2. Mammals. I. Title.
QL86.L48 1984 599′.05 83–19527
ISBN 0–582–46056–5

Set in 10/11 pt. Linotron 202. Bembo
Printed in Great Britain by
The Bath Press, Avon

Contents

List of tables

Foreword

BUCKINGHAM PALACE.

The introduction of wild species of animals, or plants for that matter, comes under the general description of "things that seemed to be a good idea at the time". It is a pity that the consequences of such introductions were not known at the time. It might have saved some terrible ecological disasters. It might also have persuaded breeders, collectors and zoos to be a great deal more careful about preventing escapes.

This book is full of useful and interesting information and, hopefully, it may yet prevent further disasters.

Philip

1985

Preface

The object of this book is to describe when, where, why, how and by whom the various alien mammals now living in wild state throughout the world were introduced, how they subsequently became naturalized, and what effects – for good or ill – they have had on the native biota, and vice versa. The criteria for inclusion of a species are that it should have been imported from its natural range to a new country or region either deliberately or accidentally by human agency, and that it should currently be established in the wild in self-maintaining and self-perpetuating populations unsupported by, and independent of, man.

Each species account is a monograph on an individual mammal: at the end of some of them are brief details of the artificial distribution of some allied, related or associated species. In general, only species which have become successfully naturalized are included. Exceptions have been made in the case of species which have failed to become established in some regions but have succeeded elsewhere, and when initial failures have been followed by subsequent successes. Natural immigrants from one country to another have normally been included only if they have been augmented by deliberate or accidental introductions by man, as in the case of the nine-banded armadillo (*Dasypus novemcinctus*) in Florida, USA, or in the case of an established alien self-colonizing a new country (e.g. the coypu (*Myocastor coypus*), which invaded Turkey from the Soviet Union). Translocations and/or transplantations of a species from one part of a country or its natural range to another, and from the mainland to offshore islands or vice versa, and acclimatized species (e.g. the reindeer (*Rangifer tarandus*) in Scotland) have in general only been included if the species has been successfully naturalized elsewhere.

Terminology

The definitions (with sources) of various terms used in the text are as follows:

Naturalization: 'The introduction (of animals and plants) to places where they are not indigenous, but in which they may flourish under the same conditions as those which are native.'[1] More particularly, 'The establishment of self-regenerating populations [unsupported by and independent of man] of an introduced species or race in a free-living state in the wild.'[2]

Acclimatization: 'Grown or become habituated to a new climate.'[3] Used to describe an exotic species living in the wild in an alien environment or climate with the support of and dependent (e.g. for food) on man.

Feral: 'Animals or plants that have lapsed into a wild from a domesticated condition.'[4] It is thus, strictly speaking, incorrect to refer to a *wild* (as opposed to a *domestic*) animal which has escaped from captivity and has become naturalized as 'feral'. The term is not, as suggested by De Vos and Petrides,[5] synonymous with 'wild'.

Introduction: 'The deliberate or accidental release of animals or plants of a species or race into an area in which it has not occurred in historical times.'[6] 'The movement by man of live organisms to a new location outside their recent natural geographic range.'[7]

Reintroduction: 'The deliberate or accidental release of a species or race into an area in which it was indigenous in historical times.'[8] 'The act of returning a species to an area within its original geographic range from which it has disappeared.'[9]

Transplantation/Translocation/Stocking: The deliberate or accidental movement by man of a species or race from one area where it is established (as either native or alien) to another area within the same range.

Restocking: 'The deliberate or accidental release of a species or race into an area in which it is already present.'[10]

Established: 'Set up on a secure or permanent basis.'[11] A successfully introduced species or race. Synonymous with 'naturalized'.

Exotic/Alien: 'A species native to an area located outside of, or foreign to, the area under review.'[12] 'An introduced species.'[13]

Adventive: An introduced species or race which is as yet unestablished.

Native/Indigene: 'An individual or population which is a member of the natural biotic community.'[14]

Classification and sequence

The classification and sequence followed is that of Corbet and Hill (1980) in *A world list of mammalian species*, which is itself based on Simpson (1945) and McKenna (1975).

Distribution of species

Zoogeographically the world was divided in 1858 by P. L. Sclater into six great faunal regions: the **Palaearctic** (Europe, northern Africa south to the mid-Sahara, and northern Asia); the **Ethiopian** (Africa south of the Palaearctic and southern Arabia); the **Oriental** (tropical Asia and western Indonesia); the **Nearctic** (North America south to central Mexico); the **Neotropical** (central and South America, southern Mexico and the Antilles); and the **Australasian** (Australia, Tasmania, Papua New Guinea and some Pacific and eastern Indonesian islands). The division between the Australasian and Oriental regions is a deep trough in the ocean bed between Papua New Guinea and Borneo; this division is defined by (Alfred Russel) Wallace's Line, which extends south of the Philippines between Borneo and the Celebes and between the islands of Bali and Lombok. The Palaearctic and Nearctic regions together form the **Holarctic** region. The tropical region (not one of the faunal regions) lies between the tropics of Cancer and Capricorn (i.e. between 23°N and S of the equator).

Geographically the world is divided into six continents which are politically subdivided into countries or nations: the continent of **Europe** includes Turkey-in-Europe and European USSR; **Asia** includes the remainder of Turkey, Israel, the Soviet Far East and Arabia; **Africa** includes Madagascar (the Malagasy Republic), the Seychelles, the Mascarenhas, Réunion, Mauritius and St Helena; **North America** includes Mexico, Central America, the Caribbean territories, the West Indies, the Hawaiian Islands and Bermuda; **South America** comprises all land south of Central America; **Australasia** (and **Oceania**) include the Antarctic territories, the Malay archipelago, the East Indies, Indonesia, Borneo, Melanesia, Papua New Guinea, Polynesia, Micronesia, New Caledonia, the New Hebrides, the Solomon Islands, Fiji, the Philippines, the Celebes, the Moluccas, Tahiti, Tonga and other oceanic island groups.[15]

The system adopted here is based on these geographical and political boundaries rather than on the less-precise division into faunal regions, with the following exceptions: Australasia and Oceania have been separated – the former comprising principally Australia, Tasmania, New Zealand and Papua New Guinea (together with their associated territories and offlying islands), and to the latter have been assigned most of the distant offshore island groups, including Hawaii.

Nomenclature of domesticated species

The use of formal binominal names for domesticated animals has for long been a subject of contention among taxonomists. Since those species discussed here are living in a feral state it has been considered justifiable to treat them on the same footing as their wild congeners.

Feral species, rats and mice

Through human agency the distribution of some feral domestic mammals and such commensals of man as rats (*Rattus rattus* and *R. norvegicus*) and house mice (*Mus musculus*) is virtually cosmopolitan. In these cases only the ecologically and/or historically more interesting and important introductions are discussed in the text. Introductions of domestic species from one country to another are only mentioned if feral populations have resulted.

Old, new and alternative geographic names

Where these apply the old or principal name is normally used with the new or secondary name in parentheses.

British county names

In 1974 the names and boundaries of a number of counties in England and Wales, and in 1975 of some in Scotland, were changed. Since many of the events described herein antedate these changes, and as it would be confusing to use both old and new names, the old names and boundaries have been adhered to throughout.

Distribution maps

These are intended to give a general indication only of the natural and naturalized distribution (both of which are in some cases discontinuous) of each species. They are based largely on Corbet's *The mammals of the Palaearctic region: a taxonomic review* (1978), Hall's *The mammals of North America* (1981) and Marlow's *Marsupials of Australia* (1983). Species for which maps are omitted are regarded as virtually cosmopolitan.

Naturalized British mammals

More detailed information on these will be found in Lever's (1977) *The naturalized animals of the British Isles*, which also discusses birds, fish, reptiles and amphibians naturalized in Britain.

Errors and omissions

The author would welcome information regarding any errors and omissions in the text.

NOTES

1. *Oxford English dictionary*.
2. Linn (1979), derived from Boitani (1976).
3. See note 1.
4. Ibid.
5. 1966; 114.
6. See note 2.
7. See note 5.
8. See note 2.
9. See note 5.
10. See note 2.
11. See note 1.
12. See note 5.
13. See note 2.
14. See note 5.
15. Based on the *Universal decimal classification*, 4th international edn (Publication No. 179 British Standards Institution, London, 1972), Gr. 9.

Acknowledgements

Without the generous co-operation of a host of correspondents throughout the world this book could never have been written. For their assistance and patience in answering my requests for information I am greatly indebted to the following:

Dr R. J. van Aarde (University of Pretoria, South Africa); Mr A. A. Adcock (Arapawa Island Wildlife Trust, New Zealand); Mr R. Albignac (Laboratoire de Zoologie, Madagascar); Mr G. L. H. Alderson (Rare Breeds Survival Trust, UK); Dr F. Alvarez (Estacion Biologica de Doñana, Spain); Mr B. Andrej (University of Ljubljani, Yugoslavia); Dr V. Anna (Janus Pannonius Múzeum, Hungary); Dr R. H. Ansell (late Chief Veterinary Officer, southern Sudan); Dr I. A. E. Atkinson (Department of Scientific & Industrial Research, New Zealand).

Mr M. E. Ball (Nature Conservancy Council, UK); Mr B. Barnett (University of California, USA); Dr K. Bauer (Naturhistorisches Museum, Vienna, Austria); Mr D. Baxter (Texas Parks & Wildlife Department, USA); Dr B. D. Bell (Victoria University of Wellington, New Zealand); Mr J. Bensusan (Gibraltar Museum); Professor R. C. Bigalke (University of Stellenbosch, South Africa); Señor J. Black (Charles Darwin Research Station, Galápagos Islands); Mr I. M. Bodea (Directa Generala a Silviculturii, Romania); Major B. D. M. Booth (UK); Dr W. R. P. Bourne (University of Aberdeen, Scotland); Professor C. H. Brooke (Portland State University, USA); Mr R. K. Brooke (University of Cape Town, South Africa); Mr J. F. Bugalho (Director of the Game Department, Lisbon, Portugal); Dr G. Bump (late US Department of the Interior); Mr J. A. Burton (Fauna & Flora Preservation Society, UK); Mr A. D. Butcher (Melbourne Zoological Gardens, Australia).

Dr L. Calvopiña (Charles Darwin Research Station, Galápagos Islands); Mr N. Campbell (Nature Conservancy Council, UK); Señor J. Cardona (Puerto Rico); Miss L. Cayot (Syracuse University, USA); Dr D. I. & Mrs N. G. Chapman (England); Professor Cheng Tso-Hsin (Academia Sinica, Peking, China); Dr D. A. Clark (Universidad de Costa Rica); Mr B. des Clers (International Foundation for the Conservation of Game, Paris, France); Mr N. S. Coad (Department of Lands & Survey, New Zealand); Professor B. E. Coblentz (Oregon State University, USA); Professor Y. Coineau (Muséum National d'Histoire Naturelle, Paris, France); Mr G. T. Corley Smith (Charles Darwin Foundation for the Galápagos); Mr M. Coulter (Charles Darwin Research Station, Galápagos Islands); Mrs M. Cox (Smithsonian Institution, Washington, USA); Dr K. Curry-Lindahl (Unesco, Sweden); Dr S. de Crombrugghe (Group de Travail pour l'Étude de l'Équilibre Forêt-Gibier, Belgium).

Señor C. F. P. Del Prado (Ministerio de Agriculture, Peru); Dr J. Dorst (Muséum National d'Histoire

Naturelle, Paris, France); Dr D. Duffy (University of Cape Town, South Africa).

Mrs Sólveig Einarsdottir (Iceland); Mr A. Elkadiri (Ministère de l'Agriculture, Morocco).

Mr J. A. Fa (Gibraltar); Dr C. J. Feare (Ministry of Agriculture, Fisheries & Food, England); Mr C. Ferriera (Department of Nature Conservation, South Africa); Dr. O. C. Filho (Instituto Oswaldo Cruz, Brazil); Mr R. S. R Fitter (Fauna & Flora Preservation Society, UK); Dr J. E. C. Flux (Department of Scientific & Industrial Research, New Zealand); Dr L. Fowler (University of Florida, USA); Dr H. J. Frith (Commonwealth Scientific & Industrial Research Organization, Australia); Dr P. J. Fullagar (Commonwealth Scientific & Industrial Research Organization, Australia).

Señor J. M. Gallardo (Museo Argentino de Ciencias Naturales, Argentina); Mr P. Goriut (International Council for Bird Preservation, Cambridge, England); Mr D. Green (University of Texas, USA); Dr J. C. Greig (Department of Nature & Environmental Conservation, South Africa); Mr J. P. Gudjonsson (Iceland).

Dr O. Hamann (University of Copenhagen, Denmark); Dr D. Harmel (Texas Parks & Wildlife Department, USA); Professor V. G. Heptner (Moscow University, USSR); Mr S. Hess (late Department of the Environment, Iran); Mr J. E. Hill (British Museum (Natural History), London); Miss A. Hogan (Bermuda); Dr M. W. Holdgate (Department of the Environment, UK); Dr R. Holmes (Massey University, New Zealand); Mr W. S. Huey (New Mexico Department of Game & Fish, USA); Dr S. R. Humphrey (Florida State Museum, USA); Sir Rex Hunt (Falkland Islands).

Dr G. Ilany (Nature Reserves Authority, Israel); Dr C. Imboden (International Council for Bird Preservation, Cambridge, England); Dr T. Ito (Tohoku University, Japan).

Professor Z. Jaczewski (Polish Academy of Sciences, Poland); Dr G. W. Johnstone (Department of Science & Technology, Australia).

Dr H. W. Kale (Florida Audubon Society, USA); Mr R. Van den Kerchove (Parc à Gibier, Hosingen, Luxembourg); Dr C. M. King (Department of Scientific & Industrial Research, New Zealand); Mr R. J. Knowles (British Museum (Natural History), London); Dr M. Konecny (University of Florida, USA); Dr F. Köster (Charles Darwin Research Station, Galápagos Islands); Dr P. Kramer (Charles Darwin Research Station, Galápagos Islands); Dr H. Kruuk (Natural Environment Research Council, UK).

Dr R. Laws (British Antarctic Survey); Dr J. N. Layne (Archbold Biological Station, Florida, USA); Dr G. A. Letts (Northern Territory of Australia Legislative Assembly); Dr H. W. Levi (Harvard University, USA, who generously presented me with his collected bibliography of introduced and transplanted species); Dr R. Liversidge (McGregor Museum, Kimberley, South Africa); Mr P. H. Lloyd (Department of Nature & Environmental Conservation, South Africa); Mr J. L. Long (Agriculture Protection Board of Western Austra-lai); Mr Lu Tai-Chun (Academia Sinica, Peking, China).

Professor R. D. Martin (University College, London, England); Mr I. L. Mason (Food & Agriculture Organization of the United Nations, Rome, Italy); Mr F. Marseu (Ministry of Agriculture, Fisheries & Forests, Fiji); Dr C. McFarland (Costa Rica); Dr L. F. G. McIntyre (Sugar Industry Research Institute, Mauritius); Professor T. L. McKnight (University of California, USA); Señor A. O. Martinez (Dirección Forestal, Parques y Fauna, Uruguay); Mr M. B. Mecke (US Department of the Interior); Mr J. C. G. Millar (Department of Nature & Environmental Conservation, South Africa); Professor V. Mimra (Research Institute of Foresty & Gamekeeping, Czechoslovakia); Dr P. D. Moehlman (University of Florida, USA).

Mrs J. Nishida (Hawaii Audubon Society, USA).

Dr W. L. A. Oliver (Jersey Wildlife Preservation Trust); Dr D. Orwin (Lincoln College, New Zealand); Mr A. W. Owadally (Forestry Service, Mauritius); Professor O. T. Owre (University of Miami, USA).

Mr J. A. Pedersen (University of Oslo, Norway); Mr J. C. Pernetta (University of Papua New Guinea); Drs F. and J. J. Petter (Muséum National d'Histoire Naturelle, Paris, France); Dr J. Pollock (England); Dr G. R. Potts (Game Conservancy, UK).

Dr C. W. Ramsey (Texas A & M University, USA); Señor J. R. Rau (Instituto de Zoologia, Valdivia, Chile); Dr H. F. Recher (Australian Museum); Dr R. Reynold (Charles Darwin Research Station, Galápagos Islands); Mrs D. Reynolds (Charles Darwin Research Station, Galápagos Islands); Mr M. G. Ridpath (Commonwealth Scientific & Industrial Research Organization, Australia); Mrs B. Rowe (Arapawa Island Wildlife Trust, New Zealand); Professor R. L. Rudd (University of California, USA); Dr M. R. Rudge (Department of Scientific & Industrial Research, New Zealand).

Mr D. Schierbaum (New York State Conservation Department, USA); Dr R. Schloeth (Commission Fédérale du Parc National, Suisse); Mr C. Schreiner (Y-O Ranch, Texas, USA); Sir Peter Scott (World Wildlife Fund); Mr D. Sideridis (Forestry Department, Ministry of National Economy, Greece); Dr A Siefke (Ernst-Moritz-Arndt Universität, Griefswald, GDR); Professor W. R. Siegfried (University of Cape Town, South Africa); Dr T Simkin (National Museum of Natural History, Washington, USA); Dr C. D. Simpson (Texas Tech University, USA); Professor J. D. Skinner (University of Pretoria, South Africa); Mr G. W. Smith (Department of Recreation & Conservation, British Columbia, Canada); Dr R. H. N. Smithers (University of Pretoria, South Africa); Mr W. A. Snyder (New Mexico Department of Game & Fish, USA); Mr N. Stronach (Balamuck Station, Papua New Guinea); Mr C. T. Stuart (Department of Nature & Environmental Conservation, South Africa); Dr R. W. Summers (Ministry of Agriculture, Fisheries of Food, England).

Mr R. H. Taylor (Department of Scientific & Industrial Research, New Zealand); Dr J. G. Teer (Rob & Bessie Welder Wildlife Foundation, Texas, USA); Dr T. C. Telfer (Hawaii Department of Land & Natural Resources, USA); Professor J. C. Thibault (France); Dr S. Thórisson (Museum of Natural History, Reykjavik, Iceland); Miss J. Thornback (International Union for the Conservation of Nature & Natural Resources); Mr B. Tinnock (Lincoln College, New Zealand); Dr P. Q. Tomich (State of Hawaii Department of Health, USA); Miss A. Todd (British Antarcitc Survey); Miss J. Tucker (International Union for the Conservation of Nature & Natural Resources); Dr D. G. Tulloch (Commonwealth Scientific & Industrial Research Organization, Australia).

Mr M. N. P. Utsi (Reindeer Council of the UK).

Dr R. E. Vaughan (Mauritius); Dr J. Vincent (National Parks, Game & Fish Preservation Board, South Africa); Dr A. de Vos (late University of Waterloo, Canada); Dr T. de Vries (Universidad Catolica, Ecuador).

Professor F. R. Walther (Texas A & M University, USA); Mr D. Watling (World Wildlife Fund, Indonesia); Mr A. R. Werner (England); Dr W. A. Weyland (Rijksinstitut voor Natuurbeheer, Arnhem, the Netherlands); Mr A. H. Whitaker (Department of Scientific & Industrial Research, New Zealand); Dr R. S. Whitley (Grasslands Trials Unit, Falkland Islands); Mr C. Winegarner (Archbold Biological Station, Florida, USA); Dr D. B. Wingate (Department of Agriculture & Fisheries, Bermuda); Professor J. M. Winterbottom (University of Cape Town, South Africa); Dr J. G. Williams (late Coryndon Museum, Nairobi, Kenya); Mr L. E. Williams (Florida Game & Fresh Water Fish Commission, USA); Mr L. L. Williamson (Wildlife Management Institute, Washington, DC, USA); Professor K. Wodzicki (Victoria University of Wellington, New Zealand); Mr J. H. Wood Roberts (Rare Breeds Survival Trust, UK).

Dr D. W. Yalden (University of Manchester, England).

For their patience and diligence in tracing innumerable references for me I am grateful to the staff of the various libraries in which I conducted my researches. In particular my thanks are due to the following members of the British Museum (Natural History): Mr M. J. Rowlands; Mr A. P. Harvey; Miss S. A. Angel; Mrs C. Comben; Miss P. Cook; Mrs A. Datta; Miss N. French; Miss P. Gilbert; Mr M. R. Halliday; Miss D. M. Hills; Miss A. Jackson; Miss J. Jeffrey; Miss P. D. Jenkins; Miss A. Lum; Mrs K. Martin (who also researched on my behalf); Miss A. Pope; Miss M. Seely; and Mr N. Thomson. The staff of the following British libraries were also of great assistance: Commonwealth Institute of Entomology; Game Conservancy; Linnean Society of London; Ministry of Agriculture, Fisheries & Food; Nature Conservancy Council; Royal College of Veterinary Surgeons; Royal Geographical Society; Science Reference; and Zoological Society of London. I have also to thank Dr M. Boyd of the Science Reference Library for helping me carry out a computer-search of *Biological Abstracts*.

I am most grateful to Dr G. B. Corbet and Dr J. Clutton-Brock of the British Museum (Natural History) and to Mr H. V. Thompson, Dr L. M. Gosling and Mr S. Baker of the Ministry of Agriculture, Fisheries & Food who kindly read and annotated the typescript for me. For drawing my attention to various errors of omission and commission and for their comments in general I am greatly indebted to them all.

I should also like to thank the staff at Longman for their help and advice throughout; Miss S. Bunney for her patience and diligence in bringing order to my typescript and for compiling the index; Dr H. Moore for his work on the distribution maps; Mrs N. Hansell for her drawings; and last, but by no means least, Mrs R. Rugman for her patience in deciphering and typing my well-nigh illegible manuscript.

To H. R. H. The Duke of Edinburgh I extend my grateful thanks for kindly contributing the Foreword to this book.

Christopher Lever

Windsor, Berkshire
January 1985

Introduction

Man is an inveterate and an incorrigible meddler, never content to leave anything as he finds it but always seeking to alter and – as he sees it – to improve. In no field is this more true than in that of the animal and vegetable kingdoms.

One of the ways in which man has sought to modify the natural environment is by the introduction of animals (principally mammals and birds) and plants throughout the world. Of the former, mammals have been introduced to new regions by human agency with a variety of motives: for sport; for sentimental or nostalgic reasons; as an aesthetic amenity; as a potential source of food; as fur-bearers; and as a form of biological control of pest species. Some have been released simply out of curiosity to see what would become of them in a new environment, whereas others have escaped from captivity or domesticity: three have used man as an unwitting means of transportation.

In the case of mammals released for sporting purposes, the introduction is made, as Bates (1956) points out, 'to compensate for alleged deficiencies in the local fauna; but the deficiency may be apparent [rather than real] because man is hankering after some particular kind of sport remembered from "back home" and is unwilling to adapt his habits to the sporting possibilities of the new fauna'. An example is the release of European rabbits (*Oryctolagus cuniculus*) by the early colonists in Australia, despite the abundance of marsupial game. The liberation in the United States of different species of ungulates is an instance of the domestic importation of exotic mammals to fill an imagined, rather than a real, need for big game animals. On the other hand, the introduction of various deer to New Zealand (where the sole native mammals are two species of bat and the only indigenous quadrupeds are reptiles and amphibians) and to Australia (where of the 18 orders of land mammals only 4 – the Monotremata, Marsupialia, Chiroptera and Rodentia – are represented) filled a genuine need.

Few, if any, mammals have been introduced for purely sentimental, nostalgic or aesthetic reasons. These – and the provision of an additional food supply – have usually been the concomitants of releases for sporting purposes. Where a naturalized mammal is regarded solely as an aesthetic amenity, it has usually become established as the result of escapes from captivity, as, for example, in the case of the red-necked or Bennett's wallaby (*Macropus rufogriseus*) in England.

Mammals introduced solely to provide a potential source of food have in most cases been domestic species (e.g. goats, *Capra hircus*), released on offshore islands for the benefit of whalers, sealers, buccaneers and ship-wrecked mariners.

One of the most successfully introduced fur-bearing mammals is the muskrat (*Ondatra zibethicus*). First imported from North America to Czechoslovakia in 1905, it has spread more widely and more rapidly than any other mammal deliberately released by man; as Elton (1958) points out, 'in zoogeographical terminology a purely Nearctic species has become Holarctic within half a century'.[1] The muskrat has one other zoogeographical distinction; having become successfully established in the wild in the British Isles (mainly in England) in the 1920s and 1930s as the result of escapes from fur-farms, it remains, nearly half a century later, the only naturalized alien vertebrate to have been successfully eliminated (by the mid-1930s) in Britain and, indeed, one of the few in the world.

The classic and most notorious example of a mammal introduced as a form of biological control of an existing pest species is the small Indian mongoose (*Herpestes auropunctatus*) in the West Indies. First imported to Jamaica in 1872 to control rats which were laying waste the sugar-cane plantations, it achieved its objective so effectively that before long it was forced to turn for food to terrestrial endemic reptiles, mammals and birds, exterminating some and gravely endangering others. There can be little doubt that the mongoose in the Caribbean has devastated more indigenous animals than any other species deliberately introduced by man anywhere in the world. As Bates (1956) has rightly said, 'Now the mongoose itself is regarded as a first-class pest everywhere that it has been introduced'.

Mammals released simply out of curiosity to see what would become of them in an alien environment (a peculiarly irresponsible reason but one which is probably a factor in many introductions) include the North American grey squirrel (*Sciurus carolinensis*), which has become a pest species in Britain and, to a lesser extent, in South Africa.

Several mammals have become naturalized in alien countries as the result of escaping from captivity or domesticity: an example of the former is the Chinese or Reeves's muntjac (*Muntiacus reevesi*) in England, and of the latter the Arabian camel (*Camelus dromedarius*) in Australia. Three human inquilines, the black and brown rats and the house mouse, have made use of man as an unwitting means of their conveyance around the world – frequently as shipborne stowaways.

Various unfortunate consequences may attend the naturalization of animals in new environments: of these, the most important are the transmission of parasites and diseases; damage to human food and economic resources; disturbance of the ecosystem; interspecific competition with indigenous species; predation; and morphological and/or genetic changes in native populations.

As De Vos and Petrides (1966) point out,[2] most diseases are likely to have more serious effects on hosts which have not been previously exposed to them than on their original hosts, because the former have not had the opportunity to develop any in-built immunity; such diseases may prove difficult or in some cases even impossible to eradicate. Although in time natural selection tends to result in an accommodation between a pathogen and its host (as in the case of myxomatosis among European rabbits in Britain and Australia), a new host may become endangered or even exterminated before that occurs. When an alien and an indigene compete for the same ecological niche, the host which originally introduced the disease may partially or completely displace the native species.

Epizootic diseases transmitted through the introduction of mammals and which primarily attack domestic livestock include rinderpest, foot-and-mouth (epizootic aphthae), bovine tuberculosis, mange, brucellosis, leptospirosis, and swine fever. Among diseases which most seriously affect man are histoplasmosis, cryptococcal meningitis, toxoplasmosis, encephalitis, infectious equine encephalomyelitis or 'sleeping sickness' and, most serious of all, rabies or hydrophobia, the last of which can be transmitted by all homoiotherms, although only very rarely by birds.

Parasites introduced by alien mammals include the tick *Trombicula akamushii*, which harbours the rickettsial disease, scrub typhus (spread by rats, mice and other rodents); the Indian rat flea (*Xenopsylla cheopis*), which is responsible for the Black

Death or bubonic plague; a flea, *Orchopeas howardi*, by the grey squirrel; a louse, *Pitrufquenia coypus*, by the coypu; the cattle tick (*Boophilus microplus*); and the buffalo fly (*Siphona exigua*) and buffalo louse (*Haematopinus tuberculatus*) in Australia. At least two parasitic worms – a cestode and a trematode – were introduced with muskrats to Britain, and three nematodes and two cestodes by rabbits to New Zealand.

The most outstanding example of a naturalized mammal depleting man's food resources – both through direct consumption and through reducing the amount of fodder available for domestic livestock – is the rabbit in the British Isles, Australia and New Zealand. In New Zealand the Australian brush-tailed possum (*Trichosurus vulpecula*), introduced for its fur and now a serious silvicultural and horticultural pest, has upset the delicate ecological balance of New Zealand's fragile ecosystem. The latter is also an important reservoir for tuberculosis.

Naturalized mammals frequently compete – principally for food, shelter and habitat – with indigenous species (especially closely related ones). Although the degree to which it has occurred is debatable, there has undoubtedly been some interspecific competition for territorial possession between grey and red squirrels (*Sciurus vulgaris*) in Britain. That arch predator, the voracious North American mink (*Mustela vison*), has, *pace* some theoreticians, caused havoc among many native mammals and birds in the British Isles and continental Europe. In *The red book: wildlife in danger,* Fisher, Simon and Vincent (1969) state that since AD 1600 animals (mainly mammals) introduced around the world have been responsible for some 23 per cent of mammalian and 19 per cent of avian extinctions.

Introduced species can also be responsible for genetic and/or morphological changes in indigenous populations. Although natural selection usually favours native genotypes, continuous infiltration or introgression of an alien's genes into an indigenous population can eventually have an effect, which may be either beneficial or detrimental. The Kamchatka sable (*Martes zibellina kamtschadalica*), translocated to western Siberia, has interbred with the native form (*M. zibellina*) to produce offspring with a more luxuriant pelt.[3] German red deer (*Cervus elaphus*) transplanted into Norway, on the other hand, are believed to have been responsible for the virtual extinction of the species there, probably because the German race is less hardy.

Extensions of a species' distribution sometimes result in changes that lead to the development of new subspecies, such as have occurred in the house mouse in the Faeroe Islands[4] in the North Atlantic and on Macquarie and Marion islands and on South Georgia in the South Atlantic. Sometimes these changes have been deliberately engineered by man, as in the case of unsuccessful attempts in the Soviet Union to produce a 'frost-resistant' strain of coypu.[5] An introduced species will frequently change its feeding habits in its new range; the raccoon-dog (*Nyctereutes procyonoides*), which in the Caucasus preys on game birds, hares and domestic poultry, subsists in its Far Eastern home largely on a diet of fish and crabs. Aliens all too often cause damage which is unknown in their native range; an example is the grey squirrel which is regarded as a harmless species in its North American homeland.

The benefits derived from the presence of naturalized mammals include the provision of (new) game species, further sources of food and an economically valuable extension of a country's natural resources. Although it is possible to argue for the conservation of at least some naturalized animals because of their undoubted economic importance, it is more difficult to justify their conservation on purely scientific grounds, unless the animal concerned is a relic population of an endangered species which has been introduced for the express purpose of conservation. There is, indeed, an ideological contradiction between on the one hand attempting to conserve natural ecosystems and native species, and on the other advocating the conservation of aliens which are, in general, ecologically undesirable.

The introduction or reintroduction of an endangered species can be a valuable tool in the hands of the conservationist. As the Working Group on Introductions of the UK Committee for International Nature Conservation states in its report *Wildlife introductions to Great Britain*:

> By it ecosystems can be enriched in species and this restoration of the full ecosystem complement might have an important influence on ecosystem functioning. It has been suggested for example that poor regeneration in British woods may be related to heavy grazing of seed and seedlings by herbivore populations which are abnormally large because of low predator popu-

lations [the brown bear (*Ursus arctos*), the wolf (*Canis lupus*) and the wild boar (*Sus scrofa*) were formerly natives in Britain]. Rabbits before myxomatosis certainly played a crucial role in maintaining vast acreages of downland, heathland and other plagioclimax ecosystems vital to many native species [e.g. the large blue butterfly (*Maculinea arion*) which has recently become extinct in Britain[6]] and now in their absence reverting to scrub.

Moreover, by dispersing the population of a species introductions or reintroductions can also lessen the risk of extinction by catastrophic changes in the original range. The Parma wallaby [*Macropus parma*] population deliberately introduced to Kawau Island, New Zealand is for example now the mainstay of the species. Paradoxically this is also the case with the coypu, the subject of an eradication programme as a pest in this country [the UK] and almost everywhere it has been introduced [although not in Louisiana, USA], yet now threatened by poaching for food [and by excessive harvesting for fur] in its native South America.

Finally increased diversity is generated by introductions and reintroductions and high diversity is commonly regarded as a conservation objective on both amenity and scientific grounds.[7]

The reintroduction to the wild of endangered species bred in captivity is increasingly becoming accepted as the only justification for keeping wild animals in confinement. A classic case among mammals is that of the 10 Arabian oryx (*Oryx leucoryx*), bred in captivity in the Phoenix and San Diego zoos in the United States and in the Sultanate of Oman, which were successfully reintroduced in the Jiddat-al-Harasis in Oman and in Jordan in 1982 and 1984.[8] The umbrella pretexts of 'education' and 'scientific' research are no longer – if, indeed, they ever were – valid.

As Bates (1956) points out, 'It is not always easy to get a vertebrate species established in a new region, even though the new environment may seem quite suitable . . . over a period of seventy-one years many unsuccessful attempts were made to introduce rabbits into Australia.'

De Vos and Petrides (1966), quoting Dorst (1964), explain the successful introduction of animals in the following three ways although, as they point out, many other complex factors are frequently involved.

1. In case of a vacant and suitable ecological niche, the introduced species will occupy it to become well established. It may even proliferate to pest proportions.
2. Where a closely related autochthonous species is present in an ecological niche, an introduced species may be more aggressive and successful in competition so that the native species is gradually reduced in numbers and may be eliminated.
3. Also the introduction of predatory species may threaten a predator–prey balance, and modify this to the detriment of native species.

Factors which increase the likelihood of success of an introduced species include a congenial climate; a suitable habitat; an abundant supply of acceptable food; freedom from predators; an absence of competition from more aggressive native species; the ability to reproduce rapidly; the absence of, or abandonment of, the instinct to migrate; a large enough initial stock; and a degree of adaptability. It is seldom that all these factors occur in the case of any one species, although as De Vos and Petrides (1966) point out, 'In general, the more generalized the characteristics of a vertebrate species, particularly in its food habits, the better are its chances of establishment.' When enough of the above factors do occur a species' establishment frequently follows a typical sigmoid curve, as described by Elton (1927) in his classic *Animal ecology*:

> When an animal spreads rapidly upon being introduced into a new country, there is usually a definite sequence of events, which is rather characteristic. At first it may be almost unnoticed for some time, or else is highly prized as forming a link with the home country. . . . The next stage is that the animal may suddenly appear in the dimensions of a plague, often accompanied by a migration, as when huge armies of rats marched over New Zealand in the early days. . . . Finally, after a good many years there is often a natural dying down of the plague. This is in most cases not due to the direct efforts of mankind in killing off the pests, but appears rather to be due to the animal striking a sort of balance with its new surroundings, and acquiring a set of checks which act fairly efficiently.[9]

The sequence of events following the successful introduction of an exotic species into an alien environment may thus be summarized as follows.

1. **Release**, followed by a marked reduction in stock due to predation or death by natural causes.
2. **Breeding**, followed – as the result of the appearance of adaptive changes in the behaviour and ecology of survivors – by a rapid increase in the population (described by Elton as an 'ecological explosion') to the maximum numbers that the colonized area will support.
3. **Establishment**, followed by contraction in

numbers and distribution until a level is reached at which both become stabilized.

It is axiomatic that alien species have their greatest impact on insular biota. As the report by the UK Working Group on Introductions states:

> Species evolved in isolation from one another to fill similar niches are potentially direct competitors should they be brought together, and one will usually (perhaps always) prove the stronger. Island species are particularly vulnerable for they are rarely equipped to contend with either direct competitors evolved in the more populous and harder school of continental evolution, or with predators or parasites absent from their own impoverished floras and faunas.
>
> Introduced species released from competition, predation or other limiting factors of their native range can also spread rapidly. The realised range, or niche width, of a species in its indigenous ecosystems may be, because of these factors, much narrower than its potential range in their absence.[10]

In his classic *The naturalization of animals and plants in New Zealand* Thomson (1922) wrote:

> In regard to any natural enemy it is, of course, absolutely certain that it cannot exterminate, but can only keep in check, the animal it is intended to cope with. If it does more, then its own means of livelihood are imperilled, or it has to find other victims.[11]

In other words, a predator is controlled by the amount of available prey and not vice versa. Although this is certainly true in the case of continental land masses, it does not necessarily apply when an alien predator is introduced to an offshore island. With particular reference to rats and the mongoose in Jamaica (where the former were exterminated after the introduction of the latter), Lewis (1953) wrote:

> The introduction of any animal into a new habitat involves certain important and highly complicated biological principles and interrelationships. In any long established and undisturbed area the native plants and animals exist in fairly constant proportions – a natural balanced condition. The balance of populations fluctuates slightly from year to year and may, over a period of years, gradually change; but if a highly prolific predatory animal is introduced into a community, which will provide no natural enemies, it is obvious that the natural balance of the community will be greatly upset.[12]

Research on the extinction of indigenous birds on small island groups and their replacement by exotics suggests that frequently native birds are not lost as a direct result of interspecific competition with adventives, but rather that the latter take over only after the former have been decimated by introduced mammalian predators, disease or despoilation of their habitat by man or other (e.g. herbivorous) mammals.[13]

Over 30 years ago Westerskov (1953) suggested various points for consideration when contemplating the introduction of an animal to a new environment.

1. Whether the new species would be of value for sport, food, its pelt, as a form of biological control of an existing pest, or on amenity or aesthetic grounds.
2. Whether the new species would be, or could in any way become, a pest, e.g. by causing soil erosion and agricultural, horticultural, sivicultural, piscicultural or other economic loss, or by competing with and/or destroying native fauna and flora.
3. Whether a vacant ecological niche exists for the new species to fill.
4. Whether the climate and habitat are suitable.
5. Whether there is an abundant supply of suitable foods.

Since Westerskov's paper three major policy manifestos for introducing animals (and plants) have been published. In chronological order these are: *Problems of species introductions* (Anon 1968), *Re-introductions: techniques and ethics* (Boitani 1976) and *Wildlife introductions to Great Britain* (Linn 1979). A summary of their recommendations is as follows:

1. The risk of accidental introductions should be minimized and, when these occur, the reasons should be determined.
2. Reintroductions are in general more acceptable than new introductions, provided that:
   (a) they are undertaken with individuals of a taxon as close as possible to that of the original stock, and that
   (b) their removal does not endanger the survival of the donor population;
   (c) the causes of extinction are known and no longer exist, and that the reason was not that the species was a pest or dangerous;
   (d) suitable food and habitat are available; and
   (e) the species is unlikely to become a pest in the existing demographic conditions.

3. The introduction of a new species should only be permitted when:
   (a) the factors governing its home distribution and population are properly understood;
   (b) no stock of a desirable and comparable native species already exists;
   (c) it seems likely to be an economic, sporting and/or aesthetic asset and ecologically and economically innoxious;
   (d) the stock to be used is free from diseases and parasites;
   (e) it would not compete with native species, other than pests;
   (f) it could, should the need arise, be eradicated or at least controlled;
   (g) its biology (especially its rate of spread and ecological – and socio-economic – impact) is carefully studied both before and after introduction;
   (h) a suitable and, if possible, vacant ecological niche exists;
   (i) the climate and habitat are suitable; and
   (j) there is an abundant supply of acceptable food.
4. If suitable enclosed experimental areas are available, where a new species can be controlled and monitored prior to release, these should be utilized.
5. The agreement of the local human population should, if possible, be obtained, and adequate funds should be available as compensation for any economic loss incurred.
6. Where necessary, relevant legislation (e.g. for the protection of the introduced species) should be enacted before the release.
7. The control of the exportation of native species for introduction elsewhere (as exists in the United States) should become universal.

The overall policy guideline for introductions of alien species around the world – both animals and plants – is encapsulated by the declaration in Article 6.8 of *Wildlife introductions to Great Britain* (retaining the original italics):

> Because of their dubious benefits to wildlife interests and our inability to predict the effects of introduced species on our native flora and fauna and ecosystems, *we accept the provision in Article 11.2 b of the Draft Convention on the Conservation of European Wildlife and Habitats 'to strictly control the introduction of non-native species'*, and *suggest that the basic premise from a conservation viewpoint should be that deliberate introductions of alien species with the intent to naturalise them in native ecosystems are for the most part undesirable.*[14]

NOTES

Based on a paper by the author, 'Introductions: reasons, results and criteria – with special reference to mammals', delivered at the British Herpetological Society, Fauna & Flora Preservation Society and Mammal Society Joint Symposium on Introduced Species, held at the Zoological Society of London on 27 November 1982.

1. p. 26.
2. pp. 116–17.
3. W. Lindemann, *Transactions of the 8th North American Wildlife Conference* 62–6 (1956).
4. Bourlière (1954).
5. Davis (1964).
6. See p. 395.
7. Linn (1979: 10–11); see also A. D. Bradshaw, *Proceedings of the Royal Society* **197**: 77 (1977).
8. See p. 274.
9. pp. 111–12.
10. Linn (1979: 8–9).
11. p. 73.
12. p. 172.
13. J. M Diamond & C. R. Veitch, *Science* **211**: 499 (1981).
14. Linn (1979: 17).

Brush-tailed Possum

(*Trichosurus vulpecula*)

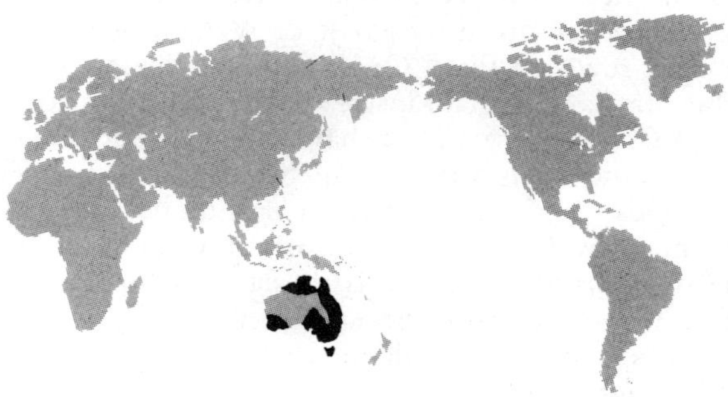

NATURAL DISTRIBUTION
Principally eastern and northern Australia and Tasmania.

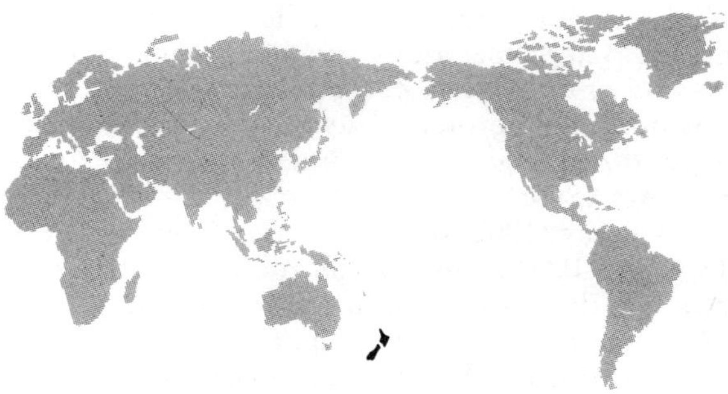

NATURALIZED DISTRIBUTION
Australasia: New Zealand.

AUSTRALASIA

New Zealand[1]

According to the Annual Report for 1892 of the Wellington Acclimatization Society: 'These animals were first liberated in the bush behind South Riverton in 1858 by Mr Basstian. Some years after, one or two opossums[2] escaped from confinement in the same neighbourhood. In 1889 they were found to have increased enormously.'

Mr T. D. Pearce of Invercargill told Thomson (1922) that: 'The opossums in Southland were liberated between 1865 and 1868 in the Longwoods by Mr Christopher Basstian, who brought them from Victoria or Tasmania.' Further information given to Thomson by Mr J. L. Watson of Invercargill revealed that: 'One pair were liberated by the late Captain Hankinson at Waldeck, Riverton, in 1875 or 1876. They have increased marvellously and are plentiful in the South Longwoods.'

Pracy (1962), however, records that recent information[3] from a descendant of Captain John Howell claims that it was the latter who was in fact responsible for the earliest introductions of possums to New Zealand, at Riverton between 1837 and 1840.

Table 1 gives particulars of the major importations of possums into New Zealand. Thus, between 1837 (or 1858) and about 1920 well over 460 possums were released in New Zealand. Originally imported from Victoria (greys) and especially Tasmania (larger and superior-pelted blacks) as a potentially valuable fur-bearer, most of the later liberations resulted from translocations within New Zealand. In the early years possums were protected under the Protection of Certain Animals Act 1861 and the Animal Protection Act 1880.

As Pracy (1962) points out, the period between 1890 and 1900 is historically the most interesting phase in the naturalization of the possum in New Zealand, in that it is highlighted by the activity of the various acclimatization societies both in regard to peak importations and also to the artificial translocation of New Zealand-bred stock. The promise of a lucrative fur trade inspired around 90 releases of possums during this period, which were first controlled by the Animals Protection Amendment of 1895.

The period between 1900 and 1910 showed a reduction in the number of releases, although two were made by the Department of Tourist and Health Resorts in Rotorua and Kumara.

The decade between 1910 and 1920 was marked by disputes between the acclimatization societies (who wanted further releases) and the New Zealand Fruitgrowers' Federation who argued that these should be banned and that, wherever possible, possums should be destroyed. As one irate orchardist wrote to Thomson:

> If you want to see how opossums and fruit trees thrive together. take a run down to Motutapu. Opossums you will see, but it will need a guide to show you where the fruit trees were planted. I have several acres in orchard, which today is free from opossums, and needs only the regular care to combat moth, scale,

Table 1 Principal introductions of possums into New Zealand

Date	*Quantity*	*Place of release*	*Introducer*
?1837–40	?	Riverton	Capt. J. Howell
1858	?	S. Riverton	Mr C. Basstian
1865	21 pairs	Canterbury	N. Canterbury AS
1869–74	?	?	Auckland AS
1870	?	Kawau Island, Hauraki Gulf	Sir George Grey (1812–98); Governor and later Prime Minister
1875–6	1 pair	Waldeck, Riverton	Capt. Hankinson
1875–7	9	?	Auckland AS
c. 1877	1 pair	Sanatorium Hill, Waikato	Mr Thornton
1890	12	Gisborne	Gisborne AS
1890	?	Auckland Islands	Southland AS
1890	15	Stewart Island	Southland AS
1890	24	Lake Te Anau	Southland AS
1890	44	West Coast Sounds	Southland AS
1890	24	N. and S. Canterbury	Canterbury AS
1891	11 pairs	Nelson	Nelson AS
1891	?	Otago	Otago AS
1892	?	Central Range	Wellington AS
1892	19	Paraparaumu	Wellington AS
1893	10	Kapiti Island, Cook Strait	Wellington AS
1893	10*	Wainui-o-mata	Wellington AS
1893	72	West Coast Sounds	Capt. Bollons, NZGS *Hinemoa*
1894	36	28 (from Victoria) West Coast Sounds; 8 (from Tasmania) Catlins	Otago AS
1895	12	Catlins	Ogato AS
1896	?	Taranaki	Wellington AS
1902	6 pairs	Hampden	Waitaki AS
1905	20	Kumara, Westland	Government Tourist and Health Resorts Department
1905	20	Rotorua	Government Tourist and Health Resorts Department
1915	14	Karangahake (Kumara and Mt Tuhua)	Auckland AS
c. 1920	40	Coromandel	Sir Hubert Ostler

Source: After Wodzicki (1950)
* Cost 22*s*. 4*d*. a pair. AS, Acclimatization Society

> scab, mildew, blight, dieback, fungus, beech, collar rot, birds, rabbits, picnickers and small boys. These I can manage to fix during the daylight, but cannot see why a set of cranks, who have nothing of their own to destroy, should compel me to sit up at night to shoot further vermin. If I am counted out in the assumption, will some 'boobs' join me in bringing pressure on the Government for the introduction of rattlesnakes, tigers and other interesting subjects, because the rattle and claws are beautiful, and the meat would compete with the local bully?[4]

Fruit-farmers eventually took matters into their own hands, and began destroying possums in large numbers. To combat this slaughter the acclimatization societies persuaded the government to declare, by an Order in Council of 1911, that possums were 'imported game' within the meaning of the Animals Protection Act 1908, thus making their future destruction illegal.

As early as 1912, when 60,000 skins were harvested in the Catlins district alone, it was clear that possum-trapping could be a financially viable venture: as a result, following representations to the government from trappers, farmers and settlers citing damage to crops, orchards and forestry, all protection for possums was removed in the same year, but was reinstated in certain specified areas of bush in 1913 as a result of representations from the acclimatization societies. From 1915 to 1924 importations of possums were confined to pets, and after the latter year seem to have ceased altogether.[5]

In 1916 the Department of Internal Affairs opened an enquiry into the ecological status of possums in New Zealand. Having, as might be expected, received conflicting and contradictory statements from pastoralists and the acclimatization societies (the Auckland Society said: 'We shall be doing a great service to the country in stocking these large areas [of bush] with this valuable and harmless animal'), in 1919 the government invited Professor H. B. Kirk to undertake a comprehensive survey of the problem. His report (Kirk 1920) found, firstly, that 'The damage to New Zealand forests is negligible and is far outweighed by the advantages that already accrue; on the other hand, the damage to orchards and gardens is indisputable', and, secondly, that 'Opossums may, in my opinion, with advantage be liberated in all forest districts except where the forest is fringed by orchards or has plantations of imported tree species in the neighbourhood'.

The results of Kirk's report were far-reaching, and strongly influenced subsequent legislation and government policy. From 1922, however, in spite of Kirk, the Department of Internal Affairs turned down further repeated applications for permission to release more possums. Thus all liberations after 1922 (and many between 1910 and 1920) were strictly illegal; this did not, however, prevent a number from being made.

In the late 1930s and early 1940s, when damage first became apparent in commercially valuable exotic tree plantations, the New Zealand Forestry Service, which had hitherto been in favour of possums, now declared against them. This, however, did not prevent the North Island Acclimatization Council from recommending that 'grey' possums (*T. vulpecula*) and bobucks or mountain possums (*T. caninus*) from eastern Australia be imported, due to the depreciating value of black (Tasmanian), brown and red-brown pelts and the increasing demand for grey. Fortunately, in 1949 an amendment to earlier legislation on possums cancelled all restrictions on controlling them and made harbouring or liberating them subject to heavy penalties.

'By 1890', as Thomson (1922) records, 'these animals had increased to a great extent in the forest covering the southern Longwood Range.' By 1923 possums were abundant in the country around New Plymouth, on Stewart Island and on the eastern side of Lake Kaniere on the west coast, and also occurred around Warkworth north of Auckland. Their dispersal owed more to repeated translocations of New Zealand-bred stock than to natural spread, which was on the whole slow. By 1947, for example, they had only extended their range from Warkworth to north of Kaiwaka, a distance of only 40 to 50 km in 24 years. Their spread in extensive and continuous areas of bush was more rapid, but by the late 1940s, according to Wodzicki (1950), they had also spread into the environs – where they were already becoming a nuisance – of cities such as Auckland, New Plymouth, Wellington and Invercargill. By this time possums were present throughout North Island except in Northland, around Lake Taupo and between there and Rotorua, extending from Auckland and the Bay of Plenty of Taranaki, and occurring in the south on both sides of the main divide.

In South Island possums were abundant on the

west coast and in Southland, but were more scattered from north Canterbury to Dunedin, and absent from central Marlborough and Otago. By 1947 they were established at Doughboy Bay on Stewart Island and had also spread to around Milford Sound.

Although it had originally been believed that possums followed the bushline, Wodzicki reported them from an altitude of 1,400 m on Mount Holdsworth, in the Tararua Ranges, and in open tussock grassland. Precipitation apparently had no bearing on their rate of dispersal, since they thrived equally in areas of heavy rainfall on the west coast and on Mount Egmont, and also in the drier districts of Canterbury and Poverty Bay. It is, however, believed that the grey phase fares better in dry and open forests whereas the black form favours wet sclerophyll forestland. The availability of cover and adequate supplies of food seem to be their most important limiting factors. Their few failures, such as in Fiordland and South Westland, and on Auckland Island, were attributed by Pracy (1962) to too small an initial stock, the lack of suitably dry nesting material and the advent of exceptionally cold and damp weather before they had had a chance of becoming properly acclimatized.

As Pracy points out, the history of the possum's naturalization in New Zealand has been marked by controversy, even among biologists, which still continues. Dr L. Cockayne and Professor Kirk envisaged a thriving fur trade yielding an annual income of some £200,000, but although they recognized damage to forests caused by deer, they apparently did not identify that caused by possums – especially to exotic varieties – and failed to appreciate the interaction in woodland between the two species. They even gave approval to a suggestion by the Otago Acclimatization Society that possums should be released in forestland on both sides of the Alpine Range.

Most scientists are, however, now agreed that the possum in New Zealand is a major pest: in native bush (which, having evolved in isolation over millions of years in an environment devoid of browsers, has developed no innate resistance to them) they feed on various parts of five finger (*Pseudopanax arboreus*), titoki (*Alectryon excelsus*), makomaka (*Avistotelia racemosa*), karaka (*Corynocarpus laevigata*), houhou (*Schefflera digitata*), mahoe (*Melicytus ramiflorus*), broadleaf (*Griselinia lucida*), konini (*Fuschia excorticata*), matipo (*Pittosporum eugenioides*), kohekohe (*Dysoxylum spectabile*), miro and nikau fruits (*Rhopalostylis sapida*), toro (*Suttonia salicina*), ratas (*Metrosideros umbellata* and *M. robusta*), tree-ferns, the bark of *Coprosma talpida*, kamahi (*Weinmannia racemosa*), broadgum (*Panax*), maple (*Pittosporum*), supplejack (*Rhipogonum*), and on several other species. Possum numbers often increase sharply once the ground cover has been opened up (resulting in drier conditions and easier access) by the browsing of deer, and the resultant defoliation of the subclimax brush canopy caused by arboreal possums is often difficult to detect from ground level. Much harm is also done by bark-stripping. It has been suggested that the killing of dominant trees by possums improves the habitat for deer by opening up glades on the forest floor.

The damage possums cause to orchard, farm and garden crops and exotic trees is considerable: in orchards they eat apples, peaches, nectarines and other stone fruits; in gardens they consume the buds and shoots of roses and flowering shrubs, peas, beetroots, turnips and other vegetables, and on farmland lucerne and clover. They also cause power failures on lines carried on unprotected poles. Their most serious damage on agricultural land is caused by the breaking of leading shoots on pines and the defoliation of sapling willows and poplars planted to discourage erosion.

Possums in New Zealand have few if any predators other than man. In spite of unlimited trapping since 1949 and the introduction of a bounty in 1951 (which, having proved ineffective, was withdrawn in 1961), possums are now probably the most widely distributed alien mammal in New Zealand, and have colonized virtually all suitable habitats on both the main islands, on Stewart Island and on some offshore islands (e.g. Kapiti, Codfish and Chatham) from the seacoasts to the subalpine uplands. On Chatham Island forest degeneration caused by possums (and cattle, sheep and pigs) has gravely endangered the Chatham Island taiko or magenta petrel (*Pterodroma magentae*). Population densities of possums vary between 7 or 8 per hectare in lowland mixed podocarp–broadleafed forest to 30 per hectare in rata–kamahi forest in Westland. Although as long ago as 1963 over 1 million skins were being harvested annually (and the figure for 1978–9 was 2.62 million yielding a net return of $A13,533,000), the damage possums cause in New Zealand far outweighs the revenue derived from the sale of their pelts.

In 1967 it was discovered that up to 30 per cent of possums on the west coast of South Island were infected with tuberculosis,[6] which is known to have been present in cattle in New Zealand for many years. It seems certain, therefore, that possums originally contracted the disease from cattle and that they now provide a reservoir of cross-infection which is greatly hampering the bovine tuberculosis eradication programme. Tuberculosis, almost certainly contracted from possums, is known also to occur to a lesser extent in most, if not all, alien mammals in New Zealand, being, perhaps, especially prevalent in hedgehogs (*Erinaceus europaeus*). Discovery of the disease has intensified efforts to eradicate possums in both main islands, largely with carrots poisoned with compound 1080 (sodium monofluoroacetate) dropped from light aircraft.

Wodzicki (1950) also records the unsuccessful introduction to New Zealand of the following Australian marsupials (excluding Macropodidae): two common ringtails (*Pseudocheirus peregrinus*) by the Canterbury Acclimatization Society in 1867; two marsupial cats (*Dasyurus* sp.) by the Canterbury Acclimatization Society in 1868, and an unknown number of 'short-faced' (short-nosed) bandicoots (*Isoodon* sp.) by the Auckland Acclimatization Society in 1873.

The **opossum** *Didelphis marsupialis* (known locally as the 'manicou') – a native of eastern Mexico to northern Argentina and Trinidad – was probably introduced to Grenada, the larger Grenadines, St Vincent, St Lucia, Martinique and Dominica. It is believed to be partially responsible for the disappearance from Dominica of the bridled quail-dove (*Geotrygon mystacea*).

It seems possible that the **opossum** *Marmosa chapmani* on Grenada, Carriacou and Islet Ronde in the Grenadines is not indigenous there but is an alien introduced from Trinidad.

The **sugar glider** (*Petaurus breviceps*), a native of northern and eastern Australia and Papua New Guinea, has been successfully introduced to the island of Tasmania.

NOTES

1. For much of the information on possums in New Zealand I am indebted to Wodzicki (1950) and Pracy (1962).
2. Australian/Papua New Guinean species are usually known as possums; New World species are referred to as opossums.
3. M. Rodrigue, *Manawatu Evening Standard* (6 February 1959).
4. Thomson (1922: 32–3).
5. Roots (1976), however, claims that possums were introduced to Chatham Island in 1928.
6. The possum is also the only reported wildlife reservoir in the world of *Leptospira balcanica* according to S. C. Hathaway, *Proceedings of the 1st symposium on marsupials in New Zealand, 1977* (Zoology Publications from Victoria University of Wellington No. 74, 1981).

Tammar Wallaby

(Macropus eugenii)

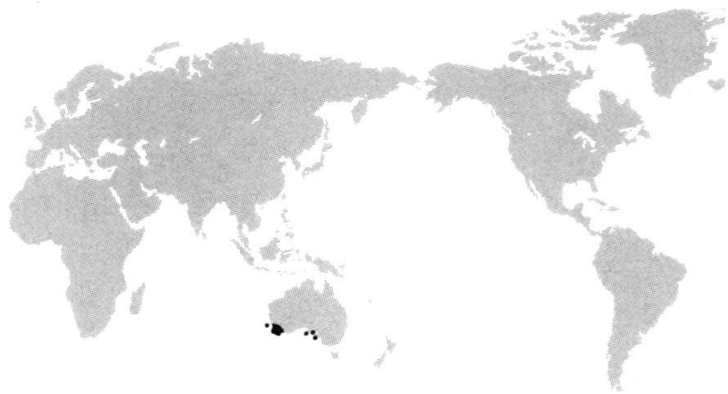

NATURAL DISTRIBUTION
Parts of southwestern and southern Australia.

NATURALIZED DISTRIBUTION
Australasia:New Zealand.

AUSTRALASIA

New Zealand

Tammar or dama wallabies occur in New Zealand on the 2,100-ha Kawau Island in Hauraki Gulf, where they were introduced in about 1870 by the island's owner, Sir George Grey (Governor of New Zealand 1843–53 and 1861–7, and Prime Minister 1877–9), and in the region of Rotorua on North Island. According to the Forest Service and Wildlife Branch of the Department of Internal Affairs they were turned down in about 1912 at the southern end of Lake Okareka by the late Mr H. R. Benn; Mr W. Parkes, however, a close friend of Mr Benn, told Wodzicki and Flux (1967a) that he knew of no wallabies having been released by Mr Benn. Their origin around Rotorua thus remains obscure, but it is known that at least by 1930 they were sufficiently well established in that region to permit shooting. L. T. Pracy of the New Zealand Forest Service believes that a number were translocated in 1939 from Kawau to Rotorua, where by 1946 they were distributed over an area of approximately 10 by 20 km to the north and west of Lakes Tarawera and Okataina; 20 years later they had expanded their range south and east to between west of Lake Rerewhakaaitu and Rotorua and covered an area of around 20 by 30 km. By the 1970s they had spread east and north of Rotorua and to Lake Rerewhakaaitu and the northern boundary of Kaingaroa State Forest. It has been suggested that there is at least one other species in addition to the tammar in the Rotorua region.

According to Wodzicki and Flux (1967a) the Rotorua tammars are considerably larger and heavier than those on Kawau Island, which suggests that different subspecies may be involved; if so this would tend to cast doubt on Pracy's belief referred to above.

Around Rotorua tammar wallabies generally inhabit undeveloped areas of native grass, scrub, and fern, and to a limited extent cutover logged indigenous forest, from which they emerge at night to feed on nearby paddocks. Damage in the Rotorua area is generally reported to be small, though locally serious destruction of crops in the ecotone between pastoral land and forest and scrub sometimes occurs. On Kawau, where they are said to damage newly planted pines, they are the most numerous species present, especially at the southern end of the island and in the central clearings where the vegetation is more sparse; in 1966 Wodzicki counted 49 tammars (and/or parmas[1]) on Kawau out of a total of 59 wallabies of 4 species present, and 73 tammar skulls out of a total of 118.

NOTE

1. See p. 15.

Parma Wallaby
(Macropus parma)

NATURAL DISTRIBUTION
Very locally in parts of southeastern Australia.

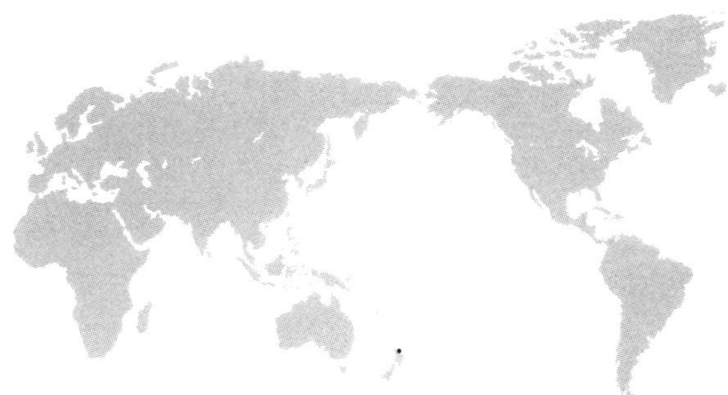

NATURALIZED DISTRIBUTION
Australasia: New Zealand.

AUSTRALASIA

New Zealand

In Australia parma wallabies, having for long been considered extinct, are now believed to be confined – due largely to predation by introduced foxes (*Vulpes vulpes*) and competition for grazing and shelter from sheep – to two or three colonies in New South Wales. Following a suggestion in 1958 that the parma might have been one of the species turned down on Kawau Island by Sir George Grey in about 1870, it was rediscovered there in considerable numbers in 1966 by Wodzicki and Flux (1967b). Parma wallabies, which have been fully protected since 1968, have now been recorded from most parts of Kawau. On the northern half of the island, which is covered with thick scrub (mostly *Leptospermum ericoides* with a small amount of *L. scoparium*), they are more abundant (25 out of 31 skulls found in 1966) than the tammar wallaby (*M. eugenii*).[1] In the south and centre of the island Wodzicki found 49 parmas (and/or tammars) out of a total of 59 live wallabies (the two other species identified being black-tailed (*Wallabia bicolor*) and brush-tailed (*Petrogale penicillata*) rock wallabies), and 40 out of a total of 118 skulls. They spend the daylight hours in dense scrub, only emerging in the evenings to feed. Since the area covered by scrub has expanded in recent years, Wodzicki and Flux considered that parmas have probably increased in numbers at the expense of tammars, and that, following the extensive control of wallabies (mainly tammars) in plantations in the south of the island, parmas could well soon become the commonest species on Kawau. A number of individuals have recently been exported to Australia in an attempt to re-establish the species in its natural range.

The Rabbit Destruction Council's Annual Report for 1962, quoted by Wodzicki and Flux (1967a), suggested that 'It is . . . a pity that a vigorous campaign of destruction is not waged against wallabies wherever they are [on Kawau], in an endeavour to eliminate this pest from the country entirely'. Had this indiscriminate policy been implemented, the world might well have lost the parma wallaby which, as Wodzicki and Flux (1976b) point out, is 'of almost as much interest in Australia as the takahe [*Notornis mantelli*] in New Zealand'.

NOTE

1. See p. 13.

Red-necked or Bennett's Wallaby

(*Macropus rufogriseus*)

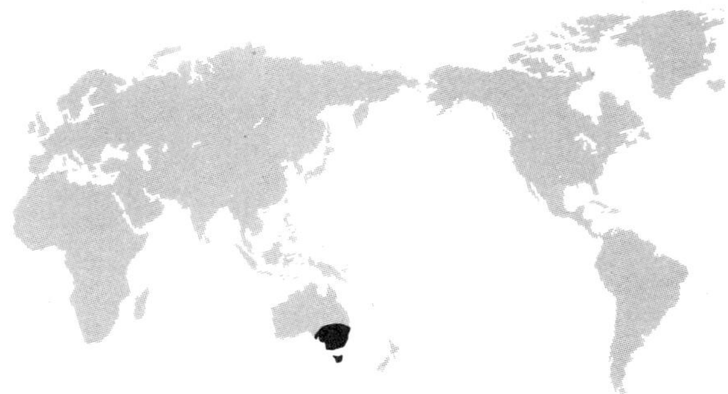

NATURAL DISTRIBUTION
Eastern and southeastern Australia from central Queensland to South Australia, and Tasmania.

NATURALIZED DISTRIBUTION
Europe: British Isles (England);[1] Australasia: New Zealand.[1]

EUROPE

British Isles (England)[2]

In the mid-1930s the late Lieutenant-Colonel (then Captain) Henry Courtney Brocklehurst established a small private menagerie in the grounds of Roaches House near Leek in Staffordshire, England. On 11 June 1936 he obtained a pair of Bennett's wallabies for his collection from Whipsnade Zoo (where the animals are still maintained in a semi-wild state); a single young was born in March 1937, and two more – one of which was drowned in the following October – in March

1938; a fourth wallaby was born in May 1939. These five animals escaped from their enclosure in 1939 or 1940 – when because of the war it proved no longer possible to continue to feed them or to maintain the fences of their enclosures – and were the ancestors of the stock of wild wallabies which have been established since then in what is now the Peak District National Park. Here they live on heather moorland and in thick scrub north-west of Leek, and in woodlands near Hoo Moor in neighbouring Derbyshire some 16 km to the north. They eat mainly heather (*Calluna*), and also grass (usually *Festuca*), young bracken-shoots (*Pteridium*), pine-scrub and birch-scrub and bilberries (*Vaccinium*).

The earliest published reference to wallabies in the Peak District appeared in a letter to the *Buxton Advertiser* on 16 March 1951; it reported that a single animal had been seen near Errwood Hall – 8 km north of the source at Roaches House – in 1940, and that in 1951 another had been observed at Hindlow Quarry near Buxton, which died two days later after being caught, exhausted in the snow, at Harpur Hill Quarry, 10 km north-east of Roaches. In 1944 a lone wallaby was seen at Barmoor Clough near Chapel-en-le-Frith, 19 km north-east of the original source. A letter in the *Buxton Advertiser* on 16 July 1954 reported that a wallaby had been seen earlier in the year in Burbage near Buxton, 11 km north-north-east of Roaches. In the following year one was seen near the Yorkshire Bridge by the Ladybower Dam, 29 km north-east of Roaches. The 16 November 1956 issue of the *Buxton Advertiser* referred to a wallaby which had spent most of the previous summer in Chapel-en-le-Frith. In the winter of 1963 a wallaby was captured near Sheen – 10 km east of Roaches – but again died, exhausted, in the snow. The *Daily Telegraph* of 25 July 1970 reported that a wallaby had been seen earlier in the month near Werrington, 14 km south-south-west of source. It is interesting to note that, with one exception, all these records are from the north and east of the original source at Roaches House.

The hard winter of 1947 probably checked the increase in the wallabies' population: by the early 1960s, however, their numbers were estimated to have reached 40–50, and one observer reported seeing seven bolted by terriers from cover in quick succession. The exceptionally severe winter of 1962–3 again caused serious casualties; several wallabies were found dead in the following spring – six or seven being reported from one small area; a nucleus, however, managed to survive, and four or five were seen together in the spring of 1965.

Between 1967 and 1970 Dr D. W. Yalden and G. R. Hosey of the University of Manchester paid a total of 35 visits to the wallabies' territory, where they recorded 34 separate sightings on only 14 occasions; in 1970 the total population was estimated at around a dozen.

The Peak District wallabies appear to be divided into two subcolonies, which seem to have little if any contact. In 1971–2 the total population was estimated at between 14 and 20, and continued slowly to increase to 1977. In August of the following year, when the population was judged to have fallen to the 1971–2 level, a much publicized release of a young male was made by Riber Castle Zoo in conjunction with Granada Television: the consequences, if any, of this irresponsible and unnecessary action, are unknown. In October 1978 Dr Yalden counted a minimum of 18 wallabies in the Peak District National Park.

In 1977, after the death of Colonel Brocklehurst's brother, Sir Philip Brocklehurst, the 459 ha occupied by the wallabies were sold for sheep-farming and were overstocked with 2,000 ewes. Competition from the sheep for grazing, browsing and – equally important – for shelter, coupled with the hard winter of 1978–9, reduced the population still further, until extinction seemed inevitable. After representations from a number of interested parties the Peak Park Planning Board intervened and purchased for £185,000 around 400 ha of the wallabies' habitat in the south-west of the National Park,[3] where, after the exceptionally severe winter of 1981–2, Dr Yalden estimated the population to number about 15, including two or three pouch young.

A second, but so far less well-documented colony of Bennett's wallabies is established in north-central Sussex. The source of this colony is presumed to be Leonardslee Park, Lower Beeding, near Horsham, where some were introduced around 1908 by Sir Edmund Loder, Bt.

In the early 1940s numerous reports were received of a small but apparently thriving colony in the Ashdown Forest/St Leonard's Forest district. In September 1965 a feeding wallaby was disturbed in a cornfield at Laybrook Farm, Thakeham, some 13 km south-west of Leonardslee.

No wallabies were reported between 1965 and 1968, which prompted Sir Giles Loder, Sir Edmund's grandson, to declare, somewhat rashly as it turned out, that 'conditions now make it impossible for abnormal [sic] animals to survive in a feral state even if the natural habitat is suitable'.[4] In the following year no less than five reports of wallabies were received from the north and south of Leonardslee: from St Leonard's Forest, Horsham, 6 km from source; from Worth Forest, Crawley (10 km); from Turner's Hill and Crawley Down (16 km); and from Felbridge, East Grinstead (21 km). In 1970 a wallaby was found dead on a railway line at Christ's Hospital, Horsham, 8 km north-west of Leonardslee, and a live individual was later observed in Lower Beeding. The expansion of Crawley New Town provides an effective barrier to further extension of range of this colony to the north as does Horsham to the west.

Although wallabies have been living in the wild for more than 40 years in both the Peak District and Sussex they have not so far spread far from their original homes. The reasons for this are probably twofold: first, the suitability of their existing habitat and an abundant food supply which, combined with their small population and basically sedentary nature, renders further colonization unnecessary; and second, their vulnerability when attempting to cross roads and railway lines, which have been the source of a number of deaths in the past.

Cover, shelter, security from disturbance and a continuing adequate food supply are the wallabies' principal requirements. Provided that these can be maintained the future in England (where the climate is not dissimilar to that of their Tasmanian homeland) of these harmless and interesting marsupials should be assured.

A small colony of Bennett's wallabies on Herm in the Channel Islands in the 1890s was exterminated, according to Roots (1976), to provide food for British troops during the First World War. Roots also says that a number introduced into a 100-ha forest near Bonn, West Germany, in 1887 increased before being killed off by poachers. Mitchell (1983) records that in about 1975 two pairs were released on Inchconnachan in Loch Lomond, Scotland by the Countess of Arran. The first of several animals arrived on the mainland in 1979 (exactly how is uncertain), and in December 1982 one was seen as far south as Balloch Park at the southern end of the loch.

AUSTRALASIA

New Zealand[5]

Thomson's (1922) record of the introduction to New Zealand of the black-tailed or swamp wallaby (*Wallabia bicolor*) clearly refers to *rufogriseus*.

According to Thomson, A. M. Johnson imported some Bennett's wallabies from Tasmania (for shooting, fur and meat) on behalf of the Canterbury Acclimatization Society in 1867; in the same year three were established in the gardens of the Auckland Acclimatization Society, who acquired a fourth in 1874; what became of them is unknown.

In April 1870 a Christchurch newspaper claimed that 'The merit of the introduction into Canterbury province of the brush Kangaroo [*rufogriseus*] is due to Captain Thomson, and for the thriving condition of those in the Society's gardens, their adaptability to the province has been proved, whilst their increase has been such as to now render their liberation desirable in suitable localities'.[6] It may have been a pair of these which Wodzicki (1950) records as having been released on Mount Nimrod, Timaru by the South Canterbury Acclimatization Society in 1903.

In about 1874 Mr Michael Studholme released some Bennett's wallabies (which he had either imported direct from Tasmania or had purchased from the Canterbury Acclimatization Society) on his sheep station at Te Waimate. In February 1916 Mr E. C. Studholme informed Thomson that:

> I can just remember seeing them turned loose here, two does and one buck being the number liberated. They gradually spread along the adjoining hills [on either side of Waimate Gorge], and are now to be found as far north as Bluecliffs [30 km away, and west to the Hakataramea watershed beyond Kaiwarua]. It is very hard to estimate the number there are at the present time, but it is quite safe to say there are thousands of them. Parties which go out shooting have killed as many as seventy in a day or two. They live chiefly in the bush, scrub and fern about the gullies and gorges,

coming out in the evenings to feed in the open ground. Their food chiefly consists of grass, but they are very hard on certain trees, barking many of them, particularly the Ohaus or five-leaf (*Panax arboreum*). I believe they are not a wallaby, but scrub-kangaroo, as they are quite large, some of the old bucks weighing over 60 lbs [27 kg].[7]

In about 1948 a number of Bennett's wallabies were released at the head of Quartz Creek on Mount Burke Station, between Lakes Hawea and Wanaka. In 1964 about 50 were seen on the north-west face of Mount Maude, and in the late 1960s a few are said to have been transferred from Waimate to Hawea.

According to the Annual Report of the Rabbit Destruction Council for 1964, quoted by Wodzicki and Flux (1967a), wallabies had by 1947 'increased to such an alarming extent that some method of control was necessary. From 1947–56 cullers employed by the Wildlife Branch of the Department of Internal Affairs destroyed 68,608. This method of destruction did not even cope with the natural increase of the pest and had the undesirable effect of spreading the animals over a more extensive area'. The Noxious Animals Act 1956 passed responsibility for the control of wallabies to the New Zealand Forestry Service, with whom it remained until the Rabbits Amendment Act 1959 empowered rabbit boards to destroy wallabies, and a special Wallaby Destruction Committee was established at Waimate. The Annual Reports for 1964–6 of the Rabbit Destruction Council, quoted by Wodzicki and Flux, reveal that shooting and the dropping of poisoned baits from the air reduced the estimated population of 500,000 to 1,000,000 wallabies on 765,000 ha to 3,000–4,000 by 1965 and to 1,000–2,000 by the following year.

A survey in 1946 revealed that wallabies were well distributed between north Pareora and Waihao Forks; they were increasing at Otaio Gorge, Albury, Bluecliffs, Mount Nimrod and Waihao Forks, were stable at Waihaorunga and decreasing at Cave. They are currently thinly but fairly widely spread mainly over an area of about 75 by 40 km west and north of Waimate, and also at Hawea. The largest and most extensive population is spread over around 400,000 ha of predominantly tussock grassland in the Hunter Hills (the majority occupying about a quarter of that area), with gullies thinly covered with bush and scrub, mainly at altitudes of between 300 and 600 m. The animals cause a considerable amount of damage to agricultural crops and by browsing on shrubs and young plantations of exotic trees, so their continued control is essential. They are the only wallaby on New Zealand's South Island and are not found on North Island.

NOTES

1. The subspecies of the red-necked wallaby in England and New Zealand is Bennett's wallaby (*M. r. bennetti*).
2. For much of the information on wallabies in England I am indebted to Yalden & Hosey (1971).
3. C. Lever, *Daily Telegraph* (24 March 1980).
4. *Sussex Mammal Report* (1968).
5. For much of the information on wallabies in New Zealand I am indebted to Wodzicki & Flux (1967a).
6. Quoted by Thomson (1922).
7. Thomson (1922: 26).

Brush-tailed Rock Wallaby

(*Petrogale penicillata*)

NATURAL DISTRIBUTION
Parts of eastern Australia.

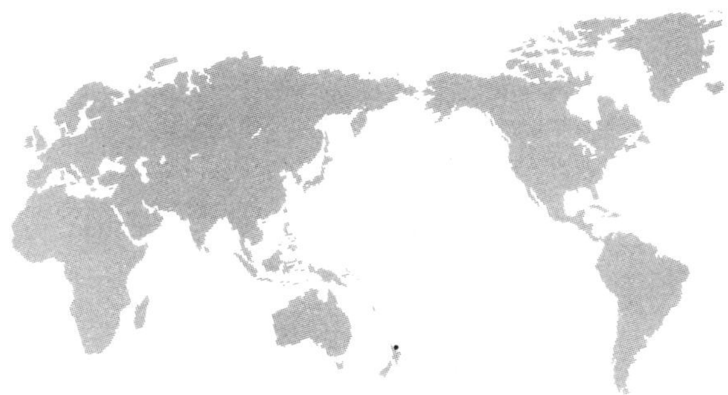

NATURALIZED DISTRIBUTION
Australasia: New Zealand; Oceania: Hawaii.

AUSTRALASIA

New Zealand

Thomson (1922) records that in 1873 the Auckland Acclimatization Society received a rock-wallaby '*Petrogale xanthopus*?' (the yellow-footed or ring-tailed rock wallaby of eastern and south-eastern Australia) from Sir James Fergusson; what became of it is unknown. At about the same time Mr John Reed (or Reid) released some 'small brown rock wallabies' on Motutapu, from where they spread across a narrow isthmus to Rangitoto Island, where by 1912 they were apparently numerous. As brush-tailed rock wallabies also occur on Kawau, it seems certain that they were one of the several marsupials turned down there in about 1870 by Sir George Grey.

Rock wallabies are very agile and can climb up to 4 m from the ground on sloping trees, though they feed mostly in grassy clearings on cliff tops. They appear to be most abundant on Motutapu; on Kawau they are reported to be scarce (in 1966 Wodzicki saw 6 out of a total of 59 live wallabies on the island and found only 2 *penicillata* skulls among a total of 118) but occur wherever there are cliffs; on Rangitoto, which is mostly recent scoria with a fragmentary vegetation, they are not confined to cliffs but are found in small numbers all over the island. There is also said to be a mainland colony of rock wallabies (? sp.) in Northland.

OCEANIA

Hawaii

On 18 August 1916 Mr Richard H. Trent of Honolulu purchased a pair of brush-tailed rock wallabies (the female with joey in pouch) for $100 from a dealer, Mr Ellis J. Joseph, who had acquired a number from the Sydney (Australia) Zoo, which he was shipping to the American mainland. The wallabies escaped from Trent's private menagerie on Alewa Heights on the following day and were chased by a dog which succeeded in killing the joey. Little more was heard of the adults, until the *Pacific Commercial Advertiser* of 29 March 1921 announced that 'The wallaby family has increased in number from the original couple . . . it is not believed that there are more than 50 today'. By the 1970s about 100 were established on Oahu, mostly on the sheer south-facing Kalihi cliffs. They are not currently (1983) open to public hunting.

In about 1870 Sir George Grey released some **swamp** or **black-tailed wallabies** (*Wallabia bicolor*), natives of eastern Australia, on Kawau Island north of the Hauraki Gulf, where they are known as 'wallaroos'. Today they occur over most of Kawau in damp or scrubby areas, but are nowhere abundant. Of 59 wallabies seen alive in the southern and central parts of the island in May 1966 only 4 were *bicolor*, and of 118 skulls collected in the following September only 3 belonged to this species.

As Wodzicki and Flux (1967a) point out, Thomson's (1922) records of the introduction of *bicolor* to the mainland of New Zealand clearly refer to *rufogriseus*.[1]

At about the same time that Sir George Grey released swamp wallabies on Kawau Island he also introduced a number of **black-striped wallabies** (*Macropus dorsalis*). This species, which is also a native of eastern Australia, is now considered to be very rare (there have been no definite records since 1954) if not extinct on Kawau, and occurs nowhere else in New Zealand.

Wodzicki (1950) records the following unsuccessful introductions of Macropodidae to New Zealand: three kangaroos (*Macropus* sp.) on Dunrobin Station by Mr Christopher Basstian in 1868 and three in the same year on Bluff Hills by the Southland Acclimatization Society; 'roan wallaroos (*Osphranter erubescens*)' between 1860 and 1870 on Kawau by Sir George Grey; and some 'long nosed rat kangaroos' (potoroos – *Potorous* (*apicalis*) *trydactylus*) by the Auckland Acclimatization Society in 1867.

Of approximately a dozen species of marsupials introduced to New Zealand since 1837 only the brush-tailed possum (*Trichosurus vulpecula*)[2] and six of the wallabies have ever become successfully established. Ambiguous early accounts by Thomson (1922) and others about which species were introduced and where they came from has resulted in considerable confusion in the past over which ones are actually established, and may well have been responsible for the delayed discovery of *Macropus parma*.[3] Thus Thomson (1922) wrote:

> About 1870 Sir George Grey introduced a number of species of marsupials into the island of Kawau, and among them was a wallaby which increased in an almost incredible manner . . . even in Sir George Grey's time, as many as two hundred wallabies would be killed in a Battue. Mr Cheeseman tells me that when the island of Kawau was sold, the new owners encouraged shooting parties to go down – indeed contracts were let to kill the marsupials of the island – and

the slaughter was great. There are still a few left about Kawau, not more than a dozen of two, according to Colonel Boscawen.[4]

This account suggests that Thomson considered the wallabies on Kawau (and those at Waimate, Hawea and Rotorua) to be all one species (*Wallabia bicolor*), instead of six separate ones. Of these, the two established on the mainland, *Macropus rufogriseus* and *M. eugenii*, have not spread far from their original points of liberation, and only one, *rufogriseus*, has ever reached high-enough numbers to become economically embarrassing by damaging crops and providing competition for food and shelter for sheep.

NOTES

1. See p. 18.
2. See p. 7.
3. See p. 14.
4. pp. 26–7.

Nine-banded Armadillo

(*Dasypus novemcinctus*)

NATURAL DISTRIBUTION
From Uruguay and northern Argentina north through Central America and parts of Mexico to the southeastern USA.

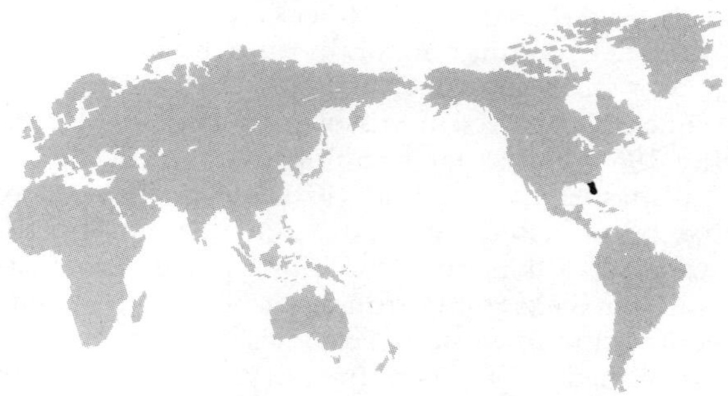

NATURALIZED DISTRIBUTION
North America: USA.

NORTH AMERICA

USA

The nine-banded armadillo[1] is the only living edentate to have reached the United States, where the subspecies established is *mexicanus* (= *texanus*). Probably for a variety of reasons (including the advance of civilization with its accompanying destruction of large predators and land-clearance for agriculture, and climatic change in the late 1880s) it seems to have been slowly extending its range north from Mexico since the middle of the last century.

The earliest reference to the creature in the United States appears to be that in John James Audubon and Rev. John Bachman's *Quadrupeds of North America* (Volume III, 1854) where it is described as existing 'in the southern portion of Texas . . . along the lower reaches of the Rio Grande'. H. H. Bailey in 1905 indicated that armadillos had pushed north and west, had ascended the Pecos River valley, and had crossed the southeastern border of New Mexico.[2] Although not officially reported from Louisiana until 1926, it seems probable that armadillos had in fact arrived there by about 1914. By 1943 they had crossed over to the eastern side of the southern Mississippi. The armadillo apparently entered eastern Oklahoma from Arkansas (where it was fairly well established in the south-west by 1944, having probably arrived from both Texas and Louisiana) in about 1936, travelling via the Canadian / Arkansas / Red River systems. It has since extended its range naturally into Kansas, where it was first recorded in about 1943, and subsequently appeared in Missouri (by 1947), Colorado (by 1966) and as far into northwestern Florida as Tallahassee by the mid-1970s. Since then there has been a continued extension of range northwards in Oklahoma and northwards and eastwards in Texas and throughout Florida, but some contraction in western Texas. Armadillos are well established in west-central Arkansas, and from the south-east of that state they have spread up the Arkansas River and by 1972 had travelled up the Mississippi to Tennessee. Their mid-1970s natural distribution included two-thirds of Mississippi, one-third of Alabama and the western end of the Florida panhandle. As Layne (1976) has written, 'The history of the armadillo in the United States is one of the most striking examples of the natural invasion of an area by a species in the annals of biogeography.'

Although armadillos invaded western Florida by a natural extension of range from Alabama, their presence in peninsula Florida is the result of human interference. Bailey (1924) has written:

> I learned [in 1922] that during the late war a marine had been stationed at the canal locks near Hileah, and that he had brought a pair of armadillos from Texas when he came. These he liberated on leaving his station after the war. In February, 1924, I learned of a female with four young being killed, and I am thus inclined to think they are increasing in numbers.

Two years earlier, in 1922, an armadillo was killed by a dog near Miami in Dade County, and two more, which were believed to have escaped several months earlier from near Cocoa, were seen at Williams Point, Brevard County. In about 1924 the animals already in the wild were reinforced by the escape of others during a storm from the Cocoa Zoo in Brevard County.[3]

In 1934 an armadillo was killed at Flagler Beach, Flagler County, and two years later another, having escaped from an over-turned circus truck belonging to the Scholtz Field Zoo, Daytona, was recaptured at Titusville in Brevard County.[4] By 1938 armadillos had spread to within 10 km south of Indian River City, Brevard County, and to near Crescent City in Putnam County: three years later the animals were being reported from Pinemount, Suwannee County, and in 1943 an individual was observed in South Jacksonville, Duval County. All these records, with the exception of the one from Pinemount, occurred east of the St John's River.

By 1943 armadillos were apparently well established in Brevard, Volusia, Flagler and Putnam counties, east of the St John's River. Six years later they had extended their range to include St John's, Indian River, St Lucie, Martin, Okeechobee, Osceola, Polk, Orange, Seminole and Lake counties; reports of individual sightings from Hamilton, Alachua, Marion, Sumter, Pasco, Manatee, Hardee, De Soto, Lee, Hendry, Palm Beach and Broward counties suggest a diffusion from the eastern regions of the State.

By 1951 armadillos had invaded several offshore islands, including Merritt Island and another to the south of Cape Canaveral; scattered individuals were observed in Pinellas, Hillsborough, Hernando, Citrus, Levy, Dixie and Taylor counties on the west coast, and in Wakulla County in the panhandle. By this time, therefore, armadillos had colonized, within a period of only 20 to 30 years, most of Florida except the Big Cypress Swamp and the Everglades of the southwest, and possibly the western panhandle.

The original introduction of armadillos to Miami during the First World War apparently was unsuccessful, and the animals first became established in Florida in the northern east-coast district, probably as a result of the escapes from the Cocoa Zoo around 1924, reinforced by an introduction said to have been made near New Smyrna in Volusia County in the early 1930s. By 1972 their main range in Florida covered the entire peninsula south to beyond Palm Beach County except for the Everglades (where, even in the extreme south, occasional sightings have been reported in recent years), extending westward in the panhandle to beyond the Aucilla River. They are thus now one of the most widespread and abundant mammals in the state, and the gap between the animals invading naturally from the west and the introduced population in the east is closing rapidly. From Florida armadillos have travelled north through Nassau County to colonize much of southeastern Georgia and have even crossed the border into South Carolina. Individuals reported from North Carolina, Washington DC, Delaware and Decatur near Atlanta, Georgia, were probably imported.

As elsewhere, armadillos in Florida occur in a wide variety of habitats including turkey-oak and longleaf pine, rosemary scrub, rough pasturage, orange groves, tung (*Aleurites fordii*) orchards, gardens, pastoral land and residential areas. They show an apparent preference for a damp environment: especially favoured are moist hammocks[5] and prairies, and improved pastures interspersed with cabbage palm and live oak hammocks. They feed mainly on insects, millipedes, snails, earthworms and other invertebrates, and occasionally on lizards, snakes, reptiles and amphibians.

Although sometimes preyed on by pumas (*Felis concolor*), black bears (*Ursus americanus*), domestic dogs, alligators (*Alligator*) and other large predators, the armadillo's principal enemy is man, who annually kills large numbers both accidentally in cars and deliberately by shooting. The *Orlando Sentinel* of 18 March 1951 summarized popular sentiment as follows:

> This destructive creature has grown in numbers to such an extent that [orange] grove owners are up in arms in protest against the armadillo depredations in the groves, where their burrowing activities fill up the ditches with their excavations. Nurserymen will tell you that a single armadillo that gains entrance to a slat house or beds where tender plants are coming to maturity [can cause] thousands of dollars damage in a single night. Hunters claim that the armadillo is very devastating to quail eggs.

In spite of widespread belief, the last of these charges is not confirmed by positive data. Armadillo burrows, on the other hand, are a serious hazard to horses on ranchland, to roadsides, dykes and to the banks of watercourses. Although armadillo depredation on ground-nesting birds is negligible, they are known to prey on some of Florida's rare endemic insectivorous reptiles such as the Florida scrub lizard (*Sceloporus woodi*), the sand skink (*Neoseps reynoldsi*), the worm lizard (*Rhineura floridana*) and the crowned snake (*Tantilla relicta*). Armadillos have also been found to be vectors of Chagas' disease, which can be fatal to man. On the other hand, the discovery in the 1970s that many armadillos are infected with *Mycobacterium leprae*, which causes human leprosy, provided a regular source of supply of the bacillae from which human vaccine can be prepared.

Also on the credit side is the fact that armadillos undoubtedly consume large quantities of injurious insects. Their burrows provide shelter for many indigenous animals and they are themselves of value in scientific and medical experimental research. As well as being excellent eating for man, they may also be an important source of food for the larger mammalian carnivores and for necrophagivorous birds such as vultures (Cathartidae) and the red-throated caracara (*Daptrius americanus*), whose original food resources have been diminished by habitat destruction.

The principal factors limiting the extension of the armadillo's range in the United States are prolonged droughts and cold weather; the latter – due to the animal's imperfect temperature control and non-hibernation – sometimes decimates the population.

The **six-banded armadillo** or **peludo** (*Euphractus sexcinctus*), a native from Uruguay to Surinam east of the Andes, and the **small armadillo** or **pichis** (*Zaedyus pichiy*) of the Argentinian pampas, have both become established – as escaped pets – in parts of central Chile; the latter, which was kept in captivity before 1847, is said to be especially common in the Province of Nuble.

NOTES

1. The name 'armadillo' is derived from the Spanish, in references to the animal's armour-like cladding.
2. *Biological survey of Texas* (1905).
3. *Cocoa Tribune* **20** (10 December 1936).
4. For the history of the early spread of armadillos in Florida I am indebted to Sherman (1943), Newman (1949) and Neill (1952).
5. An area with deep and rich soil with hardwood vegetation.

European Hedgehog

(*Erinaceus europaeus*)

NATURAL DISTRIBUTION
Western Europe from the Mediterranean coast to central Sweden and Finland, east to a line from about the Oder to the Adriatic.

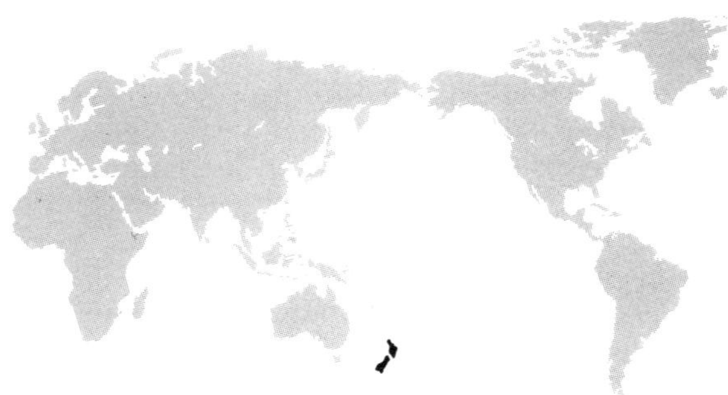

NATURALIZED DISTRIBUTION
Australasia: New Zealand.

AUSTRALASIA

New Zealand[1]

The hedgehog is the only large insectivore to have become widely established (throughout New Zealand) outside its natural range.

In 1870 Mr D. Robb, purser of the *Hydaspes*, presented a pair to the Canterbury Acclimatization Society who, in the following year, also received a further one from a Mr Nottidge; all three died. The earliest successful introduction (to South Island) took place in 1885 when three (two males and a female, the sole survivors of an original shipment of 100 from England to the Otago Acclimatization Society) were placed in a suburban garden; within eight weeks the female had died, and the two males had been released or had escaped. Others must have been imported in the late 1880s, as by about 1890 hedgehogs were being reported from Sawyer's Bay, Dunedin, and by the turn of the century from the Bay of Plenty.

In 1892 or 1894 (accounts differ) Mr Peter Cunningham of Merivale, Christchurch, received 12 hedgehogs from England in exchange for a like number of weka rails (*Gallirallus australis*). On their first night in the colony the hedgehogs escaped from a pigeon-house in which they had been placed, and within a few years were well established in the Christchurch area of Canterbury. Outside Canterbury, hedgehogs were first seen in South Island at Maungati in 1907; Peel Forest, 1910; Bluecliffs, 1915–16; Kaikoura, 1933; and Darfield in 1937.

Early translocations in various centres on the east coast of South Island account for much of the spread of the hedgehog up to the late 1940s. Large areas of tussock-country appear to have been, as on the east coast of North Island, an important factor in the extension of range. This belief is supported by a comparison of the spread of early releases in Nelson and on the west coast with more recent ones in the Marlborough district. By the late 1940s only the central area of Marlbrough, the west coast south of Hokitika and north of Westport, and bush and alpine districts remained uncolonized.

The earliest recorded introduction of hedgehogs to North Island (where the animals had first been observed about three years previously) was by Mr W. W. Smith, who in 1913 released two pairs from England in the Public Park, New Plymouth, where they settled down well and rapidly increased.

As recently as the late 1940s the central part of North Island and some areas north of Auckland and around the Bay of Plenty still remained uncolonized. In much of North Island hedgehogs appear to have spread from centres of earlier translocations, such as Wellington, Manawatu, Taranaki, Waikato and Auckland. Colonization of Northland, on the other hand, appears to have resulted from later liberations.

By the mid-1940s hedgehogs had found their way to a number of offshore islands, such as Waiheke off Auckland, the Chatham group and Stewart Island, on the last of which they had, within the space of four or five years, colonized all the cleared land around Horseshoe and Halfmoon bays.

The natural extension of range of hedgehogs in New Zealand has been comparatively slow, and numerous translocations have played a major role in the animal's spread and current wide distribution in both North Island (where it is today particularly abundant around Palmerston North and Auckland) and South Island.

According to Wodzicki, climate has had little effect on the hedgehog's distribution in New Zealand, the most important single factor being the availability of suitable habitats: those most favoured are settled suburban districts with gardens, orchards and farmland, followed by scrub, bush, and secondary growth forest, up to altitudes (in Otago) of 2,000 m; on occasion, even coastal sand-dunes and open tussock-country have been occupied.

Hedgehogs were imported to New Zealand as a form of pest control – a task which they perform admirably; they eat many injurious insects (especially beetles) as well as slugs, snails, earwigs, millipedes, moth larvae, grubs, woodlice, worms, frogs, mice, and young rats and rabbits. Unfortunately they also damage young seedlings and orchard and vegetable crops, and eat the eggs and kill the fledglings of various ground-nesting birds, such as weka and banded rails, quail, duck, chukar partridges, pheasants, skylarks and domestic fowl.

Today, hedgehogs are widespread and common throughout New Zealand, only some high mountain ranges and large forested areas remaining uncolonized; they are more abundant than in the British Isles, and occupy a wider spectrum of habitats than in mainland Europe. Many are infected with tuberculosis, almost certainly contracted from possums (*Trichosurus vulpecula*).[2]

Although the hedgehog is a native of the British mainland its presence on the following offshore islands (and probably on many others also) is the result of human agency: Shetland (mainland), Unst, Yell, Foula, Fetlar, Muckle Roe, Bressay, Whalsay, East and West Burra, Ventray, Orkney (mainland), and Canna. Its presence in Ireland may likewise be due to introduction by man.

The principal range of the **Algerian hedgehog** (*Erinaceus algirus*) is north-west Africa, from where it has been introduced to Malta, the Balearic Islands, south and southwestern coasts of France, parts of Spain, and possibly to the Canary Islands.

Tail-less tenrecs (*Tenrec ecaudatus*), natives of Madagascar (the Malagasy Republic), are said to have been introduced to control insects to Mauritius and to the islands of Mahé and Silhouette in the Seychelles where, with the help of the introduced African frog (*Rana mascariensis*), they are reported to have been responsible for the extinction of three endemic frogs – *Nesomantis thomasetti, Sooglossus sechellensis* and *S. gardinieri.*

The **house shrew** (*Suncus murinus*) – a native of much of the Oriental region – has been introduced to many islands and isolated sea-ports from the Persian Gulf and the Red Sea (including Aden, Bahrain, Basra and Suez) to south-east Asia. In Japan it is found on the island of Fukue (Goto), in Nagasaki and Kagoshima on Kyushu, and throughout the nearby Ryukyu Islands. House shrews have been established since before 1912 on Dyer Island, South Africa, and also occur on the islands of Pemba and Zanzibar, Tanzania. **Greater** and **lesser white-toothed shrews** (*Crocidura russula* and *C. suaveolens*) were introduced, possibly by Neolithic people, to the Channel Islands and Isles of Scilly, England. The **masked shrew** (*Suncus cinereus*), which feeds on the cocoons of the larch sawfly, was introduced to Newfoundland, Canada, from New Brunswick in 1958, where within a few years it was spreading rapidly. **Shrews** (*Crocidura* and *Suncus*) may have been introduced to some of the Moluccas in Indonesia, and they occur on most or all of the Lesser Sunda Islands east to Timor; at least one (*Suncus*) often lives in houses and, according to Carter, Hill and Tate (1945) in *Mammals of the Pacific world*, 'is carried about on native boats like the house mouse'.

The **Siberian mole** (*Talpa altaica*) was extensively transplanted – presumably as a source of fur – in western Siberia between 1929 and 1942.

NOTES

1. For much of the history of the hedgehog in New Zealand I am undebted to Wodzicki (1950).
2. See p. 11.

Crab-eating Macaque

(*Macaca fascicularis*)

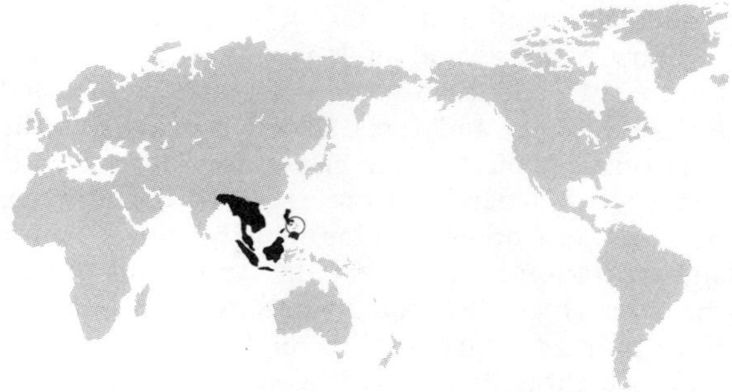

NATURAL DISTRIBUTION
Southeastern Asia from northern India to Burma, Thailand, Vietnam, Cambodia, Malaysia, Singapore, the Philippines and Indonesia.

NATURALIZED DISTRIBUTION
Asia: Micronesia (Angaur Island); Africa: Mauritius; North America: USA.

ASIA

Micronesia (Angaur Island, Palau)

Crab-eating macaques are believed to have been introduced to the tiny island of Angaur in the Palau group by German phosphate miners between 1900 and 1914. Poirier and Smith (1974), who carried out a survey of the population in 1973, were told by an old man that he distinctly recalled a pair imported by German miners, and that, as far as he knew, no more were subsequently introduced.

During the period of Japanese occupancy between 1914 and 1945 some macaques were killed for food, sport and research. Over the years the environment on Angaur has undergone a drastic alteration, due largely to mining by the Germans, building and other development by the Japanese and saturation bombing by the Americans during the war which devastated the rainforest areas, all but obliterating arboreal cover for the monkeys, who are thought to have taken refuge in caves. The annual rainfall on Angaur is high – 350 cm a year being not unusual.

Since the war two major typhoons – the most recent in 1964 – further reduced the rainforest cover. Such is the macaques' hardiness and adaptability to new conditions, however, that in 1973 Poirier and Smith estimated the population to number between 480 and 600 – all reputedly the descendants of the single original pair.

The macaques' preferred habitat is secondary-growth forest; they sleep in riverine trees and seldom travel more than 1,000 m to forage. They are primarily vegetarian, but probably also consume termites, dragonflies and grasshoppers. Poirier and Smith observed no crab-eating, which is said to take place only in coastal populations who dig out and eat crabs on the sandy beaches.

According to Roots (1976), crab-eating macaques have also been introduced to Celebes (Sulawesi) and the Lesser Sunda Islands.

AFRICA

Mauritius

Crab-eating macaques introduced to Mauritius were in 1979 estimated to number between 12,000 and 15,000. They pose a serious threat to several endemic and endangered birds: e.g. the Mauritius pink pigeon (*Columba mayeri*) which, although Roots wrote in 1976 that it had been exterminated, had in fact a population of about 50 in 1980; the Mauritius echo parakeet (*Psittacula echo*); the Mauritius kestrel (*Falco punctatus*) – of which only about 20 remain; the Mauritius cuckoo-shrike (*Coracinia typica*); the Mauritius fody (*Foudia rubra*); a weaver; and a flycatcher – known locally as the *coq-du-bois*. With some of these birds they compete for food and on all they sometimes prey. Crab-eating macaques are, in addition, said to have been partially responsible (with man) for the extermination of the dodo in about 1681 and of the Mauritius blue pigeon in 1826.

NORTH AMERICA

USA

On several occasions free-ranging colonies of crab-eating macaques (and rhesus macaques (*M. mulatta*)[1]) have been established in North (and Central) America for tourism and medical and behavioural research. One such colony of *M. fascicularis* is established as a tourist attraction in the Monkey Jungle near Miami, Florida.

NOTE

1. See p. 31.

Rhesus Macaque
(*Macaca mulatta*)

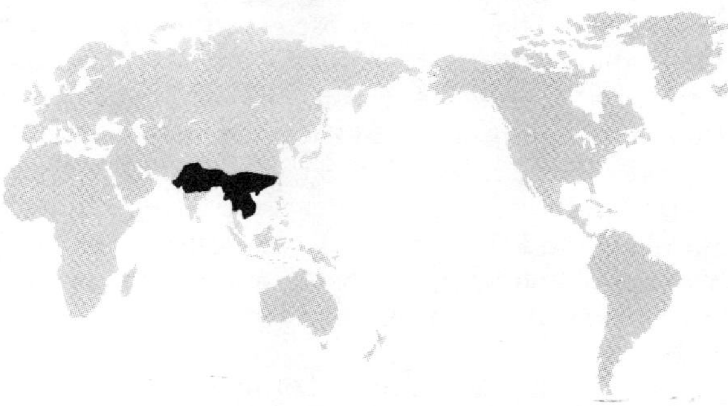

NATURAL DISTRIBUTION
From eastern Afghanistan through northern India to southern and eastern China.

NATURALIZED DISTRIBUTION
Asia: China; North America: USA, West Indies (Puerto Rico); South America: Brazil.

ASIA

China

An isolated colony, probably the result of an introduction by man, exists near Peking (Beijing) in northern China.

NORTH AMERICA

USA

In 1933 two pairs of rhesus macaques were imported from India to Silver Springs, Florida to provide authenticity during the filming of *Tarzan Finds Boy*. They were subsequently released on the right bank of the Silver Run River, which they later crossed. A census in 1968 indicated a total of 78 animals divided approximately equally between both sides of the river. By 1983 there were about 65 rhesus macaques and squirrel monkeys (*Saimiri* sp.)[1] on the south bank and around 200 on the north bank, and the population was believed to be doubling every five or six years.

The monkeys are now a familiar sight to game wardens, fishermen and visitors in the dense forest and marshland bordering the Oklawaha River, which joins the Silver Run about 6.5 km downstream from Silver Springs; reports range from 17.5 km north of the confluence of the two rivers to around 28 km to the south. Most of the sightings occur within a few hundred metres of the Oklawaha's banks, but at least one individual has been reported 8 km away at Fort McCoy. In spite of the fact that several have been shot, these two scattered colonies appear to be thriving, and provide a valuable tourist attraction. Because of the slow reproductive rate of the higher primates they are unlikely to increase or spread quickly, and pose little if any economic or ecological threat. For food and shelter they appear to favour white ash, maple and cypress trees; they also eat the leaves of poison ivy, wild plums, holly berries, elderberry blossom and insects.

In 1947 the March of Dimes Foundation established the Okite Farm for polio research near Blufftown in South Carolina: the farm held a stock of monkeys which varied between 3,000 and 7,000 – mostly rhesus, but also crab-eating (*Macaca fascicularis*),[2] baboons (*Papio* sp.) and spider monkeys (*Ateles* sp.). None ever became established locally, but in 1956–7 a colony of rhesus is said to have existed on Hilton Head Island, some 16 km offshore from Blufftown; none remained in 1974.[3] In the late 1970s a population of free-ranging rhesus macaques was established on Loggerhead Key, Florida, where they were bred for cancer research.

West Indies

A free-ranging population of rhesus macaques, introduced for behavioural studies, is established at the Caribbean Primate Research Center on Puerto Rico. During the Second World War a large number were released for pathological research on Morrillo del Diablo Key near the Isle of Pines, Cuba. After the abandonment of the project in the 1950s some of the monkeys swam to the Isle of Pines in search of further supplies of food, where they failed to become established.

SOUTH AMERICA

Brazil

In the 1940s about 300 rhesus macaques were released for research into yellow fever on Ilha do Pinheiro in Guanabara Bay, Rio de Janeiro; in 1974 around 100 remained at large.

Japanese macaques(*Macaca fuscata*) are established at La Moca, Texas, USA. **Formosan (Taiwan) macaques** (*M. cyclopis*), which escaped from a zoo in about 1942–3, are well established on Oshima (and on another island near Shikine in the Izu group) south of Tokyo, Japan, where they may also exist on the mainland at Kiyozumi, Chiba. Macaques (? sp.) are also said to have been introduced to Batjan Island in the Moluccas and to Timor. **Celebes macaques** (*M. nigra*) have been introduced to the island of Amboina (Ambon) 800 km to the east, and **pig-tail macaques** (*M. nemestrina*) to the Andaman Islands in the Indian Ocean. In January 1978 some **bear** or **stump-tailed macaques** (*M. arctoides*) were landed on Hall's Island in Harrington Sound, Bermuda, for behavioural studies and scientific research by Rutgers University, USA; in May 1983 the free-ranging population numbered 10.

NOTES

1. See p. 37.
2. See p. 29.
3. Hausfater & Walker (1974).

Barbary Ape

(*Macaca sylvanus*)

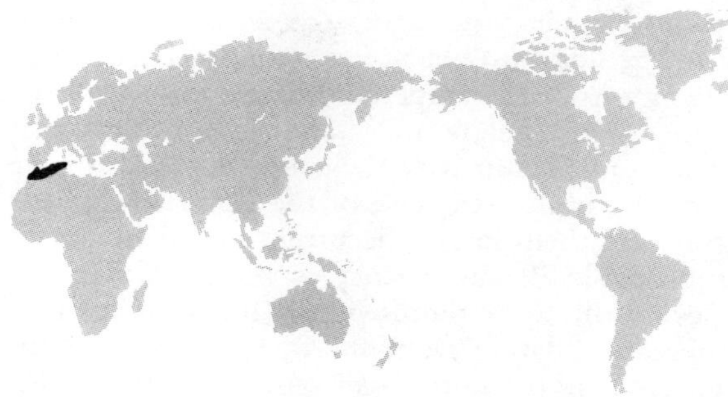

NATURAL DISTRIBUTION
Morocco and northern Algeria.

NATURALIZED DISTRIBUTION
Europe: Gibraltar, ? France.

EUROPE

Gibraltar[1]

The Barbary apes or macaques[2] on Gibraltar are the oldest established colony of free-ranging monkeys in Europe. Exactly when they were first introduced to the Rock is uncertain, but they are known to have resided there for at least the past 240 years.

Two theories have been put forward to account for their presence on the Rock. From fossil deposits it is known that monkeys attributable to the genus *Macaca* occurred in Europe north of the Alps (including parts of southern and eastern England) from the early Pliocene (about 5 million years ago) to the last interglacial of the Pleistocene (about 150,000 years ago.) Zeuner points out that from geological evidence it is known that monkeys returned to temperate Europe in each of the three interglacial periods, and he considers that there is no reason why they should not also have come back after the end of the last glaciation or to regard their natural occurrence on Gibraltar as improbable. If this theory is accepted then the apes on the Rock represent a remnant indigenous population dating from the Pleistocene. The absence of any definite fossil remains in the otherwise rich deposits on the Rock, however, makes this hypothesis unlikely.

The second, and more likely, supposition is that the apes were introduced by man; although they are not referred to on the Rock in Phoenician, Roman or Moorish times, the most credible – though admittedly speculative – theory is that they were originally imported from North Africa by the Moors who, under the Saracen Tariq, captured and fortified the Rock in AD 710–711. Since, however, no mention of them is made during the period of Spanish occupancy between 1492 and 1704, it is possible that the Moorish importation died out, subsequently being replaced by new stock.[3]

The earliest positive accounts of monkeys on Gibraltar date back only to 1740,[4] when mention is made of a large importation of apes 'and other game from Barbary', and a poll tax was levied on 'apes, Moors, Jews and other aliens'. An entry in the governor's journal dated 27 November 1749 granted permission to three members of the garrison – Sir George Sutton, Major Lambert and Captain Morgan – to shoot game on the hill 'they taking care to supply it with game in proportion to the quantity they destroy'. Just over a hundred years later, in 1856, when the population had fallen to only four or five, a Garrison Order prohibited the killing of the monkeys for any reason; nevertheless, seven years later only three remained on the Rock.

To save the apes from extinction on Gibraltar the then Governor, General Sir William Codrington, imported a fresh supply from North Africa, as is shown by the following extract from his journal on 26 May 1865: 'Turned out four apes, wild from Barbary, two males and two females, all young.'[5] (The *Gibraltar Directory* of 1898 records the presence of a dozen apes on the Rock in 1864; this is presumably either an error or includes some retained in captivity.) In June 1875 an official complaint was lodged that up to 60 apes were stealing fruit from the trees, tearing stones from the garrison's walls, and breaking wooden railings and roof gutters from the house of a naval officer, and even boarding ships in the harbour. A colonel in the Royal Engineers submitted an official claim to the Exchequer alleging that a group of apes stole his fruit, dug up his potatoes, frightened his children, tore his trousers and slept in his bed! These depredations became more frequent as increased building on the lower parts of the Rock forced the apes to the less-hospitable regions of the Upper Rock, where vegetation was sparse.

In March 1880 the British Admiralty forbade the destruction of the entire population – though the governor was empowered to issue a permit allowing a maximum of three to be killed 'to commence with' – and a warden was employed to try to keep the animals away from residential areas. Towards the end of 1893 Mr P. L. Sclater read a communication from the governor to the Zoological Society of London reporting a further increase in the Rock's ape population, which in 1900 was claimed to have reached a total of 130.

Probably because food (principally corms of the dwarf palm (*Chamaerops humilis*), roots, nuts, berries, fruit and leaves) on the Upper Rock was insufficient to sustain such a large population (around 200 in 1910), the apes began once again to raid the 16 ha of cultivated gardens, vineyards and orchards of the Lower Rock. In 1913 an unsuccessful attempt was made, by appointing a master gunner of the Royal Garrison Artillery to provide them with food, to enduce the apes to keep to the Upper Rock.

Complaints of depredations continued to be received and a proposal was put forward to reduce the population to 12. In 1915 an 'Officer in Charge of Apes' was appointed to supervise, feed and control the population – an appointment now automatically assumed by the officer commanding the Gibraltar Regiment. An increase in the feeding allowance, together with the enactment of an ordinance in December 1918 which prohibited 'encouraging Rock Apes to come down from the Upper Rock or feeding them at any place other than' at the Queen's Gate, did little to put an end to the problem. In 1923 permission was sought, and fortunately refused, to transfer the remaining apes on the Rock to North Africa.

Between 1920 and 1939 the population fluctuated considerably (to a low of only four in 1924), possibly through intertroop competition, but three years later had risen again to 27, and an unsuccessful attempt was made, on instructions from the governor, to reduce the population to 10. In 1928 a male was captured and sent to the London Zoo, and others were subsequently despatched to Bristol and other British zoos and to Dehiwela in Ceylon (Sri Lanka).

By 1931 only 10 apes remained on the Rock, and no breeding took place in either that or the following year. After consultations with the Zoological Society of London the then governor, General Sir Alexander Godley, arranged for the importation of two young animals from North Africa in 1931 followed by a further five from Tangier, Morocco, in the following year. In spite of these injections of new blood the population only numbered 11 on the outbreak of war in 1939, and by 1943 had fallen to only 7, one of which – a young female – had been imported from North Africa in the preceding year.

Mindful, perhaps, of the tradition that if the apes ever died out the British would leave the Rock, the British Prime Minister, Mr Winston Churchill, gave instructions for the population to be reinforced, and between 1942 and 1945 a male and six females were imported from North Africa. Despite several deaths from pneumonia during the hard winter of 1946, which reduced the population to 16, numbers had increased after a further two years to 24. The new population was split, as it is today, into two troops, centred on Queen's Gate and Middle Hill. Both forage for roots and berries on the Upper Rock, but are also fed twice daily, at 0900 hours and 1600 hours, by the army, which reduces raids on inhabited areas to a minimum. Each animal is named, and a roll-call is taken at every assembly.

Between 1948 and the present day the population has fluctuated between 24 and 40. The Queen's Gate troop is an important tourist attraction drawing an audience of up to 1,000 visitors a day. The extent to which human influence and manipulation (by culling aggressive and adventurous individuals) are altering the characteristics of Barbary apes on the Rock is important, since a balance is desirable between their conservation and the maintenance of their genetic integrity. With births exceeding deaths annually, some culling is essential; between 1949 and 1980, 76 (31 per cent) of young of the year were exported, and a further 10 were culled. As a gene bank the Rock contingent may, however, be declining in importance, as a result of breeding successes in captivity in two 11- and 20-ha parks in France – the Montagne des Singes, Kintzheim, and the Parc de Rocamadour – and one at Salem in West Germany started by Baron Gilbert de Turkheim, where a combined total of around 200 monkeys are born each year. This may result in the necessity of more extensive culling in Gibraltar in the future.

NOTES

1. For much of the history of Barbary apes on Gibraltar I am indebted to Zeuner (1952) and Fa (1981).
2. Although normally referred to as an 'ape', *M. sylvanus* is in fact a true tail-less monkey – a macaque. It is the only one of 13 species to occur outside Asia.
3. A third theory – that the apes came from Africa by way of an underground tunnel which emerged in St Michael's Cave on the Upper Rock – can, perhaps, be discounted!
4. F. E Zeuner, on the other hand, considers that 'It is certain that apes were present on the Rock when Gibraltar was captured by the British in 1704'.
5. Anon. Sketches of natural history in Gibraltar. *The Field* 10 July: 79 (1880).

Green Monkey

(*Cercopithecus aethiops*)

NATURAL DISTRIBUTION
From Senegal and Somalia to South Africa.

NATURALIZED DISTRIBUTED
North America: West Indies (St Kitts, Nevis, Barbados).

NORTH AMERICA

West Indies

Neither R. P. J. B. Du Tertre, in his *Histoire générale des îles de S. Christophe, de la Guadeloupe, de la Martinique, et autres dans l'Amerique* (1654), nor C. C. Rochefort, in *Histoire naturelle et morale des îles de l'Amerique* (1658),[1] make any reference to the existence of green monkeys – also known as savannah, vervet or grivet monkeys – in the West Indies. J. B. Labat, however, in his *Nouveau voyage aux îles de l'Amerique* (1722),[2] describes how green monkeys – introduced as pets to St Kitts and Nevis by the French – escaped and became established in the wild during the Anglo-French wars of the second half of the seventeenth century.

Although R. Ligon, in *A true and exact history of the Island of Barbados* (1673), makes no mention of green monkeys on Barbados, G. Hughes in *The natural history of Barbados* (1750) wrote:

> they are not very numerous in this Island; they chiefly reside in inaccessible Gullies; especially where there are many Fruit trees. The great mischief they do to the neighbouring Planters is digging out of the Earth their Yams and Potatoes, and sometimes breaking and carrying off a great many ripe Sugar-canes. As a Law of this Island provides a Praemium for destroying these, as well as racoons [*sic*], they yearly rather decrease than multiply.

On St Kitts green monkeys have been particularly damaging to the island's birdlife, and are believed to have exterminated the St Kitts bullfinch (*Loxigilla portoricensis grandis*), a subspecies of the Puerto Rican form. They seem fortunately not to have affected the Lesser Antillean bullfinch (*L. noctis*) on Barbados.

Mona monkeys (*Cercopithecus mona*) were introduced from Cameroun in West Africa to Grenada and St Kitts in the Lesser Antilles, possibly by nineteenth century slave-traders. On the former they became established in small numbers in the dense inland primeval forests of Grand Etang.

Squirrel monkeys (*Saimiri* sp.) from Central or South America have been introduced as a tourist attraction to the same area of the Silver Run River around Silver Springs in Florida as the rhesus macaque (*Macaca mulatta*).[3] They too have moved downstream from Silver Springs to the Oklawaha River, where they appear to occupy the same range as *mulatta*. Although Roots (1976) claims that squirrel monkeys of Costa Rica and Panama are thought to have been introduced by man, the red-backed squirrel monkey (*S. oerstedi*) is in fact indigenous to the region. The common squirrel monkey (*S. sciureus*) ranges from Columbia to the basin of the Amazon.

The **brown pale-fronted capuchin** (*Cebus albifrons*), a native of Venezuela and the upper Amazon, has been imported to the island of Trinidad (and also, according to Roots (1976), to Ecuador); he also claims that **langurs** (*Presbytis* sp.) of south-east Asia have been introduced to Lombok Island, Indonesia.

Monkeys (? sp.) – known locally as 'Nanukrawa' and said to have been introduced by the Japanese to New Britain, Papua New Guinea, during the Second World War as a potential source of food – are established in the vicinity of Aipati, Aimaga and Siac villages on the limestone plateau country of Cape Gloucester.

In June 1972 one male and three female **chimpanzees** (*Pan troglodytes*) were released on Bear Island near Ossabaw Island off the coast of Georgia, USA, to assess the feasibility of establishing a free-ranging and breeding colony of the species in subtropical North America. Following the death of two of the females a further four (one of which bred successfully) were introduced, and three years later the troop was still in existence.

Between 1975 and 1977 a free-ranging colony of **gibbons** (*Hylobates* sp.) was established on Hall's Island in Harrington Sound, Bermuda, for scientific and behavioural studies by the International Psychiatric Research Foundation and the Rockman Research Institute, USA.

Three monkeys (? sp.) introduced to the island of Floreana in the Galápagos group in the 1930s soon disappeared.

NOTES

1. Translated by J. Davies as *The history of the Caribby-Islands* (1666).
2. Vol. 5, pp. 7–8.
3. See p. 32.

Red Fox
(*Vulpes vulpes*)

NATURAL DISTRIBUTION
Most of the Palaearctic region from North Africa, Arabia, central India and Vietnam in the south, north to the Arctic coast, and North America south to northern Florida and New Mexico.

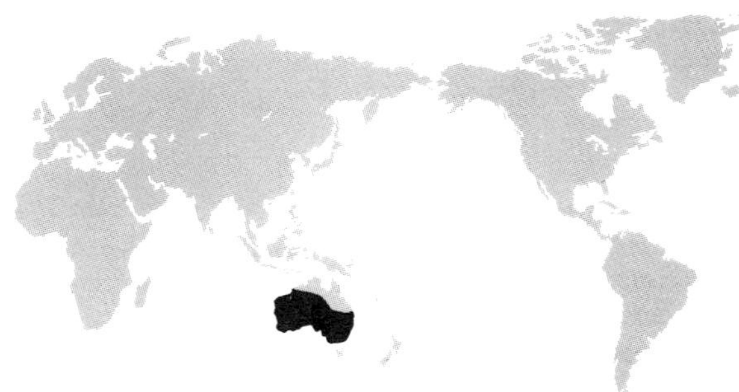

NATURALIZED DISTRIBUTION
Australasia: Australia. (Reinforced in the British Isles from continental Europe; 'silver' colour phase imported to Finland; introduced from England to the USA to establish populations in new areas.)

AUSTRALASIA

Australia[1]

Foxes were imported to Australia from England in the nineteenth century, mainly by hunt clubs, for the express purpose of hunting, and not for rabbit control. The earliest release seems to have taken place on the property of Mr T. H. Pyke near Keilor, Melbourne, Victoria, in 1845. In

November 1890 the Sydney *Echo* described the safe arrival of two dog foxes (out of a shipment of six) from England in 1855, and in the *Australasian* of 25 April 1868 Gideon Long – probably referring to one dog fox and two vixens brought out from England by Mr George Watson on the *Suffolk* in 1864 and subsequently released by the Melbourne Hunt Club – expressed the hope that 'They may help in due time to keep in check the kangaroo as well as the pheasants and the rabbits'. In 1868 the Melbourne Hunt released another fox to join those already in the wild, and more were subsequently turned down by the Flemington Hunt.

In July 1869 Mr Arthur Bean returned to Adelaide, South Australia, from England on the *Pekina* with two fox-cubs, which the captain described as 'regular bull-terrier foxes'. Bean was no hunting man, and refused to give them to the Adelaide Hunt. What he did with his cubs is unknown, but in October 1888 four foxes and 'a startling number of tracks' were seen along the Coorong.

None of these early introductions was apparently very successful, and it seems likely that the fox population of Australia today is mainly descended from two subsequent liberations – both in Victoria. In 1871 'a well-known sportsman and acclimatiser' – a Dr King – released a pair from England 20 km from Ballarat, and in the early 1870s Mr Thomas Chirnside – one of Australia's most successful pioneer pastoralists who arrived in the colony at the age of 20 in 1839 – turned down several on Point Cook, from where they spread by 1878 to the property of a Mr Phelps – 'a good game preserver' – at Laverton. From there they moved on to Braybrook and the Little River at Bird Rock. By 1880 they had reached the north shore of Corio Bay, 64 km from Keilor, and shortly afterwards joined forces with those released at Ballarat by Dr King. In the same year Mr Robert Hood of Merrang reported in the *Warrnambool Standard* the sighting of foxes on the Hopkins River, 96 km west of Ballarat:

> As my brother and I were beating some long grass on the banks of the Hopkins for rabbits, I noticed the dogs feathering and hunting differently from what they do on either hare or rabbit. I called to my brother 'There is something strange here'. Just then he saw something stealing away up the river's bank. I have heard rumours of their being seen before in this neighbourhood. I hope there will be no mercy shown them.

Thus, by about 1880 foxes had spread over some 13,000 km^2 of Victoria and were established in parts of South Australia. The history of their dispersal over the rest of the country is uncertain: dates are available of when they were first noticed in some districts, but there is no evidence to show whether they arrived by natural expansion of range or were translocated by man – possibly to control rabbits. Thus, in 1886, they were seen at Bendigo in Victoria, and seven years later a bounty scheme was announced by the shires of Euroa, Benalla and Shepparton. In the same year (1893) foxes had moved north to the Murray River, where they were first reported at Bundalong, 16 km east of the present site of Yarrawonga. A few months later Mr E. D. O'Dwyer of Rockwell, Savernake, some 32 km to the north-west, trapped one of the first foxes in New South Wales (whose pelt was displayed in the Momalong Hotel in Berrigan) and by 1900 they had reached Cobar and Armidale in the same state.

By 1911 foxes were reported to be established in southern Queensland; eight years later they extended from the McPherson Ranges and the Albert River in the south to the Burnett at Bundaberg on the coast and as far west as Charleville. In 1920 one was seen at Longreach, around 1,600 km north of Melbourne, and by 1933 they had reached Julia Creek, some 320 km to the north-west, which remains approximately their present northeasterly limit.

In Western Australia foxes were first reported near and to the west of Eucla in 1911–12, and within a further four years they had penetrated 160 km west of the South Australian border. In 1916 one was seen at Esperance and another at Sandstone, 400 km north-west of Kalgoorlie, having apparently followed much the same line of invasion from Victoria via South Asutralia as the rabbit some 17 years earlier. By 1925 foxes were distributed along the coastline of Western Australia north to Geraldton but had not spread very far inland; only between Mingenew through Geraldton to Northampton were they to be found in any numbers.

After 1925 the spread of foxes in Western Australia – especially in pastoral regions – grew apace; by 1931 they occurred in nearly every district except Marble Bar (which they reached in the following year) and the Kimberleys. By 1934 they were well established from the extreme south-west (where they were first seen at Manjimup) to

Derby in the north, whence they spread inland to the east Kimberleys by 1943, and to north of the Fitzroy River in the west Kimberleys by the early 1950s. In December 1968 one was trapped on Lissadell Station in the east Kimberleys, but in general foxes have not flourished in the Kimberleys as elsewhere in Western Australia, where they are now otherwise widely distributed. The decline of the brush-tailed or black-flanked western rock wallaby (*Petrogale penicillata lateralis*), the crescent nail-tailed wallaby (*Onychogalea lunata*), last recorded in 1930 in Western Australia, and the endemic mallee fowl (*Leipoa ocellata*), have all been attributed to predation by foxes and feral cats.

In about 1890 two British army officers stationed at Anglesea Barracks in Hobart, Tasmania, imported a pair of foxes from England for hunting: they were, however, fortunately destroyed by the authorities before they could be released, and there are no foxes in Tasmania today.

Foxes are currently found throughout mainland Australia except in northerly parts of Queensland and the Northern Territory. They occur in all habitats except dense tropical rainforests, and at all altitudes. In spite of the fact that Australia's native animals are ill-adapted to predation from a new carnivore, the effect foxes have had on the country's wildlife is uncertain. They prey on, among other species, superb and Prince Albert's lyrebirds (*Menura novaehollandiae* and *M. alberti*), brush turkeys (*Alectura lathami*), Australian bustards (*Ardeotis australis*), pectoral quail (*Coturnix pectoralis*), brown or swamp quail (*Synoicus ypsilophorus*), curlews (*Numenius* spp.), spotted nightjars (*Eurostopodus guttatus*), Richard's pipits (*Anthus novaeseelandiae*), singing bush-larks (*Mirafra javanica*), plovers (*Charadrius* spp.), red-necked avocets (*Recurvirostra novaehollandiae*), Australian stilts (*Himantopus leucocephalus*), brolgas (*Grus rubicunda*), the eggs of emus (*Dromaius novaehollandiae*) and black swans (*Cygnus atratus*), marsupial mice (Dasyuridae), bandicoots (*Perameles* and *Isoodon* spp.), the duck-billed platypus (*Ornithorhynchus anatinus*), boodies (*Bettongia lesueurii*), rabbits (*Oryctolagus cuniculus*), small reptiles and amphibians, and small invertebrates. Foxes may, indeed, have had a more damaging affect on Australia's fauna than even the introduced dingo (*Canis familiaris dingo*);[2] yet as Frith (1979) points out, in some places, such as the woodlands of the southeast and the northeast highlands of New South Wales, both foxes and the smaller native mammals are abundant. It seems possible that foxes are a greater threat to man's domestic stock than to Australia's wild fauna. Analysis of stomach contents suggests that sheep (as carrion), poultry, rabbits and invertebrates form foxes' staple diet. Opinion is divided on their role as predators of newborn lambs; some foxes undoubtedly do kill lambs (and some surplus killing is known to occur), but most appear to eat mainly already dead lambs and afterbirth. Research has shown that foxes are the intermediate hosts of tapeworms, which have caused the rejection of much of Australia's exported meat.

Were it not for an Act of 1867 which prohibited their importation, it is likely that foxes would have had a similar affect on the wildlife and domestic stock of New Zealand as they have had in Australia.

Although the red fox is indigenous in the British Isles, the population has from time to time (certainly until at least 60 years ago) been reinforced for hunting purposes by importations from continental Europe. Writing in 1884 E. N. Buxton refers to foxes from Spain being turned down in Epping Forest, Essex. In *The influence of man on animal life in Scotland* Ritchie (1920) wrote: 'in several districts of South Britain, foxes, and badgers also, are actually imported from the Continent and from Scotland, and are set free to establish themselves in the hope that the hunt may benefit thereby'. This is reiterated by Harrison Matthews (1952) who, in *British mammals*, wrote that 'Importations of foxes have been made from the continent into various parts of England, and also from one district to another within the country, especially from those parts of Scotland where foxes are vermin', but he does not make clear to what period he is referring.

Silver foxes – a colour phase of the red fox – were in 1938 imported into Finland from Alaska in the expectation that they would interbreed with the type species and thus improve fur quality.

According to Gilmore (1946) 'There is strong evidence that the native American red fox [*V. fulva*] did not range into the middle eastern United States from the north in aboriginal times, and that the present red fox of the central eastern states is a direct descendant of the [conspecific] European red fox (*Vulpes vulpes*), which was introduced between 1650 and 1750 from England for fox hunting'.[3] According to the *Maryland Conservationist*,[4] the *Centreville Record* claimed that in about 1650 the master of an English tobacco schooner imported the first European foxes (eight pairs, from Liverpool) to North America, which were released near the coast of Maryland, where they multiplied 'marvellously' but did not spread into Virginia until the hard winter of 1679–80. Lloyd (1980), however, in *The red fox*, says that foxes were turned down – not to supplement native stock but to establish populations in new areas – in Maryland in the mid-eighteenth century, and subsequently on Long Island and in New Jersey, Virginia and other eastern states, where they became established. As woodlands were cleared for agriculture the English foxes spread west until they encountered the native *fulva*, probably south of the Great Lakes. Hall (1981) suggests as 'a tempting speculation' the reason for the subsequent movement further westward of the red fox:

> On the Midgrass and Shortgrass Prairies of the High Plains wolves were abundant as long as natural mortality in the bison herds provided most of the wolves' food. When the wolf population declined, the coyote population increased. When the coyote population decreased in the 1900s the red fox moved westward.[5]

It seems likely that foxes from England have had some influence on the genotype which exists in the eastern states today. Red foxes now occur all over North America except for southern Florida, most of Arizona, and a band running approximately north–south from Alberta through much of Montana south to northwestern Texas and Mexico.

Arctic foxes (*Alopex lagopus*) of the Eurasian and North American tundra were originally translocated to the Aleutian Islands[6] off the southern coast of Alaska for fur-farming in 1886 by the Russian–American Company. Pairs were landed on many islands – especially those with large colonies of seabirds to provide abundant food – where they were left to breed, surplus stock being subsequently culled. By 1925 foxes had been planted on no fewer than 77 islands, and by 1936 more than 25,600 pelts had been harvested. The economic depression of the 1930s and the outbreak of the Second World War in 1939 together effectively destroyed the market for furs. No longer controlled by trapping, foxes in the Aleutians rapidly increased, and the populations of many endemic birds were either exterminated or endangered.

Large colonies of ancient murrelets (*Synthliborampus antiquus*) and Cassin's auklets (*Ptychoramphys aleuticus*) disappeared from Sanak Island; storm petrels (*Oceanodroma* spp.) were eliminated on Salt and Ilak islands; Cassin's auklets became extinct on Keegalod and Adugak islands; whiskered auklets (*Aethia pygmaea*) were eliminated on the Near islands; on St Lawrence Island crested auklets (*A. cristatella*), parakeet auklets (*Cyclorrhynchos psittacula*), horned puffins (*Fratercula corniculata*) and tufted puffins (*Lunda cirrhata*) are heavily preyed upon by foxes; on Big Koniuji Island 13 foxes slaughtered a total of 858 auklets and puffins in a period of 12 weeks; a recent survey of more than 100 fox-infested islands south of the Alaskan peninsula revealed a complete absence of nocturnal seabirds. The most serious ecological problem in the Aleutians, however, is the near extinction, mainly through predation by foxes, of the endangered and distinctive Aleutian Canada goose (*Branta canadensis leucopareia*), whose breeding range is now confined to the 1,720-ha island of Buldir.

In the absence of native terrestrial predators, and in an attempt to restore the islands' ecosystems to their original state, a proposal was put forward in 1980 to introduce to the Aleutians sterile red foxes – which are competitively superior in sympatric mainland habitats – as a form of biological control.

Arctic foxes have also been introduced to improve the fur harvest on the Kola Peninsula in the northwestern Soviet Union; on several of the Kuril Islands between Hokkaido and Kamchatka; on the Commander (Komandorskie) Islands between Kamchatka and the Aleutians; and on the Pribilof Islands in the Bering Sea.

On the Falkland Islands the **wolf** *Dusicyon australis*, which was not in fact a wolf although it bore a closer resemblance to one than did any other member of the genus, was exterminated by fur-trappers and sheep-farmers between 1873 and 1876. Charles Darwin, who visited the islands in 1833 and 1834 and referred to it as *Canis antarctica*, reported that it was the only native land mammal, and that it preyed largely on birds. It seems to have been either a relic of a domesticated form of *Dusicyon* or a feral hybrid which evolved from a cross between a domestic *Canis* and a *Dusicyon* species. It is improbable that a single mammal (and one which was a large carnivore) could have survived on the islands throughout the Pleistocene in a periglacial region. Since the shortest distance between the South American mainland and the islands (nearly 500 km) is too great to make emigration by 'rafting' seem likely, *D. australis* was almost certainly introduced to the Falklands, where it became known as the 'warrah', during the early Holocene by South American Indians – probably from the Chilean side of Tierra del Fuego, since all the early archaeological sites known on the continent's southern tip occur in that country.[7] The Falklands are thus the only subantarctic islands whose fauna have evolved in the presence of a long-established mammalian predator.

In the late 1940s the **Argentine grey fox** (*Dusicyon griseus* (*fulvipes*)) – also known as the Patagonian fox – was released on Tierra del Fuego to reduce the numbers of introduced rabbits; predation on the ruddy-headed goose (*Chloëphaga rubidiceps*) has been high, and the increase in the fox population has been accompanied by a decrease in that of the goose, due apparently to their similar preference for grassland habitats. In the early 1930s an *émigré* Scottish shepherd, Mr John Hamilton, released some Argentine grey foxes on Weddell, Statts and Beaver islands off the west coast of West Falkland: today they still occur on these three islands and also on Tea Island in the Weddell group, as well as on River Island near Pebble and on Split near Roy cove. According to Dr R. S. Whitley the foxes have a marked influence on some species of birds, in particular on the upland goose (*Chloëphaga picta*), which on these islands nests only on isolated outcrops which foxes cannot reach. Outside the nesting season the foxes feed largely on limpets, crabs and fish – especially 'dogfish' (*Harpagifer hispinis*) – and in autumn on 'diddle dee' berries. Some sheep-farmers are said to have sustained losses due to fox predation during the lambing season.

NOTES

1. For much of the history of foxes in Australia I am indebted to Rolls (1969) and Frith (1979).
2. See p. 47.
3. See also S. N. Rhoads, *The mammals of Pennsylvania and New Jersey* (privately published, 1903).
4. Vol. 7, No. 2.
5. Vol. 2, p. 938.
6. For much of the information on Arctic foxes in the Aleutian Islands I am indebted to E. W. West & R. L. Rudd ('Biological control of Aleutian Island Arctic fox', preprint).
7. J. Clutton-Brock, *Science* **197**: 1340–2 (1977).

Raccoon-dog

(*Nyctereutes procyonoides*)

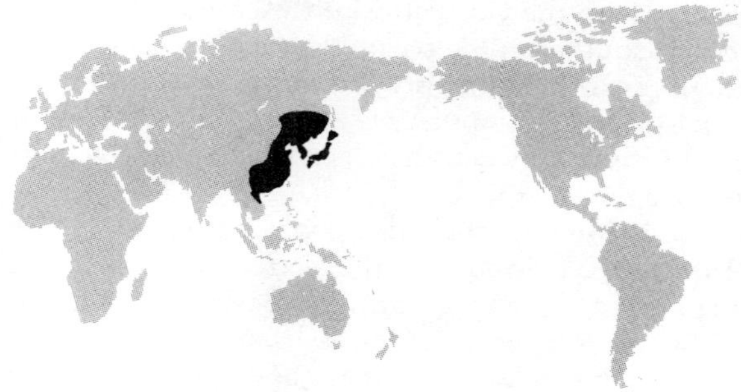

NATURAL DISTRIBUTION
Eastern Asian woodlands from the River Amur to Yunnan and North Vietnam, west to Shansi and east Szechuan, and all the main islands of Japan where, however, it is declining.

NATURALIZED DISTRIBUTION
Europe: USSR, Finland, Sweden, Romania, Poland, Hungary, Germany, Austria, Czechoslovakia, Bulgaria, Switzerland, the Netherlands. (Appearance in Mongolia in the early 1960s was probably the result of natural extension of range from western China.)

EUROPE

USSR

Raccoon-dogs of the Siberian race *ussuriensis* were first imported from the Far East to European Russia in 1928 or 1929, when 415 pregnant females were introduced to the mountain forests of the Teberda Wildlife Reserve in Transcaucasia, to the Gumistinskii Reserve in Abkhazia and to southern Ossetia and Karatalinia; in general these introductions proved unsuccessful, but a small remnant population lingers on in the lowland forests between the Terek and Sulak rivers in the Dagestan ASSR.

In 1934–5 raccoon-dogs were released in the Tatar ASSR and in the latter year also in Leningrad Province and in Bashkiria, and in 1948 in the Mari, Chuvash and Mordvinian ASSRs. From these centres they spread into the neighbouring Ul'yanovsk and Kuibyshev regions where, except locally, they survive only in small numbers. In Kalinin Province, where they were introduced in 1934, they multiplied so quickly that by 1946 the annual harvest of pelts was 150 per cent above that of the Far East.

In the North Caucasus raccoon-dogs were first turned down in 1934, when 1,147 gravid females were released in mountain woodlands bordering the lower reaches of the Volga, Terek and Kuma rivers, and along the Don and its tributaries: a further introduction was made in Ryazan Province in 1936. Raccoon-dogs are now widely distributed from the Volga and the Don to the Caucasus and the Black Sea.

In the Ukraine, where they are almost ubiquitous and cause a certain amount of damage to nesting waterfowl and vineyards in Transcarpathia, raccoon-dogs were first released in 1936.

The Astrakhan Region, where by 1937–9 raccoon-dogs were firmly entrenched and increasing in the Volga Delta, received three importations totalling 324 animals between 1936 and 1939. The value of their pelts does not compensate for the damage they cause to wildlife nor for the reservoir they provide for rabies, which has destroyed large numbers of livestock – mainly cattle and horses – in the area, where by 1966 raccoon-dogs were the dominant wild animal.

Although raccoon-dogs introduced to the Orgeyev Region of Moldavia between 1949 and 1954 initially increased and spread, few survive there today.

Seventy-eight raccoon-dogs liberated in the nut plantations of Arkit in Kirgizia in 1934 all soon disappeared. Ten years later the Kolkhozes freed nine males and six females on the banks of the Issyk-Kul', where, although they established themselves in *Hippophae* thickets, they increased to only about 100–150 individuals. Thirty turned down in 1950 in Susamyr all died in the ensuing hard winter.

Raccoon-dogs have become most successfully naturalized in Belorussia, where 100 released in 1963 in the Lyuban', Parichi and Chechersk districts multiplied rapidly and are currently established throughout the region except in Chechersk. By the mid-1970s the population in Estonia was estimated to number more than 7,000.

Between 1929 and 1967 a total of 8,651 raccoon-dogs were liberated in European Russia, where they are now widespread in a variety of habitats north to around 63° N and south to the Caucasus. In central Russia they produce pelts of some commercial value which, in the colder climate of Siberia, are, surprisingly, said to be of poor quality. In the Far East raccoon-dogs eat mainly small vertebrates – especially fish – and crabs, as well as vegetable matter, including fruit and acorns. In European Russia, where they compete for food with superior-pelted native mustelids – and indeed throughout Europe – raccoon-dogs have largely altered their diet to game birds, hares and domestic poultry.

Finland; Sweden; Romania; Poland; Hungary; Germany; Austria; Czechoslovakia; Bulgaria; Switzerland; Netherlands

Raccoon-dogs first arrived in Finland in 1935 (one account says not until the early 1950s) following a rapid extension of range from European Russia. Most of the early records were from eastern Kymi in south-east Finland, which raccoon-dogs reached by way of the Karelian Isthmus, and in Kainuu and Kuusamo in the east of the country, where they arrived via Russian east Karelia.

As the result of dispersals along the Baltic coast, raccoon-dogs appeared first in southern Sweden in 1945, and subsequently in Romania (1951), in Poland (where they advanced at an average of 135 and a maximum of 300 km a year) in 1955, in Czechoslovakia in 1959 (having travelled north from Poland and south-east from Romania), in Hungary and East Germany in 1961 (though they were rumoured to be in northeastern Carpathia shortly after the war), in West Germany and Austria in 1962 (though there were unconfirmed reports of them in Soltau in Lower Saxony and in the Trauchgau area as early as 1960), and in Bulgaria in 1967.

During the next decade almost the whole of West Germany was overrun, though Turkey, Greece and Yugoslavia in the south-east, Denmark and Norway in the north and France and Benelux in the west had not yet been invaded. Since then raccoon-dogs have spread steadily westwards as far as Switzerland and the Low Countries and north into central Sweden. The preferred avenues of expansion are along waterways affording plenty of cover, and mountainous regions tend to be avoided. Siivonen (1953) believes that the raccoon-dog has travelled westwards in Europe because it was 'originally a member of the fauna of the coastal region of the Pacific Ocean . . . the strong west-bound and north-west-bound distribution . . . may be a sign of its tendency to spread to the Atlantic coast, influenced by the Gulf Stream'.

In 1980 a fur farmer in Ontario, Canada imported 140 raccoon-dogs from Europe. Were any to escape and become established in the wild they could prove a serious threat to native wildlife in North America.

In the early 1960s raccoon-dogs first appeared (probably as the result of an extension of range from western China) in the Peoples' Republic of Mongolia near Lake Buir and Mount Choibalsan.

Dingo
(Canis familiaris dingo)

NATURAL DISTRIBUTION
Originally probably India and southeastern Asia.

NATURALIZED DISTRIBUTION
Australasia: Australia.

AUSTRALASIA

Australia

For many years controversy raged over the ancestry and origin in Australia of the dingo or warrigal.[1] Early authorities maintained that it was indigenous, citing fossil remains found with those of now extinct marsupials (e.g. the large herbivorous *Diprotodon* and the carnivorous *Thylacoleo*) as evidence of its existence in the Pleistocene: this association, however, is now known to be due to redistribution of earth deposits. Another theory suggested that the dingo arrived in Australia from Antarctica via the Bass Strait land-bridge between Tasmania and the Australian mainland, which disappeared around 12,000–13,500 years ago; the absence of dingoes in Tasmania, however, discounts this suggestion. A claim has also been made that dingoes arrived via the Torres Strait land-bridge between Papua New Guinea and Queensland, which was inundated by the sea about 6,500–8,000 years ago; the dissimilarity between the dingo and the wild dogs of Papua New Guinea, coupled with the absence in Australia of other Asiatic placental mammals, makes this theory, too, untenable. In any event, the Torres Strait land-bridge would have been valueless unless similar links had existed between Papua New Guinea and the Australian-Malayan islands and between them and the Indo-Malayan region.

The mammalian fauna of Australia is characterized by a comparative paucity of terrestrial placentals (bats and rodents being the only representatives of this group) and both are believed to have arrived adventitiously after the separation of Australia from the Asian mainland during the Cretaceous period. Apart from the dingo (and such later introductions as domestic dogs, cats and foxes) there are not – and never have been – any terrestial non-marsupial carnivores in Australia from which the dingo could have evolved.

It therefore seems clear that the dingo was introduced as a domestic animal to Australia by Aboriginal voyagers from southeastern Asia, perhaps some 3,000–7,000 years ago, and that it subsequently became feral; its most likely ancestor is the plains wolf of India (*Canis lupus pallipes*). The true dingo, which is morphologically similar to the Indian 'pariah' dogs and the 'singing' dogs of Papua New Guinea[2] and is probably descended from a very small founder population, is a relict which for thousands of years survived in isolation as probably the only 'pure-bred' domestic dog in the world, without the introgression of genes from any other breed until the arrival of the first European settlers' dogs in the late eighteenth century. Professor N. W. G. Macintosh (quoted by Rolls 1969), who examined and dated a 3,000-year old skeleton (the earliest then discovered) found at Fromm's Landing on the lower Murray River, said it had 'precisely the same proportions as those of present-day dingoes'.

The earliest European reference to dingoes comes from the Dutch navigator Jan Carstensz who noted their tracks while exploring the eastern shore of the Gulf of Carpentaria in the *Pera* and *Arnheim* in April 1623. In January 1688 the English privateer William Dampier in the *Cygnet* observed the track of a 'Beast as big as a great Mastiff-Dog' on the north-west coast, where in September 1699 – while conducting a governmental survey in the *Roebuck* – 'my men then saw two or three Beasts like hungry Wolves, lean like so many Skeletons, being nothing but skin and bones'.[3] Referring to the Aboriginals of Cooktown, on the east coast of the York Peninsula, Captain James Cook, who later observed Aboriginal women suckling dingo pups, noted in 1770 that 'Tame Animals they have none but Dogs . . .'.[4] The first settlers at Port Jackson saw dingoes on their arrival in 1788, and in the following year John Hunter (later vice-admiral, and governor of New South Wales 1795–1801) shipped a half-grown dingo on his vessel the *Sirius* as a gift to a Dutch official at the Cape of Good Hope.

Until the importation by European settlers of domestic stock which provided easy prey, dingoes probably never reached sufficient numbers seriously to reduce the populations of any indigenous species. Thereafter, however, they increased considerably, and in many districts were soon taking a heavy toll of sheep, calves, and wild animals. By the middle 1860s, following more than half a century of persecution by man, few dingoes remained south of the Murrumbidgee River, as a result of which the populations of kangaroos, wallabies, pademelons (*Thylogale*), rat-kangaroos, bilbies (*Macrotis lagotis*) and bandicoots noticeably increased. When sheep runs began generally to be fenced and active day-to-day shepherding ceased, the dingo population staged a recovery, with a concomitant decline in other wildlife communities: the extinction on the Australian mainland of the thylacine or Tasmanian wolf (*Thylacinus cynocephalus*) and the Tasmanian devil (*Sarcophilus harrisi*) has been attributed to predation by dingoes.

Today, although they feed largely on kangaroos, wallabies and rats, dingoes – which are scavengers as well as predators – are still, through their depredation of domestic stock, an economic pest in many pastoral regions. They have few, if any, predators apart from man, by whom they are poisoned with compound 1080 (sodium monofluoroacetate), shot and trapped. So-called 'dingo-proof' fences have been erected (e.g. along the border between New South Wales and Queensland) but are only partially effective.

Dingoes, which never bark like a domestic dog but only yelp and howl, do, however, interbreed freely with feral dogs, the hybrid offspring usually resembling dingoes, with short pricked ears, a bushy tail, and a brownish, reddish or yellow (occasionally melanistic, leucistic or mottled) pelage. The general appearance of most dingoes is distinctly 'foxy' – indeed they were much hunted by the early colonists before the introduction in the 1870s by Thomas Chirnside of the European red fox (*Vulpes vulpes*),[5] for which, according to John Gould, they provided 'no mean substitute'.

Dingoes are today widely distributed throughout the mainland of Australia in a variety of habitats (eucalyptus forest, scrub, bush, desert, and savannah comprising mallee, mulga and spinifex grassland) except for parts of central New South Wales and the mid-coastal regions of Western Australia. Many are almost certainly hybrids with feral domestic dogs. It is probable that only in remote areas of the outback and in parts of the Northern Territory do pure-bred dingoes survive.

NOTES

1. An Aboriginal word sometimes also used (especially in the outback) to denote any wild creature.
2. See p. 50.
3. *Voyage round the world* (London, 1697).
4. J. C. Beaglehole, ed., *The journals of Captain James Cook: the voyage of the Endeavour, 1768–1771* (Cambridge University Press for the Hakluyt Society, London, 1955).
5. See p. 39.

Domestic Dog
(Canis familiaris)

NATURAL DISTRIBUTION
Originally descended from various races of the wolf (*Canis lupus*), which ranges over parts of North America, northern Europe and Asia (and possibly also from the golden jackal (*C. aureus*) of Africa, southeastern Europe and India).

NATURALIZED DISTRIBUTION
Now widespread as a result of domestication probably more than 10,000 years ago in Palaeolithic times.

The following are some of the more significant examples of the presence throughout the world of feral dogs, and the effects they have on their ecosystems.

EUROPE

Italy

At the time of writing (1983) the largest population of feral dogs in Europe (believed to number upwards of 80,000) is said to be established in Italy where – hunting principally at night in packs 20–30 strong – they prey mainly on sheep, cattle and deer.

ASIA

India

The 'untouchable' pariah dogs of India are regarded as valuable scavengers in and around human habitation. They are also, however, notorious vectors of rabies.

NORTH AMERICA

Bermuda

Until their removal from the Castle Harbour Islands National Park feral dogs were the principal predators in Bermuda of nesting white-tailed tropic birds (*Phaethon lepturus*).

West Indies

On the islands of Cayman Brac and Little Cayman, mid-way between Jamaica and Cuba, feral dogs have greatly reduced the populations of the ring-tailed iguana (*Cyclura carinata*); on Great Inagua in the Bahamas they are responsible for the extermination of two subspecies of snakes found only there and on Sheep Cay. Feral dogs in Cuba are said to be descended from those introduced by the early Spanish conquistadors under the command of Diego Velazquez in 1511. They hunt in packs, are very fierce, and are reputed sometimes even to attack a solitary man.

AUSTRALASIA

Australia

In all probability domestic dogs arrived in Australia with the first colonists from England in 1788. In Tasmania dogs were originally bred for kangaroo-hunting, but were allowed to run free when the value of skins made dog-breeding no longer economic; as early as 1830 packs of roaming feral dogs are said to have made the streets of Hobart unsafe for pedestrians at night.

Following two affrays between European settlers and Aborigines near Campbelltown, south of Sydney, in 1827, a contemporary newspaper, quoted by Rolls (1969), reported that 'dogs are roving masterless in the bush, killing and scattering sheep, and as their destruction is attended with so much difficulty, nay almost impossibility, as they shun the approach of white men, it is not easy to say when tranquility will be restored to the surrounding flocks'; they or their progeny were shortly said to be pestering horsemen in Sydney even during the hours of daylight.

Today, feral dogs exist in most parts of Australia, though their exact distribution is difficult to determine; this is partly due to the frequent problem of differentiating between feral dogs and dingoes (*C. f. dingo*),[1] since the two animals are known to hybridize freely in the wild. Feral dogs and dingoes are together the most significant predators (89 per cent) of domestic livestock – principally sheep – in Australia.

New Zealand

Predation by feral dogs is believed to have been partly responsible for the extinction in New Zealand (in 1902) of the Auckland Islands merganser (*Mergus australis*).

Papua New Guinea

The 'singing dogs' of the Highlands of Papua New Guinea (so named from their eerie, communal howling) may, like the true dingo, be descended unmixed from a prehistoric population.

OCEANIA

Hawaii

Domestic dogs were probably originally introduced to Hawaii by early Polynesian voyagers. Following the importation of sheep, cattle, pigs, goats and poultry by European settlers, feral dogs became established on all those islands where suitable prey occurred, and even now constitute a threat to livestock, wildlife and even, on occasion, to man himself. Packs of feral dogs are partially responsible for the reduction in numbers on Maui of the Hawaiian race of the dark-rumped petrel (*Pterodroma phaeopygia sandwichensis*); on Maui and Hawaii of the endemic Hawaiian goose or nene (*Branta sandvicensis*); and on Kaui of the endemic Hawaiian duck or koloa (*Anas platyrhynchos wyvilliana*). On Kure Atoll, one of the few breeding grounds of the endangered Hawaiian monk seal (*Monachus schauinslandi*), feral dogs have been eliminated since August 1976; they have yet, however, to be removed from other rookeries, where they take a heavy toll of newborn pups.

Galápagos Islands[2]

It has for long been suggested that domestic dogs were first released in the Galápagos – on Isla Santiago (James) – in 1685 by the Viceroy of Peru, to exterminate the goats which were said to have been previously introduced by English privateers, who careened and revictualled their vessels at a semi-permanent base which they established at James Bay.[3] There is, however, no firm evidence that either animal was introduced to the Galápagos in the seventeenth (or indeed eighteenth) century. Confusion may have arisen with the mastiff dogs released by the Spaniards on the Juan Fernandez Islands off the coast of Chile in 1686 to control the goats introduced in 1572. In any case, it seems hardly likely, in view of the abundance of easily obtainable food in the form of tortoises, turtles and ground-nesting birds and their eggs, that the need to introduce any new animals to the Galápagos would have arisen.

As with other domestic stock, dogs were probably first imported to the islands by early settlers in the nineteenth century. General José Villamil founded the first colony on Floreana (Charles) in 1832; temporary settlements were subsequently established on a number of other islands – the earliest permanent ones being those founded on San Cristóbal (Chatham) by Señor Manuel Cobos in 1869 and on southern Isabela (Albemarle) by Señor Antonio Gil in about 1893.[4] Until their extermination in the early 1970s (mainly by poisoning donkey carcases as bait with compound 1080 (sodium monofluoroacetate) distributed by the Health Authority for rat control) feral dogs, abandoned when the settlements were deserted, survived continuously on both the two former islands – on Floreana since 1846[5] and on San Cristóbal since before 1856.[6]

In 1851 Villamil landed on Isabela with a pair of great danes which, since they did not accompany him back to San Cristóbal, must have died, escaped or been released.[7] When Dr Simeon Habel visited Isabela in 1868 he reported that 'the men told me there were many wild cats, dogs, and also asses on Albemarle. The two latter are also to be seen on Indefatigable [Santa Cruz] and Chatham Islands'.

For five years after he settled on southern Isabela in about 1893 Antonio Gil mined sulphur from the caldera (crater) of Volcan Sierra Negra, which he transported by donkey to a small community on the south coast. When the Hopkins and Stanford expedition visited this village a year after Gil's arrival they reported considerable destruction of eggs of the giant tortoise (*Testudo* (*Geochelone*) *elephantopus*) in the area by feral dogs.[8] In 1902 a garrison – named Villamil in memory of the general – was formed in the port settlement, and Gil moved to the highlands to farm near the new community of Santo Tomás (named after the founder of the islands, Fray Tomás de Berlanga, Bishop of Panama), where within four years a population of 200 people was established, many of whom hunted the feral cattle and donkeys which roamed the grasslands surrounding the caldera wall.[9] In the same year (1906) the schooner *Academy* reported the presence of feral dogs in the pampas above Santo Tomás as well as along the coast several kilometres from Villamil.[10] The feral cattle and donkeys that wandered over the slopes of Sierra Negra also attracted hunting parties from Manuel Cobos' settlement on San Cristóbal; these frequently brought dogs – mainly pointers – with them to help find their prey. Several bitches gave birth to litters of pups while on Isabela, and many of the latter were left behind to fend for themselves when their owners returned to San Cristóbal.

In 1915 N. G. Martínez described the packs of feral dogs which preyed on cattle around Santo Tomás as a 'terrible plague'.[11] In 1930 a borzoi bitch and a male Alsatian (German Shepherd) and male Pharoah hound were imported to Santo Tomás by a Spaniard, Señor Emilio Pozo, who remained on the island for two years before returning to Guayaquil in Ecuador. Whether the dogs or any of their progeny were left behind on Isabela is unrecorded.

Feral dogs were first observed on the coast north of Sierra Negra in 1934 by a party of American scientists studying marine iguanas (*Amblyrhynchus cristatus*) at a small cove west of Elizabeth Bay.[12]

On southern Isabela feral dogs have been present continuously – albeit in some places intermittently – since before 1868; Kruuk (1979) estimated the population to number between 200 and 500 (probably nearer the lower figure), and at the time of writing (1983) an eradication campaign (mainly by poisoning with compound 1080), started in 1981 under the joint auspices of the Galápagos National Park Service, the Charles Darwin Research Station and the Frankfurt Zoological Society, has probably reduced it to half that figure.

The Norwegian Ulva expedition found feral dogs on Santa Cruz in about 1925, which proved a nuisance to early Norwegian settlers until 1928.

Thereafter they declined, and seem to have disappeared entirely between 1930 and 1935. Their reappearance in the late 1960s or early 1970s has been attributed by Kruuk (1979) to various possible sources: two or three animals which vanished near Puerto Nunez during a goat hunt in 1968 or 1969; two 'Dobermann pinscher type' dogs which were deliberately released on the south coast in 1973; and/or a pregnant bitch which strayed from her owner while hunting on the south coast near Tortuga Bay, and was later seen with a litter of six pups. Between 1975 and 1977 the entire population of land iguanas (*Conolophus subcristatus*) on Santa Cruz – apart from a few which were rescued by the Charles Darwin Research Station – were eliminated by feral dogs. Kruuk estimated the population of dogs on the island in 1979 at between 25 and 70.

At present, feral dogs on southern Isabela occur mainly on the coast from Elizabeth Bay on the Perry Isthmus to Punta Cristóbal, and in the west on Cerro Azul – at 1,689 m, the island's second highest volcano. A few are also found around Cartago Bay on the eastern side of the isthmus. Feral dog populations on Isabela are constantly subject to reinforcement by domestic (mainly hunting) dogs from the villages of Santo Tomás and Villamil; these animals, before they become truly feral, pass through a free-ranging and mainly scavenging intermediate 'pariah' stage during which they are based on Volcan Sierra Negra. Barnett and Rudd,[13] who draw a clear distinction between 'pariah' and feral packs, suggest the use of a chemical vasectomizing agent (chlorhexidine digluconate) for domestic dogs to prevent further recruitment of the feral population.

The discovery in 1979 that feral goats had managed to cross the seemingly impassable larva fields of the Perry Isthmus, which divides Isabela in two, has led to the fear that dogs might follow their example; this would seriously threaten resident populations of giant tortoises and land iguanas – which have been seriously affected elsewhere – and breeding colonies of Galápagos flightless cormorants (*Nannopterum harrisi*) and blue-footed boobies (*Sula nebouxii*).

The dogs of coastal southern Isabela stand between 50 and 70 cm high at the shoulder and measure about 100 cm in length, excluding the tail. Most are short-haired with noticeably large ears (adaptations for heat dissipation) and are white in colour with brown or black spots. Those at the cooler and moister higher altitude of Cerro Azul are similar in appearance, but have longer hair and smaller ears.

As elsewhere, feral dogs in the Galápagos seem to be opportunistic predators, eating whatever is most readily available. Barnett and Rudd found the prey of coastal dogs to be divided between marine iguanas (35 per cent), Galápagos penguins (*Spheniscus mendiculus*) (32 per cent), juvenile Californian sealions (*Zalophus californianus*) and South American fur-seals (*Arctocephalus australis*) (8 per cent), Audubon's shearwaters (*Puffinus l'herminieri*) (7 per cent), brown pelicans (*Pelecanus occidentalis*) and blue-footed boobies (6 per cent), and traces of dogs, cats and black rats (*Rattus rattus*) (2 per cent). Some observers also report predation of nesting dark-rumped petrels (*Pterodroma phaeopygia phaeopygia*), of the American race of the greater flamingo (*Phoenicopterus ruber*) and of swallow-tailed gulls (*Creagrus furcatus*). On the other hand the accusation by V. Ziswiler[14] that feral dogs have 'denaturalized' the waved or Galápagos albatross (*Diomedea irrorata*) and have contributed to the decline of the endangered Galṕagos hawk (*Buteo galapagoensis*) can be discounted; the former, as Thornton (1971) points out, breeds only on Hood (Española), an uninhabited island free from dogs, and the latter often nests in trees.

The predominance of small prey in the diet of feral dogs removes the need for hunting in packs, and Barnett and Rudd rarely observed groups of more than three individuals, although Kruuk (1979) reported packs of between four and eight. Kruuk and Snell (1981), whose findings were corroborated by Barnett and Rudd, suggested that in the region of Caleta Webb alone feral dogs annually consume 27 per cent of the population of marine iguanas, thus gravely endangering their future. The breeding range of the Galápagos penguin is confined to the northern coast of Isabela and coastal Fernandina (Narborough), where the cool water of the Humboldt current flows through the Bolivar Strait[15]: increasing predation by dogs on nesting colonies of penguins, which are extremely sensitive to such pressure, could endanger their survival in the Galápagos.

On Cerro Azul the population of 2,000 or so feral cattle is a major food resource for feral dogs and, when they occurred in substantial numbers, feral goats were also probably a preferred prey. Barnett and Rudd found that of the dog faeces which they examined 50 per cent contained the remains of cattle: the balance of the diet included coleopterans (14 per cent), orthopterans (11 per cent), ferns and grasses (8 per cent), cannibalized dogs (6 per cent), and traces of cats, small ground finches (*Geospiza fuliginosa*) and black rats (8 per cent). Surprisingly few of the cattle – mostly calves – are killed directly, the usual method employed by the dogs being harassment of their prey, on whose carcasses they subsequently scavenge. Barnett and Rudd explained this strategy by the reduced risk of injury to, and smaller expenditure of energy by, the dogs, or by the modification of behaviour patterns and/or morphology resulting from nearly 16,000 years of domestication[16], which, they suggest, can transform a predator into a scavenger. In view of the killing by feral and domestic dogs of livestock (e.g. sheep, and other species throughout the world) the former explanation seems the more likely.

Feral dogs in the Galápagos – where at the time of writing (1983) they occur only on southern Isabela and Santa Cruz – have seriously affected some populations of marine iguanas, land iguanas, the recruitment of giant tortoises and local colonies of Galápagos penguins and fur-seals. If dogs can be eradicated, these species might recover, and it could also become feasible to release land iguanas (which are currently being successfully reared in captivity at the Charles Darwin Research Station on Santa Cruz) on that island, as they have already been at their ancestral home near Cartago Bay on eastern Isabela. On the other hand, the elimination of feral dogs could lead to an increase in the number of feral cats (which prey, *inter alia*, on marine iguana and giant tortoise hatchlings, and on various seabirds), and also in that of feral cattle and goats which, through overgrazing, degrade the vegetation and accelerate soil erosion. The eradication of cats might in turn result in an increase in the number of rats, which are held to be largely responsible for the decline of dark-rumped petrels on Santa Cruz, Isabela, San Cristóbal and Floreana. It thus seems vital that the successful elimination of dogs in the Galápagos should be followed by concerted campaigns against other introduced pests.

Other oceanic islands

Feral dogs are believed to have been partly responsible for endangering the survival of the Marianas race of La Pérouse's megapode (*Megapodius lapérouse lapérouse*), the Ryukyu rabbit (*Pentalagus furnessi*) on the islands of Amami Oshima and Toku-no-Oshima south of Kyushu, Japan, and the mysterious kagu (*Rhynochetos jubatus*) of New Caledonia. As a result of predation by dogs (helped by cats, foxes and black rats) the distinctive and rare Aleutian Canada goose (*Branta canadensis leucopareia*) now only nests on the 1,720-ha island of Buldir. Feral dogs have occurred on Tristan da Cunha since about 1817 (six years after it was first colonized by settlers from Massachusetts, USA); on Kerguelen they were established from 1903 to 1928, on Heard Island (south-east of Kerguelen) between 1947 and 1955, on Macquarie Island from shortly after its discovery in 1810 until about 1820, on Chatham Island, and at one time (though no longer) on South Georgia, Inaccessible, Gough, Campbell, Auckland, Enderby and Robinson Crusoe (Masatierra) and Alejandro Selkirk (Masafuera) islands in the Juan Fernandez group.

In 1929 **coyotes** (*Canis latrans*) – natives of much of North America apart from the southeastern United States – were transplanted to Florida for hunting; they are at present said to be most abundant in the Lake Wales region of Polk County.

J. W. Fitzpatrick recorded[17] that in the 1860s about a dozen jackals from India (presumably **golden jackals** (*Canis aureus*) were released for hunting in South Australia, where they apparently disappeared.

NOTES

1. See p. 48.
2. For much of the information on feral dogs in the Galápagos I am indebted to Kruuk (1979a), Kruuk & Snell (1981) and B. D. Barnett & A. L. Rudd ('Feral dogs in the Galápagos Islands – impact and control', preprint).

3. e.g. V. W. von Hagen, *Ecuador and the Galápagos Islands* (1949); T. Heyerdahl & A. Skjölsvold, *American Antiquarian* **22**: 1–71 (1956); Thornton (1971); Roots (1976).
4. For a more detailed account of the colonization of the Galápagos Islands see p. 359.
5. Slevin (1959).
6. H. Melville, 'The encantadas' in *Four short novels* (1856).
7. Naveda (1950).
8. Heller (1903).
9. Slevin (1959).
10. Slevin (1931).
11. *Impresíons de un viaje* (Talleres de Policía Nacional, Quito, Ecuador, 1915).
12. W. A. Robinson, *Voyage to Galápagos* (Jonathan Cape, London, 1936).
13. See under note 2.
14. *Exinct and vanishing animals: a biology of extinctions and survival* (Longmans, London; Springer-Verlag, New York, 1967).
15. G. T. Houvenaghel, in R. Boje & M. Tomczak eds, *Upwelling eeosystems* (Springer Verlag, Berlin and New York, 1978), pp. 181–200.
16. I. L. Brisbin, *ASB Bulletin* **21**: 3–8 (1974) and *American Kennel Club Gazette* **93**: 22–9 (1976).
17. *Irish sport and sportsmen* (1878).

Raccoon
(*Procyon lotor*)

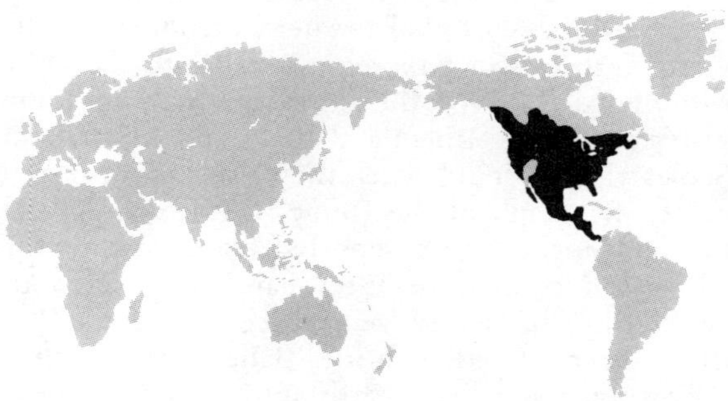

NATURAL DISTRIBUTION
Most of North and Central America from southern Canada to Panama; absent from the tundra and boreal forest zone.

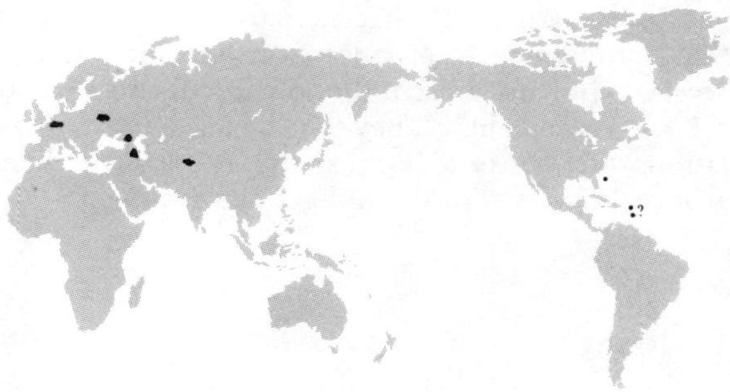

NATURALIZED DISTRIBUTION
Europe: Germany, France, the Netherlands; Eurasia: USSR; North America: West Indies (?Barbados, ?Guadeloupe, Grand Bahama).

EUROPE

Germany; France; Netherlands

In 1934 two pairs of raccoons were released into the wild near the Edersee area of Kreis, Frankenberg, Hessen, in what is now West Germany. They are said to have been reinforced in the wild after the Second World War by animals that escaped from American servicemen who had imported them as pets from the United States. Their descendants soon began to spread, initially mainly on the Eifel Plateau in Hessen and in eastern North Rhine-Westphalia, later extending northwards into Lower Saxony and east into the Harz Mountains. In 1964 they first appeared in some eastern parts of Rheinland-Pfalz and in 1970 in Schleswig-Holstein, when they also first occurred on the left bank of the Rhine in Rheinland-Pfalz but had not yet been reported from west of the Rhine in North Rhine-Westphalia or in Saarland.

South of the River Main raccoons were still at this time confined to north Baden, from where they spread northeastwards into East Germany, probably via Thuringia. A discrete colony east of Berlin is derived from escapes from captivity in 1945. By 1970 there were estimated to be around 20,000 raccoons in Hessen alone, and within a further decade much of both East and West Germany had been overrun. They are shot and trapped extensively on both sides of the border.

From Germany, adjacent parts of France and the Netherlands were colonized, and raccoons in western Europe are continuing rapidly to extend their range. Where they occur in dense populations they pose a serious threat – due to their arboreal habits – to the indigenous fauna.

EURASIA

USSR

In 1929 a pair of raccoons were imported to Russia from North America by the zoo in Tashkent. Attempts to naturalize the species began in 1936, the nucleus comprising 22 animals (one report says 20–15 males and 5 females) born in the zoo which were released in the walnut (*Juglans* sp.) forests near Gavinskii in the Fergana Valley in Kirgizia, Turkestan. These were supplemented by the liberation of at least 10–12 in the same area in the following year: they soon began breeding in the wild and within a few years had spread up to 40–50 km from their point of release.

In the autumn of 1952 a further 33 raccoons – 16 males and 17 females – were turned down in the neighbouring forests of Kara-Aminskii. Although by 1954 the population only numbered an estimated 300 – due probably to exceptionally heavy snowfalls in each of the preceding winters – raccoons today are well established over an area of 12,000–15,000 ha in the nut forests of Arslanbob, the western Achinsk District, the Dzhalalabad Region and, to a lesser extent, in the Bazar-Kurgan Region.

In 1941 21 raccoons – 11 males and 10 females – imported from breeders in western Europe (probably Germany), were released in the Ismail Region of Azerbaijan. Here they adapted well, and were soon breeding in the eastern parts of the Zakataly-Nukha Valley; from Ismail (and later from nearby Kutkashen) some were subsequently translocated to nut forests in the Khachmas Plain, in Talysh, in Terek, in Sulak and along various tributaries of the Kuban. Everywhere they thrived and bred, and the commercial harvesting of pelts began in 1954. By the end of the decade they had spread still further into parts of Vartaschen, and currently occupy a territory of around 1,000 km^2. Their further extension of range is limited on the north by the slopes of the Great Caucasian Range and to the south by the Steppe Plateau.

In 1954 52 raccoons – followed by a further 76 four years later – were turned down at three localities in Belorussia – in the mixed oak and hornbeam forests of Gomel and Brest and in the valley of the Pripet River. Here they spread slowly and are still common only locally.

In 1953 43 raccoons were introduced to the mountain forests of the Pskem River valley in Uzbekistan, where by the following year they were established and breeding.

In 1937 two pairs obtained from the Moscow Zoo were unsuccessfully released on Petrov Island in the Maritime Territory. By 1954 a further 486 raccoons had been translocated from Transcaucasia to both coniferous and broadleaved forests in the mountain river valleys of the Maritime Territory; nowhere were they ever abundant and, since their mortality exceeds their birthrate, they are currently declining.

By 1966, when raccoons were established in parts of Azerbaijan, Dagestanskaya, Krasnodarskiy Kray, Kabardino-Balkarskaya, White Russia, Kirgizia, Uzbekistan, Primorskiy Kray and Stavropol'skiy Kray, there had been 26 introductions to the Soviet Union involving a total of 1,222 animals, of which 504 had been liberated in the Caucasus, 98 in Central Asia, 130 in Belorussia and 490 in the Far East. Most were translocations from the Zakataly-Nuka Valley in Azerbaijan, supplemented by 38 from west European (probably German) breeders and 26 from Soviet zoos.

Almost everywhere in the Soviet Union raccoons adapted well to their new environment, failing to become well established only on Petrov Island in the Maritime Territory and in the Kabardino-Balkarian ASSR. They have been most successful in broadleafed hardwood nut-forests up to an altitude of 3,500 m, and in orchards with plenty of hollow trees (which provide adequate dens) situated in plains with marshes and still waters – areas with more varied habitats and fast-flowing mountain streams, which often dry up in summer, being apparently less suitable. Raccoons are most abundant in Transcaucasia and Uzbekistan, and especially in the Zakataly-Nukha Valley, in the nut-forests along the lower regions of river valleys of Kuba-Khachmas in Azerbaijan, on the Sulak Plain in Dagestan and in young mixed woodlands on the right bank of the Pripet River in Belorussia.

Apart from suitability of climate and adaptability to different habitats, coupled with the absence of important predators or competitors, the principal reason for the successful naturalization of raccoons in the Soviet Union has been that, being polyphagus, they have an abundant and varied supply of foods, on which they feed in small and approximately equal quantities. Altogether they are known to eat over 250 species of fauna and flora – the latter from forest canopy to ground level. Even in years of crop failure, when a preferred food is unavailable, raccoons are, unlike, for example, badgers (*Meles meles*) and brown bears (*Ursus arctos*), able to lay down a store of fat for hibernation. Plant foods include green briar berries (*Smilax excelsa*), hazelnuts (*Corylus avellana*), acorns of *Quercus longipes*, Virginia creeper (*Vitis sylvestris*), and cherries and plums (*Prunus* spp.). Animal food comprises injurious insects (including click beetles (Elateridae), weevils (Curculionidae), leaf beetles (Chrysomelidae), longhorn beetles (Cerambycidae), crickets (Gryllidae) and water beetles (Dytiscidae)), reptiles and amphibians, molluscs, crawfish (Decapoda), small rodents, birds and their eggs and, to a lesser extent, fish.

To a limited extent raccoons in the Soviet Union are preyed on by yellow-throated martens (*Martes flavigula*), jackals (*Canis aureus*), wolves (*C. lupus*), lynx (*Felis lynx*) and red foxes (*Vulpes vulpes*), and in times of shortage may compete for food with foxes, wild boar (*Sus scrofa*), bears, edible dormice (*Glis glis*) and beech martens (*Martes foina*). Their main limiting factor is their inability to withstand sub-zero temperatures for more than around nine months of the year, snow-cover for more than about 18 weeks a year, or temperatures below −42 °C.[1]

NORTH AMERICA

West Indies

G. Hughes, in his *The natural history of Barbados* (1750), refers to a 'Praemium' on the destruction of raccoons, which may have been introduced to the island by man, as they may also have been on Guadeloupe. In 1932 or 1933 Mr Jack Morris released a pair of *P. lotor elucus* (and a pair of opossums which soon disappeared) from Florida near West End Settlement on Grand Bahama where they thrived and bred, before long becoming a pest to peanut-farmers and corn-farmers, subsequently spreading even into the pine-forest zone. Raccoons of the subspecies *P. l. maynardi* were introduced to New Providence in the Bahamas from North America possibly before 1874.

In about 1905 an unsuccessful attempt was made by the New Zealand Government (through the agency of T.E. Donne) to establish a pair of raccoons in the wild at Rotorua.

Southern coatis (*Nasua nasua*) are established on Robinson Crusoe (Masatierra) in the Juan Fernandez archipelago off the coast of Chile, to which they were introduced from the South American mainland (probably, on the evidence of the small size of their molar teeth, Venezuela or Ecuador) by *'seudonaturalistas'* after the islands were declared a national park in 1935. The population, which in 1976 was estimated to number between 2,500 and 5,000, has, in the absence of controlling agents, devastated the island's vegetation and avifauna.[2] An endemic hummingbird, the Fernandez firecrown (*Sephanoides fernandensis fernandensis*), however, is still numerous despite heavy predation.[3] (The coatis on Robinson Crusoe Island have been mistakenly referred to as kinkajous (*Potos flavus*) – to which they are closely related – which are said to have been introduced 'for the purpose of eating the rats which exist there, but today are a serious danger for birds, destroying them'.[4] Kinkajous are, in fact, largely frugivorous.)

NOTES

1. Lavrov & Pokrovsky (1967).
2. C. M. Pizarro, 'El Archipiélago de Juan Fernández y la conservación de sus recursos naturales renovables', *Bolletin de la Academia de Ciencias, Instituto de Chile* **1** (2): 83–103 (1969) (reprinted in *Museo Nacional de Historia Natural*, Serie Educative No. 9 (Santiago 1974)). See also *Plan de Manejo Parque Nacional Juan Fernández* (Chilian Corporacion Nacional Forestal (and FAO), 1976).
3. W. R. P. Bourne (1983b and personal communication by letter, 1983). I am also grateful to Dr Bourne for drawing my attention to the references in note 2, and to Dr J. Clutton-Brock, Miss D. M. Hills and Miss S. Angel of the British Museum (Natural History) for confirmation of the identity of this species.
4. Quoted by De Vos, Manville & Van Gelder (1956: 177) as personal communication from L. E. Peña.

Weasel, Stoat, Polecat, Ferret, Polecat-ferret

(Mustela nivalis, Mustela erminea, Mustela putorius, Mustela furo, Mustela putorius furo)

Note: Natural distribution shown is composite range.

NATURAL DISTRIBUTION

M. nivalis: most of the Palaearctic and Nearctic regions from the Arctic coasts south to North Africa, Afghanistan, central China, Nebraska and Missouri. *M. erminea*: Eurasia and the Nearctic Region from the coast of the Arctic southwards to the Pyrenees, Alps, Kashmir, Japan, California and Virginia. *M. putorius*: Europe from the Atlantic coast to the Urals, north to south-eastern Norway, southern Finland and Sweden, south to the Mediterranean and the Black Sea.

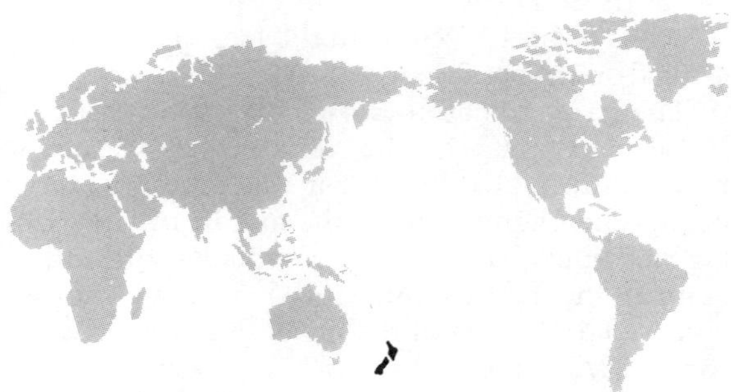

NATURALIZED DISTRIBUTION

Australasia: New Zealand.

EUROPE

British Isles

The polecat-ferret is, as its name implies, a cross between the wild (European) polecat and the domestic ferret. There has been – and indeed still is – much controversy among systematists over the origin of the domestic ferret; all agree that it is a domesticated albino descendant of the polecat, but whether its ancestor is *M. putorius* (with which it is fully interfertile) or the (Asiatic) steppe polecat (*M. eversmanni*) remains uncertain. These two wild forms have been regarded in the past as conspecific, but current opinion in eastern Europe and Russia, where they appear to be sympatric, is that they are separate species.

The ferret was kept in domestication by the ancient Greeks and Romans: the Greek historian Strabo (born *c.* 63 BC) states in his *Geography* that it was imported to Spain from North Africa, and the Roman, Pliny (*c.* AD 23–79), describes in his *Naturalis historia*, how, under the name *viverra*, it was used for hunting rabbits. (The English name 'ferret' is not, as might perhaps be expected, derived from *viverra*, but from the French *furet*, which is itself derived from the Latin *fur*, meaning a thief or robber.)

The ferret has been known in domestication in Britain since at least the late thirteenth century; it is therefore possible that it was first introduced during the reign of one of the middle or late Plantagenet kings, to help in providing sport and to control the rabbits introduced a century or so earlier. The capture of rabbits with ferrets at Waleton in 1272 is mentioned by J. E. Thorold Rogers in his *History of agriculture and prices in England* (Vol. I, 1866), and a decade later Richard le Forester was paid 3*s*.6*d*. for catching rabbits with ferrets for the King at Rhuddlan Castle in Flintshire, north Wales.

Feral polecat-ferrets in the British Isles are the result either of escaped albino ferrets breeding with wild polecats, or are escaped dark or particoloured animals which man has produced deliberately by crossing the domestic albino ferret with the wild polecat.

On the island of Mull off the west coast of Scotland, where the polecat has never been indigenous, ferrets and polecats were both kept in domestication in about 1933–4; they soon escaped and interbred freely in the wild, before long becoming pests throughout the island where they preyed on rabbits, ground-nesting wild birds and domestic poultry. The well-known Scottish naturalist and author, Seton Gordon, recalls that polecat-ferrets were set free on the island of Harris in the Outer Hebrides to control rabbits where, in Mr Gordon's opinion, they were responsible for exterminating the ptarmigan (*Lagopus mutus*) on the Hill of Clisham – its last haunt in the Outer Hebrides. From Harris they spread to the northern part of the island (known as Lewis), where they are said to be especially common around Uig. There are also thriving populations on the Scottish islands of Arran and Bute, on the Isle of Man in the Irish Sea and on the Isle of Anglesey off the coast of north Wales, as well as in Renfrewshire and parts of Yorkshire.

A native of the British and Irish mainland, the stoat occurs on the main island in the Shetlands as an introduction by man.

Netherlands

During a period of afforestation between 1910 and 1930 pines, alders and oaks were planted on 600 ha (about 5 per cent) of the island of Terschelling off the coast of Friesland. Channels cut to drain the plantations provided ideal breeding sites for water voles (*Arvicola terrestris*), and after about 1920 the increased population began to cause serious damage to trees and gardens on the island. In 1931, after a bounty system, mouse typhoid and poison had all proved ineffective, 102–104 weasels and 6–9 stoats were imported from the mainland as control agents. Other mammalian prey species available on Terschelling included common shrews (*Sorex araneus*) and pygmy shrews (*S. minutus*), hedgehogs (*Erinaceus europaeus*), long-tailed field or wood mice (*Apodemus sylvaticus*), house mice (*Mus musculus*), brown rats (*Rattus norvegicus*) and rabbits (*Oryctolagus cuniculus*). By 1934 the weasels had, surprisingly in view of the size of the original stock, vanished. The stoats, however, thrived, and by the same year had increased to at least 180 individuals; by 1937, when the water voles had been successfully exterminated (and the numbers of some nesting waders and waterfowl had noticeably declined), the stoats had established a fluctuating but secure population.

Feral ferrets are established in the wild in many places in continental Europe, being especially common, according to Roots (1976), in Sicily and Sardinia. The presence of weasels on some Mediterranean islands and in the Azores is probably due to introductions by man.

ASIA

Japan

For a period in the 1930s polecats (? sp.) are said to have been established in the wild in the Tôhoku district of Japan.

AUSTRALASIA

New Zealand

Referring to weasels, stoats and ferrets, Thomson (1922) wrote: 'Nothing in connection with the naturalization of wild animals in New Zealand has caused so much heart-burning and controversy as the introduction of these bloodthirsty creatures.'[1]

As the introduced rabbit increased to plague proportions in New Zealand it was natural that thought should be given to some form of controlling agent. Thomson records that 'one authority actually proposed to introduce Arctic foxes, because their fur would be so valuable. When it was pointed out to him that they would probably prefer lamb to rabbit, he replied that as they did not know anything about lambs in their native haunts it was improbable that they would take to eating them in New Zealand[!]'[2] Fortunately, as Thomson points out, the proposal was rejected.

Instead, in 1867, the Canterbury Acclimatization Society imported five ferrets, followed by a further one in the following year: these were not released but were allowed to breed in captivity, their offspring being sold to private landowners. In 1873 the society had a stock of half-a-dozen in their Christchurch gardens, and six years later five (presumably from this source) were released in the Conway River area north of Christchurch.

The proliferating rabbit population led to increasing pressure on the government by harassed sheep-farmers. Between March 1882 and July 1883 the agent-general shipped 1,217 ferrets from London in 32 consignments, of which only 376 (at a cost of £3.2*s*.6*d*. each) survived the voyage. In the following year a further 3,041, imported from Lincolnshire, England, were released in Marlborough and (400) in Otago.

A year latter two shipments of 202 weasels and 60 stoats arrived from London, of which 183 weasels and 55 stoats were safely landed. Of these, 67 weasels were liberated on a peninsula of 3,200 ha on Lake Wanaka where they successfully reduçed, but did not succeed in exterminating, the rabbit population; 28 weasels and 6 stoats were released on the shore of Lake Wakatipu, 15 weasels near the Waiau River in Southland and 8 stoats at Ashburton; the remainder were sold in Wellington, Christchurch and Dunedin. (There is a further but unsubstantiated report that an additional 300 stoats and weasels were released north of the Waiau in 1885.)

In the following year the government imported a further two consignments of 82 stoats and 126 weasels, which were distributed equally in the Wilkin River district, around the Makarora at the head of Lake Ohau, and in the vicinity of the Waitaki River. In the same year 32 stoats and 116 weasels were turned down by the government between Marlborough and West Wairarapa, and a further 55 stoats and 167 weasels were released privately at Riddiford's Station in West Wairarapa where, although rabbits were still abundant, ferrets were firmly established.

At a meeting at Masterton in 1886 to discuss the administration of the Rabbit Act, a resolution was passed 'that the introduction of ferrets, stoats and weasels in large numbers is the only means by which the rabbit pest can be successfully put an end to'. This view was strongly opposed by the various acclimatization societies who, since they had been responsible for the introduction of the rabbits, were not surprisingly overruled. Indeed, until the practice was abandoned in 1897, ferrets were bred in captivity in enormous numbers by the Department of Agriculture for subsequent release into the wild, and all mustelids in New Zealand were legally protected until 1907. As Thomson (1922) wrote, 'the mischief had been done, stoats and weasels are common in nearly every part of New Zealand and in some parts are enormously abundant'.[3]

By the late 1940s, according to Wodzicki (1950), stoats, weasels and ferrets were collectively widely distributed in New Zealand, being absent only from Chatham and Stewart islands and a few other small offshore islets.

Stoats were especially widespread, occurring in a variety of habitats which included paddocks, plantations, bush, orchards, gardens, dried-up riverbeds, sand-dunes and native forests; they had even occupied bushland on the wet west coast from Greymouth to Jackson Bay. Although they were most plentiful in regions of high rabbit density, they were also numerous in such comparatively rabbit-free districts as Warkworth, North Auckland, eastern parts of the Bay of Plenty and Gisborne.

Weasels were more locally distributed; in North Island, according to Wodzicki (1950) quoting F. E. Wilkin, they were plentiful by 1930 on the north side of Doubtless Bay in North Auckland. On the Waimea Plains in Nelson weasels increased considerably after their release in about 1890, but apparently decreased after the arrival of the stoat. E. F. Stead, quoted by Wodzicki, considered that weasels were local and rare in the Canterbury area in 1948, and Wodzicki found them to have been absent for some years in the Kurow, Queenstown, Oamaru and Lawrence districts.

Ferrets seem to have always had a restricted range in New Zealand. Wodzicki reported them to be rare or absent in the North Auckland peninsula from Kaipara harbour northwards, in the eastern part of the Bay of Plenty, on the east coast generally and in Poverty Bay, in large areas of Taranaki, in west Nelson and along the whole of the western coast of South Island. Elsewhere they were locally distributed on both main islands, preferring dry bush, paddocks, dried-up riverbeds and open tussock land: being poor tree-climbers their range is apparently largely governed by the distribution of rabbits.

New Zealand's alien mustelids failed to achieve more than a limited control over the country's introduced rabbit population; instead, they found the native birds to be much easier prey, since they were not pre-adapted to avoid predators as were the rabbits[4] and other – all introduced – mammalian species such as hedgehogs,[5] house mice,[6] black rats (*Rattus rattus*),[7] brown rats (of which stoats are the principal predators),[8] kiore or Polynesian rats (*R. exulans*),[9] brown hares (*Lepus capensis*),[10] and brush-tailed possums (*Trichosurus vulpecula*).[11] Species which declined dramatically, or in some cases were exterminated, through the attentions of introduced mustelids and other predators include weka rails (*Gallirallus australis*), banded rails (*Rallus philippensis*), introduced California quails (*Lophortyx californicus*), takahes (*Notornis mantelli*) (which are presently confined to parts of the Murchison and Kepler mountains in South Island's Lake District), kiwis (*Apteryx* spp.), two parrots (kakapos (*Strigops habroptilus*) and keas (*Nestor notabilis*)), laughing owls (*Sceloglaux albifacies*), blue or mountain ducks (*Hymenolaimus malacorhynchos*), paradise shelducks (*Tadorna variegata*), New Zealand scaup or black teal (*Aythya novaeseelandiae*) and tuis (*Prosthemadera novaeseelandiae*); the eggs and chicks of albatrosses (*Diomedea* spp.), petrels (Procellariidae), little blue penguins (*Eudyptula minor*), saddlebacks (*Creadion carunculatus*), robin-flycatchers (*Petroica* spp.), bellbirds (*Anthornis melanura*), pigeons (Columbidae), gulls (*Larus* spp.), and, on occasion, the eggs and nestlings of black swans (*Cygnus atratus*) are also sometimes taken. Between 1894 and 1900, in an attempt to save them from predation by stoats, Mr Richard Henry translocated 400 kakapos from Fiordland to some small islands in Dusky Sound; unfortunately his efforts were unrewarded, since the predators later gained access to the islands.

Today stoats are widely distributed in considerable numbers throughout New Zealand, being especially common in native forests, including all the national parks; these are among the most important reserves for the future conservation of the remaining native avifauna, and the reduction or extermination in them of stoats is being attempted by the National Parks Authority. Elsewhere, stoats help to keep in check rabbits, rats and mice, but also prey – as described above – on native wild birds, on domestic poultry and on game stocks; they are even accused of sometimes attacking sickly lambs and piglets.

As a result of its declining population the weasel is now of little economic or ecological significance in New Zealand. Feral ferrets occur most commonly on pastoral farmland; they are locally abundant and potential pests elsewhere (e.g. at the colony of breeding albatrosses at Taiaroa Head near Dunedin, and of the endangered black-winged stilts (*Himantopus novaeseelandiae*) in the Mackenzie Basin – their sole remaining breeding site). Birds of prey are the main predators of mustelids in New Zealand.

King and Moors (1979) have sought to explain the successful co-existence of stoats and weasels where they overlap in their native range, and the failure of the latter on Terschelling Island and its comparative lack of success in New Zealand. They hypothesize that the larger stoat is more of a generalist than the weasel and dominates the latter in interspecific encounters. Stoats, on the other hand, are unable to penetrate the runways and nests of small rodents, and are limited by delayed implantation to the production of a single litter per season. Weasels eat mainly small rodents, which they are more efficient than stoats in exploiting. They have a high potential rate of increase and can achieve dense populations wherever or whenever small rodents are abundant: their dependence on unstable populations of the latter, however, makes them liable to local extinction.

The biology of the two species therefore favours the stoat in interspecific competition with the weasel, whereas the latter has superiority in exploitation; this results in an unstable balance of competitive advantage determined by the environment. Where local prey populations fluctuate asyncronously neither species, except temporarily and locally, can gain complete superiority over the other. They co-exist because, in a sufficiently diverse environment, weasels can avoid confrontations with stoats, which can themselves avoid dependence on a single prey resource.

As King and Moors point out, since there were no voles (*Microtus* or *Clethrionomys*) on Terschelling Island or in New Zealand, both introduced populations of weasels were at a considerable disadvantage, and neither were likely to flourish. Whereas weasels died out on Terschelling within four years of their importation, however, they still persist in scattered localities in New Zealand after nearly a century. The two main islands of the latter are very large and have a diversity of habitats in comparison with Terschelling, thus giving weasels the environmental heterogenity they require to avoid confrontation with stoats and to operate their fugitive foraging strategy. The size distribution of prey on Terschelling is more favourable for weasels than it is in New Zealand, which suggests that they might have survived, albeit in reduced numbers, on Terschelling were it not for interference from stoats. Terschelling (110 km^2) is less than one-third of the size of the smallest British island on which both species satisfactorily co-exist.

The **Siberian weasel** or **kolinsky** (*Mustela sibirica*) was introduced to the island of Hokkaido, Japan, before 1901, where it has multiplied and interbred with the indigenous Japanese weasel *M. s. itatchi*. In 1933 *M. sibirica* (or *M. s. itatchi*) was introduced to two small islands, Rishiri and Rebun, off the northern tip of Hokkaido, and also to South Saghalien; 15 years later the same species was released in Okujirijima to control brown rats and the large Japanese field mouse (*Apodemus speciosus ainu*). Japanese weasels have also been introduced, with partial success, to combat rats on the Ryukyu Islands, where, however, they have also preyed on small indigenous mammals and birds. The World Health Organization has recently been investigating the possibility of using Japanese weasels to control introduced rats on other Pacific islands, where, by gnawing coconuts, they help to provide ideal breeding sites for the mosquito vectors of dengue and Bancroft's filaria, which causes elephantiasis in man.

Korean weasels (*M. s. coreana*) were introduced, probably to fur-farms, in Japan shortly after 1930. Escaped animals were soon established in the wild in some places (e.g. Kobe City and near Osaka), co-existing with *M. s. itatchi*, whereas in others (e.g. near Akashi, in some parts of Chugoku (southwestern Hondo) and in Tokushima City, Shikoku) the alien species replaced the native one.

Between 1929 and 1942 attempts were made to naturalize the (Asiatic) **steppe polecat** (*Mustela eversmanni*) in western Siberia.

The **Kamchatka sable** (*Martes zibellina kamtschadalica*) has also been translocated to western Siberia where, by interbreeding with the indigenous Siberian sable (*M. zibellina*), a heavier and more valuable pelted animal has resulted. **Siberian sables** have themselves been translocated to Khabarovsk Kray on the Sea of Okhotsk.

Bobak marmots (*Marmota (baibacina/himalayana/sibirica) bobak*) were introduced to western Siberia between 1929 and 1942.

The **striped skunk** (*Mephitis mephitis*) and the **sea otter** (*Enhydra lutris*) have both been unsuccessfully introduced from North America to the Soviet Union.

According to Strange (1972) **otters** 'from the channels of Tierra del Fuego' (possibly *Lutra felina* or *L. provocax*) and **skunks** (perhaps *Conepatus humboldti*, a species of hog-nosed skunk from Patagonia) were landed on a number of islets off West Falkland by an *émigré* Scottish shepherd, Mr John Hamilton, in the early 1930s. Of these, only the otter may remain; in 1942 one was shot at East Bay, Fox Bay, and in 1965 spraints[12] were found in a creek on the south-east coast of East Falkland. So far unidentified tracks observed on some beaches on other islands may have been made by otters, but this is uncertain. Since 1976 possible sightings have been reported from Weddell Island and from the West End of Pebble Island. Recently some children gave convincing descriptions to Dr R. S. Whitley of two animals which they had seen on Sea Lion Island off the east coast of East Falkland.

NOTES

1. p. 70.
2. p. 71.
3. p. 72.
4. See p. 408.
5. See p. 26.
6. See p. 368.
7. See p. 357.
8. See p. 352.
9. See p. 345.
10. See p. 389.
11. See p. 7.
12. The droppings of otters.

North American Mink

(*Mustela vison*)

NATURAL DISTRIBUTION
Most of North America apart from islands within the Arctic Circle, southwestern parts of the USA and Mexico.

NATURALIZED DISTRIBUTION
Europe: British Isles, Iceland, Norway, Sweden, Finland, Denmark (including the Faeroe Islands), the Netherlands, France, Spain; Eurasia: USSR. (North America: Newfoundland.)

EUROPE

British Isles

North American mink were first imported to fur-farms in Britain in 1929, and escapes were soon being reported from places as far apart as Lancashire, Warwickshire and Berwickshire in Scotland. During the 1950s the industry began to expand considerably, and large numbers of mink (up to 700 in a single consignment) were imported direct from America (where ranching in captivity began in 1871) or via Scandinavia.

The earliest unconfirmed reports of escaped mink breeding in the wild came shortly after the Second World War from Lancashire, where there were a large number of fur-farms. Although considerable depredations were reported by local poultry farmers the animals did not apparently become established.

In about 1953 around a dozen mink escaped from a farm at Bovey Tracey in Devon and they, and possibly their offspring, established a colony on the River Teign. In 1956 a female with young was observed in an old mill leat nearby, and in the following year it was officially announced that the animals were now breeding in the wild. At about the same time they were reported from Teignmouth, and also from the River Avon in Hampshire. A year later a pregnant female containing six young was trapped on the River Deben in east Suffolk, and reports were received of mink in the wild in Yorkshire. In the late 1950s two mink – killed in separate areas well away from mink-farms – were identified in Scotland.

During the late 1950s and early 1960s mink in the wild began to expand their range considerably: in England they were reported from the Devon Avon; from a number of stretches of the Hampshire and Wiltshire Avon and from the rivers Test and Itchen in the same counties; from the Stour in Dorset; from the Avon in Gloucestershire; and from the Wharfe and Lune in Lancashire and Yorkshire. In Wales mink established themselves on a stretch of the River Teifi on the Carmarthenshire/Cardiganshire border, in west Pembrokeshire, and on the rivers East Cleddau and Taf on the Pembrokeshire/Carmarthenshire border. In Scotland mink were reported on the Tay and Dee in Perthshire and Aberdeenshire.

In the middle and late 1960s mink increased in numbers in Devon, Lancashire, Wiltshire, Hampshire, Dorset and west Wales, and first appeared in Sussex (on the rivers Ouse and Cuckmere), and in Buckinghamshire, Berkshire, Somerset, Kent, Northamptonshire and Northumberland. By the late 1960s they were being reported from the rivers Taw, Dart, Walkham, Exe, Axe, Coly, Culm, Tavy, Mole, Torridge, East and West Webburn and Yeo in the west of England, and from much of Dartmoor. In 1967 mink were observed on the Tiddy, Hayle, Looe, and Tressillian in Cornwall.

By 1963 only Aberdeenshire, Perthshire and Kirkcudbrightshire were heavily infested in Scotland; by the late 1970s, however, mink were widespread, being especially numerous in Ayrshire, Stirlingshire, Banffshire, Midlothian, Roxburghshire, Peebleshire, Lanarkshire, Kincardineshire, Fife, Selkirkshire, Dumfriesshire and Morayshire.

In Ireland the first fur-farms were set up between 1950 and 1953, and by 1960 some 40 breeders were established. Most of the mink imported into Ireland came from Scandinavia, supplemented by stock from Britain and North America.

The earliest report of mink established in the wild in Ireland came in 1961 when 30 escaped from a farm near Omagh, County Tyrone, and colonized a length of the River Strule; by 1967 they had spread to the Mourne and Glenelly, and by the following year to Victoria Bridge, 24 km west of their point of escape. First reports of mink in other counties during the 1960s came from Dublin (1962), Armagh, Down and Leix (1965), Antrim and Wicklow (1966), Kerry and Kildare (1967), and Fermanagh and Meath (1968).

Despite a campaign by the Ministry of Agriculture between 1965 and 1970 to eradicate them, by the late 1970s mink were widespread throughout much of England, Scotland and Wales and were locally common in Ireland. In England they occur principally on the rivers Exe and Teign (Devon); on the Avon, Wylye and Beaulieu (Hampshire and Wiltshire); on the Stour (Dorset); on the Wharfe, Wyre, Ribble and Lune (Yorkshire and Lancashire); and on the Cuckmere and Ouse in Sussex. In Wales they are present in significant numbers on the Teifi (Carmarthenshire and Cardiganshire); on the Cych, Ceri and Dulas; and on the East and West Cleddau and Taf in Pembrokeshire and Carmarthenshire.

In Scotland mink are found mainly on the Tweed, Gala Water, Teviot and Kale Water (Roxburghshire, Selkirkshire and Berwickshire); on the Forth and Teith (Perthshire and Stirlingshire); at Boquhan (Forth), Blane, Earlsburn Reservoir, Balfron (Selkirkshire and Stirlingshire); on the Tay, Almond and Earn (Perthshire); on the Dee, Don, Devron and north and south Ugie and Cruden Water (Banffshire and Aberdeenshire); on the Spey (Banffshire and Morayshire); on the Doon (Ayrshire); on Urr Water, Dalbeattie (Kirkudbrightshire); and around Dalkeith, Gorebridge and Penicuick in Midlothian. They are widespread on the island of Harris and Lewis in the Outer Hebrides, and are said to have occurred on North Uist and on Arran.

In Ireland mink are breeding in some numbers in County Tyrone, and have spread throughout the watershed of the rivers Dodder and Liffey in County Dublin. They are also established in parts of Fermanagh, south-west Donegal, Antrim, Kerry and around Strangford Lough in County Down.

Mink are the only alien non-domesticated carnivore to have become firmly established in Britain in historic times. Escaped ranch mink usually survive well in the wild, being mostly a cross between the large and fertile 'Alaska' strain and the superior-pelted 'Quebec' strain. Ranch mink come in a variety of colour mutants (e.g. beige, blue, cream and white) known in the trade variously as Aleutian, Palomino, Pastel, Pearl, Platinum and Topaz. The offspring of escaped ranch mink usually revert to the standard dark-brown coloration.

One of the principal reasons for the success of the mink in becoming established in Britain is that it has filled a broadly based feeding niche previously unoccupied by any other carnivore. Because of the comprehensive nature of the mink's diet, which consists in Britain of approximately equal proportions of mammals, birds and fish, it is not a serious competitor for food with any other species.

By extrapolating to the rest of Britain findings in a small area of the west of England, there has recently been an unfortunate attempt by some academic scientists[1] to minimize the depredations caused by mink (whose presence they have even suggested in print should be 'enjoyed'); practical experience from other parts of the British Isles, however, confirms that the damage mink cause to wildlife, to avicultural collections, to domestic poultry and to fish-breeders can be extremely grave.

Most seriously at risk are undoubtedly ground-nesting birds – especially colonially breeding species. Although coots (*Fulica atra*) and moorhens (*Gallinula chloropus*) form the bulk of the mink's avian prey, many other species are also taken. The little grebe or dabchick (*Podiceps ruficollis*) has entirely disappeared from many mink-infested waters, and several species of ducks have been put at risk in various parts of Britain. At Duddington Loch in Scotland, for example, not only were a considerable number of different species of wildfowl destroyed before mink were identified as the culprits, but their presence is believed also to have inhibited migratory wildfowl from visiting the loch. Predation by mink in both lentic and lotic fisheries has in several instances caused economic problems. Only in its consumption of mammalian prey, which consists largely of rabbits, rats and grey squirrels (*Sciurus carolinensis*), does the mink display any redeeming dietary feature.

The precise results of interspecific competition between the alien mink and the now gravely endangered and almost entirely piscivorous native otter (*Lutra lutra*) is difficult to assess. The decline of the otter has coincided very largely with the spread of the mink, and it seems not unreasonable to suppose that these fluctuations may be connected. At the very least it is probable that mink have colonized those waters vacated by otters, and that the former have prevented the latter, which are notoriously shy animals, from becoming re-established in their previous haunts, much as the alien grey squirrel[2] did to the native red squirrel (*Sciurus vulgaris*) in the 1920s. There is also evidence both that otters will not breed in waters where mink abound and also of a correlation between the absence of otters and the presence of mink in some areas through competition for food; this competition may not be excessive but it nevertheless exists. Mink, for example, may catch small fish before they become large enough to attract the attention of otters, and as their diet is much more broadly based are better able to survive than otters when fish are scarce. It has also been suggested that escaped mink may have been one of the causes of the spread of canine distemper in British otters.

Mink depredations on domestic poultry, avicultural collections and commercial fish-farms, hatcheries and stews have caused considerable losses and made necessary the expenditure of large sums in anti-mink precautions.

In an attempt to achieve some degree of control over the population and spread of mink in Britain, the government in 1962 took action through the Mink (Importation and Keeping) Order, made under the Destructive Imported Animals Act 1932, which banned the importation and keeping of mink except under licence. In 1965 a similar order was made by the Irish government; three years later this prohibition was extended also to Ulster. Efforts were intensified to limit mink to those areas in Britain where they were already well established, and to attempt to prevent their further encroachment into previously uncolonized districts. As a result, the total population of mink in Britain is probably fairly stable. In North America adults regularly migrate over considerable distances in winter and spring during the mating season, while the young disperse in the autumn, and it is probable that the same occurs in Britain; their distribution, therefore, continues to increase.

Undoubtedly the greatest potential threat posed by mink in Britain is the risk of their gaining access to hitherto uninfested predator-free offshore islands with large populations of colonially nesting seabirds. In 1977 permission was sought to establish a mink-farm on the island of Westray in Orkney, Scotland, which not only supports the second largest colony of seabirds in the British Isles (more than a quarter of a million birds of 19 different species) but also a small endemic mammal, the Orkney vole (*Microtus arvalis orcadensis*).[3] This application was regarded by conservationists as a test case, and its refusal should ensure that insular species in Britain, where mink are not already present, remain free from the depredations of this voracious alien carnivore.

Iceland

American mink were first introduced to Iceland for fur-farming in 1930; escapes soon took place, and by 1937 (if not before) populations had become established in the wild. By the early 1950s wild mink were distributed throughout south-western Iceland from Mýrdalur in the south to the Breidafjördur area in the west, occurring both coastally and on rivers and lakes inland; by the mid-1970s the whole of the mainland had been overrun.

Much has been written about the alleged depredations by mink in Iceland on wild birds, domestic poultry, salmon, trout and even sheep, but opinions vary as to their culpability. Although there is definite proof of many attacks by mink on farmers' poultry stocks, accusations of havoc caused to wildlife are largely based on circumstantial evidence.

In 1974, J. Fjeldså conducted a census of breeding waterfowl on lakes Mývatn and Vikingavatn in north-east Iceland.[4] In 1961 the population of ducks on the former amounted to an estimated 50,000; this had fallen by 1970 to around 30,000, and had declined still further to about 15,500 by 1974. Fjeldså considered that although a change in habitat selection had contributed to this decline it had been caused at least in part by mink predation. Mink were found to occur principally along the shoreline and on offshore islets, where they were said to prey on such hole-nesting species as Barrow's goldeneye (*Bucephala islandica*), and also on gadwall (*Anas strepera*), wigeon (*A. penelope*) and scaup (*Aythya marila*). The lower incidence of mink in marshy and swampy areas was held to account for the lesser decline of such species as teal (*Anas crecca*), mallard (*A. platyrhynchos*), pintail (*A. acuta*), tufted duck (*Aythya fuligula*), long-tailed duck (*Clangula hyemalis*), scoters (*Melanitta nigra*) and mergansers (*Mergus merganser*). Where they occurred in large numbers mink were considered to be more serious predators than arctic skuas (*Stercorarius parasiticus*), great black-backed gulls (*Larus marinus*) or ravens (*Corvus corax*).

In the early 1970s 10–20 per cent of ducks found dead were believed to have been killed by mink; by 1974, when mink were said still to the increasing, Fjeldså estimated that the figure had risen to 40–49 per cent. In addition, habitat disturbance by mink was believed to give rise to mass displacement of nesting species and, in the same way that eider ducks (*Somateria mollissima*) and long-tailed ducks in the Arctic are inhibited from nesting by the presence of Arctic foxes (*Alopex lagopus*), to result in many non-breeders.

Since Fjeldå's survey the rate of decline of duck populations on Mývatn and Vikingavatn has been questioned, as has the cause of the displacement of breeding birds which has undoubtedly occurred. The species which have apparently suffered most from mink predation in Iceland are waders, puffins (*Fratercula arctica*) and black guillemots (*Cepphus grylle*), though again it is uncertain whether mink have been the main cause of declining populations and dispersals or whether the latter are due to other causes. Although Sven-Axel Bengtson has claimed that mink are 'now a menace to waterfowl in many places in Iceland,'[5] further research is clearly necessary before mink can finally be acquitted or condemned.

Since 1939 the Icelandic Government has paid a bounty on mink killed, but this has not prevented a continual increase in the population and its spread throughout the country.

Norway

Mink were imported from North America to fur-farms in Norway in the late 1920s, and were first recorded in the wild in about 1930. By 1948 wild mink were established over much of southern Norway – especially in eastern districts ('herreds') – and occurred regularly along parts of the Atlantic west coast and to a lesser extent on the south coast bordering the Skagerrak. In 1949 mink first appeared north of the Trondheimsfjord, and by the middle of the following decade they were apparently distributed continuously throughout most of the country to the south. In the west they were particularly common in Rogaland and Romsdal; along the south coast they occurred regularly in parts of Vest-Agder and occasionally elsewhere. North of the Trondheimsfjord mink were distributed over much of Nord-Tröndelag, and were sometimes seen as far north as Bardu: today they remain more abundant in the south of the country than in the north.

As in Iceland much of the evidence of mink predation on nesting seabirds in Norway is circumstantial. In Møre and Romsdal in western Norway they have been accused (together with white-tailed or sea eagles (*Haliaeetus albicilla*), eagle owls (*Bubo bubo*), red foxes (*Vulpes vulpes*) and man) of heavy predation on gull and tern colonies, and on black guillemots which have greatly decreased (to an estimated 700–800 pairs) since about 1950. Serious damage to colonies of shags (*Phalacrocorax aristotelis*), eider ducks, great black-backed gulls, common gulls (*Larus canus*), common and arctic terns (*Sterna hirundo* and *S. paradisaea*), black guillemots and puffins in the western islands has also been attributed to predation by mink. At Stavanger mink are said to be responsible for the devastation of seabird colonies on islets 2–3 km offshore (they have been recorded as making sea passages of up to 5 km), and to have attacked swimming eider; in 1976 on Hekkingen, Troms, in northern Norway and in 1972 on Røst, where kittiwake (*Larus tridactylus*) chicks were found decapitated in their cliff-side nests, mink are said to have been the culprits. Again, as in Iceland, more research into mink predation is necessary to establish the degree of their guilt in the devastation of Norwegian seabirds.

Sweden

Accounts differ as to whether the farming of American mink in Sweden was begun in 1925 or 1928.[6] However that may be, by 1937 41,365 mink were distributed unevenly throughout the country among 627 farms – totals which by the outbreak of war two years later had risen to 107,232 and 1,224 respectively. Although a severe fall in the price of fur, coupled with the rationing of food and conscription for military service, resulted in a contraction in the fur-farming industry during hostilities, by 1945 Sweden had 714 farms holding a total population of 98,507 mink.

The earliest record of free-living mink in Sweden was near Limedsforsen in Kopparbergs *län* (district) in 1928. In the following year mink were observed in the wild near Ullånger, Västernorrlands *län*, and escapes were reported from around Anundsjö some 50 km to the north.

In the early 1930s two mass escapes occurred which were probably largely responsible for the establishment of a viable population of wild mink in Sweden. In the autumn of 1932 about 20 escaped from a farm at Bjurholm, Västerbotten *län*, and between 1933 and 1934 a similar number escaped from a farm at Hjärtum near the River Göta älv in Göteborgs and Bohus *län*.

Subsequent escapes occurred in two main areas: the northern one extended over the coastal districts of mid-Norrland, while that in the south included the regions around the extensive system of large eutrophic lakes in the plains of Gotaland; the latter area consisted primarily of those districts south-west of Lake Vänern, and around lakes Takern and Hjälmaren.

In the 1930s, therefore, mink were only locally distributed in the wild in Sweden and the tendency towards dispersal was slight, probably because the population density was insufficient to make expansion of range necessary.

During the Second World War the population of wild mink increased considerably, though the rate of spread was still slow. Large numbers were trapped, especially in Västernorrlands *län*; many of the watercourses in this district (and in Gävleborgs *län*) debouch from Jämtlands *län* where the catches of mink – mostly in the east – noticeably increased, indicating a dispersal westwards.

In 1941–2 mink were trapped in the river system of the Dalälven in western Kopparbergs *län* and in the northern valley of the River Klarälven in Värmlands *län*, these animals almost certainly originated in Norway (where the fur-farming industry suffered during the war from much the same problems as in Sweden), some of whose rivers run into that country.

During the war years (when predation on domestic fowl, waterbirds and small game was first reported) the number of mink trapped in southern Sweden also increased, and there is believed to have been some dispersal from the region of the River Göta älv. On the basis of trapping figures lakes Hjälmaren and Mälaren in Örebro *län* appear to have supported the highest populations, with probably some dispersal from the former to the latter.

In the early post-war years, from 1946 to 1949, the number of fur-farms in several districts – especially in Blekinge *län* – multiplied considerably. The largest uninfested areas were to be found in the extreme north and south of the country: elsewhere dispersal now occurred polycentrically, and subsequent spread can only be traced by the trapping figures for the various districts.

Between 1949 and 1954 wild mink populations continued to increase and to extend their range. During the trapping season of 1952–3 mink were captured in all *län*, the heaviest catches being recorded in Blekinge, Västernorrlands, Jämtlands and around Lake Vänern in Orebro *län*. In southern Sweden the main centre of dispersal (into Kronobergs) was along the larger watercourses of Blekinge which flow from north to south. From southern Älvsborgs, where the rivers run from north-east to south-west, mink spread into Hallands *län*.

Between 1954 and 1959 the annual catch of wild mink in Sweden more than doubled from 6,600 to 14,000: during this period the south of the country reported catches above the national average, suggesting both an increase in the population and an extension of range. In northern Sweden the catch increased markedly in Västernorrlands and Västerbottens – in the latter as the result of continued northward expansion. Göteborgs and Bohus also reported an increase in the number of mink trapped, probably as a result of colonization of the archipelago which supplies a rich source of food and adequate shelter. A large increase in the catch of mink also occurred in Kronobergs where the abundance of watercourses provide an ideal habitat.

The number of mink trapped between 1959 and 1964 remained at much the same level as during the previous five-year period; the highest figures were reported from two coastal *län* – Blekinge, and Göteborgs and Bohus – with a tendency towards a continued population increase in the south, where one of the last areas to be invaded (in 1963) was the Island of Öland. Only the mountainous regions in the far north remained uninfested.

The principal reasons for the success of the mink in becoming successfully established in Sweden (and, indeed, elsewhere in Europe) are the similarity of the climate with that of its North American homeland, an ample supply of suitable habitats, an absence of important predators, and – perhaps most important of all – the fact that mink are catholic facultative feeders, adapting their diet to the seasonal availability of prey: in summer, when their activity renders them most vulnerable, crayfish are a staple food; fish are eaten mainly in winter and spring, birds when on migration or nesting, and rodents throughout the year.

Much has been written about the interaction between mink and the mainly piscivorous otter.[7] In southern Sweden the food of mink and otters overlaps by as much as 60–70 per cent, though they tend to take fish of differing sizes. In general there appears to be an inverse correlation between the density of otters and mink due to habitat preference and specific interference. Although in some places relatively dense otter populations may keep those of mink low, in most areas mink appear to have an adverse effect on otter populations because of competition for food – especially in winter and spring – which results in the confinement of otters to optimal habitats.

There is also some evidence to suggest that otters are reluctant to breed where mink abound. In the Altai Mountains of the Soviet Union mink are held to be responsible for driving otters from fast-flowing montane streams to the more sluggish lowland rivers, where they are more at risk from pollution as well as from man and other predators.

In the 1960s and 1970s Dr Rune Gerell of the University of Lund made a study of the food habits of mink in Sweden.[8] In the west coast Lysekil Archipelago he found that eider ducks and puffins had deserted their nesting sites – the latter from their only locality in Sweden – none of which had been recolonized. Between 1972 and 1974 Gerell found that eider ducks and gulls initially suffered severe predation after the arrival of mink, but that subsequently a balance between them was established – indeed a number of young mink were believed to have themselves fallen prey to gulls. Hole-nesting species were found to be most at risk, black guillemots and razorbills (*Alca torda*) suffering a decline in numbers during the 1960s in colonies on the east coast of Sweden.

Finland

Mink for ranching were first imported to Finland[9] in the late 1920s, but it was not until after the Second World War that the industry became firmly established: by 1964 some 1.5 million mink were distributed among more than 2,000 farms, most of which were situated in coastal districts.

Between 1951 and 1964 a total of 734 reports were received of mink established in the wild, of which 570 (78 per cent) were from communes within 50 km of the coast. The largest areas of infestation were in the province of Vaasa and in the southwestern corner of the province of Oulu to Kalajoki, where mink extended inland in the communes of Ilmajoki, Nurmo and Lapua – in the last up to 100 km from the sea; it was believed that this region of Ostrobothnia contained more mink than the whole of the rest of Finland.

Another extensive area of colonization occurred around Turku and Naantali in the south-west and on the inner islands of the Turku Archipelago: here, infestation extended to communes south-east of Turku and to Hanko, Tammisaari and neighbouring islands in the province of Uusimaa. Reports were also received indicating the presence of mink around Rauma and Pori.

In the early 1960s mink increased rapidly in the Åland Islands and first appeared in communes around Porvoo and on nearby islands off the coast of the Gulf of Finland. Inland, mink were well established solely in eastern Lapland and in communes around Jyväskylä; elsewhere they occurred only sporadically.

Since the 1960s mink have continued to expand their range in Finland, where they have no serious predators, and are themselves reported to take a heavy toll of gamebirds and small mammals; the European mink (*M. lutreola*) has been legally protected in Finland since 1945, but is declining largely through interspecific competition from its more aggressive relation. As Westmann points out, even had there been no farms in Finland for mink to escape from, colonization would almost certainly have taken place by invaders from Norway and Sweden.

Denmark

In Denmark about one million mink are kept in captivity; between 1,000 and 3,000 are reportedly shot and trapped annually in the wild, where they are known only to occur in the Vejlerne – a large marshy tract which is the most important wetland area for birds in the country, but where so far only limited damage to colonies of black-headed gulls (*Larus ridibundus*) has occurred.

On the Faeroe Islands mink-farming commenced during the 1940s, and by 1960 one ranch contained between 6,000 and 7,000 breeding females: although this farm was well fenced a number of mink managed to escape after a river nearby overflowed its banks; one was seen killing domestic poultry 2 km away and was eventually found in a den with the remains of two brown trout (*Salmo trutta*).

Netherlands[10]

Exactly when mink farming began in the Netherlands in uncertain, but individuals which had presumably escaped were first observed in the wild in the mid-1950s; free-living populations were not established for some years thereafter.

As in North America mink in the Netherlands prey on muskrats (*Ondatra zibethicus*),[11] and are frequently caught by muskrat-trappers. In the last decade the number of mink taken in places also infested by muskrats in the province of Noord-Brabant increased from 1 in 1971 and 19 in 1974 to 50 in 1979, the last figure representing almost half the total for the whole country. As muskrats expanded their range (which in 1979 covered around 75 per cent of the country) mink did the same, and only the peatlands of Utrecht, Zuid-Holland, Noord-Holland and Friesland remained comparatively uninfested by both animals.

Although mink were believed to have been breeding in the wild in the Netherlands for several years previously – especially in eastern Brabant (Asten and Deurne) and in Zeeuws-Vlaadesen (Zuiddorpe) – breeding was only confirmed as recently as 1979 when young were observed in an old muskrat nest on a small island in a fish pond at Helenaveen in Noord-Brabant.

France

American mink occur in northwestern France – the only country in western Europe possessing a remnant population of the European mink – where concern has been expressed for the survival of the latter in the face of interspecific competition from its North American congener.

Spain

American mink that escaped from fur-farms in about 1982 are believed to be responsible for the decline in numbers of the Pyrenean desman (*Galemys pyrenaicus*) – a species included in *The IUCN mammal red data book* – and for the extinction of the subspecies *G. p. rufulus* in mountain streams of the Spanish Central Pyrenees.

Elsewhere in Europe American mink have been farmed commercially in Germany (by 1926), Switzerland, Belgium, Austria, Italy, Czechoslovakia, Poland, Bulgaria, Hungary and Romania, in some, if not all, of which feral populations have from time to time become established.

EURASIA

USSR

American mink were first introduced to the Soviet Union for farming in captivity in 1928, although large-scale importations for release into the wild did not commence until 1933. In that year a single shipment was made to Karelia, where the species is now the region's most economically important fur-bearer.

In the Volga-Kama Territory 178 mink were released into the wild in the Tatar ASSR in 1934, 72 in Bashkiria in 1935, 192 in Mari in 1948 and around 150 in the adjacent Chelyabinsk Region in 1960–2 where they thrived and soon became the most important fur-bearer of the area, spreading from Tatar to the Ul'yanovsk and Knibyshev regions.

In 1956 94 mink (28 males and 66 females) were translocated from the Choya area of the Gorno-Altai Region to the nut-forests of the south and the valleys of the Kara-Ungur and Kugart rivers in Kirgizia. They soon began to spread and – reinforced by 11 pairs obtained from a *sovkhoz* (breeding establishment) at Frunze which were turned out in the valley of the Sary-Kamysh River in 1961 – had, within five or six years, reached densities which in some places justified the commencement of commercial trapping. In the Sary–Chelek Reserve mink were introduced primarily to control rodents – especially the Turkestan rat (*Rattus (turkestanicus) rattoides*)) and the long-tailed field or wood mouse (*Apodemus sylvaticus*) – which are serious pests in the nut-growing and fruit-growing districts of southern Kirgizia.

In 1962 100 pairs of mink from the *sovkhoz* at Frunze were released in the Susamyr forests, in the floodland of the upper reaches of the Susamyr River and along the Karakolka River, where they became established.

In western Siberia naturalized American mink are somewhat patchily distributed. They have succeeded best in the Altai Mountains and in the Kemerovo Region; in the former 443 were released between 1937 and 1940 supplemented by a further 133 in 1950. Unfortunately they compete with several native mustelids – especially Siberian weasels or kolinskys (*Mustela sibirica*) – and in some places have displaced otters from the fast-flowing mountain streams into the larger and more torpid lowland rivers where they are more vulnerable to predation and pollution. In the taiga the population was high and widely distributed as early as 1966.

American mink were first translocated (from the Anyui and other rivers) to the Soviet Far East in 1936, when 3,040 were turned down in no fewer than 52 localities in the Khabarovsk and Maritime territories. In the latter, where introductions continued until well into the 1960s, 961 were released in the Sankhobe River area of the Sikhote-Alin Reserve. Within 30 years they had become widely distributed over an area of 2,800 km^2 in both territories, where they found abundant food and shelter, few predators, but several competitors for food, such as Siberian sable (*Martes zibellina*), raccoon-dogs (*Nyctereutes procyonoides*),[12] badgers (*Meles meles*), stoats (*Mustela erminea*) and Siberian weasels. Trapping began within five years of the initial releases.

Between 1938 and 1961 a further 4,162 mink were translocated to the Soviet Far East – 2,094 to the Khabarovsk Territory, 946 to the Maritime Territory, 539 to the Amur Region, 310 to Magadan and 273 to Sakhalin – plus more to Kamchatka; almost everywhere they settled well and became established. In Kamchatka mink escaped from a *sovkhoz* at Avacha and from another at Ust'Kamchatsk, where they colonized parts of the valley of the Avacha River and compete for food with the native sable. On Sakhalin Island 79 mink translocated from the Khabarovsk Territory to the valley of the River Poronai adapted well and by 1960 – with the aid of additional escapes from farms – had colonized some 2,500 km^2. Trapping began in 1961.

In 1950 37 mink from Lithuania were translocated to the Utena District where, surprisingly, they failed to establish themselves. Seventy-six from the Tatar ASSR released in the Anikshchyai and Ukmerge districts fared better, adapting well to the new habitat and soon breeding. Forty-four mink turned down in the Arkhangelsk Region in 1947 multiplied rapidly and before long had colonized several river valleys. Between 1953 and 1958 895 mink (725 between 1954 and 1957) were distributed among 12 lakes and rivers in Belorussia where they soon became established.

By 1948 over 3,700 American mink had been released at over 50 sites in the Murmansk Region: large numbers had also been turned down in the Karelo-Finnish SSR; the Kalinin and Voronezh regions; the Tatar, Mari and Bashkir ASSRs; the Georgian and Azerbaijan SSRs; the Sverdlovsk, Omsk, Novosibirisk and Kemerov regions; the Altai Territory; the Gorno-Altai Autonomous Region; the Krasnoyarsk Territory; the Irkutsk Region; the Buryat-Mongolian ASSR; the Chita Region; the Khabarovsk Territory; the Amur Region; and the Maritime Territory.

By 1962 the total had risen to 16,709 mink released at nearly 200 places in six republics of the Soviet Union – Azerbaijan, Georgia, Belorussia, Kirgizia, Ditovskaya and the Russian Union of Federated Soviet Republics (RSFSR). Most were turned down in the last of these, where they are currently at large in four krays (regions), 8 ASSRs (autonomous republics) and 20 oblasts (provinces). They have also become established on at least two islands – Sakhalin and Urup.

The results of releases show a rather uneven pattern, and in many areas mink are established only in non-commercially viable numbers; in most places, however, they have adapted well to new conditions, and their range and population are still increasing. In general they have fared best in Siberia and in the Far East – especially in Primor'ye, Priamur'ye and in the Altai – and are economically most valuable in the Altai, Krasnodar, Khabarovsk and Primorsk krays, in Bashkiria and in the oblast of Kemerovo. In the Soviet Union as a whole *M. vison* and the European mink (*M. lutreola*) each provide about 50 per cent of the annual harvest of wild pelts.

In the Soviet Union mink feed largely on rodents – mainly bank voles (*Clethrionomys glareolus*), grey red-backed voles (*C. rufocanus*), water voles (*Arvicola terrestris*), economically valuable introduced muskrats,[13] young Arctic hares (*Lepus timidus*) and young brown hares (*L. capensis*), house mice (*Mus musculus*) and brown rats (*Rattus norvegicus*) – as well as on fish (e.g. large pike and catfish), frogs (Ranidae), freshwater crayfish (Isopoda), ground-nesting birds and aquatic insects. Their principal predators are wolves (*Canis lupus*), red foxes, in some places otters, lynx (*Felix lynx*), wolverines (*Gulo gulo*), yellow-throated martens (*Martes flavigula*), and diurnal and nocturnal (e.g. eagle owl) raptors.

It is ironic that, as Novikov (1962) points out in his *Carnivorous mammals of the fauna of the U.S.S.R.*, 'in regions where the American mink has become acclimatized, it is displacing the small European form'.

For many years American mink were reared in captivity on the island of Hokkaido, Japan, but none are known to have escaped or become established in the wild.

Mink appear first to have been imported to insular Newfoundland[14] from mainland Nova Scotia in 1934 to stock a farm in the south-east near St John's and others at Springdale in the north-west. By 1952 about 70 farms had been established which within two years had increased to about 100 – mainly in the Avalon and Bonavista peninsulas, and in Notre Dame Bay – with a total population of between 50,000 and 60,000, yielding 25,000 to 35,000 pelts per annum. The last of these farms became defunct in 1971 by which time, however, the descendants of earlier escapes – supplemented by two deliberate releases by the Canadian government in 1948 (18 on Chapel Island in Notre Dame Bay and 13 on Swale Island in Newman Sound, Bonavista Bay) – were firmly established.

The first wild-trapped mink were caught near Springfield during the 1938–9 season, and by 1944 mink were also being captured in the wild around St John's. In the late 1940s and early 1950s large numbers of mink escaped from ranches in the Corner Brook area, where they soon became established. No mass escapes were reported from the Bonavista Peninsula where mink were first trapped in the wild in 1961. Seven years later mink first appeared in the Musgrave Harbour district, probably having made the sea-crossing from Chapel or Swale islands.

By the early 1970s mink occupied most of the suitable habitats in Newfoundland and were continuing to increase and spread into hitherto unoccupied districts; only the southern half of the Burin Peninsula, the northern half of the Great Northern Peninsula and the mountainous west coast regions remained relatively uninfested. Their success has been due to a combination of an amicable climate, suitable habitats, an abundance of food, and an absence of important predators other than man who annually harvests an average of over 1,000 pelts.

In 1959 and 1960 the Falkland Islands Company imported 96 mink in two shipments (which were to be fed on the losses and culls from sheep flocks) as the founder-stock of a commercial fur-farming venture which, perhaps fortunately, failed shortly thereafter.

After an application to legalize mink-farming in New Zealand, the country's Wildlife Service issued a strongly worded warning in 1980 of the devastation to New Zealand's already impoverised fauna which would almost certainly result from the granting of such an ill-considered request.

In 1956 three mink – a male and two females, believed to be *M. lutreola* rather than *M. vison* – were released on l'Île du Chat, Kerguelen to control the population of introduced rabbits, but by 1963 they had died out.

The **Siberian mink** (*Mustela lutreola sibiricus*) has been translocated within the Soviet Union to Central Russia.

NOTES

1. Linn & Chanin (1978a,b) and Chanin & Linn (1980).
2. See p. 323.
3. See p. 372.
4. Fjeldså (1975).
5. *Oikos* **23**: 35–58 (1972).
6. For much of the information on the spread of mink in Sweden I am indebted to Gerell (1967, 1968).
7. E.g. S. Erlinge, *Svensk Jakt* **98**: 563–6 (1960); *Skånes Naturskyddsförenings Arsberättelse* **50**: 133–46 (1963); *Oikos* **18**: 186–209 (1967), **19**: 81–98, 259–70 (1968), **20**: 1–7 (1969), **23**: 327–35 (1972); *Viltrevy* **4**: 343–71 (1967). See also R. Gerell, *Oikos* **18**: 233–46 (1967), **20**: 451–60 (1969), **21**: 160–73 (1970) and Gerell (1967, 1968).
8. See under note 7.
9. For much of the information on mink in Finland I am indebted to Westmann (1963).
10. For much of the information on mink in the Netherlands I am indebted to Litjens (1980a).
11. See p. 336.
12. See p. 44.
13. See p. 339.
14. For the history of mink in Newfoundland I am indebted to Northcott, Payne & Mercer (1974).

Small Indian Mongoose

(*Herpestes auropunctatus*)

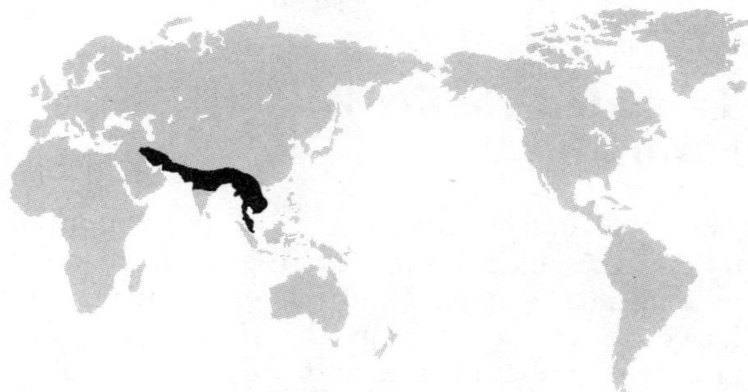

NATURAL DISTRIBUTION
Northern Arabia east through Iraq, Iran, Afghanistan and Kashmir; northern India east to Nepal, Assam and Burma; Thailand, Indo-China and Malaysia.

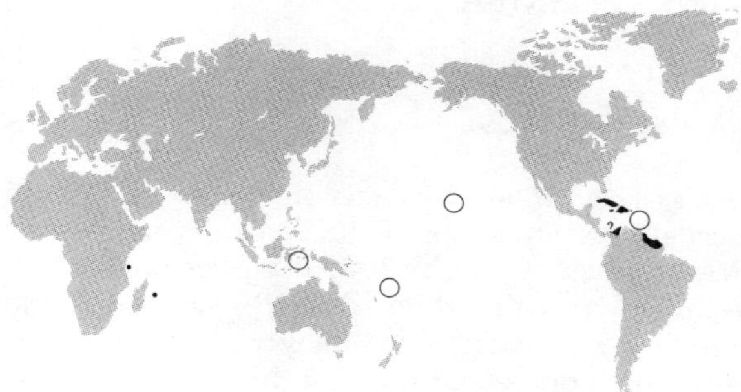

NATURALIZED DISTRIBUTION
Asia: Indonesia (Ambon); Africa: Tanzania (Mafia Island), Mauritius; North America: many islands in the West Indies; South America: Surinam, Guyana, French Guiana, ?Colombia; Oceania: Hawaii, Fiji.

ASIA

Indonesia (Ambon)

According to Haltenorth and Diller (1980), in their *Field guide to the mammals of Africa, including Madagascar*, the small Indian mongoose has been introduced to the island of Ambon (Amboina) off the southwestern corner of Seram (Ceram).

AFRICA

Tanzania (Mafia Island); Mauritius

The small Indian mongoose has been introduced to and is established on Mafia Island off the coast of Tanzania, and on Mauritius where it was released to control rats in the sugar-cane plantations. On the latter it also preys on the olive white eye (*Zosterops chloronothus*) and the Mauritius fody (*Foudia rubra*), and has been largely responsible for confining the introduced guineafowl (*Numida meleagris*) to the Black River Gorges area.

NORTH AMERICA

West Indies

As early as 1654 black rats (*Rattus rattus*) were a serious pest on some of the French islands in the West Indies, where by 1789 it was estimated that – together with brown rats (*R. norvegicus*) and grey or 'cane-piece' rats – they were annually destroying a quarter or more of the Jamaican sugar-cane crop.

In an attempt to combat these depredations Sir Charles Price, Bt, introduced some ferrets (*Mustela furo*) from England in about 1750 which were, however, apparently rendered ineffective by attacks from chigoes (*Tunga penetrans*), a form of burrowing flea also known as chiggers or jiggers. A year or so later Price apparently made a second attempt with some unknown species of mustelid from South America, which was called by the natives the 'Charley Price Rat', but this, too, was unsuccessful.

In 1762 a planter, Thomas Raffle,[1] introduced from Cuba an ant (*Formica omnivora*), which preyed fairly successfully on young rats and became known locally as the 'Tom Raffle' ant. Unfortunately the ants themselves soon reached pest proportions, and in an only partially successful attempt to control them and the rats another planter, Mr Anthony Davis, in 1844 introduced to Jamaica from Barbados (to which they had been imported from Cayenne via Martinique) the giant toad (*Bufo marinus*).

As early as 1814, in *Hortus Jamaicensis*, John Lunan was advocating the introduction of mongooses from India which 'might extirpate the whole race of the vermin'. Although at least half-a-dozen unsuccessful attempts were made to establish the animals in Jamaica before 1872 (most of which had been bred in captivity in England and were said to be 'literally afraid of a rat'), it was not until that year that Mr W. Bancroft Espeut – prompted by a suggestion from his wife and having unfortunately overcome initial government opposition by 'repeated and urgent solicitation' – made the first successful introduction to his Spring Garden estate at Portland:

> . . . on the 13th February, by the East-Indian ship 'Merchantmen', I received [from Calcutta] 9 of these animals, 4 males and 5 females, one large with young. I paid for them £9, 'in reimbursement of cost of attending the procuring and transmitting'. I let these nine animals out in four lots, three pairs and one lot of three . . . Within a few months young ones were seen about; and in less than six months there was evidence, clear and certain, that the rats were much less destructive than had ever been known . . . [by 1875] I enjoyed relief and immunity . . . within three years neighbouring estates found a similar benefit; and some of my brother sugar-planters . . . began to buy all they could procure from the natives, who, setting traps on my lands, stole all the Mungooses [*sic*] they could obtain, and sold them at prices ranging from £1 at first down to 5s each . . . By this means, and naturally, the Mungoos has now become general all over the island; and the beneficial results of the introduction of this useful animal may safely be taken as exceeding £150,000 a year.[2]

As a result of its spread in Jamaica the mongoose became known to the natives as 'Massa Espeut's Ratta'. 'I question much', Espeut continues, 'if such enormous benefit has ever resulted from the introduction and acclimatization of any one animal, as that which has attended the Mungoos in Jamaica and the West Indies; and I marvel that Australia and New Zealand do not obtain this useful animal in order to destroy the plague of Rabbits in those countries.' Since mongooses were already preying on other species of wildlife apart from rats, and were attacking the islanders' domestic stock, the 'marvel' is that Espeut could write with such naivety and complacency. Mr D. Morris, the Director in 1882 of the Jamaican Public Parks and Gardens, was more perspicacious: 'The introduction and complete naturalization of an animal possessing such strong predatory habits and remarkable powers of reproduction, must have an important influence on all indigenous and introduced animals capable of being affected by it.'[3]

Espeut claims to have exported mongooses from Jamaica to Cuba, Puerto Rico, Grenada, Barbados, Trinidad[4] (where five were despatched to Mr Eugene Lange at Santa Cruz in about 1875), and elsewhere. In Cuba, where they spread only slowly due to unfavourable climatic and ecological conditions, they were not well established until 1886, and even by 1929 only occupied some 2,600 km^2 around Havana: today they are still confined to the western regions of the island. In Puerto Rico mongooses were established by about 1877 on the coast at Arecibo, San Juan, Fajardo, Arroyo, Ponce and Mayaguez, and inland at Utuado and Adjuntas. Barbados was colonized between 1877 and 1879, and the Dominican Republic, Hispaniola, before 1895.

According to a letter in the *St Croix Avis* of 20 February 1884 the mongoose was introduced to the island from Jamaica earlier in that year (some accounts say 1882), where it was soon established. Other introductions in the Virgin Islands were made before 1898 to Vieques and St Thomas, and to St John, Tortola and Guana. Elsewhere mongooses were successfully introduced to Martinique (and subsequently to Long Island) and St Lucia to control the venomous pit viper or *fer-de-lance* (*Bothrops atrox*) of tropical South America – which had possibly itself been introduced to the islands – and also to St Vincent, Marie Galante, Guadeloupe (and possibly to La Désirade), Antigua, St Martin, St Kitts and Nevis. Two attempts were made in the 1880s to establish the mongoose on the island of Dominica; both were unsuccessful, possibly due to predation by the boa or *tête-du-chien* (*Constrictor c. orophias*), but more probably because of the island's high precipitation. The mongoose's present distribution in the West Indies – which includes nearly all the larger islands such as Cuba, Jamaica, Hispaniola, Puerto Rico, Trinidad, St Croix, Buck, Vieques, St Thomas, St John, Tortola, St Kitts, Nevis, Antigua, St Martin, Guadeloupe, Marie Galante, probably La Désirade, Martinique, St Lucia, St Vincent, Barbados and Grenada, but excludes Dominica, Culebra, Saba, St Eustatius, Montserrat, Barbuda, Carriacou and several small islets and cays such as Protestant and Green which offer sanctuary to species decimated by mongooses elsewhere – had been achieved by around the turn of the century, by which time most authorities (in Jamaica by as early as 1883) had legislated against its importation.

In the *US Department of Agriculture Yearbook for 1898*, T. S. Palmer summed up the activities to date of the mongoose in Jamaica:

> Its omnivorous habits became more and more apparent as the rats diminished. It destroyed young pigs, kids, lambs, kittens, puppies, the native 'cony' or capromys, poultry, game, birds which nest on or near the ground, eggs, snakes, ground lizards, frogs, turtles' eggs and land crabs. It was also known to eat ripe bananas, pineapples, young corn, avocado pears, sweet potatoes, coconuts, and other fruits. Toward the close of the second decade the mongoose, originally considered very beneficial, came to be regarded as the greatest pest ever introduced into the island. Poultry and domesticated animals suffered from its depredations, and the short-tailed capromys, *Capromys brachyurus*, which was formerly numerous, became almost extinct except in some of the mountainous districts. The ground dove, *Columbigallina passerina*, and the quail dove, *Geotrygon montana*, became rare, and the introduced bobwhite, or quail, was almost exterminated. The peculiar Jamaica petrel, *Aestrelata caribboea*, which nested in the mountains of the island, likewise became almost exterminated. Snakes, represented by at least five species, all harmless, and lizards, including about twenty species, were greatly diminished in numbers. The same thing was true of the land and freshwater tortoises and the marine turtle, *Chelone viridis*, which formerly laid its eggs in abundance in the loose sand on the north coast. The destruction of insectivirous [*sic*] birds, snakes, and lizards was followed by an increase in several injurious insects, particularly ticks, which became a serious pest, and a Coccid moth, the larvae of which bore into the pimento trees. In 1890[5] a Commission was appointed by the government to consider whether measures should be taken to reduce the number of the animals, and the evidence collected showed conclusively that the evil results of the introduction of the mongoose far out-weighed the benefits to the sugar and coffee plantations.[6]

In 1918 C. B. Williams[7] analysed the stomach contents of over 180 mongooses in Trinidad; as a result of his investigations he estimated that the food of one mongoose over a period of three months amounted to 26 rats, over 500 grasshoppers, 14 birds, 17 lizards, 18 snakes and 30 frogs or toads (including *Bufo marinus*, to whose venom the mongoose appears to be resistant or immune).

As in the case of feral cats in New Zealand the havoc wrought by mongooses on wildlife in the Caribbean reads like a tale of disaster. Among mammals by about 1880 they had exterminated the Jamaican rice rat (*Oryzomys antillarum*), by about 1897 the St Vincent rice rat (*O. victus*), and subsequently the Antillean and St Lucia giant rice rats (*Megalomys desmaresti* and *M. luciae*) on Martinique and St Lucia. On St Croix they attack and kill fawns several days old and more than twice their own weight (700 g) of the Virginia or white-tailed deer (*Odocoileus virginianus*), introduced to the island from North America in about 1790.[8] On Grenada the introduced agouti was nearly exterminated by mongooses as long ago as 1910. By 1972 Brown's hutia (*Capromys brownii*) was confined – through predation by mongooses and man – to the south-east portion of Jamaica, where according to Mittermeir (1972) it appears to have 'struck an equilibrium with the mongoose, as have a number of other Jamaican animals'. Other Capromyidae endangered by mongoose predation include the bushy-tailed or black-tailed hutia (*C. melanurus*) and the dwarf hutia (*C. nanus*) – both of Cuba – and the Hispaniolan hutia (*Plagiodontia aedium* (*hylaeum*)).[9] Mongooses are believed to be at least partially responsible for the decline of the Cuban and Haitian solenodons (*Solenodon cubanus* and *S. paradoxus*). On Trinidad they have endangered or exterminated the water vole (*Nectomys palmipes*), the rice rat (*Oryzomys brevicauda*), the spiny pocket mouse (*Heteromys anomalus*) and the spiny rat (*Loucheres guianae*).

The part played by mongooses in the decline of many species of Caribbean birds is confused by predation also by black rats and opossums (*Didelphis marsupialis insularis*), locally known as manicous. According to J. Bond[10] mongooses were probably responsible for the extirpation on Jamaica by 1881 of the uniform crake (*Amaurolimnas concolor*) – known locally as the water partridge – which remains widespread in tropical America. By 1905 the ruddy quail dove (*Geotrygon montana*) was extinct on St Vincent and had become rare on Grenada, where five years later the island's dove (*Leptotila wellsi*) was also very uncommon. At one time the bridled quail dove (*G. mystacea*), which later staged a recovery, was nearly exterminated on the island of St Croix. Other birds to have suffered include the introduced bobwhite quail (*Colinus virginianus*); the black-capped petrel or diablotin (*Pterodroma hasitata*); two subspecies – *guadeloupensis* on Marie Galante and *amaura* on St Kitts, Nevis and Antigua – of the burrowing owl (*Athene cunicularia*), both late last century; the rufous nightjar (*Caprimulgus rufus otiosus*) which may be conspecific with Chuck Will's widow (*C. carolinensis*), on St Lucia; the Puerto Rico whip-poor-will (*C. vociferus noctitherus*); house wrens (*Troglodytes aedon* (*musculus*)) in the Lesser Antilles, except on Dominica (where there are no mongooses) and Grenada; the pearly-eyed thrasher (*Margarops fuscatus densirostris*) on Guadeloupe, Martinique and St Lucia; the white-breasted thrasher (*Cinclocerthia brachyurus*); and Semper's warbler (*Leucopeza semperi*) on St Lucia. Bond suggests that the guineafowl, which was introduced in 1508 from west Africa to the West Indies where it is still widely kept in domestication, would probably have become feral on many islands were it not for the mongoose. In 1977 Mrs A. S. de Dod told J. L. Long that in the Dominican Republic feral guineafowl were rapidly decreasing due to over-shooting and mongoose predation.[11]

According to Barbour (1930), by 1929 the following valuable insectivorous lizards and skinks had been seriously affected by mongoose predation: *Ameiva polops* (extinct on St Croix but thrived on Protestant and Green Cays where mongooses are absent); *A. exsul* (extinct on St Thomas – where its main prey, larvae of *Phyllophaga*, reached plague proportions – but common on Culebra where mongooses do not occur); *A. aquilinia* (extinct on St Vincent); other species of the genus *Ameiva* – e.g. *cineracea* and *dorsalis* – which formerly occurred as distinct species on most of the larger islands, were extinct on Nevis, Guadeloupe and St Lucia, and were virtually confined to settled districts in St Kitts and Antigua. On St Eustatius, Montserrat and Dominica (where mongooses were absent) the lizards were still common; *Mabuya sloanii* (extinct on St Thomas and St Croix, and rare or extinct in Puerto Rico); *M. mabouia* (extinct on Martinique); *M. luciae* (extinct on St Lucia); *M. aenea* (extinct on St Vincent and Grenada, and rare on Trinidad); *M. lanceolata* (extinct on Barbados); and *Gymnophthalmus pleii* (extinct on Martinique and St Lucia). Harmless or useful snakes depleted by mongooses included *Alsophis rufiventris* (extinct on St Kitts and Nevis, but common on Saba and St Eustatius where the mongoose is absent); *A. antillensis* (extinct on St Thomas); *A. sanctae-crucis* (extinct on St Croix); *A. leucomelas* (extinct on Guadeloupe and Marie Galante); *Drymobius boddaerti* (extinct on St Vincent and rare on Grenada); *Leimadophis exiguus* (extinct on St Thomas); *L. cursor* (extinct on Martinique); *L. boulengeri* (extinct on St Lucia); *L. perfuscus* (extinct on Barbados); and *L. melanotus* (extinct on Grenada but common on Trinidad and the mainland).

Since Barbour's time the blind snake *Typhlops richardi* has become very rare (or extinct) and the iguana (*Iguana iguana*) greatly depleted on St Croix; on Trinidad the ground lizard *Ameiva surinamensis*, the matte, teguexin, teju or jacuaru (*Tupinambis nigro-punctatus*), the slippery back lizards *Mabuya aurata* and *M. agilis*, ground lizards of the genus *Cnemidophorus*, and the snakes *Spilotes pullatus*, *Coluber corais* and *Oxyrhopus coronatus* have all been much reduced or extinguished. In Jamaica the iguana *Cyclura collei* became extinct (or nearly so) due to overhunting of the adults by man and predation of the eggs and young by mongooses (a remnant population survived on Goat Island until the arrival of mongooses in about 1943); adults of the Jamaican boa (*Epicrates subflavus*) were killed by man, and the young by mongooses; the skinks *Diploglossus occiduus* and *Mabuya spilonotus* were both probably exterminated by mongooses, as was the sluggish and diurnal ground snake *Alsophis ater*. Elsewhere *A. rijgersmaei, Iguana iguana rhinolopha* and land crabs have suffered severe depredation; on some sugar plantations the introduced toad *Bufo marinus* and piping frogs of the genus *Leptodactylus* have also been depleted. The destruction of many insectivorous vertebrates is alleged to have resulted in an increase on many islands of insect pests. The eggs and young of the endangered green turtle (*Chelonia mydas*), and probably those of the brown pelican (*Pelecanus occidentalis*) and the magnificent frigate bird (*Fregata magnificens*) – especially on Buck Island off St Croix – also suffer from mongoose predation. Since the last war the Cuban whistling duck (*Dendrocygna arborea*) has declined throughout its range, which includes Cuba, Hispaniola, Puerto Rico and the Virgin Islands and used also to include Jamaica, due to a combination of overshooting, the collection by man of eggs for food, and mongoose predation. On some islands the slaughter of domestic stock – particularly poultry – has had serious economic consequences.

In 1952 two pairs of mongooses were introduced (ostensibly for pest-control purposes) to the 182-ha Buck Island. Ten years later the island was designated a National Monument by the US government, and a campaign designed to eradicate the mongooses, which had already caused several species of nesting seabirds to desert the island, was begun by the National Park Service. By 1972 the campaign was abandoned as a failure, and the reintroduced ground lizard *Ameiva polops* was itself found to have been exterminated; now officially listed as endangered by the US Fish and Wildlife Service, a remnant population estimated to number around 300 survives only on the mongoose-free 1.6-ha Protestant and 4.4-ha Green Cays off St Croix.

It is hardly surprising that Westermann (1953a) should have written: 'The introduction of the Indian mongoose into the Caribbean islands is an example of irresponsible interference in the natural balance, as notorious as the unfortunate importation of European rabbits and foxes into Australia.'

In 1950 it was discovered that in Puerto Rico the introduced mongoose was an important reservoir and vector of the rabies virus; this was apparently the first time that rabies had been reported in a member of the genus *Herpestes*. According to Tierkel et al. (1952), there were only 21 cases of rabies in Puerto Rico (in dogs and domestic stock) between 1911 and 1923, and from the latter year until 1950 the island was considered to be one of the comparatively few rabies-free areas in the world (Pimental 1955a). Taylor (1965) states that on Grenada 'The number of cases of human infection with the disease through the bite of a mongoose is few but there have been numerous reports of domestic animals (donkeys and cattle) being bitten and subsequently dying of rabies-like symptoms.'

The only benefits the mongoose conveys in compensation for the havoc it wreaks in the Caribbean – where it is regarded with favour only by some planters – is the possible destruction of some injurious insects and the control of rats and mice. In respect of rats, Myers (1931) wrote:

It has been suggested [by Williams (1918) et al.] that

the damage done by the mongoose to birds has been offset by its destruction of rats, which would ordinarily destroy many tree-nesting birds beyond the reach of the mongoose. But it has been otherwise pointed out [by Barbour (1930) et al.] that the latter has made rats essentially arboreal animals [in both the West Indies and Hawaii], thus driving them further into the haunts of birds. It may be mentioned, however, that the black rat, which was introduced into New Zealand in the early days of settlement, eventually became almost wholly arboreal there, following on the later importation of its arch-competitor, the brown rat, which is now the domestic rat. It seems probable to me that the same thing has happened in the West Indies, [and Hawaii] and would have happened even without the intervention of the mongoose.

As Westermann (1953a) points out, arboreal rats are also a common feature on mongoose-free islands. By preying on terrestrial brown rats the only partially arboreal mongoose favours the normally inferior but largely arboreal black rat – which lives and even breeds in coconut trees – in its interspecific competition with the former. Sometimes only descending to the ground at night to feed, the black rat frequently escapes the worst attentions of the diurnal mongoose. The situation is much the same also in Hawaii.

The mongoose has thrived in the West Indies principally because it is virtually free from predators, has an ample supply of easily available prey species, and lives in an ideal climate with an abundance of suitable habitats. Its principal enemy is man, by whom it is trapped, shot, hunted with dogs and – in spite of its high degree of resistance – poisoned with strong doses of warfarin, zinc phosphide, arsenic trioxide, compound 1080 (sodium monofluoroacetate) and strychnine. Mongooses are also preyed upon by boa constrictors; as mentioned above their failure to become established on the island of Dominica may have been partially due to predation by the island's arboreal *tête-du-chien*, which also kills mongooses (and opossums) on St Lucia; on Trinidad mongooses are preyed upon by *Constrictor constrictor* and by *Boa cookii*, and on Grenada by the arboreal *B. c. grenadensis* and *B. hortulana*.

There can be little doubt that the mongoose in the Caribbean has helped to endanger or exterminate more species of mammals, birds and reptiles than any other animal deliberately introduced by man anywhere in the world. It provides the classic example *par excellence* of the havoc which can result from the attempted biological control of an already established species by the introduction of an alien outside its natural range.

In February 1892 a plan was put forward by the US Department of Agriculture to import mongooses (species unspecified, but probably *auropunctatus* from the Caribbean) to some western states to try to control fossorial pocket gophers (Geomyidae) whose burrows were a serious hazard on cattle ranches. According to Palmer (1899) only 'the most strenuous efforts' of those with a knowledge of mongoose biology prevented its implementation.

Exactly 60 years later G. A. Seaman concluded an address at the North American Wildlife Conference with these sobering words: 'I wish to beg most seriously and sincerely one thing, and that is: that under no circumstances the mongoose be considered for introduction into the American Continent.'[12] Yet in 1978 it was revealed that in the previous year a young female Indian mongoose had been trapped on Dodge Island, Port of Miami, Florida, where two had been killed eight weeks previously. Whether these were deliberately introduced from the West Indies or arrived accidentally in consignments of fruit and vegetables is uncertain, but since the climate of the Gulf Coast states appears ideal for mongooses clearly the utmost vigilance must be exercised to prevent a repetition of the occurrence, which could prove an economic and faunal disaster to the southeastern United States.

SOUTH AMERICA

Surinam; Guyana; French Guiana; Colombia

From Barbados the mongoose was introduced to Surinam in 1900,[13] and shortly afterwards to Guyana and French Guiana. According to Roots (1976), 80 mongooses were released in northern Colombia in 1951.[14] In the first three countries – where they face competition from and are preyed on by native predators – they are apparently established mainly in settled coastal areas, from where their expansion is inhibited by wide river systems and vast areas of dense tropical rainforest.

OCEANIA

Hawaii

According to Palmer (1899), some large East Indian mongooses – presumably the Indian grey mongoose *Herpestes edwardsi* – were released in about 1881 on sugar plantations near Hamakua on Hawaii to try to control black and brown rats and the Polynesian rat (*R. exulans hawaiensis*) which had become serious economic pests, but they soon disappeared. A few months later a few pairs of *H. auropunctatus* were imported from Calcutta, but nearly all were drowned while being offloaded at Hilo. In 1883 the Hilo Planters' Association raised $1,100 and commissioned Mr J. Tucker to purchase some mongooses from Jamaica; on 30 September the *City of New York* arrived in Hawaii with 11 cases containing 72[15] 'Javan Golden Brown Mongooses' (in fact *H. auropunctatus*), which were at once released on the coast between Hilo and Hamakua; two years later they were reinforced by a second consignment of 215 imported from Jamaica by Mr Joseph Marsden of New Orleans. The offspring from these two introductions, supplemented by further importations from the West Indies, were subsequently liberated on the islands of Maui, Molokai and Oahu. A proposed introduction to Kauai was foiled when a crate containing some mongooses fell (or was deliberately pushed) overboard before it could be landed. 'It is interesting to note', wrote A. J. Berger in 1972, 'that Kauai, the only one of the main islands on which the mongoose has not been introduced, is the only island which still has all the endemic birds known to have occurred there since 1778.'[16] Until the introduction of effective poisoning in 1938, plantations on Kauai suffered worse depredations from rats than those on mongoose-infested islands.

As in the Caribbean a certain measure of rat-control was soon achieved, and even now mongooses are looked upon favourably by some plantation owners. In 1924 Pemberton (1925) analysed the scats of 356 mongooses in Hawaii: he found that 88 per cent contained the remains of rodents, of which 56 per cent were composed exclusively of rodents and 36 per cent of a mixture of rodents and insects; the balance of 12 per cent consisted solely of insect remains. By about 1910, however, mongooses were beginning to turn their attention to members of the Hawaiian avifauna. As Kramer (1971) wrote in *Hawaiian land mammals*, the mongoose must be charged with an immense amount of destruction to certain species of ground-nesting birds'. Most of these are colonially nesting species which are especially vulnerable to attacks by predators. Climatic conditions, habitat and food requirements are probably more important factors in the survival of noncolonial ground-nesting birds than is predation by mongooses.

On Molokai mongooses have been accused of the extermination of the dark-rumped petrel (*Pterodroma phaeopygia sandwichensis*), and on Maui with that of Newell's shearwater (*Puffinus newelli*). On Hawaii the endemic goose or nene (*Branta sandvicensis*) was reduced from an estimated population of 25,000 in the late eighteenth century to under 50 in 1948 through excessive shooting, habitat destruction and predation by introduced rats, cats and mongooses. Thanks to a captive breeding programme at the Pohakuloa Game Farm on Hawaii by the State Division of Fish and Game and by the Wildfowl Trust in Britain the nene is now re-establishing itself in Hawaii.

In recent years mongooses have become naturalized on the island of Kauai (which had hitherto been uninfested) where they threaten the endemic Hawaiian duck or koloa (*Anas platyrhnchos wyvilliana*) in what is now the bird's last haunt.[17] Other species to have suffered severely from predation by mongooses include the introduced ring-necked pheasant (*Phasianus torquatus*) and the peaceful dove (*Geophila tranquilla*). Although Hinton and Dunn (1967) state that mongooses do not appear to prey on red-footed boobies (*Sula sula rubripes*), Roots (1976) claims that they have been seen climbing trees and taking young from their nests. On the other hand, the endemic short-eared owl (*Asio flammeus sandwichensis*) and the introduced skylark – both ground-nesters – appear to be comparatively free from mongoose molestation.

As in the West Indies the mongoose in Hawaii must be regarded as a reservoir and potential vector of rabies, an outbreak of which occurred on Oahu in 1967. According to Roots (1976), mongooses on Hawaii are also infected – probably by the brown rat – with the protozoan *Leptospira icterohaemorrhagiae*, which is the cause of Weil's disease in man. In addition they are hosts to the roundworm *Trichinella spiralis*, which could be passed on to man in the form of trichinosis by pork from pigs which have scavenged on infected mongoose carcasses.

The only positive benefit conveyed by mongooses in Hawaii is, as in the West Indies, a measure of rat control. Hinton and Dunn (1967) suggest that by keeping down the rat population mongooses may have reduced to some extent the amount of bubonic plague or Black Death in the islands – caused by *Pasteurella pestis* and transmitted by the rat flea *Xenopsylla cheopis* – which first appeared in Hilo on Hawaii in 1900, but this must be conjecture. The same authors also surmise that by spreading the seeds of such fruits as guavas and balsam apples through undigested seeds passed in their faeces mongooses may actually benefit some frugivorous species of birds, but this again is supposition. What is certain is that the ratio of good done by killing rats to the harm caused by destroying birds is greater in Hawaii than in the West Indies.

Today mongooses occur on Hawaii, Maui, Molokai, Oahu, Kauai and – since about 1965 – on Mokuoloe (Coconut Island) in Kaneohe Bay, Oahu, and on Ford Island in Pearl Harbor, on all of which they are now widespread and firmly established. On Hawaii and Maui they are found from sea-level to an altitude of over 3,000 m, though mostly below 700 m: they prefer warm and humid regions with mixed natural vegetation and plenty of shelter – especially grasslands, scrub and pineapple, sugar-cane and coffee plantations – but in places are even found on or near the coast, where they feed on crabs and stranded fish.

Fiji

In the same year that they were successfully introduced to Hawaii (1883), small Indian mongooses were released to control rats in sugar-cane plantations in Fiji,[18] where they are today common throughout Viti Levu and Vanua Levu. They also occur on Beqa and Yanuca off the former island and on Rabi off the latter. Although they have probably caused a decline in various species of ground-nesting birds, Pernetta and Watling (1978) attribute the possible extinction in Fiji of the barred-winged or Woodford's rail (*Nesoclopeus poecilopterus woodfordi*), the grass owl (*Tyto capensis*) and the lesser wandering whistling tree duck (*Dendrocygna arcuata pygmaea*),[19] and the current rarity of the red-throated lorikeet (*Charmosyna amabilis*), the pink-billed parrot-finch (*Erythrura kleinschmidti*), the long-legged warbler (*Trichocichla rufa*) and the banded iguana (*Brachylophus fasciatus*) more to habitat destruction and human interference than to mongoose predation. Roots (1976) claims that mongooses have seriously reduced the populations of three other (unspecified) rails and the snake *Ogmodon vitianus*. According to Gibbons (1984), mongooses have seriously affected such ground-nesting birds as the banded rail (*Rallus philippensis*) and some (mainly diurnal and only partially arboreal) reptiles, including the skinks *Emoia nigra* and *E. cyanura*. All these species are rare or absent on Viti Levu and Vanua Levu but relatively common elsewhere.

A few years after Espeut's unfortunate suggestion quoted above, some **mongooses** (probably *Herpestes auropunctatus* but possibly either *edwardsi* or *smithi*) were released in northern Queensland, Australia by the Victoria Sugar Company, the Hamleigh Company and the Gairloch Company, to combat rats which had followed a plague of locusts in the sugar-cane fields. Fortunately they soon all disappeared.[20]

In 1910 the **Indian grey mongoose** (*Herpestes edwardsi*) was introduced to some of the Ryukyu (Loochoo) islands in the East China Sea and to Tonaki, Japan (where it now preys on domestic poultry) to control rats and the venomous pit vipers *Trimeresurus flavoviridis* and *T. okinavensis*. It was also subsequently introduced to a number of islands in Malaysia and to Mauritius; on the latter it is said to have been partially responsible for decimating partridges, quail, black-naped hares (*Lepus nigricollis*), and Timor deer (*Cervus timorensis*) introduced from Djakarta, Java, in 1639. Fifteen grey mongooses released on a farm in Southland, New Zealand in 1870 fortunately failed to become established. In the 1960s some grey mongooses[21] were released around Monte Circeo, 100 km south of Rome, Italy, to control the venomous asp viper (*Vipera aspis*).

The **Egyptian mongoose** (*Herpestes ichneumon*) of Africa and Asia Minor, which occurs in southern Spain and Portugal probably as the result of an early introduction, has more recently been released to control rats and the highly venomous nose-horned viper (*Vipera ammodytes*) on the island of Mljet between Korčula and Dubrovnik off the Dalmatian coast of Yugoslavia where it has now taken to killing domestic poultry. It is also said to occur in Madagascar (the Malagasy Republic).

The **banded mongoose** (*Mungos mungo*) is alleged to have been introduced in about 1935 from the African mainland to the island of Zanzibar.

The **small Indian civet** or **rasse** (*Viverricula indica*) has been introduced to the Philippines, Comoros, Madagascar (where it is believed to have arrived with the early Indo-Malayan colonists), Zanzibar, Pemba, Socotra, and possibly to the mainland of Tanzania. The **large Indian civet** (*Viverra zibetha*) has been introduced to the Andaman Islands where it is locally abundant.

Two other civets – the **Malaysian** (*Viverra tangalunga*) and the **common palm civet** or **Toddy cat** (*Paradoxurus hermaphroditus (philippinensis)*) – occur on some of the Celebes and Moluccas, the latter even on the Kei and Aru isles and on some or all of the Lesser Sunda Islands east to Timor. They were probably transported by man who sold them for their musk and later released them.

In both ancient and modern times **masked palm civets** (*Paguma larvata*) have been introduced, probably from Formosa or southern China, to zoos and private menageries in several parts of Hondo, Japan. Escaped animals have occurred in the wild in Nagano, Yamanashi and Shizuoka in central Hondo, and in Shikoku and Shinano. They appear to have had no detrimental affect on any native species.

The general distribution of the **genet** (*Genetta genetta*) is Israel, Arabia and most of Africa; the absence of fossil records and of related species in Europe, coupled with a plausible motive (the control of rodent pests) for its introduction to Europe, suggests that its presence in Spain, Portugal, the Balearic Islands and southern and central France may well be the result of importations by man.

NOTES

1. Not, as stated by Espeut (1882), Sir Stamford Raffles, who was not born until 1781.
2. Espeut (1882). (Several subsequent accounts incorrectly say £45,000, and Roots (1976) says £10,000.)
3. Quoted by Espeut (1882).
4. Unsubstantiated claims have been made that mongooses were successfully introduced into Trinidad in about 1870.
5. *Journal of the Institute of Jamaica* **2:** 273–5 (1896).
6. Palmer (1899).
7. *Bulletin of the Agricultural Department of Trinidad and Tobago* **17**: 167–86 (1918).
8. Seaman & Randall (1962).
9. Some taxonomists regard this as two separate species, respectively Cuvier's and the Dominican hutia.
10. *Birds of the West Indies* (Collins, London, 1960).
11. J. L. Long, *Introduced birds of the world* (A. H. & A. W. Reed, Sydney, 1981).
12. Seaman (1952).
13. A. M. Husson, *Lutra* **2**(1): 12–13 (1960).
14. Hinton & Dunn (1967), in their comprehensive monograph *Mongooses*, make no reference to this introduction.
15. Laycock (1966) – confusing this with the second consignment – says 215.
16. *Hawaiian birdlife* (University Press, Honolulu, 1972).
17. J. Kear & G. Williams, *Wildfowl* **29**: 5–21 (1978).
18. Palmer (1899) says that an earlier introduction in about 1870 failed to become established.
19. But cf. P. Scott, *Key to the wildfowl of the world* (Wildfowl Trust, Slimbridge, 1957).
20. *Victorian Naturalist* **62**: 205 (1946). According to Palmer (1899) over 100 were also unsuccessfully released near the Murray River in New South Wales, as well as others in New Zealand (see *Final report of the Royal Commission of Inquiry into the Extermination of Rabbits in Australasia*, 9 (1890)).
21. Erroneously described by Hinton & Dunn (1967), Roots (1976) and others as the Egyptian mongoose (*H. ichneumon*).

Domestic Cat

(Felis catus)

NATURAL DISTRIBUTION
The domestic cat is derived from the European and Asian wild cat (*F. silvestris silvestris*), which ranges from western Europe as far east as central India and western China, and from the African and Arabian subspecies *F. s. libyca*.

NATURALIZED DISTRIBUTION
Now virtually world-wide as the result of early domestication – possibly in prehistoric times.

The following are some examples only of the presence throughout the world of feral cats, and of the effects they have on their ecosystems.

EUROPE

British Isles

Over the years many domestic cats have been deliberately turned out by their owners or have escaped to lead a feral existence. There have also been a number of deliberate introductions to various parts of the British Isles.

In 1930 about a dozen were set free on St Kilda off the north-west coast of Scotland, but by the following year only three remained; they are believed to have contributed to the decline and eventual extinction of the St Kilda house mouse (*Mus musculus muralis*).

In the 1890s cats were released on the island of Noss in the Shetlands to control rats, and subsequently to South Havra and to Holm of Melby in the same group. In Ireland some 36 cats were turned down on Great Saltee Island off Ballyteige Bay in County Wexford in 1950 as a form of rabbit control.

Few urban and rural areas in Britain (and in continental Europe) are today without their complement of feral cats, many of which exist independently of man.

ASIA

Japan

Cats said to have been left behind on the island of Tori Shima, one of seven volcanic peaks south of Honshu, when it was evacuated by troops stationed there during the Second World War, seriously reduced the numbers of the bush warbler *Cettia canturians panafidinicus*; further south, on Iwo Jimo, feral cats, assisted by introduced rats, exterminated the white-browed rail (*Poliolimnas cinereus brevipes*).

AFRICA

Mauritius

Cats on Mauritius have, with the help of rats and the introduced crab-eating macaque (*Macaca fascicularis*),[1] been responsible for decimating the Mauritius cuckoo-shrike (*Coracinia typica*), the Mauritius fody (*Foudia rubra*), a weaver and a flycatcher known locally as the *coq-du-bois*. They are believed also to be partially responsible (with man) for the extinction of the dodo by about 1681.

South Africa

Cats were landed on Dassen Island, north of Cape Town, in the second half of the nineteenth century, where their principal prey is introduced feral rabbits (*Oryctolagus cuniculus*). In March 1981 the population was between 65 and 85.

NORTH AMERICA

USA; Canada; Mexico

Presnall (1958) reported only one area of serious conflict in the United States between feral cats and native species; on Anacapa Island, in the Channel Islands National Monument off California, they were reported to prey on a nesting colony of brown pelicans (*Pelecanus occidentalis*). Since then cats are believed to have exterminated Leach's storm petrel (*Oceanodroma leucorhoa kaedingi*), Bewick's wren (*Thryomanes bewickii*) and the towee (*Pipilio* sp.) on the Mexican island of Guadaloupe, 260 km west of California; the rock wren (*Salpinctes absoletus exsul*) on another Mexican offshore islet, San Benedicto in the Revilla Gigedo group; and the Socorro dove (*Zenaida graysoni*) in the same archipelago, 525 km west of the Mexican mainland. With the help of pigs and dogs, feral cats are responsible for endangering in Texas Attwater's prairie chicken (*Tympanuchus cupido attwateri*). On Sable Island off the coast of Nova Scotia, Canada, cats introduced in about 1880 exterminated a population of rabbits which had themselves been introduced half a century earlier.

West Indies

Wherever they have become established in the Caribbean (as on other insular groups) feral cats have devastated the native wildlife. On Swan Island, off the coast of Honduras, the loss of an endemic thrush and the lizard *Tiaporus fuliginosus* has been attributed to feline predation, as has that of the St Vincent island house wren (*Troglodytes aedon musicus*) in the Windward Islands. Feral cats left behind to fend for themselves when the lighthouse on Navassa Island, between Jamaica and Hispaniola, was no longer manned exterminated at least seven terrestrial reptiles. Several lizards on Beata Island, off the south coast of the Dominican Republic, have been eradicated (with some help from dogs and goats) by feral cats. The Jamaican boa, which was comparatively safe from the introduced small Indian mongoose (*Herpestes auropunctatus*),[2] which is only semi-arboreal, has suffered greatly from predation by cats, as has the island's black rail (*Laterallus jamaicensis jamaicensis*); elsewhere cats (helped by rats and mongooses) have been partly responsible for endangering the Haitian and Cuban solenodons (*Solenodon paradoxus* and *S. cubanus*) – the only two existing members of their family – and two skinks, *Diploglossus occiduus* and *Mabuya spilonotus*.

AUSTRALASIA

Australia

Domestic cats almost certainly arrived in Australia in 1788 with the first settlers from England; within a century – if not before – they were feral in all suitable habitats throughout the colony; in 1896 the Calvert Scientific Exploration Expedition found one in the desert nearly 650 km from the nearest human habitation.

In Western Australia Long (1972) refers to the only recorded deliberate release of cats in the state when the colonial secretary ordered the liberation of 200 in about 1899 between Eyre Patch and Mount Ragged on the south coast, in a vain attempt to prevent the incursion of rabbits from South Australia. By 1900 feral cats in Western Australia had reached as far east as the South Australia border, and in some places were sufficiently common to be eaten by the Aborigines. In 1908 particularly large numbers were reported from around Murchison. E. Troughton, on a collecting trip in the southwest of the State in 1921, reported (1938) that feral cats were abundant and were preying on possums (Phalangeridae), including the rare honey possum (*Tarsipes spencerae*), numbats (*Myrmecobius fasciatus*), planigales (*Planigale* spp.) and marsupial mice (Dasyuridae). They are also held to have been responsible for the near-extinction of the ground parrot (*Pezoporus wallicus*), the Western Australian noisy scrub bird (*Atrichornis clamosus*), and (with the introduced red fox, *Vulpes vulpes*) for the decline of the brush-tailed or black-flanked western rock wallaby (*Petrogale penicillata lateralis*). In many districts the reduction of small insectivorous marsupials, which had evolved in a relatively predator-free environment, resulted in a proliferation of insect pests.

Frith (1979) reported feral cats in such remote areas as the Gibson Desert (Western Australia), the Arnhem Land escarpment (Northern Territory), and monsoon forest edge near the Jardine River in the Cape York Peninsula in Queensland. Feral cats are today widely distributed throughout Australia, where their numbers are continually augmented by the addition of new blood.

New Zealand

Domestic cats were probably first brought to New Zealand by visiting sealers and whalers when they began to establish shore stations towards the end of the eighteenth century. When the Russian commander Bellinghausen visited Macquarie Island and Enderby Island in the Auckland group in 1820, he found a number of feral cats which he believed had been left behind by early seal-hunters in about 1812, and where their descendants remain. Captain Musgrave, master of the wrecked schooner *Grafton*, who was a castaway on Auckland Island in 1864, trapped a cat (the descendant of a number landed in about 1840) which 'soon cleared the hut of mice, which were dreadfully common'.

Thomson (1922) recorded that in 1868 H. H. Travers found Chatham Island overrun with cats which had already killed many of the indigenous birds, and that in 1911 F. A. D. Cox claimed that on Mangere Island in the Chatham group cats liberated to kill rabbits had 'exterminated the small native birds' – presumably the endemic Chatham Island fernbird (*Sphenoeacus rufescens*) and Chatham Island robin-flycatcher (*Petroica traversi*). On the main island in the Chatham group the endemic (and now extinct) banded rail (*Cabalus dieffenbachi*) – known to the islanders as *meriki* – was very scarce as early as 1850 as a result of feline predation, as was the closely related *C. modestus* on Mangere.[3] Originally established on Pitt, Mangere and Little Mangere in the Chatham group the endemic robin-flycatcher disappeared from the two former islands soon after the cats' arrival, surviving only on Little Mangere after the cats had been removed. Here it was further endangered by the rapidly deteriorating habitat, and was saved from extinction (if only temporarily) by the removal of the surviving population to the regenerating scrub forest on Mangere Island. The latest (1977) census revealed a population of three pairs plus a single male. Feral cats on Chatham Island may also be jeopardizing the survival of the recently (1978) rediscovered and endangered endemic taiko or magenta petrel (*Pterodroma magentae*).

Ernest Dieffenbach in *Travels in New Zealand* (1843) refers to the presence of feral cats on the mainland of North Island in southern Auckland, and their spread thereafter appears to have coincided with that of early settlements and exploration and, from the early 1870s, with that of the introduced rabbit.[4] The population was (and still is) constantly reinforced by fresh escapes, and also in the past by deliberate releases to control rabbits.

By the late 1940s Wodzicki (1950) found feral cats widely distributed on both main islands and on Auckland Island; on Campbell Island (where they were introduced after 1916, probably to control brown rats (*Rattus norvegicus*)[5], and were said to be declining); on Stewart Island (where the population was of long-standing); on Herekopare Island; on Little Barrier Island; and on Macquarie Island.

Feral cats have had a more devastating affect on the wildlife of New Zealand and its offshore islands than on the whole of the rest of the world: the list of species they have exterminated or endangered reads like a roll-call of avian disaster.

On the two main islands they have wreaked havoc on terns and gulls (Laridae), penguins (Spheniscidae), shearwaters and petrels (Procellariidae), kiwis (*Apteryx* spp.), the rare shore plover or tuturuato (*Charadrius novaeseelandiae*), wekas (*Gallirallus australis*), pheasants and quail (Phasianidae), collared grey fantail-flycatchers (*Rhipidura fuliginosa*), paradise or New Zealand shelducks (*Tadorna variegata*), parakeets or kakarikis (Psittacidae), bellbirds (*Anthornis melanura*) and bush wrens (*Xenicus longipes*). In the Mackenzie Basin, its only remaining nesting place, feral cats (and introduced ferrets, *Mustela furo*) have put at risk the rare black-winged stilt (*Himantopus novaeseelandiae*).

As mentioned above, cats have exterminated the Chatham Island banded rail, robin-flycatcher and fernbird, and also the saddleback (*Creadion carunculatus rufusater*) on Little Barrier Island where they prey as well on the kiore rat (*Rattus exulans*), on the black or Parkinson's petrel (*Procellaria parkinsoni*)[6] and on Cook's petrel (*Pterodroma cookii*).

On Stewart Island, subantarctic snipe (*Coenocorypha aucklandica iredalei*), bush wrens, South Island kokakos (*Callaeas cinerea cinerea*), South Island saddlebacks (*Creadion carunculatus carunculatus*), yellowheads (*Mohoua ochrocephala*), robins (*Petroica australis rakiura*), tuis (*Prosthemadera novaeseelandiae*) and fernbirds (*Bowdleria punctata stewartiana*) have all been eradicated, and feral cats are also known to prey and/or scavenge on the endangered kakapo (*Strigops habroptilus*), which was rediscovered on Stewart Island as recently as 1977. Feral cats on Auckland Island had by 1974 seriously endangered the endemic Auckland Island flightless teal (*Anas aucklandica aucklandica*) (one of the two flightless forms of the brown teal), which is now largely confined to six of the smaller outer islands in the group, and are believed to have been partially responsible for the extinction (in 1902) of the Auckland Islands merganser (*Mergus australis*).

On Campbell Island, 240 km southeast of the Aucklands, cats (and brown rats) are believed to have endangered the survival of the second flightless form of the brown teal, the Campbell Island flightless teal (*Anas aucklandica nesiotis*), which is now confined to the jagged rocks of tiny offlying Dent Island. Introduced to Herekopare Island sometime after 1911 (possibly not until 1931) cats had, by 1943, reduced the avian population from an estimated 400,000 to a few thousand, and had exterminated six breeding species. The cats themselves were finally eradicated in 1970.

The arrival of rabbits on Macquarie Island in 1880 (or possibly earlier) resulted in a dramatic surge in the number of cats, introduced in about 1812, and flightless weka rails (*Gallirallus australis scotti*) which had been introduced from New Zealand in 1867, 1872 and 1879 (possibly as early as about 1830) as a source of food for sailors: the cats began to prey less on wekas, which themselves killed young rabbits. The increased population of cats and wekas, assisted by black rats (*Rattus rattus*), between them succeeded in exterminating the island's endemic parakeet (*Cyanoramphus novaeseelandiae erythrotis*) between 1880 and 1891 and the banded rail (*Rallus philippensis macquariensis*) by 1894. Although on Macquarie today cats subsist largely on introduced rabbits, they have (with the help of black rats and wekas) greatly reduced the populations of Antarctic terns (*Sterna vittata*), grey-backed storm petrels (*Garrodia nereis*), white-headed petrels (*Pterodroma lessoni*), common diving petrels (*Pelecanoides urinatrix*), blue petrels (*Halobaena caerulea*), sooty shearwaters (*Puffinus griseus*),

Antarctic or dove prions (*Pachyptila desolata*), fairy prions (*P. turtur*) and thin-billed prions (*P. belcheri*), and have exterminated the island's population of grey petrels (*Adamastor cinereus*). They also scavenge on the carcases of dead southern elephant seals (*Mirounga leonina*) and on penguins. It has been estimated that on Macquarie an individual cat kills approximately 213 birds per year. Based on the body weight of prey required by a cat per day, it has been further calculated that approximately 57,000 rabbits, 46,000 Antarctic prions and 11,000 white-headed petrels are killed annually by cats on Macquarie.[7] The population, whose main limiting factor is the availability of food in winter, was in 1977 estimated to number between 250 and 350.

On Stephen's Island in the Cook Strait cats belonging to the lighthouse-keeper not only eradicated a rare endemic reptile, the tuatara (*Sphenodon punctatus*), but also, in the same year (1894) that it was identified by Lord Rothschild as a species new to science, the island's endemic bush wren (*Xenicus lyalli*).[8]

Cats in New Zealand also sometimes prey on sickly lambs and even on calves. The only benefit they convey in return is to kill small numbers of introduced stoats (*Mustela erminea*), rabbits and rats.

OCEANIA

Fiji

Cats were first introduced to Fiji in the early nineteenth century, and by the 1870s had become so numerous in the interior of Viti Levu that they were being hunted by the Fijians for food and trophies. Since then the population on Viti Levu and Vanua Levu appears to have declined, probably because prey in the form of birds, iguanas and other lizards has also diminished. Feral cats are still, however, established on most inhabited islands, including Taveuni, Yaqaga, Ovalau, Kadavu, Matacawalevu, Yadua, Galito, Malolo Laila, and possibly Vatua Vara. Gibbons (1984) considers that feral cats are the most important single factor in the decline of *Brachylophus* iguanas and several other lizards (e.g. *Emoia* skinks) in Fiji.

Hawaii

Feral cats have been established on all the main islands in Hawaii since the days of European settlement in the early nineteenth century, and were common in forestland on Lanai and Oahu by 1892. In the crater of the dormant volcano, Haleakala, on Maui (the species' only substantial colony) and on Mauna Kea on Hawaii, they prey on the eggs and nestlings of the Hawaiian race of the dark-rumped petrel (*Pterodroma phaeopygia sandwichensis*); on Maui and Hawaii they are predators of the endemic nene or Hawaiian goose (*Branta sandvicensis*); and on Kaui of the endemic Hawaiian duck or koloa (*Anas platyrhynchos wyvilliana*). Feral cats also prey on Captain King's 'perroquet' – a honeycreeper – the ou (*Psittirostra psittacea*).

Galápagos Islands[9]

Cats were probably introduced to the Galápagos by eighteenth-century sealers, whalers and buccaneers who visited the archipelago to careen and reprovision their vessels. They were – and frequently still are – an integral part of many a ship's company, and what easier way (other than by drowning) of getting rid of unwanted kittens than to put them ashore to fend for themselves. These early feral animals would have been reinforced with new blood when the first permanent settlements were founded – on Floreana (Charles) in 1832, on San Cristóbal (Chatham) in 1869 and on southern Isabela (Albemarle) in about 1893.[10]

When Dr Simeon Habel visited the Galápagos in 1868 he reported that 'At the spring (at Tagus Cove) I saw a wild cat come down to drink; and the men told me there were many wild cats, dogs and also asses on Albemarle.'

From Floreana, San Cristóbal and Isabela cats were presumably later transported to Santa Cruz (Indefatigable) and to Santiago (James), and these remain the only islands currently supporting feral populations.

As elsewhere, feral cats in the Galápagos are opportunistic predators, and appear to change their diet seasonally in response to the availability of fresh water. In times of drought they eat predominantly grasshoppers (*Schistocercus*) which enable them to survive without incurring a water debt from a high protein diet. At other times they prey mainly on black rats – of which they are the principal natural controlling agent – and birds such as the local race of the endangered dark-rumped petrel (*Pterodroma phaeopygia phaeopygia*), greater frigate birds (*Fregata minor*), blue-footed boobies (*Sula nebouxii*), brown pelicans, Darwin's ground finches (*Geospiza* spp.), cuckoos and Galápagos mockingbirds (*Nesominus trifasciatus*). Cats may have been responsible for exterminating on Floreana the island's subspecies of mockingbird, which is now confined to a single cat-free offlying islet. On the other hand the accusation by V. Ziswiler[11] that cats (and dogs) have 'denaturalized' the endangered waved or Galápagos albatross (*Diomedea irrorata*) cannot be sustained since, as Thornton (1971) points out, that species nests only on the small uninhabited island of Española (Hood), from which both predators are absent. Cats also kill Sally Lightfoot crabs (*Grapsus grapsus*), the hatchlings of marine iguanas (*Amblyrynchus cristatus*) and giant tortoises (*Testudo* (*Geochelone*) *elephantopus*), snakes (*Dromicus* sp.), lava lizards (*Tropidurus* spp.), crustaceans (mainly a large intertidal isopod), centipedes and other small invertebrates. In some localities lava lizards may be the principal prey of feral cats, in which case the latter could be exerting an influence on the former's evolution. According to Laurie (1983), predation by cats is largely responsible for the low densities and imbalanced populations of marine iguanas on those islands where the predators occur. In addition cats scavenge on larger animals such as Californian sealions (*Zalophus californianus*), South American fur-seals (*Arctocephalus australis*) and fish, and in turn are preyed on by the endemic Galápagos hawk (*Buteo galapagoensis*), which takes kittens up to 8 to 10 weeks old, and occasionally by feral dogs.

Kerguelen

Fairly widespread on Kerguelen in the early nineteenth century, feral cats had disappeared by the early 1900s (possibly by around 1850) only to be reintroduced in 1952 near Port aux Français on Grande Terre. By 1960 trapping had greatly reduced the population, but a relaxation of the campaign allowed it to build up again to its former level, and despite intensive shooting it is still increasing. The cats' diet on Kerguelen is similar to that on Macquarie Island; rabbits (introduced in 1874) are eaten at all seasons, and birds comprise most of the rest of their food: the principal species to suffer are soft-plumaged petrels (*Pterodroma mollis*), white-headed petrels, grey-faced petrels (*P. macroptera*), blue petrels (*Halobaena caerulea*), Antarctic or dove prions, common petrels and South Georgian diving petrels (*Pelecanoides georgicus*).

A campaign designed to eliminate cats from Kerguelen has been in operation since 1973, and resulted in the shooting of between 447 and 685 annually up to 1977. Until 1975 the cats were confined to Péninsule Courbet on the eastern side of Grande Terre, but in that year faeces were discovered to the south and west. Clearly, therefore, shooting has been ineffective in preventing their dispersal into new areas, and is at best only slowing up their rate of expansion. Biological control, as on Marion Island (see below), if considered socially acceptable, seems the most effective method of eliminating cats (and other pests) from small islands.

Crozet Archipelago (Île Aux Cochons)

On Hog Island in the Crozet Archipelago (east of Kerguelen in the southern Indian Ocean), where cats were introduced in about 1887, their population is currently estimated at about 400 in summer and 100 in winter. Although some rabbits are eaten in winter, introduced house mice form their staple diet. Their preferred avian prey seems to be Salvin's prions (*Pachyptila vittata salvini*), and other seabirds such as Cape petrels (*Daption capenses*), common and South Georgian diving petrels, black-faced sheathbills (*Chionis minor*), kelp gulls (*Larus dominicanus*), Kerguelen terns (*Sterna virgata*), and introduced Kerguelen pintail (*Anas acuta eatoni*).

Aldabra Island; Îles Glorieuses

On Aldabra, 320 km north-west of Madagascar (Malagasy Republic), feral cats are held to have been responsible for the extermination before

1898 of the endemic flightless rail (*Rougeticus aldabranus*). On Îles Glorieuses, 200 km to the south-east, the avifauna has also been devastated by cats.[12]

Marion Island

In 1949, two years after Marion Island (some 1,900 km south-east of Cape Town in the Prince Edward group) had been annexed by South Africa as a weather station, five domestic cats were introduced to control an outbreak of house mice in the station's foodstore. By 1965 the former had spread to most parts of the island, and within a further decade had multiplied to about 2,200, increasing two years later to around 6,200; having achieved the purpose for which they were introduced, they began to prey on nesting seabirds such as Salvin's prion, the soft-plumaged petrel, the Kerguelen petrel (*Pterodroma brevirostris*), the grey-faced petrel and the white-chinned petrel (*Procellaria aequinoctialis*),[13] killing an estimated 600,000 annually and endangering several species. They also scavenge on penguins, seals and young albatross chicks. In an attempt to control the population 100 cats deliberately infected with feline panleucopaenia virus (FPL) were in 1977 airlifted by the South African government (which was roundly condemned by the country's Society for the Prevention of Cruelty to Animals) to Marion, where by 1980 the cats' winter population of some 2,000–3,000 had been reduced by about 60 per cent. At the time of writing (1983) there was no sign of the development of any acquired immunity.

Seychelles

On Frégate Island in the Seychelles feral cats (and rats) have endangered the sole surviving population of the endemic Seychelles magpie robin (*Copsychus sechellarum*) or *pie chanteuse* as it is known locally, and the islands' endemic paradise flycatcher or *veuve* (*Terpsiphone corvina*). In an attempt to save these species the International Council for Bird Preservation has recently succeeded in eradicating cats on Frégate. Feral cats are also present on North and South Islands of the Farquhar Atoll and on Eagle (Remire) Island in the Amirantes.

Other oceanic islands

On Tristan da Cunha (since about 1817), Nouvelle Amsterdam (since the early nineteenth century), Île St Paul (from the early nineteenth century until before 1874), Ascension, Tahiti, New Caledonia, Choiseul (Solomons), Raoul (Kermadecs), the Aleutians, the Falklands, South Georgia, Juan Fernandez and on other oceanic island groups, nesting birds and other animals have suffered grave predation by feral domestic cats: especially affected have been frigate birds (*Fregata aquila*) and sooty terns (*Sterna fuscata*) on Ascension; the rare Tristan starchy (*Nesocichla eremita eremita*) and the Tristan rail (*Porphyriornis nesiotis*); other rails on Tahiti and New Caledonia, and the crested pigeon (*Ocyphaps lophotes*) on Choiseul Island. With the help of foxes and black rats, cats have confined the nesting range of the rare Aleutian Canada goose (*Branta canadensis leucopareia*) to the 1,720-ha island of Buldir. A recent report suggests that cats (assisted by rats) may be destroying a large colony of sooty terns at Denham Bay on Raoul Island in the Kermadec group,[14] where they are believed previously to have been responsible for eradicating the Kermadec parakeet (*Cyanoramphus novaeseelandiae cyanurus*) and an indigenous fruit pigeon. Endangered birds are particularly vulnerable on those islands where introduced resident prey species provide an alternative food source for feral cats (and other introduced predators) at seasons of the year when the endangered bird species may be absent.

A number of **jaguarundis** (*Felis yagouarundi*) – natives of central and northern South America – are believed to have been deliberately released at Chiefland and Hillsborough River State Park in Florida, USA, shortly before 1942. A decade later they occurred from Dunellon through Chiefland to Jena in eastern Marion County, at the southern end of the central Florida ridge in Osceola, Polk and Highlands counties, in the Everglades National Park and perhaps in Hillsborough County. They are currently established in parts of central Florida, where they cause a considerable amount of trouble to local poultry farmers. Their preferred habitat is thick cover in the vicinity of water.

Ocelots (*Felis pardalis*), which occur naturally from Arizona, USA, southwards to northern Argentina, are believed to have escaped from captivity and to exist in the wild in scattered regions of Florida, especially around Miami.

In May 1983 three captive-bred **lynx** (*Felis lynx*) from Czechoslovakia were reintroduced to the Vosges Mountains of Alsace, France, from which they had been absent for over a century. In the following September a further pair from Riber Zoo, Matlock, Derbyshire, England, were set free in the same region.[15]

NOTES

1. See p. 30.
2. See p. 79.
3. H. O. Forbes, *Ibis* **6**: 531–3 (1893).
4. See p. 406.
5. See p. 352.
6. M. J. Imber, 'The black petrel', *Wildlife – A Review* **6**: 37–9 (1975).
7. Van Aarde (1980) and Jones (1977). See also E. Jones, *Notornis* **27** (1): 11–20 (1980).
8. The holotype of the Stephen's Island wren (of which only about 20 specimens – all killed by the lighthouse-keeper's cat – were ever discovered) is in the American Museum of Natural History in New York.
9. For much of the data on feral cats in the Galápagos I am indebted to Dr Michael J. Konecny of the University of Florida, USA.
10. For a more detailed history of the settlement of the islands see p. 359.
11. *Extinct and vanishing animals* (Longmans, London; Springer Verlag, New York, 1967).
12. W. L. Abbott, *Proceedings of the US National Museum* **16:** 762, 764 (1894).
13. Anderson & Condy (1974).
14. R. H. Taylor, *Notornis* **26**: 199–202 (1979).
15. *World Wildlife News* 27 (Autumn 1983).

Domestic Ass or Donkey
(*Equus asinus*)

Note: Only present (reduced) natural distribution of wild ancestor is shown.

NATURAL DISTRIBUTION

The ancestor of the domestic ass or donkey,[1] the African ass (*E. africanus*), was originally a native of northwestern and northeastern Africa, including Morocco, Algeria, Tunisia, Ethiopia, Somalia and the Sudan, and possibly of parts of western Asia. It is now confined to the Danakil region of Ethiopia and to parts of Somalia.

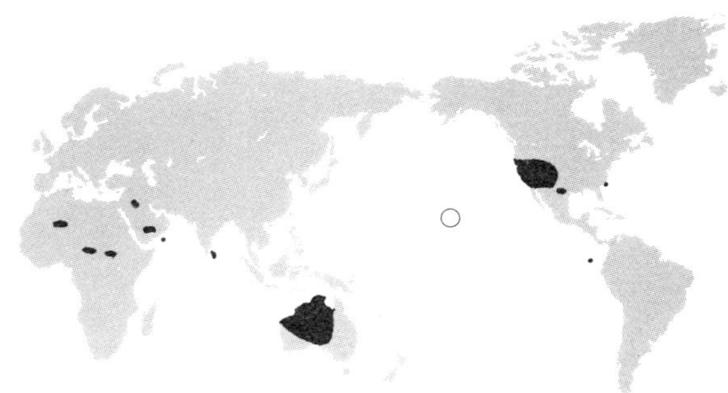

Note: Only principal feral populations mentioned in the text are shown.

NATURALIZED DISTRIBUTION

Asia: Saudi Arabia, Sri Lanka, Yemen (Socotra); Africa: Algeria, Sudan, Chad; North America: USA; Australasia: Australia; Oceania: Hawaii, Galápagos, ? other oceanic islands.

ASIA

Saudi Arabia

Feral donkeys which roam the desert regions of Saudi Arabia compete for grazing with herds of domestic camels.

Sri Lanka

A feral population of blackish-brown pygmy donkeys with white muzzles in the north-west of the island is strictly protected by the Department of Wildlife Conservation.

Yemen (Socotra)

Feral donkeys on the island of Socotra in the Indian Ocean may be derived from domesticated animals originally introduced by the ancient Egyptians.

AFRICA

Algeria; Sudan; Chad

Feral donkeys – probably descended from domesticated stock – roam the Ahaggar (Hoggar) high plateau in the Algerian central Sahara, where they are sometimes hunted for sport by the Tuareg. Free-ranging donkeys in the Sudan and in mountainous regions of the eastern Sahara, such as Tibesti in Chad, are more likely to be the descendants of feral domestic rather than of indigenous stock.

NORTH AMERICA

Canada

According to G. Sagard-Théodat, in his *Histoire du Canada* (1866), the first donkeys in Canada were a pair brought to New France by the Récollet Fathers in 1615. More were imported in 1671 by Jean Talon who had been appointed the first intendant when New France became a crown colony in 1665. So far as is known none ever became established in the wild.

USA[2]

Domestic donkeys (known in the United States as 'burros') were introduced to the New World by seventeenth-century Spanish conquistadors, by whom they were soon being bred on stud-farms in Mexico, and probably also in Hispaniola, Cuba and elsewhere.

'These hardy and obstinate asses', as McKnight (1958) graphically describes them, were soon being widely used as pack animals by prospectors, miners, stockmen, explorers and shepherds in both North and South America; although by the nineteenth century they were in use virtually throughout the west of the United States, they were most valuable in the dry rugged terrain with sparse vegetation of northern Mexico and the southwestern United States – especially in Arizona, California, Colorado, Nevada, New Mexico and Utah.

As with other forms of domestic livestock, burros became feral as the result of escaping from, or being released by, their owners. Although this undoubtedly first occurred in the early years of their history in North America, where feral burros were soon widely distributed throughout the western states, McKnight believed that because of their value to European settlers and Indians they were not well established until the nineteenth century, when the advent of the railways, better roads and the decline in the mining boom reduced their usefulness. Their success in the wild, despite their fairly sedentary nature, has been largely due to the wide area over which they became feral, coupled with a relatively high breeding rate, the existence of few significant predators (though coyotes (*Canis latrans*), bobcats (*Felis rufus*) and golden eagles (*Aquila chrysaetos*) kill some young and possibly a few sickly adults), and their ability to survive in a rocky, arid and inhospitable terrain.

By the late nineteenth century large numbers of feral burros were established in many parts of the south-west; the population fluctuated (but generally increased) until the 1920s and 1930s, when more stringent control measures halted the advance.

By the late 1950s feral burros still occurred – albeit in a restricted range – in all the western states apart from Washington and Montana. The largest numbers were established in Arizona, California, Nevada and New Mexico with smaller populations in Colorado, Idaho, Oregon, Texas, Utah and Wyoming.

Table 2 gives details of the main concentrations of burros in the United States. McKnight (1964) estimated the total population at between 5,500 and 13,000. He found the principal concentrations to be in the rugged mountainous terrain and associated desert basins of Inyo and San Bernardino counties of southeastern California; along the lower reaches of the Colorado River from the Grand Canyon to Mexico; and in parts of western New Mexico. In Arizona feral burros were largely confined to the Colorado River area (especially in the Grand Canyon) in the west. Because of strict control measures burros were restricted to the extreme south of Nevada.

'Probably the hardiest and most adaptable of all the feral exotics', wrote McKnight (1964), 'is the "wild" burro . . . the burro is without peer when it comes to eking out an existence on the barest of forage in the hottest and driest of lands [being able to withstand up to 30 per cent dehydration], and yet he is able to survive a protracted, cold, snowy winter as well.' It is this adaptability, coupled with a catholic diet, which is largely responsible for the continued survival of feral burros in the fragile ecosystem of the arid west. Nowhere is this more typified than in the 800,000 ha of the Death Valley National Monument (much of which lies more than 65 m below sea-level and where in summer the temperature can reach 55 °C) between the Panamint and Armagosa mountains on the California/Nevada border which supports over 2,100 burros – the densest concentration in the United States. Throughout the west overgrazing and overbrowsing, together with the fouling of meagre desert water resources, have reduced food and water supplies for native wildlife and domestic livestock and cover for ground-nesting birds, and cause serious soil erosion. The mammalian species most affected seem to be mule deer (*Odocoileus hemionus*), pronghorn antelope (*Antilocapra americana*) and, especially, desert bighorn sheep (*Ovis canadensis*) – the last of which occupies much the same range as, and favours a similar habitat to, the burro. Interspecific competition between burro and bighorn is likely to favour the former over the latter due to the burro's more facultative feeding habits; because of a lesser susceptibility to diseases

Table 2 Main concentrations of feral burros in the United States

State	*Distribution*
Arizona	Grand Canyon, south of Colorado River; Black Mountains, near Lake Mead; lakes Mohave and Havasu; Bill Williams River drainage ranges; Kofa, Chocolate, Trigo and Castle Dome mountains west to the Colorado River; Wickenburg; Castle Hot Springs; Mohawk Mountains; Organ Pipe Cactus National Monument; Black and Galiuro mountains north-west of Tucson
California	On the west side of Death Valley in the Panamint Mountains; Saline Valley in Inyo County; Coso and Argus mountains; parts of San Bernardino County; near Ravendale in the Smoke Creek Desert; parts of the southern Sierra Nevada; Avawatz, Clark, Chemehuevi and Whipple mountains; southeastern Imperial County; Merced River valley between Bagby and Hornitos; Eureka Valley
Colorado	Fort Garland in San Luis Valley
Georgia	Ossabaw Island
Idaho	Owyhee Canyon in the south-west
Nevada	Colorado River ranges especially between Havasu Creek on the east and Pierce Ferry on the south and west; south-west of Goldfield; Wassuk Range; Smoke Creek Desert; Virgin Valley in Charles Sheldon National Antelope Range
New Mexico	Luera and Elk mountains; Copper Canyon; Bandelier National Monument; Mesa Prieta east of the Puerco River; Tularosa Basin
Texas	Big Bend National Park
Utah	Confluence of Colorado and San Juan rivers
Wyoming	Vermilion Creek/Salt Wells area in south-west

Source: After McKnight (1958, 1964), who obtained his data from a questionnaire/survey

and endoparasites; because of its higher rate of reproduction and survival of young; and especially because of its bolder and more aggressive nature.

'The native desert bighorn sheep in both Death Valley and the Grand Canyon', said a recent report, 'are doomed unless something is done to eliminate the wild burros.' Their intrusion into the Cottonwood Mountains, in Butte Valley, and at Eagle Springs in the Panamint Mountains has resulted in the virtual disappearance of sheep: in the first half of the last century some 4,800 bighorns ranged over 3,600 km^2; by the mid-1970s only 520 remained on less than 1,000 km^2. The state Department of Fish and Game, in a report on the *Status of feral burros in California*, said:

> On the lower western slopes of the Panamints and in Butte Valley, the better browses have been killed out, and low value plants such as white bursage are nearly gone. The burros are now heavily utilizing creosote bush. The foliage of this plant is rarely taken by any animal. Utilization of creosote bush reflects a very severely depleted range condition. Old timers tell of former high numbers of bighorn sheep in the Panamints. At present a very low population exists . . . from examination of the serious damage wrought to the range by burros, it seems safe to assume that they have been the chief cause in the large reduction in bighorn numbers in the area. The most serious aspect of range plant depletion in desert areas is the extremely slow recovery rate of native vegetation. Even under complete protection from foraging burros, the desert vegetation might require several decades to return to its original condition.

The Wild Free-Roaming Horse and Burro Act of 1971 (enacted in response to public opposition to their slaughter) gave the two animals protection on Bureau of Land Management and US Forest Service land, claiming that both animals were 'living symbols of the historic and pioneer spirit of the West' and, quite incorrectly, that they were an 'integral part of the natural system of the public lands'. A census in 1976 (when limited culling was again allowed) revealed about 3,700 burros in California, 2,668 in Arizona and 1,360 in other southwestern states – a total of over 7,700; other estimates judge the countrywide population at nearly double this figure. The consumption of forage by burros has been calculated at some 10 t per day – a figure which is expected to double by the mid-1980s.

Between 1924 and 1968 the National Parks Service (which was exempted from the 1971 Act) shot 6,300 burros in the Grand Canyon and Death Valley, of which over 4,000 were killed in the latter where the population was maintained at around 700. By 1981 the programme organized by the Fund for Animals to airlift burros out of the Grand Canyon – as an alternative to shooting them – had been completed at a cost of $500,000, and 200 of the 557 removed in nets slung under helicopters from the 500,000-ha national park had been allotted under the 'Adopt-A-Burro' programme. A similar scheme in Death Valley resulted in the removal of 451 burros. If desert bighorn sheep are to survive, the adoption of live burros or legal culling will have to be continued; the future of both species is largely in the hands of man.

AUSTRALASIA

Australia[3]

The Scottish-born landowner, Sir Thomas Elder, imported the first donkeys to Australia in 1866 or 1886. Until the arrival of motorized transport in the 1930s donkeys were extensively used – in teams of up to 80 – in hauling freight and supplies during the early years of settlement and development between outback sheep- and cattle-stations, remote towns and isolated mines, and railheads or ports, and also in the construction of the Overland Telegraph and the Transcontinental Railway. Between about 1900 and 1920 (in the Kimberleys in Western Australia and in northern Queensland until after the Second World War) donkey- and mule-studs[4] were established on many stations from Queensland and New South Wales to Western Australia.

Although sporadic sightings of escaped or released feral donkeys were made as early as the 1900s, established groups (possibly from teams travelling up from South Australia) were first reported in the 1920s in the Channel country of Queensland – especially in the Georgina drainage – by Henry Lamond. Many more became feral in the 1930s, when their usefulness decreased. Where their numbers were not controlled the population increased explosively: in the Mc-

Arthur River area of the Northern Territory, for example, 40 freed in 1936 had within 30 years multiplied to 1,500.

Since the early days of pastoral development in the north of Western Australia, the Kimberleys have been the site of the largest concentration of donkeys – both domestic and feral – on the continent; in this region horses have proved an unreliable means of transportation due largely to an endemic but non-infectious disease – 'walkabout' or Kimberley horse disease – caused by the ingestion of poisonous shrubs which do not, however, affect donkeys or mules.

According to Long (1972), donkeys were first used as pack or draught animals in Western Australia in the late nineteenth century to haul equipment and supplies from Carnarvon to the Kimberleys. In 1894 there were some 131 donkeys and mules in the colony.

In 1904 the government of Western Australia imported a number of donkeys from Sumatra and Mauritius; the bulk of those in use, however, were home-bred animals descended from Sir Thomas Elder's importation.

As the donkey was superseded by motor transport in the 1930s large numbers became established in the wild. To combat their depredations they were officially declared 'vermin' at Nullagine and Hall's Creek in 1949, at Port Hedland, Coolgardie and Dundas in 1953, in the West Kimberleys in 1954, at Marble Bar in 1957, and at Wyndham and Meekatharra in 1958.

By the late 1960s Long records that feral donkeys were widely distributed in the Kimberleys, Pilbara, North West and Goldfields districts. In the Kimberleys they ranged from the border with the Northern Territory to Derby, being especially numerous between the East Kimberleys and the Northern Territory. In 1957 the Agriculture Protection Board of Western Australia estimated that there were upwards of 100,000 donkeys in this area; during the next decade between 90,000 and 100,000 were destroyed, but by 1970 the overall population had only been reduced by some 10,000–20,000. Around 20,000 were judged to be established on Bulloo Downs and Prairie Downs, where 1,000 were once counted within 15 km of a single waterhole. In 1958 there were said to be 4,000 in the Savory Creek area east of Mundiwindi, and in 1960 2,000 on Maroonah Station and a like number around Mundiwindi. Some stations claimed to support over 10,000 feral donkeys (more than their entire stock of domestic animals) and herds of up to 500 were reported. By 1968 donkeys had been eliminated on a number of stations in the East Kimberleys but were still abundant in parts of the outback.

In the Pilbara and North West districts Long reported that donkeys ranged from the De Grey River nearly as far as Onslow, and southwards through the Upper Gascoyne region. In the Goldfields Long found donkeys to be widely dispersed from Lake Wells in the north to Norseman in the south, and west through Bullfinch to Boodarockin: they also occurred in small and scattered groups around Meekatharra and on the Boolardy Station in the Murchison. In 1960 there were estimated to be about 3,000 in the vicinity of Meekatharra.

Serious efforts to control donkeys in the Kimberleys began in 1959, and within three years 30,000 had been accounted for; by 1965 40,000 had been destroyed and the population had been halved. In the late 1970s helicopters began to be used to hunt donkeys, and now 600–700 can be shot in a single day; pet-food companies with refrigerated trucks garner up to 10 t of meat an hour. In spite of this slaughter tens of thousands of feral donkeys remain in the Kimberleys today.

The Victoria River district of the Northern Territory, east of the Kimberleys, is another region where 'walkabout' disease has limited the employment of horses. In the late nineteenth and early twentieth centuries donkeys regularly hauled freight south from Darwin to this region, and from the Victoria River depot near Timber Creek to Victoria River Downs and Wave Hill. In the early 1900s several stations in the area set up their own stud-farms.

As in the adjacent Kimberleys the feral population in the Northern Territory exploded during the 1930s. In 1942 several hundred were mustered and redomesticated for service with the allied forces in New Guinea. In 1956, when the population in the Victoria River district was estimated to number around 150,000, attempts were made to reduce the herds by shooting; no accurate figure exists for the total number destroyed, but by the mid-1970s the largest station had removed at least 28,000, and the total may have been double that figure.

The report of the Board of Inquiry into Feral Animals in the Northern Territory, published in 1979, recommended that a survey should be undertaken by the Northern Territory Parks and Wildlife Commission, in conjunction with the Primary Industry Division, to assess more accurately the population and range of feral donkeys in the territory, to establish the degree of environmental damage they cause, and to enquire into possible markets for their meat and hides.

Although not as widely distributed in Australia as feral horses, donkeys are found over much of the interior and north of the continent. They occur most commonly in all the farming regions of the Northern Territory and in the northern two-thirds of Western Australia, and in lesser numbers in the drier regions of South Australia and in parts of western and northern Queensland. The largest concentrations are established in the Kimberleys of Western Australia and in the Victoria River Downs region of the Northern Territory. Although in the West Kimberleys numbers have decreased in recent years, in the East Kimberleys – especially in the Dixon and Durak ranges – feral donkeys are still abundant. In the Victoria River Downs district the population, though also high, is unevenly distributed. Elsewhere, lesser numbers exist on both sides of the Roper River and near the McArthur River in the Northern Territory; south and east of Port Hedland and in various localities south of Onslow in Western Australia; west of Lake Eyre in South Australia; and between Alice Springs and the South Australia/Northern Territory border. The total population is divided between states as shown in Table 3.

Table 3 Distribution of feral donkeys in Australia

State	*Estimated population*
Western Australia	40,000–70,000
Northern Territory	20,000–40,000
South Australia	5,000–10,000
Queensland	1,000–5,000
Total	66,000–125,000

Source: From McKnight (1976)

Donkeys have been successful colonists in Australia for a variety of reasons: they are more resistant to the heat and aridity of semi-desert regions than any mammal apart from the introduced camel (*Camelus dromedarius*),[5] and are well adapted to life in a barren and inhospitable terrain; their facultative feeding habits enable them to eat almost any available plant; they will dig with their hooves to a depth of 10–12 cm for moisture; and can tolerate brackish water. Being of a gregarious nature, donkeys usually assemble in herds of between 150 and 200, but small groups, pairs and even lone individuals also occur. Although their preferred habitat is riverine pastures with a variety of forage, they are well able to exist also in rugged hill country.

As in the United States, feral donkeys in Australia consume forage and water intended for domestic stock, and in the Kimberleys and Victoria River Downs district have been largely responsible for the degeneration of large areas of pasturage and for soil erosion. They break down fences, overturn water-troughs, foul natural water supplies and stampede and even on occasion attack domestic cattle, to whom they may also pose a threat as vectors of parasites and diseases: in the Northern Territory donkeys are commonly infested with intestinal nematodes which might affect domestic horses. Because they are easier to catch and redomesticate than horses donkeys are still sometimes used for packing or hauling, for the breeding of mules and for riding by Aborigines. Their skins are sold for hide and their meat for pet-food.

Although McKnight (1976) considered that the range and population of feral donkeys would decrease rapidly where they were numerous (and less so where they were uncommon) as settlement and agriculture spread, and in spite of the fact that between 1979 and 1981 70,000 were killed, some authorities believe that the position regarding distribution and numbers remains only static.

OCEANIA

Hawaii

Tomich (1969), in *Mammals in Hawaii*, says that the earliest donkeys in the Hawaiian group were imported from England to the island of Oahu on board the *Active* by the first British consul, Richard Charlton. Andrew Bloxam, naturalist on the *Blonde* during her voyage to the Sandwich Islands in 1824–5, visited the *Active* when she called at Valparaiso in Chile and recorded that 'On February 6, 1825, Captain Charlton sailed for the Hawaiian Kingdom, with a most extraordinary cargo on board, consisting of donkeys and Jews' harps'.

By the 1970s feral donkeys survived in small numbers only in the lava fields of the Kaupulehu –Kiholo area of the North Kona district, on the McCandless Ranch in the South Kona district, on the Huehue Ranch (between 50 and 100), and on the Hualalai Ranch on the main island of Hawaii.

Mules are known to have been present on Hawaii before 1851, although whether they were imported or home-bred is uncertain. By the 1970s a few remained in the upper Halawa Valley on Molokai, in the Waimanu Valley on Hawaii, in the upper Hamakua irrigation system on Kohala Mountain and in the Waipio Valley. The ecological impact of feral donkeys and mules in the Hawaiian Islands seems to be generally insignificant.

Galápagos Islands[6]

Donkeys (and other domestic animals) were probably introduced to the Galápagos Islands by the first governor, General José Villamil, who annexed the archipelago in the name of Ecuador in 1832 and established a colony on Isla Floreana (Charles). Temporary bases were subsequently set up on other islands, but the first permanent settlements were those established on San Cristóbal (Chatham) in 1869 and on southern Isabela (Albemarle) in about 1893, to both of which donkeys were probably imported.[7]

The earliest written reference to donkeys in the Galápagos is by Dr Simeon Habel who visited the islands in 1868. 'At the spring (at Tagus Cove)', he wrote, 'I saw a wild cat come down to drink; and the men told me there were many wild cats, dogs, and also asses on Albemarle. The two latter are also to be seen on Indefatigable [Santa Cruz] and Chatham Islands, wild horses also being found on the last named.'[8] O. Salvin, who visited the Galapágos in HMS *Petrel* in July 1875, reported that 'of animals introduced, it was estimated that there existed . . . of donkeys, 2000 on Charles, some on Chatham, large numbers on James [Santiago], and 25 or 30 on Albemarle'.[9] It is possible that the donkeys seen by Habel and Salvin on Isabela were descended from some which Villamil may have left behind when he landed on the island in 1851.[10]

Donkeys were probably taken to unsettled parts of the islands (e.g. Volcan Alcedo and Volcan Sierra Negra on Isabela) by early miners, such as Señor Antonio Gil, to transport sulphur and drums of tortoise-oil from the interior, where the animals were slaughtered, to the coastal settlements.[11] It is likely that from time to time some managed to escape and became established in the islands' inland regions.

Today both dry and moist zones of all five major islands – Floreana, Santa Cruz, Santiago, San Cristóbal and Isabela – support herds of feral donkeys; none of the insular populations has ever been eradicated and the animals have never occurred on any of the smaller islands. No official census of the population has ever been made, but Fowler (1983) has estimated that between 500 and 700 live in herds up to 30 strong on Volcan Alcedo and some 100–300 on Volcan Sierra Negra – both on Isabela; around 300 are established on San Cristóbal, 200–300 on Santa Cruz, 500–700 on Santiago, and 2,000–5,000 (estimates vary) on Floreana – giving a total population for the five islands of between 3,600 and 7,300. Except where fresh water is not readily available the various populations are, in the absence of natural predators or control by the National Park Service (who are, rightly, more concerned with eradicating rats, dogs and goats), probably increasing. Fowler found a considerable number of *burritos* (foals) on Alcedo after a wet season.

Compared with other introduced species donkeys have a relatively insignificant effect on the Galápagos ecosystem. On Volcan Alcedo Fowler found that they trample and destroy some nests of the giant tortoise (*Testudo* (*Geochelone*) *elephantopus*). There are two principal nesting sites for tortoises on Alcedo – one on the northern and one on the southern caldera (crater) floor. Fowler found that out of 58 nests on the northern floor, 9 (15 per cent) were damaged or destroyed by donkeys, while on the southern floor 9 out of 28 nests (32 per cent) were damaged and 4 (14 per cent) destroyed; the hatching rate of damaged nests in the south was only 36 per cent compared with 60–65 per cent for undisturbed clutches.

There is a considerable dietary overlap between donkeys and tortoises on Alcedo: grasses such as

Paspalum sp. and *Panicum* sp., several species of *Sida*, a Malvaceae (especially *S. rhombifolia*), *Hyptis rhomboidea, Rubiaceae* sp. and *Ipomea* sp. are favourite foods for both animals; it is possible that, in the dry season when food reasources are in short supply, donkeys limit the amount of forage available for tortoises, although the latter can withstand both food and water stress for several months. No permanent supply of fresh water exists on Volcan Alcedo. Both donkeys and tortoises are dependent on the *garua* (a mist formed by low moisture-laden clouds) which condenses on trees, from where it sometimes drips to the ground to form muddy pools. Donkeys also at times lick moisture from mosses and lichens. (Fowler records that up to 5 gallons (nearly 23 l) of water can be collected from a single tree overnight.) There is evidence that on Isabela donkeys also trample on the nests and burrows of the endemic Galápagos land iguanas (*Conolophus subcristatus*).

The principal harm done by donkeys to the islands' flora is their destruction of much of the archipelago's endemic cactus *Opuntia insularis*. On Floreana they also browse heavily on *Scalesia* and on the island's endemic *Alternanthera nesiotes* and trample on the burrows of nesting dark-rumped petrels (*Pterodroma phaeopygia*). By their overgrazing they have also adversely affected the herb cover, and by trampling have opened up the shrub layer, thereby accelerating soil erosion: by thus creating fresh grazing they may, on the other hand, be indirectly benefiting the giant tortoises.

Other oceanic islands

A population of some 60 donkeys remains on Ascension Island in the South Atlantic, where they were abandoned by British marines in the nineteenth century. In the Chagos archipelago in the Indian Ocean more than 100 donkeys are said to survive in the uninhabited part of Diego Garcia, with smaller numbers on Boddam Island in Saloman Atoll, on Coin de Mire Island in Pecos Banhos Atoll and elsewhere in the Chagos group, where they were abandoned when the human inhabitants were evacuated in the early 1970s.[12] Donkeys have also been introduced to Uapou in the Marquesas, to Tristan da Cunha (in about 1860), to Walpole Island in New Caledonia, to Macquarie Island and to Robinson Crusoe (Masatierra) in the Juan Fernandez Islands off the coast of Chile. Between 1949 and 1953 mules were established on Îles Kerguelen.

NOTES

1. The wild African ass was domesticated in Egypt and Mesopotamia at least by 3400 BC.
2. For information on donkeys in the United States I am indebted to McKnight (1958, 1964).
3. For much of the history of feral donkeys in Australia I am indebted to McKnight (1976).
4. Mules are the offspring of mares and jack asses.
5. See p. 135.
6. For much of the information on feral donkeys in the Galápagos Islands I am indebted to Fowler (1983). See also Fowler (1979, 1980).
7. For a more detailed history of the settlement of the archipelago see p. 359.
8. Quoted by Salvin (1876). (I am grateful to Dr Lynn Fowler *per* Miss Gayle Davis of the Charles Darwin Research Station for drawing my attention to this reference.)
9. Salvin (1876: 456).
10. Naveda (1950).
11. See van Denburgh (1914).
12. M. J. Howells (letter to W. R. P. Bourne (1982) and 'The birds of Diego Garcia', in press for *Sea Swallow* **33**).

Domestic Horse

(*Equus caballus*)

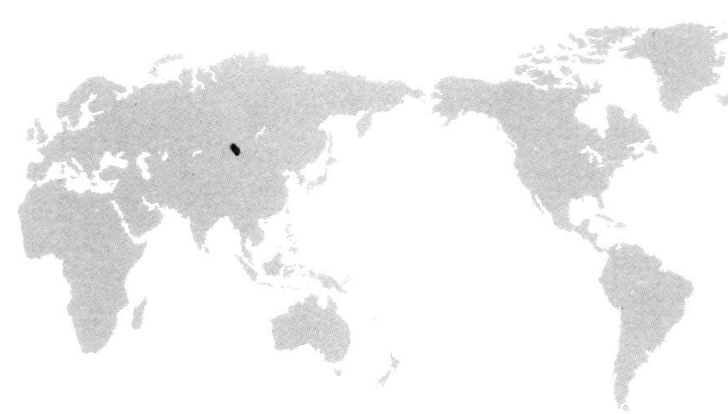

Note: Only present (reduced) natural distribution of wild ancestor is shown.

NATURAL DISTRIBUTION

The ancestor of the domestic horse, the wild or woodland horse or tarpan (*E. ferus*), ranged from eastern Europe through the steppes to Mongolia[1]: it is now confined (in the form of Przewalski's horse, *E. f. przewalskii*) to a small area on the border between southwestern Mongolia and Sinkiang in northwestern China.

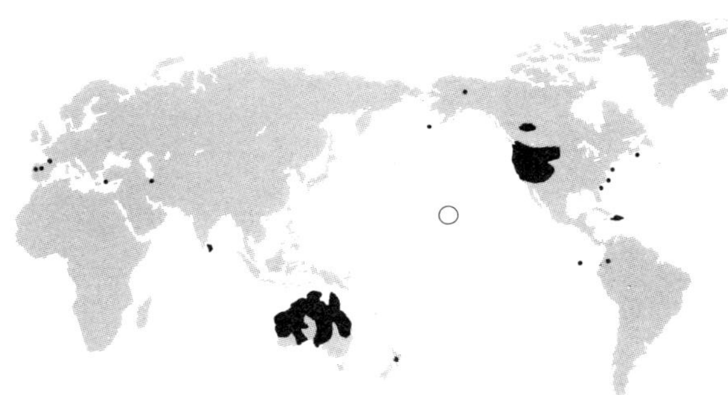

Note: Only principal feral populations mentioned in the text are shown.

NATURALIZED DISTRIBUTION

Europe: France, Greece, Portugal, Spain; Asia: Sri Lanka, Iran; North America: USA, Canada, West Indies (Hispaniola); South America: Colombia; Australasia: Australia, New Zealand; Oceania: Hawaii, Galápagos and various other oceanic islands.

EUROPE

France[2]

A primitive breed of pony known as the Potiok survives in small numbers in a semi-feral state in the Basque country of southwestern France.

Greece

On the island of Skyros in the Sporades in the Aegean Sea lives a stock of tiny delicately boned semi-feral horses of great antiquity, which stand only a little over a metre (9 to 11 hands) high at the shoulder.[3]

Portugal

A feral herd of about 30 Garrano Luso-Galiziano (Lusitanian–Galician) ponies survives in the Peneda-Gerês National Park in northwestern Portugal. One or two are removed annually for cross-breeding with domestic stock.

Spain

About 500 semi-feral horses roam the Sueve Mountains of Asturias in northwestern Spain. Of these, around 50 are small ponies – standing about 120 cm at the withers – known as Asturcones. In Galicia to the south-west considerable numbers of Galician ponies – only fractionally larger than Asturcones – inhabit mountainous regions in the provinces of Lugo and Pontevedra. There are three geographical races – those of La Grova, Lugo and Coruña. About 6,700 are slaughtered annually for meat. They are heavily preyed on by wolves (*Canis lupus*), which in, for example, 1971 (when the total population numbered 21,700) accounted for 1,069.

ASIA

Sri Lanka

A population of small feral horses (known as Mannar ponies) in the north-west of the island, whose colour ranges from dark brown to buff with a white forehead patch, is strictly protected.

Iran

In April 1965 a population of semi-feral miniature 'Caspian' horses was rediscovered in northern Iran on the southern shores of the Caspian Sea; this ancient race bears a close resemblance to artistic representations of small horses in Persia in the second and first millennia BC.

NORTH AMERICA

USA; Canada

Palaeontological evidence shows that several members of the family Equidae were present in North America up to the late Pleistocene;[4] all, however, except those which spread to Asia and thence to Europe via the Bering Strait land-bridge,[5] became extinct well over 8,000 years ago.

'Probably the most striking and best known of the feral animals on this continent', wrote Mcknight (1964), 'is the horse . . . symbol of the frontier, freedom and the pioneering spirit . . . more so than any other species, the feral horse has had a history that can be traced.'

As exploration and settlement in Mexico (where Eurasian horses were first introduced in 1519; see below) extended, horses gradually spread north towards the United States. In the southeastern states they first appeared with the disastrous expedition, led originally by Pánfilo de Narváez and subsequently by Cabeza de Vaca, which from 1528 to 1536 explored the Gulf coast country between Florida and Texas, and later with the expedition of 1539–43 led by Ferdinando de Soto which explored the region between Florida, Arkansas and Louisiana. In the south-west horses were first used by Francisco Coronado during his exploring expedition of 1540. Although horses were lost or abandoned by all three expeditions, most authorities agree that none became established in the wild. Subsequent explorers (e.g. Antonio de Espejo in 1582–3 and Juan de Oñate in 1598) reported that Indians in the south-west appeared terrified at the sight of horses, which suggests that they had had no previous contact with them. Indeed, the first feral 'mustangs'[6] in the southwestern states (probably in Texas and New Mexico) are believed to have been animals which escaped from mission-stations founded by Oñate around the turn of the century.[7]

In spite of scanty water supplies, poor grazing, an inhospitable terrain and predation by wolves, coyotes (*C. latrans*) and cougars or pumas (*Felis concolor*), feral herds – augmented by further escapes from miners, prospectors, ranchers, settlers, explorers and other sources – rapidly increased and extended their range in the American West.

Although some writers have suggested that American Indians acquired their first horses by recapturing feral mustangs, most authorities agree with Denhardt (1947) in *The horse of the Americas* that:

> The natives obtained their original horses, and always by far the greatest number, from the Spaniards or neighboring tribes and not from the wild herds. The Indians had mounts by the time the wild herds dotted the plains, and they always preferred domesticated animals to the *mesteños*. Mustangs were hard to catch and, once caught, harder to tame.

The tribes nearest to Mexico (in Texas and New Mexico) acquired their horses first. McKnight (1959) describes how:

> Apaches and Comanches were particularly active, attacking Spanish settlements before the end of the seventeenth century, capturing and freeing many horses. They transmitted horses to other tribes, as did Spanish and French traders. Thus horses spread through the Southwest to Navajos, Zuñis, Utes and other tribes; into the southern plains to Pawnees, Osages, Kansas, Wichitas, and others; into the lower Mississippi Valley to Caddoes, Hasinais, and others; into the northern plains to the Blackfeet and Shoshonis; to California and Pacific Northwest tribes; and even to Canadian Blackfeet by the 1750s.[8]

In general, horses dispersed from Texas and New Mexico more rapidly northwards – and to a lesser degree to the northwest – than they did due west. Nonetheless there were, by the mid-nineteenth century, large herds of feral horses in the San Joaquin Valley (and doubtless elsewhere) in California.

The results of this dispersal were fivefold: with greater mobility intertribal warfare, buffalo-hunting and the range of nomadic tribes rapidly increased; horse-rustling became a test of manhood for young braves; and further escapes greatly increased the distribution of feral herds throughout much of the American West and parts of southwestern Canada. Herds of feral horses – believed to be derived from animals belonging to Indians, who began travelling north from Santa Fé in about 1600, arriving in the Flathead area in about 1700 – were well established in the hinterland of British Columbia when the first European settlers arrived a hundred years later. Their numbers were considerably augmented after the start of ranching in the 1840s.

The first horses in eastern Canada were imported to New France (Quebec) from France by Poutrincourt in 1610. In 1665 Jean Talon, who was appointed the first intendant when New France became a crown colony, brought over a further 12 mares and 2 stallions, which were followed by additional shipments until 1671 when because of the cost involved (120 livres per mare and 200 livres per stallion, plus the expense of food and transport), Louis XIV decreed that importations should cease. The Indians were much astonished by these 'Moose of France' which were distributed among local farmers from whom, inevitably, some escaped to found feral herds.[9]

By the late eighteenth century feral horses had spread from the Rio Grande in the south to the Athabaska River in Alberta in the north, and from the Mississippi west to the Pacific coast. Although it is known that the largest number of feral horses (sometimes in herds several thousand strong)[10] occurred at this time in Colorado, New Mexico, Oklahoma and Texas (especially between the Rio Grande and the Palo Duro canyon),[11] no accurate estimate was ever made of the population: the most likely total was probably between two and five million.

As the settlers' wagon-trains rolled westwards feral horses declined in some areas and increased in others. Dense settlement usually drove those horses which were not recaptured or killed into the back country, whereas in sparsely populated districts feral stallions often lured domestic mares into the wild. In addition, the widespread custom of permitting cow ponies to roam freely in winter on the open range annually augmented the feral herds. The fencing of rangeland after the end of the Civil War in 1865 effectively eliminated feral horses from much of the Great Plains (which had previously been their principal stronghold) and reduced their distribution in other parts of the west, where, however, several million still survived, mainly in western Texas and eastern Montana: the establishment of ranches for horses, cattle and sheep in some western states (especially in Arizona, western Colorado, Nevada, eastern Oregon, Utah and Washington) in the 1870s and 1880s (when the total population is said to have been around 3.5 million), increased pressure on feral herds, but also resulted in the loss of valuable domestic horses which were often allowed to be free-ranging. At times, as for example in the 1890s, when there was a surplus of cow ponies, unwanted stock were deliberately released to join their brethren in the wild.[12] By the end of the century the majority of feral horses occurred west of the Rocky Mountains.

The outbreak of the Boer War in South Africa in 1899 greatly increased the demand for horses, and large numbers were purchased (at between $2 and $5 each) in the United States, Canada and Mexico by the British army for use as draught or pack animals or as cavalry remounts. Mortality during the voyage across the Atlantic was high (it was calculated that some 75 per cent of horses shipped from New Orleans – the main port of embarkation – died *en route* to South Africa), and most of the survivors soon succumbed to disease or were killed.[13] The demand for horses thus continued unabated until the conclusion of hostilities in 1902, resulting in a great depletion of the feral herds.

In some states (in eastern Oregon, for example) the US army released 'a number of well-bred stud horses' with an eye to future military recruitment,[14] and during the First World War many thousands of feral horses were re-captured in Oregon's Lake and Harney counties (and elsewhere) for use as cavalry remounts and for draught purposes with the allied forces, at an average price of $30–40 a head.[15]

In the early 1920s the abandonment of many farms (especially in eastern and central Montana) by homesteaders discouraged by poor soil, pests, drought and the fall in the price of wheat, made more land available for feral horses whose herds, augmented by abandoned domestic stock, rapidly increased and spread.

In the mid-1920s, chicken-feed processing – especially in California – opened up a market for horse-flesh, which by 1924 was also being used in the production of food for fur-farms and domestic pets. In the late 1920s the demand for horse-meat for human consumption also increased apace, and several hundred thousand horses were slaughtered – especially in Oregon and Washington – for shipment to Scandinavia, the Netherlands and elsewhere in Europe.[16] Finally, as the price for cotton fell, plantations in the southeastern states reverted from mechanical- to horsepower. 'The combined effect of these four factors', claimed McKnight (1959), 'caused a marked decrease in the feral-horse population, a decrease that has generally continued until the present day.'[17]

The enactment of the Taylor Grazing Act in 1924 (whereby grazing rights on certain designated range-lands under the jurisdiction of the Bureau of Land Management were henceforth to be rented to ranchers) increased once again the pressure on feral horses and accelerated their decline.

By the 1930s, however, horses were still widespread – albeit in diminished numbers – on land controlled by the Forest Service, the Bureau of Land Management and the Bureau of Indian Affairs, and all three bodies undertook intensive eradication programmes, resulting in the removal of tens of thousands of feral mustangs. McKnight (1959) gives the following examples: 10,000 removed from the Navajo Indian Reservation in a round up described by Wyman (1945) in *The wild horse of the West* as 'one of the largest in the history of the Southwest'; more than 10,000 slaughtered in the early 1930s on the San Carlos Indian Reservation in Arizona following an outbreak of dourine disease; around 1,000 killed in Sitgreaves National Forest between 1935 and 1942; several thousand removed from a single county in Montana in the 1930s; and 4,000 taken in the space of two years from Malheur County in Oregon.

In western Canada intensive efforts to eradicate feral mustangs began in British Columbia as early as 1924, and by the early 1960s BC Forest Service records show that well over 15,000 had been shot and a further 1,500 corralled and removed. Although small numbers of feral horses had been established in Alberta for many years, the peak population (which was never large) was not reached in that province until the 1940s or 1950s.

After the Second World War, during which mustangs were left in relative peace, due to lack of demand by the military and the absence of manpower, renewed efforts were made to eliminate them on now economically valuable ranges. In the mid-1940s and early 1950s well over 100,000 feral horses were removed from rangelands in western states – the majority from Nevada. Although Wyman (1945) points out that as early as the 1920s light aircraft were employed in Oregon and Wyoming to 'spook' horses from inaccessible areas onto the open plains where they could be more easily corralled, it was not until 1946–7 that aeroplanes were extensively used in roundups. The majority of horses captured were slaughtered for pet-food and to supply fur-farms, the remainder being used on ranches or in rodeos. The success of these roundups (which resulted by the late 1950s in the survival of only a few remnant herds) aroused considerable public opposition (originating in Nevada), which eventually secured a ban in most states on the use of mechanical aids in capturing feral horses. In response to continued opposition from the public the Wild Free-Roaming Horse and Burro Act of 1971 gave the two animals protection on land under the jurisdiction of the US Forest Service and the Bureau of Land Management.

McKnight (1959) summed up the status of feral horses in North America as follows:

1. The 'wild horse' of the West is a feral exotic, descended from strayed domestic stock, but adapting with rather remarkable adroitness to an unfettered existence in a sometimes inhospitable environment.
2. The horses have limited economic value. Presumably as a result of scanty grazing and inbreeding, they tend to be runty, big-headed, coarse and weak but have agility and stamina. Occasional individuals are handsome and well built, but these are the exceptions rather than the rule. In general the horses bear little resemblance to the famed mustang of yore: indeed, their most common appellation in the West today is 'broomtail'. When captured, they are difficult to break. The combination of inferior physique and difficult domestication means that their principal economic destination is either the rodeo or the cannery.
3. The feral herds tend to be deleterious to stock raising. They trample spring grass, pollute watering places, sometimes injure cattle, entice domestic horses with 'the call of the wild', and compete with domestic stock for forage and water.
4. They provide an interesting addition to the native fauna in many parts of the West, particularly in areas where native hoofed animals are scarce or completely lacking.[18]

Table 4 gives details of the distribution of feral horses in North America.

Sable Island, Nova Scotia is the only breeding place in the world of the rare Ipswich sparrow (*Passerculus princeps*).[19] The fragile scrub vegetation in which the Ipswich sparrow nests (comprising marram grass, beach pea, blueberry, rose, mayberry, winterberry, juniper and crowberry) depends for its survival on the stability of the sand-dunes, which are being progressively eroded by overgrazing and trampling by feral horses.

Thus 13 western states in the USA and Canada's two western provinces supported between them in the mid-1960s some 18,000 to 34,000 feral horses, and a further few hundred semi-feral animals occurred on some offshore islands on the Atlantic seaboard. The largest concentrations were to be found in Nevada, especially in Elko county in the north-east (near those established in the Owyhee drainage basin of Idaho and Oregon); in the north-west around the Smoke Creek Desert which adjoins California's Modoc and Lassen counties (which in the east were also densely occupied); and in various ranges and basins in west-central Nevada, extending as far west as the White and Excelsior mountains of California. Large numbers also occurred on Indian rservations such as the Mescalero Apache (north-east of Alamogordo) and various Pueblo reservations of the upper Rio Grande Valley in New Mexico; in the Southern Ute Reservation in southwestern Colorado; in the Warm Springs Reservation north of Bend in Oregon; in the Yakima, Colville and Spokane reservations in Washington; in the Uintah and Ouray Reservation in northeastern Utah; and in the Fort Apache Reservation in the Mogollon Rim country of eastern Arizona. Other regions in the western United States well stocked with feral horses included Montana, the Tularosa basin of southern New

Table 4 Distribution of feral horses in North America

State/ Province	*Regions*	*Total population*
United States		
Alaska	Big Delta area and Unalaska Island	25–50
Arizona	Indian reservations of: Hualapai, Havasupai, Fort Apache, Gila Bend, Colorado River and Papago; Sacaton, Galiuro, Cerbat, Aquarius, Black, Chocolate, Castle Dome, Tucson and Tortolita mountains; Sitgreaves National Forest; Gila and Bill Williams rivers; Palomas Plain; Organ Pipe Cactus National Monument	500–1,000
California	Eastern Lassen and Modoc counties; Devil's Garden, Modoc County; Goosenest District, Klamath National Forest; east and north-east of Mono Lake in Excelsior Mountains; Mono and Inyo counties, White Mountains; Coso, Tehachapi and Providence mountains; Saline Valley; Hunter Mountain region of Death Valley National Monument; Palm Canyon drainage of Riverside County	800–1,800
Colorado	Moffat, Rio Blanco, Garfield and Mesa counties in north-west; between Piceance and Douglas creeks south of White River; Book Cliffs; southern and western Montezuma County; Disappointment Creek drainage of Dolores and San Miguel counties	500–1,500
Idaho	Owyhee County; east of Mountain Home, Elmore County; east of Weiser, Washington County; Butte, Custer and Lemhi counties	500–2,500
Georgia	Semi-feral bands on Little St Simons and Big and Little Cumberland islands	400–500
Maryland/ Virginia	Semi-feral on Assateague and Chincoteague islands	?
Montana	Fort Peck Reservoir; southern Powder River and Rosebud counties; Pryor Mountains in southeastern Carbon County; south-west of Dillon in Rocky Hills range; west of Missoula and south of Frenchtown; east of Ravalli in Lake and Missoula counties; Kootenai National Forest in Lincoln County	500–800
Nevada	Elko, northern Eureka and Lander counties in the north-east; Humbolt and Washoe counties, and central and southern Nevada	5,000–7,000
New Mexico	Mescalero Apache and Pueblo Indian reservations; Tularosa Basin; Carson National Forest in Rio Arriba County; north-west of Albuquerque on Black Mesa and north in the Jemez Mountains; on Mount Taylor, north-east of Grants; near Carracas Mesa, west of Dulce; east of Raton near Berila Mesa; east of Las Vegas along the Canadian River; central Socorro and southeastern Otero counties; Gila National Forest	3,000–5,000
North Dakota	Badlands country of Little Missouri River in Billings and McKenzie counties	Up to 25
North Carolina	Semi-feral on Ocracoke, Shakleford and Cedar islands	?
Oregon	Lake, Harney and Malheur counties in south-east; Wallowa, Baker, Union and Umatilla counties in north-east; between the main and south branches of the John Day River in the Malheur National Forest; Prineville; Warm Springs Indian Reservation; Applegate River in Rogue River National Forest in southwestern Jackson County	2,000–4,000

Continued

Table 4 *(continued)*

State/ Province	*Regions*	*Total population*
United States *(cont.)*		
Utah	Box Elder, Western Tooele, Juab, western Millard, Beaver, western Iron and northern Washington counties; east of Antimony; Uintah River drainage of Uintah and Ouray Indian Reservation; Ashley National Forest; near Colorado border around Dead Man Bench; Hill and Willow creeks watersheds of Uncompahgre Addition of Uintah and Ouary Indian Reservation; Book Cliffs; southern Emery, eastern Wayne and eastern Garfield counties; Sooliers Mesa in San Juan County	1,200–2,500
Washington	Yakima, Colville and Spokane Indian reservations	500–2,000
Wyoming	Fire Hole County, south of Rock Springs; north of Wamsutter in the Red Desert; south of Sweetwater River in northern Fremont County; Powder River drainage in Sheridan, Campbell and Johnson counties; between Thermopolis and Meeteetse in Hot Springs County; in Sand and Fifteen Mile Draw areas of Washakie and Hot Springs counties	300–800
Canada		
Alberta	Between the Highwood and Athabaska rivers in the foothills of the Rocky Mountains; in the Peace River country along the border with British Columbia; north-east of Edmonton and north of railway line to Bonnyville	1,500–2,000
British Columbia	Drainages of Upper Fraser and Thompson rivers	1,000–2,000
Nova Scotia	Sable Island (150 km off-shore)	200–300

Source: After McKnight (1959 and 1964)

Mexico; between the Yampa and Little Snake rivers in northwestern Colorado, and neighbouring regions of Utah and Wyoming; parts of Tooele and Juab counties in western Utah; and some stretches of the Salmon River drainage basin in Idaho.

Large numbers were also to be found in the foothills of the northern Rocky Mountains between the Highwood and Athabaska rivers in Alberta, and in parts of the upper Frazer and Thompson river drainages of British Columbia. By 1980, however, it was estimated that in the United States only about 16,000–17,000 (one report says as many as 45,000) feral horses survived in 11 states.

As might be expected of an animal with such a widespread distribution, the feral horse in North America occupies a very varied habitat. It is mostly found in rough and rugged country, in the south-west favouring slopes and canyons of the desert mountains, alluvial fans and even the semi-desert plains. Mustangs prefer grassland regions, sagebrush associations and piñon–juniper scrubland of the west to forested areas. They are found in a variety of climatic conditions from extreme cold to moderate heat, but avoid the desiccating temperatures of the southwestern deserts. Although at their zenith they gathered in immense herds they now usually associate in small groups numbering from 5 to 15, roaming as widely as the availability of food and water and human settlement will permit.

Feral horses have undoubtedly been responsible, through overgrazing, for much range degeneration in the past, and still cause limited damage in places where they remain abundant. The few benefits that they confer, reported by McKnight (1964), include the spreading of undigested seeds through their faeces; ice-breaking at domestic stocks' watering places and keeping cattle-trails free from snow in winter; their aesthetic value; and their consumption of dead grass in winter, which assists spring growth.

Feral horses in North America today have few if any significant predators, their main controlling agent being man. McKnight (1964) considered that:

> The future for feral horses in Anglo-America is speculative, and in part dependent on legislation. In general it can be expected that the rapid decrease in their numbers of recent years will slow down markedly . . . The

decline will probably continue, but at a much slower pace, and in some areas increases in their numbers are to be expected.[20]

West Indies (Hispaniola)

The Genoese Christopher Columbus introduced the first (Spanish) horses to the New World when, during his second voyage of discovery, he landed between one and three dozen on Hispaniola in December 1493. For the next 30 years more horses were brought from Spain by nearly every ship bound for Hispaniola, which soon became a major horse-breeding centre in the New World. The French corsair brothers, Jean and Raoul Parmentier, saw considerable numbers of feral horses near Santo Domingo (in what is now the Dominican Republic) when they visited Hispaniola in 1526. In 1785 Sanchez Valverde wrote that since 1725 the Dominicans had hunted horses which they sold for a 'vile price'. A few still remain today in the Massif de la Selle near the Haiti–Dominican border – principally in the Savane Bourrique ('Jackass Meadow') region.

In the years following Columbus's voyages more horses were imported to the New World by the British, French, Portuguese and Dutch, and by 1511 stud-farms on both privately owned and government *estancias* had been established in Puerto Rico, Jamaica and Cuba. From these four islands horses were transshipped to other islands in the Caribbean, to Mexico, to British colonies in North America and to Central America and Peru.

Horses were first introduced to Mexico by Hernando Cortes during his expedition of 1519; since, however, according to Cabrera (1945) in his *Caballos de América,* most, if not all, of Cortes' 16 horses died within two years, future Mexican herds (both domestic and feral), which were soon widespread, originated in animals imported from the islands mentioned above.

SOUTH AMERICA

Colombia

Small dun-coloured horses with thick, longish coats are established at altitudes of 3,300–4,500 m in the *páramos* (high plateaux) of the Sierra Nevada de Santa Marta. They are believed to be descended from introductions made by the Spanish conquistadors.

AUSTRALASIA

Australia[21]

Exactly when horses were first introduced to Australia is uncertain, but they were undoubtedly imported during the early years of settlement in the late eighteenth century.

Rolls (1969) records that in about 1794 Private James Brumby arrived in Sydney with the New South Wales Corps, where he later settled and began breeding horses on a land-grant near Windsor. In 1804 he accompanied Colonel William Paterson on his colonizing expedition to the Tamar Valley in Tasmania and subsequently, on Paterson's appointment as lieutenant-governor, to Port Dalrymple. Before leaving New South Wales Brumby abandoned his stock, which became the first feral horses in Australia, where they are commonly known as 'brumbies'.[22]

As farming began to spread in New South Wales in the early nineteenth century more and more horses escaped from pastoralists and from travelling drovers or were deliberately turned loose to fend for themselves. By the 1830s so-called 'bush-horses' were common in the hill country around Sydney, from where they spread north, south and west with the expansion of pastoralism.

As McKnight points out, it is possible to correlate the early spread of feral horses with that of human settlement and the establishment of cattle stations – in southern and northern Queensland in the 1850s and 1860s respectively, in the Barkly Tableland in the 1870s and in the Top End (both in the Northern Territory) in the 1880s, and in the Kimberleys in the north-east of Western Australia in the 1890s. The development of herds of feral horses in Victoria, South Australia and elsewhere in Western Australia followed – as did the spread of farming – a more irregular pattern. Towards the end of the century the largest concentrations are said to have occurred between the Wilson and Cooper rivers in southwestern

Queensland where, as on the Barcoo River, they had been long established. A cattle-drover named McGillevery, on an overland drive from the Gulf of Carpentaria to Adelaide in 1871, reported large numbers at the confluence of the Thompson and Barcoo rivers north-east of the present township of Windorah.

From the late 1860s to about 1890 the feral horse was a major pest in parts of Victoria and New South Wales; he was, as a contemporary account describes him, 'a very weed among animals . . . the squatters shoot him off in organized battues to prevent his lawless depredations upon unfenced sheep walks'.

In the second half of the nineteenth century a market developed for horse-hide and hair; in 1869 hides were fetching 4*s*. each and hair 1*s*. 6*d*. a pound in Sydney, and large numbers of feral horses to meet this demand were rounded up at waterholes such as that at Wagga Wagga in southern New South Wales, hunters receiving 2*s*. to 2*s*. 6*d*. for each animal. Around the turn of the century some stations, especially in Queensland and the Northern Territory, found that the market for horses was more profitable than that for cattle; large numbers were bred for use as cavalry remounts by the British army in Egypt and India. As the army became increasingly mechanized between the two World Wars this demand declined, and many horses were given their freedom. Nevertheless Rolls (1969) records that in the 1930s two hunters shot 4,000 in a single year at Innamincka on North West Cooper creek in northeastern South Australia, and that from 1945 to at least the late 1960s between 5,000 and 7,000 were killed annually on the adjacent Coongie outstation. One hunter shot 700 in a single night, and the slaughter of 300–400 was commonplace.

The principal factors governing the existence of feral horses in Australia are hunting by man, climatic conditions, and the degree of recruitment by further escapes. In especially wet seasons, when food and water are abundant, foaling sometimes reaches 100 per cent, resulting in an expansion of range, often into hitherto unoccupied territory. During times of drought feral horses (in common with other species) tend to decline, and the reduction in the number of available watering places makes them easier to capture. On the other hand drought also causes the abandonment of domestic horses which help to swell the feral herds.

The present distribution and population are probably less than at any time in the twentieth century, and are likely to decline still further as human settlement spreads into the outback and more intensive farming methods increase.

Feral horses can still be found in Australia today in most of the more sparsely settled regions apart from desert areas. In a few places sizeable herds occur, but most associate in small and scattered bands. The largest concentrations are found in semi-arid and subhumid hills and plains and, especially, in monsoonal woodland and bush in the Top End of the Northern Territory. Although an apparently comparable habitat exists in the Cape York Peninsula of Queensland, it supports (for some unknown reason) only small numbers of feral horses; their relative absence in the Kimberley district of Western Australia can, on the other hand, be attributed to the presence there of the endemic and fatal 'walkabout' or Kimberley disease,[23] although large numbers were reported in the East Kimberleys in 1959.

McKnight (1976) found that feral horses were also common in other cattle-ranching areas, though they were mostly widely scattered in comparatively small numbers. As well as in the Top End he reported feral horses to be most frequently encountered in the central part of the lowlands of the Gulf of Carpentaria between Normanton in Queensland and Borroloola in the Northern Territory; in a crescent west of Alice Springs in central Australia through the Amadens basin and Stuart Bluff Range; in a broad band in southwestern Queensland and northeastern South Australia, especially in the drainages of Cooper Creek and the Bulloo River; and in central Western Australia between (and including) the Pilbara and Goldfields districts.

In Western Australia Long (1972) considers that some herds of feral horses have probably been established since the early colonial era. In pastoral regions they have been reported on many stations from Nullagine in the north to Esperance in the south. The largest area of distribution is from Juna Downs and Mount Vernon east to the Canning Stock Route, where there are said to have been between 4,000 and 5,000 in 1960. Shire Councils' estimates for the same year reported 500 in Meekatharra and near Carnegie, 200–300 in Norseman, 150–200 at Shark Bay, and 'a few' at Leonora, Upper Gascoyne, Sandstone and at Granite Peak in the Wiluna area.

In the south-west of Western Australia Long reported that in 1960 feral horses occurred in coastal areas near Esperance (including Cape Le Grande, Pink Lake and Fleming Grove); about 100 were judged to occupy the coastal regions between the Moore River, Cockleshell Gulley, Green Head and Eneabba. In the extreme south-west of the state small numbers were reported from the western parts of the Brookton and Pingelly shires; north of Wandering; in the Boddington shire; north-east of Collie; and south-east of Boyup Brook at Tone River. In 1969 a number were seen near the coast south of Jerramungup.

On the sheep stations of Western Australia, South Australia and New South Wales feral horses occur only in the more remote regions because sheep-runs require fewer horses than cattle-stations and because, being generally smaller, they tend to be more securely fenced.

McKnight (1976) refers to A. B. ('Banjo') Patterson's poem *The man from Snowy River,* which describes the mustering of a large herd of feral horses in southeastern New South Wales around the turn of the century. In the 1950s there were said to be 'about a thousand brumbies within riding distance of Charlotte Pass'[24] in the Snowy Mountains, and as recently as 25 April 1966 the *Sydney Morning Herald* was able to refer to problems caused by 'mobs of wild horses' in Shoalhaven Shire between the Snowy Mountains and the coast. Other scattered bands occurred in mountainous districts in eastern Victoria (especially along the Mitta Mitta River and the New South Wales border); in the Barrington Tops region of east-central New South Wales and in a number of hilly areas of east-coastal Queensland, including one district within 50 km of Brisbane. The total population is divided between states as shown in Table 5.

According to Frith (1979) feral horses are 'most numerous in the mountains of the south-east, in the south-west of Queensland, and in the vast areas of broken country in the north-west of the continent'. In *Feral animals in the Northern Territory: report of the Board of Inquiry*, the population in the territory was estimated to number between 20,000 and 50,000, most of which were concentrated on the Carmour Plain, Murganella, parts of the Roper, McKinlay, Margaret, Daly and Victoria river catchments, together with the central fringe of the Gulf of Carpentaria and the western McDonnell Ranges in the Alice Springs district (Letts, Bassingthwaighte & de Vos 1979).

Table 5 Distribution of feral horses in Australia

State	*Estimated population*
Northern Territory	40,000–60,000
Queensland	40,000–60,000
Western Australia	30,000–50,000
South Australia	10,000–20,000
New South Wales	5,000–10,000
Victoria	3,000–5,000
Total	128,000–205,000

Source: From McKnight (1976)

In the absence of interference by man or drought feral horses prefer to associate in small groups consisting of one stallion and from six to a dozen or so mares – colts being excluded at the age of between a year and 18 months. Occasionally more heterosexual herds of between 40 and 70 (and even up to 200) are observed. Feral horses are strongly territorial, with a home range of between 50 and 150 km^2. Even when drought and famine threaten their survival they seem reluctant to travel far in search of fresh supplies. Although they need water daily in hot weather, in cooler conditions they drink only every other day or so. They are less catholic in their diet than either donkeys or goats, always seeking the best forage (such as saltbush, grasses, edible mulga and other acacias) even when no water is nearby. Dingoes (*Canis familiaris dingo*) may kill a few sickly foals, but their main causes of death are 'birdsville', 'walkabout' and other diseases, poisonous plants,[25] falling and bogging, prolonged drought and shortage of food, but most of all predation by man.

The problems caused by feral horses include the consumption of food and water intended for cattle and sheep, the fouling and poaching of watering places, the interference with domestic stock which results in stampedes (especially during mustering), the damage they cause to fences and water-troughs, and the luring into the wild by feral stallions of domestic mares. The benefits they confer in return are few: small numbers are redomesticated for riding or as draught animals; their meat is used for pet-food and fish-bait and, abroad, for human consumption; there is also a market for their hides and for the hair from their manes and tails.

Feral horses are captured by rounding them up – with motor-cycles, trucks and light aircraft – into trap yards sited near watering places, or are shot by hunters. As human settlement advances into the outback and as husbandry becomes more intensive the range available for feral horses – and thus their distribution and numbers – is likely to decrease. In places where the climate and habitat is suitable for horses but not for domestic stock, and where they are unmolested, horses will continue to exist. Their long-term survival (as with so many species) is largely in the hands of man.

A small breed of feral horse known as the Timor pony (standing some 12 hands or 1.22 m high) is established in the Northern Territory of Australia. Exactly when Timor ponies were first imported is uncertain, but since it is known that water buffaloes[26] and banteng or Bali cattle[27] were introduced from Koepang on the island of Timor to early settlements at Fort Dundas on Melville Island and at Raffles Bay and Port Essington between 1826 and 1840, it seems not unreasonable to suppose that ponies were brought in at the same time. T. B. Wilson, in his *Narrative of a voyage around the world* (1835), records that a 'Timor horse' and a mare in foal were left behind at Fort Wellington when it was abandoned in 1829, and G. W. Earl, in *Enterprise in tropical Australia* (1846), states that horses had previously been shipped to the Victoria Settlement from Timor; in 1922 John Lewis recalled that there had been 'numerous Timor ponies at the old settlement' in the 1870s and that his party 'captured and broke in many and found them useful for riding'.[28] None of these authors, however, mentions exactly when the animals were first introduced. Frith (1979) records that 'the garrison at Port Essington in 1838 found feral Timor ponies and these were assumed to be the progeny of animals released at Raffles Bay, when that settlement had imported 'stock' from Timor. There were further importations to Port Essington during the course of that settlement.'

D. Johnson on his expedition to Arnhem Land in 1948 reported large numbers west of Port Essington.[29] G. A. Letts, leader of an Animal Industry Branch expedition to the Cobourg Peninsula in 1960, recorded small groups of one stallion and three or four mares across the length of the peninsula from Blue Mud Bay to Port Essington and Mount Bedwell; he estimated that there were up to 300 on the peninsula and several hundred more on the adjacent Malay Bay and Murganella Plain, where they interbred with their larger feral brethren.[30] In the following year A. Newsome of the Commonwealth Scientific and Industrial Research Organization saw small bands at Danger Point.

Timor ponies, which are of a slender and graceful appearance, occur in a variety of colours – bay, brown, grey, chestnut and cream – usually with long manes and tails. Although by grazing the common sedge (*Fimbristylis cymosa*) they add to the erosion of the sandy plains and also foul waterholes, because of their small numbers the overall effect is probably minimal. McKnight estimated the population in 1976 at between several hundred and a thousand, and Letts, Bassingthwaighte and de Vos (1979) state that a Primary Industry Division survey in 1978 suggested a population of around 440. Timor ponies are found throughout the Cobourg Peninsula, especially around Vashon Head, as far east as Malay Bay and Murganella. They also occur on Croker Island off the north coast of the peninsula where they have been redomesticated.

Between 1975 and 1978, on the pretext of habitat conservation, the Wildlife Service shot 150 Timor ponies on the Cobourg Peninsula. A moratorium on further shooting was then declared, which the Report of the Board of Inquiry in 1979 strongly endorsed: it also recommended the adoption of a suggestion, first put forward as long ago as 1846 by Captain John Lort Stokes, that a substantial fence, some 11 km in length, should be erected across the neck of the peninsula to prevent the entrance or exit of Timor ponies and other feral ungulates; the report considered that Timor ponies (and banteng[31]) should be relocated outside the peninsula, with a small herd only – as a tourist attraction – being retained within the Cobourg National Park.

New Zealand

Horses were first introduced to New Zealand in the early nineteenth century by European settlers, from whom some soon escaped into the wild. They were imported to the Chatham Islands in the 1840s and remained common there in unsettled districts until at least 1868, and probably until much later. They were first reported in the Hawke's Bay region prior to 1877.

Thomson (1922) reported their presence on Mount Tarawera, around the base of Ruapehu, in many localities on the North Island volcanic plateau, and possibly still on the Chatham Islands' tableland. At one time they were numerous in the bush country of Hokianga and along the west coast of the Auckland Peninsula.

Wodzicki (1950) found that by the late 1940s, although a few still remained in the back country of the Bay of Plenty, most were concentrated in the central part of North Island – principally in the Kaingaroa State Forest from where their range extended in a narrow strip as far south as the headwaters of the Mohaka River. Another strip continued around the southeastern shore of Lake Taupo to join a triangular extension roughly bounded by lines joining Tokaanu, Taumarunui and Taihape. Horses were also said to occur along the northwestern shore of Lake Taupo. They occupied mainly open country on the Central Plateau surrounded by native forests and mountain ranges which formed natural barriers to any possible spread. They were the size of large ponies and, especially in winter, were generally in poor condition; they caused a certain amount of damage by browsing young leading shoots when the Kaingaroa Forest was being established, and were of little if any compensatory economic value.

At the present time feral horses in New Zealand are confined to some 24,000 ha of the Southern Kaimanawa Range where they roam freely on land owned by the army, the Ohinewairua Station, the New Zealand Forest Service and Mr R. A. L. Batley of the Homestead, Moawhango, Taihape. The population is derived from nineteenth century liberations and escapes augmented by releases of redundant cavalry horses from the Waiouru army depot prior to 1940, and more recently by the addition of a number of Exmoor and Welsh pony stallions from Britain. As the result of increasing pressure on the population through illegal hunting, the New Zealand Forest Service in April 1978 formed the Kaimanawa Wild Horse Preservation Committee; the results of a survey in February 1979 showed a total population of around 174, of which about 30 are established on Mr Batley's Motumatai Private Scenic Reserve.[32]

OCEANIA

Hawaii

On 20 May 1803 the merchantman *Lelia Bird*, under the command of Captain William Shaler, called at San Borgia, California, where the captain was presented by a missionary with a stallion and a mare in foal. A second mare, also in foal, was subsequently purchased at Cape San Lucas. On 21 June the *Lelia Bird* dropped anchor in Kealakekua Bay, Hawaii, and on the following day one of the mares was landed at Tooagah Bay: 'this', according to H. W. S. Cleveland in *Voyages of a merchant navigator of the days that are past* (1886), 'was the first horse ever seen in Hawaii, and it excited great astonishment among the natives.' Captain Shaler then proceeded on to the island of Maui, where the remaining mare and the stallion were landed, probably at Lahaina, as a gift for King Kamehameha I who, however, 'only remarked that he could not perceive that their ability to carry a man quickly from one place to another would be sufficient compensation for the great amount of food they would necessarily require'.[33]

Andrew Bloxam, naturalist on the *Blonde* during her voyage from England to the Sandwich Islands in 1824–5, noted that on the island of Oahu 'cattle, goats and horses were very plentiful' and that in the south of Maui there were 'a few horses, but none in the neighbourhood of Lahaina'.

By the 1970s feral horses in Hawaii survived only in Waimea Canyon on the island of Kauai.

Galápagos Islands

As with other domestic livestock, horses were probably introduced to Isla Floreana (Charles) by the first colonists under the command of General José Villamil of Ecuador in 1832. Subsequently permanent settlements were established on San Cristóbal (Chatham) in 1869 and on southern Isabela (Albemarle) in about 1893.[34] 'Wild horses' were observed on San Cristóbal by Dr Simeon Habel when he landed on that island in 1868. In July 1875, however, when HMS *Petrel* visited the archipelago, Salvin (1876) reported that 'of animals introduced, it was estimated that . . . of horses, 150 existed on Charles, but none elsewhere'. Today, over a century later, feral horses are known to be established only on Volcan Sierra Negra on southern Isabela and in the highlands of San Cristóbal. Because they are ecologically unimportant compared with other feral ungulates in the archipelago they have been little studied, and their total population is uncertain. A number of herds believed to contain between them several hundred individuals roam over the caldera (crater) of Sierra Negra, where they are subject to limited exploitation by the inhabitants of Villamil and Santo Tomás. Since, however, very few giant tortoises (*Testudo (Geochelone) elephantopus*) remain on the volcano, whose vegetation is changed almost irretrievably from its pristine condition, and as the horses are prevented by lava flows from spreading to Volcan Alcedo (where feral donkeys[35] are established) or to Cerro Azul, there are at present no plans for their eradication.

Other oceanic islands

Horses have also been introduced to the island of Fatuhiva in the Marquesas; to Raivavae, Tubuai and possibly to Rurutu in the Austral and Rapa islands; to the island of Eua, Tonga and to Easter Island; possibly to Walpole in New Caledonia; to Maron, Luf and Manus islands, Papua New Guinea; to Macquarie Island; to Robinson Crusoe (Masatierra) in the Juan Fernandez group; to Auckland Island (first in about 1850); to the Agalégas; and to Tristan da Cunha. For some years after 1905 horses were established on South Georgia in the South Atlantic, and after 1950 on Kerguelen in the southern Indian Ocean.

According to Corbet and Southern (1977):

Wild horses were present in Britain during the last glaciation and as late as the Boreal phase of the postglacial period (i.e. around 7,000–10,000 years ago), but there is no good evidence that they survived beyond that time. Domesticated horses have been present since the Neolithic period (about 5,000 years ago).

There are no truly feral populations [in Britain] at present. Several primitive breeds live in a semi-wild state, e.g. in the New Forest (Hampshire), Dartmoor and Exmoor (Devon), Wales, the Lake District, Northumberland, Shetland, the Hebrides and western Ireland. It is frequently claimed that some or all of these are the direct descendants of indigenous wild horses but there is no evidence to support this and they are much more likely to be descended from domestic stock brought to Britain at any time since the Neolithic period.[36]

As previously described, horses were first introduced to South America in the early sixteenth century when importations from Hispaniola, Puerto Rico, Cuba and Jamaica were made to Peru; more were imported to Peru by the Spanish conquistador Francisco Pizarro in 1527 or 1531.

The French explorer Louis Antoine de Bougainville introduced three horses (as well as pigs and cattle) to East Falkland in 1764 where by the 1840s – when between 2,000 and 3,000 were running wild – their grazing had eliminated much of the tussock grass (*Poa flabellata*); the erosion which followed removed most of the shelter and nesting-sites for such seabirds as the Magellan penguin (*Spheniscus magellanicus*), petrels, wildfowl and waders. No feral horses remain in the Falklands today.

NOTES

1. The horse is generally believed to have been first domesticated in Asia between 5000 and 2000 BC According to Corbet (1978), it seems probable that the 'wild' horses which survived in Poland until about 1812 and in the Ukraine until around 1850 were in the main geniune remnants of the indigenous species. The eastern plains race, Przewalski's horse (*E. f. przewalskii*), survived in the wild until quite recently from the Altai Mountains in western Mongolia into Dzungaria in northern Sinkiang, China. Some still remain in the Takhin Shar-nuru massif of the Gobi-Altai region of southwestern Mongolia, and many nearly or entirely pure-bred individuals exist in captivity. Attempts have been made to reproduce the tarpan by selective breeding of primitive strains of domestic horses (which

occur in various parts of Europe, e.g. the Pyrenees, parts of Germany and the Camargue in southern France) with Przewalski's horses, and a semi-feral herd of this kind exists in Poland.

2. For information on feral horses in France, Portugal and Spain I am indebted to Mason (1979).
3. J. Coulentianou, *The Ark* **9**: 454–9 (1982).
4. For much of the history of feral horses in North America I am indebted to McKnight (1959, 1964).
5. See p. 144.
6. 'Mustang' is an Anglicized corruption of the Castilian *mesteño* meaning an animal belonging to *la mesta* (everyone and nobody, i.e. strictly ownerless). (See Cabrera (1945: 167.) Virtually synonymous terms include *cimarron* in Hispaniola, *bagual* in Argentina and *mostrenco* in Venezuela.
7. M. L. Moorhead, *New Mexico Historical Review* **32**: 107 (1957); Cabrera (1945: 166); and Denhardt (1947: 103).
8. See also D. E. Worcester, *Pacific Historical Review* **14**: 409–17 (1945) and *New Mexico Historical Review* **20**: 1–13 (1945).
9. Saunders (1935).
10. J. A Bursey, *New Mexico* **11**: 10–12, 36–7 (1933); and W. H. Chamberlin, *New Mexico Historical Review* **20**: 14–57, 144–80, 239–68, 336–57 (1945).
11. Denhardt (1947: 101).
12. D. Taylor, *Western Livestock Journal* **35**(45): 40–2; **35** (52): 80–2 (1957).
13. W. C. Barnes, *American Forest and Forest Life* **30**: 643–8 (1924).
14. McKnight (1964: 7).
15. Wyman (1945: 124).
16. Wyman (1945: 212).
17. p. 515.
18. pp. 516–19.
19. So-named in the *American Naturalist* in 1872 by C. J. Maynard because the holotype was collected at Ipswich in Massachusetts. The species winters on the Atlantic seaboard principally from Massachusetts south to Cumberland Island in southern Georgia.
20. p. 20.
21. For the history of feral horses in Australia I am indebted to McKnight (1976) and Rolls (1969).
22. In the *Australian Encylopaedia* he is promoted to Major William Brumby, and possible geographic and Aboriginal derivations of the term are suggested.
23. See also p. 96.
24. Fitzpatrick (1950: 33).
25. See also p. 96.
26. See p. 250.
27. See p. 258.
28. *Fought and won* (W. K. Thomas, Adelaide, 1922).
29. *Records of the American–Australian Scientific Expedition to Arnheim Land; Zoology* **4**: 427–515 (1964).
30. Letts (1964).
31. See p. 259.
32. Mason (1979).
33. pp. 96–8.
34. For a more detailed account of the human settlement of the Galápagos see p. 359.
35. See p. 98.
36. p. 408.

Wild Boar and Domestic Pig
(*Sus scrofa* and *Sus domestica*)

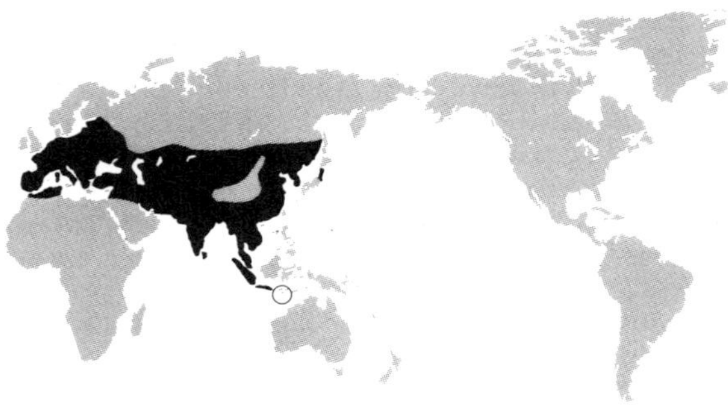

Note: Only present (reduced) natural distribution is shown.

NATURAL DISTRIBUTION
Wild boar (from which are descended the domestic pig) range from Europe and North Africa to southeastern Siberia, eastwards to Sri Lanka, Java, Honshu, Taiwan and Korea.

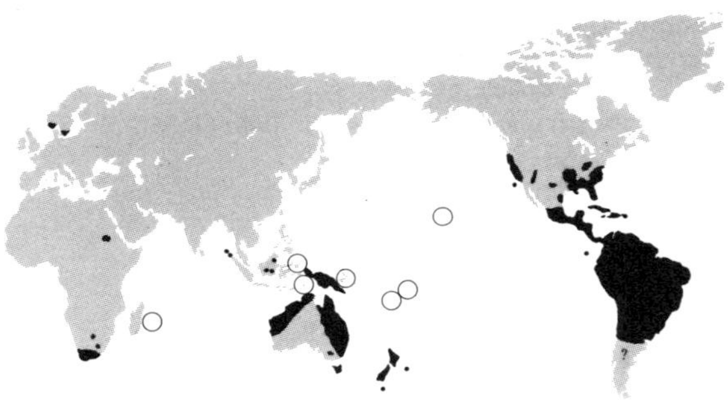

Note: Only principal feral populations mentioned in the text are shown. Distribution in Central and South America is uncertain and discontinuous.

NATURALIZED DISTRIBUTION
Europe: Norway and Sweden (south); Africa: South Africa, Sudan; North America: USA, West Indies (Hispaniola, Puerto Rico, Mona Island, Jamaica, Cuba, Barbuda, Virgin Islands) and much of Central America; South America; Australasia: Australia, New Zealand, many islands in Indonesia; Oceania: Hawaii, Galàpagos, Fiji, Mauritius, various other oceanic islands.

EUROPE

Wild boar have been reintroduced outside their present reduced European range in parts of southern Sweden and Norway where, as in England, Scotland, Wales (but not Ireland) and Denmark they were at one time indigenous but are now extinct. In England wild boar died out in the seventeenth century, and in Scotland and Wales probably considerably later. Subsequent attempts to reintroduce them in Britain, mainly in enclosed parks, all failed. In 1822 the Earl of Fife, on the advice of the Margrave of Anspach, turned down some wild boar on his unenclosed Mar deer forest in Aberdeenshire, Scotland, where they soon disappeared.

AFRICA

South Africa; Sudan

As early as 1925 domestic pigs (described as 'common long-snouted "boervarks"') were introduced to forestry plantations near George and at Stellenbosch in Western Cape Province, South Africa to control the larvae of the pine tree emperor moth (*Nudaurelia cytherea*) which were defoliating the trees. According to a Department of Forestry report, quoted by Siegfried (1962), some 'Austrian Boars' were imported later in the same decade in an attempt to breed a pig which would be more successful in rooting out the moth's pupae.

In the 1930s the Department of Forestry liberated some 'Bavarian' wild boars in the Kluitjieskraal Forest plantation for the same purpose. Today, the progeny of feral domestic pigs and wild boar (which freely interbreed) are found in many (mainly exotic) plantations of the southwestern Cape, from which they sometimes emerge to rootle among agricultural crops.

Feral domestic pigs have also been reported from several areas of dense indigenous scrub and alien prickly pear cacti in Eastern Cape Province, as well as from the Transvaal and Natal and possibly elsewhere in South Africa, where their ecological impact appears to be minimal.

Feral pigs are also established in a number of regions in the Sudan.

NORTH AMERICA

USA[1]

Domestic pigs were almost certainly first introduced to North America by Christopher Columbus during his second voyage of discovery in 1493. The earliest documentary evidence of importations, however, refers to those made to Florida by Ferdinando de Soto during his exploration of that state, Arkansas and Louisiana between 1539 and 1543. When French colonists attempted to settle in Florida in 1560 they were supplied by the Indians with pork from feral herds which were allowed to roam free.[2] More introductions followed to North and South Carolina, Georgia and elsewhere, in the seventeenth and eighteenth centuries, and feral populations became widespread in several southeastern states. In 1697 Jonathan Dickinson reported that Indians hunted deer and pigs on what was probably St Simons Island, Georgia. Pigs were first released in California by the Spanish in 1769 when they established a refitting base at San Diego for galleons from Manila, and by the Russians at Fort Ross in Sonoma County perhaps as early as 1812. More escaped during the gold rush of the 1850s, and their descendants were being hunted by the 1880s. It was not, however, until the latter decade that the first European wild boar were introduced to the United States when Mr E. H. Litchfield turned some down in Litchfield Park in the state of New York, where they survived until the 1920s.

According to McKnight (1964), the most important factor in the establishment and spread of feral pigs was probably the practice of 'free-range'. From the early days of settlement in the southeastern states – from Florida to eastern Texas and from Virginia to southern Missouri – there were many areas where cattle, pigs and other domestic stock were permitted to roam freely on open unfenced range. Concomitant with this practice was the establishment of so-called 'hog-claims' whereby, in Florida for example, a farmer could turn out a boar and two sows and be entitled to all the progeny produced in that area; such claims could be bought, sold, bartered, or inherited. The custom of free range combined with that of 'hog-claims' provided a virtually unlimited reservoir of potential feral animals, which was supplemented by those which escaped from fenced enclosures or were abandoned by their owners. When the practice of allowing domestic pigs to range free became illegal in the 1950s, large numbers were left behind in the coastal plains and swamps along many major river drainages from South Carolina to Texas.

'By far the most numerous of the feral ungulates in Anglo-America', wrote McKnight, 'is the feral hog [which] is so widespread and numerous in the southeastern and southwestern states as to be numbered in the hundreds of thousands.'

The Smoky Mountains of North Carolina and Tennessee (which now form the Great Smoky Mountains National Park) were largely owned in the early 1900s by such corporations as the Great Smoky Mountains Land & Timber Company which, on 24 February 1908, sold a vast tract of land to the Whiting Manufacturing Company, owned by two Englishmen, the brothers W. S. and Frank Whiting. To reward their financial adviser, George Gordon Moore of New York, who had enabled them to raise a loan of some $2 million, the brothers gave him permission to establish a game reserve on 650 ha of forest around the 1,800 m Hooper Bald[3] in the Snowbird Mountains of Graham County in western North Carolina, in part of what is now the Cherokee National Forest.

In 1909 Moore constructed a road up the mountainside to the summit of the bald, where two years later he built a lavish clubhouse and fenced in a 'boar lot' of over 200 ha. In April 1912 the first consignment of game from Europe (from the Harz Mountains in northern Germany) for Moore's reserve, including 'fourteen Russian wild hogs', arrived in the township of Murphy, from where three days later they were laboriously transported by ox-cart to Hooper Bald, where they were released in their enclosure.[4] According to a contemporary account: 'as the beady-eyed tuskers whooshed from the crates, some of the mountaineers went shinnying up trees like scared squirrels. They had been used to wild mountain razorbacks and acorn-splitter hogs all their lives but these wild pigs from abroad were something different.'[5]

In charge of the clubhouse and reserve Moore appointed a local homesteader, Garland 'Cotton' McGuire, to whom eventually Moore presented the deeds of the entire reserve. During a 'pig-sticking' and shooting hunt organized by McGuire in about 1920 (when only two animals were killed) about 100 managed to escape, and by the early 1930s were established in the surrounding Appalachians in North Carolina and Tennessee, where they interbred with feral domestic pigs and rapidly increased until checked by an outbreak of hog cholera in 1932. They did not cross the Little Tennessee River into the Great Smoky Mountains National Park – where they are now a serious pest – until the 1940s, but thereafter increased and spread rapidly, until by 1955 they were said to number about 500 and by the mid-1970s had multiplied to around 2,000: they now occupy an area of some 1,500 km^2. In the Appalachians, where the annual harvest is in excess of 120,000, hunting is controlled by the North Carolina Wildlife Resources Commission and the Tennessee Game and Fish Commission.

Following their natural dispersal into Tennessee (where they are now established in most of the 120,000 ha of the Ocoee, Hiwassee and Tellico districts of the Cherokee National Forest), wild boar were stocked on the 32,000-ha Catoosa Wildlife Management Area in Morgan and Cumberland counties of the Cumberland Plateau by the National Park Service in the early 1960s, and on the Anderson Tully Wildlife Management Area in the Mississippi River bottomlands as recently as 1979. In 1971 30 were translocated from the southern Appalachians of Tennessee to a private ranch in Boone County in southwestern West Virginia, where their descendants now occupy an area of some 115 km^2, and others have been acquired by the Florida Game and Freshwater Fish Commission. By 1975 wild pigs were present in parts of 66 of Florida's 67 counties: the highest populations are in large forested areas, usually managed for timber or pulpwood, with dense understories for shelter. Levy, Dixie, Lafayette and Taylor counties – which are dominated by slash pine flatwoods interspersed with cypress swamps, hydric oak hammocks and bayheads, and coastal salt marshes – are the most densely populated. Elsewhere in Florida relatively high populations are protected on private property, state land, wildlife management areas and national refuges.[6]

In 1924 some wild boar were recaptured on the slopes of Hooper Bald and transhipped to the San Franciscito Ranch in California, near the northern end of the Monterey Division of the Los Padres National Forest in the Carmel region, where by 1929 they had become established on national forest land. New blood, in the form of two dozen yearlings from Hooper Bald, was introduced to the Carmelo Creek watershed in 1932 to provide game for commercial hunting after the end of the deer-shooting season.[7] Wild boar now range over 100 km south from the Carmel Valley through the Santa Lucia Mountains of Monterey and San Luis Obispo County as far as Nacimiento Lake. In 1966 the population was estimated to number about 3,000; a few were also established on the Channel Islands off Los Angeles.

In 1928 some wild boar were imported from Europe by the Powderhorn Ranch on the Gulf coast of Texas, where later in the same year they escaped into the wild. More, from the southern Appalachians, were translocated to the St Charles Ranch (now the Aransas National Wildlife Refuge) in eastern Texas by Mr Leroy G. Denman in the early 1930s. In Texas wild boar are also currently established on the Edwards Plateau where they damage fencing and occasionally kill young sheep and goats: they also destroy the nests and eggs of wild turkeys (*Meleagris gallopavo*), compete for food with the native collared peccary (*Tayassu tajacu*), and cause serious damage to farm crops. Their population in Texas, which in 1967 was estimated at around 10,000, has since then greatly increased.

In 1938 a hurricane destroyed the fences of some enclosures at Corbin's Park in Sullivan County, New Hampshire, allowing some wild boar (which had been held in the park since 1889) to escape. The present population (the only pure-bred wild boar one in the United States) is said to amount to less than 40. Small numbers of European wild boar have from time to time been released on coastal islands of Georgia and North Carolina to add new blood to feral domestic herds. Illegal stocking by private citizens in Tennessee, North Carolina and – especially – in California has also assisted the species' spread.

Outside California and New Hampshire feral pigs (whose varied ancestry is reflected in their appearance) occur over an area of more than 108,000 km^2 of swamps, deltas, riverbeds and dense 'piney woods' of the Lower Coastal Plain of the Gulf and Atlantic coasts, from Virginia to Texas, where their population by 1964 was estimated at 1.5 million, of which some 67 per cent was in Texas and Florida. Other large concentrations were found in montane pine-forests of the southern Appalachians of Alabama, Georgia, North Carolina and Tennessee; in the Ozark-Ouachita uplands of Arkansas, Oklahoma and Missouri, and in the hill country of mid-Texas, where the total combined population numbered many thousands. In western states up to 10,000 feral pigs were unevenly distributed from southern New Mexico to southwestern Oregon. Even with 90 per cent piglet mortality it has been estimated that unless controlled the population could increase at an annual rate of 33 per cent. In West Virginia, North Carolina, Tennessee, California and parts of Florida feral pigs are accorded game animal status, and the annual harvest (most of which comes from the last two states) averages over 100,000 animals.

Table 6 gives particulars of the distribution of feral pigs in the United States. The range of feral pigs in the country is principally determined by climatic conditions and land usage. Prolonged frosts limit their ability to grub for roots and tubers, though montane populations, whose autumn and winter diet is composed mainly of nuts, are not affected. Since adequate cover is an essential requirement minimal land clearance for agriculture or settlement is an important factor. Their principal predator is man, by whom they are highly regarded as a game animal to be shot, trapped or hunted with hounds. Bobcats (*Felix rufus floridanus*), black bears (*Ursus americanus*) and foxes (*Vulpes fulva*) kill some young. Feral pigs tend to associate in sounders of between half-a-dozen to a dozen, though solitary mature boars are not uncommon.

In the Great Smoky Mountains National Park the rooting activities of feral pigs and wild boar have severely damaged the herbaceous forest understorey of laurel thickets – known locally as 'laurel hells'. In winter pigs tend to root in low successional tulip (*Liriodendron tulipifera*) forest and tulip silverbell (*Halesia carolina*) forest. At higher altitudes they grub around shrub-thickets on grassy balds. In the northern hardwood and grey beech (*Fagus grandifolia*) forests their activities cause considerable damage in late spring and early summer. In pig-infested grey beech stands understorey cover ranges from only 2–15 per cent compared with between 80 per cent and 100 per cent in similar but uninfested plantations.

Among the species most seriously affected are the tubers and shoots of wapota (*Sagittaria platy-*

Table 6 Distribution of feral pigs in the United States

State	*Distribution*
Alabama	Widespread apart from the east; especially common near the Gulf Coast and in the Alabama and Tombigbee river drainages
Arizona	Below the Hoover Dam on the Colorado River; in the Dragoon, Whetstone and eastern slopes of the Chiricahua Mountains; in the southeastern corner of the Grand Canyon National Park; and in the San Pedro Valley, Cochise County
Arkansas	Widespread, especially in the bottomlands of the Mississippi and White rivers
California	Northern coastal ranges and the Sacramento Valley; southern coastal ranges between Monterey and Santa Barbara; the bottomland of the Colorado River; marshes at the southern extremity of the Salton Sea; Afton Canyon of the Mohave River; Santa Catalina, Santa Cruz and Santa Rosa islands
Florida	Widespread in lightly settled regions
Georgia	Widespread in maritime areas and on some coastal islands, and scattered in hilly districts of the north
Louisiana	Widespread in suitable localities
Mississippi	Widespread, especially in the south and west
Missouri	Scattered throughout the south-east
New Mexico	Peloncillo Mountains, Hidalgo County
North Carolina	Marshes of the est and hill country of the west
Oklahoma	Hills in the east
Oregon	In the south-west, especially in the drainage of the Rogue River
South Carolina	Mostly in the east
Tennessee	Montane regions of the east and the Mississippi bottomland
Texas	'Piney woods' area of the east and hill country of central Texas; near the Gulf Coast, particularly around Corpus Christi
Virginia	Coastal region and hills of the hinterland
West Virginia	Widespread but unevenly distributed

Source: After McKnight (1964)

phylla), the long-leaved pine (*Pinus palustris*) and bulbs and rhizomes of *Lilium superbum*, *Dioscorea batatus*, *Medeola virginiana*, *Claytonia virginica*, *Stachys clingmanii* and *Aristolochia* spp. Depredations to natural plant communities are most noticeable in the Great Smoky Mountains National Park, in the Cumberland Island National Seashore in Georgia and in the Gulf Islands National Seashore of Mississippi, where rootling by pigs tends to favour alien species (which are used to competition from pigs at home) over natives who are not so accustomed.

Competition for food – especially nuts in the hardwood forests of California and the south-east – is the main adverse influence exerted on native wildlife; white-tailed deer (*Odocoileus virginianus*), wild turkeys, squirrels (Sciuridae), black bears and racoons (*Procyon lotor*) are the species most seriously affected, particularly in Alabama, Arkansas, Florida, South Carolina, Tennessee and Texas. On the other hand the pigs' rooting activities tend to make more food available for these and other species, so the overall impact may not be as great as is sometimes imagined. A more direct, if proportionately less important, influence is their omnivorous diet which includes small mammals (sometimes as carrion), reptiles, amphibians, ground-nesting birds and their eggs, and invertebrates. In the coastal marshes of Mississippi, where their rootling undermines levees and causes flooding, pigs both compete for food with and kill muskrats (*Ondatra zibethicus*). Elsewhere, damage to fencing and water supplies (by wallowing) and to agricultural, horticultural and silvicultural crops (especially seedling trees) has been reported. Pigs also provide a reservoir for such diseases as cholera, swine brucellosis (*Brucella suis*), trichinosis, foot-and-mouth, African swine fever, pseudorabies and bovine tuberulosis.[8]

West Indies

Domestic pigs were introduced to Hispaniola shortly after its discovery by Columbus in 1492; they were reported to be well established in the wild and numerous by the Spanish historian/naturalist Oviedo y Valdès in 1535. Today they frequent arid limestone regions with sparse soil and xerophytic brush vegetation, especially in the Baradères Peninsula and in the Tapion de Papaye region of the north-west; the south coast near Grand-Gozier and Anse-à-Pitre near the Haiti/Dominican border; Morne Bienac near Gonaives; the western portion of the Massif de la Hotte; and pine forests in lightly settled areas of the high Massif de la Selle. In appearance they resemble 'Appalachian razorbacks', being mainly black in colour, slender, rangy and having long tapering snouts.

On Mona Island (between Puerto Rico – where they also occur – and Hispaniola) predation by feral pigs on the nests of ground iguanas (*Cyclura stejnegeri*) in the island's Forest Reserve averages 25 per cent, rising to 100 per cent in very dry years. In Cuba feral *scrofa*, introduced by early Spanish colonists, live sympatrically with collared peccaries and white-lipped peccaries (*Tayassu pecari*) imported more recently for hunting purposes. Feral pigs in Jamaica (partly descended from those introduced by the Spanish in the late fifteenth century and partly from domestic stock abandoned by slaves following the abolition of slavery in 1838) show a remarkable adaptability to habitat variations and a high degree of tolerance to extreme climatic conditions, surviving from xerophytic forest-scrub, brackish marshes and mangrove swamps at sea-level on the arid south coast, to wet montane forest at altitudes of over 2,000 m in the cooler east. Feral pigs also occur in the West Indies on a number of the Virgin Islands and on Barbuda in the Leeward Islands. On some islands they are important predators of iguana eggs.

CENTRAL AND SOUTH AMERICA

By the late sixteenth century Spanish colonial settlements were well established in Mexico, some Central American countries, Peru and Chile, and the Portuguese had colonized parts of Brazil. The widespread practice of permitting domestic pigs to free-range inevitably led to the early formation of feral populations, and their progeny now occur over much of Central and South America. The descendants of pigs introduced between 200 and 400 years ago are hunted for meat and for the tusks of the boars in the Llanos Orientales of Colombia. In Argentina feral pigs hybridize with European wild boar which were introduced to his large *estancia* in La Pampa by Señor Pedro Luro in 1906. Jacob le Maitre and Willem Cornelis Schouten found large numbers of feral pigs (and goats) on the Juan Fernandez Islands off Santiago, Chile, when they landed to replenish their ships in 1616. In Brazil feral pigs are established in parts of the Matto Grosso on the Bolivian border. Señor Paulo Negueira Neto of the Brazilian Ministry of the Interior informed Tisdell (1982) that there are said to be European wild boar in the states of Santa Catarina and Paraná, and that in the latter they interbreed with feral domestic pigs in the Pon ta Grossa region.

AUSTRALASIA

Australia[9]

As McKnight (1976) remarks, feral pigs are the most detested and numerous of all Australia's escaped domestic livestock; their total population greatly exceeds that of all other ungulates combined.

It was at one time generally believed that feral pigs in Australia today are descended from some introduced by Captain Cook during his first voyage of discovery – hence the nickname 'Captain Cookers'. The release was commonly held to have been made near Cooktown on the Endeavour River on the east coast of northern Queensland where Cook careened his vessel in 1770. There is, however, no evidence in Cook's meticulous journal or log (or elsewhere) to support the belief that such a liberation ever in fact took place; the error seems to have arisen through confusion with the release of pigs which Cook did make in New Zealand in 1773. During his third voyage Cook obtained a pair of pigs at Cape Town in 1776, which were later released near Adventure Bay on the island of Bruni off Tasmania, which Cook mistakenly believed was part of the mainland of Australia; these pigs and their progeny were, however, all soon killed off by the local Aborigines.

'Hogs' (49) and 'pigs' (25) were among the livestock imported to Australia by the early colonists in the First Fleet of 1788, and many more (some allegedly from Siam (Thailand)) were brought in on subsequent occasions. These were at first allowed to range free around the settlement at Sydney Cove, but as early as 17 February 1795 permission was granted to kill any pigs found on private land.[10] At one time pigs were allowed to roam at will over a small island (possibly Philip Island) near Norfolk Island, but this practice was eventually abandoned.

In 1803 the Frenchman Nicolas Baudin released some pigs on Kangaroo Island, South Australia, as a source of food for sealers and shipwrecked sailors; within 20 years they were sufficiently abundant for one of the island's inlets to be named Hog Bay. In succeeding years more were liberated on Tasmania and on Flinders and French islands and other islands in the Bass Strait.[11] In *Discoveries in Australia . . .* (1846) Captain John Lort Stokes, commander of HMS *Beagle* during her voyage of 1837–43, records that he saw some pigs on an unnamed island in the Bass Strait, and landed a sow to join a pair he found on Swan Island in the Banks Strait off the northeastern tip of Tasmania. Joseph Beete Jukes, naturalist on HMS *Fly* during her survey of the northeastern coast of Australia in 1842–6, tells in his *Narrative . . .* (1847) how he released a pair of pigs on an island off the Queensland coast which, with their offspring, he shot in the following year.

In northern Australia pigs were both imported from Indonesia and translocated from settlements in the south. At the first post, established in 1824 at Fort Dundas on Melville Island off the coast of the Northern Territory, there was a herd of 54 pigs by 1826: whether these originated in Indonesia or were translocations from the south is uncertain.

In 1827 Fort Dundas was abandoned in favour of a new settlement (Fort Wellington) at Raffles Bay on the Cobourg Peninsula, some 30 km east of Port Essington, and a shipment of 20 pigs arrived from Koepang, Timor. When, two years later, Fort Wellington was in turn deserted most of the stock – including some pigs – were turned loose to fend for themselves. Captain (later Sir) James Bremer, who established the Victoria Settlement at Port Essington in 1838, arranged for both the translocation of pigs from Sydney and also for at least one consignment of 60 from Kissa (Kisar) Island north-east of Timor. At most of these settlements pigs were permitted 'to wander at liberty to a distance from the swamps and forts'. Some no doubt escaped into the bush where the population was subsequently reinforced by abandoned stock when the settlements were deserted. In 1866 the schooner *Beatrice* brought some pigs from Timor to a surveying party at Escape Cliffs west of Darwin, where a member of the group subsequently reported that 'a considerable number of pigs and boars were at the same time let go. The bulk of these crossed over the Adelaide River at the Narrows.'

According to K. Dahl,[12] feral pigs in the Northern Territory in 1894–6 were generally referred to as 'Chinese pigs' and were believed to be descended from importations made by coolies from China during the gold rush of the 1850s. Pullar (1950) received an unconfirmed report of an introduction of pigs from New Guinea to the Daintree River in northeastern Queensland before 1900.

As settlers spread from coastal areas they took domestic pigs with them into the outback, where they were again permitted to free-range. P. Cunningham in *Two years in New South Wales* (1827) observed that 'they [pigs] are allowed to run in the bush during the day . . .', and J. Henderson claimed in *Excursions and adventures in New South Wales* (1851) that 'pigs thrive and breed readily . . . they must, however, be allowed to run loose'. Thus until about the middle of the nineteenth century (or shortly thereafter) pigs were kept in a semi-feral state around the early settlements. When, after about 1865, properties began to be fenced and large runs and stations were subdivided, pigs were gradually abandoned by their owners and allowed to become truly feral. Most of the pigs in Queensland and the Northern Territory and on the upper Darling and around the confluence of the Lachlan and Murrumbidgee rivers in New South Wales, on Flinders and Kangaroo islands and in the Darling ranges of Western Australia are believed to have originated before about 1870. The population on Kangaroo Island was augmented around the turn of the century by farmers who released their stock after a religious revivalist convinced them that they were 'unclean'. The earliest documentary references to truly feral pigs on the mainland of Australia, traced by Pullar, were by H. Finch-Hatton in *Advance Australia* (1885) on the Dawson River in east-central Queensland and by Dahl *In savage Australia* (1926) on the Adelaide River south of Darwin in the Northern Territory and at Hill Station near Broome in the north-east of Western Australia in 1894–6.

In suitable habitats, such as swamps and the marshland of river systems and heavily timbered country, feral pigs spread widely and rapidly –

between 1913 and 1921, for example, throughout the drainages of the Georgina and Diamantina rivers in the Queen's Channel country. By about 1950 Pullar found that in northeastern Australia the population was more or less contiguous throughout most of Queensland and New South Wales (including the Darling River) south to the Tilpa and its tributaries from the Warrego to the Bogan; feral pigs also occurred in isolated swampy areas around the junction of the Lachlan and Murrumbidgee rivers; in the Nandewar ranges; on Campbell Island near Barham on the Murray River; at Morisset near Newcastle; in state forestland at Bago near Tumbarumba; near the Queensland seaboard on the Burrum and Mary rivers; at Wide Bay and at Yabba on the Brisbane River; and on Prince of Wales, Hammond, Curtis, Facing and Moreton islands. In Queensland feral pigs were thus already established virtually everywhere apart from the south-west region of low precipitation, densely settled parts of the south-east, the relatively tree-less grassy plains of the Mitchell River and the western slopes of coastal ranges.

In Western Australia, where as early as 1911 feral pigs occurred on the De Grey, Gascoyne and Fitzroy rivers and around Beagle Bay, Pullar reported that isolated colonies existed at the mouth of the Forrest River and in coastal marshes to the north; around the mouth of the Isdel River and along the upper reaches of the Fitzroy River; on the De Grey River and its tributary the Shaw; on the coast near Northampton; at the mouth of the Hill River; on the Wooroloo Brook near the Avon River; in the Darling Range in the south-west; and in portions of the neighbouring coastal marshes. Long (1972) added the following localities in Western Australia for the 1950s: in the vicinity of the Serpentine, Brunswick, Harris, Williams and Hotham rivers, and along the south-west coast between Mandurah and Bunbury. Subsequently small numbers were reported in hilly and swampy country near Waroona, Harvey, Collie, Drakesbrook, Serpentine/Jarrahdale and Dwellingup.

In the Northern Territory pigs were found to occur along the coast and river systems of eastern Arnhem Land and on the Cobourg Peninsula between the King and Daly rivers and in the swamplands near Maranboy. By 1964 they had moved north into Arnhem Land along the Roper River.

In South Australia Pullar found that the only extant colony was established west of the Middle and Harriet rivers at the western end of Kangaroo Island. In the Bass Strait Flinders Island supported small populations, but there were apparently none on Tasmania. In Victoria the last big colony – in the Otway Forest on the headwaters of the Gellibrand River – was exterminated by bush fires in 1938–9.

Since 1950 there has been a continual fluctuation in the population, which has seldom remained static for more than a few years, but little apparent overall change in general distribution. The *Agricultural Gazette of New South Wales* in 1955 reported that in the north and centre of the state '. . . wild pigs are more or less confined to certain areas, such as the Walgett and Morree districts and marshes on the Macquarie and Lachlan rivers'. A succession of mild winters allowed them to spread into the Cobar and Hillston districts and inland from the Darling River near Bourke.

In Western Australia Long reported that by the early 1970s feral pigs were established in many areas of the Kimberleys, Pilbara, north-west, Lower Murchison and Geraldton as well as in the south-west. In the Kimberleys they occupied a number of river systems between Derby and Wyndham, and were present until at least 1962 along the Fitzroy, Isdell, Forrest, Drysdale, Charnley, Robinson, Lennard and Margaret rivers – albeit in small numbers. In the Pilbara and north-west a few were reported on the De Grey, Shaw and Fortescue rivers, and also along the Hutt River, near Nanson and on the Murchison and Bowes rivers near Northampton. In 1968 a sounder of about 20 animals was first seen at Swan View, and in the following year some appeared in the John Forrest National Park.

McKnight (1976) found that in Queensland feral pigs are established principally along the Condamine/Balonne, Warrego, Paroo, Bulloo, Thomson, Diamantina, Cloncurry/Flinders, Gilbert and upper Burdekin river systems – especially in the sheep country of southern and central Queensland. In New South Wales they occur mainly in northern, western and central regions, particularly in the drainages of the Darling River and along the Lachlan River and its Willandra billabong.[13] Isolated communities exist along the Murrumbidgee and Murray rivers and in various river basins in eastern New South Wales. In the Northern Territory most suitable coastal habitats have been colonized from the Moyle River on the west coast, along the Daly basin extending northwards and eastwards to the coast and into Arnhem Land, as well as (according to McKnight) such offlying islands as Peron, Bathurst, Croker, the Goulbourns and Elcho. Smaller intrusions have occurred in the Roper River valley and in the northern catchment of the Victoria River.

In Western Australia feral pigs are fairly widespread in the Kimberleys of the north-east – especially along the Fitzroy River – and also further north along the Isdell and Charnley rivers; elsewhere in the state they are only lightly scattered, mainly along the Fortescue River in the north-west and in the vicinity of the Harris, Williams, Collie, and Harvey rivers in the south-west. Several very small and scattered populations occur in South Australia and Victoria, and a few also in Tasmania. There are two long-established colonies on Kangaroo Island off the coast of South Australia and on Flinders Island in the Bass Strait. The total population in Australia is divided between states as shown in Table 7.

Table 7 Distribution of feral pigs in Australia

State	*Estimated population*
Queensland	300,000–1,000,000
New South Wales	100,000–300,000
Northern Territory	50,000–100,000
Western Australia	15,000–40,000
Victoria	5,000–15,000
South Australia (including Kangaroo Island)	2,000–10,000
Flinders Island (including Tasmania)	1,000–5,000
Total	473,000–1,470,000

Source: From McKnight (1976)

In 1980, however, D. M. Flynn[14] estimated the population of feral pigs in Australia to be: Queensland, 5 million; New South Wales, 3–5 million; Western Australia, 500,000; Northern Territory, 200,000; and Victoria, South Australia, Tasmania and other offshore islands, 2,000 – a grand total of between about 8.75 and 10.75 million. The wide variation in the above totals (Flynn's may well be a considerable overestimation) is due to the violent fluctuations that occur in the population – largely through a high degree of fecundity coupled with a short gestation period of only 12 weeks, countered by droughts, floods (which sometimes drown large numbers) and hunting by man. Especially in Queensland and New South Wales numbers tend to vary enormously, and in some places denser settlement accompanied by more intensive husbandry may result in local extermination. An apparent increase in the Kimberleys of Western Australia and in the Top End of the Northern Territory in recent years (to a total in the latter estimated by the Primary Industry Division at between 100,000 and 200,000) may be due to previously incomplete reports.

Feral pigs in Australia are all descended from escaped domestic animals augmented by the deliberate release of prime boars in an attempt to improve the feral stock. There have been no introductions in Australia of wild boar such as have taken place in the United States and South Africa.

Australia's feral pigs are extremely variable in appearance. The predominant coloration is blackish or piebald, but rusty-red, fawn, dun, roan, white, spotted, striped or mixed individuals are not uncommon. Pullar found two extreme varieties – with many intermediate forms – which for convenience he dubbed 'early type' and 'recent type'. The former type, which occurs principally in the Top End of the Northern Territory and in the rainforests of the Cape York Peninsula of Queensland, is lean, muscular, small in size, black or red in colour, with a massive head and shoulders, a long snout and a prominent backbone (hence the term 'razor-back') and very small hind legs; in general conformation it is not unlike a Tamworth pig – the most primitive domestic breed and the closest in appearance to the wild ancestral type; an adult boar weighs about 65 kg. The latter type resembles an 'unimproved' domestic animal – again with a large head and well-developed shoulders – but having a broad back and with hind-legs and fore-legs which are less disproportionate. The usual colour is blackish or 'foxy', but paler or mixed colours are also quite common; an adult boar may weigh up to 135 kg. The canine tusks (especially the lower ones) of adult males are long (up to 300 mm) and well developed. Pullar believed that 'early types' are direct descendants of animals which escaped or were released in the early years of settlement (between about 1830 and 1880), whereas the 'recent type' are the offspring of later recruitment.

Feral pigs in Australia occur in a variety of habitats including coastal sand-dunes, marshes, mangrove swamps, forest, scrub, savannah, rough hills, grassy floodplains and, especially, rivers and watercourses – hence the term 'watercourser' used in some parts of Queensland. Their prime requirements are food, water and cover, for the last of which they favour lignum scrub (*Muehlenbeckia cuninghamii*). They are truly omnivorous and will consume any available edible matter, including leaves of such species as the *Pandanus* palm, fruit, roots, rhizomes and tubers of nut grass (*Cuperus rotundus*), bullrush (*Typha angustifolia*), *Phragmites* spp., lignum, and other species, as well as grubs and beetles. Unlike other ungulates they also kill (and eat as carrion) small mammals, reptiles, amphibians, shellfish, and ground-nesting birds. They are gregarious (usually associating in sounders of between 6 and 20 or so) and mainly crepuscular, and will range up to 15 or more km a day in search of food; in some areas they appear to be partially migratory. In the tropics salt-water crocodiles (*Crocodilus porosus*) and pythons (*Python amethystinus*) are their main predators; elsewhere red foxes (*Vulpes vulpes*), dingoes (*Canis familiaris dingo*), feral dogs and wedge-tailed eagles (*Aquila audax*) prey on piglets and sick or injured adults. Their most important controlling factors, however, are prolonged droughts, poisoning (with strychnine, arsenic, phosphorus and compound 1080) and hunting by man (especially in southern Queensland and in New South Wales) with rifles, bows and arrows, hounds and, occasionally, by 'pig-sticking'. Unlike other feral animals they are very prolific (with one or two litters of up to a dozen or more young per year), and Tisdell (1982) considered that in view of the paucity of non-human predators an annual population increase of some 60 per cent could be expected. Feral pigs are extremely pugnacious, and solitary boars and sows with young may well attack without provocation.

Feral pigs are regarded by Australians as their country's most pernicious pest since they not only compete with domestic livestock for land usage but also prey on foals, kids, lambs, calves, bogged or injured adults and poultry. In some places – especially on ecologically important offshore islands – they are held to have been responsible for the extermination of various species of ground-nesting birds. Other charges levelled against feral pigs include damage to vermin-proof and stock fencing, and, especially in artesian and subartesian areas, to water troughs, dams, tanks, drains, embankments and roads; wallowing in and fouling natural watering holes where domestic stock may become bogged; the acceleration of erosion and the creation of hazards to both motor vehicles and horses by their rootling activities; depredations on growing root, cereal and vegetable crops (particularly in parts of Queensland, New South Wales and Victoria) and on stored grain and hay; and the consumption and trampling of sugar-cane in parts of Queensland. Feral pigs are also important vectors of disease; as early as the 1920s an outbreak of anthrax in New South Wales was traced to feral pigs, and more recently they have been shown to have a high incidence of spirochaetal granuloma, 'Queensland fever', leptospirosis and melioidosis. Pullar considered that the dissemination by pigs of such diseases as foot-and-mouth, swine fever, Aujeszky's disease and rinderpest was a distinct possibility. They are also regarded with concern as potential reservoirs of tuberculosis: McKnight (1976) reported that postmortem examinations in the early 1960s revealed that 85 per cent of 260 samples and 46 per cent of 113 samples in the early 1970s had tuberculosis-like lesions; many of these were confirmed as *Mycobacterium tuberculosis* var. *bovis*.

Tisdell (1982), who suggested that 'Possibly no other vertebrate pest causes so much damage now that rabbits are not as great a problem as in the past', estimated that the loss to agriculture caused by feral pigs in Australia is in the region of $A70–80 million a year (mostly in Queensland and New South Wales), of which over half is accounted for by a reduction in grain (mainly wheat) production, with significant losses also being incurred by livestock farmers through predation of lambs and the reduction of the stock-carrying capacity of pasturage. Lesser damage is suffered by oil seeds (including soya bean) and sugar-cane crops (respectively 2 and 1 per cent (20,000 t) of the total harvest), but even so this amounts to $A1.4 million and $A500,000 per annum respectively.

On the credit side their carrion-eating habits give feral pigs some value as scavengers, and they also to some extent help to control blowflies, rats, rabbits and venomous snakes, which they kill by trampling with their sharp hooves. Pullar noted that during a plague of rats in Queensland in the 1940s thousands were destroyed by feral pigs. In some areas adults killed in the wild, and in particular feral piglets fattened in captivity, are a useful source of pork. Especially in Queensland and New South Wales and on Flinders Island they are hunted for sport, which provides both recreation and revenue in the form of licence fees for landowners. McKnight suggested that their rooting may act as a form of cultivation to assist plant growth, and that by providing an alternative prey for dingoes and eagles piglets may help to protect the young of domestic stock; these benefits, if indeed they occur, must be of only marginal importance, and are far outweighed by the damage pigs cause.

Tisdell estimated that in New South Wales alone sport-hunting accounts for at least 150,000 feral pigs a year, and that in the country as a whole between 100,000 and 200,000 sportsmen annually kill in the region of half a million pigs. At present professional hunters take only around 60,000 a year, which yield 2,000 t of meat for export – mainly to France, West Germany and Japan, with lesser amounts to the United States, the Netherlands, Hong Kong and Bahrain – to the value of approximately $A8 million, but this is expected to rise within the next few years to about 500,000. The maximum sustainable annual yield for the whole of Australia has been calculated to be at least 2.5 million.

New Zealand[15]

The first pigs to land in New Zealand were two young ones presented to a Maori chieftain at Doubtless Bay by de Surville of the *Saint Jean Baptiste* in December 1769: they are believed to have been the first animals introduced to New Zealand since the Maoris brought with them from Hawaiki (probably the island of Savaii in western Samoa) their domestic dogs and the kiore or Polynesian rat (*Rattus exulans*)[16] some 400 or more years earlier: what became of them is unknown.

Four years later, in 1773, Captain James Cook's second expedition was responsible for the first release of pigs into the wild – hence the appellation 'Captain Cookers'. (In 1847 G. F. Angas claimed that Spanish adventurers were the first to liberate pigs in New Zealand.[17]) On 2 June Cook records that while they were in Queen Charlotte Sound in South Island Captain Tobias Furneaux of the *Adventure*:

> Put on shore, in Cannibal Cove, a boar and two breeding sows, so that we have reason to hope this country will, in time, be stocked with the animals, if they are not destroyed by the natives before they become wild, for, afterwards, they will be in no danger.[18]

George Forster, Cook's naturalist, states in his *Journal* that 'they were turned into the woods to range at their own pleasure'.[19]

In October of the following year (1774) Cook 'went to the place where I left the hogs and fowls; but saw no vestiges of them, nor of any body having been there since . . . As we were coming away Mr Forster thought he heard the squeaking of a pig in the woods . . . probably they may have been those I left with them when last here'.[20] The earlier part of this statement is confirmed by Forster who says that he

> took the opportunity to visit the innermost recesses of West Bay, in order to be convinced, if possible, whether there was any probability that the hogs, brought thither about a year before, would ever stock those wild woods with numerous breeds. We came to the spot where we had left them, but saw not the least vestiges of their having been on the beach, nor did it appear that any of the natives had visited this remote place; from whence we had room to hope, that the animals had retreated into the thickest part of the woods.[21]

Later in the month one of Cook's party saw a large black boar on Long Island in Queen Charlotte's Sound which Cook

> thought might be one of those which Captain Furneaux left behind, and had been brought over to this isle by those who had it in keeping. Since they did not destroy those hogs when first in their possession, we cannot suppose they will do it now, so that there is little fear but that this country will, in time, be stocked with these animals, both in a wild and domestic state.[22]

A month later, on 2 November, Cook gave some pigs (which he had brought from Tonga or Tahiti) to Maoris who came out by canoe to the *Resolution* while she was off Cape Kidnappers, south of Napier, North Island. Six days later he landed another pair nearby, 'so that it is hardly possible all the methods I have taken to stock this country with these animals should fail'.[23]

When Cook visited Queen Charlotte Sound again (where he presented another pair of pigs to a local Maori chief) on his third voyage in February 1777 he reported that:

the animals which Captain Furneaux sent on shore here, and which soon after fell into the hands of the natives, I was now told were all dead; I was afterwards informed by the two youths who went away with us that Tiratou, a popular chief amongst them, had one of the sows in his possession but I could get no intelligence about the fate of those I had left in West Bay, and in Cannibal Cove, when I was here in the course of my last voyage. I have, at different times, left in New Zealand not less than ten or a dozen hogs, besides those put on shore by Captain Furneaux. It will be a little extraordinary, therefore, if this race should not increase and be preserved here, either in a wild, or in a domestic state, or both.[24]

When Captain Raven landed the first sealers at Dusky Bay in 1792 he is believed to have offloaded some pigs with them. Captain Philip Gidley King, while he was lieutenant-governor of New South Wales, presented 10 young sows and 2 boars from Norfolk Island (where he had recently been governor and commandant) to the Maori chiefs Tooka and Woodoo at the Bay of Islands, North Island, when a number of flax-dressers were repatriated while he was visiting New Zealand in 1793. Of these, however, Captain Bamford of the *Fancy* subsequently reported that all but one had perished: apparently they were mistaken by the Maoris for horses (which they had seen on Cook's vessels) and two were killed by riding, while a third was summarily executed after trespassing on a sacred burial ground. (According to another account these pigs were released and preserved by Tooka and Woodoo on an island off North Cape.)

In succeeding years many more pigs were landed in New Zealand, from which (and possibly from Cook's earlier introductions) the feral pigs encountered by early settlers at the beginning of the nineteenth century were probably descended. The crew of the English brig *Tula* found feral pigs on Cornwallis Island in 1831, and in the same year the *Argo* from New South Wales transported 30 pigs from the Cook Strait area of South Island back to her home port of Sydney – a practice which was subsequently repeated on many occasions by other ships. A few years later Captain Jay of the American vessel *Mary Mitchell* found that pigs were highly valued by the Maoris, in one of whose encampments, according to Clark (1949), he saw 'four sows, two with litters of piglets, and two boars'. 'Gone-native' European whalers and sealers – known as 'Pakeha-Maori' – in the Cloudy Bay area employed themselves in the off-season in planting potatoes and hunting pigs.

Edward Shortland, an early protagonist of the Maoris, in his *Southern districts of New Zealand* (1851), found that by 1840 pigs were widely established throughout New Zealand; Marlborough Sound on the Banks Peninsula, parts of northern Westland and the hinterland of the present township of Dunedin were 'thick with pigs'; he reported that in acquiring a whaler in 1843 the Maoris paid part of the purchase price with 41 pigs, which were valued at about 12*s*. each. Thomson (1922) records that in 1840 a Major Bunbury, in his report on the proclamation of Stewart Island as a Crown possession, said of Paterson Inlet that 'the Europeans there employ themselves in the culture of pigs and fowls'. Two years later Captain William Smith, chief surveyor of the Wellington Land Company, wrote of Otago that 'at Tautuku Bay is a great deal of land cultivated by a number of industrious men who have many pigs and goats'. In *Contributions to the early history of New Zealand* (1898), T. M. Hocken recounted how in 1845 two sailors who had jumped ship sold some pigs which they had captured in the bush to whalers at Otago Heads where, as at other whaling-stations, scavenging pigs were frequently observed.

There were, however, as yet few feral pigs in the Nelson area; Dumont D'Urville, for example, in his *Voyage de la corvette l'Astrolabe* (1830–5), said that he had seen none in Tasman Bay in 1827, and John Deans, in *Pioneers of Canterbury: Deans' letters 1840–54* (1937), writing of Nelson in 1842, claimed that 'the pigs are all extirpated hereabouts long ago'.[25] Ernest Dieffenbach, on the other hand, described how in 1839 'the natives had come from Cannibal Cove to catch pigs, which overrun the island' [of Matuaru],[26] but a few miles to the east; in the same year both Shortland and Edward Jerningham Wakefield observed large numbers around Port Underwood and Queen Charlotte Sound.

According to Thomson 'the abundance of wild pigs in the country was of great value to explorers, particularly to prospectors, and also to shepherds, miners and back-block settlers' – as well as to the native Maoris who valued them for food, barter and plunder, and allowed them to run semi-wild (as did some European settlers) around their encampments. Dieffenbach observed that the Maoris 'have great quantities of pigs, which have run wild, but are easily caught by dogs'.[27] By about 1840 *Te poaka* ('the porker') was a feature on many Maori *tanguis*.

Angas reported[28] that by the early 1840s feral pigs (augmented by escapes of free-ranging domestic animals) were abundant – mainly around abandoned Maori *pas* and in woodland where fern-roots were said to give their flesh the flavour of veal – between the Waikato River and the Bay of Plenty in North Island. So-called 'pork-traders' used to travel up river to obtain from the Maoris pigs with which to supply the inhabitants of *pakeha* settlements and sealers, whalers and ships on the coast.

Dr Monro, who accompanied T. Tuckett on his expedition through Otago in 1844, wrote of the hill country south-west of Saddle Hill that:'There is a famous cover for pigs, too, between the upper part of the Teiari [Taieri] Valley and the sea . . . the whalers come up the river in their boats and kill great numbers of pigs here; as the Maoris told us.' According to Robert Gillies, 'In 1848 [the year of the settlement of Otago] wild pigs were very common on the site of Dunedin'[29]; six years later he and some companions killed 70 behind Flagstaff Hill in the space of two days. Writing of this period James Drummond said that 'they multiplied astonishingly and enormous numbers assembled in uninhabited valleys far from the settlements'.[30]

Summing up the position to date F. von Hochsetter, in *New Zealand: its physical geography, geology and natural history* (1867), wrote:

> Cattle and swine run wild in various districts of the islands, and it is astonishing to what numbers the wild pigs are multiplying. They find an excellent and everywhere plentiful food in the fern-roots, which formerly served the Maoris as a chief article of food. They retire shyly from the immediate vicinity of the settlements, because the settlers hunt them down energetically; but they congregate in the yet uninhabited valleys in enormous numbers. The Wangapeka Valley in the Province of Nelson I saw for miles up and down literally ploughed up by thousands of wild pigs. They are nearly all black. . . . three men in twenty months, upon an area of 250,000 acres [100,000 ha], killed not less than 25,000 of them . . . where the wild pigs are very numerous, they do a great deal of damage to sheep-breeding.[31]

In some areas – presumably as a result of seldom if ever encountering man – feral pigs appear to have been remarkably tame: writing of the hill country of Marlborough Aston (1915) recounted how 'two sows, in response to our grunts . . . came up close and smelt us. My companion made a grab at one leg, and pig and man went rolling down the hill together'.

On distant offshore islands pigs were first landed by Captain A. Bristow in 1807 on the main island in the Auckland group, where they were described in 1840 by Mr (later Sir) Joseph Hooker (naturalist and assistant-surgeon on the *Erebus* during Sir James Clark Ross's Antarctic expedition of 1839–43, and subsequently Director of the Botanic Gardens at Kew, England) to be abundant; Hooker reported that they seemed to feed mostly on *Pleurophyllum criniferum*, while the *Erebus*'s chief surgeon, McCormick, observed that they also ate *Stilbocarpa polaris*. A decade later Captain Enderby found them still present in large numbers, yet Captain Musgrave, whose ship was wrecked on Auckland Island in 1865, was unable to find any with which to feed the survivors of his crew. In 1842 pigs were introduced to Enderby Island (where they had died out by 1887) in the Auckland group as a potential source of food for shipwrecked sailors. These introductions led Edgar R. Waite to say that:

> There can be small doubt that the introduction of pigs to the Auckland Islands has resulted in considerable havoc among the ground-nesting birds, by destroying both eggs and young . . . native plants are also suffering . . . *Bulbinella* and *Pleurophyllum* evidently being favourites.[32]

On Arapawa Island in the Cook Strait small pigs of the early 'Berkshire' variety (with mottled sandy or cream and black coloration, unmarked legs, a long straight tail terminating in a tassel, pricked ears and a long snout) were first introduced in around 1850 when, with the decline in their profession, whalers became small farmers. Their descendants remain on the island to this day.

Three pigs released on Campbell Island, south-east of the Aucklands, in 1865 by a Captain Norman apparently did not long survive.

According to a report made to Wodzicki (1950), as late as the 1880s the Waimate plains on the mainland were over-run with pigs, which were 'killed almost on the doorsteps of the farmhouses'. Further evidence of their abundance is that the poles of the first telegraph line between Wellington and New Plymouth had to be protected from rubbing by pigs. Shortly thereafter 1,200 pigs were slaughtered in the course of a single drive in south Taranaki.

In 1922 Thomson wrote that 'at the present time they [pigs] are still common in nearly all bush country which is not too near settlement', where they found an abundance of both food and shelter.

Wodzicki reported that by the late 1940s pigs were fairly widespread on both main islands, and on such offshore ones as Arapawa, Mayor, Chatham, Pitt, D'Urville and Great Barrier: on Stewart Island a few survived in the north-west between the Freshwater and Ruggedy rivers, but the population on Poor Knights Island had been exterminated by the Department of Internal Affairs in 1936. On the subantarctic islands of Enderby, Adams, Rose, Campbell and Macquarie (where they were introduced at an unknown date) pigs had long been extinct, but they survived in considerable numbers on the main island of the Auckland group, where, assisted by feral cats, they have been responsible for a decline in the endemic Auckland Island flightless teal (*Anas aucklandica aucklandica*); they also on occasion eat yellow-eyed penguins (*Megadyptes antipodes*) and the eggs and young of albatrosses and shags, and dig up the burrows of nesting petrels.

On South Island, which was less densely populated than North Island, the largest numbers occurred in parts of Marlborough, north Canterbury and the eastern regions of Nelson Province. In the south of South Island Wodzicki found that the main areas of infestation were centred near Herbert in north Otago, in the Catlins district and in the Longwood and Princess ranges in Southland. The central areas of South Island supported small numbers of pigs, with fewer still in Westland.

Wodzicki reported that although in many places the distribution of pigs coincided with forestland, most of the heavier infestations occurred on the forest edge and in areas of secondary growth. In some parts of Nelson, for example, pigs were found to be especially common where hill pastures were reverting to fern and scrubland, and near Dunedin they occurred only on ferny and scrubby areas. In the Catlins district there were smaller numbers of pigs than might have been expected from the amount of bush and secondary growth.

In North Island Wodzicki found that the heaviest infestations occurred on the Coromandel Peninsula, in southwestern Waikato, in large portions of eastern Taranaki, in the Kaingaroa State Forest, and in several places east of the main divide from East Cape to Wairarapa. In spite of apparently suitable habitat there were few pigs in the Rimutakas, and none in west Taranaki, the Tongariro National Park, Manawatu, and east Hawke's Bay, probably as the result of closer settlement and the concomitant reduction of favourable habitat. Northland, north of Auckland, supported only small numbers, in spite of an abundance of secondary native bush.

Wodzicki found few pigs in pastoral areas and above the bush line (about 1,300 m) where there is no cover and little food; most live in the forest ecotone and in scrub, fern, tussock and marginal farming areas. (Their exceptional appearance in the Kaingaroa State Forest – where they caused a considerable amount of damage – was, Wodzicki considered, explained by an abundant food supply associated with the establishment of pine plantations, and protection from hunting.) Favourite habitats include damp, fertile woodlands, where succulent fern roots flourish and strong secondary growth inhibits agriculture: here slow forest regeneration takes place through the bracken fern which is one of the pigs' preferred foods. Leached forest soil, on the other hand, where regeneration tends to occur through manuka, is less suitable for pigs, and few are found in regions of such secondary growth – as, for example, in northern Auckland.

Feral pigs in New Zealand have caused a great deal of damage to the country's native fauna and flora. According to Kirk (1896) they were responsible for the extinction on the mainland of a rare endemic reptile, the tuatara (*Sphenodon punctatus*), and for the considerable reduction of the population of kakapos (*Strigops habroptilus*), and other ground-nesting birds. By way of compensation pigs also root out and kill large numbers of young rabbits. On the Auckland Islands they wrought havoc among the eggs and fledglings of nesting colonies of seabirds.[33] Kirk added that the formerly abundant endemic orchid *Gastrodia cuninghamii* had been seriously depleted by pigs and rats. On the Auckland group the greater part of feral pigs' food consisted of *Pleurophyllum criniferum*, *Stilbocarpa polaris* and *Bulbinella*.[34] On the Chatham Islands feral pigs nearly succeeded in exterminating two endemic plants – the Chatham Island lily (*Myosotidium nobile*) and, by grubbing up its succulent tap root, *Aciphylla traversii*. Other plants adversely affected by pigs in New Zealand included, according to Thomson (1922), two spear grasses *A. squarrosa* and *A. colensoi* (whose succulent and aromatic roots were grubbed up), *Gaya lyallii*, the bracken fern *Pteris aquilinia* var. *esculentum* (whose starchy rhizomes were relished by pigs who thus helped farmers by delaying forest regeneration), *Gastrodia sesamoides*, *Marattia fraxinea*, and the umbellifers *Ligusticum* and *Angelica*. L. B. Moore and L. M. Cranwell summed up the position when they wrote that feral pigs are 'a constant menace, destroying the natural balance between the vegetation and its environmental factors'.[35].

In exotic plantations – especially in their early stages when Douglas firs (*Pseudotsuga douglasii*) and Corsican pines (*Pinus laricio*) were particularly affected – and on agricultural land Wodzicki considered that feral pigs were only exceeded as a menace by alien goats and deer. By rooting for food they also damage forest roads and firebreaks. On farmland, which they frequent especially in winter, breeding sows in need of calcium are responsible for considerable losses among newborn lambs, and on occasion even kill sickly adult sheep.

Many feral pigs in New Zealand bear characteristics of the domestic Tamworth, Berkshire and Hampshire breeds. Wodzicki found that an average adult boar weighted between about 70 and 120 kg and stood nearly a metre high; individuals of as much as 180 kg or more and up to 2 m in length have been recorded. Between 6 and 10 piglets (of which a maximum of half-a-dozen are usually reared successfully to maturity) may be born at any season, though most sows farrow in spring and early summer. Feral pigs' replacement rate is high, and even heavily depleted populations can be restored to their former levels in under a year. Disease is apparently not an important mortality factor.

Since Thomson's time the reclamation of scrubland and cutover forest for agricultural purposes has destroyed much of feral pigs' habitat and has reduced their range. According to Whitaker and Rudge (1976) they are, however, still widely distributed on both main islands, albeit generally in small numbers. A survey undertaken in 1971 showed pigs to be present on 40 per cent of North Island and 20 per cent of South Island, but that they were abundant in only some 7 per cent of their range.[36] They still occur on several offshore islands, including Arapawa (where they may closely resemble the original Oxford Sandy and Black), Great Barrier, Mayor, D'Urville, the Chatham group and on Auckland Island in the subantarctic, but have disappeared from Stewart Island. They are most abundant in native forests, pine plantations, marginal farmland, and in fern, scrub, bush and tussock country, where they frequently travel considerable distances in search of food. Their rootling activities have in places played a significant part in rainforest degredation and in the acceleration of erosion. They are now, however, generally regarded as of only minor importance in protection forests, and only a local problem in commercial woodlands and on agricultural land. There is some difference in coloration and conformation between discrete populations, and also a suggestion of a reversion to the ancestral 'Tamworth'-like wild form.

Feral pigs in New Zealand are of value only as a sporting asset; they are controlled mainly by shooting, poisoning and trapping. Of these, shooting – by professional hunters and, to a lesser extent, by amateur sportsmen – has been the most effective method. Tisdell (1982) was unable to estimate even the approximate number of feral pigs killed annually in New Zealand, but found that between 1967 and 1979 a total of over 5,000 t of meat to the value of over $NZ4.5million was exported – as from Australia, mainly to France, West Germany and Japan, with lesser quantities being sent to the United States, the Netherlands, Hong Kong and Bahrain.

Indonesia

Groves (1981) has traced the distribution of feral pigs of three different origins in the Indonesian region. He has argued convincingly that the Celebes wild boar or warty pig (*S. celebensis*) has been truly domesticated, and that feral populations in the Lesser Sunda Islands, the Moluccas and Sula Islands (Halmahera, Flores, Timor and Simaleue, and probably also in Pulau Saleyer, Pulau Peneng and possibly in Pulau Nias) should be ascribed to this species.

S. celebensis has also been introduced, for hunting in a semi-feral state, on the island of Buru. Another 'warty' pig, the bearded pig (*S. barbatus*) of Malaya, Sumatra and Borneo, has been introduced to the Philippines via Palawan, and a third, the Javan warty pig (*S. verrucosus*) – the type race of which may now be extinct in the wild – to Papua New Guinea via the Lesser Sunda and Maluku islands. Hybridization between 'warty' pigs and *S. scrofa* has occurred widely.

Feral pigs of *S. scrofa* ancestry (*vittatus* and *cristatus/moupiensis* types) occur on Tinian (in the Marianas), on Sanana (in the Sulu Islands), on Sumba, Babi, Sumbawa, Banyak, Pulu Nias, Flores (where they have apparently not hybridized with *celebensis*), Enggano (to which, according to local tradition, they swam ashore from a shipwreck in the late nineteenth century), Siberut, the Admiralty Islands, the Lesser Sunda and Celebes islands, Borneo, Papua New Guinea and on Espiritu Santo in Vanuatu in the New Hebrides.

On Papua New Guinea (where they have been accused of introducing the tick *Amblyomma cyprium cyprium*) and in Salawatti, the Aru Islands, the Kei Islands, Goram, Seram, Ambon, Ternate, Morotai and Bacan Groves believes that feral pigs are hybrids between *S. celebensis* and *S. s. vittatus*.

To account for the feral distribution of *celebensis* and *scrofa* in Indonesia Groves has hypothesized that the former were domesticated on Sulawesi (their original home) in the early Holocene, and that they were carried by early Bugis and Macassans settlers or traders to the Moluccas and Lesser Sunda and Sula islands, where they were traded or released for food or for hunting. Subsequently, trade with western Indonesia or with Malayan seafaring peoples resulted in the introduction of *S. s. vittatus* which, for whatever reasons, proved superior as a domesticate and replaced *celebensis* except in a few isolated places; the latter was then allowed to revert to its original wild state or, as in Papua New Guinea and elsewhere, the two species hybridized.

The origin of feral pigs on the Andaman (and Nicobar) islands is uncertain, but they were probably introduced as domestic stock by early negrito Onges and Jarawar tribal settlers (in whose middens their remains have been found) from the south-east Asian mainland. Abdulali (1962) has shown that there are two distinct and apparently stable types in the Andamans – a long-snouted form from Little Andaman and a short-snouted one from Middle Andaman.

OCEANIA

Hawaii

Pigs (possibly of south-east Asian origin and similar in appearance to the Eurasian wild boar) are believed to have been brought to Hawaii from Tahiti by early Polynesian colonists in about AD 1000 as a source of food and for use in divination. Although they are said to have been already common at the time, Cook presented a boar and a sow to the islanders of Niihau (or Kauai) on 2 February 1778. Soon after Cook's arrival pigs were being reared in captivity in Hawaii for compounding rents, for religious sacrifice, for witchcraft and, after the lifting of King Kamehameha I's *kapu* (taboo) in 1819, for food. Some, inevitably, escaped from domestication to become established in the wild, and in the 1820s small numbers were exported as foundation stock for farms in the Pacific north-west of the United States. A century later some of their descendants made the return journey from California to Hawaii.

The early Polynesian pig gradually became assimilated, through interbreeding, with later European arrivals, which present-day animals most closely resemble. Feral pigs, known locally as *pua'a*, are now generally distributed in upland forests on Hawaii, Molokai, Oahu, Kauai and Niihau, and on Maui especially in the Kahakuloa region: on Kahoolawe they died out shortly after 1841, on Laysan before 1910, on Sand Island in the Midway Atoll group around 1915, and on Lanai during the 1930s.

In some native forested areas (such as the Haleakala National Park on Maui), where they occur in high densities and root up seedlings of koa (*Acacia koa*) and ohia (*Metrosideros collina*) trees, feral pigs are significant silvicultural pests, and watersheds and open rangeland at times also suffer from their rootling activities. Damage to montane forestland has endangered a number of endemic passerines; among the rarest of these is Bishop's o'o (*Moho bishopi*), a honeyeater which feeds largely on the nectar of lobelias whose roots pigs dig up; this little bird was last sighted on Molokai in 1904, but in 1982 was rediscovered on the nearby island of Maui. The decline to near-extinction of the endemic Hawaiian goose or nene (*Branta sandvicensis*) and the dark-rumped petrel (*Pterodroma phaeopygia*) has been largely attributed to predation of the eggs and young by feral pigs. For many years they are known to have been reservoirs for brucellosis, leptospirosis and trichinosis. In several areas they are shot for food and sport, and hunting is their main controlling factor. In 1975 their population was estimated at around 80,000 (possibly the highest of any state in the USA) of which some 10,000 are culled annually.

Galápagos Islands

Together with other domestic livestock pigs were probably first introduced to the Galápagos in 1832 when General José Villamil established a colony and raised the flag of Ecuador on the island of Floreana (Charles). More were doubtless imported when San Cristóbal (Chatham) was first permanently settled in 1869 and southern Isabela (Albemarle) in about 1893, and some probably escaped shortly thereafter to found feral populations.[37]

O. Salvin, who visited the islands in July 1875 on board HMS *Petrel*, reported that:

> of animals introduced, it was estimated that . . . pigs existed in large numbers on Charles, Chatham and James [Santiago], but none on Albemarle. [It is possible that the pigs on Santiago were the descendants of some which escaped when English privateers established a semi-permanent base on the island a century earlier.] Pigs and goats do not seem to have been turned out on the Galapagos so soon after their discovery [by Fray Tomás de Berlanga, Bishop of Panama, in 1535] as was usually the case with other islands. I find no mention of pigs in the writings of the earlier voyagers; but Mr Darwin says that on Charles Island there were many wild pigs and goats. This was in 1835. . . . Dr Habel [who landed on the archipelago in 1868] . . . speaks of wild cattle and swine on Charles Island.[38]

Rollo H. Beck, a member of the California Academy of Sciences' expedition to the Galápagos in 1905, refers to the presence of feral pigs on both Santa Cruz (Indefatigable) and Santiago. The animals on the former were reinforced in about 1927 by a disillusioned Norwegian pig farmer who abandoned his stock, which was said to consist of an 'improved' white breed from the South American mainland and black feral animals from Floreana,[39] where they were extensively hunted in the 1930s and 1940s.[40] Today feral pigs occur on all five main islands – Floreana (around 1,000), Santa Cruz (over 5,000), Isabela (over 5,000), Santiago (over 20,000) and San Cristóbal – a total of well over 30,000 animals.

Pigs in the Galápagos consume a wide variety of foods. They uproot the nests of the endemic giant tortoises (*Testudo* (*Geochelone*) *elephantopus*) and those of the East Pacific green turtles (*Chelonia mydas agassizii*), and eat their eggs and hatchlings. On Quinta Playa beach on southern Isabela about 30 per cent of nests of the latter species are attacked – mainly either when the eggs are fresh or when the hatchling turtles are making their way to the surface.[41] On Espumilla beach on Santiago nearly every nest is attacked, usually within a few days of laying. The eggs and hatchlings of both land iguanas (*Conolophus subcristatus*) and marine iguanas (*Amblyrynchus cristatus*) are rooted out from their burrows and eaten, as, most probably, are adults of the latter species. Snakes (*Dromicus* sp.) are also sometimes killed.

Feral pigs are serious predators of the adults, eggs and fledglings of the rare and endemic dark-rumped petrel (*Pterodroma phaeopygia phaeopygia*) – to such an extent that on Santa Cruz in 1935, during the petrel's breeding season, pork became so tainted by the smell and taste of the birds that it was virtually inedible.[42] The American race of the greater flamingo (*Phoenicopterus ruber*) and the endemic Galápagos dove (*Zenaida galapagoensis*) also suffer from predation; only in lagoons protected by mudflats too soft for feral pigs (and dogs) to cross have flamingos been breeding successfully in recent years, and several vulnerable nesting sites have been totally abandoned.[43] Werner (1977) believed that predation by feral pigs is one of the reasons for the continuing decline in the Galápagos of colonies of nesting seabirds.

On the island of Santiago, which has the densest population of pigs, they (and feral goats) have seriously damaged the native vegetation, which in the highlands has been reduced to a stunted growth, thus accelerating erosion.[44] The principal climax tree of the upland zone, *Zanthoxylum fagara*, which supports epiphytic mosses, provides shade and intercepts the damp air of the *garua* (wet) season – thereby providing favourable conditions for a luxuriant understorey – has been widely destroyed, thus creating dustbowls and hastening soil run-off during the rainy season. Elsewhere, five species of trees and shrubs (*Alternanthera echinocephala*, *Gossypium barbadense*, *Waltheria reticulata*, *Scalesia atractyloides* and *S. darwinii*), three of which were endemic, are believed to have disappeared.[45]

In the *Miconia* belt on Santa Cruz pigs eat mainly the rhizomes of the fern *Pteris aquilina* and the berries of *Miconia*. In the lowlands they feed largely on the fruit of *Psidium galapageium* in season, and at other times on herbs, grasses and the larvae of stag beetles which live in the bark of *Scalesia*.[46]

In recent years the National Park Service has been endeavouring to eliminate pigs from the breeding sites of giant tortoises. During the nesting season stone walls are erected in an attempt to exclude both pigs and feral dogs from those breeding areas where eradication has so far proved impossible.

Fiji

Pigs were introduced to Fiji before 1300 BC by early Polynesian and Melanesian settlers. They are abundant in forests and grassland on all the main islands (including Viti Levu, Vanua Levu, Taveuni, Yaqaga, Ovalu, Kadavu, Yadua and perhaps Matacawalevu), where they subsist mainly on fruit and roots. According to Gibbons (1984), the effect they have on *Brachylophus* iguanas is uncertain, but is probably greater than is generally appreciated.

Mauritius

As early as 1512 the Portuguese navigator Pedro Mascarenhas released 'hogs, goats and fowls' on the island of Mauritius in the Mascarenes, where in the late seventeenth century the two mammals were a major factor in the extinction of the dodo. Today, feral pigs, helped by other naturalized mammals, have exterminated or endangered much of the island's endemic avifauna and some of its herpetofauna, and their rootling activities have assisted in the spread of exotic plants such as guava (*Psidium* spp.) and brambles (*Rubus* spp.) which are stifling the little remaining native vegetation.

Other oceanic islands

Pigs were released, as a source of food for nineteenth century sealers and whalers, on the French subantarctic territories of Île Amsterdam, Île Saint Paul (where they were abundant by 1823) and Île aux Cochons (Pig Island) in the Crozet archipelago. Captain C. M. Goodridge. who was shipwrecked on 'Pig Island' in the cutter *Princess of Wales* in 1820, wrote in his *Narrative of a voyage to the South Seas . . .* (1834) that feral pigs were already 'very numerous and very ferocious', and that they fed mainly on the roots of tussock-grass and seabirds – especially penguins. Sir James Clark Ross, who commanded a British expedition to the Antarctic in 1839–43, recorded in his *Voyage to the South Seas* (1847) that in 1840 'Pig Island' was 'so overrun with these animals that you could hardly land for them . . . in less than six years, they had increased in an almost incredible manner, although great numbers are every year killed by the sealers, not only for their present subsistance, but salted down for supplies on their voyage to and from the Cape'. In view of Goodridge's sighting in 1820, Ross's implication that pigs were first introduced in 1834 is clearly incorrect. Surprisingly, in view of their previous abundance, feral pigs appear to have died out on Île aux Cochons by around 1860, having presumably succumbed to starvation, disease and/or overhunting.

At one time or another feral pigs have also been established in the Seychelles; on L'île du Nord in the Agalégas; on Inaccessible Island (from sometime between 1882 and 1905 until the 1960s); Tristan da Cunha (before 1810); Raoul Island in the Kermadecs (by 1837); Kerguelen (in the nineteenth century); Île del Coco, Costa Rica; possibly on Masatierra and Masafuera in the Juan

Fernandez Islands, Chile; Eiao and Fatuhiva in the Marquesas; Raivavae, Tubuai and possibly Rurutu in the Austral or Tubuai Islands; Meetia (Mehetia), Tahiti and Huahine in the Society group; Suvarov (Suwarrow) and Tongareva atolls in the Cook Islands; Tofua, Tonga; Hull (Orona) atoll and Sydney (Manra) in the Phoenix group; Malden atoll in the Equatorial (Line) Islands; Aguijan in the Marianas; Palau (Babeldaob) in the Carolines; Ponape; Kosrae (Kusaie); and Walpole, New Caledonia. On Guam, the Solomon Islands, Santa Cruz and the New Hebrides in the south-west Pacific (as in Papua New Guinea) feral pigs are believed to be responsible for the introduction of the tick *Amblyomma cyprium cyprium*. On many oceanic islands feral pigs have degraded the vegetation, resulting in soil erosion, and have devastated colonies of ground-nesting birds.

Pigs were released on Bermuda as a future source of meat for shipwrecked mariners when (or shortly after) the islands were discovered by the Spanish explorer Juan de Bermudez in 1515. When the English Admiral, Sir George Somers, was wrecked on the Bermudas *en route* to Virginia in 1609, he recorded that 'our people in the Bermudas found such abundance of hogs that for nine months' space [the duration of their sojourn in the islands] they plentifully sufficed, and yet the number seemed not diminished'.[47] Despite their abundance feral pigs appear to have been exterminated (or redomesticated) by the early colonists between about 1620 and 1630.

When pigs were first released on Bermuda by the Spanish in the early sixteenth century the population of the endemic Bermuda petrel or cahow (*Pterodroma cahow*) is believed to have been in excess of one million. By the time of British settlement a century later its numbers had fallen to around 100,000 due to predation of its nesting-burrows by feral pigs; the human inhabitants, who slaughtered large numbers for food, were responsible for its near-extinction shortly thereafter.

Pigs (together with goats and cattle) were first introduced to New France in Canada from the mother-country in 1541 by Jacques Cartier, whose crew, due to the prolongation of the voyage, were, according to the contemporaneous Richard Hakluyt in his *Principall navigations, voiages and discoveries of the English nation* (1589) 'constrained to water [them] with Sider and other drinke'. In the early years of the following decade Portuguese sailors landed pigs (and cattle) on Sable Island off the coast of Nova Scotia as a source of supply for castaways and fishermen on the Grand Banks, where they survived for more than two centuries.[48] The English navigator Sir Humprey Gilbert is believed to have revictualled his vessels on Sable Island after his colonization of Newfoundland in 1583. When, 15 years later, the Marquis de la Roche abandoned the colonists on Sable Island to their fate they were able to survive only because of the existence of these feral livestock.

Marc Lescarbot, in his *History of New France*, claims to have imported some pigs in 1607, and Samuel Champlain records that the Recollet Fathers brought with them 'a few pigs' as well as other domestic stock when they arrived in Quebec in 1615; in 1634 Jesuit missionaries imported to New France two sows and eight piglets which, together with their cattle, were allowed to run free, causing havoc to their crops; they subsequently transported pigs, cattle and poultry in birch-bark canoes up the Ottawa River to the Huron Indians of Sainte-Marie. Whether any of these early importations ever became established in the wild is uncertain, but so far as is known there are no feral pigs in Canada today.

In 1764 the French explorer Louis Antoine de Bougainville introduced domestic pigs (and cattle and horses) to the Falkland Islands, where, by the 1840s, their grazing and rootling had destroyed much of the tussock grass (*Poa flabellata*) on East Falkland; the resulting erosion eliminated most of the shelter and nesting-sites of such seabirds as the Magellan penguin (*Spheniscus magellanicus*), petrels, wildfowl and waders.

Commodore John Byron, the leader of an English expedition to the islands, landed a number of domestic pigs at Port Egmont on Saunders Island in 1765, where hunting their feral descendants became a popular pastime. To provide emergency food supplies nineteenth-century sealers and whalers landed pigs on Beaver, Bleaker, New, Quaker, Speedwell and West Point islands. How numerous feral pigs became is uncertain, but in 1838 some 3,000 skins were exported. According to Strange (1972), the last introduction of pigs was to Jack's Island off Albemarle on West Falkland in the 1920s, and none has been reported since 1952.

In 1930 the **white-lipped peccary** (*Tayassu pecari*), a native from Mexico to Paraguay, and the **collared peccary** (*T. tajacu*), which is found from the southern United States to Patagonia, were both introduced for hunting purposes to the extreme west of Pinar del Río and the Sierra Cristal, Oriente, Cuba, where they live sympatrically with feral *scrofa* landed by early Spanish colonists. In 1972 a pair of the latter species were released by the Countess of Arran on Inchconnachan in Loch Lomond, Scotland. One disappeared in the following year and the other was removed in 1974.

According to Oliver (1984), the presence of the **bush pig** (*Potamochoerus larvatus*, a race of *P. porcus*) on Mayotte in the Comoro Islands and on Madagascar (where it is the only terrestrial mammal, apart from the recently extinct *Hippopotamus lemerlei*, that is congeneric with any African form) is probably due to human intervention. Bush pigs currently pose a serious threat to the gravely endangered Malagasy tortoise (*Geochelone yniphora*) in northwestern Madagascar.

Dammerman (1929) believed that the unbalanced and impoverished nature of the mammalian fauna of the island of Buru suggests that the presence there of the **babirusa** (*Babyrousa babyrussa*) (*babi* = pig, *rusa* = deer), a native of the Celebes and Togean islands, is the result of human intervention. In concurring with this view Groves (1980) points out that the close similarity between the Buru and Sula islands' forms suggests either a very recent spread from the former to the latter or a human introduction to the Sula group, since the babirusa appears never to have existed on the island of Peleng – an essential stepping-stone between the Celebes and Sula had it extended its range without the help of man.

NOTES

1. For much of the history of feral pigs in the United States I am indebted to McKnight (1964).
2. Hanson & Karstad (1959).
3. A 'bald' is a treeless area on montane forest in the Great Smokies.
4. Some authorities state that Moore was the head of a syndicate of English sportsmen who were jointly responsible for introducing wild boar.
5. Quoted by Laycock (1966).
6. W. B. Frankenberger & R. C. Belden, unpublished report, Federal Aid to Wildlife Restoration Program: Florida Pittman–Robertson Project W-41-R (1976).
7. Some authorities say that Moore was the owner of the San Franciscito Ranch; others suggest that it was Moore who dispatched the boar from Hooper Bald.
8. For many years badgers (*Meles meles*) in parts of the west of England and brush-tailed possums (*Trichosurus vulpecula*) (see p. 11), in New Zealand are also known to have been reservoirs for bovine tuberculosis.
9. For the history of feral pigs in Australia I am indebted to Pullar (1950, 1953) and McKnight (1976).
10. W. A. N. Robertson, *Report of the Australian Association for the Advancement of Science* **21**: 295(1932).
11. E. H. Hallack, *Kangaroo Island: 'Adelaide's sanatorium'* (W. K. Thomas, Adelaide, 1905); A. E. Martin, *Twelve hundred and more place names in South Australia, Western Australia and the Northern Territory* (NSW Bookstall Co., Sydney, 1943).
12. *In savage Australia* (Philip Alan, London, 1926).
13. A branch of an Australian river which, instead of returning to the main stream like an anabranch, comes to a dead end.
14. *Report on vertebrate pests and exotic animal diseases* (Australian Bureau of Animal Health, Canberra, 1980) (quoted by Tisdell 1982).
15. For the early history of pigs in New Zealand I am indebted to Thomson (1922), Clark (1949) and Wodzicki (1950).
16. See p. 345.
17. *Savage life and scenes in New Zealand* (1847).
18. Cook, *A voyage towards the South Pole and round the World . . . (in HMS Resolution and Adventure), 1772–1775* (London, 1777), Vol. 1, p, 123.
19. Ibid., Vol. 1, p. 221.
20. Ibid., Vol. 2, p. 154.
21. Ibid., Vol. 2, p. 467.
22. Ibid., Vol. 2, p. 155.
23. Ibid., Vol. 2, p. 159.
24. Cook, *A voyage to the Pacific Ocean . . . (in HMS Resolution and Discovery), 1776–80* (London, 1784), Vol. 1, p. 131.
25. pp. 49–50.
26. *Travels in New Zealand* (1843), Vol. 1, p. 30.
27. Ibid.
28. See note 17.

29. *Transactions of the New Zealand Institute* **10**: 306–24 (1877).
30. Ibid. **39**: 503–8 (1906).
31. p. 162.
32. *Proceedings of the Zoological Society of London* 857 (1897).
33. C. Chilton, *Philosophical Institute of Canterbury, Christchurch* (1909).
34. Cockayne (1904).
35. *Records of the Auckland Institute and Museum* **1**: 219–30 (1934).
36. J. T. Martin, *Journal of Mammalogy* **56**: 914–15 (1975).
37. For a more detailed history of the settlement of the archipelago see p. 359.
38. Salvin (1877).
39. Kastdalen (1982).
40. M. Wittmer, *Floreana* (1961).
41. Green (1979).
42. M. P. Harris, *Condor* **72**: 76–84 (1970).
43. Eckhardt (1972).
44. *Charles Darwin Research Station Scientific and Conservation Report* No. 13 (1968).
45. See note 42.
46. Kastdalen (1982).
47. Quoted by J. H. Lefroy, *Memorials of the discovery and early settlement of the Bermudas or Somers Islands 1515–1685* (1877–9), Vols 1 and 2.
48. G. Patterson, *Transactions of the Royal Society of Canada* **11**: 28 (1894) and **11**: 133–4 (1897); see also Saunders (1935).

Arabian Camel
(*Camelus dromedarius*)

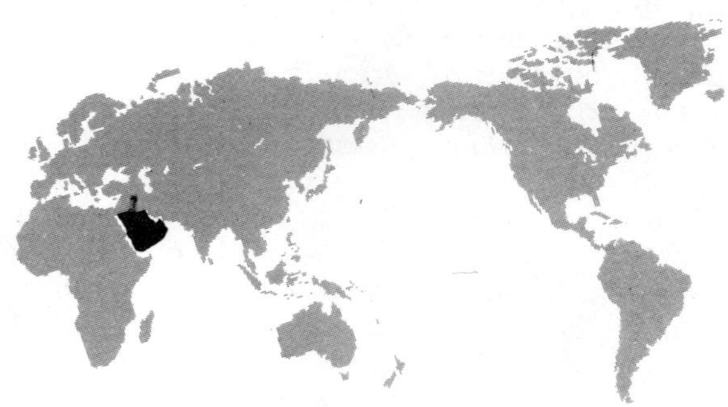

Note: Original distribution in Arabia uncertain.

NATURAL DISTRIBUTION
Originally confined to Arabia,[1] this species – known also as the one-humped camel or dromedary[2] – now occurs only in domestication[3] (mainly in northern and northeastern Africa, the Middle East and Central Asia) and *ferae naturae* in Australia.

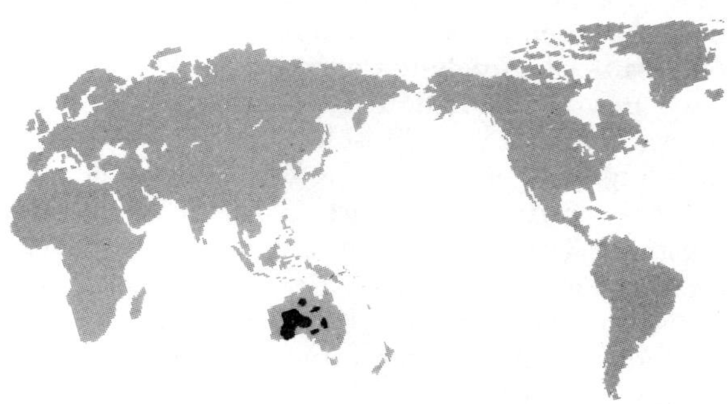

NATURALIZED DISTRIBUTION
Australasia: Australia. (Feral for a short time in the USA and Canada in the mid-nineteenth century.)

AUSTRALASIA

Australia[4]

The continent of Australia possesses, after the African Sahara, the second largest more or less continuous expanse of desert (comprising the Great Sandy, Gibson, Great Victoria and Simpson) and semi-desert land in the world; it is therefore appropriate that the country is the home of the only feral camels currently in existence.

The earliest written references to the possibility of introducing camels to Australia date from the 1830s, some 50 years after the establishment of the first European settlement but only a quarter of a century after the first crossing of the Blue Mountains in New South Wales. As the result of enquiries launched by a government official of that colony in 1836, a report[5] was laid before Governor Burke in January of the following year which suggested that although the acquisition of camels in the upper Ganges Valley for shipment to Sydney via Bombay or Calcutta would present little difficulty, the engagement of drivers might be less easy. Three years later, in September 1839, the *Sydney Herald* (later the *Sydney Morning Herald*) published an editorial claiming that camels were 'admirably adapted to the climate and soil of New South Wales'.

In South Australia the Governor, George Gawler, recommended as early as August 1840 (only four years after the establishment of the first European settlement in the colony) the purchase of two pairs of camels for purposes of exploration. Two months later the first (one-humped) camel[6] to reach Australia – the sole survivor of an original shipment of six or nine from the Canary Islands – arrived at Port Adelaide in South Australia. For six years this animal was employed on the Horrocks' Penwortham sheep station near Clare; in July 1846 John Horrocks and five companions set off with the camel and five horses to explore the uncharted country west of Lake Torrens; six weeks later, near Lake Dutton northwest of Port Augusta, the unfortunate camel, having unwittingly caused Horrocks accidentally to discharge his gun, blowing off two fingers and knocking out several teeth, was summarily despatched.

The second importation of camels to Australia took place two months after the first, in December 1840, when a pair (from an unknown source) arrived on the *Calcutta* in Hobart, Tasmania, from where they were shipped to Melbourne, where they were exhibited by a Mr J. Ardlie and a young camel – the first in Australia – was born. Late in the following year they were offered for sale in Sydney, but finding no purchasers were overlanded back to Melbourne where, according to the *Sydney Morning Herald* of 19 January 1842, 'not all the eloquence of the auctioneer could coax a bidder to support him'. They were subsequently acquired for £75 each by the Governor of New South Wales, Sir George Gipps, by whom they were returned to Sydney and exhibited in the Domain. While the camels were in Hobart, Governor Franklin of Tasmania wrote to Gawler to suggest that E.J. Eyre should take camels with him on his expedition to Western Australia; this proposal came rather late in the day, since Eyre had departed from Adelaide some six months previously.

According to the Perth *Inquirer* of 11 November 1846, the government of Western Australia unsuccessfully offered a reward of £100 to anyone importing a pregnant female camel to the colony.

Little further interest in introducing camels appears to have been evinced by New South Wales, but the *South Australia Register* – the colony's most influential newspaper – was an enthusiastic proponent, extolling the animals' virtues and deploring the lack of governmental support.

The third importation of camels to Australia was, however, delayed until June 1860, when the *Chinsurah* from Karachi in British India landed 24 camels and three so-called 'Afghan' camel-drivers under the command of Dost Mahomet (with one British ex-soldier with experience of the animals) at Melbourne in the charge of Mr George Landells, for use on the proposed transcontinental expedition led by Burke and Wills and sponsored by the Royal Society of Victoria. Subsequently a further six camels, landed at Melbourne from 'overseas', were acquired by the Victoria Exploration Committee.

On 20 August 1860 the expedition – consisting of 20 men, 26 camels and 23 horses – with accompanying wagons and carts – set out from Melbourne. In the following year a small group of four men, six camels and one horse reached the estuary of the Flinders River on the north coast of Queensland. During the return journey the horse and two of the camels were eaten, two more camels (one of which was recaptured 14 years later) were abandoned and the remaining two shot when they became bogged, before three of the men (including Burke and Wills) died at Cooper's Creek. John McKinlay's unsuccessful relief expedition (which included four camels, two of which were provided by the Victoria Exploration Committee) confirmed the value of camels in the desert regions of Australia.

In 1857 Mr (later Sir) Thomas Elder – a Scottish-born South Australia pastoralist and landowner – made a pilgrimage by camel from Cairo to Jerusalem. Clearly impressed by the animals' capabilities in desert-like country, in 1862 Elder, founder of the firm of Elder & Co., authorized Mr Samuel J. Stuckey (at his own prior suggestion) to travel to Melbourne to discuss with Landells the possibility of importing further camels from India. From Melbourne Stuckey travelled on to Bombay and thence to 'Kurrachee' (Karachi); here, although the acquisition of camels proved no problem, the provision of transportation seemed impossible, and he was forced to return to Adelaide empty-handed. Late in 1865 Stuckey returned to Karachi, this time succeeding in loading 124 camels (together with other stock) and 31 drivers on board a specially chartered vessel, the *Blackwell*, which berthed a month later, in early January 1866, in Port Augusta, where the surviving 121 camels and other stock were offloaded. This shipment was said to comprise approximately equal numbers of light riding 'Mekraua' camels, medium-weight hill camels from 'Scinde' (Sind) and heavy-weights from Kandahar. According to the *South Australia Register* of 27 January, Elder & Co. were hoping for a wool-clip from the 38 Kandahars averaging 25–30 lb (11–14 kg) per beast at 8*d.* per lb. The handlers who accompanied the camels to Australia were universally known as 'Afghans' or 'Ghans', although some were Egyptians, Persians, Indians or Turks; most came from Rajasthan, Baluchistan or from what is now West Pakistan.

Fifty-nine of Stuckey's camels died from mange (*Sarcoptes scabiei* var. *cameli*) on the 400-km journey on foot from Port Augusta to the Elder property at Umberatana; many others recovered after liberal applications of 'Stockholm tar', but the problem persisted as the animals' only serious recurring disease in Australia. After a short period of acclimatization at Umberatana the surviving camels were transferred to the Beltana Station, 65 km to the south-west and 350 km due north of Adelaide where, under the management of Mr N. E. Phillipson, the first camel stud in Australia was established.

During the next 70 years camels from India were imported to Australia on a number of occasions, but home-bred animals from Beltana were generally held to be of the best quality. In the opinion of the Chief Inspector of Rabbits for Western Australia:

> The finest camels I have seen in Australia were those bred by the late Sir Thomas Elder, of South Australia. The light and heavy breeds were kept separate and the bulls were selected with the greatest care. So well and favourably known were Sir Thomas Elder's camels that he could always command from twenty to thirty pounds more than the price of imported stock.[7]

The station at Beltana later became the base of the camel-carting company of Elder & Stuckey, which at first operated only between Port Augusta and Lake Hope, but subsequently expanded its range to include much of central Australia. From Beltana camels were dispersed eastward into New South Wales, north-east into Queensland, north into the Northern Territory and westwards into Western Australia. Although most of the early shipments of camels from India were disembarked at Port Augusta in South Australia, from the late 1880s they began increasingly to arrive at ports in Western Australia – at Albany, Carnarvon, Port Hedland, Derby, Wyndham and especially at Fremantle, where late in 1887 the *Abergeldie* and *Yeoman* arrived from India with over 300 camels each; few found purchasers in Fremantle, and most were shipped onward to Port Augusta. Once they had overcome their initial fear and suspicion, Australian Aborigines proved capable camel handlers and invaluable assistants to 'Afghans' and Europeans.

Although Burke and Wills, and their would-be rescuer, McKinlay, were the pioneers in the use of camels for exploring and surveying expeditions, Major (later Colonel) Peter Egerton Warburton was the first to depend exclusively on camels, when, during his expedition of 1872–4 from Adelaide to Roebourne between the Fortescue and De Grey rivers in Western Australia, he was accompanied by 30 provided by Beltana. Later in the 1870s Ernest Giles relied solely on Elder's Beltana camels during his various exploring expeditions in southern and western Australia, especially during his double crossing of the continent in 1875–6 from Port Augusta to Perth and back. At least three other pioneers of Australian exploration – W.C. Gosse, John Ross and W.H. Tietkins – made use of Beltana-bred camels which, by the late 1870s or early 1880s, had come to be regarded as the standard means of transport in the country's desert-like regions. Travelling in the Simpson Desert in 1883 Charles Winnecke covered 445 km with camels provided by Elder, each carrying a load of 315 kg; for 16 days the camels travelled without water and then, after a single day's rest, continued for a further 240 km, again with no water. During one stage of the Elder Scientific Exploring Expedition of 1891–2, organized by the Royal Geographical Society to survey the Great Victoria and Gibson deserts between western South Australia and the Indian Ocean, and equipped and provided with camels by Elder, 40 of the animals covered 820 km in 34 days on some 18 l of water each. It is little wonder that in the outback camels were regarded as 'more indispensable than damper, bully beef, or blackfellow'.[8]

After Elder's major importation in 1866 no more camels were brought in to Australia until the 1880s, when a number were imported by Western Australia and, especially, by South Australia: the records of the South Australia Department of Agriculture reveal that six camels from India were imported by the colony in 1883, 661 in 1884, 53 in 1885 and 422 in 1887. Most – even including those destined for other Australian colonies – were shipped from Karachi to Port Augusta in South Australia, a voyage of between 4 and 6 weeks. Among the major importers were W.R. Cave & Co., George Wills & Co., Akbar Kushar, Abdul Wadi (who once brought in 500 in a single shipment) and Tagh Mahomet – the last three presumably being the descendants of earlier 'Afghan' herders.

Large consignments continued to arrive during the 1890s, when gold strikes in Western Australia opened up new avenues for camel-transport to and from Kalgoorlie through Murchison and Pilbara north-west to the Kimberleys. Between 1894 and 1897 more than 6,600 camels were shipped from India direct to Western Australia, with a corresponding reduction in importations to South Australia.

After the turn of the century (when the 'colonies' became 'states') the introduction of camels to Australia began gradually to decline. The last major importation, which was also numerically the largest single shipment on record, was undertaken as a speculative venture by a private citizen who borrowed capital from an insurance company to finance the purchase and transport of camels to Pilbara, north of the Hamersley Range in Western Australia. On 21 March 1907, over 500 camels from Bombay were offloaded from the *Century* in Port Hedland harbour. On arrival in Port Hedland from Perth the state pathologist, Mr (later Sir John) Cleland found many of the new arrivals to be infected with both mange and the trypanosomes of Surra disease (*Trypanosoma evansi*);[9] the mangey camels were treated with 'Stockholm tar', those infected with surra were destroyed, and the remainder were quarantined at Carlindie Station, 80 km from Port Hedland, for a year before being released for service.

Between 1840 and 1907 (most from the 1880s onwards) a total of between 10,000 and 20,000 camels were imported to Australia from India, of which perhaps 50–65 per cent were landed in South Australia.

Following Phillipson's success in camel-breeding at Beltana, a number of other (often less well organized) privately owned studs were set up in South Australia, Western Australia, Queensland and New South Wales. Depots were also instituted for training, breeding and depasturing government-owned camels.

The total of domesticated camels in use in Australia at any one time is impossible to assess accurately; the population probably exceeded 5,000 by the middle 1890s, may have doubled by around the turn of the century, and had probably doubled again to a peak of 20,000 by the early 1920s. Much of this growth in the population owed more to natural increase rather than to repeated importations; most domesticated camels were used in Western Australia and South Australia, with smaller numbers in Queensland and New South Wales, a few hundred only in the Northern Territory and virtually none in Victoria.

The uses made of domesticated camels in Australia – apart from exploring and surveying – were many and various. Undoubtedly the most important was in mining; from Southern Cross in the south of Western Australia north to Hall's Creek in the Kimberleys, and from Milparinka in New South Wales to Chillagoe in Queensland – over half a continent – camels were used to transport stores, supplies, provisions and machinery from railhead to mine and ore on the return journey.

Other uses to which camels were put included riding by the Northern Territory Camel Patrols; packing (a single camel could carry a load of 180–270 kg or more 15–30 km in a day); hauling (when a team of up to two dozen camels could transport half a tonne of freight 24–26 km a day); the transport of wool-clips from sheep stations to market; earth-moving during the construction of dams and water-tanks; drainage; borehole-drilling for water; and ploughing. The Number 1 Rabbit Proof Fence in Western Australia in the 1880s; the Canning Stock Route from Hall's Creek to Wiluna in Western Australia (1906–8); the Transcontinental Railway from Port Augusta, South Australia, to Kalgoorlie, Western Australia (1912–17); the Central Australian Railway from Alice Springs in the Northern Territory to Port Augusta (1879–1929); the Queensland Border Fence in the 1880s; the Overland Telegraph Line between Darwin in the Northern Territory and Adelaide in South Australia (1870–2); and the Birdsville Track – all were constructed with the aid of Australia's 'ships of the desert'. At one period camels were in use as a means of transport in Australia from the Atherton Tableland in the north-east to the district of Wimmera in Victoria in the south-east, south-west to Perth and Albany and as far north-west as the northern Kimberleys. Only occasionally were they either milked or eaten.

The advent of motor vehicles for transporting goods was the first and most serious blow to the employment of domesticated camels in Australia. A declining demand for public works construction resulted in the closure of many governmental camel depots. Until the late 1930s, however, camels continued to be used for exploration and surveying in the outback (the last major expedition to make use of them being C.T. Madigan's during his west to east crossing of the Simpson Desert in 1939) and for police work in parts of South Australia, Western Australia and in the Northern Territory. Camels were last employed by regular police patrols at Marree in northern South Australia in 1949, and occasionally until as late as 1955–6; they were used for carrying mails in a few remote regions in northern South Australia until 1942; at isolated mines until the middle to late 1940s (the last being a uranium mine near Mount Painter in the Flinders Range in South Australia); and in repairing vermin-proof fencing until the early 1950s. The last government camel depot in South Australia – at Mulloorina – closed in 1938, and the last in Western Australia – at Jiggalong – a decade later. The domesticated camel in Australia was a long time a-dying.

Exactly when feral camels first became established in Australia is uncertain; a few escaped as early as the time of the Burke and Wills expedition in 1860, and they were doubtless joined by further small numbers until the gradual decline of domesticated animals in the early 1920s led to greater numbers becoming established in the wild. Since their meat and hides were of little value few camels were killed when they ceased to be of use as beasts of burden, the majority being released to fend for themselves – a role to which they readily adapted. As McKnight (1976) says:

> Dromedaries can find feed even when other animals cannot, because they are tall enough to browse the high scrub of the outback. Their watering needs are minimal. Their resistance to temperature extremes is legendary. Apart from flies and mange (two not inconsiderable exceptions), they seem to be unbothered by insects and disease. Except under unusual circumstances, they are invulnerable to Australian predators. Apart from two factors – long-term drought and man – living is easy for feral camels in Australia.

From the late 1920s to the mid-1930s the feral camel population in Australia markedly increased. They were established, apart from in the northern and western Kimberleys, in the eastern two-thirds of Western Australia; throughout much of the southern half of the Northern Territory; in South Australia north of the 32nd parallel; west of the Darling River in New South Wales; and in the western portion of Queensland from the New South Wales border nearly to the Gulf. During this period the overall population was variously estimated at between a few thousand and a quarter of a million. McKnight suggests that 'the maximum population was reached at some time in the late 1930s or early 1940s and may have totalled anything between 30,000 and thrice that number'.

After the Second World War the numbers of feral camels in Australia began gradually to decline as the result of increasing land clearance by man and his greater interest in desert water resources. As the human population began to expand into the more remote outback regions, the pressure on feral camels, as on other species of wildlife, began greatly to increase.

From time to time feral camels have been redomesticated by Aborigines for transportation purposes, especially in northwestern South Australia and in the south-west of the Northern Territory. McKnight estimated that in the late 1960s about 400 redomesticated camels were owned by Aborigines and a further 90 or so by pastoralists.

'The dominant element in the contemporary camel picture', wrote McKnight (1969), 'is the vast remnant of now feral animals, now somewhat decreased in numbers, but more noticeable than ever before.' In many places feral camels have been established in the wild for several generations – normally in small groups, but sometimes in herds of 25–50 or even 300–500 in times of drought – over 960,000 km^2 of the Australian outback. Because of the immensity and remoteness of their habitat they are, however, seldom seen by man; indeed, in 1966 McKnight 'criss-crossed a large area between the Macdonnell and James Ranges south of Alice Springs . . . fresh camel tracks were commonplace, but no animals were seen. Two days later a systematic search of the area in a light aircraft revealed a complete absence of camels'.

Australia's feral camels occupy a variety of habitats, ranging from lightly settled arid regions of sandy ridges, where domestic livestock are unable to survive but whose sparse but succulent vegetation provides ideal camel fodder, to featureless limestone plains, rugged hills and deserts. Whenever possible they avoid hard rocky areas, whether gibber[10] or mountain ridges, whose hard surfaces their soft feet cannot withstand. They are largely nomadic, ranging about widely over hundreds of thousands of square kilometres and covering 50–70 km a day where suitable food is plentiful; during the winter rut adult bulls assemble harems of 10–20 cows – for the rest of the year the bulls remain in all-male groups (mobs) or with their cows and calves.

The feral camel in Australia is a facultative herbivore; according to H.H. Finlayson:

> Probably the most valuable trait of the camel is the extraordinary catholicity of his taste in food plants, for there is scarcely any plant, small or great, in the Centre, which he does not lay under toll with evident relish. Some of them are singularly unattractive to outward seeming like the spiny Bassias and Solanums, and many are utterly nauseous to the human palate and shunned even by goats. A notable example is the woody, bright orange fruit of the white wood, *Atalaya* sp., which is intolerably bitter but which the camels will seek out and eagerly devour. For the most part he is a browser. His great height and flexible neck are valuable specializations for seizing and dragging down the twigs and bushy tops of small trees and large shrubs; and, quantitatively, the almost ubiquitous arborescent mulga is his most valuable supply.[11]

The camel's preferred foods are the leaves of eucalypts and acacias; mulga (*Acacia aneura*); saltbush (*Atriplex* sp.) and the succulent parakeelya (*Calandrinia balonensis*). Camels can exist for many days without drinking except in times of serious drought, when they gather at water-troughs intended for domestic stock.

As a result of the camel's wide selection of forage considerable numbers die from eating poisonous plants, prominent among which are emu bush (*Eremophilia maculata*), ironwood (*Acacia excelsa*), gidyea (*A. camebagei*) and gastrolobium (*Gastrolobium grandiflorum*). Camels appear to suffer from few diseases apart from mange. Introduced red foxes (*Vulpes vulpes*), dingoes (*Canis familiaris dingo*) and wedge-tailed eagles (*Aquila audax*) are believed to take some sickly calves, and broken limbs, bogging, snakes and poison in dingo bait account for small numbers. The camel's main controlling agents, however, are recurrent prolonged droughts and predation by man, whether as outraged pastoralist or tourist 'sportsman'.

Feral camels can be extremely destructive to the water-troughs used for domestic stock and also to fencing, thus allowing into pastures predators such as dingoes and foxes, and the egress of stock. They foul natural waterholes and destroy scrub that provides shelter for sheep, which they are sometimes accused of kicking and biting to death. Domestic stock in general, and horses in particular, seem to have an inherent fear of camels. The main use to which their carcasses is put after being shot is as bait for dingoes – especially the fatty hump which dingoes like and which decomposes only slowly even in hot sun. Some have been rounded up and sold to menageries, zoos and circuses (both in Australia and abroad) and for use in medical and scientific research.

In 1967 McKnight (1969) considered that 'From information at hand, I believe that there may be 15,000 to 20,000 feral camels existing in Australia today. Probably 40 per cent of the total [6,000–8,000] is found in Western Australia, perhaps 30 per cent [4,500–6,000] in the Northern Territory, some 20 per cent [3,000–4,000] in

South Australia, and the remainder [1,500–2,000] in Queensland.' The largest single concentration occurs in the Simpson Desert south of Alice Springs in the south of the Northern Territory below the twenty-second parallel – especially in the south-west, south of the Palmer River and between Lake Amadeus and the Petermann Ranges – and in northern South Australia. A second major population centre is in the north-west of the country from the Fortescue and De Grey drainage basins eastwards into the Great Sandy Desert, while a third occurs to the west and south in the Gibson Desert and a fourth in the western area of the Nullarbor Plain. A sizeable population in the country around Durham Downs between Cooper's Creek and Lake Yamma Yamma in southwestern Queensland is apparently separated from that in the Simpson Desert by over 300 km of camel-free country. Smaller outlying concentrations occur in the middle and upper drainages of the Ord River in the eastern Kimberleys, and on Victoria River Downs, Humbert River, Waterloo and Mistake Creek stations in the neighbouring fringe of the Northern Territory; in the Dobbyn/Kajabbi region north-west of Cloncurry and north-east of Mount Isa in northwestern Queensland; between the Barkly Tableland and the Simpson Desert in the Lucy/Manners creeks region of the eastern Northern Territory; south of Lake Yamma Yamma in southwestern Queensland; in the Lake Gairdner–Wirraminna area north of the Eyre Peninsula in South Australia; and in a small coastal district near Yalata Mission at the head of the Great Bight in South Australia.

The feral camels of Australia are unique, and their impact – if any – on the biota of the outback is not yet fully understood. McKnight believed that they were slowly declining, especially in Western Australia where they are regarded as 'vermin' when they trespass on sheep stations, and foresaw a big decline in the western Nullarbor Plain (where many were shot in the 1970s to save vermin-proof fencing erected to protect sheep during a period of intensified farming) and probably also around Pilbara due to an increase in both mining and sheep-farming activities. In the Northern Territory, south and south-west of Alice Springs, McKnight believed that increasing pressure would be put on camels by encroaching human settlement. In the Simpson Desert, south-east of Alice Springs, the camel population fluctuates according to the amount of rainfall and the degree of slaughter by farmers on nearby cattle stations. In South Australia the camel's position may be relatively stable, except where competition with sheep arises or where they occur in large numbers, as on the fringe of the Simpson Desert. In Queensland and New South Wales (if any survive in the latter) their continued existence depends largely on the tolerance of local pastoralists, doggers and vermin controllers.

As McKnight (1969) points out, the exploitation of Australia's feral camels has, principally because of their nomadic way of life, been little studied. Camel blood shows some apparent important similarities to that of some merino sheep; if these impart a special resistance to body desiccation in conditions of great heat, it might be possible, through artificial selection, to breed new varieties of heat-resistant merinos.

There could well be a market in Australia for pet-food and hide, and a small export market already exists; the report (Letts, Bassingthwaighte & de Vos 1979) on feral animals in the Northern Territory says that live camels from Alice Springs are exported for racing and for meat to the Middle East at prices up to $A1,000 each, f.o.b. Enquiries have also been received from Ethiopia, Libya, Mozambique, India and Saudi Arabia. Unfortunately, catching camels at watering places – the easiest and cheapest way to obtain them in quantity – is only possible in exceptionally dry years. The cost of capturing camels individually in desert areas is high; prices for unbroken camels on stations range from $A150–200 each, and up to $A300 delivered to Alice Springs. To these sums must be added the cost of food and transport, so the net profit is not high.

The camel farm at Alice Springs has proved a considerable tourist attraction; there could well be scope for a moderate expansion of this sort of enterprise and for the greater use of camels in some desert national parks.

Australia's feral camels, where they are currently established, seem unlikely to pose a significant threat to any of the country's native fauna and flora. Should they spread out from their present range, however, their impact on the environment could substantially increase. As the Board of Inquiry's Report recommended, 'Some further studies would be useful on populations, distribution and movement, habits and habitat'.

As early as 1836 Major George H. Crossman – later Deputy Quarter-Master General of the US Army – conceived the idea of importing camels to carry supplies to military frontier outposts in the southwestern United States.[12] On 19 May 1855 Major Henry C. Wayne, having obtained the support of Senator Jefferson Davis of Mississippi, sailed for Europe, returning to Texas a year later, on 14 May 1856, with a shipment of 34 camels, one of which had been born during the voyage. In the following year a further consignment of 44 camels arrived safely in Texas; these two shipments were the foundation stock of the US army's first and last camel herd at Camp Verde, 100 km north-west of San Antonio.

Arabian camels first entered into military service at El Paso and Bowie, Texas. In the four years preceding the start of the Civil War in 1861 camels were in regular service around Camp Verde, and between San Antonio and the Gulf Coast. In the autumn of 1856 the War Department ordered a survey of a new wagon road between Fort Defiance in New Mexico and the Colorado River on the Californian border. The 25 camels used during the two-year expedition of 1857–8 proved superior in staying power to the horses and mules also employed. They were, however, generally distrusted by the military, muleteers and civilians alike.

Soon after the outbreak of the Civil War a number of camels were captured by the Confederates, some of which were retained for use by their Post Office Department while the remainder were turned loose in the desert. Union forces caught three camels in Arkansas (where a number were established in the wild in the mid-1870s near the Gila River) and sold them by auction. At the end of the war in 1865 the surviving 66 camels at Camp Verde were sold for $31 each to a travelling entertainer for exhibition in Mexico.

The War Department's camels were not the only ones introduced into Texas in the mid-nineteenth century; in 1858 a number arrived from a British vessel at the Watson Ranch near Houston, where they remained for several years.

Fifteen Bactrian or two-humped camels (*C. bactrianus*) from Central Asia (where, in parts of China and Mongolia, around 1,000 still exist in the wild today) were imported from Manchuria in 1860 by a San Francisco company for hauling salt across the 320 km of desert separating Esmeralda County from the Washoe Silver Mill. When a more convenient source of salt was discovered the camels became redundant, the last one hauling cord wood in Virginia City until 1876. An ambitious plan for camel caravans between Texas and California likewise failed.

The advent of the railways, coupled with the animals' own unpredictability and unpopularity, finally put an end to the commercial exploitation of camels in the American south-west. Just when the last feral camels died out in the United States is uncertain, but they were never established widely or in large numbers.

In 1862 Mr John C. Calbreath of Lillooet, British Columbia, Canada, on behalf of a syndicate headed by Mr Frank Laumeister, purchased, for $300 each, 22 or 23 Bactrian camels in San Francisco for hauling freight in the goldfields of the Cariboo. The camels arrived at Port Douglas in May in a barge towed by the sternwheeler *Flying Dutchman*. The 'Dromedary Express', as it was misnamed, which it was hoped would earn its proprietors $60,000 in the first season, proved a dismal failure; quite apart from the uncompromising behaviour of the camels, it was soon discovered that their hooves were no match for the rocky terrain of the Fraser Valley and the Cariboo. In 1864 the experiment was abandoned; some camels were returned to the coast for disposal, others were killed for meat or died in winter storms, while a few were turned out in the Lac La Hache area and near Westwold, where they survived in the wild for several years. The last Canadian camel is said to have died near Grande Prairie in the northern Okanagan in 1905.

Camels have been introduced to several other countries apart from Australia, the United States and Canada, but in none have feral populations ever become established.

In Europe, Ferdinand II de Medici, who ruled Tuscany from 1627–70, imported some from Tunisia to San Rossore near Pisa in 1622, where their descendants survived until the outbreak of the last war. The Muslims introduced camels to southern Spain, especially around Huelva; a breeding herd, derived from Canary Islands stock, was established on the royal estate at Aranjuez from 1786 until the start of the Spanish

Civil War; 30 introduced to Barcelona in 1831 by the Captain-General of Catalonia did not long survive. Robert Swinhoe, in his *Narratives of the North China campaign of 1860* (1861), even went so far as to suggest that 'they would form an acceptable addition to the beasts of burden at home [in England], where in flat portions of the country they might be usefully and economically employed'! Unsuccessful attempts were also made towards the end of the nineteenth century to acclimatize two-humped camels in Russia, Poland, Portugal, Sicily, Macedonia and Bulgaria.

In Asia, camels introduced to Java soon died from an infection of the liver. The most successful introduction of camels outside their natural range, apart from Australia, was made from Morocco to the Canary Islands off the coast of West Africa by Jean de Béthencourt between 1402 (when he annexed the islands for Spain) and 1406. Camels have been exported to a number of other countries from the Canaries, where they continue to be used on agricultural land and as beasts of burden.

The British, Portuguese and Germans introduced camels with some success during the last century to their possessions in southwestern Africa (e.g. the Kalahari Desert): the German stock came from the Canaries and later from North Africa, while some of the British animals were obtained in Somalia. More recently, Arabian camels have been introduced to a number of large fenced private farms in the Orange Free State in South Africa.

In the Caribbean unsuccessful attempts were made over a period of half a century to acclimatize camels in Jamaica. In Cuba 70 camels employed in the copper mines near Santiago in 1841 were subsequently used in the sugar-cane plantations.

In South America Juan de Reynaga imported some camels in the mid-sixteenth century from the Canary Islands to Peru; the experiment was, however, soon terminated by royal decree, on the grounds that the cheapness of camel-drawn transport would interfere with the profits obtained from slave labour.

In 1856 the Brazilian government asked Isidore Geoffroy Saint-Hilaire, biologist of the Société Impériale Zoologique d'Acclimatation, to obtain some camels for introduction to the dry northern province of Ceara. On 23 July 1859 the *Splendide* landed 14 camels from Algeria in Brazil, but the experiment was not successful. Finally, an unsuccessful attempt was made by the father-in-law of Simon Bolivar to acclimatize camels in Venezuela.

In about 1845 Richard Clement Moody (appointed first governor of the Falkland Islands in 1841), having noted the similarity between the vegetation and climatic conditions of the islands and those of eastern Patagonia, suggested that an introduction of South American **guanacos** (*Lama guanicoe*), which occur from Peru south to Tierra del Fuego, for their wool and skins might be 'worthy of the attentive consideration of the enterprising settler'. The animals were first introduced in 1862, and Strange (1972) refers to an interesting report of 1871 when the Duke of Edinburgh, second son of Queen Victoria and Prince Albert, visited the islands. 'On Monday Captain Packe organized an expedition for the Prince to Mare Harbour fifty miles [80 km] distant from the settlement to the southward. Here His Royal Highness remained until the following Wednesday evening shooting Guanacos, wild geese, duck and snipe, etc.'

In the early 1930s an *émigré* Scottish shepherd, Mr John Hamilton, landed a small number of guanacos on Statts Island off Weddell Island, where they increased to a herd of several hundred. In 1956 and 1959, as the result of severe overgrazing, nearly 400 had to be shot to save the remainder from starvation. In 1970, when a single animal (which subsequently became aggressive and had to be shot) was transferred to West Point Island, the herd numbered about 60. Since then it has been subject to considerable fluctuations in numbers (probably as the result of inbreeding and periodic food shortage), ranging from around 300 down to the current (1983) total of between 150 and 200. The Falkland Islands population (probably the largest free-ranging one in the world, since most of the mainland herds are semi-domesticated) provides a valuable gene bank of a species which is declining in its natural range.[13]

NOTES

1. Fossil records show that the early evolution of the camel family took place entirely in North America. The earliest known ancestor, *Protylopus*, which was

little bigger than a hare and which dates from the late Eocene, subsequently evolved during the Pleistocene (around 500,000 years ago) into the genus *Camelus*. During one of the following Ice Ages the Bering Straits between Alaska and Siberia froze over, enabling early camels to cross into Asia, from where they travelled via the valley of the Ganges in India west across the Sahara Desert and North Africa to the Atlantic coast, and south into northern Tanzania.

2. From the Greek *dromos* meaning a road; strictly applicable only to the riding/racing variety.
3. In Arabia since at least the sixth century BC.
4. For the history of the camel in Australia I am indebted to McKnight (1969, 1976).
5. 37/1255 (Reg.) 2 February 1837, Archives of New South Wales, Mitchell Library, Sydney.
6. Only very small numbers of Bactrian or two-humped camels were imported to Australia: Philipson (1888–9) claimed that of around 6,000 camels in the country in 1895 only five vere Bactrians. In *The red centre: man and beast in the heart of Australia* (Angus & Robertson, Sydney 1952), H. H. Finlayson records that a pair of Bactrians were used for a number of years at Mount Elba in South Australia, and subsequently died in the Adelaide Zoo. In the early 1900s a station on the Paroo River in northwestern New South Wales kept a large male Bactrian camel for breeding purposes, but few offspring were produced. In about 1941 a hybrid (one-humped × two-humped) camel – bred by an 'Afghan' in Kalgoorlie – was employed in the goldfields of Western Australia, where it was said to be an excellent draught animal. (In spite of the fact that two years previously Peru had banned the exportation of alpacas in 1847, attempts were made to import some from that country to Australia. Ten years later a small number were shipped from England to South Australia, and in the following year (1858) one was despatched to Victoria and another to South Australia; later in the same year the government of New South Wales imported, via Sydney, 280 alpacas to a property near Goulburn, where by 1880 all had been killed by heat and drought.)
7. 'Camel breeding in Australia', *Western Australia Journal of Agriculture* (1908) (quoted by McKnight 1969).
8. B. Threadgill, *South Australian exploration 1856–1880* (Public Library, Museuum & Art Gallery of South Australia, Adelaide, 1922).
9. A protozoan remittent fever, dangerous to domestic cattle, which was believed to be the most fatal of all camel diseases, and was endemic, and frequently epidemic, in parts of western India.
10. A desert area whose surface is covered with rocky stones and boulders.
11. Finlayson, *The red centre*, op cit. (note 6).
12. For information on camels in the United States I am indebted to L.B. Lesley, *Uncle Sam's camels* (Harvard University Press, Cambridge, Mass., 1929).
13. R. Perry, *Country Life* **171**: 420–2 (1982) and Dr R. S. Whitley (who also supplied information on other naturalized mammals in the Falkland Islands), personal communication (1983).

Chinese Water Deer

(*Hydropotes inermis*)

NATURAL DISTRIBUTION
Yangtze Valley in central China and northeastern China and Korea.

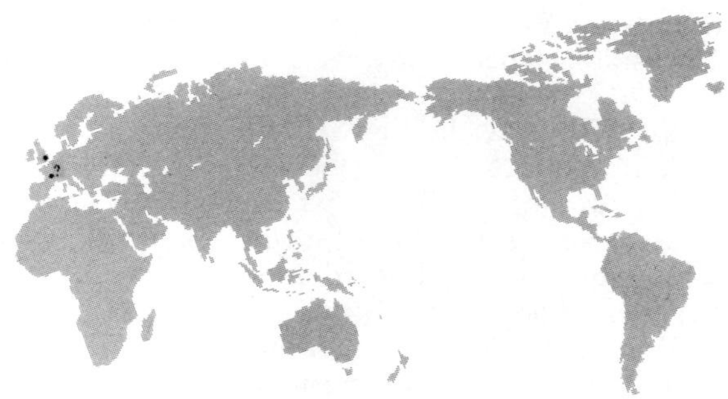

NATURALIZED DISTRIBUTION
Europe: British Isles (England), France.

EUROPE

British Isles (England)

Around the turn of the century the eleventh Duke of Bedford made an unsuccessful attempt to establish a pair of Chinese water deer in the park of Woburn Abbey in Bedfordshire. A few years later the duke made a second attempt to introduce these deer to his menagerie which proved more successful, and thanks to their fecundity a breeding herd was soon established. During the First World War a few managed to escape from the park, and became naturalized in the surrounding woodland, mainly to the south and east of Woburn; here their numbers were subsequently augmented by some animals which escaped from nearby Whipsnade Zoo, which the duke had presented with 32 water deer between 1929 and 1931. In the 1930s and 1940s water deer from Woburn were sent to private collections at Cobham in Surrey, and at Leckford Abbas and Farleigh Wallop in Hampshire from which a number managed to escape but where they never became established in the wild.

In 1945 water deer were reported for the first time in Buckinghamshire, and in 1950 from Northamptonshire and Oxfordshire. In the latter year two pairs were despatched from Woburn to a private menagerie in Shropshire, from which some escaped in 1956 and became established in the surrounding countryside; this population has apparently since died out. Also in 1950 six water deer from Woburn escaped from a collection in Yorkshire, where, however, they did not long survive.

In 1970 came the first record of water deer in Cambridgeshire near the border with west Norfolk and Lincolnshire. Two years later several were observed in eastern Norfolk, and they were said to be 'well established and self-perpetuating' between Woburn and Whipsnade.

By about 1978 Chinese water deer were naturalized in modest numbers in suitable habitats in parts of Bedfordshire, Buckinghamshire, Cambridgeshire, Norfolk and possibly Northamptonshire, and outliers have been reported from time to time in neighbouring counties. There has been little expansion of range or numbers in recent years.

An adult water deer buck may stand 60 cm high and weigh up to 14 kg; the doe is slightly smaller. Like the muntjac (*Muntiacus reevesi*),[1] water deer are a territorial rather than a herd species, and are seldom seen other than singly or in pairs. In England they frequent woodland and grassy downland in addition to the marshes, reed-beds and swamps which they favour in the east. They are of particular interest for two reasons: neither the buck nor the doe grows any antlers (their scientific name means 'weaponless water-drinker'), but the former possesses 7–8-cm long canines which protrude from the upper jaw; and, unusually among Cervidae, as many as six fawns may be produced at parturition. These two characteristics suggest that the species is the survivor of a very ancient breed.

The reasons for the failure of the Chinese water deer – despite its prolificness – to become better established in England are probably twofold; the fawns weigh barely 450 g at birth, and are thus vulnerable to a variety of both avian and mammalian predators; and the heavy coat, which is essential to its survival in the Far East, renders it susceptible to dehydration and heat-exhaustion in the comparatively mild English climate. So far water deer are not sufficiently widespread or numerous for any damage to have been reported.

France

Between about 1960 and 1965 a number of Chinese water deer were released for sporting purposes near Solignac-le-Vigen, a few kilometres south of Limoges. They adapted well to the lightly wooded country, frequenting principally the damper areas where they were often unnoticed, and caused little if any agricultural damage. Within a few years one was killed some 80 km to the south-west of their point of introduction. In the past Chinese water deer have been maintained in extensive enclosures in the zoo at Clères north of Rouen.

Bentley (1978) records that in 1868 a pair of 'paddy-field deer' – presumably *inermis* – were breeding in the Royal Park in Melbourne, Australia from which, however, there is no record of any being released. He also states, without mentioning a specific date, that 'water deer were said to be established somewhere on the South Australian coast . . . it was said that a ship which was carrying some of these deer was wrecked somewhere off Yorke Peninsula and that the animals had been released or had otherwise found their way ashore'; they were reported to be established 'in difficult swampy country'.

NOTE

1. p. 147.

Chinese or Reeves's Muntjac

(*Muntiacus reevesi*)

NATURAL DISTRIBUTION
Densely wooded hilly regions of southern and eastern China and Taiwan.

NATURALIZED DISTRIBUTION
Europe: British Isles (England).

EUROPE

British Isles (England)

Two species of muntjac (also sometimes known as 'barking' deer from their distinctive call) – the Chinese or Reeves's and the Indian (*M. muntjak*) – have been introduced to England; both have escaped from captivity and may have interbred in the wild. Recent research, however, suggests that if hybridization did occur the resulting offspring are likely to have been sterile. From the measurement of skulls and jaws and chromosome studies, it is concluded that muntjac in England today are the Chinese or Reeves's variety.[1]

In about 1900 the eleventh Duke of Bedford introduced (possibly from France) some Indian muntjac to his seat, Woburn Abbey in Bedfordshire, where for a number of years they thrived both inside and outside the park. They were subsequently augmented by introductions of the less-aggressive Chinese or Reeves's muntjac, which also flourished within the confines of the park and in the surrounding woodland. During the 1930s the population in the wild was supplemented by escapers from nearby Whipsnade Zoo, which between 1929 and 1931 had acquired 18 Indian muntjac from Woburn, and by others (of unknown species) from a private collection in the neighbouring county of Hertfordshire. Since then muntjac have considerably extended their range into parts of eastern, central, southern and south-western England, and are continuing to spread. Table 8 gives the earliest recorded dates for muntjac in English counties.

Muntjac are small deer; an adult buck stands some 45 cm high, weighs around 14 kg and grows rudimentary (6–8 cm long) antlers from bony pedicles extending from frontal bones on the skull (hence the alternative name of 'rib-faced' deer), and long (up to 2.5 cm) upper canine teeth. In both sexes the skull has deep depressions between the frontal bones for the suborbital glands.

Like Chinese water deer (*Hydropotes inermis*),[2] muntjac have a lengthy ancestry and are territorial rather than gregarious animals, usually occurring singly or in pairs. They are essentially deciduous or mixed forest dwellers, and require plenty of cover through which they run with deceptive speed in a characteristic hunched position.

Table 8 Earliest records for muntjac in England

Date	*County*	*Locality*
1922	Bedfordshire	Wrest Park
*c.*1925	Hertfordshire	Ashridge Park
1933–4	Buckinghamshire/ Northamptonshire border	Brackley, Salcey Forest, Yardley Chase
1934	Warwickshire	Ryton Woods, near Kineton
1941	Staffordshire	Needwood Forest
1940 or 1941	Suffolk	Parham
1941 or 1942	Essex	Near Colchester
Before 1948	Derbyshire	Matlock
1949	Hampshire	New Forest
1951	Norfolk	Walsingham
1952	Oxfordshire	Stoke Lyne, near Bicester
1953	Leicestershire	Charnwood Forest
1955	Berkshire	Twyford
1958	Middlesex	Enfield
1958	Gloucestershire	Withington Woods
Before 1959	Huntingdonshire	?
Late 1950s	Cambridgeshire	Hatley Wood
Late 1950s	Kent	Lullingstone Castle Park, near Dartford
1962	Monmouthshire	Llanvair Grange, Abergavenny
1966	Surrey	Barnes Common
1966	Sussex	Various localities
1966	Wiltshire	Various localities
1967	Devon	Widworthy near Honiton
1968	Dorset	Studland, Rempston, Purbeck
1969	Denbighshire (Wales)	Pontfadog, Ceiriog Valley
1971	Rutland	Rutland Wood
1972	Cornwall	Mithian near Truro
1973	Somerset	Wanstrow, Frome
1973	Yorkshire	Flamborough

Source: After Lever (1977)

Muntjac are currently widely distributed in suitable woodlands in parts of the southern half of England, especially in Bedfordshire, Buckinghamshire, Hertfordshire, Northamptonshire, and Oxfordshire. Much of their expansion of range has taken place since about 1950 and they are continuing to spread. Their success as colonists has been attributed to a number of reasons: being shy and unobtrusive and causing only limited damage they draw little attention to themselves, and their tiny antlers are a meagre trophy for sportsmen; their territorial organization, which tends to banish young males, has encouraged their expansion, and the fact that they can breed throughout the year helps to ensure the survival of a large number of fawns. Although they cause minor damage to young hardwood plantations by browsing and to unfenced gardens, there is no evidence of agricultural depredations, and muntjac in England are economically unimportant.

Chinese muntjac were introduced to France in about 1890 but in spite of assertions to the contrary they never succeeded in becoming established in the wild. They have, however, been maintained on a number of large private estates and in some zoological parks – in the latter especially at Clères, north of Rouen.

NOTES

1. D.I. Chapman & N.G. Chapman, *Journal of Natural History* **16**: 381–7 (1982).
2. See p. 145.

Axis, Chital or Spotted Deer

(Cervus axis)

NATURAL DISTRIBUTION
India, Nepal and Sri Lanka.

Note: Naturalized distribution in Brazil uncertain.

NATURALIZED DISTRIBUTION
Europe: USSR, Yugoslavia; North America: USA; South America: Argentina, Brazil, Uruguay; Australasia: Australia, Andaman Islands; Oceania: Hawaii.

EUROPE

USSR

In 1939 39 axis deer were released in the Teberda ravine in the Caucasus as a source of supply of antlers for which there was believed to be a developing market. Within three years the population had risen to 150, but by the end of the Second World War had been halved through indiscriminate slaughter by German troops. By 1956 a small herd which had been translocated from the Altyagach Region in Azerbaidjan two years earlier had increased to about one hundred, but was later reduced by half due to foot-and-mouth disease.

In 1953 16 axis hinds and 4 stags were transferred from a *sovkhoz* (stud farm) in the Ussuri Territory to the Yerevan Zoo in Armenia. After a period of acclimatization in the zoo they were released on 25 May 1954 in the Khosrov State Forest in the Vedi district, where they disappeared, due, it was believed, to predation by wolves (*Canis lupus*) and lynx (*Felis lynx*). A second introduction in 1958 of 14 hinds and 5 stags proved more successful, and by April 1961 the population had increased to 58.

Before the Second World War axis deer were kept in captivity in Askania Nova in the Ukraine, where, by the end of hostilities, a herd of some 200 was established at Burkuty. From there, axis deer were released in hunting reserves in the Kiev, Cherkassy, Vinnitsa and Kherson regions in 1956–7, and subsequently in the Dnepropetrovsk and Volyn regions. Most of the reserves were enclosed, but the deer in the Chernomorskii Reserve, in the Samara pine forest, in the Dnepropetrovsk Region and elsewhere were allowed to roam free.

Two dozen axis deer released in forests in Lithuania in 1954 adapted well to their new environment, and by 1961 had increased to 67. In May 1959, September 1960 and October 1961, 22, 50 and 25 axis deer, respectively, were released in woodlands in central Kodry in Moldavia.

Yugoslavia

Axis deer have been introduced to several other regions in Europe, including (unsuccessfully) to France in about 1890; they are successfully established on the Istra Peninsula and its offlying island of Brionski Otaki in the Adriatic Sea in north-western Yugoslavia.

NORTH AMERICA

USA

Axis deer are currently by far the most numerous exotic ungulate in Texas. They were originally introduced for hunting in the 1930s and early 1940s to the King Ranch – at the time the largest in the world – and the Bear Creek Ranch by Mr Richard Friedrich. Since then they have been extensively and repeatedly stocked on numerous ranches (especially on the Y-O Ranch near Mountain Home and on the Rickenbacker and Greenwood Valley ranches) – mainly on the Edwards Plateau, in the Rio Grande and on the Coastal Plains. Several herds of over 500 animals are known to exist and at least two, one of which is on the Powderhorn Ranch near Port O'Connor on the coast, number around 1,200. In 1964 the total population was estimated at around 2,200: by 1966 it had increased to 6,450, by 1971 to 11,171, and by 1974 to 19,581 , of which some 6,085 were free-ranging. In 1979, when the population was said to be 22,799, axis deer were present in 64 counties in Texas; 138 ranches in 59 counties totalling 153,293 ha contained axis deer within deer-proof fences. Free-ranging animals occurred in 20 counties with a total population estimated at 7,887; the counties holding the largest numbers include Bandera, Bexar, Blanco, Calhoun, Kendall, Kenedy, Kerr, LaSalle, Live Oak, Robertson and Real – principally on the Edwards Plateau, the South Texas Plains, and in the Gulf Prairies and Marshes where a large population exists in predominantly low (up to 2-m high) live oak (*Quercus virginiana*) interspersed with open grasslands.

Seasonal patterns of precipitation and temperature in most of the Coastal Plains, the South Texas brush country and the arid Edwards Plateau are largely similar to those of the axis deer's Indian homeland. Although primarily catholic grazers rather than browsers, their willingness to change to forbs (weeds and broadleaved herbaceous plants) and browse with a low nutritive value enables them to maintain condition during periods of drought: at other times the availability of water is an essential requirement. Axis deer are not fastidious in their choices of cover; oak mottes[1] on the Coastal Plains and wooded drainages on the Edwards Plateau are used for shelter from sun, wind and rain.

Being primarily grazers, axis deer compete for food mainly with domestic cattle and sheep; only when grass is in short supply and they turn to forbs and woody browse do axis come into conflict for forage with other herbivores such as the native white-tailed deer (*Odocoileus virginianus*), and exotics such as the sambar deer (*Cervus unicolor*),[2] fallow deer (*Cervus dama*),[3] sika deer (*Cervus nippon*),[4] blackbuck (*Antilope cervicapra*),[5] nilgai (*Boselaphus tragocamelus*)[6] and mouflon sheep (*Ovis musimon*)[7]: competition with other herbivores for cover, of which there is an abundance in the form of live oak, mesquite (*Prosopis juliflora*) and juniper (*Juniperus* spp.), appears to be minimal. Feral wild boar (*Sus scrofa*),[8] on the other hand, are avoided by axis deer who seem unable to tolerate their presence. Low temperatures accompanied by heavy snowfalls when food is scarce during the calving season (mainly February and March) seem to be the species' main limiting factors. Axis deer are high regarded as game animals in Texas both for their venison and also because velvet-free hard-antlered stags can be found throughout the year.

In about 1940 some axis deer were released on the Ottinger Ranch near Drake's Bay on the Point Reyes Peninsula in Marin County north of San Francisco, California, where there were around 400 in 1977. Others, said to be established near San Simeon in San Luis Obispo County, are descended from some which escaped from the Rancho Piedra Blanca (Hearst Ranch). There are also a number of axis deer in Volusia County, Florida, where several escaped from a private collection in the 1930s.

SOUTH AMERICA

Argentina

In 1906, 1910 and in about 1932 several shipments of axis deer (the earliest by Señor Pedro Luro to his *estancia* in the southern pampas) were made from India to Argentina, where considerable numbers are now established in Buenos Aires, Santa Fé, Neuquen, Rio Negro and possibly other provinces.

Brazil

Axis deer are established in the wild in several widely distributed localities in Brazil.

Uruguay

In the 1930s Señor Aaron de Anchorena enclosed some axis (and fallow) deer on his 500-ha *estancia* of La Barra de San Juan near the mouth of the San Juan River in the Department of Colonia, some 200 km from Montevideo. Here, and on the edge of scrubland on the banks of the Negro River to the north in the Department of Soriano, axis deer are now widely established in the wild.

AUSTRALASIA

Australia[9]

Axis deer were the first members of the genus to be introduced to Australia. As early as 1803 (possibly even earlier) a number were imported by Mr John Harris, who had arrived in the colony as a surgeon in the New South Wales Corps in 1790; within a decade of their arrival a herd of some 400 was established in Harris Park near Blackwattle Swamp at Parramatta (then known as Rose Hill, and the first settlement outside Sydney), from where a number escaped into the surrounding bush; according to the New South Wales National Parks and Wildlife Service others were deliberately freed at St Mary's, a western suburb of Sydney; although they have not been officially sighted since 1835, Frith (1979) considered that some may still survive in the gorges and deep valleys of the Blue Mountains. It is possible that other liberations were made from Harris' herd at Parramatta.

In 1829 (the same year in which fallow deer were introduced to the island[10]) Captains Edward Dumaresq and Anthony Fenn Kemp of Cross Marsh, Kempton, imported some axis deer to Longford, Tasmania where, although they did not survive in the wild for long, they may have hybridized with the fallow deer which still exist in considerable numbers.

In Victoria three axis deer (presumably a stag and two hinds) were turned out in 1862 on Sugarloaf Hill near Yan Yean (Kinglake), and others were freed 'near the source of the Yarra River'; in the following year four are recorded as having been released at Longerenong near Horsham, where there were also probably earlier, and certainly later, introductions (e.g. five at Yering in 1870). Within five years axis deer were apparently well established in this locality and had extended their range 70 km from the point of release, occupying an area 100 km in diameter. This population, however, eventually died out, and it seems unlikely that any axis deer survive in the state today.

In 1865, 1867 and 1872 the Queensland Acclimatization Society attempted to naturalize axis deer near Toowoomba in the Darling Downs area west of Brisbane; although in 1869 the society was able to report that the deer were increasing on Mr A. Weinholt's property at Maryvale, and by 1876 complaints were being received of damage to crops, the deer apparently did not survive for long.

In 1886 (not 1866 as stated by Whitehead (1972)) the vessel *Quetta*, on which Mr William Hann (the owner of a property at Maryvale[11] on the Burdekin River near Charters Towers in northern Queensland) was a passenger, called at Ceylon (Sri Lanka) *en route* from England, where an axis hind was presented to the master, Captain Jarvis E. Withers, who passed it on to Hann on the ship's arrival at the port of Townsville. A few years later two stags and a second hind were imported by Hann, and from these two pairs (not a stag and two hinds as stated by Whitehead (1972)) – presumably of the form *C. a ceylonensis* – are descended the considerable population currently established in and around the Maryvale Creek. Although the main concentration is still to be found within 25 km of the point of release, the deer are apparently spreading slowly northwards, and small parties have been recorded up to 65 km from Maryvale, with occasional sightings from as far as 560 km.

Axis deer in Australia associate in herds of between 20 and 100 individuals. They are primarily a woodland species, and lie up during the day in dense riverine thickets, from which they emerge at night to feed. Although several authors (e.g. Bentley (1978) and Whitehead (1972)) refer to the dingo (*Canis familiaris dingo*) as an important predator, an animal from south-east Asia where there are many large carnivores is, as Frith (1979) points out, unlikely to be seriously troubled by dingoes once its population has become firmly entrenched. Severe droughts, coupled with a botanically impoverished habitat, are more likely reasons for the species' slow rate of expansion; the existence of more suitable habitats in the Burdekin valley and near the coast led Frith to suggest that the axis population in Queensland, which in 1956 was estimated to number around 600, may well 'increase its rate of expansion in the future if it is not seriously interfered with by man'. Despite Whitehead's (1972) claim that a few axis deer were 'still surviving in the vicinity of Toowoomba', none has been reliably reported in the area for over a century.

Andaman Islands; Java; Nicobar Islands; Papua New Guinea

After the First World War axis deer were landed on the Andaman Islands in the Bay of Bengal to provide food for timber workers. By the end of the Second World War they had reached pest proportions and, by browsing on seedlings, were inhibiting forest regeneration: in 1952–3 two female leopards (*Panthera pardus*) were released as a form of biological control, but this was clearly ineffective, as axis deer are still locally abundant in the Andamans. According to Bentley (1978) axis deer were introduced unsuccessfully to Java and to the Nicobar Islands north of the Andamans in 1846. The same author records that they 'are reliably reported in the Madang area [of Papua New Guinea] and photographs of skins of deer shot in the district confirm this'. The deer are said to have been imported by German rubber planters around the turn of the century, and to have inhabited mostly virgin tropical rainforest and dense rubber (*Ficus elastica*) plantations, from which they emerged at night to feed. In 1975, however, Dr E. Lingren informed N. G. and D. I. Chapman[12] that axis deer, if they ever existed in Papua New Guinea, were now extinct.

OCEANIA

Hawaii[13]

There are several versions of exactly how axis deer, which are known in the islands as *kia*, were introduced to Hawaii; all, however, agree that only one shipment of eight deer was ever made, from which the whole of the present population is descended.

The most likely version is that in 1867 four pairs, obtained from the region of the upper Ganges River in India, were shipped from Calcutta on a ship bound for Hong Kong; an unknown number succumbed to the rigours of the voyage, and were replaced with others by a Mr Magniac of Jardine Matheson & Co. of Hong Kong. The eight deer were forwarded from Hong Kong as a gift for Lot Kamehameha – who reigned as King Kamehameha V – on board the *Loch na Garr*, which dropped anchor off Honolulu in December 1867. One of the stags died during the voyage, but a calf was born in Honolulu Harbor. The deer were probably presented to the king by his physician/botanist, Dr William Hillebrand, who obtained them while on a visit to Calcutta and arranged for their shipment through the consulate in Hong Kong. (Another, but, according to Tomich (1969), less likely version, is that the deer were obtained from the Imperial Japanese deer park and presented to Kamehameha by the Hawaiian Consul to Japan.)

In January 1868 the eight deer were transferred on the king's private yacht *Ka Ma I Ie* from Oahu to his estate on Molokai, where the first herd was established on the forested slopes above Kaunakakai. Within 20 years the population on Molokai had increased to around 1,000, and by 1898, when they were said to number between 6,000 and 7,500, they had become a serious agricultural and ranching pest. In 1900 the Molokai Ranch Company, in an attempt to eradicate the deer, engaged the services of two professional hunters from California, who were so successful that by 1903 only a few axis remained in the native kiawe (algaroba/mesquite; *Prosopis chilensis*) forest.

Soon after their arrival on Molokai several axis deer were returned to Oahu, where prior to 1898 a herd, which later died out, became established on Diamond Head. Another colony, originating in a stag and two hinds which escaped from captivity in about 1910, formed in the Moanalua Valley, where by 1938 the population had multiplied to around 1,000. Two years later, when these deer too had become serious pests, the owners of the valley – who had hitherto protected them – were forced to remove 300; by 1962 only about 25 remained, and by the late 1970s no more than 20 still survived.

In 1954 a pair of axis deer were released at the Kaneohe Marine Corps Air Station on the windward side of Oahu, where they remained for several months.

In 1920 Mr George Munro landed 12 axis deer from Molokai on Lanai; here they became established on the Palawai Plateau, where they were soon causing damage to the native vegetation and to pineapple plantations. In the late 1920s over 300 were shot, and by the end of the Second World War only a few remained; since then they have been maintained for licenced hunting.

In September 1959 two stags and three hinds were released in the Red Hill area on Maui, and in the following July one stag and two hinds were turned out on the Kaonoulu Ranch on the same island, where by 1968 the population had increased to between 85 and 90. Following the discovery of tuberculosis in axis deer on Molokai, a proposal in 1969–70 to release some in the vast semi-arid parkland forests on the main island of Hawaii aroused bitter opposition.

Since 1959 the population on Lanai and, since 1956, most of that on Molokai, have been under the control of the State Division of Fish and Game. On both these islands axis deer are well distributed in all but urban and intensively cultivated regions, being especially common on the latter on the leeward coast between Laau Point and the headlands of the Halawa Valley. On Oahu a small relict herd survives in the brushland above Pearl Harbor. On Maui the population on privately owned ranchland in the saddle between West Maui and Haleakala is believed to be slowly increasing.

With the exception of those in the high (900–1,300 m) rainforests of eastern Molokai (where the precipitation averages over 500 cm per annum), axis deer in Hawaii live in areas where the annual rainfall is between 50 and 150 cm. The rainforest is composed mainly of mixed fern and ohia (*Metrosideros* sp.) trees, with both native and exotic grasses. The lower parts of the deers' habitat, below 500 m, is largely semi-desert in nature with some open savannah-type acacia forest – primarily kiawe with an understorey of lantana (*Camara lantana*) and ilima (*Sida fallax*), and grasses such as the native pili or tanglehead (*Heteropogon contortus*). On Lanai, where their principal range is semi-desert, the brushland is relatively open, the predominant ground cover consisting of pili grass, kiawe trees, klu or huisache (*Acacia farnesiana*) and haole koa bushes or white popinac leadtree (*Leucaena glauca*). The deers' favourite habitat is in or near the dense kiawe forests which cover the narrow coastal plain adjacent to the brushland, where their principal foods are kiawe (especially when the seeds fall), huisache and white popinac leadtree. In much the same habitat on Molokai kiawe and lantana furnish the greater part of their diet.

In the late 1970s some 3,000 of the islands' total population of around 4,000 axis deer (a fall of 1,000–2,000 from that of the mid-1960s) were concentrated on Molokai where, as a result of hunting pressure, the deer lie up in thick forest during the day from which they emerge at night to feed. Writing in 1969 Tomich considered that the future of axis deer as an important game species seemed assured on Lanai, Molokai and possibly on Maui. The fact that all the axis deer in Hawaii (where they are regarded as genetically equal to those in their native homeland) are descended from a single introduction of eight animals well over a century ago, 'should', wrote Graf and Nichols (1967), 'provide some interesting food for thought for those sportsmen who persistently contend that inbreeding degrades our game species'. A contributory factor in the successful naturalization of axis deer in Hawaii may be that, at between 19° and 22° N, the islands lie at the same latitude as that of the deers' native range in central India.

Axis deer have been introduced for sporting purposes to several large (85,000 ha or more) fenced private ranches in the Western Cape, South Africa.

In 1867 Mr J. A. Ewen of Dunedin in New Zealand shipped seven axis deer from Melbourne (where they had been imported from Calcutta) for Mr Gordon Rich of the Otago Acclimatization Society, by whom they were released in the Goodwood Bush area between Oamarn and Palmerston South. Within a decade they are believed to have increased to around 100, but were soon causing so much damage to agricultural crops that by 1890–5 they had been exterminated. Writing about a pair in the wild in 1875, Donne (1924), in *The game animals of New Zealand*, records that the stag, 'having developed the pernicious habit of butting pedestrians, was shot'.

In 1893 the Zoological Society of Calcutta presented a pair of axis deer to the Wellington Acclimatization Society, who turned them out in the government's Game and Bird Sanctuary on Kapiti Island (2,000 ha) in Cook Strait, where by 1907 they had only increased by one.

In 1902 the Otago Acclimatization Society imported three axis deer from India, which were placed in quarantine on Quail Island; what subsequently became of them is unknown.

In 1906 Donne, on behalf of the Government Tourist Department, ordered two stags and eight hinds from the German-born animal dealer Carl Jamrach in London who was, however, able to despatch on the SS *Papanui* only the two stags and four of the hinds 'in charge of a man named H. Page', which arrived in Wellington on 8 August 1907. One was kept in the Wellington Zoo while the remaining five were released in the 25,000 ha of the Tongariro National Park, south of Lake Taupo.

On 22 October 1907 six more axis deer were ordered by the New Zealand government from Jamrach, who on 12 June of the following year shipped them from London by the SS *Rimutaka* which arrived in Wellington on 30 July: the deer were placed in quarantine for eight weeks on Soames Island from where they were released at Dusky Sound in the Fiordland National Park. (One account says that one of the deer died on the voyage and that the surviving five were liberated in 1909.)

None of these plantings seems to have met with lasting success, and since no axis deer have been seen in New Zealand since about 1920–30 the species must be presumed to be extinct.

NOTES

1. Clumps of trees in prairie country.
2. See p. 223.
3. See p. 176.
4. See p. 213.
5. See p. 273.
6. See p. 247.
7. See p. 311.

8. See p. 117.
9. For much of the information on the history of axis deer in Australia I am indebted to Bentley (1978) and Frith (1979).
10. See p. 182.
11. Several authorities appear to have failed to distinguish between Maryvale, Toowoomba, and Maryvale, Charters Towers.
12. Chapman & Chapman (1980: 68).
13. For much of the information on the history of axis deer in Hawaii I am indebted to Graf & Nichols (1967), Tomich (1969) and Kramer (1971).

Fallow Deer

(*Cervus dama*)

NATURAL DISTRIBUTION

Originally the Mediterranean region (probably including North Africa) eastwards to southern Iran. (For original natural distribution see text.)

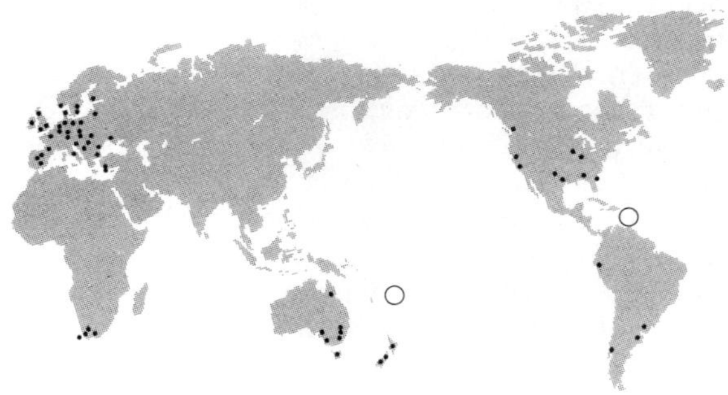

Note: After Chapman and Chapman (1980).

NATURALIZED DISTRIBUTION[1]

Europe: British Isles, Austria, Belgium, Bulgaria, Czechoslovakia, Denmark, Finland, France, Germany (East and West), Greece (island of Rhodes), Hungary, Italy, the Netherlands, Norway, Poland, ?Portugal, Romania, Spain, Sweden, USSR, Yugoslavia; Africa: South Africa; North America: USA, Canada, West Indies (Lesser Antilles (Barbuda and Guana)); South America: Argentina, Chile, Peru, Uruguay; Australasia: Australia, New Zealand; Oceania: Fiji.

EUROPE

British Isles

England

Although fallow deer are known to have been present in England during the Last Interglacial period of the late Pleistocene, around 150,000 years ago, no remains have been found since the end of the Ice Age around 10,000 years ago, and the present stock is descended from subsequent reintroductions. Although Chapman and Chapman (1975, 1980) suggest that fallow were probably imported by the Normans in the eleventh century, a much earlier reintroduction, possibly by the Romans, has been suggested (Lever 1977).

By the thirteenth century the pursuit of fallow deer for sport had become a popular pastime of the nobility, and vast tracks of woodland in England were enclosed for this purpose. In the county of Essex alone, for example, it is known that between 1200 and 1390 more than 130 such enclosures were in existence.

By the early seventeenth century there were more than 700 deer parks containing fallow in England; during the Civil War (1642–9) the fences of many parks fell into disrepair, resulting in the escape of numerous fallow into the surrounding countryside. The eighteenth century saw a revival of the interest in deer parks, and by the end of the following century there were nearly 400 in England holding an estimated 71,000 fallow dear. The two World Wars of the present century resulted in the break-up of many parks, and the escape of more fallow into the surrounding woodlands. Although fallow still occur in forests where the species has been wild for centuries, the majority of the present population is established in coniferous, deciduous and mixed woodlands in the vicinity of former parks – especially those abandoned during the present century.

Fallow are today the most widespread of the four alien deer naturalized in England; they occur in all the counties south of a line from the Humber to the Severn, as well as in several to the north. In the following survey the probable sources of current populations of wild fallow deer in individual counties are given (with dates where known) in brackets.

In west Kent fallow are found from Eynsford and Sevenoaks to as far east as Maidstone and Chatham [Knole, Surrenden-Dering, Chilham, Mersham, Hatch and Waldershare parks]. In east Sussex they are well established from the Northam/Robertsbridge/Battle area westward to beyond Ashdown Forest;[2] in west Sussex they occur from St Leonard's Forest near Horsham to the South Downs; north of the Downs fallow are established near Midhurst, and west of the Rother Valley to the Surrey and Hampshire borders [Ashburnham, Britling, Buckhurst, Arundel, Petworth, Parham and Cowdray parks].[3]

Throughout much of Hampshire fallow deer are widespread; they occur in the coniferous and mixed plantations of Bramshill Forest in the north-east, around Petersfield, near the coast at Emsworth, in the oak, beech and coniferous woodlands of the Forest of Bere in the south-east, in Harewood Forest near Andover, around Hursley and Farley Mount near Winchester, and west of the River Test towards Salisbury [Aldermaston, Hackwood, Hurstbourne, Hursley and Uppark parks]; the largest population (about 1,000) lives in the south-west in the 27,000 ha of the New Forest. One-third of this forest, created by William the Conqueror in 1079, comprises deciduous and coniferous plantations; the remainder, which consists of natural deciduous woodland – principally beech (*Fagus sylvatica*) and pedunculate or common oak (*Quercus robur*) and sessile oak (*Q. petraea*) with an understorey of holly (*Ilex aquifolium*) – moorland, heath, grassland and agricultural land, is the most extensive surviving area of semi-natural vegetation in lowland Britain.

In Dorset fallow deer are present in the east on the border of the New Forest, between Lulworth Cove and Wool in the south-east, between Maiden Newton and Bridport in the south-west and in the 2,000 ha of mixed hardwoods and softwoods of Cranborne Chase in the north-east [Sherborne, Hooke, Melbury, Hyde House and possibly Charlborough parks].[4]

In Devon fallow occur intermittently from the Dorset border to the Otter Valley.[5] West of the River Exe they are established in some numbers in the woodlands and adjacent farmland of the Haldon Hills.[6] On the eastern edge of Dartmoor fallow live in the Teign Valley (especially between Dunsford and Fingle Bridge)[7] in plantations of Douglas fir (*Pseudotsuga menziesii*), Norway spruce (*Picea abies*) and Japanese larch (*Larix leptolepis*), as well as in deciduous woodlands of birch (*Betula pendula*), rowan (*Sorbus aucuparia*), oak, hazel (*Corylus avellana*) and beech, with an understorey mainly of holly, bramble, grasses and thyme (*Thymus drucei*). In south Devon fallow occur in Plym Forest north-east of Plymouth [Shute, Bicton, Blatchford, Cready, Whiddon and Werrington parks].[8]

In Cornwall small numbers of fallow deer are found on Torpoint [Mount Edgecumbe], to the south of Bodmin Moor, near Lostwithiel and Falmouth, and around Truro. In northwestern Somerset there are fallow in and around the coniferous plantations on Croydon Hill south of Dunster, and around Birds Hill, Combe Sydenham [Dunster, Nettlecombe, Combe Sydenham and Hatch Court parks]. Fallow are widely distributed in suitable woodlands in Gloucestershire, especially in the 13,000 ha of largely mixed hardwoods (with some conifer plantations) of the Forest of Dean, where their main haunts are Highmeadow Woods west of Coleford and near Speech House west of Cinderford [Lydney, Over, Great Barrington, Brockhampton and other parks]. Across the River Severn in Monmouthshire fallow are established around Monmouth, on both sides of the River Wye, especially in Beaulieu and Lady Park woods and in Redding Enclosure towards Staunton in the east, and at Hendre and White Hill in the west [probably Wyastone Leys and Hendre parks].[9] In Wiltshire fallow occur in the oak, beech and mixed woodlands of Savernake Forest north-west towards Marlborough, and to the east of Salisbury. In Surrey a few occur in the vicinity of Farnham [Witley and Farnham parks]. In Berkshire fallow are present in small numbers as far east as Maidenhead Thicket and northwards to the border with Oxfordshire [Windsor, Hall Place, Aldermaston, Hampstead, Welford and Littlecote parks].

In Oxfordshire fallow deer are widely distributed, especially in the Woodstock/Charlbury area, in Bernwood Forest east of Oxford and in the conifer and beech woods around Stonor, north of Henley-on-Thames, and east of Chinnor and Aston Rowant in the Chiltern Hills on the border with Buckinghamshire [Ditchley, Cornbury, Great Barrington, Glympton, Shotover, Stonor and Nuneham parks], where they occur especially on the Hertfordshire and Bedfordshire borders [Ashridge and other parks]. Large numbers are established in the 1,600 ha of principally deciduous woodland (mostly beech, oak, ash (*Fraxinus excelsior*), sweet chestnut (*Castanea sativa*) and wild cherry (*Prunus avium*)), commons and scrub (largely bracken, gorse and birch) at Ashridge in Hertfordshire where they also occur in smaller numbers around Watford, Cuffley, Northaw and Hitchin, and between Nuthampstead and Bishop's Stortford on the Essex border [Woodhall (1916), Hatfield, Ashridge (1928) and King's Walden parks].

In East Anglia fallow deer are widespread in suitable habitats; in Essex they are largely confined to the north and west of the county, being particularly abundant in Epping Forest (where they have been established since the fifteenth century), Navestockside, Mill Green, Roxwell, Hatfield Forest, Strethall, Markshall and Clacton-on-Sea [Langleys, Easton Lodge, Dunmow, Chelmsford and Weald parks].[10] In west Suffolk fallow occur in the mainly coniferous King's Forest; in deciduous woodlands to the south-east around Saxham, Whepstead, Lawshall, Long Melford, Hawkedon and Cavendish near Bury St Edmunds, and westwards through Dalham, Ousden and Lidgate near the border with Cambridgeshire.[11] Fallow deer are established in the southern half of east Suffolk through Felsham, Chelsworth, Ashbocking, Helmingham, south of Ipswich, and between the Orwell and Deben rivers to the mainly coniferous Tunstall and Rendlesham forests [Woolverstone, Henham (1914), Livermere Hall and Ickworth parks]. In Norfolk fallow occur principally in the west around King's Lynn and Sandringham. A few are also found from Cromer on the north coast south to Norwich [Blickling, Dudwick (12 black deer in 1942), Sandringham, Houghton and Holkham parks].[12]

Although much of Cambridgeshire and Lincolnshire are composed of flat, tree-less expanses of intensively cultivated farmland which are inimical to deer, fallow are found in some localities in both counties. In Cambridgeshire these include Gamlinghay in the south-west and the Soke of Peterborough in the north-west – especially in the triangle formed by Peterborough, Stamford and Wansford-in-England. In Lincolnshire fallow have occurred for many years in the Stamford/Bourne/Kirkby Underwood/Grantham districts, and recently have moved into Linwood and Willingham forests and to near Wragby and Woodhall Spa west of the Wolds in central Lincolnshire [Grimsthorpe, Burghley and Brocklesbury parks].[13] In 1977 some fallow were released on the 100-ha Read's Island in the River Humber west of Barton, where their present status is uncertain.

In parts of some Midland counties fallow deer have been established in the wild for several hundred years. In Northamptonshire they are found between Oundle and Corby, in the mixed woodland of Rockingham Forest (where they have lived since at least the early thirteenth century) and in Whittlewood Forest in the south [Milton, Fawsley and other parks]. In Nottinghamshire they occur in Sherwood Forest (where they were established by the thirteenth century) and in the Underwood/Annesley area [Welbeck, Thoresby and Rufford parks]. The 2,400 ha of mixed woodland, heath and scrub which constitute Cannock Chase (between Birmingham and Stoke-on-Trent) in Staffordshire have provided a home for fallow deer since at least 1271; there have been fallow in Needwood Forest near Cheadle since before 1313, and others are established around Abbots Bromley. In Leicestershire fallow are found in the deciduous woodland of Charnwood Forest [Staunton Harold, Calke Abbey and Donington Hall parks], between Pickworth and South Witham (especially in Morkery Wood) in Rutland [Exton and other parks],[14] and around Spondon and Matlock in Derbyshire. In Warwickshire they occur around Edge Hill south-east of Stratford-upon-Avon [Ragley Hall and possibly Charlecote, Alscot and Ettington parks]. In Worcestershire fallow are found in the woodland, pastures and scrub on Bredon Hill [Hagley Park], and in Shropshire in the oak and beech woodlands of Wyre Forest,[15] where they have been present since the thirteenth century [Mawley Hall (*c.* 1880), Cleobury, Mortimer, Longnor, Loton and Attingham parks]. In Herefordshire fallow are found principally near Hereford, Leominster and Ross-on-Wye, and in the 3,600 ha of mainly coniferous plantations – intermixed with some deciduous woodland – of Mortimer Forest south-west of Ludlow, where a few with long 'shaggy' hair first appeared in the 1960s [Moccas, Garnstone (black variety), Hampton Court, Ludford, Haye and Moor parks].[16]

The north of England supports much smaller numbers of fallow deer than do the Midlands and the south, but a few herds are scattered as far north as the Yorkshire/Co. Durham border and Cumberland and Westmorland. In Yorkshire fallow occur north-west of Barnsley;[17] in Stang Forest south of Barnard Castle; in the Swale Valley west of Richmond; near Swinton Park, Masham and West Tanfield; in the Laver Valley; in woodlands in Nidderdale north-west of Harrogate; near Knaresborough east of Harrogate;[18] and south-east of Wakefield [Duncombe, Studley Royal, Allerton, Nostell Priory, and possibly Castle Howard and Scampston parks]. In Cumberland and Westmorland fallow are found in ash and other deciduous woodlands near Arnside in the south of the Lake District near the border with Lancashire [Dalemain and Lowther parks].[19] In Northumberland a small herd of fallow is present near Alnwick.

Scotland

In Scotland there is a tradition that monks first brought fallow deer to Kildalton on the Isle of Islay in about AD 900. They are first referred to on the mainland in 1283 when a herd was established in the King's Park at Stirling. The present population of between 2,000 and 3,000 is largely confined to the Lowland and Border counties and to three islands in the Inner Hebrides. Small relict populations may still survive in parts of the Highlands.

In the Borders fallow are established in Kirkudbrightshire, Dumfriesshire and, to a lesser extent, in Wigtownshire, to which some were reintroduced in 1971. In Kirkudbrightshire they are found near Newton Stewart at the southern extremity of the 6,200 ha of mixed deciduous and sessile oak woodland and softwood plantations which form Galloway Forest. In Dumfriesshire the 200 or so fallow in the woods around Raehills, near Moffat, are the descendants of some introduced in 1780.

Further north fallow occur around the shores of Loch Lomond in Stirlingshire and Dunbartonshire as well as on three islands in the loch itself – Inchlonaig, Inchcailloch and Inchmurrin. East of Loch Lomond, straddling the Stirlingshire/Perthshire border, lies the 17,000 ha Queen Elizabeth Forest Park (comprising the forests of Rowardennan, Loch Ard and Achray) in which a small population of fallow has been established since the early 1970s. The stronghold of fallow deer in Perthshire lies to the east of the River Tay between Dunkeld and Ballinluig; smaller numbers also occur between Blairgowrie and Dunkeld. On the west coast a few fallow are believed to be established at Inverary in Argyllshire and on the Mull of Kintyre.

Despite the claim, by Chapman and Chapman (1980), that 'the Dunkeld populations are probably the most northerly ones established in Scotland', fallow are still reported from time to time in a number of Highland counties: these include

Banffshire (Arndilly House woods, Craigellachie); Inverness-shire (Balmacaan, Loch Ness); Morayshire (between Fochabers and Craigellachie in the Spey Valley); Ross and Cromarty (Achnalt/Garve; Corriemoillie; Kinlochluichart; Lochrosque; Scatwell); and near Dunrobin in Sutherland.[20]

Although fallow have occurred in the past on no fewer than 11 islands off the west coast of Scotland they currently survive on only three – Mull (around Loch Ba', Glenforsa and Ben More), Islay (more than 50 near Kildalton) and Scarba, where 21 were counted in 1974.

Wales

Fallow deer have been established in Wales for well over 700 years, and by the sixteenth century they are known to have been running free in the Royal Forest of Snowdon in Caernarvonshire.

In the south-east they occur in Margam Forest near Port Talbot in Glamorganshire, where pine plantations predominate among larch, oak and other deciduous woodlands. In the south-west they are found in the forests of Brechfa (mostly conifers with some oaks) and Caeo (mostly oak, beech and sycamore (*Acer pseudoplatanus*)) between Llandilo in Carmarthenshire [Golden Grove Park] and Lampeter in Cardiganshire; small numbers also occur in the woods south of Carmarthen. In Montgomeryshire fallow are established on Breidden Hill, Welshpool, and the Long Mountain [Powis Castle (1947) and Loton parks], and in the native hardwood and coniferous plantations of the Wye Valley in Brecknockshire and Radnorshire along the Monmouthshire and Herefordshire borders.

In Denbighshire in north Wales fallow occur in the east to the south of Wrexham and between Denbigh and Colwyn Bay on the north coast [Wynnstay Park]. In Merionethshire a mass escape of 90 fallow from Nannau Park near Dollgellau in 1963 resulted in the formation of a number of wild herds,[21] some of which moved northwards to the mainly coniferous plantations of Coed-y-Brenin Forest, while others colonized oak and softwood plantations on the slopes of the Dovey Valley. In Caenarvonshire fallow occur near Bangor on the Menai Straits; across the straits on the Isle of Anglesey a small number are established on the sand dunes and in the conifer plantations of Newborough Warren on the west coast around Bodorgan near Aberfraw.

Ireland

In Ireland (where in 1970 the total population in the wild was judged to be between 2,500 and 3,000) fallow deer were probably first introduced for hunting by the Normans in the eleventh century. Today they are established in at least four of the six counties in Northern Ireland – Tyrone, Armagh, Down and Antrim. The largest concentration occurs near the River Blackwater between Glaslough (in County Monaghan in the Republic) and Caledon in County Tyrone, near the border of County Armagh. In County Down fallow are found on the Clandeboye estate between Bangor on the north coast and Newtownards at the northern end of Strangford Lough. To the north in County Antrim fallow are more widely distributed near Randalstown north of Lough Neagh [Shane's Castle Park (*c.*1933; dark form)]. Although fallow are believed to be absent from County Londonderry, they may still occur around Colebroke, Florence Court and Crom Castle in County Fermanagh [Florence Court, Crom Castle and Kiltiernay House parks].[22]

In the Republic of Ireland fallow deer are widespread throughout much of County Waterford except in the extreme east, occurring principally in woodlands south of the River Suir near Kilsheelan and Clonmel, along the River Nier, north of Cappoquin and near Villierstown [Curraghmore Park, *c.*1909]. Fallow are also widely distributed in County Tipperary, where they inhabit plantations on the slopes of the Knockmealdown Mountains near Clogheen and Ballyporeen, the Glen of Atherlow in the Galty Mountains, and smaller woods around Bansha, Dundrum, Rear Cross, Newport, the Silvermines Mountains and near Kilcooley Abbey in the east [Ballinacourty, Dundrum House (*c.* 1918), Gurteen (1890s) and Castle Lough (*c.*1920) parks]. A few occur south of Tallow on the eastern border of County Cork.

North of the River Shannon in County Clare fallow are found in the Bunratty/Tulla area and on both sides of the Clare/Galway border near Lough Cutra, Dromoland Castle and Raheen parks; in the east of Galway they are also established in woodlands near Derrybrien on the Slieve Aughty Mountains and at Kylebrack and Woodford. In 1970, in Portumna Forest on the northern shore of Lough Derg, no fewer than 370 fallow were estimated to live in an area of no more than 560 ha.[23] Further north fallow also occur in smaller numbers around Ahascragh, on both banks of the River Clonbrock and near Ballygar.

In the centre of the Republic, in counties Offaly and Leix, fallow are widely distributed; in the former they occur mainly near Birr Castle and Clonfinlough and around Tullamore [Charleville, Kinitty Castle, Bin Castle, Knockdin Castle and Thomastoun parks], and in the latter around Clonaslee, the Slieve Bloom Mountains, Camross, Montrath, Emo, Ballyfin, Ballybritas and Stradbally [Emo Park]. Eastwards, in County Wicklow, fallow have been established in the Wicklow Mountains since 1244[24] and on the western forested slopes they range from the region of Kiltegan northwards to the Lackan reservoir, eastwards towards Delgany and south through plantations around Carrick Mountain to the coast at Wicklow [Powerscourt and Ballycurry parks]. Further south still fallow are found south of Aughrim and across the border into County Wexford.

In 1888 Count Considine released some fallow deer on Lambay Island, 25 km north-east of Dublin, where their descendants still remain; in 1974 Lord Revelstoke told Chapman and Chapman (1980) that the herd, maintained at about 30 and mostly of the black variety, roam freely over the 1,000-ha island.

In County Westmeath fallow are found north and south of Mullingar. Further west, in County Roscommon, a herd is established near Boyle [Rockingham Park, *c.*1910], and in County Mayo on the west coast another is present on the slopes of Partry Mountain near Lough Mask. To the north-east, in County Sligo, fallow occur in woodlands between Collooney and Lough Gill and around Slish Wood [Marktree Castle Park]. In the north of County Monaghan fallow are found near Glaslough and in the south between Rockcorry and Cootehill near the border of County Cavan [Hilton Park].

From time to time fallow are reported from counties Kildare, Meath, Longford, Leitrim, Wexford, Carlow and Kilkenny, but nowhere with regularity. They are entirely absent from counties Donegal, Kerry and Limerick, in all of which, however, they have occurred in the past; this suggests that although still widely distributed there may have been some overall reduction in range in recent years.

The damage caused by fallow deer in the British Isles – limited to browsing seedlings and occasionally larger trees and fraying[25] saplings – is of little economic significance.

Austria

In the 1920s and 1930s fallow deer were established in the wild in Austria in the Leithagebrige in northern Burgenland and in the Dunkelsteiner Wald,[26] but by the late 1940s both these populations had been exterminated.[27] Today the species occurs in the wild in only three localities. Since 1932, when fallow from Lüneberg Heath in (West) Germany were introduced, they have been established at Anthering near the River Salzach north of Salzburg close to the German border; further releases of fallow from Czechoslovakia were made in 1943. The deer roam over approximately 1,000 ha of moist woodland (composed mainly of willow (*Salix* sp.), oak (*Quercus robur*), lime (*Tilia* sp.),popular (*Populus* sp.), alder (*Alnus* sp.), elm (*Ulmus* sp.), ash (*Fraxinus* sp.) and Norway spruce (*Picea abies*)) interspersed with pastureland in which rape (*Brassica napus*) and oats (*Avena orientalis*) have been planted especially for the deer. Fallow are also present near Horn, about 65 km north-west of Vienna, where around 60 were counted in 1971.[28] A similar number also occur in woodland (largely Norway spruce, European larch (*Larix decidua*) and pines (*Pinus mugo* and *P. cembra* with some ash and sycamore (*Acer pseudoplatanus*) and meadowland near Mautern in Steiermark in central Austria.

Belgium

In about 1850 fallow deer were introduced to the 1,500-ha royal forest of Ciergnon west of Rochefort in the Ardennes; this herd – the only one in Belgium – has recently been maintained at under 100 head.

Bulgaria[29]

In Bulgaria, where in 1973 the total population was said to number about 1,000, fallow are widely distributed – especially in the east.

The earliest introduction was probably in 1904

when a buck and two does, imported from Germany, were enclosed in the Kricim hunting reserve near the River Maritsa east of Plovdiv; they were subsequently augmented by further importations whose progeny were later liberated in other localities. A herd of about 80 is still enclosed at Kricim – all other Bulgarian populations being free-ranging.

The largest concentration (said to number about 350 in 1969) is probably that established in 3,000 ha of deciduous forest at Voden near Kubrat, north of Razgrad where the plain of the Danube widens into the Deli Orman Plateau. From here, fallow have spread north as far as the Romanian border on the Danube, west towards the River Lom and eastwards to the north of Novi Pazar almost as far as Shabla near the Black Sea.

In 1970 more than 250 fallow deer were estimated to live in 900 ha of the Yambolski-orman Forest near Yambol, some 145 km south of Voden. To the east, near the coast of the Black Sea, fallow are found in the 2,700 ha of the Ropotamo Reserve in the Stranja Mountains west of Mitschurin and south of Burgas.

South of the Balkan Mountains in central Bulgaria, fallow occur in the woodland and scrub of the Stara Zagora Plain. Further south there are some to the north of the River Arda and near the Studen Kladenez Dam in the hills between Kurdzhali and Khaskova, while to the west of Stara Zagora fallow live near Pazardzhik on the River Maritsa.

In the west of the country fallow are found in the 10,000 ha of mainly pine forest and pasture which form the Vitoscha Nature Reserve south of Sofia.

Czechoslovakia

According to Nožička (1966), fallow deer were being kept in captivity in Czechoslovakia as early as 1465, although apparently none were released into the wild, where they are now widespread, until last century. In 1974 there were thought to be 2,200 at liberty with a further 2,900 emparked, which by 1977 had risen to a combined total of around 7,000.[30]

The largest concentrations of wild fallow are in Western Czechoslovakia (Bohemia), especially south-west of Plzeň near the border with West Germany where three populations (near Kydne, west of Plzeň and on the River Mže near Střibo) occur from west of the River Radbuza to east of the Úhlava. South of Prague a population is established in the region of Dobřis. In northern Bohemia fallow occur around Mělnik; near Varnsdorf close to the border with East Germany; to the east around Jan Nisou near the Polish border, and to the south near Dvůr Kralové. In eastern Bohemia two more populations are established to the south-east of Hradec Kralové and east of the Elbe. Southern Bohemia supports a single population in the Luznice Valley between Tábor and Bechyně.

In mid-Czechoslovakia (Moravia) fallow deer occur around Moravski Budějovice; near Blanska, Olomouc and east from Přerov between Brno and Ostrava; to the east of Stražnice; and near the Austrian border between the confluence of the Thaya and Morava rivers south of Břeclav.

The largest population (about 1,000) of wild fallow in eastern Slovakia lives in the Malé Karpaty Hills north of Bratislava; two more groups are established near Trencin and Zilina to the north-east between the rivers Váh and Nitra; further east fallow are established near Levice, east of the River Hron, and in the 12,000-ha hunting reserves near Topolčany. The only fallow in lowland Czechoslovakia (below 2,000 m) are found south of Břevclav.

Denmark

Having become extinct as a native species during the last (Würm) glaciation, fallow deer were reintroduced to Denmark in the thirteenth century. Documents of 1231 in the reign of King Valdemar II (1202–41) mention the presence of fallow on no fewer than 16 small islands.[31] Today, although fallow are widely distributed in the wild, their population is exceeded by that of enclosed deer; out of a total of nearly 4,000 Danish fallow in 1970 more than half were in captive herds, of which that in the Jaegersborg Deer Park north of Copenhagen on Zealand is the largest, holding about a third of the country's entire population.[32] Between the end of the Second World War and 1970 free-living fallow died out on the island of Samsø, at Kastrup (Zealand) and at Rosenvold and Sostrup (Jutland). Apart from the disappearance of fallow from Løvenholm east of Auning in eastern Jutland, their present distribution is much the same as it was in 1970.

The largest concentration of fallow deer in Denmark (around 400) inhabits the 5,000 ha of coniferous, beech and other deciduous plantations and scrub of the Grib Skov Forest between Hillerød and Esrum in northern Zealand.[33] South-west of this forest about 100 fallow are established in woodlands on the Bognaes Peninsula in the southern Roskilde Fjord. Between Ringsted and Slagelse in central Zealand a small number of fallow (about 10 in 1970) has lived in woodland near Sorø Academy since 1965; they probably arrived from Naesbyholm to the south, where there is at present a herd of around 80. South-east of Naesbyholm about 70 fallow live in the arable farmland, deciduous woodland (including beech (*Fagus* sp.) birch (*Betula* sp.) and oak (*Quercus* sp.)) and coniferous plantations at Gisselfeld, from where some have dispersed eastwards to the marsh-type habitat of Holmegaards Mose. Near Faske, to the east of Gisselfeld, about 20 fallow occur at Lystrup, from where they often wander eastwards to Vemmetofte. Formerly, fallow were abundant in this region (sometimes in herds 60–70 strong), particularly in the vicinity of Gisselfeld, Lyrup, Vemmetofte, Bregentved, Jomfuens, Egede, Beldringe and Knuds Skov.

Between December 1977 and February 1979 (when 64 were counted) some fallow were landed on three occasions on the 70-ha island of Hasselø in the Kattegat, some 30 km off the north coast of Zealand; it is planned eventually to maintain this herd at between 150 and 300 by culling surplus beasts for venison.

In 1970 four herds of wild fallow were established on the island of Lolland;[34] east of Maribo some two dozen animals occurred outside the Krenkerup Park, from which their ancestors are believed to have escaped. About 100 fallow lived in the woods between Bremersvold and Kjaerstrup east of Rødby, and a like number inhabited the beech woods and conifer plantations to the north of Christianssaede, where the nearby corn and sugar-beet fields provided forage. On the coast north-east of Maribo, between Oreby and Berritgaard, a further two dozen were present in 1970.

Fallow are known to have occurred on the island of Langeland in the time of King Valdemar II, and the present population may well be their direct descendants, augmented by further introductions; today, fallow are found in mainly deciduous woodlands (beech, oak, lime (*Tilia* sp.), alder (*Alnus* sp.), and hazel (*Corylus* sp.) with some conifers and agricultural land around Tranekaer in the centre of the island and to the south near Hjortholm Manor.

The island of Funen also supports two populations of fallow deer – on the 100-km^2 Revnholt estate north-east of Ringe, and on the 480 ha of the Wedellsborg estate near Husby on the west coast, where the deer grow antlers of exceptional quality. At the end of the Second World War both wild and emparked fallow occurred at Revnholt, but since 1950, after the break up of the park, all the fallow (in the autumn of 1978 more than 100) have been free-ranging. The mixed hardwood and coniferous plantations with a dense understorey provide the deer with an abundance of food and shelter.

About 100 fallow live on the 100-ha island of Romsø off the north-east coast of Funen: on the 250-ha island of Aebelø off the north coast an unsuccessful attempt was once made to eradicate the resident fallow[35] which, by 1970, were said to number about 250.[36]

A few small discrete populations of fallow occur – mostly in the east – in Jutland. One is established in mixed deciduous and softwood plantations in the north at Børglum south of Glimsholt. Until at least 1970 a small herd, dating from the 1920s, was to be found in the Emmedsbo plantations near the coast, 15 km north-west of Grenaa. In 1970 about 30 fallow occurred at Palsgard, also near the coast, 20 km south-east of Horsens. In 1876 fallow were introduced for hunting to the 300-ha island of Livø in Limfjorden in northern Jutland, where they established themselves in mixed softwood and ancient oak plantations and on heathland (mainly juniper (*Juniperus commune*) and heather (*Calluna vulgaris*)) on the north coast: new blood was introduced in 1957 and again in 1975 (a buck and two does), but by the spring of 1978 only three – two does and one fawn – remained.

Finland

A herd of fallow deer at Kytäjä near Hyvinkää some 50 km north of Helsinki, descended from some introduced from central Europe in 1938[37] and said to number about 50 in 1973, is believed to be the only free-ranging one in Finland. At 60° 38′ N it is also the most northerly population of fallow in the world.

France

Although the overall distribution of fallow deer in France seems to have contracted during the last 50 years, they are currently present in small numbers in Aisne, Aube, Calvados, Oise, Yvelines, Essonne, Haute Marne, Val de Marne, Bas Rhin, Haut Rhin and Moselle in the north, and in Landes and Tarn in the south: the majority of the population, believed to number only around 700 in 1974, is in the *départements* of Bas Rhin, Haut Rhin and Moselle in Alsace-Lorraine; most of the remainder occur north of Paris in Aisne, Oise and Somme.

Fallow deer, which have probably occurred in Alsace-Lorraine since at least the sixteenth century,[38] are currently found in some 20,000 ha of land along the River Ill (a tributary of the Rhine) between Erstein and Guemar south of Strasbourg. The bulk of the population occurs in the Illwald Forest with smaller colonies in the Marckolsheim-Schoenau Forest north of Ried and, since 1973, in parts of La Harth.[39] The area occupied by the deer – around 9,500 ha – is composed of woodland and water-meadows; the former contains 2,800 ha of hardwoods with oaks (*Quercus robur* and *Q. sessiliflora*), ash (*Fraxinus elatior*), poplars (*Populus* spp.) and Norway maple (*Acer platanoides*) predominating, and with an understorey for cover of hazel (*Corylus avellana*), privet (*Ligustrum europaeum*), dogwood (*Cornus sanguinae*), hawthorn (*Crataegus oxycantha*) and blackthorn (*Prunus spinosa*). The Illwald Forest accounts for more than half this area, the remainder consisting of small woods scattered among the meadows of the Reid Centre-Alsace. The population in this region was variously estimated in 1975 to be 228 and more than 500.

In the 1860s some fallow deer were introduced to Rouvray Forest near Orival north of Arras;[40] here they thrived and spread towards Rouen,[41] Le Havre and into the Hallates region; those now living in Delville Wood east of Longueval, about 30 km north-east of Amiens in the *département* of Somme, may either be direct descendants of this nineteenth century introduction or of more recent park escapes. Similarly, fallow believed to inhabit some small woods around Ivry-le-Temple between Pontoise and Beauvais south-west of Amiens in the *département* of Oise, could be descended from some known to have been in the nearby forest of Gisors in the last century[42] or from park escapes.

Between Reims and St Quentin north-east of Paris in the *département* of Ainse a small herd of escaped fallow inhabit the Samoussy Forest near Laon. Before the First World War the remnant of the ancient oak and beech royal hunting ground of Compiegne held a stock of fallow deer which had, however, all been killed by around 1920. J. L. Reed[43] mentioned the presence of a few fallow in this forest, where their present status is unknown.

Germany (East)[44]

Free-ranging fallow deer are widely distributed in East Germany – especially in the central and northern regions – and appear to be increasing. In 1977 the total population was estimated at 11,600 and more than 5,000 were shot, compared with barely 300 twenty years earlier.

In the far north several hundred fallow are established on the eastern half of the Baltic island of Rügen and on the Darss-Zingst Peninsula to the west in Rostock. To the south-east a few occur on the island of Usedom close to the Polish border and between Greifswald, Anklam and Uechkermünde.

The largest area of more or less contiguous distribution lies to the south of Rügen and is bounded by Demmin, Neubrandenburg, Prenzlau, Eberswalde, Oranienburg, Neuruppin, Malchow Lübe and Güstrow: fallow were first introduced to this region in about 1755, and there are now about 50 separate populations, a quarter of which consist of several hundred animals. In the province of Neubrandenburg alone the population in 1977 was judged to number 3,800. Herds of fallow also occur in the north around Schönberg and between Bützow and the River Recknitz south of Rostock. Most of this area consists of Scots pine (*Pinus sylvestris*) and beech (*Fagus sylvatica*) woodlands bordering agricultural land.

In the Hagenow/Ludwigslust area to the west nearly 1,000 fallow are established close to the West German border, while to the south a small number are found in the vicinity of Osterburg and in the Altmark region west of Stendal, where the species was first introduced in 1713. The eastern portion of the province of Magdeburg near the Potsdam border west of Berlin supports an estimated population of over 2,000 fallow.

The second largest area of nearly continuous distribution is in the centre of the country south and west of Berlin, and extends from Potsdam to Fürstenwalde, Lübben, Dessau, Zerbst and Brandenburg, and north-west as far as Havelberg. In 1977 the province of Potsdam was estimated to hold more than 2,250 fallow. To the south-west fallow occur in the principally coniferous forests between Halberstadt and Oschersleben.

In southern East Germany smaller populations of fallow are found around Sondershausen near Erfurt, north of Görlitz on the Polish border, to the south of Forst, between the rivers Nysa and Spree and around Cottus. Between 1960 and 1970 fallow began to colonize areas near Jena, Werdau, Grimma, Oschatz and Torgau, where they have subsequently been reinforced by the introduction of new blood.

Germany (West)

As in East Germany fallow deer are widespread in the Federal Republic – mainly in the north and west. Although they have been held in parks since at least the sixteenth century, the present wild population is derived from comparatively recent escapes augmented by some deliberate introductions. In 1967 the total population was considered to be in excess 20,000 – the majority in Schleswig-Holstein, Lower Saxony and Westphalia;[45] in the hunting season of 1976–7 over half that number were shot, of which 84 per cent were killed in those three states.

In Westphalia[46] fallow were first reported in the wild in 1883 in oak woods near Kottenforst, having escaped from a deer park at Kommende-Muffendorf. In the 1930s some fallow escaped from captivity near the spa of Bad Dridburg and established themselves around Höxter west of the River Weser; eight years later a number escaped from an enclosure (abandoned after the Second World War) about 40 km north-west of Höxter near Detmold on the fringe of the Teutoburg Forest; in both these districts their descendants still survive. Fallow presently established in the vicinity of Dülmen near the River Alme north-east of Büren originated from some deliberately released from a neighbouring park in about 1890. Fallow also occur in the Sauerland Hills near Wittgenstein in southeastern Westphalia, and in the lowlands of the north-west in the Münster region.

Although the afforested area of Schleswig-Holstein is meagre compared with much of the country, it was the first state in Germany in which fallow deer (from Denmark) were released, in about 1800. Today they are firmly established in several localities, especially in the north and east around Flensburg (near the Danish border), Eckernförde, Rendsburg and Oldenburg, and in the south in Bad Segeberg, Steinberg and Stormarn.[47] The descendants of 10 fallow released near Schleswig in 1939 have become established discontinuously north to Satrup, south to Dannewerk, east to Kappeln and west to Wittbek, and are continuing to expand their range; by 1969 the population had increased to about 670. South-east of Schleswig fallow are established in beech and mixed woodland, pastures and arable fields near Salzau between Kiel and the Selenter See, and they have occurred to the south near Plön since well before 1885–6, when a census revealed the presence of more than 800.

In Lower Saxony also fallow have been increasing their population and distribution since the end of the Second World War; herds are now widely scattered from around Goslar in the foothills of the Harz Mountains through the Deister Hills 40 km south-west of Hanover[48] in the south-east to the plains of the north-west, where they are especially common west of the River Elbe, in East Friesland and around Oldenburg. South of Hamburg and between the rivers Elbe and Aller fallow are also established on Lüneberg Heath, especially in a nature reserve of 20,000 ha of heather and juniper heathland interspersed with oak, birch and coniferous woodlands, between Undelok and Wilseder.[49] Further east fallow are found in several places south of Dömitz near the Elbe on the East German border. The estimate of 7,000 fallow in Lower Saxony in 1967 may have been overly conservative.

In Hesse the majority of fallow occur in the south, especially around Erbach near the boundary of Baden-Württemberg, around Darmstadt south of Frankfurt-am-Main and in the Taunus Hills near Wiesbaden west of Frankfurt. Smaller numbers are found near Ziegenhain.

In Baden-Württemberg fallow are also established principally in the south, around Donaueschingen and Stockach on either side of the Danube south towards Lake Constance on the Swiss border; a few also occur in the north-east near Crailsheim.

The stronghold of fallow in Bavaria is between Dillingen and Donauwörth on both sides of the Danube and about 40 km to the east at Stammham near Ingolstadt; smaller numbers also occur in the north near Schweinfurt.

In the Rhineland-Palatinate fallow are widespread but nowhere abundant; they are found principally near Koblenz west of the Rhine, between Saarburg and the Luxembourg border, around Birkenfeld and near Germersheim on the Rhine.

Greece

T. de Heldreich, in *La faune de Grèce* (1878), referred to the presence of a small number of fallow deer on the Greek mainland in the mountainous Manina Forest north-west of Lake Trikhonis between the River Akheloos and Catouna in Akarnania, where he doubted that they would long survive: in this belief he was correct, and there have been no wild fallow on the mainland of Greece for many years.

On the Greek island of Rhodes in the Aegean Sea off the south-west coast of Turkey, fallow have existed for well over 600 years. T. R. B. Dicks[50] records that they were imported (traditionally from Asia Minor to eradicate snakes) by the Knights Hospitallers of St John of Jerusalem who conquered the island in 1309, and protected a part of it as a hunting preserve. The deer were exterminated during the period of Turkish domination (1522–1912) but were subsequently reintroduced by the Italians during the early years of their rule (1912–45). The present population (estimated in 1973 to number between 300 and 400) lives mainly on the forested slopes of the Profita Hills, where the main tree species are pines (*Pinus brutia* and *P. halepensis*), cypresses (*Cupressus sempervirens* and *C. horizontalis*), oaks (*Quercus aigelops, Q. coccifera* and *Q. infectoria*) and junipers (*Juniperus macrocarpa* and *J. phoenica*). Since the nineteenth century, when bronze statues of a buck and a doe were erected on either side of Mandraki harbour, fallow deer have been regarded as the official emblem of the island of Rhodes.

Fallow deer were present on the island of Cyprus from about 7,000 years ago until they were hunted to extinction in medieval times. In an attempt to re-establish them on the island the Cypriot forest authority acquired a pair from the Bonn Zoo in West Germany in 1980 followed by a further three from the Bern Zoo, Switzerland in 1982.

Hungary

Near Kecskemét and Szolnok east of Budapest some 600 fallow deer are established in agricultural land and woodland (mainly acacia (*Robinia pseudocacia*), with some oaks (*Quercus* spp.), poplars (*Populus* spp.) and pines (*Pinus* spp.) and a wealth of privet (*Ligustrum vulgare*), hawthorn (*Crataegus* sp.) and ground elder (*Aegopodium* sp.) for browse and cover) which form the 15,000-ha Pusztavacs hunting preserve; they were first introduced to this region in the late nineteenth century and now more than 200 are culled annually.

To the east fallow occur in the 8,000 ha of low-lying flat and fertile woodland near Gyula between the River Körös and the Romanian border. On the grassy prairies (*puszta*) of the Great Plains to the north fallow live around Taktakenez and Taktaharkany on either side of the River Tisza.

Between Budapest and Komáron west of the Danube fallow are established in ancient woodlands near Gyarmapuszta and Telki on the Veŕtes plateau. In the south-west about 300 fallow live in La'bod near the Yugoslavian border.

Today, Hungarian fallow, which are believed to total about 4,500 head, are renowned among sportsmen the world over for the high quality of their trophy antlers,[51] the best of which come from the 7,000-ha forest and hunting reserve (mainly Turkey oak (*Quercus cerris*) and hornbeam (*Carpinus betulus*)) of Gyulaj near Tamási, some 50 km south of Lake Balaton.

Italy

Most of the localities holding fallow deer in Italy (where the total population in 1973 was estimated to be no more than 1,000) lie between Bologna and Rome. North of the latter, near Terni in Umbria, fallow are established in some private hunting preserves and in state-owned forests on the slopes of Monte Peglia. West of Umbria fallow inhabit some of the mixed hardwood and softwood plantations of the Foreste Demaniah Camaldoni, 80 km from Arezzo in the extensive plains of the Maremma in Tuscany and Latium. Near Pisa fallow are found in the Migliarino hunting reserve; to the south in the state-owned forest of Feniglia in Grosseto, and in the hunting preserve of Sant' Agostino in Viterbo. They also occur in state-owned forests near Pistoia, Lucca and Florence on the slopes of the Appenine Hills, in woodlands near Forli, and on Mount Catria, 80 km east of Ancona on the Adriatic.

In the north-east some of the fallow released in northwestern Yugoslavia in 1971 crossed the border into Italy, and are now established in several districts within 25 km of Trieste, where their numbers are limited by heavy poaching.[52] In the far south a few also live in the dense state-owned oak, holly (*Ilex* sp.), laurel (*Prunus* sp.) and myrtle (*Myrica* sp.) forests of Sila on tableland near Catanzaro in Calabria.

In *I quadrupedi di Sardegna* (1774), F. Cetti referred to the presence of fallow deer on the island of Sardinia; they were said to be especially abundant on the Plain of Sindia and were widely distributed elsewhere, at least 3,000 being killed annually. As recently as the mid-1940s between 150 and 300 still remained, and in the 1950s 30 to 50 were regularly seen in the Castiadas Hills east of Cagliari on the south coast. The removal of legal protection resulted in their extinction on the island before 1976.[53] A small number of fallow imported from Tuscany and Yugoslavia, are currently enclosed at Marganai near Iglesias and near Nuoro, could provide the nucleus of a future wild population. Exactly when fallow became extinct on the island of Sicily is uncertain.

Netherlands[54]

Fallow deer may already have been established in parks in the Netherlands when, in the early seventeenth century, Prince Maurice of Nassau imported 100 from England which he liberated on sand-dunes near The Hague; because of the damage they caused to the vegetation they were all killed off 20 years later. In 1647 William II released some more at Hof te Dieren near Rheden which, by the end of the following century, had also all been shot.

A hundred years later, in about 1880, William III enclosed some fallow in the royal forest at Het Loo, where they were released in 1912. This forest, which is now fenced, is divided by the Apeldoorn/Amersfoort road; the larger section of 6,000 ha supports about 200 fallow (probably the largest woodland population in the country) but the smaller 3,000-ha portion contains only about 20. By around 1915 fallow from Het Loo had colonized a forest a few kilometres to the northwest near Elspeet. In 1973 around 50 fallow lived in the 300-ha enclosed Noorderheide portion of the forest and a further 80 in the 1,200-ha Elspeterbos/Elspeterstruiken sections.

In 1915 fallow were turned down in the 2,100 ha of the Deelerwoud north of Arnhem, where by the late 1930s the population had increased to around 600 but has since declined to only 200. In 1945 some deer managed to escape into the 5,000 ha of the Hoge Veluwe Forest, where the last one was shot about nine years later.

To the west, north of Ede, fallow have lived since before the outbreak of the Second World War in the De Valouwe Forest, where around two dozen are at present established in a 500-ha enclosure. Nearby, fallow which escaped from parks in 1963 and 1968 have settled in the vicinity of Leersum. The inland forest populations of fallow in the Netherlands live at altitudes of between 100 and 200 m.

A number of free-ranging herds of fallow occur on the sand-dunes of the west coast: in Zeeland some have been established since an escape from a neighbouring enclosure in 1968; on the island of Walcheren a small herd roams over 350 ha of marram grass (*Ammophila arenaria*) dunes and 50 ha of conifer plantations near Oranjezon between Oostkapelle and Domburg; in the late 1960s some escaped fallow became established near Haamstede on the island of Schouwen. North of Leiden fallow (said to number about 20 in 1974) have been living in the 1,200 ha of Kennemerduinen since some escaped from a nearby park in the late 1950s; in 1972 a few spread onto coastal dunes to the south-west, which lie up to 20 m below sea-level.

Norway

The exact date when fallow deer were first introduced to Norway is uncertain. As early as 1305–7 some hides were shipped from Norway to England, but Chapman and Chapman (1980) suggest that these may have been in transit from Denmark. In 1645 fallow are known to have been enclosed in a deer park on the island of Kalvö in Högsfjorden, on the west coast due east of Stavanger, from where in 1674 they were moved east to Larvik on the Skagerrak. Others must at some time have been landed on the island of Malmö at the eastern entrance to Larvik Fjord, since it is recorded that in 1705–6 wolves (*Canis lupus*) killed off the last of the population. In *The natural history of Norway* (1755), E. Pontoppidan refers to predation by wolves on 'the common deer' by which, Chapman and Chapman (1980) consider, he probably meant fallow; he goes on to say that in former times these deer were more abundant in the vicinity of Trondheim and Bergen, but that recently many had been devoured by wolves, although a few remained on the mainland and on some offshore islands.

Fallow were released in Norway on a number of occasions in the early 1900s: three from Denmark – presumably a buck and two does – were landed on Rauø Island in Oslo Fjord where, although they had multiplied to 50 by 1911, they subsequently died out; similarly some translocated from the island of Hankö to Hurdalen in 1903 soon disappeared. Those landed on Skorpø Island north of Stavanger in the following year swam to the neighbouring larger island of Strønen, from where five years later a buck crossed to the nearby island of Røttingen;[55] a few years later the herd on Strønen ceased to exist.

The only extant wild population of fallow deer in Norway – descended from seven imported from Denmark in 1901–2 – lives on the well-wooded 6.75-km^2 island of Hankö at the eastern entrance to Oslo Fjord. By 1936 the deer had increased to some 300 but were greatly reduced during the Second World War; in 1968 a herd of about 100 remained, which four years later had declined – presumably through shooting – to around 40. This area has perhaps the mildest climate in Norway; much of the remainder of the country lies under deep snow for lengthy periods in winter, which fallow apparently cannot tolerate.

Poland

Fallow deer, which were probably first introduced to Poland during the seventeenth century, are now widely distributed, especially in the west, and, as the result of more recent introductions, are currently also becoming established in some central and eastern regions.[56] In 1972 the total population was estimated to be about 2,600 spread over at least 30 of the country's 49 *voivodeships* (provinces).

In the east fallow occur on the East Polish Plateau, or Lublin tableland, in the upland region of Łysogóry in the Kielce/Radom/Tomaszów Mazowieki triangle of the Little Polish Plateau, and to the north of Lódź. Further north they are found in many districts between Bydgoszcz in the east and Olsztyn north of Warsaw in the west, but are absent from the extreme north-east (where the climate is more severe than in the rest of the country) in the vicinity of Bialystok and the Mazurian lakeland. East of Warsaw fallow are established between Wegrów and Minsk Mazowiecki and eastwards past Siedlce. In the west they are scattered in woodlands from near Poznań almost as far north as Szczecinek and south towards Wrocław. South-west of Poznań fallow occur in several localities beyond the River Odra, especially between Zielona Góra and Wegliniec near the border with East Germany. They are also found south-east of Wrocław in woodlands to the east and south of Opole eastwards to beyond Katowice (near the Czechoslovakian border) around Pszyzyna. In the *voivodeship* of Szczecin in the north-west fallow are known to occur only between Chojna and the River Odra on the East German border.

Most of the habitat occupied by fallow deer in Poland is mixed woodland on the edge of pastures and arable land below 200 m. Of this area some 35 per cent is pine / oak association in which Scots pine (*Pinus sylvestris*), pedunculate or common oak (*Quercus robur*) and sessile oak (*Q. petraea*) predominate, with lesser numbers of silver birch (*Betula verrucosa*), beech (*Fagus sylvatica*), alder buckthorn (*Frangula alnus*), Norway spruce (*Picea abies*) and *Padus* sp. An additional 29 per cent consists of pine/bilberry woodland dominated by Scots pine and bilberry (*Vaccinium myrtillus*), with some birch, larch (*Larix* sp.), beech, juniper, rowan (*Sorbus aucuparia*) and hornbeam (*Carpinus* sp.). A further 26 per cent comprises oak / hornbeam forest with some stands of pine, beech, birch, small-leaved lime (*Tilia cordata*), aspen (*Populus tremula*) and hazel (*Corylus avellana*). Four per cent of the territory occupied by follow is wetland forest, in which alder (*Alnus glutinosa*) predominates, the remaining 6 per cent being agricultural land.[57]

Portugal

If fallow deer are established in the wild in Portugal, which is doubtful, they occur only in a few localities and in very small numbers. Interlopers from across the border in Spain are sighted from time to time to the north and south of Barrancos, further north near Campo Maior and Castelo de Vide, and towards the Sierra de las Mesas north of the River Tagus.

Romania[58]

In Romania fallow deer are widely distributed, especially on the plateaux and in the peripheral plains (which together comprise two-thirds of the country) and in the Carpathian foothills. In 1975 the total population was in the region of 3,800 – an increase of about 1,100 on that of 1968.

The largest concentrations (believed to total about 600) are found on the Tisza Plains in the provinces of Crisana and Maramures in the north-west, where the forest steppe vegetation consists largely of open grassland and oak woods.

On the Moldavian Plateau in the north-east fallow occur near Suceava, Iasi and Bacau and, further south, around Galati. A population of around 200 is believed to be established in the tableland and oak woods of the northerly portion of the Dobrogea Plain in the south-east between the Danube and the Black Sea.

Fallow occur in several districts in the Great Wallachian or Romanian Plain in the south of the country. A population of some 300 was estimated by Cotta[59] to be present to the north of Bucharest with a small colony further north in the vicinity of Polesti: about 200 were reported in the Arges region in the centre of the plain, while the Oltenia district in the west supported a population of nearly 500. Although the northern part of the plain is covered by woodland – in which oak predominates – much of the remainder is arable farmland.

In central Romania fallow live in oak and beech forests (with some cleared areas) around Cluj and Tirgu Mures, and further south towards Brasov near the River Otbul on the tableland, hills and plains of the Transylvanian Plateau. They are also established in beech forests at Hunedoara in the foothills of the southern Carpathians.

Spain

Chapman and Buck in 1893[60] reported the presence of wild fallow deer in Spain only in the vicinity of some royal estates outside Madrid, and Cabrera (1914) found them solely in the western province of Cáceres. Today, they are still neither abundant nor widespread, but appear to be increasing and expanding their range.[61]

In the province of Oviedo in the north-west a herd is established at Sueve, while another occurs in the north-east at Quinto Real in Navarra. In central Spain fallow are found in Serrania de Cuenca and in the south at Lugar Nuevo and Cazorla-Segura in Jaen.

In 1917[62] some fallow were released by the Duke of Tarifa in the Coto Doñana region between Seville and the Gulf of Cadiz in the south-western province of Huelva, where by 1924 they were being hunted by King Alfonso XIII.[63] The present population (estimated to number over 1,000) on the Biological Station Reserve is almost certainly the largest in Spain.[64] The reserve, which is less than 100 m above sea-level, consists of 45,000 ha of coastal sand-dunes, water-meadows, woodland and scrub, and the marshes and mudflats of the delta of the River Guadalquiver, of which the deer mainly frequent the marshes and meadowland and cork oak (*Quercus suber*) woods between the marshland and scrub. The marshes – Las Marismas – are subject to considerable seasonal variation. In winter 40,000 ha become submerged with only a few small islets and mudflats appearing above the surface. In spring, as the waters recede, the area becomes densely carpeted with sedges. Although the chain of lagoons in the woodlands and scrub beyond the marsh retain their water through the hot, dry summer, the rest of the area – apart from some pools and mudholes – becomes a desert of baked and cracked soil, which the onset of the autumnal rains turns into a morass. The meadows which border the marshland are flooded in winter but dry out in summer, and it is here, in the reedbeds, that the fallow deer give birth.

Attempts to naturalize fallow on land north of Coto Doñana failed because of poaching.[65]

Sweden

Fallow deer were first introduced to Sweden in the sixteenth century; the exact date is uncertain but those imported from England in 1579 by King John III were not the earliest. They now occur, mostly below 200 m, in the southern *län* (provinces) of Skåne, Halland, Småland, Västergötland, Dalsland, Östergotland, Södermanland, Närke, Västmanland and Uppland,[66] where in 1971 the total population was estimated to be between 4,500 and 5,000, of which some 1,100 are shot annually. Vagrants are occasionally reported from further north in Värmland and Dalarnia.

In the extreme south, in Skåne, fallow are widely distributed in a triangle formed by Svelala, Sjöbö and Lake Vomb (especially at Häckberga and Övedskloster), from north-west of Lake Vomb as far as Angelholm on the west coast (especially in the vicinity of Wrams-Gunnarstorp), and east of Lake Ring where they are particularly abundant at Maltesholm between Horby and the east coast. The habitat occupied by the deer is more or less the same in all these regions – largely mixed woodland near agricultural crops (e.g. rye , potatoes, rape, carrots, corn, beets and legumes). The forested areas are variable in composition, and include Scots pine (*Pinus sylvestris*) plantations (where *Deschampsia flexuosa, Carex arenaria* and *Rubus idaeus* form much of the ground cover); stands of Norway spruce (*Picea abies*); mixed spruce/beech woods; and deciduous trees such as beech (*Fagus sylvatica*), ash (*Fraxinus excelsior*), hazel (*Corylus avellana*), hawthorn (*Craetagus* sp.), birch (*Betula* sp.) and oaks (*Quercus robur* and *Q. borealis*).[67]

Elsewhere in southern Sweden small colonies of fallow are widely scattered. In Västergötland they occur mainly around Lake Mjörn north-east of Göteberg, south of Lake Vänern (and on the island of Lurö), especially in the vicinity of Gössater. There are two main concentrations of fallow in Östergötland – one to the north-east of Lake Åscunden and the other between Lake Vättern and Norrkopping north of Lake Roxen. In Närke fallow deer are established principally between the north end of Lake Vättern and Lake Hjälmaren. South-east of the latter they occur in Södermanland, and further east on the island of Lidingo in the Baltic Sea near Stockholm. In Uppland fallow are found around Vasteras, south of Sigtuna and near Vaxholm (all just north of Lake Malaren), and north of Uppsala.[68] At the southern end of the 1,400-km^2 island of Öland off the south-east coast of Sweden in the Baltic a herd of fallow (maintained at around 150 by culling between 60 and 70 annually) are held in enclosed oak/birch woodland in the 1,000-ha Ottenby Nature Reserve.

USSR

At present, fallow deer occur in the wild in the Soviet Union in deciduous woodlands and forest and grassland steppes in a small number of localities, mostly below 200 m in the comparatively temperate south-west, and probably also in Lithuania.

Fallow deer are known to have been living in the forests of Lithuania and Belo-Russia as early as the sixteenth century,[69] and by the beginning of the seventeenth century they were reported to be abundant on the Ukrainian steppes.[70] A hundred years later they were established in Czar Alexander I's hunting preserve in the Białowieża Forest (in which oak (*Quercus* sp.), ash (*Fraxinus excelsior*) and hornbeam (*Carpinus* sp.) are the dominant trees), where by 1857 the population was estimated to be about 7,000, with a further 7,000 red deer (*Cervus elaphus*), 5,000 roe deer (*Capreo-*

lus capreolus), and 1,900 European bison (*Bison bonasus*);[71] in spite of further introductions, by 1920–30 fallow in this forest had become extinct.[72]

In the seventeenth century park-bred fallow were planted in a number of forests in Lithuania which by the 1950s held half the Soviet Union's total complement of the species (barely 200), and where a remnant stock probably still remains. Fallow bred in Lithuania were also translocated to the Gatchina hunting preserve south of Leningrad and to the forest of Belo-Russia, where fallow from Germany were also released in 1932 in the vicinity of Voloshin/Ivenez/Molodetscho near Minsk;[73] by 1974, however, none remained.

A herd formerly established on the island of Khortiza in the River Dnieper near Zaporozhye in the Ukraine had similarly disappeared by 1974, although further south the population on the Biryuchiy Peninsula in the Sea of Azov mustered about a hundred. The 867-ha Askania-Nova Reserve on the wide plains between the rivers Dnieper and Molochnaya – where fallow have been maintained since 1918 – holds a stock in an extensive enclosure on the steppe pastures.[74] North-west of Khakov, near the Ukrainian border on the southern edge of the central Russian uplands, lie the 20,000 ha of mixed (mainly deciduous) woodlands which form the Donbass forests of Tepensky, Majaksky, Nickolsky and Khakov, where fallow were first reported in about 1960: whether any remain there now is uncertain, as is the outcome of a proposal in 1966 to liberate fallow in the deciduous woodlands of the Carpathian foothills in the Ukraine.

A herd of about 40 fallow (in 1974) in woods near Kishinev in Russian Moldavia near the Romanian border may be descended from an introduction in 1961–2.[75]

In 1888, on the order of Czar Alexander III, 54 fallow were turned out on mountain slopes of the Borhomi hunting preserve in Georgia, 150 km west of Tbilissi in the southern Caucasian foothills. In the following two years more were released, and by 1918 a population of several hundred had become established in a 500-ha enclosure in the reserve where by 1920, however, following the revolution, the entire stock was slaughtered by the local peasantry.[76] Subsequent attempts to naturalize fallow in the Caucasus have all been unsuccessful.[77]

Yugoslavia

Although most of the relatively few wild fallow deer at present in Yugoslavia occur in the mainly beech and oak woods on the floodplains of the larger rivers in the north and east of the country (mostly below 200 m and seldom above 500 m), successful experimental introductions to hunting areas in Montenegro in the central-west could eventually lead to a wider distribution.[78]

About 80 km north-west of Zagreb in Slovenia in northwestern Yugoslavia, not far from the Austrian border, fallow are established on the gentle and lightly wooded slopes of the Pohorje Hills (the most westerly extension of the Alps) between Kozina and Sezane. A second population occurs in the Velenje and Posavje hunting preserves which straddle the River Sava between Trbovje and Litija east of Ljubljana, and a third in the comparatively flat Slavonia region of Croatia. Another herd (of about 30) lives around Bosanski Novi and Dubica south-east of Zagreb near the River Una, with more north-west of Belgrade in the Fruska Gora Hills which lie, mostly below 500 m, between the Danube and the Sava from Novi Sad and Sremska Mitrovica westwards to Kunjecvia in the region of Vinkovci. In 1958 some fallow were released near Lipovica in Serbia, about 20 km south-west of the capital, where there is now a herd of around 70. In 1965 10 fallow were introduced to the Majdan-Kucajna Reserve some 160 km south-east of Belgrade, where there is currently a small enclosed herd: in the same year a like number were liberated in the fenced Boranja Reserve near the River Drina south of Loznica. North of Gornji Milanovac and between Majdan-Kucajna and Boranja some fallow were introduced in 1961 to an enclosed reserve at Rudnik, where within four years they had multiplied to about 70.[79] In 1971 others were released near the Italian border, which they subsequently crossed to become established around Trieste.

Fallow deer also occur on the offshore islands of Cres (in the Gulf of Kvarner) and Brioni (only 6 km long, off Pula between the gulfs of Kvarner and Venice); fallow from the latter island, which supports a population of some 500, have been translocated to most places on the mainland where the species now occurs.

AFRICA

Libya; Tunisia; Algeria; Egypt

In 1906 J. G Millais reported that until late last century fallow deer were established in the wild in parts of Libya and Tunisia, and in coastal woods around La Calle near Tripoli on the Algerian/Tunisian border.[80] The origin of these deer is uncertain; L. Joleand,[81] who records their introduction to Egypt in the sixteenth century, has suggested that they may have been imported by early Phoenician or Roman traders.

Madagascar

In 1932 some fallow deer from Czechoslovakia were released on a high wooded plateau in the Massif de l'Ankaratra near the forestry station of Mandjakatompo, 60 km south of Tananarive, where they subsequently suffered from severe poaching. Since then much of the plateau has been deforested and has become pseudo-steppe with *Aristida rufescens* as the dominant grass. Thirty years after their introduction, J. Vincent[82] reported that the deer appeared to have moved into the dense rainforest to the east; since then this region too has suffered from deforestation, and, despite Whitehead's (1972a) claim that 'Fallow deer have been satisfactorily introduced', they are now believed to be extinct on the island.

South Africa

In South Africa fallow deer are found in the wild in Cape Province, the Orange Free State and the Transvaal; in Natal there are only two small enclosed herds – one in highland sourveldt in the south-west and the other on the border of valley bushveldt and coastal hinterland.

J. N. Thompson[83] has recorded that as early as 1869 fallow deer were being kept in the park of Newlands House, Cape Town. When Newlands were sold by the government the herd of 100 deer were acquired for £200 by Mr Duncan Baxter of Lourensford, Somerset West. In 1897 Cecil Rhodes introduced some fallow to his estate at Groote Schuur on the slopes of Table Mountain, from where some were subsequently translocated to Lourensford. Large numbers of fallow deer have since been, and are still being, distributed as breeding stock to many estates throughout South Africa.

Most fallow deer in South Africa occur in Cape Province, where by 1970 they were widely scattered in at least 32 of the 113 districts, from Cape Town in the south-west to Mafeking in the north-east. Table 9 gives particulars of their numbers and distribution. In 1963 three fallow deer were transported from Groote Schuur to Robben Island (site of a notorious prison camp) 8 km off the coast in Table Bay and some 10 km north of Cape Town; by 1977 the population had multiplied to around 40.

About half the localities in which fallow deer occur are in the Karoo, which is largely semi-desert with a low and irregular rainfall, where the vegetation is mainly composed of open, stunted xerophytic shrubs, with patches of woodland along watercourses. Most of the other localities are situated either in temperate grasslands or in the southwestern Cape, which has damp, mild winters and hot, dry summers which support a fynbos vegetation of sclerophyllous evergreen shrubs.

Probably the largest herd of fallow deer in Cape Province (estimated in 1974 to number about 350) occurs in oak and pine plantations and open grassland on the lower slopes of the Devil's Peak at Groote Schuur. Another large population of around 200 in the Middleburg district is descended from nine bucks and three does released on the Culmstock Estate in 1908; this population occupies an area of shrubby vegetation on brackish soil (known as brakveld), low-lying ground which is flooded in the rainy season (vleis), and dense riverine scrub.

In the Orange Free State fallow deer are known to occur in at least two (Harrismith and Heilbron) of the 49 districts, but are in fact probably distributed far more widely. The earliest introduction seems to have been that made by Sir Percy FitzPatrick to Buckland Downs, Harrismith, in the early years of the present century; here they flourished, and had increased to at least 183 by 1972.[84] Three fallow (presumably a buck and two does) were also released before the First World War by the Vereeniging estates 60 km

Table 9 Distribution of fallow deer in South Africa

District	*Minimum population*
Albany	4
Albert (Burgersdorp)	45
Bedford	124
Bredasdorp	91
Britstown	1
Caledon	7
Calvinia	1
Cape Town	2
Colesburg	55
Cradock	4
Fort Beaufort	3
Hanover	1
Hay	1
Hopetown	15
Kimberley	2
Laingsburg	?
Middleburg	160
Mafeking	1
Molteno	24
Paarl	56
Philipstown	4
Piketberg	1
Queenstown	1
Somerset East	95
Stellenbosch	2
Sterkstroom	8
Steynsburg	2
Sutherland	150
Tarkastad	1
Vryburg	5
Wodehouse	1
Worcester	2
Total	869

Source: After Chapman and Chapman (1980) via R. C. Bigalke

south of Johannesburg on the Transvaal border, where by 1937 they had multiplied to about 50[85] and had increased still further to some 250 by 1978: from here fallow have been translocated to other parts of South Africa. In 1910 some were released on Mr Charles Newberry's Pyrnnsberg Estate between Gumtree and Clocolan in the Ficksburg district on the Lesotho border, and at least 40 were present in 1937;[86] their current status, however, is uncertain. Fallow deer have also been introduced to the Franklin Game Reserve in Blomfontein.[87] All these areas are situated in the largely treeless and undulating grassy plains of the highveld with heavy rainfall in summer and drought and severe frosts in winter.

In the early 1900s a Mr Fick despatched some springbok (*Antidorcas marsupialis*) to the Zoological Society of London, receiving in exchange for his Palmietfontein Estate near Volksrust in the Transvaal several fallow deer, which had increased to more than 350 by 1967, and from where between 80 and 100 are annually translocated to other estates in South Africa.[88] Fallow are also said to have occurred at Bryanston near Johannesburg until at least 1973.

Although some debarking of fruit-trees and vines by fallow deer has been reported, their ecological impact in South Africa appears to be minimal. In recent years interest has been evinced in breeding fallow in captivity for their venison.

NORTH AMERICA

USA

Although the earliest recorded importation of fallow deer (from England) to the United States took place in 1878,[89] most releases have been made during the present century with home-bred rather than European stock. According to Presnall (1958), fallow have been more widely liberated in the United States than any other alien ungulate. Although in some places they have failed to become established, they now occur in the wild on the mainland in at least eight states, from California in the west to Georgia in the east and from Texas in the south to Nebraska in the north, as well as on one east-coast offshore island. Proposals for further releases in some states (the subject of much controversy) are still being considered.

In 1925 a dozen fallow from a captive herd at Prattville were freed (but failed to survive) in the Black Warrior National Park in Wilson and Laurence counties, Alabama.[90] At about the same time a fenced herd was formed at Miller's Ferry in Wilcox County, from which in the early 1930s about 20 managed to escape; by the 1960s they had increased more than tenfold and had spread up to 32 km from their point of escape – a rate of dispersal of about 1 km per annum. During the last 20 years there seems to have been little further expansion of range in Wilcox and Dallas counties, where the deer live on the alluvial river plains of the Alabama River – an area of dark lime or basic soil formerly submerged by salt water. Much of the land is used for agriculture (especially for grazing and the cultivation of such crops as soya beans, cotton and cereals) and silviculture; the dominant native trees include shingle oak (*Quercus imbricaria*), overcup oak (*Q. lyrata*), other oak species, nutmeg hickory (*Carya myristicaeformis*) and red cedar (*Juniperus virginiana*).

Some of the 30 white fallow deer which escaped in Mobile County in 1946 may have survived until at least 1965; others which escaped at Chapman in Butler County soon disappeared.[91]

The largest population of fallow deer in California lives on the Point Reyes Peninsula in Marin County, some 50 km north-west of San Francisco. Three-quarters of the deer released, which came from the San Francisco Zoo, were the white variety; one account states that 12 were introduced as early as 1930. In 1942 15 deer, all of which were white, were freed on the Ottinger Ranch near Drake's Bay, followed by a further 11 five years later and two more in 1954; shooting began in 1956. Although some normally pelaged, black and menil individuals are present, white animals still predominate. A census carried out by the National Park Service in March 1977 revealed the presence of more than 500 fallow (as well as black-tailed or mule deer, *Odocoileus hemionus*), domestic cattle and 400 axis deer (*Cervus axis*),[92] ranging over 6,400 ha of woods, scrub and grassland on the peninsula. In the last of these, which is the animals' favourite habitat, *Bromus rigidus*, *B. mollis*, *Festuca dertonensis* and *Holcus lanatus* are the principal species grazed by the deer, which also browse Scotch broom (*Cytisus scoparius*), blackberry (*Rubus vitifolius*), *Ceanothus thyrsiflorus*, willow (*Salix* sp.) and red alder.[93]

In 1949 51 white fallow (one account says 20 as early as 1937) from the Rancho Piedra Blanca (Hearst Ranch) at San Simeon in San Luis Obispo County were turned out on the Ridgewood Ranch near Ukiah south of Willets, Mendocino County, in northwestern California, where they became firmly established within 11 km of their point of release, and by the mid-1960s had multiplied to between 250 and 350. In spite of both legal and illegal shooting, at least 130 were in 1970 judged to roam over 7,300 ha of oak woods and grassland radiating out from the rolling hills and ridges of the Walker Valley, a few kilometres north of Lake Mendicino – a habitat which they share with native mule deer. The dominant grasses are *B. mollis*, slender wild oat (*Avena barbata*) and dogtail (*Cynosurus echinatus*); the most abundant trees are garry oak and California black oak (*Quercus kelloggii*), with lesser numbers of valley oak (*Q. lobata*), interior live oak (*Q. wislizenii*) and blue oak (*Q. douglasii*); a dense understorey of madrone and manzanita (*Arctostaphylos manzanita*) provides cover for the deer. In the late 1970s a few fallow were released from the Hopland Field Station of the University of California, also in Mendocino County, where their present status is uncertain.

In 1934 two pairs of fallow were turned loose in the Mount Madonna County Park in Santa Clara County at the southern end of San Francisco Bay where, as the result of annual culling, there was a herd of only 20 in 1976. On the southwestern side of the bay a second herd (numbering about 50 in 1965 but now of unknown size) is established on the Pomponio Ranch in the adjacent county of San Mateo.

In about 1976 a herd of around 30 fallow escaped from a ranch 12 km south of Red Bluff in Tehama County, where they spread out along the Sacramento River between Tehama and the confluence of the Thomes Creek; in spite of poaching a few may have managed to survive.

In 1967 and again in 1970 some white fallow were released on Lone Pine Ranch in southwestern Trinity County, where in 1971 the herd numbered between 20 and 30. The 85 fallow which escaped from the Rancho Piedra Blanca in 1953 (where they had been maintained since the 1930s) were less successful; within a decade only a solitary individual remained. Similarly, of the 13 white fallow released in 1960 on the Butts Ranch near Call Mountain in San Benito County only two does remained 10 years later; a dozen freed near Yreka in Siskiyou County in the mid-1950s had all vanished a decade later.[94]

In 1918 20 fallow deer (and some native white-tailed deer (*Odocoileus virginianus*)) were released in an area known locally as the Land Between the Lakes between the Barkley and Kentucky lakes in western Kentucky, where the population in

1975 – believed to be the largest individual one in North America – was estimated at some 700–800, with an even greater number of white-tailed deer.[95] This area of 68,850 ha, which is currently being developed for educational, sporting and recreational purposes, consists largely of rolling meadowland, wooded uplands and steep hillsides which rise to an altitude of 170 m. The commonest tree species are oaks – shingle oak, cherrybark oak (*Quercus falcata* var. *pagodaefolia*), white oak (*Q. alba*), scarlet oak (*Q. coccinea*), post oak (*Q. stellata*), and northern red oak (*Q. rubra*) – flowering dogwood (*Cornus florida*), black walnut (*Juglans nigra*) and hickory (*Carya* sp.). The pastures contain an abundance of fescues such as *Festuca elatior*, other grasses, legumes such as white clover (*Trifolium repens*) and perennial herbs.

In the early 1970s a small number of fallow were introduced to the Fort Knox – and possibly also to the Fort Campbell – military reservations and in the Grayson Lake Wildlife Area in Carter County. In the two army reservations they are now believed to be extinct; their present status in the wildlife area is uncertain.

In 1939 20 white fallow (followed by a further 53 in 1940) from Pioneer Park, Lincoln, were released on the Ray Hall Ranch in the Beaver River Valley in Boone County, Nebraska, 16 km north-west of Petersburg – a region of extensive grass- and cornfields with a cover of willows, cottonwood (*Populus palmeri*) and other species.[96] By the end of the Second World War fallow had spread 3 km east and 25 km north-west of their point of release; although a decade later unconfirmed sightings were reported from the adjacent counties of Madison, Holt and Hall, their overall distribution had increased but little, remaining centred on Boone and Wheeler counties. Between 1957 and 1965 the population fluctuated between 105 and 157; between 1966 and 1973 culling – to avoid conflict with agricultural interests and competition with white-tailed and mule deer – had reduced the herd to between 50 and 60.

The earliest release of fallow deer in Texas took place on 20 November 1930 when Mr LeRoy G. Denman introduced some to his ranch on the Black Jack Peninsula near Austwell on the central Texas coast. Between that date and 31 August 1936 a total of 500–600 (obtained from dealers in Alabama, Wisconsin (Milwaukee) and California (the San Diego Zoo)) were released, of which by the end of 1938 less than two dozen remained.[97] In the mid- to late-1930s others were turned out on the King Ranch, which stretched eastwards from the Rio Grande to the Gulf of Mexico on the South Texas Plains, on Mr Richard Friedrich's Bear Creek Ranch in central Texas, on the Rickenbacker Ranch and elsewhere in the state.

In 1964 there were 220 fallow deer in Texas; two years later the total had doubled to 445 and by 1971 had risen to 2,617[98] and by 1974 to 4,483, only 60 of which, however, were free-ranging. By 1979, when the total had increased to 7,922, 121 ranches in 61 counties reported the presence of fallow deer on a total of 126,128 ha of deer-proof fenced rangeland. Free-ranging herds totalling 2,508 deer were established in 14 counties – especially in Blanco, Kerr, Menard, Real, Robertson and San Saba – mostly on the Edwards Plateau in central Texas, which in 1974 was estimated to hold over 61 per cent of the state's total population. Fears have been expressed that with free-ranging alien sika (*Cervus nippon*) and axis deer, fallow could provide unwelcome competition for native white-tailed deer.[99]

The hill country of the Edwards Plateau – 9.7 million ha in extent – rises to an altitude of over 900 m and supports a vegetation of grassland and open savannah. The former consists mainly of needle grass (*Aristida* sp.), bluestems (*Andropogon* spp.), grammas (*Bouteloua* sp.), Texas winter grass (*Stipa leucotrichia*), buffalo grass (*Buchloe dactyloides*) and curly mesquite (*Hilaria belangeni*). On the rocky outcrops grow live oak (*Quercus virginiana*), shinnery oak (*Q. breviloba*), juniper (*Juniperus ashei*) and mestuite (*Prosopis chilensis*). The principal land use is for the raising of beef cattle, sheep and angora goats, with some cultivation at lower altitudes and near settlements.

A herd of fallow deer in the Sacramento Mountains in east central New Mexico is descended from some which escaped from captivity; the size of the present population is unknown.

In addition to the mainland populations, in the early 1950s some fallow deer were landed on the temperate and lowlying island of Little St Simons off Brunswick in the state of Georgia, where by 1974 a population of between 500 and 600 ranged over 4,000 ha of live oak and pine woodland with an understorey of palmetto (*Sabal* sp.) and myrtle (*Myrica* sp.), and salt marshes. Although the deer mainly graze on the marshland, acorns are also an important constituent of their diet. Some fallow released in 1957 near Tocoa in the Lower Mountains on the mainland of Georgia were exterminated within a year by feral dogs and poachers.

In about 1932, and again in 1968–9, fallow deer were landed on the island of Martha's Vineyard 8 km off Cape Cod on the coast of Massachusetts.[100] By 1976 only a small number, of the black phase, were believed to remain (in the West Tisbury and Gay Head districts), and three years later Mr R. E. Woodruff informed Chapman and Chapman (1980) that fallow on the island were now extinct. Fallow were also landed in 1938 on Nantucket Island east of Martha's Vineyard,[101] but none have been known there for more than 30 years.

From time to time fallow have also been introduced elsewhere in the United States; a number released nead Blufftown, Indiana and in Louisiana (for hunting by archers) were illegally shot before they could become established. Similarly, fallow released on the Prodenville Estate, Michigan, in spite of Whitehead's (1972) claim that 'they have done well there', in fact did not long survive. Between the 1920s and 1940s fallow were released in Conejos, Larimer and Rio Blanco counties, Colorado, where by the late 1950s Presnall (1958) estimated there were around 75; this population is now apparently extinct. Between 1920 and 1945 fallow were released in Worcester and Talbot counties in Maryland, where Presnall thinks that a few may have survived until the late 1950s. At one time there was also a herd of fallow on Mills Island in Chincoteague Bay,[102] but by at least 1974 all three populations appear to have died out. Presnall refers to the presence of fallow deer in the vicinity of Camp Gruber in Muskogee County, Oklahoma, but these had disappeared before 1976. A solitary buck shot near Natchez Trace in Tennessee in 1968 was probably either a vagrant from Kentucky or had escaped from an enclosure near Jackson.

Canada

British Columbia is the only province in Canada to which fallow deer have been introduced. They were first released in about 1895 by Mr Irving H. Wheatcroft on James Island at the southern end of the Straits of Georgia, where they were soon so well established that periodic culling became necessary. In 1974 Mr J. H. Todd told Chapman and Chapman (1980) that he believed that fallow (from the Duke of Devonshire's Chatsworth Estate in Derbyshire, England) had been landed on James Island in 1907. In 1931, 1934 and 1935 some were translocated by the Game Commission from James to Saltspring and Pender islands and to the Alberni district on Vancouver Island. In about 1954 a number swam the 2 km from James to Sidney Island, and others have occassionally made the crossing to the Saanich Peninsula.[103] No fallow have been reported from Pender Island since 1930 or 1940 (accounts differ), and although a solitary buck is said to have been observed bear Alberni in 1952 there have been no subsequent sightings there.

Fallow at present occur in British Columbia only on James, Saltspring and Sidney islands, a few kilometres north of the United States border. Of these Saltspring, the largest with 18,000 ha of mountains, forest and lakes, supports the smallest population numbering only between 10 and 20. A herd of around 125 lives on the smallest, James Island, with an area of no more than 280 ha. The largest population (believed to number several hundred in November 1977) is established on the 713 ha of woods, pastures and rocky outcrops of Sidney Island, where the dominant trees include Douglas fir (*Pseudotsuga menziesii*), red alder (*Alnus rubra*), broadleaf maple (*Acer macrophyllum*), garry oak (*Quercus garryana*) and madrone (*Arbutus menziesii*), with ground cover of salal (*Gaultheria* sp.),ferns, stinging nettles (*Urtica dioica*), grasses and mosses. Fallow deer (and other introduced mammals, such as mule or black-tailed deer (*Odocoileus hemionus*),[104] and birds) have become naturalized on Sidney only since freshwater-holes were provided on the island. The climate of the straits is uniformly mild.

West Indies

Fallow deer are established in the wild on two of the Leeward Islands – Barbuda and Guana – in the Lesser Antilles.

It seems probable that fallow were first landed on Barbuda (situated 40 km north of Antigua with an area of 155 km^2) in the early years of the eighteenth century by the Codrington family of Dodington, Gloucestershire, England, to whom the island had been assigned in 1680, and by whom it was retained for over 200 years as a kind of feudal fief. Sir Hans Sloane mentions in his *Voyage to the islands of Madera, Barbadoes, Nieves, St Christopher's and Jamaica* (Vol. II, 1725) that fallow deer were sometimes imported to Jamaica; their place of origin is not stated but it may well have been Barbuda, since a letter written from Antigua in 1740[105] to Lady Codrington in England refers to the presence of 'about a thousand head of deer' on Barbuda, some of which were netted and shipped to Antigua where their venison fetched 2*s*.6*d*. a pound. In another letter, dated 18 November 1834[106] to Sir Christopher Codrington, mention is made of a sample skin despatched from Barbuda to England (and of 13 others sent in 1831), which, since they were riddled with shot, resulted in a net loss of 4*s*.7*d*. In 1857 Charles Darwin presented the skins of two normally coloured and one black fallow to the British Museum (Natural History)[107] with the information that the species was 'Domesticated and naturalized about 100–150 years ago in the Island of Barbuda'. Both H. N. Coleridge, in *Six months in the West Indies in 1825* (1826), and T. Southey, in *Chronological history of the West Indies* (Vol. III, 1827), mention the existence of deer on Barbuda, but neither names the species; Millais[108] refers to a 'recent' introduction of fallow to the West Indies, but unfortunately fails to name the island.

For many years it was believed that the only deer on Barbuda were white-tailed from North America.[109] In 1974 Mr J. L. Robinson reported the presence of fallow on the island to Chapman and Chapman (1980); four years later Mr David Watters confirmed that fallow deer of at least two colour forms – common and black – are indeed established on Barbuda. They live mainly in xerophytic, evergreen woodland and scrub, especially in the so-called Highlands which rise to 34 m above sea-level.[110] The dominant evergreen tree in the Highlands, which due to their inaccessibility are seldom visited by man, is the white cedar (*Chamaecyparis sphaeroidea*); other species include whitewood (*Bucida buceras*), satinwood (*Chloroxylon swietenia*) and a few deciduous trees. The scrub consists mainly of thorn-trees, acacia, hogwood (*Comocladia dodonaea*), cankerbury (*Solanum racemosum* var. *igneum*), black sage (*Lantana camara*) and spiny succulents.[111] Fallow also on occasion visit the coastal swamps and grassland.

The 4-km^2 low-lying island of Guana (known also as Guiana, Iguiana and Iguana), whose highest point is only 15 m above sea-level, lies southwest of Crabbs Peninsula barely 100 m off the north-west coast of Antigua. Although part of the coastline is fringed with mangrove swamps, most of the interior consists of xerophytic trees and scrub. As on Barbuda the Codrington family is believed to have introduced the original stock of fallow to Guana in the eighteenth century where, again as on Barbuda, they were for long believed to be white-tailed deer – even by the present owners of the island, the Fuller family; in 1978, however, Mr Morris Nedd, who had experience of the fallow on Barbuda, identified the deer on Guana also as fallow.

SOUTH AMERICA

Argentina

Fallow deer occur in two separate regions – about 1,000 km apart – in Argentina; one lies in the Pampas zone in the east in the province of Buenos Aires, while the other is situated to the southwest in the largely coniferous forests and colder, higher, wetter and more rugged country of the Lakes region in the foothills of the Andes.

In the former region, fallow are found in an area bounded by Bahia Blanca in the south, Mar del Plata in the east and Buenos Aires in the north, with occasional reports of sightings north of Rosario in the province of Santa Fé. Although fallow were first introduced to the Park Pereyra Iraolo shortly after the turn of the century, the present population is mostly descended from stock imported from Spain and Poland in 1905 by Señor Ernesto Tornquist, who enclosed them on his *estancia* in the Sierra de la Ventana north of Bahia Blanca. Some 25 years later his sons released the deer into the surrounding hills, which rise to little more than 1,200 m, where they have flourished ever since and where they now roam over 300,000 ha between Sierra de la Ventana Cura-Malal, Saavedra and Tres Picos. On 40 ranches (with a combined area of 195,000 ha) in the first two districts a survey in 1972 revealed the presence of some 5,600 fallow, together with 3,800 feral goats (*Capra hircus*),[112] 6,000 guanacos (*Lama guanicoe*), and some feral horses (*Equus caballus*).[113] Within this area about 200 fallow are established in a wildlife reserve. For much of the year the deer wander on the open undulating plains and in the foothills to an altitude of about 400 m above sea-level, where, although an abundance of grasses and herbs provides nutritious fodder, there are few trees or shrubs for shelter.

To the north and north-east fallow are widespread between Daireaux, Olavarria, Bolivar, Tapalque, General Alvear and 25 de Mayo eastwards to Tandil, Ayacucho, Loberia, General Pueyrredon and Mar del Plata on the coast. In 1906 fallow were turned out on Señor Pedro Luro's 50,000-ha Huetal *estancia* near the townships of 25 de Mayo and General Alvear, where within half a century the population was said to number in excess of 20,000. In about 1958 4,400 deer were shot and the remainder dispersed onto the surrounding flat and treeless cattle ranches, where at present between 10,000 and 15,000 fallow live in creeks and on the edges of grass-fringed lagoons.

South of the Sierra de la Ventana fallow are established in the region of Villarino and on the Verde Peninsula in Bahia Blanca bay. Two hundred kilometres further south a small population lives on Isla Gamba north of Bahia San Blas.

The largest number of fallow (estimated in 1977 at around 3,000) in western Argentina lives at an altitude of around 760 m on Isla Victoria in Lake Naheul Huapi in the Parque Nacional of the province of Neuquen, close to the Chilean border. The island is timbered with false beech (*Nothofagus* sp.), bamboo (*Bambusa* sp.) and conifers such as alerce (*Fitzroya patagonica*). A smaller population inhabits the Huemel *estancia* on the shores of the lake. Both colonies are descended from an introduction from Uruguay in the 1930s or 1940s by Señor Aaron de Anchorena (see below under Uruguay).

Since the late 1930s fallow have also been living 40 km east of San Carlos de Bariloche on the southern shore of Lake Nahuel Huapi, on the *estancia* of La Primavera at Lago Trafal near Confluencia on the Rio Negro border, where there were thought to be around 180 in 1977. Further north, at San Martin de los Andes at the eastern end of Lake Lacar, a herd of about 40 fallow live in the Parque Diana, and more occur 40 km further north at Junin de los Andes.

Chile

In 1887 some fallow from Europe (possible Spain) were released in the national park near Lota south of Concepción, from where they were subsequently translocated to a reserve at El Escuadron a few kilometres to the north near Coronel, where some managed to escape into the nearby Nahuelbuta Mountains. Those which remained were later shipped to the Isla Santa Maria in the Golfo de Arauco, where they flourished and after 15 years had increased to about 100. During a round-up a few years later many fell to their death from the cliffs. Some of the survivors were returned to the mainland on ranches in the hills near Collipulli, between Osorno and Concepción,[114] where none are now believed to remain.

In 1925 some fallow imported from the Hamburg Zoo were established in a fenced enclosure on a ranch between Temuco and Puerto Moutt in the province of Valparaiso (where around 30 still survive), and from where in later years deer were taken to stock other enclosures in Bió-Bió and Osorno.

Although most fallow in Chile are enclosed, there are at least two free-ranging colonies: one is established on the 2,500 ha of the Isla Altué-Huapi in Lake Ranco near Rio Bueno in the province of Valdivia (an area of high precipitation), some 150 km north-west of Lake Nahuel Huapi. Seventy-five kilometres due south in Osorno Province an island in Lake Rupanco in 1975 supported a total of 200 red deer (*Cervus elaphus*),[115] fallow and mouflon (*Ovis musimon*).[116] Both these islands are largely covered by forest, in which olivillo (*Aextoxicum punctatum*), ulmo (*Eucryphia cordifolia*) and *Laurelia philippiana* predominate.

Peru

In 1948 some fallow deer were introduced to the Huacraruco Ranch in the San Juan region of Cajamarca Province in the northern Peru, from where they were subsequently released in the nearby basin of the Chicamo River where they became firmly established. They appeared to weather successfully the 1963–4 outbreak of foot-and-mouth disease which seriously affected the population of native white-tailed deer (*Odocoileus virginianus*), but since then have declined as the result of illegal hunting to around two dozen in 1974; some are to be maintained as a tourist attraction on the adjacent Sunchabamba Game Reserve.

Uruguay

In the 1930s Señor Aaron de Anchorena (see above under Argentina) introduced axis (*Cervus axis*) and fallow deer to a 500-ha enclosure on his low-lying *estancia* (less than 130 m above sea-level) of La Barra de San Juan near the mouth of the San Juan River in Colonia, some 200 km west of Montevideo, from where some were taken to found successful high-altitude colonies at Lake Nahuel Huapi in Argentina. Small numbers of fallow (as well as a larger population of axis) are currently established in the wild in the densely wooded riverine grassland near their former enclosure. In 1978 Señor A. O. Martinez informed Chapman and Chapman (1980) that at least one fallow buck was consorting with a small herd of axis deer on the edge of scrubland on the Rio Negro north of Colonia in Soriano.

AUSTRALASIA

Australia[117]

In the second half of the nineteenth century large numbers of fallow deer were imported to Australia by private landowners and various acclimatization societies, both for aesthetic reasons and as a potential source of sport; after an initial period of increase following their introduction they have decreased in most areas, partially as a result of rural settlement. Although they have adapted well to a wide range of habitats and climatic conditions in Australia, fallow have not penetrated densely forested montane regions, preferring instead open lowland woodlands. Today, they are found in parts of New South Wales, Queensland, South Australia, Victoria and, especially, on Tasmania.

New South Wales probably has the largest population of fallow deer on the Australian mainland. In July 1862 Mr E. Cox transferred six from his brother's park in Tasmania to his property at Mulgoa, which by late 1864 had increased threefold. In 1886 a pair were released by the Lamb family of Rooty Hill on the eastern side of Lake George, north-east of Canberra, followed at a later date by others at nearby Bungendore. By the outbreak of the First World War they had multiplied to some 200 and had extended their range eastwards to Braidwood, north-west to Goulburn and south as far as Queanbeyan.[118] The population in the hills around Lake George, which is believed to number in excess of 1,000 and which ranges within a radius of 80 km from the point of release, is probably the largest and most stable on the mainland of Australia. A second, and smaller, herd is established near Goulburn.

At various dates up to 1924 (when half a dozen were released by Mr Colin Campbell in Ranger's Valley) fallow were turned down in the New England range north of Glen Innes in northeastern New South Wales, where they became established on both sides of the Severn River in an area bounded by Glen Innes, Emmaville and Deepwater.[119] In recent years new blood has been introduced and the deer are now culled annually. The habitat of both the Lake George and Glen Innes regions consists of open park-like grassland and woodland dotted with eucalyptus trees interspersed with yellow box (*Eucalyptus melliodora*), New England peppermint (*E. nova-anglica*) and other shrubs.

Fallow deer introduced to Bungowanah Park, Albury, in about 1884 by Sir Rupert Clarke, crossed the Murray River (which forms the state boundary in that area) into Victoria. Others released in about 1914 or in the 1920s in the Delegate/Bombala region in southeastern New South Wales became established in the Snowy Mountains, from where they too spread into Victoria. A small herd also occurs near Nowra, 130 km south-west of Sydney, and another (which is descended from seven released at Port Hacking in the 1880s and is of the melanistic form) exists in the Sutherland district south of the Royal National Park, some 60 km south of Sydney.

The largest concentration of fallow deer in Queensland is probably that established in tropical monsoonal woodland of the Atherton Plateau south-west of Cairns in the north of the state. Two populations also occur on the Darling

Downs in the south-east; here the descendants of three bucks and seven does sent from Tasmania to Bowen Park by Mr John Bisdee in 1865 and released at Maryvale in 1870[120] are established in the rugged mountainous country of the Upper Freestone Valley in the Maryvale State Forest, where the main tree species include narrow-leaved ironbark (*Eucalyptus crebra*), blue gum (*E. teretecornis*) and ribbon gum (*E. viminalis*). Ninety kilometres to the north-west fallow were released in 1890 on the Pikedale Station west of Stanhope; their descendants survive in the Pikedale/Jubbinbar Mount area. Fallow deer introduced in the 1870s at Westbrook near Toowoomba, and six turned out at Canning Downs near Warwick (both on the Darling Downs) in 1872, have since died out, the latter being killed because of alleged damage to crops.

Although fallow were released in at least eight localities in South Australia (mostly around Adelaide) loss of habitat to human settlement and agriculture have contributed to their present confinement to only four or five areas in the south-east of the state. (Whitehead's map (1972) erroneously shows them to be established only in the south-west.) Before 1895 fallow were released by Mr John Riddoch in Yallum Park in the south of the state, from where some were subsequently translocated to nearby Nangwarry and Penola; escapes from this area were the ancestors of the present population in scrub and pine plantations in the Mount Gambier/Millicent/Naracoorte district. Small numbers are also reported in native woodland (dominated by brown stringybark (*E. baxteri*)) in the region of Furner.

Around the turn of the century a number of fallow deer were translocated to Bungaree (about 170 km north-east of Adelaide), and by a Mr Gilbert to Angaston and Pewsey Vale, respectively 55 and 45 km north-east of Adelaide, from all of which some managed to escape into the surrounding countryside; all three populations are now extinct.

In about 1917 fallow deer were transported from Pewsey Vale to a 4-ha enclosure at Mr E. A. Brooks's Buckland Park station, Two Wells, near Port Gawler, from where some were later released on the adjacent maritime land, where they found abundant cover from eucalypts, banksias and acacias, and browsed and grazed on silver banksia (*Banksia marginata*), myrtle or scrub wattle (*Acacia myrtifolia*) and tussock grass (*Poa australis*). By the end of the Second World War the population had multiplied to around 1,500, a figure which (in spite of increased sporting shooting) was maintained for a further 20 years, since when commercial shooting has resulted in a considerable decline. This population is established in a uncharacteristic habitat – the estuary of the Gawler River, which is covered with samphire (*Arthrocnemum* spp.) and mangroves (*Avicennia marina*), backed by a zone of lignum (*Melothria micrantha*) which provides shelter for the deer. From Buckland Park (where a small herd survives) some fallow have been taken to Ashbourne, 45 km south of Adelaide, as the nucleus stock of a deer-ranching enterprise.

In the early 1900s some fallow deer were imported to the Rosebank Station near Mount Pleasant, from which a number escaped; their descendants survive in the extensive state-owned pine (pincipally Monterey pine (*Pinus radiata*)) plantations in the Mount Crawford/Springton area, some 45 km north-east of Adelaide.

In 1949 nine fallow deer were released in privately owned woodland in the Flinders Range north-east of Port Augusta; although they were said to have increased to about 100 by 1978, in the following year they were reported to have disappeared.

In Victoria fallow deer were first introduced for hunting to Mr Thomas Chirnside's Werribee Estate as early as the 1840s. By the 1870s, when the herd had increased to more than 700, deer drives were being held to reduce the rapidly growing population, none of which, however, remain there today. In the following decade Mr Paul de Castella obtained a buck and two does from Sir Richard Fry in Tasmania which he released near Lilydale and Healesville in the Yarra Valley, some 40 km north-east of Melbourne, where they multiplied prodigiously on the river flats and open wooded foothills of the Warburton ranges. Although protection was suspended immediately after the First World War, by 1922 there were said still to be thousands of fallow in the district. Today, only remnant populations survive around Marrondah Dam near Healesville, where the main woodland species are narrow-leaf peppermint (*Eucalyptus radiata*), messmate (*E. obliqua*), blackwood (*Acacia melanoxylon*), silver wattle (*E. dealbata*), hop bitter-pea (*Daviesia latifolia*), forest wiregrass (*Tetrarrhena juncea)* and bracken fern. Occasional reports are received from other places in the eucalyptus belt, especially from Kinglake where narrow-leaf peppermint, broad-leaf peppermint (*E. dives*), brown stringybark, red stringybark (*E. macrorrhyncha*) and shrubs and tussock grass are abundant, but where ground cover is generally sparse.

On 2 September 1851 the Melbourne *Argus* carried the following notice: 'Fallow Deer. It cannot be too generally known that three of these beautiful creatures (two bucks and a doe) are for sale at Kirk's Bazaar.' On 19 September the newspaper reported that the deer had been bought for breeding purposes and would be released on Phillip Island where 'the interesting experiment of the introduction of fallow deer into Victoria will be tested under very favourable conditions'. It seems probable that these deer were the nucleus of the herds started on the island in 1860 by Mr John McHaffie (where they caused considerable damage to agricultural crops) and subsequently between the Bass and Powlett rivers, all of which are now extinct.

In the late ninteenth century fallow were turned out by Captain W. H. Bacchus in the Blackwood ranges, some 25 km north of the township of Bacchus Marsh, where a few may still survive. Occasional sightings of fallow are made in the Grampian Hills in western Victoria, and small numbers may possibly exist in the Wimmera ranges to the north, where they were first introduced in 1862.

On the Victoria/New South Wales border fallow deer, which are reported from time to time in open woodland on the Corowa/Albury/Lake Hume stretch of the Murray River (where black box (*Eucalyptus largiflorens*) and river redgum (*E. camaldulensis*) predominate), are derived from those released by Sir Rupert Clarke near Albury in New South Wales in about 1884. Eastwards, where fallow were at one time established, 19 were released in late 1978 in the Railway Block of the Koetong plantation on the Koetong Plateau in the Tallangata Forest, where the principal tree species are Monterey pines and native hardwoods.

Fallow in the Snowy Mountains of eastern Victoria are descended from plantings just before or after the First World War in the Delegate/Bombala area of New South Wales. Until at least the mid-1970s occasional sightings of fallow deer made around Casterton on the Victoria/South Australia border were the result of a release in the latter state. On 1,600-ha Sunday Island in Corner Inlet north of Wilson's Promontory a few semi-domesticated fallow coexist with hog deer (*Cervus porcinus*).[121]

In 1912 52 fallow deer (from the herd kept at Concord, New South Wales, by Miss Edith Walker, and probably the largest single consignment ever made in Australia) were introduced to Port Essington on the Cobourg Peninsula in the Northern Territory, together with four barasingha or swamp deer (*Cervus duvauceli*) and three sambar (*C. unicolor*);[122] the deer were shipped on the SS *Mataram* from Sydney in the care of Mr Dudley Le Souef, director of the Melbourne Zoo. Today, only the sambar survive.

Queen Victoria presented some fallow (and red) deer to Western Australia, where they were kept in the Zoological Gardens in South Perth. In 1899 a pair of their offspring were released in the Cape Leeuwin area, followed by more in subsequent years, and others were turned down near Gingin north of Perth. In neither of these localities did fallow long survive.

The largest numbers of fallow deer in Australia today occur on the island of Tasmania. Although it is generally agreed that most of the early importations were made by private landowners before the formation of the Tasmania Acclimatization Society, opinions differ regarding the exact year of the first introduction. In 1826 a meeting was convened in Hobart to discuss the importation of game animals from England. Some authors suggest that fallow were introduced by Captains Edward Dumaresq and Anthony Fenn Kemp in the same year (1829) as axis deer (*Cervus axis*);[123] Allison and Coombes (1969) say that Mr James Cox of Campbell Town imported the first fallow deer to Tasmania from England in 1834 (followed by a single buck in 1845), whereas Bentley (1978) suggests that the earliest introduction may have been in 1836 by Mr John Bisdee of Hutton Park, Green Ponds, north of Hobart, who imported fallow from England for hunting in succeeding years. However that may be, by 1863 a number of plantings had been made and the Acclimatization Society of Victoria's Annual Report for that year claimed that 'between six and eight hundred deer are running wild in Tasmania'.

A century later the number of fallow – the only free-ranging deer in Tasmania – was causing concern to farmers and horticulturists, and by 1972 the population was said to be at least 7,000–8,000. Fallow deer in Tasmania occupy an area of around 400,000 ha in the east and central Midlands, where their range is divided by the north/south Midland Highway. To the east fallow are scattered within the area east of Launceston, Campbell Town and Ross, bounded approximately by Mathinna, Lake Leake and Tooms Lake. In this area they are not distributed continuously, being largely absent from Conara Junction eastwards along the South Esk Valley, and especially to the north and south of Avoca. West of the Midland Highway most of the deer lie within an area approximately bounded by Oatlands, Bothwell and Poatina; a discrete population occurs between the Macquarie River and the Highway around Epping, to the east of Poatina.[124]

Fallow deer in Tasmania occupy dry eucalyptus woodland with ground cover of bracken (*Pteridium aquilinum*) on the ecotone of sclerophyll forests – a range which they share, probably advantageously, with domestic cattle and sheep. Fallow deer shooting is of considerable economic importance in Tasmania.

New Zealand

After red deer (*Cervus elaphus*)[125] fallow deer are the most widely distributed deer in New Zealand; Table 10 gives details of the principal introductions, most of which were made for sporting purposes. Although most releases were successful – an abundance of food, an absence of predators, a favourable climate and protection from shooting until 1890 enabling fallow to colonize woodland, grassland and scrub below 600 m – their rate of dispersal (on average less than 1 km a year) has been relatively slow in comparison with that of other alien ungulates; this has resulted in a number of scattered populations located near the original point of release. By the 1920s fallow had achieved such high concentrations that in some places they were officially declared vermin, and shooting restrictions were lifted; the bucks were, and indeed still are, much prized by stalkers for their trophy antlers. In many areas numbers remained high – sometimes resulting in environmental damage – until the 1950s; since then the population has declined following the establishment of a lucrative export market for venison to West Germany and the Netherlands.

On North Island fallow are well established in several localities. On the southern headland of the west side of Kaipara harbour fallow mostly live in sand-dunes and scrub where the manuka or tea-tree (*Leptospermum scoparium*) predominates. To the south small herds occur on the South Head of Manakau harbour as far as the estuary of the Waikato River. Until at least 1977 fallow were present in the Puriri/Matatoke area at the southern end of the Coromandel Peninsula, where they are now, however, believed to be extinct. Further south fallow occur at Maungakawa (Te-Tapui) between Cambridge and Matamata, and to the east another herd is established south of Tu Puke, mid-way between Lake Rototua and the coast.

The largest population of fallow deer in New Zealand, which has a high proportion of black individuals, is found in the south-west corner of North Island north of Wanganui. Here the deer range over nearly 2,400 km^2 of land – rising in places to an altitude of almost 1,000 m – between the Whangaehu River in the east and the Waitotara in the west, divided between privately owned farmland (where the deer sometimes conflict with agricultural interests) and state forestry plantations, which also on occasion suffer depredation. The fallow favour for shelter dense manuka scrub, gorse (*Ulex* sp.) and kiekie (*Freycinetia banksii*) in the ecotone between farmland and state-owned native forests (the latter comprising mainly an association of hardwoods and conifers) in which the deer are subject to heavy culling by stalkers, with no close season. In this area fallow are also being experimentally farmed for venison on land considered unsuitable for domestic stock.

The only offshore islands on which fallow deer currently occur in New Zealand are Rangitoto and Kaikoura (and possibly on Kawau) in North Island's Hauraki Gulf.

Most of the widely scattered regions in South Island where fallow are found are at higher altitudes than are those in North Island. In the north-west Nelson State Forest Park a scattered population occurs in the Mount Arthur area, where its range is bounded by the Mount Arthur and Pikirunga ranges in the east, by the Leslie and Peel streams in the south and west, and by the Cobb and Takaka rivers in the west and north. Here, both fallow and red deer live principally in native forest where red beech (*Nothofagus fusca*) – in places mixed with silver beech (*N. menziesii*) –

Table 10 Principal introductions of fallow deer to New Zealand

Year	*Number*	*Locality*	*Introduced by*	*Source*
1862	Unknown; unsuccessful	Flaxbourne, Ward, Marlborough	Nelson AS	Tasmania
1864	1 male; 2 females	Aniseed Valley, Nelson	Nelson AS	Richmond Park, Surrey England
1867	? (unsuccessful)	Morven, Otago	Otago AS	?
1869	12	Ranch of Mr David McKellar of Brookside, near Tapanui, Blue Mountains	Otago AS	England
1870	1 (cream or white; ? albino)	Blue Mountains, Tapanui	Otago AS	via *City of Dunedin* (? from Australia)
1870	?	Kawau Island	Sir George Grey	? England
?1870	?	Motutapu Island, Bay of Plenty and Rangitoto Island	? Sir George Grey	? England
1870s	?	Southland	?	Tasmania
1871	4	Culverden Estate	Canterbury AS	Mr John Bisdee, Tasmania
1877	28* (survivors of shipment of 30 or 33)	Maungakawa and Matamata Ranges (18), and near the Wanganui River (10)	Auckland and Wellington AS (per Mr Falconer Larkworthy)	From London via *Thurland Castle* on 22 December 1876.† Arrived Auckland May 1877; 22 from Mr Jeremiah Colman of Carshalton Park, Surrey, the remainder from Richmond Park
*c.*1887	7	Doolan Creek, Paparoa Range, Westland	?	Motatapu Island
1899	4 (unsuccessful)	Cairn Bush	Otago AS	?
1900	2 (unsuccessful; shot out when island declared a sanctuary)	Kapiti Island	Wellington AS	?
1906	3 males; 2 females	Teal Valley, Nelson	Nelson AS	Otago (Greenstone herd)
1907	1 male; 3 females	Whangamoa (Kokurua), Nelson	Nelson AS	Otago
*c.*1908	?	Barron's Flat, Upper Takaka	?	East Takaka herd
1910	2 males; 2 females	Flora Saddle, Arthur Range, Nelson	Nelson AS	Otago
?	?	South Kaipara Head	?	?
?	Survived until shot out about 1950	Selwyn or Kaikoura Island	?	?
?	?	Tauranga	?	Said to have come from Spanish Pyrenees
?	?	Lake Wakatipu	?	?
1919	1 male	?	NZ Tourist Department	Wellington Zoo
1920	1 male	?	NZ Tourist Department	Wellington Zoo

AS, Acclimatization Society

* Chapman and Chapman (1980) say that 29 survived the voyage, which started in November

† Donne (1924)

predominates; they also sometimes ascend to areas of snow-grass (*Chionochloa* sp.) above 1,400 m.

South of Nelson a few fallow deer – the descendants of those released in 1864 – still live in the head of the Aniseed Valley; during the 1920s this herd spread through the Bryant Range and merged with that established since 1906 in the Teal Valley at the northwestern end of the range.

Between Brunner and the Buller River on the west coast of Nelson a population of fallow is established in the Paparoa Range; here they live in scrub and alpine snow-grass meadows at an altitude of 1,200 m, where in some places the annual precipitation can be as high 6,000 mm.

In the centre of South Island small numbers of fallow inhabit the valley of the Opihi River near Albury, some 10 km south of Fairlie, and parts of the adjacent Two Thumbs Range. Here, where the annual rainfall averages a tenth of that on the west coast, the deer ascend to around 1,500 m.

Until extensive culling in recent years to protect newly planted saplings greatly reduced their numbers, the Blue Mountains in Otago (where they were first introduced in 1869)[126] were the main stronghold of fallow (among which were many black animals) in New Zealand. Here, they multiplied rapidly, and in the decade from 1893 some 2,000 bucks are said to have been shot.[127] In 1926 nearly 4,000 fallow were killed, and in the following year a further 2,000. In 1957–8 and again in 1961–2, some 6,000 were shot in an area of only 8,000 ha. At least 1,550 were taken in 1972, but this may be an underestimate by as much as 40 per cent. The steep and rugged area is intersected by innumerable gullies and streams. Under half is stalked – the state-owned forests of Beaumont, Rankleburn and Tapanui, which have been planted with non-indigenous trees, mainly Monterey pine but also Corsican pine (*Pinus nigra*), Ponderosa pine (*P. ponderosa*) and Douglas fir (*Pseudotsuga menziesii*). The climax forest lacks an understorey, which would provide both food and shelter, so fallow seldom venture inside for long; instead they browse in young plantations and find cover in unthinned stands. The greater part (more than 7,000 ha) is indigenous forest which includes broadleaf (*Griselinia littoralis*), putaputaweta (*Carpodetus serratus*) and fuchsia (*Fuchsia excorticata*). The damp gullies support a scrub vegetation of fern (*Blechnum* sp.), pepperwood (*Wintera colorata*) and manuka. At higher altitudes (around 1,000 m) snow tussock (*Chionochloa rigida*) is the dominant species.

Intensive shooting over many years has greatly reduced the number of fallow deer in the glacial valleys surrounding Lake Wakatipu, some 90 km north-west of the Blue Mountains – formerly one of their main haunts. Here extensive river flats surrounded by native (mainly southern beech) forest in the Capels and Greenstone valleys on the north-west side of the lake, and land on the northeastern side, provide an ideal habitat for fallow. In spring they ascend to above the tree-line in search of the new growth on mountain flax (*Phormium colensoi*).[128] Among their favourite foods in New Zealand are fivefinger (*Nothopanax* spp.), broadleaf, wineberry (*Aristotelia serrata*), the bark of stinkwood (*Coprosma foetidissima*), cabbage tree (*Cordyline australis*), buttercup (*Ranunculus* sp.), leaves of silver beech (*Nothofagus menziessi*), bush lawyer (*Rubus* sp.), *Dracophyllum* sp., putaputaweta, sedges and ferns. Relatively little damage – apart from fraying[129] – is reported in exotic woodlands; elsewhere agricultural crops, especially turnips, sometimes suffer. In general fallow show a marked preference in their choice of habitat for lowland areas such as river flats and valleys, and do not venture into subalpine scrub and tussock country to the same extent as red deer do.

Papua New Guinea

Around the turn of the century German rubber planters are said to have released fallow deer in the tropical rainforests in the Madang area of New Guinea, where by 1953 the herd was estimated to number around 150. Doubts have been raised about the correct identification of these deer which it has been impossible to resolve.[130] The only deer on the island today are Timor deer (*Cervus timorensis*).[131]

OCEANIA

Fiji[132]

In about 1880 a Captain Padd imported some fallow deer to the 770-ha island of Wakaya in the

Fijian group, where by 1929 they are said to have multiplied to around 1,000; by 1977 they had decreased to about 400. The deer live in light forest and scrub where candlenut trees (*Aleurites moluccana*), raintrees (*Albizia lebbek*), *Dysoxylum quercifolium*, *D. richii*, *Ficus* spp., mango (*Mangifera indica*), guava (*Psidium guajava*), rose-apple (*Syzygium malaccensis*), *Turnera ulmifolia, Lantana camara* and *Dendrocnide harveyi* predominate. Presumably as a result of overgrazing and overbrowsing by the deer (and by feral goats) there is, in contrast to forests on other Fijian islands, only a sparse understorey.

For many years there was confusion about the identity of the deer on Wakaya. R. A. Derrick said in 1951 that 'During the early years of the present century Red Deer were introduced, probably from New Caledonia, and they have increased to such an extent that they tend to become a nuisance'.[133] Whitehead (1972a) wrote of the deer on Wakaya that 'they would appear to be Rusa [Timor]'. From photographs taken in 1977 of the deer and of the antlers of culled animals, Chapman and Chapman (1980) were able to confirm that the deer on Wakaya are indeed fallow.

Although de Vos, Manville and van Gelder (1956) stated that 'The King Ranch [in Texas, USA] introduced [in 1926] . . . Japanese Fallow Deer', Chapman and Chapman (1980) could find no evidence of wild populations of fallow in Japan, where the species occurs only in five zoological parks. Either the importation to Texas was of captive stock or, more probably, the deer were not fallow but Japanese sika (*Cervus nippon*) which have been introduced to Texas and are natives of Japan.

According to Chapman and Chapman (1980), 'Turkey is the only Asiatic country where the European Fallow deer exists in the wild and is probably the only country where this animal has lived continuously since it evolved'. Although fallow were at one time widely distributed in Turkey – especially in the south – today only three populations are known, all in south Turkey-in-Asia between the Mediterranean Sea and the Taurus Mountains; in 1975 the total population was estimated to number only about 100.

Fallow deer currently occur in the wild in no fewer than 38 countries in all six continents, where they have adapted to a wide range of environmental and climatic conditions; they are found from 61 °N to 46 °S in temperatures between 40 °C and −9 °C; in four of Köppen's five major climatic zones; in some regions where the annual precipitation can amount to 6,000 mm, and in others where there is an average of 40 (discontinuous) days of snow cover a year (although the depth of snow may be a critical limiting factor). They occur in 8 of the 11 major vegetation types, including all 6 of the extra-tropical types. Apart from rats, mice and feral domestic animals fallow deer are the world's most widely naturalized mammal.

NOTES

1. For much of the history and distribution of fallow deer throughout the world I am indebted to the comprehensive accounts by Chapman & Chapman (1975, 1980).
2. L. R. Conisbee, *Hastings and East Sussex Naturalist* **12**: 43–6 (1977).
3. P. H. Carne, *Wild Deer of the West Sussex Downs* (West Sussex Deer Control Society, Petersfield, Hants 1970).
4. P. H. Carne, *Proceedings of the Dorset Natural History & Archaeological Society* **88**: 93–101 (1967).
5. P. H. Carne, *Journal of the British Deer Society* **1**: 42–4 (1967).
6. D. Carter, *Journal of the British Deer Society* **1**: 46 (1967).
7. W. A. Scott, *Journal of the British Deer Society* **1**: 45 (1967).
8. M. J. Penistan, *Journal of the British Deer Society* **1**: 44–5 (1967).
9. Lever (1977).
10. D. I. Chapman, *Essex Naturalist*, n.s. **1**: 1, 3–30 (1977).
11. Earl of Cranbrook and W. H. Payn, *Transactions of the Suffolk Naturalists' Society* **15**: 222 (1970).
12. J. G. Goldsmith ed., *Norfolk Bird and Mammal Report* **23**: 60–9 (1972).
13. M. Johnson, *Transactions of the Lincolnshire Naturalists' Union* **19**: 8 (1976).
14. Lever (1977).
15. N. E. Hickin, *The natural history of an English for-*

est: the wild life of Wyre (Arrow Books, London, 1971).

16. G. Springthorpe, *Journal of Zoology, London* **159**: 537 (1969).
17. J. Robinson, *Deer* **3**: 222 (1975).
18. Ibid., pp. 320–2.
19. P. Delap, in G. A. K. Hervey & J. A. G. Barnes eds, *Natural history of the Lake District* (Warne, London, 1970), pp. 176–93.
20. Lever (1977).
21. H. V. Vaughan & P. H. Carne, *Deer* **2**: 767 (1971).
22. Lever (1977).
23. F. Mulloy, *Deer* **2**: 502–4 (1970).
24. C. D. Deane, *Deer* **2**: 920–6 (1972).
25. Damage to a tree caused by deer rubbing velvet (the protective covering of skin) from their new-grown antlers.
26. R. Amon, *Die Tierwelt Niederösterreichs* (1931).
27. Kerschagl (1964).
28. F. Dobschova, *Österreichs Weidwerk* **10**: 507–13 (1972).
29. For much of the information on fallow deer in Bulgaria I am indebted to Sartorius (1970).
30. Kotrlá & Kotrylý (1977).
31. F. W. Braestrup, *Hjortebogen* (1952) and U. Møhl, 'Oversigt over dyreknogler fra Naesholm' in V. La Cour ed., *Naesholm* (Nationalmuseet, Copenhagen, 1961), pp. 395–421, 427.
32. B. Alex-Hanson, *Dansk Jagttidende* **62**: 37–8, 48–9, 63–4, 74–7, 88–91, 103–5, 118–21 (1945); **87**: 166–71 (1970).
33. Ibid. (1970).
34. Ibid.
35. I. Sporon-Fiedler, *Dansk Jagttidende* **47**: 39–43 (1930).
36. See note 32.
37. Ueckermann & Hansen (1968).
38. J. P. Stoll, *Bulletin de la Société industrielle de Mulhouse* **734**: 73–6 (1969).
39. A. Waechter, *Mammalia* **40**: 339–52 (1976).
40. H. G. de Kerville, *Bulletin de la Société des Amis des Sciences naturelles de Rouen* **2**: 117–246 (1888).
41. G. Pennetier, *Actes du Muséum d'Histoire naturelle, Rouen* **9**: 1–108 (1905).
42. See note 40.
43. *Forests of France* (Faber & Faber, London, 1954).
44. For much of the information on fallow deer in East Germany I am indebted to Siefke (1977).
45. Ueckermann & Hansen (1968).
46. Ibid.
47. Sartorius (1970) and Ueckermann & Hansen (1968).
48. L. Földes ed., *Waidwerk der Gegenwart: Natur, Mensch und Wild im Einklang* (1972).
49. M. Ross-Macdonald, *The world wildlife guide* (Threshold Books, London 1971).
50. *Rhodes* (David & Charles, Newton Abbot, 1974).
51. Sartorius (1970); see also J. Szabolcs, *A dámvad* (Mezögazdasági Kiadó, Budapest, 1975).
52. C. Calligaris, F. Perco, & F. Perco, *Ricerche di Biologia della Selvaggina* **8**, Suppl.: 155–207 (1976).
53. R. Massoli-Novelli, ibid. **7**, Suppl.: 475–90 (1976).
54. Information on fallow deer in the Netherlands is largely derived from Dr W. A. Weyland (personal communication in 1974 to Chapman & Chapman (1980)).
55. R. Collett, *Norges Hvirveldyr 1. Norges Pattedyr* (Kristiania, 1911–12).
56. A. Haber, T. Pasławski & S. Zaborowski, *Gospodarstwo Łowieckie* (1977); Z. Oko & K. Włodek, *Roczniki Akademii Rolniczej w Poznaniu*, C. Zootechnika **24**: 113–25 (1978).
57. Oko & Włodek, ibid.
58. The account of the distribution of fallow deer in Romania is based largely on V. Cotta, *Vinatorul si Pescarul sportif* **8**: 13–14 (1968).
59. Ibid.
60. A. Chapman & W. J. Buck, *Wild Spain; records of sport with rifle, rod and gun, natural history and exploration* (1893).
61. F. Hingston, *Deer* **3**: 386–7 (1975).
62. Sartorius (1970).
63. J. A. Fernandez, *Doñana, Spain's wildlife wilderness* (Taplinger Publishing, New York, 1975).
64. Alvarez, Braza & Norzagasay (1976: 12).
65. J. A. Valverde, *Archivos del Instituto de Aclimatación de Almeria* **9**: 1–168 (1960).
66. L. Siivonen, *Nordeuropas Däggdjur* (1976).
67. I. Ahlén, *Viltrevy* **3**: 177–376 (1965).
68. Sartorius (1970).
69. V. G. Heptner, A. A. Nasimovič and A. G. Bannikov, *Die Säugetiere der Sowjetunion*, Vol. 1. *Paarhufer und Unpaarhufer* (Gustav Fischer, Jena, 1966).
70. N. K. Vereschchagin, *The mammals of the Caucasus: a history of the evolution of the fauna* (Israel Program for Scientific Translations, Jerusalem, 1967).
71. K. Curry-Lindahl, *Europe: a natural history* (Random House, New York, 1964).
72. See note 69, and Lindemann (1956).
73. See note 69.
74. V. D. Treus, *Transactions of the 9th International Congress of Game Biologists, Moscow, September 1969*, 155–7 (Moscow, 1970).
75. V. D. Treus & N. V. Lobanov in Yanushevich (1966: 128–9)
76. See note 70.
77. F. F. Aliev, *Transactions of the 9th International Congress of Game Biologists, Moscow, September 1969*, 167–9 (Moscow, 1970).
78. M. Vucković, ibid. pp. 219–23 (1970).
79. D. Bojović, *Transactions of the 7th Congress of the*

International Union of Game Biologists, Beograd, September 1965, 359–64 (1967).
80. *The mammals of Great Britain and Ireland*, 3 vols (Longmans, London, 1904–6).
81. *Mémoires présentés à l'Institut d'Egypte* **27**: 1–85 (1935).
82. *Annals of the Cape Provincial Museum* **2**: 110–17 (1962).
83. *The story of a house* (Howard B. Timmins, Cape Town, 1968).
84. W. von Richter, C. D. Lynch & T. Wessels, *Orange Free State Provincial Administration, Nature Conservation Research Report* No. 1 (1972).
85. G. A. Best, F. Edmond-Blanc & R. C. Witting eds, *Rowland Ward's records of big game*, 11th edn (Rowland Ward, London, 1962).
86. Bigalke (1937b).
87. C. A. van Ee, *Annals of the Cape Provincial Museum* **2**: 53–5 (1962).
88. C. J. Uys, *Farmer's Weekly, Durban* (27 September 1967).
89. J. D. Wehausen, 'Some aspects of the natural history and ecology of fallow deer on Point Reyes Peninsula' (MSc thesis, University of California, 1973).
90. R. H. Allen, *Alabama Department of Conservation, Division of Game and Fish, State Game Management Section Bulletin* No. 6 (1965).
91. Ibid.
92. See p. 152.
93. See note 89.
94. R. M. Jurek, 'Status, ecology and behaviour of a wild population of fallow deer in Mendocino County, California, 1968–70' (MSc thesis, Humboldt State University, 1977).
95. V. A. Terpening & R. E. Hawkins, 'Ecology and population dynamics of fallow deer in Land Between the Lakes, Kentucky' (unpublished report, Cooperative Wildlife Research Laboratory, Southern Illinois University, 1975).
96. Packard (1955).
97. Halloran & Howard (1956).
98. W. E. Armstrong, 'Habitat preference of exotics' (unpublished report of Texas Parks and Wildlife Department, Project No. W-76-R-18, Job No. 18, 1975) and E. L. Young, ibid. (Project No. W-76-R-15, Job No. 18, 1972).
99. Armstrong, ibid.
100. A. J. Godin, *Wild mammals of New England* (Johns Hopkins University Press, Baltimore, 1977) and A. R. Keith, *Dukes County Intelligencer* **11** (2): 47–98 (1969).
101. A. J. Godin, ibid. and W. Uloth, *Säugetierkundliche Mitteilungen* **19**: 202–6 (1971).
102. Flyger (1959).
103. Carl & Guiguet (1972).
104. See p. 237.
105. Gloucestershire Record Office, D1610 C5.
106. Ibid., D1610 C40.
107. Ref. 1857.1.14.1–3.
108. See note 80, Vol. 3.
109. Harris (1965).
110. M. D. Brazier & J. D. Mather, *Geological Magazine* **112**: 271–82 (1975).
111. Harris (1965).
112. See p. 289.
113. These may now be extinct.
114. Schneider (1936).
115. See p. 191.
116. See p. 311.
117. For much of the history of fallow deer in Australia I am indebted to Bentley (1978).
118. Allison & Coombes (1969).
119. Ibid.
120. Roff (1960).
121. See p. 217.
122. See p. 225.
123. See p. 152.
124. J. E. Wapstra, *Fisheries and Wildlife Paper, Victoria* **8**: 23–35 (1975).
125. See p. 193.
126. Baker (1972).
127. K. Baker, 'Reproductive biology of fallow deer (*Dama dama*) in the Blue Mountains of New Zealand' (MSc thesis, University of Otago, Dunedin, 1973).
128. L. H. Harris, *Hunting fallow deer* (New Zealand Forest Service, Wellington, 1967).
129. See note 25.
130. Bentley & Downes (1968–9).
131. See p. 221.
132. R. M. Spencer, *The Field* **154**: 871 (1929).
133. *The Fiji Islands: a geographical handbook* (Government Press, Suva, 1951) (quoted by Chapman & Chapman 1980).

Red Deer
(*Cervus elaphus*)

Note: Present fragmented natural distribution is due to local extinction.

NATURAL DISTRIBUTION
The Palaearctic region from northern Britain to Manchuria and from south of the Arctic Circle in Norway (65° N) to the Himalayas and the Tunisian / Algerian border in North Africa (33° N). Fragmented distribution due to local extinction.

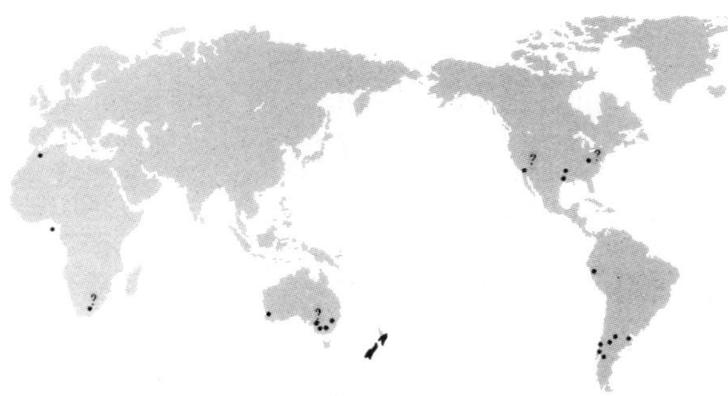

NATURALIZED DISTRIBUTION
Africa: Morocco, Fernando Poo, ? South Africa; North America: USA; South America: Argentina, Chile, Peru; Australasia: Australia, New Zealand.

AFRICA

Morocco;[1] Fernando Poo

In 1952 5 stags and 10 hinds of the Spanish race *C. e. hispanicus* – natives of central and southern Spain and southeastern Portugal – from the game reserve of Prado near Madrid were released in the vicinity of Ksar-es-Seghir on the northern coast of Morocco (between Tangier and the Spanish enclave of Ceuta) by the Spanish governor, Señor Garcia Valino. Although a number were killed by Rif poachers and others dispersed, by 1969, when they were strictly protected, they had become established in two separate herds – one of about 30, in the Ben Karrich region in the foothills of the Rif Mountains south of Tetouan, and another, of about 20, near the coast between Restingás and Mdiq. The deer adapted well to their new habitat, but there is fortunately no danger of interbreeding with the Barbary race (*C. e. barbarus*) since the sole survivors of this subspecies live more than 1,200 km to the east near the Algerian / Tunisian border.

In 1954 some red deer – presumably of the Spanish race – were introduced to the Spanish island of Fernando Poo in the Gulf of Guinea off the coast of Cameroun.

South Africa

A Mr Ohlsson is said to have imported the first red deer, from Scotland, to South Africa in the 1870s, and others were introduced to Clocolan in the Orange Free State in 1895. Several extensive (85,000 ha) fenced ranches in the Western Cape have been stocked with red deer for sporting purposes, and a few may possibly occur in the National Park near Bredasdorp.

In about 1975 some red deer escaped from a zoo or private collection near Vereeniging in the Transvaal when floods destroyed the fencing of their enclosure; in 1981 a small, but possibly developing, herd of three stags and two hinds (both of which are known to have bred) was established in several hundred hectares of oak plantation near the deers' point of escape.

NORTH AMERICA

USA

After axis deer (*Cervus axis*),[2] blackbuck (*Antilope cervicapra*),[3] mouflon (*Ovis musimon*),[4] Barbary sheep (*Ammotragus lervia*),[5] fallow deer (*C. dama*),[6] sika (*C. nippon*),[7] nilgai (*Boselaphus tragocamelus*)[8] and wild boar (*Sus scrofa*),[9] red deer are the most abundant of the many species of exotic ungulates in the state of Texas, where they were first introduced from Europe by Mr Richard Friedrich in the early 1930s. The total population, much of which is on unfenced land, has risen from 21 in 1964 to 95 on 15 ranches in 1966, 307 (1971), 407 (1974) and 434 in 1979; most of the deer are established in the hill country of the Edwards Plateau.

Red deer have also been introduced to Kentucky, where by 1950 about 120 were established in the Woodlands National Wildlife Refuge and Bernheim Forest. A small number – descended from escapes or releases from the Hearst Ranch (Rancho Piedra Blanca) – may also occur near San Simeon in San Luis Obispo County, California.

SOUTH AMERICA

Argentina

Since 1902 (in 1906 by Señor Pedro Luro to his *estancia* in La Pampa) red deer from Germany, France, the Hungarian Carpathians and Austria have been imported into Argentina. In La Pampa (where they compete for grazing with domestic cattle) they first became established in the wild in 1922. Twenty brought in the 1920s by Herr Robert Hohmann to Neuquen Province (where they come into conflict for food and territory with the native Chilean guemal (*Hippocamelus bisculus*) and southern pudu (*Pudu pudu*)) were spreading widely and rapidly by the 1940s. In 1932 Herr Carl Adolf Vogel imported a number of red deer from Austria to his estate near San Martino de los Andes in Neuquen Province, and others were introduced to ranches in Chubet Province. By the 1950s wild populations were established on the pampas at San Huberto, south of Buenos Aires, at San Martino and in the southern foothills of the Andes in the Parque Nacional Lanin and on Isla Victoria in Lake Nahuel Huapi. This lake, which is situated between the provinces of Neuquen and Rio Negro, lies at an altitude of some 760 m between the Patagonian Plateau and the Andean forest in the Parque Nacional Nahuel Huapi. The vegetation on the island (which the red deer share with fallow deer[10]) is mainly false beech (*Nothofagus* sp.), bamboos and conifers – especially alerce (*Fitzroya patagonica*). Both fallow and red deer were introduced to the island by Señor Aaron de Anchorena from his *estancia* La Barra de San Juan in the Department of Colonia in Uruguay. By the late 1970s red deer were established in the wild in considerable numbers in many widely scattered regions of Argentina.

Chile

Red deer – transplanted from Argentina and imported from Germany in the 1920s – are established in the wild in several localities between Temuco and Puerto Montt, where they are said to compete for grazing with domestic cattle. In 1975 a herd of about 200 inhabited an island in Lake Rupanco in Osorno Province, which they share with fallow deer[11] and mouflon.[12] The island is densely wooded, the principal species being olivillo (*Aextoxicum punctatum*), ulmo (*Eucryphia cordifolia*) and *Laurelia philippiana*. Another herd is established a little to the north on an island in Lake Ranco in the province of Valdivia.

Peru

In 1948 30 red deer (and several fallow) were translocated from Argentina to the Huacraruco Ranch in the district of San Juan, Cajamarca Province, in northern Peru. From here, they were subsequently released (or escaped) into the neighbouring basin of the Chicamo River. Unlike the fallow, the red deer (and native white-tailed deer *Odocoileus virginianus*, with which the red compete successfully for food and territory) were seriously affected by the 1963–4 outbreak of foot-and-mouth disease which, with predation of calves, has limited the population to between 200 and 300.

AUSTRALASIA

Australia[13]

Table 11 gives details of the principal importations of red deer into Australia. According to Rolls (1969), Captain Anthony Fenn Kemp of Cross Marsh, Tasmania, possessed some red deer in 1830 which were thriving so well that further importations were considered unnecessary. There appear to be no other references to substantiate this claim.

Rolls also refers to editorials in the Melbourne *Argus* in January and February 1860 by Edward Wilson (founder and first president in 1861 of the Victoria Acclimatization Society) stating that Queen Victoria had offered, through the Duke of Newcastle (Colonial Secretary 1859–64), to send out to Victoria some red deer from Windsor or Balmoral. The earliest, and most notable, importation of red deer into Australia arrived from Windsor on the US ship *H. B. Mildmay* for Mr Thomas Chirnside of Werribee Park, 40 km west of Melbourne, Victoria, in 1860. Two years later the Werribee herd had increased to 11, and from this source red deer were subsequently distributed to other localities in Victoria, and to Western Australia (in 1862), Queensland and, in later years, to New Zealand.

From at least the 1880s until the turn of the century, when denser human settlement drove them into heavily timbered country, red deer roamed freely over the grassy plains west of Geelong. Today the majority live in an area of around 1,000 km^2 in western Victoria from the Grampians and Ararat eastwards to Ballarat and Mount Mercer. Small groups are scattered elsewhere in the state (e.g. in the Otway Range and in western Gippsland), and especially in the Snowy Mountains region in the east near the border with New South Wales, where they are the descendants of the Quidong population (see below), which spread throughout the wooded country in the region of Brassey's Creek bounded by the McLaughlin and Snowy rivers, and around Mila.

The red deer in the generally heavily timbered Quidong region of the southeastern corner of New South Wales near the headwaters of the Snowy River (an area bounded in the east by Bombala and Delegate, in the south by the border with Victoria and in the north by Dalgety and the

Table 11 Principal importations of red deer to Australia

Date	*Number*	*Presented by*	*Presented to*	*Source*
April 1860	6	Queen Victoria and Prince Albert	Mr Thomas Chirnside, Werribee Park, Victoria	Windsor Great Park, England
1862	?	Queen Victoria*	Melbourne Hunt Club	?Windsor Park or Balmoral, Scotland
1862	?	Queen Victoria*	Western Australia	?Windsor Park or Balmoral
September 1873	2 stags; 4 hinds	Queen Victoria	Queensland AS	Windsor Park
1874†	1 stag; 2 hinds	Mr Alexander Dundas Cochrane-Baillie, MP (created Baron Lamington in 1880)	Queensland AS	?Lamington, Lanarkshire Scotland
1874	6	Duke of Richmond and Gordon	Queensland AS	?Goodwood, Sussex, England or Fochabers, Banffshire, Scotland
1888	3	Earl of Derby	Victoria AS	Knowsley Park, Lancashire, England
?1888	?	Earl of Derby	Melbourne Hunt Club	Knowsley Park

Source: After Bentley (1978)
AS, Acclimatization Society

*Bentley (1978) states that these deer were presented by Prince Albert. Whitehead (1972) – presumably following Bentley – wrote that 'two years later [i.e. in 1862] Prince Albert sent some further deer from his park at Windsor . . .': in fact the Prince Consort died in 1861

† Bentley (1978) says that 'in 1874 a further stag and two hinds were presented by the Governor of Queensland, Lord Lamington'; Whitehead (1972) – presumably again following Bentley – wrote that in about 1873 'a further presentation was made by Lord Lamington, who was governor of Queensland about that time'. The second Baron Lamington, however, was Governor of Queensland from 1895 to 1901; the deer must therefore have either been introduced at the date given by Bentley by the second baron's father, Mr Alexander Cochrane-Baillie (who was not elevated to the peerage until 1880) or by the second baron himself during the period of his governorship some 20–25 years later

Snowy River) are descended from some which escaped from Mr J. R. Logan's Aston Station between Cooma and Bombala, augmented by deliberate releases of the remainder when the property was sold in 1918. A population described by Whitehead (1972) as 'moderately abundant' and by Frith (1973) as 'light' persists in the high country of the Monaro near Delegate, from where it is spreading slowly south into the Snowy River country and eastern Victoria. According to Frith, red deer were also 'released south of Sydney in 1885 . . . [and] five Red Deer were released in the Royal National Park in 1886, but there has been no reliable sighting there since 1914'.

The six red deer presented by Queen Victoria through the Marquess of Normanby (Governor of Queensland 1871–4), which arrived in Queensland in 1873 on board the *Great Queenslander*, were transported by ox-cart to Toogoolawah where they were turned out on Mr David McConnell's station in the foothills of the Darling Downs between Esk and Cressbrook north-west of Brisbane; by 1878 they were beginning to spread, and by the 1880s considerable numbers were reported from the Black Jack, Scrub and Waterfall creeks district at the head of the Brisbane River. By 1896 the population numbered around 300 and had spread almost 100 km from the point of release.

In 1874 (but see footnote to Table 11) a further three red deer were presented to Queensland by the future Lord Lamington – probably from Scotland. Bentley (1978) refers to a report published in the now defunct journal *Wild Life*[14] that 'a stag and two does [*sic*] arrived from the Duke of Buccleuch or the Duke of Argyle [*sic*] in the following year (1874)'. Bentley considered that this report was probably authentic, since 'Queen Victoria frequently visited the Atholl Forests (the Duke of Argyle) during the Scottish stalking season'. Since Queen Victoria and the Prince Consort are not known to have visited the Duke of Argyll's Inveraray Forest or the Duke of Atholl's Atholl estate after their acquisition of Balmoral in 1852, the *Wild Life* report is either erroneous or refers to the importation from Lamington.

Around the turn of the century two pairs of red deer were landed on Hinchinbrook Island which lies off the Queensland coast 125 km north of Townsville, where they were reinforced in 1915 or 1916 by a further pair imported by Mr William Watkins on behalf of the government of Queensland. Whether any deer remain on Hinchinbrook is uncertain, but some are believed to have swum from the island to the adjacent mainland.

Bentley (1978) believed that seven red deer presented by Thomas Chirnside to the Queensland Acclimatization Society were released on Messrs. Wildash and Hutchinson's Canning Downs property near Warwick in 1903; others were freed in the Stanley Mountains during the governorship (1905–9) of Viscount Chelmsford. The present range of red deer in southeastern Queensland (which is nearly 100 km in diameter and is still increasing) is approximately bounded by a line from Esk north-east to Woodford, north to Gympie, south-west through Kingaroy to the Bunya Mountains and through Crow's Nest back to Esk. The main concentrations are in mountainous country around the head of the Brisbane and Stanley rivers and in the vicinity of Linville, Kilcoy, Monsildale, Taromeo and Cooyar creeks. The deers' habitat ranges from open, undulating pastoral country interspersed with numerous watercourses, to steeply wooded hills with thick cover of vine scrub and lantana (*Lantana* spp.). They especially favour mixed grassland, woodland and rainforest associations.

The red deer presented to Western Australia[15] by Queen Victoria in 1862 were kept in the zoological gardens in South Perth, from where their offspring were later released near Albany, at Cape Leeuwin and between Pinjarra (80 km south of Perth) and Rockingham. In 1899 a further pair were liberated at Cape Leeuwin. Four from Werribee freed on a private estate near Pinjarra in 1903 had increased to 30 by 1912: in 1908 further introductions were made in this region, and in about 1915 on the Creaton estate of Mr Douglas Paterson near Pinjarra. Although the deer turned out at Cape Leeuwin initially did well, they (and those released at Albany) died out between 1920 and 1930. The deer freed at Pinjarra in 1908 had increased to more than 150 by 1920, and in later years, as they encroached onto agricultural land, considerable numbers were culled. Between 1903 and 1906 red deer were introduced to the Porongorups, to Numarracarra near Geraldton and to Gingin. Small numbers still survive around Pinjarra, Waroona and Harvey in the extreme southwest of Western Australia. Bentley (1978) suggests that a possible reason for their decline has been their consumption of the toxic zamia palm (*Macrozamia fraseri*).

In South Australia red (and fallow) deer were introduced – probably from Werribee – to Yallum Park near Penola shortly after its acquisition in 1861 by Mr John Riddoch (the 'Squire of Penola') whose brother George later stocked his Korine Station with red deer from Yallum, which also supplied stock to stations at Penola and Nangwarry. Red deer were also liberated at Poltalloch Station near Meningie. Until the early 1900s red deer were widespread in swamplands and scrub in southeastern South Australia; since that time, especially as the result of increased human settlement and the development of land for agricultural purposes after the Second World War, there has been a steady decline in both range and population. Although Bentley (1978) believed that red deer died out in South Australia by about 1956, small numbers continued to survive until at least 1973 in scrub country around Naracoorte near the Victorian border.

Little is known of the feeding preferences of red deer in Australia and their effects – if any – on the environment and on domestic livestock. Occasional depredations are reported on maize, oats, sorghum and lucerne crops. In some districts, especially in Queensland, the deer are infested with cattle ticks (*Boophilus microplus*). The quality of heads in Australia – as in New Zealand – is, in general, poor.

New Zealand[16]

Red deer have been the most frequently introduced and are the most widely distributed large mammal in New Zealand. Logan and Harris (1967) mention some 220 separate liberations between 1851 and 1926 involving a total of over 820 animals, and there were undoubtedly many more. Table 12 gives details of the principal releases of red deer in New Zealand during that period.

According to Donne (1924), 'Lord Petre, of Thorndon Hall, Essex, at the instance of the Hon.

Table 12 Principal releases of red deer in New Zealand

Date	*Number*	*Presented by*	*Released by*	*Locality*	*Source*
North Island					
1862 or 1863	3	Queen Victoria	?	Taratai Plains, Carterton	Windsor Park, England
?1863				?Ruamahanga	
1883	11	?	Mr W. Shrimpton of Hastings	Matapiro Station, Hawke's Bay	Bushey Park, SI
1885				Mahia Peninsula	
1887 (or 1895) and 1890	3 calves	?	Wellington AS	Matthew's Station, Tararua	Martinborough herd, NI
1895–96				Wellington; Rimutaka	
1897				Tongariro National Park	
1897				Wainui-O-Mata (State Forest)	
1897	4 calves	?	Wellington AS	Galatea, Urewera	Martinborough, NI
1898	6 calves	?	Wellington AS	Wainui-O-Mata	Martinborough, NI
1899				Paraparaumu	
1899	4 calves	?	Wellington AS	Lake Waikaremoana	Martinborough, NI
1900–10	Large number of calves	?	Wellington AS	Wellington	Martinborough, NI
1900				Paraparaumu	
1901–3				Kahuterawa	
1901–2	7	?	Government Tourist Department	Lake Waikaremoana	Wairarapa
1902				Erewhon Station	
1903				Paihia	
1904				Erewhon Station	
1904				Wellington	
1904				Wainui-O-Mata	
1904				Mangatainoka	
1904				Eketahuna	
1904				Rimutaka	
1905				Lake Okareka	
1905 or 1906	3 stags; 2 hinds	?	Government Tourist Department	Tongariro National Park	Martinborough, NI
1906				Mangaterere	
1906				Ruapehu	
1906				Ruamahanga	
1906–7				Tauherenikau	
1907				Eketahuna	
1907				Matthew's Station, Tararua	
1907	9	Mr G. W. Chirnside	?	Lake Roto-Iti, Rotorua	Werribee Park, Victoria, Australia
1908				Lake Okareka	
1908				Mangaweka	
1909				Eketahuna	

Continued

Table 12 (*continued*)

Date	*Number*	*Presented by*	*Released by*	*Locality*	*Source*
North Island (*cont.*)					
1910				Ruamahanga	
1911				Matapiro Station	
1911				Ohau	
1911				Paraparaumu	
1911				Eketahuna	
1911				Waingawa	
1911				Mangaterere	
1911				Tauherenikau	
1911–12				Tokaanu	
1911–12				Wainui-O-Mata	
1912				Mount Camel	
1912				Tauranga	
1912				Galatea	
1912–14				Shannon	
1912–14				Mahia Peninsula	
1912–17				Lake Okareka	
1913				Lake Waikaremoana	
1913				Wairoa	
1913				Rimutaka	
1913				Paraparaumu	
1913				Wellington	
1913–19				Taumaranui	
1914				Rotorua	
1914–15				Te Awaite	
1914–17				Tokaanu	
1914–19				Lake Roto-Iti	
1915				Waikaremoana	
1915–16				Wairoa	
1915–19				Galatea, Urewera	
1916				Te Whaiti	
1916				Pahiatua	
1916				Rimutaka	
1916				Wellington	
1916	11	?	?	Rotorua	Paraparaumu
1916–17				Lake Okataina	
1916–17	25	Mr G. W. Chirnside, and Paraparaumu Game Farm	?	Mount Edgecumbe	Werribee and Paraparaumu
1916–17				Tora	
1917				Waimana Gorge	
1917				Matthew's Station, Tararua	
1917–18				Waiohine	
1917–19				Paraparaumu	

Continued

Table 12 *(continued)*

Date	*Number*	*Presented by*	*Released by*	*Locality*	*Source*
North Island *(cont.)*					
1918				Kahuterawa	
1918				South Kaimanawa	
1918	2 pairs	Paraparaumu Game Farm	?	Lake Waikaremoana	Paraparaumu
1918				Pahiatua	
1918–19	6 stags; 4 hinds	?	?	Mahia Peninsula	?
1919	2 pairs	Mr G. W. Chirnside	?	Lake Waikaremoana	Werribee
1919				Lake Okataina	
1919				Lake Okareka	
1919				Wainui-O-Mata	
1919				Te Whaiti	
1920				Ohau	
1920–1				Te Kuiti	
1920–1				Tiritea	
1921				Lake Okareka	
1921				Lake Roto-Iti	
1921				South Kaimanawa	
1921				Te Whaiti	
1921				Mahia Peninsula	
1921–2				Mount Edgecumbe	
1921–2				Galatea	
1922				Apiti	
1922				Pohangina	
1922				Lake Okataina	
1922				Taumaranui	
1923				Mahia Peninsula	
1923				Pahiatua	
1926				Paraparaumu	
South Island					
1851	1 pair	Lord Petre	Hon. W. E. Petre	Maitai Valley, Nelson	Thorndon Hall, Essex, England
1854	1 stag (survivor of a pair)	Prince Albert	?	Estuary of Waimea River, Nelson	Richmond Park, England
1860 or 1861	1 stag; 2 hinds	Lord Petre	Hon. W. E. Petre	Nelson Hill country	Thorndon Hall
1871	8 and 7 (all calves)	Earl of Dalhousie	Otago AS	Morven Hills, Lake Hawea and Bushey Park, Palmerston South	Invermark, Forfarshire (Angus), Scotland
1895				Hawea	
1897	9	Purchased from Mr Willoughby Bryant	Canterbury AS	Rakaia Gorge	Stoke Park, Buckinghamshire, England

Continued

Table 12 *(continued)*

Date	*Number*	*Presented by*	*Released by*	*Locality*	*Source*
South Island *(cont.)*					
1898	3	?	Wellington AS	Greymouth	Martinborough, NI
1898	2 stags	Duke of Cleveland and Lord Feversham	Otago AS	?	Raby Park, Co. Durham and Duncombe Park, Yorkshire, England
1898–9	4	?	Southland AS	Birchwood	?
1899				Nelson	
1890s				Langleydale Station	
1900				Hawea	
1900	10	?	Southland AS	Davis Flat, Lillburn Valley	Nelson
1901	3 stags; 10 hinds	Sir Rupert Clarke	Southland AS	The Monument, Lake Manpouri, near Waiau	Victoria, Australia
?1901	6	Sir Rupert Clarke	Southland AS	The Hump, Hauroko	Victoria, Australia
1901	6 calves	?	Government Tourist Department	Fresh Water River, Paterson Inlet, Stewart Island	Wairarapa
1902	12	Miss Audrey Chirnside	Government Tourist Department	Fresh Water River, Paterson Inlet, Stewart Island	Werribee
1902				Rakaia Valley	
1902				Glenorchy	
1902	6	Miss Audrey Chirnside	Otago AS	Pomahaka Valley	Werribee
1903	8	Miss Audrey Chirnside (or the Duke of Bedford)	Government Tourist Department	Lake Wakatipu	Werribee (or Woburn Abbey, Bedfordshire, England)
1903	7 or 8 calves	Miss Audrey Chirnside	Southland AS	Mount Tuhua, Westland	Werribee
1906	4 and 8	Miss Audrey or Mr G. W. Chirnside	?	Mount Tuhua and Lake Kaniere	Werribee
1908	7 (survivors of original consign-ment of 3 stags and 9 hinds)	Mr Charles W. Lucas	Canterbury AS	Poulter Valley	Warnham Park, Sussex, England
1909	8	Mr Charles W. Lucas	Canterbury AS	Poulter Valley	Warnham Park
1909			Government Tourist Department	Seaforth Valley	
1909	3	Paraparaumu Game Farm	Government Tourist Department	Dusky Sound	Paraparaumu
1910	3 young stags	Mr Charles W. Lucas	Nelson AS	Nelson	Warnham Park
1914	6	Mr Charles W. Lucas	Otago AS	Dunedin	Warnham Park

Continued

Table 12 *(continued)*

Date	*Number*	*Presented by*	*Released by*	*Locality*	*Source*
South Island *(cont.)*					
1914				Waiwera	
1917				Avondale Station	
1919–20				Pomahaka Valley	
1919–21				Waikaia	
1921				Waiwera	
1922				Nelson	
1923				Inangahua junction	
1924				Dunedin	

Sources: After Wodzicki (1950) and Logan & Harris (1967)
AS, Acclimatization Society; NI, North Island; SI, South Island

W. E. Petre [his brother], who was then a settler in Nelson, presented a stag and hind in 1850', which arrived in New Zealand in the following February. 'H. R. H. The Prince Consort', Donne continues, 'in 1853, through Lord John Russell, the Royal Ranger, presented from the Royal Park at Richmond, a stag and hind.' 'In 1862', Donne goes on, 'six red deer, two stags and four hinds . . . were presented to the people of New Zealand by the late Prince Consort.' Donne also records that 'In 1870 seventeen red deer calves were presented to the Acclimatization Society of Dunedin from Invermark Forest, Forfarshire, Scotland, by the eleventh Earl of Dalhousie'. Thomson (1922), whose claim is repeated by Clark (1949), states that Felix Wakefield, a distinguished engineer who resided in New Zealand from 1851 to 1854 and again after 1863, 'landed one stag at Nelson in 1851'. Clark also claims that a fawn was imported to Riccarton in 1853. There appear to be no other references to substantiate these suggestions.

Lord Petre's pair from Essex, released in the Maitai Valley in 1851, were thus the first red deer to be landed in New Zealand. Following the death of the hind, the stag remained for several years in the vicinity of Motueka. In 1853 the Prince Consort despatched a pair from Richmond Park on board the *Eagle*, of which the hind perished during the voyage; the stag was released on 6 September 1854 near the mouth of the Waimea River. In September 1860 a further stag and two hinds were sent by Lord Petre from Thorndon Hall on the *Donna Anita*; these were turned out in the Nelson Hill country in the following February. Their progeny spread rapidly over much of the high country in provincial Nelson and Marlborough, and later south into northern Canterbury and towards the west coast. In 1900 deer translocated from this population to Lillburn Valley west of the Waiau River formed the nucleus of a new herd in Southland.

On 1 June (or 6 July) 1862 the clipper *Triton* from London, after a voyage of 127 days, landed a pair of red deer (the survivors of an original consignment of three despatched by Mr John Morrison, the New Zealand government agent in London) in Wellington. Later in the same year the *Harwood* berthed in Lyttleton with a single hind – two others having died on the voyage from London – which was subsequently transshipped to Wellington: all these deer were presented to Sir Frederick Weld, then minister for native affairs (not 'Governor Weld' as stated by Thomson (1922) and repeated by Whitehead (1972)), by Queen Victoria, and not, as recorded by Thomson and Whitehead, by the Prince Consort who had died in the previous year.

These three deer were placed in the care of Dr Isaac Featherstone, superintendent of the province of Wellington, by whom they were confined for several months in a stable near 'Noah's Ark', Lambton Quay, from where they were subsequently transported across the Rimutaka Ranges to the Taratai Plains, where they were released on Mr Carter's run. They later crossed the Ruamahanga River and became established in the Maungaraki Ranges. According to H. Frank Wallace, 'The only stag of the trio liberated was well known. He turned nasty in his old age, and, gallantry not being his strong point, charged an unfortunate lady and compelled her to take refuge beneath a house, which, fortunately for her, was built on piles.'[17]

By 1869 this Wairarapa herd had multiplied to 'upwards of twenty'; by the end of the century, when the population (said to produce the finest heads in the country) numbered many thousands spread over hundreds of square kilometres of montane forestland, Donne (1924) records that 'on one station alone the owner made contracts for the destruction of a minimum number of three thousand deer annually for several years in succession'. Hardcastle (1906) wrote: 'The Wairarapa Forest is probably the best stocked red deer ground on the globe. On Te Awaite run bordering the East Coast, the deer may now be seen in bunches of up to a hundred head. At the beginning of last year it was estimated that there were fully 10,000 head on the station.' According to Whitehead (1972), 'no other liberations from outside were made in the Wellington district prior to 1908'; Wodzicki (1950), on the other hand, states that from 1900 to 1910 the 'Wellington Acclimatization Society made annual liberations of fawns throughout their district'.

Thomson (1922) states that in 1867 a Mr A. M. Johnson imported a single red deer stag from Hobart, Tasmania (where, according to Rolls (1969), they had been introduced before 1830), which was kept in the gardens of the Christchurch Acclimatization Society; it does not appear ever to have been released into the wild.

In 1932 A. I. McConnochie recorded that 'in 1870 seventeen calves were presented by the eleventh Earl of Dalhousie' from his Invermark Estate in Scotland to the Otago Acclimatization Society in Dunedin 'where they made themselves quite at home in their new country and rapidly developed into large herds.'[18] Eight which sailed from London on 20 October on the *City of Dunedin,* berthing at Port Chalmers towards the end of January 1871, were released on a spur near Timaru Creek in the Lindis Pass area east of Lake Hawea in the Morven Hills. The remaining nine sailed from London in the *Warrior Queen*, arriving at Port Chalmers on 29 November; the seven survivors of this shipment (which had been insured for £10 each) were freed on Mr Gordon Rich's Bushey Park Estate on the lower Shag River near Palmerston South. The deer from these two plantings (the only ones known to have come directly from Scotland) were soon well established and breeding; those in the Morven Hills – the ancestors of that region's present population – multiplied and spread over much of the country between lakes Wanaka, Hawea and Ohau, up the Hunter and Makarora rivers, across the Haast Pass into southern Westland and nearly to Mount Cook. Most of this terrain, which is steep, rugged and intersected by innumerable gullies, lies at altitudes of between 1,000 and 2,300 m and is covered with bush and open grassland. By 1908 the population was estimated by Wallace[19] to number 10,000. From Bushey Park a number of deer were subsequently translocated to the Hawke's Bay area, while by 1896 others had spread into the Horse Range where, however, they were apparently later shot out.

In 1897 the Canterbury Acclimatization Society imported nine red deer which were liberated in the Rakaia River gorge; here they rapidly increased, and by the early 1920s herds of more than 40 were commonly reported.

In 1898 Captain Montagu Cradock presented two stags – one from the Duke of Cleveland's herd in Raby Park, County Durham, the other from Lord Feversham's herd in Duncombe Park, Yorkshire – to the Otago Acclimatization Society.

On 24 November 1896 the Canterbury Acclimatization Society acquired three stags and seven hinds at a cost of £12 each from Mr Willoughby Bryant of Stoke Park, Slough, Buckinghamshire. Following the death of a hind shortly after being handed over in London to the Agent General for New Zealand, a curt telegram was despatched from the Port of London Authority demanding that he 'remove your dead deer from the dock premises, as it can not be allowed to remain here any longer'. The nine survivors left London on the SS *Waikato* on 23 February 1897, arriving in Port Chalmers on 6 May in good condition, apart from one hind which died shortly afterwards; one stag was said by a local newspaper to 'remain very wild, biting and kicking anyone attempting familiarity'. The deer were placed in quarantine for 20 weeks on Quail Island, where a calf was born. In October they were moved to Lyttleton and thence by rail to Christchurch and on to the Wilberforce River, where they were turned out on Mr William Gerard's Snowdon Station in Rakaia Gorge, where they flourished but for some years did not spread. This eventually proved to be one of the most successful introductions of red deer to New Zealand.

On 2 October 1908 Mr Charles W. Lucas despatched to New Zealand on the SS *Papanui* three stags and nine hinds from his Warnham Park herd near Horsham in Sussex; the seven deer which survived the voyage were turned out in the Poulter River region in north Canterbury. In the following September Mr Lucas sent by the SS *Turakina* eight more deer from Warnham, all of which arrived in excellent condition in Lyttleton, whence they were sent direct to the Poulter River where they were released on 24 November; by the 1920s the 15 deer from these two plantings had multiplied to a herd of several hundred.

On 30 September 1910 Mr Lucas despatched three young stags from Warnham on the *Ruapehu* to the Nelson Acclimatization Society, by whom they were subsequently released; four years later the Otago Society received six red deer from Warnham which were liberated near Dunedin.

Thomson (1922) considered that these importations account for most of the red deer in New Zealand today. From these herds the Government Tourist Department and various acclimatization societies obtained stock to found populations in many other parts of both main islands, and also on Stewart Island off the south coast of South Island, where by 1918 they were said to be numerous in woodlands in the north and west.

Red deer were initially introduced to New Zealand primarily for sporting purposes and as a source of food. In the early years their heads and body weights were in general superior to those in Scottish deer forests, and were equal to or better than those of English and European park or woodland deer: the reasons for the greater size of New Zealand red deer were attributed to the generally milder climate, more nutritious fodder and the abundance of shelter.

Apart from the three earliest introductions by private landowners most of the subsequent importations and translocations were made by the acclimatization societies of Canterbury, Otago, Southland and Wellington. According to Wodzicki (1950), some of the most regrettable liberations were those made by the New Zealand Government Tourist Department on Stewart Island in 1901–2, in the Tongariro National Park (1905 or 1906), and near the west coast sounds in 1909. Wherever they were planted the deer flourished, due, Wodzicki believed, largely to their inborn adaptability, the absence of predators and initial protection; to these reasons may be added a congenial climate, an abundance of nutritious food, a lack of competition from other species, and the early sexual maturity of hinds.

In 1924 J. Forbes[20] recorded that although the herd in Nelson had considerably expanded its range, it had not yet merged with those around Mount White and in the Rakaia Valley. The latter was still widely separated from the deer in Otago, which had themselves not joined up with the outliers of the Lake Manapouri herd in Southland. Those deer turned out in Westland and at Bushey Park had neither spread far nor become very numerous. North Island had five discrete herds most of which had not dispersed far from their points of liberation, and were apparently separated from each other by extensive tracts of unoccupied country.

By the mid-1920s most of the early plantings (both importations from abroad and translocations within the country) were expanding their range and increasing their population, although they had still not all merged, and large expanses of suitable habitats remained uncolonized. In 1930 red deer in New Zealand were officially classified as vermin, and in the following year a destruction campaign was begun by the Department of Internal Affairs.

When Wodzicki conducted his survey in the late 1940s he found that the main reservoirs of red deer in New Zealand were in relatively inaccessible valleys and mountain ranges in the north and south of South Island; these two regions were separated by only lightly populated country from south of Lake Coleridge to Lake Ohau. In North Island the species' range was less extensive; it was virtually confined to the main divide and its subsidiary ranges, with the principal concentrations in the centre of the island. During the 1930s red deer spread south and south-east of Opotiki and colonized the Kaingaroa Plains; a decade later they were being reported from around Aorangi in the Ruakumara Range, west of Tokomaru Bay. Wodzicki believed that the ease of access by man to the range occupied by red deer in North Island was the main, if not the sole, reason for the presence there of fewer and smaller herds than in South Island.

In the quarter of a century between Forbes's survey in the early 1920s and Wodzicki's in the late 1940s red deer considerably expanded their range in New Zealand; except for those in the

Horseshoe Range and on Stewart Island, where physiographic features curtailed their expansion, many of the major herds had amalgamated. Most suitable habitats – albeit some recently – had been occupied, and colonization of many of the west coast sounds and parts of the eastern coastal district in North Island was proceeding apace.

Most favoured habitats of red deer in New Zealand are, in order of preference, native forests (rather than hardwood and mixed podocarp ones), mountain tops, scrub, tussock, and grassland. Their favourite foods include *Fuchsia*, koromico (*Veronica salicifolia*), broadleaf (*Griselinia* sp.), native gum (*Panax* sp.), ribbonwood (*Gaya lyallii*), *Coprosma* spp., pepper tree (*Drimys colorata*), milk tree (*Paratrophis heterophylla*), tutu (*Coriaria* sp.), *Nothopanax colensoi* and *Pinus laricio*. Wodzicki believed that climate, cover, altitude and soil were the principal factors governing their spread. Although red deer flourish in both the dry Marlborough district and in the cold and wet of Southland their failure to colonize central Auckland and Northland is probably due to climatic conditions. Maps of their distribution largely coincide with those of bush and scrub growth; the clearance of bush – either for land development or through browsing by deer – induced the colonization of new areas. The ecotone between the limit of continuous herbaceous vegetation and scree (about 2,000 m) is the highest altitude regularly colonized. Their preference for forestland below this height may help to explain the species' westerly and northwesterly limits in North Island, where the summits of ranges in east Taranaki and south Waikato are all below the bush line. Wodzicki considered that the presence of red deer on the Kaingaroa Plains, where there are no high mountains, is explained by the existence of induced forest vegetation, open rangeland formed by thousands of hectares of fire-breaks, and by the fact that much of the region lies at altitudes of between 700 and 800 m. Distribution maps show a marked coincidence between the presence of red deer and the existence of beech forests; since beech, except saplings when food is scarce, is seldom eaten this suggests similar soil requirements, and may explain such anomalies as the absence of red deer from parts of central Westland and the Catlins region.

The economic damage caused by red deer in New Zealand is only exceeded by the havoc wrought by rabbits. In forestland their browsing and trampling effectively inhibits regeneration. As early as 1893 Walsh warned against the attempted naturalization of deer; he pointed out that New Zealand woodlands, unlike those of Europe, had never been subjected to the depredations of browsing ruminants, and stressed the interdependence of the various constituent parts of native bush affected by deer. In 1919 the Forest Department expressed its concern at the damage being caused by deer to commercial woodlands, and three years later Perham (1922) stated categorically that 'regeneration or replanting of the State's exotic plantations can never be accomplished if the presence of deer is allowed'. Thomson (1922) sounded a further warning about the potential economic threat posed by red deer, and L. Cockayne, who considered that they were the most destructive alien species in New Zealand, wrote: 'The New Zealand high-mountain forests are the finest in the temperate zones as protection forests, on account of their undergrowth and the thick water-holding bryophyte carpet, or cushions, of the floor, but these features can be entirely destroyed by deer, and with this the main value of these forests is gone.'[21]

These warnings – and those of other like-minded ecologists – went largely unheeded, since it was generally believed that deer were an important economic asset in New Zealand, and that, in any case, it would be impossible to preserve intact the country's endemic fauna and flora. V. D. Zotov et al. (1939) believed that the red deer 'has proved, so far, the most destructive animal on the Tararuas', where large quantities of woody plants were eaten and many trees severely frayed[22] between 80 and 160 cm above ground level. At high altitudes overgrazing and browsing resulted in the depletion of numbers of subalpine plants, tussock grassland and shrubs, which were replaced by low-growing bog species. Red deer were found to browse on a large variety of shrubs and seedling trees in the Tararuas (thus delaying forest regeneration), and on some species of tree ferns: their destruction through grazing of soil-retaining plants increased the risk of erosion, and the creation of tracks through the understorey and trampling helped to consolidate the forest floor.

Research has shown that all native forest in New Zealand is liable to suffer from the presence of red deer; drier woodlands with a more open undergrowth are the worst affected, while in more humid forests the greatest risk is of soil erosion. The type of damage to the vegetation has been found to correlate with its constitution, which in turn corresponds with climatic conditions, principally with temperature and precipitation.

In exotic coniferous woodlands only one – the seedling – stage in the full forest cycle (seedling regeneration, thicket, pole and mature tree) suffers severely from the presence of red deer, with some damage to the pole stage caused by fraying. In South Island as early as 1911 – and later also in North Island – damage to seedling plantations by deer was being reported on both state-owned and private land. By the late 1940s Wodzicki found that, in spite of heavy culling, such state forests as Kaingaroa, Balmoral, Hanmer and Golden Downs held substantial red deer populations which were causing considerable damage by nibbling seedlings, and lesser harm, through ring-barking caused by fraying, to pole-stage trees.

H. S. Gibbs et al. considered that 'Deer, chamois, tahr and goats . . . accelerate the rate of erosion . . . by grazing the sub-alpine and alpine plants and by destroying the undergrowth in the beech forest . . . trampling of these animals over the bare slopes also increases the movement of soil downslope, and the beaten tracks lead to the formation of gullies'.[23] W. P. Packard believed that deer were almost solely responsible for the destruction of plant cover around Lake Coleridge,[24] while Zotov (1949) considered that 'it can be readily demonstrated that these eroded areas [on Mount Holdsworth in the Tararuas] are a development of the last ten years or so as the result of trampling by deer'. As Wodzicki, however, pointed out, red deer are spread over vast areas which are subject to natural physical or accelerated erosion through interference by man; it is thus difficult to assess the precise part played in soil erosion by red deer and other alien species.

By providing competition for grazing with domestic livestock, through contributing to the acceleration of erosion by overgrazing subalpine pastures, and by their destruction of vegetative cover, deer have proved as serious a pest in New Zealand to agriculture as to silviculture. In some areas red deer feed on growing crops (especially turnips) and are accused of spreading weeds and of disturbing lambing ewes. The most important parasites for which red deer in New Zealand provide a reservoir are liver fluke (*Fasciola hepatica*), lungworm (*Dictyocaulus viviparus*), tapeworm (*Cysticercus tenuicollis*) and the cattletick (*Haemaphysalis bispinosa*).

The harm caused by red deer to New Zealand's wildlife is hard to determine. From around the turn of the century until its rediscovery by Dr G. B. Orbell in 1948 in a remote region of the Murchison Mountains in the Fiordland National Park, the flightless gallinule, the takahe (*Notornis mantelli*), was believed to be extinct. The future of its present population of about 125 is jeopardized by competition with red deer (which spread from releases near Manapouri in the takahes' range by way of the Kepler Mountains and around the head of the South Fiord of Lake Te Anau) for winter foods such as grasses (e.g. *Chionochloa conspicua*), sedges (*Uncinia affinus* and *U. clavata*), *Carex coriacea* and rushes (e.g. *Juncus gregiflorus*). Between 1948 and 1978 at least 17,000 red deer in this area were shot, and the population declined to around 815. Since 1978 there has been a tendency for the deer to desert the takahes' range in the subalpine scrub, the alpine grasslands and the climax forest, in favour of seral woodlands; this should act to the takahes' advantage.

Degredation of its habitat through trampling and browsing by red deer is said to be partially responsible for the reduction by 1980 of the total known population of the owl parrot or kakapo (*Strigops habroptilus*) to about 40, most of which are believed to be males. On the other hand, the suggestion by Roots (1976) that the reduced distribution of the endemic blue duck (*Hymenolaimus malacorhynchos*) can be attributed to the pollution of mountain streams through soil erosion caused by deer cannot be sustained; the clearance of native forests, predation by exotic mammals and the introduction of various food competitors (e.g. brown and rainbow trout (*Salmo trutta and S. gairdneri*)) are more likely reasons.[25] Hybridization with introduced wapiti (*Cervus canadensis*)[26] and sika (*C. nippon*),[27] where it occurs, is usually detrimental to those species. The clearance of dense thickets by browsing has resulted in the creation of racks[28] which have assisted in the spread of brush-tailed possums (*Trichosurus vulpecula*).[29]

The commercial value of red deer in New Zealand in some measure compensates for the damage they cause. Stalking for sport – especially by overseas sportsmen – earns foreign revenue, and the sale of venison (helped by the improved methods of recovering carcasses from remote areas, e.g. by helicopter), has resulted in a valuable export trade. Tails, antlers in velvet, testicles and sinews are much in demand in China as reputed aphrodisiacs, and the skins of embryos and newly born calves are used in the manufacture of ladies' handbags and gloves.

In about 1906–8 a large number of malformed heads and hummels[30] began to appear, especially in South Island, those from the Lake Hawea area being particularly weak, stunted, twisted and ill-formed. The government's Chief Veterinary Officer, Mr J. A. Gilruth, considered that 'their aberrant condition was congenital in origin' – an erroneous assumption which led to further importations 'in the hope', Donne (1924) wrote, 'that the introduction of new blood would, in a measure, remedy that tendency of the Hawea herd to produce aberrant antlers'. The continuing decline in the quality of New Zealand red deer heads was considered to be the result of the importation of inferior stock. Wallace,[31] on the other hand, was of the opinion that the 'environment is much more likely to be the predominating factor in determining the type of head than heredity. The effect of climate and food on the growth of a stag's horns is well known . . .'. Wallace's view was more or less paralleled by that of Wodzicki (1950), who wrote that 'owing to the absence of carnivorous enemies, deer had become too numerous for the existing food-supply, and it is considered that this, rather than some genetic factors, was responsible for the appearance of malformed heads'.

When major liberations ceased in 1926 (although illegal releases are said to continue today) red deer were established in most mountain regions from the Bay of Plenty in North Island to Southland in South Island. By the time of Wodzicki's survey in 1947 most of the larger herds had amalgamated. A decade later their range had increased still further – in North Island to the north-east of Gisborne and into the western Taranaki, in South Island into parts of central Westland and the more remote regions of Fiordland, and throughout Stewart Island. Only the north of North Island (approximately north of a line between Raglan and Tauranga) and the south-east of South Island (south of a line between Invercargill and Dunedin) remained uncolonized. Today, red deer are widely established in all suitable habitats, ranging from montane forests, scrub and grasslands to higher lowland pastures. They occur in the damp forests of Fiordland and Stewart Island in the south, in the dry beech woodlands of Canterbury and central North Island, in exotic coniferous plantations and on improved pastureland. Although they may live almost entirely in woodlands (as in continental Europe) such as the Urewera forests of central North Island, their preferred habitat remains the high country of beech forests and tussock grass in South Island, where climatic, orographic and vegetational factors combine to produce optimum conditions.

In the early days, red deer stalking in New Zealand was controlled by the various acclimatization societies, who issued licences and limited the numbers shot. Following the population explosion, and the resulting deterioration in the quality of antlers, some societies instituted culling programmes and even bounty systems. By 1927 the protection of red deer in state-owned forests was abandoned, and three years later all protection was removed. In 1931 the control of red deer became the responsibility of the Department of Internal Affairs, whose stalkers shot many thousands annually; in 1956 the Noxious Animals Act transferred this duty to the Forest Service. Between 1931 and 1968 a total of well over a million red deer were killed by government stalkers; although this resulted in the disappearance of large populations containing herds in excess of a hundred individuals, in many areas red deer are still a serious pest species. As Wodzicki (1961) has said, 'No other introduced mammal, except perhaps the rabbit, is economically as important [in New Zealand] as the red deer'.

The British race *scoticus* is confined naturally to parts of Scotland, north-west England and probably south-west Ireland. Wild populations of park breeds are established in parts of Somerset, Devon, Hampshire, Surrey, Sussex (Ashdown Forest), East Anglia (Thetford Chase), Essex, north Staffordshire, Cheshire, west Yorkshire and Durham.[32]

In continental Europe there have been many translocations of red deer within their natural range.

In the 1930s Señor Aaron de Anchorena imported a number of red deer from Europe (as well as fallow and axis deer) to his *estancia* La Barra de San Juan, 200 km west of Montevideo. Although no red deer are believed to have escaped from their 500-ha enclosure, they (and fallow and axis) from this *estancia* were used by Señor de Anchorena to stock Isla Victoria in Lake Nahuel Huapi, Argentina.

Writing of the island of Wakaya in the Fijian Islands in 1951 R. A. Derrick reported that 'During the early years of the present century Red Deer were introduced, probably from New Caledonia, and they have increased to such an extent that they tend to become a nuisance.'[33] In fact, the only deer known to occur on Wakaya are fallow deer.[34]

NOTES

1. Haltenorth & Diller (1980) and Lehmann (1969).
2. See p. 151.
3. See p. 273.
4. See p. 311.
5. See p. 307.
6. See p. 176.
7. See p. 213.
8. See p. 247
9. See p. 117.
10. See p. 178.
11. See p. 179.
12. See p. 311.
13. For much of the history of red deer in Australia I am indebted to Bentley (1978) and Frith (1979).
14. April, 179 (1949).
15. See Le Souef (1912); Kingsmill (1920); H. Colebatch, ed., *A story of 100 years; Western Australia, 1829–1929* (F. W. Simpson, Perth, 1929); Glauert (1956); Bentley (1957); Allison & Coombes (1969); and Long (1972).
16. For much of the information on red deer in New Zealand I am indebted to Thomson (1922), Donne (1924), Wodzicki (1950) and Logan & Harris (1967).
17. *Stalks abroad; being some account of the sport obtained during a two years' tour of the world* (Longmans, London, 1908).
18. *Deer forest life* (A. Maclehose, London, 1932).
19. See note 17.
20. *New Zealand deer heads* (Country Life, London, 1924).
21. *Bulletin of the New Zealand Forest Service* No. 4 (1) (1926).
22. Damage to a tree caused by deer rubbing velvet (the protective covering of skin) from their new-grown antlers.
23. *New Zealand DSIR Bulletin* No. 92 (1945).
24. *New Zealand Geographer* **3**: 19–40 (1947).
25. J. Kear & G. Williams, *Wildfowl* **29**: 16 (1978).
26. See p. 208.
27. See p. 214.
28. Paths or trails worn by deer.
29. See p. 10.
30. A stag whose antlers, due to some physiological defect, fail to grow; these animals are usually, though not invariably, heavier than average.
31. See note 17.
32. V. P. W. Lowe & A. S. Gardiner, *Journal of Zoology, London* **174**: 185–201 (1974) and V. P. W. Lowe, in Corbet & Southern (1977:411–23).
33. *The Fiji Islands: a geographical handbook* (Government Printer, Suva, 1951)
34. See p. 185.

Wapiti
(Cervus canadensis)

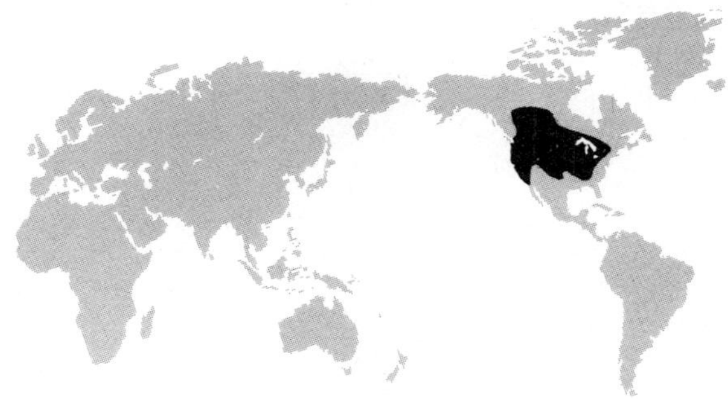

Note: Present natural distribution is much reduced from original range shown.

NATURAL DISTRIBUTION
Formerly much of North America between about 34° N and 56° N. Now, through the advance of civilization, restricted to parts of western Canada and the northwestern United States.

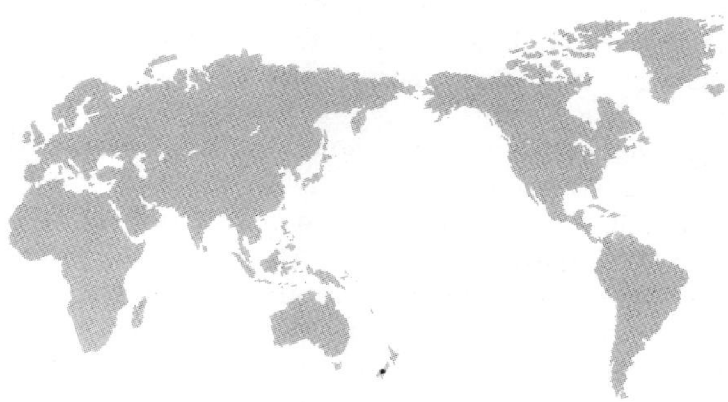

NATURALIZED DISTRIBUTION
Australasia: New Zealand. (In the British Isles, Austria and the USSR hybridization has occurred between introduced wapiti[1] and native red deer.)

EUROPE

British Isles

In an attempt to improve the quality of the heads of native red deer (*Cervus elaphus*), wapiti have from time to time been introduced to some Scottish deer forests (see Table 13). Since hybridization between the two species in New Zealand and

Table 13 Introductions of wapiti to Scotland

Date	*Number*	*Locality*	*Introduced by*	*Source*
September 1819	2 pairs	Dunkeld, Perthshire	Duke of Atholl	Lord James Murray, Datchet Park, Buckinghamshire, England
1894	1 hind	Brodick Castle and Dougarie, Isle of Arran, Buteshire	Duke of Montrose	?
*c.*1904	?	Mamore, Inverness-shire	Mr Frank Bibby	Sir Arthur Grant of Monymusk, Aberdeenshire
*c.*1912	?	Killiechonate, Inverness-shire	Sir P. Walker	Wyoming, USA
Before First World War	1 stag	Glenfinnan, Inverness-shire	Mr G. B. Bainbridge	?
Before First World War	30 +	Meoble and Lettermorar, Inverness-shire (kept in 80-ha enclosure for breeding with red deer; died out between 1914 and 1918)	Mr Walter Jones	?
Before First World War	2 stags	Cumloden, Kirkcudbrightshire	Earl of Galloway	

the Soviet Union (see below) resulted in a deterioration in heads rather than in an improvement, it is not surprising that the experiments were unsuccessful.

In about 1870 some wapiti hinds are believed to have swum from Belle Isle on Windermere, Lancashire, England, to the mainland, where they bred with native red deer on the Graythwaite Estate. In about 1859 two wapiti stags and a hind were introduced to Powerscourt, County Wicklow, Ireland, by Viscount Powerscourt, where they hybridized with red deer. In 1922 the deer at Powerscourt escaped into the Wicklow Mountains around Ballynabrocky.

Austria[2]

Several hundred wapiti were introduced to Austria by the Emperor Francis Joseph where they hybridized with the native red deer.

USSR[3]

In the nineteenth century wapiti and red deer of the Asiatic race *C. e. asiaticus (sibiricus)* were introduced by the Czars to the southern Urals, where hybridization resulted in the growth of inferior antlers.

New Zealand[4]

In the 1870s Sir George Grey (formerly Governor of New Zealand) released a pair of wapiti on Kawau Island in Hauraki Gulf where, after the death of the hind, the stag become so dangerous that he had to be shot. Some years later a second pair, from Seattle, Washington, was shipped to Dunedin by the master of a US timber vessel. During the voyage the stag died, as did the hind after giving birth in Dunedin to a male calf. This young stag eventually escaped from captivity and was subsequently shot as a 15-pointer on Flagstaff Hill.

In 1902, while on a visit to New Zealand, Mr St George Littledale of Bracknell, Berkshire, England – an experienced big game hunter – met Mr T. E. Donne of the Government Tourist Department. On his return to England in the following year Littledale approached Theodore Roosevelt – then President of the United States – who arranged for the New Zealand government to receive 10 wapiti (three stags and seven hinds) from the National Zoological Park in Washington, DC, in exchange for a number of endemic New Zealand birds, an endemic tuatara (*Sphenodon punctatus*) and some roe deer (*Capreolus capreolus*). Donne also acquired a further 10 wapiti, through Mr H. E. Richardson, from the Indian Game Reserve in Brookfield, Massachusetts; all 20 were of the Rocky Mountains subspecies *C. c. nelsoni*, and not *C. c. roosevelti* as suggested alternatively by Whitehead (1972).

Under the supervision of Mr F. Moorehouse of the New Zealand Tourist Department the deer were transferred by rail from the east coast to San Francisco, California, where at first the shipping company refused to accept them as arranged since the freight charges had not been prepaid by the New Zealand government. Moorehouse cabled Donne in New York who was eventually forced to pay the charges himself.

The wapiti, two of which died in a storm during the voyage, finally sailed from San Francisco on 11 February 1905; on their arrival in New Zealand, the 18 survivors (of which one hind died shortly after being released) were trans-shipped on the government vessel *Hinemoa* to the head of George Sound in Fiordland, South Island, where they were released on 3 March.

The wapiti were left undisturbed until 1909, when Donne reported numerous slots[5] of adults and young near the point of release. In 1921 Moorehouse, Mr J. H. Wadworth of the Southland Acclimatization Society and a professional photographer, Mr L. Hinge, visited the area to find out what success, if any, the deer had had in establishing themselves. They 'visited Lake Alice and found many tracks of deer having lately visited the area', and saw further evidence of the presence of wapiti in Caswell Sound to the south and Bligh Sound to the north of the point of release. On 22 March Moorehouse reported to the Under-Secretary of the Department of Internal Affairs that 'tracks were found leading in all directions'. At the eastern end of Lake Katherine they found 'a fine wapiti cow' (with, according to Wodzicki (1950), a calf) which Hinge managed to photograph. On returning to the western end of the lake

> Hinge found upon the shore a cast antler. Here again were seen traces where many deer had been passing to and fro in the dense bush. Returning by a different route, numerous tracks were observed . . . around the southern end of Lake Marchant [at the head of Caswell Sound] again tracks were noted, showing the wapiti had become well established and were increasing.[6]

In December of the same year the Southland Acclimatization Society despatched its head ranger, Mr C. J. Evans, to discover whether wapiti had spread to the eastern side of the divide. Near Lake Hankinson, north of the Middle Arm of Lake Te Anau, Evans 'found the first signs of wapiti. The marks were fresh . . . about two miles from Lake Hankinson on the way to Lake Thomson we got our first glimpse of one . . . in excellent condition.' Evans observed that the deer appeared to be feeding mainly on the bark of ribbonwood and black maple and on gum leaves. He reported that:

> I may venture the opinion that the valley [of the Middle and North Arms of Lake Te Anau] is carrying a herd of about twenty wapiti, but owing to the nature and feed of the country, I do not think it probable that they have extended to the high, hilly country, but have confined themselves to the valleys where feed abounds. No signs of moose [*Alces alces*[7]] were seen, nor do I think it probable that they have reached this locality, it being too far from the point of liberation . . . I have perused the Moorehouse report of last season and I am of the opinion that the main herd of wapiti is working south along the sounds, this party's first trace of them being at Lake Katherine, George Sound, and the next at Thompson Sound.[8]

Although Evans found traces of both wapiti and moose (which had been released in the area in 1910) at Dusky Sound, Moorehouse, as Banwell (1966) points out, made no mention of seeing any deer around Thompson Sound a few kilometres to the north.

In 1923 the first licences were issued for stalking wapiti, which in 1935 were open to shooting by anyone with a firearms permit. By 1924 wapiti were being shot in the Eglinton Valley, around Casewell Sound and the Middle Arm of Lake Te Anau, and in 1933 a stag was killed in the Gleisnock. The inaccessibility of their habitat, however, effectively limited the numbers shot until, in 1938, official government parties began an intensive shooting programme, which was only curtailed by the outbreak of war in the following year.

The precipitous cliffs and steep valleys of the Fiordland National Park created effective channels for the dispersal of wapiti from George Sound. The Edith and Katherine valleys at the head of the sound, close to the point of release, were soon colonized, and from there the deer spread into the Wapiti River and the Gleisnock and Stillwater valleys. The neighbouring Lugarburn was, however, not occupied until the mid-1930s, and the Midburn not for a further decade. Similarly, other valleys towards Lake Te Anau and some of the peninsula between the fiords facing the Tasman Sea were only colonized between about 1940 and 1950.

According to John K. Howard and H. Wendell Endicott,[9] the leaders of an expedition to Fiordland in 1947 from the Museum of Comparative Zoology at Harvard University, USA to compare North American and New Zealand wapiti, the breeding range of the latter was:

> Bounded on the north by a line from the north side of Bligh Sound through the Franklin Mountains to the northern side of the head of the Worsley Fiord, on the east roughly by the heads of the Worsley, North Fiord and the heads of the north-east and south-west arms of the Middle Fiord; on the south by a line running from the south side of the head of the south-west arm through the Murchison Mountains to the Tasman Sea, so as to include the south shores of Charles Sound, and on the west by the Tasman Sea.

Wapiti found outside this area were deemed by Howard and Endicott to be 'strays or wanderers, the few pioneers or "advance guards" of the expanding herd'; the principal factors limiting the deers' spread were that 'on the whole, the country is much more precipitous . . . than the country occupied by the wapiti in the United States and Canada [and] it is much more densely forested'. The New Zealand wapiti 'frequents the edges of the open slides much more so than in the States [and] on the whole he probably browses [in woodland], as distinguished from grazing [in the open], more in New Zealand than he does in the United States'. Because of this 'the capacity of the range available and the size of the herd using it' are also important factors in controlling the wapitis' distribution in New Zealand. The effects of the incursion of their territory by red deer (*Cervus elaphus*),[10] resulting in competition for both food and habitat, are aggravated by the wapitis' abandonment in New Zealand of the seasonal migration which they make in North America between summer and winter ranges.

Howard and Endicott reported that the wapitis' preferred foods at lower altitudes included broadleaf (*Griselinia* sp.), lacebark (*Hoheria glabrata*), fern (*Blechnum discolor*), fivefinger (*Nothopanax lineare*), black shrub (*Coprosma* sp.), native broom (*Carmichaelia* sp.), veronica, beech (*Nothofagus* spp.), native grasses, rimu (*Dacrydium cupressinum*) and *D. biforme* (or *bidwillii*). At higher altitudes the deer fed mainly on *Hebe* sp., mountain tutu (*Coriaria* sp.) and snowgrass (*Danthonia* sp.). The Harvard expedition concluded that 'The party did not observe any results of the feeding in this area which, in the opinion of its members, could be said to be damaging the forest. The regeneration is perfect'. (This view was at variance with other contemporary opinions.)

In their report of the expedition's findings Howard and Endicott warned against the wholesale slaughter of New Zealand wapiti which could, they suggested, provide a gene bank should the species ever become endangered in North America. Just how little was known about wapiti in New Zealand at the time is shown by the discrepancy between the various estimates of the total population. The Department of Internal Affairs put the figure at no more than 500; Howard and Endicott suggested 850; while experienced stalkers (who could, admittedly, be expected deliberately to inflate the total to ensure the continuation of their sport) claimed a total of between 1,200 and 1,500.

The wapiti enjoyed a respite from 1939 to 1952, when the region was again opened for hunting. In the late 1940s the government began determined attempts (which are still being continued) to kill off the red deer in the area, which were spreading north and were reported to be hybridizing with the wapiti. In 1948 Dr Olaus J. Murie, an authority on Rocky Mountain wapiti, conducted a detailed examination of deer in the Fiordland National Park, from which he concluded that some hybridization was undoubtedly taking place. In 1954 the region was placed under the control of the New Zealand Deerstalkers' Association who, from 1963 (when the wapiti population was estimated at around 2,000), shared management responsibilities of the 1.2 million ha with the Fiordland National Park Board.

Wapiti are now established in about a sixth of the national park from Bradshaw Sound and Coronation Peak in the south to between Sutherland and Milford sounds south of the Castle Mountain/Mount Mackenzie massif in the north (a distance of some 80 km), and 50 km inland to the western shore of the northern half of Lake Te Anau. This area includes the Franklin and Stuart ranges in the north, and the seaward side of the Murchison Mountains and foothills of the Kepler

Range in the south. Wandering stags are from time to time sighted outside this area. The species' further expansion to the south is inhibited by the firmly entrenched red deer population, and that to the north by the precipitous sides of Milford Sound and human settlement in the Cleddau Valley.

The forests of the wapitis' habitat in Fiordland are dominated by silver beech (*Nothofagus menziesii*) with lesser numbers of mountain beech (*N. solandri* var. *cliffortioides*). Rimu occurs near the coast and fiords and around low-altitude lakes, and southern rata (*Metrosideros umbellata*) on the valley slopes. Above the bushline, at around 1,000 m, there is an intermittent belt of subalpine scrub – mainly *Olearia crosby-smithiana, O. colensoi, Phormium colensoi, Neopanax colensoi, Dracophyllum* sp. and *Hebe* sp. The mountain tops are covered with snowgrasses and herbs – chiefly *Chionochloa acicularis, C. crassiuscula, C. flavescens, C. pallens* and *C. teretifolia*.

Banwell (1966) considered that 'This wapiti herd, the only one of its kind in the Southern Hemisphere, is the heritage of every New Zealander of the outdoors, be he young or old . . . it is our duty to see that they are preserved and managed in numbers compatible with their rangeland, to be maintained for future generations to enjoy'. The ecological importance and genetic value of this – admittedly unique – population hardly justifies such extravagant sentiments.

According to Dr Antoon de Vos 'there are now moves afoot to introduce Canadian wapiti to New Zealand for deer-farming purposes'.[11]

Between 1887 and 1890 wapiti were breeding in captivity in Australia in the Victoria Acclimatization Society's collection at Royal Park; whether any were ever released into the wild is unknown.

Rocky Mountain wapiti (*C. c. nelsoni*) have been transplanted into several parts of the former range of *C. c. canadensis*, especially in Colorado, Idaho, Montana, Oregon and Wyoming. The form *C. c. roosevelti* has been introduced to Afognak Island off the southern coast of Alaska.

In 1941 18 wapiti from the Wichita National Wildlife Refuge in Oklahoma, USA, were released in northern Coahuila, Mexico, where by 1943 they had all disappeared or been killed. In 1952 and 1955 further attempts were made to establish wapiti – from the Yellowstone National Park – in Coahuila, again apparently without success.

NOTES

1. Wapiti – sometimes known in North America as elk – are closely related to red deer (*Cervus elaphus*) (especially to the Asiatic form) with which they may be conspecific.
2. K. Z. Lorenz, *King Solomon's ring* (Methuen, London, 1952).
3. M. A. Romanow, *Fremde Wildarten im Urwald von Bialowies* (1932) and Lindemann (1956).
4. For much of the history of wapiti in New Zealand I am indebted to Wodzicki (1950) and Banwell (1966).
5. The tracks or footprints of deer.
6. Quoted by Banwell (1966).
7. See p. 242.
8. Quoted by Banwell (1966).
9. Quoted by Wodzicki (1950).
10. See p. 202.
11. Personal communication (1980).

Sika Deer

(Cervus nippon)

Note: Probably now extinct in North Vietnam and China (apart from Manchuria).

NATURAL DISTRIBUTION

Formerly from the Ussuri region of Siberia south through Korea and Manchuria to North Vietnam, Japan and Taiwan. Now probably extinct in North Vietnam and China (apart from Manchuria).

NATURALIZED DISTRIBUTION

Europe: British Isles, Czechoslovakia, Denmark, France, Germany, Austria, Poland, USSR; North America: USA; Australasia: New Zealand.

EUROPE

British Isles[1]

Sika deer were first introduced to England in 1860, when a pair were presented to the Zoological Society of London for their collection in Regent's Park. In the same year Viscount Powerscourt obtained a stag and three hinds (which by 1884 had multiplied to over one hundred) from the animal dealer Carl Jamrach, for his estate at Enniskerry in County Wicklow, Ireland, from where sika were subsequently transferred as breeding stock to several parks in England, Scotland and Ireland; the first of these were probably at Waddesdon in Buckinghamshire, England in about 1874 and at Tulliallan in Fife, Scotland around 1870. Since then a considerable number of deer parks have been stocked with sika, from several of which they have escaped and have become established in the surrounding countryside.

Table 14 gives the earliest dates for the introduction of sika to deer parks in English, Scottish and Irish counties.

In England sika deer are established in the wild mainly in the south, where they are most common on the heathlands of south-east Dorset in Wareham Woods as far as the limits of the Bagshot and Reading beds, and in the south of the New Forest in Hampshire. Some of the sika introduced to Brownsea Island in Poole Harbour, Dorset, in 1896 escaped by swimming to the Isle of Purbeck, where they became established; when Hyde Park, near Wareham, was requisitioned by the War Office in 1939 some sika escaped and joined up with those already on the Purbeck Peninsula, and these deer were the ancestors of the present sika population in south-east Dorset. A herd of around two dozen also survives on Lundy Island in the Bristol Channel. A pair of

Table 14 Earliest records for the introduction of sika deer to deer parks in England, Scotland and Ireland

Date	*County*	*Estate*	*Date*	*County*	*Estate*
England			**Scotland**		
*c.*1874	Buckinghamshire	Waddesdon Park (from Powerscourt, Co. Wicklow, Ireland)	*c.*1870	Fife	Tulliallan
*c.*1880	Dorset	Melbury Park	*c.*1887 or 1889	Ross and Cromarty	Achanalt, Lochrosque (from Powerscourt, Co. Wicklow)
*c.*1885	Hampshire	Hursley Park	*c.*1893	Argyllshire	Carradale (from Fawley Court, Buckinghamshire)
*c.*1890	Essex	Weald Park	Late 19th century	Sutherland	Rosehall
*c.*1890	Kent	Knole Park	*c.*1900	Inverness-shire	Glenmazeran and Glenkyllachy (from Fawley Court), and Aldourie Castle (from Rosehall)
*c.*1890	Rutland	Exton Park	1908	Peeblesshire	Dawyck Park (from Japan)
1892	Sussex	Leonardslee	*c.*1920	Caithness	Berriedale (from Welbeck Abbey, Nottinghamshire)
Late 19th century	Northamptonshire	Whittlebury Park	?	Angus	Kinnaird Castle
Late 19th century	Surrey	Park Hatch Park			
Late 19th century	Westmorland	Rigmaden Park	**Ireland**		
*c.*1900	Bedfordshire	Woburn Abbey	1860	Wicklow	Powerscourt Park
*c.*1900	Norfolk	Melton Constable Park	*c.*1865	Kerry	Kenmare (from Powerscourt)
1904–6	Yorkshire	Park Nook	1870	Fermanagh	Colebrooke Park (from Powerscourt)
Early 20th century	Gloucestershire	Great Barrington Park	1891–2	Tyrone	Baronscourt Park (from Colebrooke)
Early 20th century	Northumberland	Hulne Park	?	Down	Castlewellan Park (from Powerscourt)
Early 20th century	Nottinghamshire	Rainworth Lodge	?	Limerick	Glenstal Park (from Powerscourt)
Early 20th century	Oxfordshire	Crowsley Park, Fawley Court and Nuneham Park			
1925	Shropshire	Weston Park			
1927 or 1929	Devon	Lundy Island (from Surrenden-Dering, Kent)			
?	Wiltshire	Rushmore Park			

Source: After Lever (1977)

sika which escaped from Lord Montagu's collection at Beaulieu Abbey in Hampshire in 1904 were joined in the wild by another pair which were deliberately released in the following year; these four deer were the ancestors of the present New Forest stock which lives mainly between Beaulieu and Brockenhurst.

In southern Oxfordshire and Buckinghamshire a few sika are established in the chalkland woods (mainly beech (*Fagus sylvatica*)) of the Chiltern Hills around Henley-on-Thames and in the Hambledon Valley; small populations also exist in parts of Kent and Sussex.

In northwestern England sika are established in Pendle Forest and Ribblesdale, between Bolton-by-Bowland and Ribchester, in Yorkshire and Lancashire, where some hybridization with native red deer (*Cervus elaphus*) has occurred.

The largest Scottish population of free-ranging sika is probably that living mainly in the 1,600 ha of the Dawyck Estate in Peebleshire, from where some crossed the border into Dumfriesshire in 1971. Smaller numbers are established at Tullialan, Fife (from where some have crossed the border into Clackmannanshire), at Kinnaird, Angus and in several districts in Argyllshire.

In the Highlands sika are found on both sides of Loch Ness and the Caledonian Canal in the Great Glen, Inverness-shire: further north a few are established in parts of Ross and Cromarty, Sutherland and Caithness.

In the Republic of Ireland several hundred sika are established in the wild on the 40,500 ha of the Kenmare Estate in County Kerry, from where some have wandered over the border into County Cork. They are common in over 16,000 ha of woodlands in the eastern Wicklow Mountains (where they have hybridized with native red deer), whence some have crossed the border into County Dublin. In Northern Ireland sika are established in several districts in counties Fermanagh and Tyrone. The severe winter of 1962–3 almost exterminated sika deer in Ireland, since when they have staged a dramatic recovery and by the late 1970s were estimated to number around 2,500.

Sika have been emparked on at least two estates in Wales, but none have ever become established in the wild.

Although the sika population in the British Isles has increased considerably in recent years, especially in Scotland and Ireland, little damage has been reported and they are as yet of scant economic significance.

Czechoslovakia

Sika have been imported from the Far East to a large number of deer parks and game reserves in Czechoslovakia. As communism gained the ascendancy such symbols of decadence were abolished, and the deer were either shot or escaped. In 1977 an estimated 965 sika were established in the wild in a minimum of 11 localities in Czechoslovakia.

Denmark

Sika were first introduced to Denmark in 1900 when a pair – probably from Viscount Powerscourt in Ireland – were imported to a park at Svenstrup. Nine years later a consignment from the German animal dealer Carl Hagenbeck of Hamburg was turned out on an estate at Knuthenborg. On several occasions between 1900 and 1910 sika from Svenstrup and Knuthenborg were deliberately released, and the present wild populations in Denmark are all descended from animals bred on these two estates.

The earliest recorded free-ranging herd – of uncertain origin and date – was established on an estate at Frijsenborg; with a population in excess of 200 it is today the largest in Denmark, where there are currently at least nine separate free-ranging herds (whose future, in some cases, is in doubt) and a large number of enclosed ones.

In 1918 a stag and two hinds were transferred from Svenstrup to an estate at Katholm; their descendants, which currently number about 100, provide some of the finest trophy antlers in the country.

In 1923 two stags and eight hinds – followed two years later by a further stag and two hinds (all from Knuthenborg) – were released on the 5,000-ha state-owned Gribskov Estate in Zealand, where by 1977 a herd of around 40 was established. Three sika which escaped from a park in west Zealand in 1936 and settled on an estate at Gunderslevholm had increased to about 70 by 1977.

In 1974 Egon Bennetsen estimated the total Danish wild population at around 475, of which about 360 were in Jutland (including about 60 at Hals, Ormstrup-Tange and Sophiendal-Hemstok) and 110 in Zealand, with between three and five on Møn: at Ravenholt on Funen, where there had been a herd of about 50 in 1950, only a solitary hind remained.

Apart from two herds – at Mejlgaard and Ormstrup-Tange – probably founded by immigrants from Frijsenborg, all Danish populations were established either by deliberate releases or from park escapes. Such dispersal as occurs – led by stags – is principally in a westerly direction in central Jutland.

France

In 1890 a stag and three hinds, presented by the Emperor of Japan to M. François Sadi Carnot, President of the Third French Republic, were released in the forest of Marly in Aisne. Their success encouraged other importations – especially to Île-de-France – and sika have at various times been released in Pas-de-Calais, Ardennes, Marne, Aube, Meurthe-et-Moselle, Haut Rhin, Côtes-du-Nord, Loire-Inférieure, Haut-Vienne, Haut-Savoie, Isère, Alpes-Maritimes and Ariège. Wild populations are currently present in at least two localities in northern France and in two in the south.

Sika are also established in several places in Germany, Austria and Poland, and in parts of the Soviet Union west of the Urals.[2]

NORTH AMERICA

USA

In 1956 it was reported[3] that 'In Texas . . . the King Ranch introduced [in 1926] Japanese Fallow Deer'.[4] There is some doubt as to whether this introduction (if it indeed took place) refers to wild (Japanese) sika or to captive European fallow, which have never occurred in the wild in Japan. However that may be, sika – believed to be of the nominate race – were released on Mr Richard Friedrich's Bear Creek Ranch in central Texas in the 1930s and early 1940s; subsequent introductions to Texas such as those on the Rickenbacker Ranch near San Antonio, have included two subspecies, the Dybowski's or Pekin race (*C. n. hortulorum*) and the Formosan form (*C. n. taiouanus*). By 1964 the total Texan population numbered 634. Two years later it had risen to 875, by 1971 to 2,036 and by 1974 to 3,042; by 1979 the population had more than doubled to 6,217 in 49 counties (of which 1,236 were free-ranging in 13 counties) where 105 ranches (mainly in Blanco and Kerr counties) reported sika inside deer-proof fences on 123,943 ha. Indications are that of all exotic game animals in Texas sika have, through competition for food, had the most serious impact on native white-tailed deer (*Odocoileus virginianus*).

In about 1916 Mr Clemment Henry released from an enclosure near Cambridge, Maryland, four or five sika (believed at one time to be hog deer (*C. porcinus*)[5]) on James Island in Chesapeake Bay,[6] where they increased and spread onto neighbouring Taylor's Island and to the mainland in Dorchester County. By 1957 the deer on James Island were estimated to number about 270.[7]

In 1920 Dr Charles Law purchased (probably from Henry) two stags and three hinds which he held for several years in an enclosure near Berlin, Maryland, where three calves were born and one of the stags died. These deer were eventually sold by Law and in about 1930 were released on Assateague Island in Worcester County, Maryland, where within five years they had increased to between 100 and 125 and had spread to the Chincoteague National Wildlife Refuge in the south of the island in Virginia: in 1958 the population was estimated to number 100.[8] (Two other versions of the origin of the sika on Assateague Island have been put forward by Presnall (1958): Mr Dan Trimper of Ocean City, Maryland claimed that in 1923 Captain Will Powell gave his troop of Boy Scouts two or three pairs which were released by them at the northern end of the island; it has also been suggested that the deer swam to Assateague from a neighbouring island owned by a Mr Sam Riddle.)

On the mainland sika are largely confined to the eastern shore of Maryland, where they have spread eastward from Dorchester County across the Nanticooke River into Wicomico County, and are continuing slowly to expand their range. As in Texas they compete advantageously with native white-tailed deer.

Sika deer are also recorded as having been introduced to Michigan and Nebraska.

AUSTRALASIA

New Zealand[9]

In 1885 Mr John Bathgate of Dunedin presented three sika (presumably a stag and two hinds) to the Otago Acclimatization Society who turned them out at Otekaiki near Oamaru on the coast south of the Waitaki River in South Island; although by 1890 they were said to be increasing they eventually died out. (The original source of Mr Bathgate's stock does not appear to have been recorded.)

In July 1904 three pairs of sika – a gift to the New Zealand government from the eleventh Duke of Bedford at Woburn Abbey, Bedfordshire, England – were shipped aboard the SS *Kaikoura* in charge of Captain Reginald C. Clifford. (These were said to be the descendants of a number of the subspecies *mantchuricus* imported to England from Manchuria for the duke by Carl Hagenbeck in 1898. In a letter to T. E. Donne the duke wrote: 'If no other Japanese deer have been imported into New Zealand it will be interesting, as you will then have none but the pure-bred Mantchurian [*sic*] type.'[10] Since the present sika in New Zealand exhibit a wide range of variation in both size and coloration, either the duke was misinformed about the ancestry of his stock or there have been other and unrecorded importations of sika to New Zealand.) During the voyage two fawns were born near Wellington, only one of which, however, survived. In 1905 all seven deer were released near Merrylees (now Poronui) on Mr W. N. Smith's Taharua sheep station, 32 km south-east of Lake Taupo on the eastern edge of the Kaimanawa State Forest in North Island, where Donne anticipated that they would 'provide good sport and a choice variation of food for lucky stalkers'.

As the population increased the deer spread south along the eastern border of the Kaimanawa Forest, reaching the Otupua Valley by the 1930s and the Mohaka Valley area between the Ahimanawa and Kaweka ranges by the early 1940s, from where they dispersed through dense manuka (*Leptospermum scoparium*) scrub between the Kaweka and Black Birch ranges, south to the Burns Range between the Kawekas and Ruahines. By the late 1940s they had also spread north-west through shrublands and red beech forest into the northern section of the Kaimanawa Forest and the headwaters of streams flowing into Lake Taupo. Occasional sightings of stray stags were reported from north of Hawke's Bay. Another dispersal route was through shrublands to the edge of the Urewera Forest on the east of the developing plantations of exotic trees in the Kaingaroa Forest, which do not provide a favourable habitat. Other routes were through the Ngaruroro Valley and into the north and Taruarau valleys from the Golden Hills region, and through splinter movements from the northwesterly dispersal.

Between 1951 and 1970 the rate of spread – especially in the north-west – appears to have accelerated, possibly due to pressures exerted by hunting, logging and the development of land for agriculture and road construction. The dispersal of sika was assisted by the fact that vegetation which had been browsed by red deer,[11] to the point where it could no longer support them, retained enough nutrition to keep sika in a healthy condition; the latter could thus degrade plant communities below the level caused by the former and, except at higher altitudes, largely replaced them as they were forced to seek new territory. Where the two species occur together there seems to be little direct interspecific competition (though the smaller sika stag is more aggressive than his red counterpart), but a survey by the New Zealand Forest Service in 1963–6 revealed a degree of hybridization.

By the late 1940s Wodzicki (1950) reported that damage to young trees by browsing was causing growing concern in the northern Kaimanawas, where the effective regeneration of beech seedlings was being inhibited and excessive trampling was thinning the understorey and thus accelerating erosion.

By the early 1960s sika were established in considerable numbers over an area of some 2,600 km^2, and small scattered populations occurred over a much greater area. Today sika range over more than 11,000 km^2 of country from the Tongariro National Park and the northern Ruahines to the Kaingaroa State Forest and the Maungaharuru Range, the main concentrations being found in the northern Kaimanawa Mountains and in the Kaweka and Ahimanawa ranges. Sika have also penetrated the forests (mainly red beech with kamahi and shrub hardwoods; red and silver beech with kamahi; mountain and red beech; and black, silver and red beech with shrub hardwoods) of the southern Urewera and north-west Ruahines and those of west Taupo. They are now found on forest margins west of the Hauhangaroa and Rangitoto ranges, north and west of Lake Taupo, and seem to be slowly spreading further in those – and possibly other – directions; individual wandering stags have been reported from far outside the species' normal range (e.g. at Te Kuiti, in the Mokai Patea and Wakarara ranges and in the Wairarapa). Their preferred habitat appears to be mixed native beech (*Nothofagus* spp.) forest and native grassland and scrub, where favourite foods include fivefinger (*Nothopanax arboreum*), konini (*Fuchsia excorticata*) and karamu (*Coprosma lucida*).

Some time after 1943 some Formosan sika were imported from Formosa to the zoo on Oshima, northernmost of the Seven Islands of Izu, south of Tokyo, Japan, from where several escaped and, largely due to the absence of predators, became established in the wild; by 1950 the population numbered about 50. The deer are mainly confined to the eastern parts of Oshima which are seldom visited by man.

Kerama or black sika (*C. n. keramae*) were introduced from Japan to the Ryukyu Islands before 1757 when they are known to have been well established. The deer currently exist in very small numbers on three little islands in the group – Yakabi-Shima, Kuba-Shima and Keruma-Shima: their long isolation has resulted in the development of an insular form – the Ryuku sika – with distinctive characteristics.

Cecil Rhodes introduced sika to his estate at Groote Schuur near Cape Town in 1897, where 40 years later there were said to be about 20; there are none in the wild in South Africa today.

There are several unsubstantiated reports of sika having been introduced to Morocco; none are established in the wild there today.

Despite Whitehead's (1972) assertion that 'Sika deer have been satisfactorily introduced to the forests of Madagascar', no such introduction has ever taken place. The only alien deer known to have been introduced to the island are fallow (*Cervus dama*)[12] and Timor (*C. timorensis*).[13]

Between 1887 and 1900 the Acclimatization Society of Victoria, Australia released into the wild Japanese (*C. n. nippon*) and Formosan sika, none of which appear to have survived for long.

NOTES

1. Three races or subspecies of sika deer are represented in the wild in the British Isles: the Japanese (*C. n. nippon*), the Manchurian (*C. n. mantchuricus*) and the Formosan (*C. n. taiouanus*), all of which have been known to interbreed. In addition, herds of Kerama or black sika (*C. n. keramae*), Chinese sika (*C. n. kopschi*) and Dybowski's or Pekin sika (*C. n. hortulorum*) have been kept in captivity at Woburn, Whipsnade and Surrenden-Dering parks. In northeastern England and southeastern Ireland sika have hybridized with the native red deer (*C. elaphus*).
2. Most of the sika deer in continental Europe are believed to be the Japanese form (*C. n. nippon*), except for those in the Soviet Union, which are reputedly the Dybowski's or Pekin race (*C. n. hortulorum*).
3. De Vos, Manville & van Gelder (1956: 173).
4. See p. 186.
5. See p. 218.
6. Presnall (1958).
7. *Proceedings of the 12th Annual Conference of the Southeastern Association of Game and Fish Commissioners* (1958).
8. Flyger (1960).
9. For much of the information on sika in New Zealand I am indebted to Kiddie (1962) and Davidson (1973).
10. Donne (1924).
11. See p. 201.
12. See p. 173.
13. See p. 220.

Hog Deer
(Cervus porcinus)

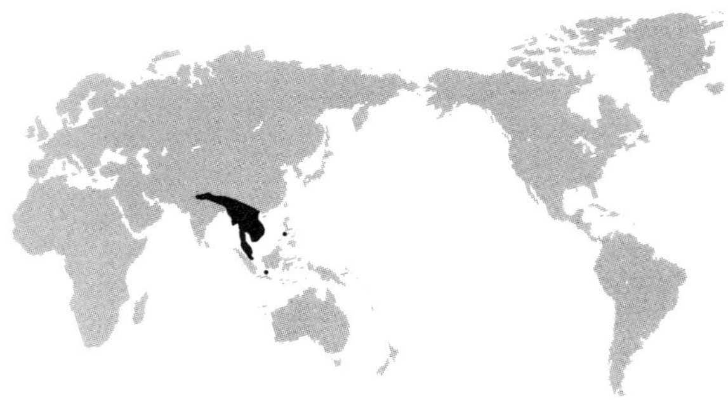

NATURAL DISTRIBUTION
Northwestern India to Indochina (including parts of Laos, Thailand, Vietnam and Cambodia), Bawean Island (Java) and Calamian Island (Philippines).

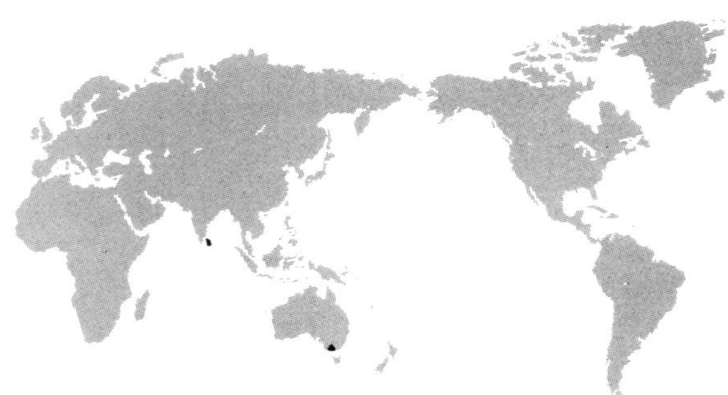

NATURALIZED DISTRIBUTION
Asia: Sri Lanka; Australasia: Australia.

ASIA

Sri Lanka

Hog deer from India were introduced to Ceylon (Sri Lanka) for hunting probably by the Portuguese in the sixteenth century, but possibly not until the seventeenth or early eighteenth century by the Dutch.

AUSTRALASIA

Australia

In 1858 (not, as stated by Whitehead (1972), in 'about 1870') the Rajah Rajendro Mullick, Mr A. Grote of Calcutta, India, and Mr C. P. Layard of Colombo, Ceylon (Sri Lanka), presented eight hog deer to the Acclimatization Society of Victoria, where they were bred in an enclosure in Melbourne. In 1866 12 were released at Cape Liptrap north-west of Wilson's Promontory. Eight deer of unspecified species mentioned in the society's Annual Report for 1870 as having been 'liberated in the bush' were hog deer turned out at Gembrook, some 55 km south-east of Melbourne.[1] Further releases are believed to have been made in succeeding years at Gembrook and on Wilson's Promontory.

Although, according to Bentley (1978), the Gembrook site was ideal for hog deer, with an abundance of swamps and river flats, they were soon all killed off by hunters. Those at Cape Liptrap and on Wilson's Promontory, on the other hand, rapidly increased, and before long had spread north-east up the coast as far as Wombat Creek between Nowa Nowa and Orbost – a distance of nearly 200 km. At apparently unrecorded dates hog deer were also landed on three or four islands near Corner Inlet – Snake (about 4,600 ha), Little Snake (about 460 ha), Sunday (about 1,600 ha) and possibly St Margaret's.

Although at one time quite common along the South Gippsland coast from Inverloch to the Gippsland lakes, by the mid-1960s – due largely to excessive shooting, poisoning and loss of habitat (especially around Norman and Oberon bays) through land clearance for agriculture and settlement – hog deer survived in considerable numbers only on Wilson's Promontory (part of which is now a national park) and the islands near Corner Inlet, with smaller populations along the lower reaches of the Tarwin River and around Lake Wellington in the Gippsland Lakes National Park. Their preferred habitat is low-lying – often swampy – densely vegetated riverine grassland, tea-tree scrub, grass-covered islands and open grass and sandy coastal plains, such as occur on either side of mounts Hunter and Roundback, at the base of Wilson's Promontory and on the Yanakie isthmus. They favour warmer areas and have not penetrated the Strzelecki Range. They both graze and browse, especially on the wattle tree (*Acacia longifolia*), but appear to cause little if any harm on cultivated land. From the quality of some of their antlers it seems likely that hog deer on Wilson's Promontory have hybridized with axis deer (*C. axis*), which have also been introduced to the national park.[2]

In 1965 Sunday Island – part of the 24,600-ha Nooramunga State Wildlife Reserve – was acquired by the Para Park Co-operative Game Reserve Society, who, in conjunction with the Victorian Fisheries and Wildlife Department, have been studying the biology of hog deer (and fallow deer (*C. dama*)[3]) on the island, where they are now protected. Snake Island, which is outside the Wildlife Reserve and National Park, and all land within a 48-km radius have also been declared a deer sanctuary.

At present, Victoria possesses the only wild population of hog deer (most of which are on Snake Island) outside the Indian subcontinent. Because of increasing pressure there on their habitat, due to the food and housing requirements of an ever-increasing human population, southeastern Victoria could become one of the last strongholds of the species in the world.

Hog deer introduced from India to Samsø Island off Denmark in 1880 and to France in about 1890 did not long survive. Although Harrison Matthews (1952) includes the species in his 'List of living British mammals',[4] it has never become established in the wild in the British Isles.

Hog deer have been introduced for sporting purposes to several large (85,000 ha or more) fenced private ranches in the Orange Free State, South Africa.

The deer on James Island in Chesapeake Bay, Maryland, USA, which were at one time believed to be hog deer, are now known to be sika deer (*C. nippon*).[5]

NOTES

1. *Victorian Historical Magazine* **36**(1) (1870).
2. On Tasmania axis have also hybridized with fallow deer.
3. See p. 182.
4. p. 388.
5. See p. 213.

Timor or Rusa Deer

(*Cervus timorensis*)

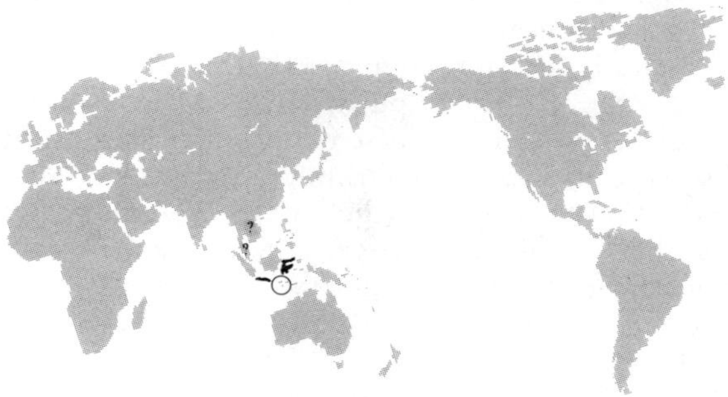

NATURAL DISTRIBUTION
Original range uncertain; the species is believed to have been widely distributed throughout the Malaysian and Indonesian archipelagos by early Malay and other voyagers. Now mainly in Java, the Celebes and the Lesser Sunda Islands.

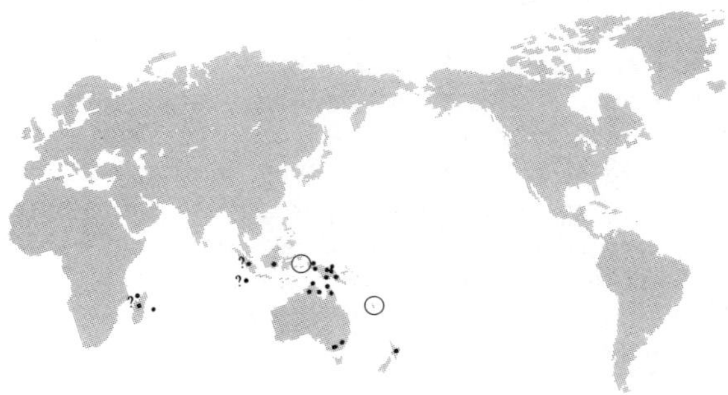

NATURALIZED DISTRIBUTION
Africa: Mauritius, Comoro Islands, ?Madagascar;. Australasia: Australia, New Zealand, Papua New Guinea, Indonesia (Borneo, Ambon, Aru, Belang-Belang, Maron and Ninigo islands, New Britain, New Caledonia, Horsburgh Island and Sumatra).

AFRICA

Mauritius; Comoro Islands; Madagascar

Deer of the subspecies *C. t. russa* from Batavia (Java) were introduced by the Dutch to the island of Mauritius as a source of fresh meat in 1639, and to Anjouan in the Comoro Islands in 1870. In the last decade attempts have been made in Mauritius, where the present population is said to number around 30,000, to increase the production of meat by experimental deer-farming in fenced enclosures. On Madagascar (the Malagasy Republic) a number of Timor deer were released experimentally in 1928 at the forestry station of Périnet east of Tananarive, where they were still apparently spreading in 1955. Since then, largely as the result of poaching, they appear to have declined, and may now even be extinct on the island.

AUSTRALASIA[1]

Australia

The Victoria Acclimatization Society's Annual Report for 1868 mentions that some 'Batavian Deer' had been imported from Java and were breeding in captivity; these were almost certainly Timor deer of the Javan subspecies *C. t. russa*. The Annual Report for 1885 refers to the presence of four in the society's collection, which by 1887 had increased to nine. The first release took place in the Dandenong Range near Gembrook south-east of Melbourne in 1890, where two years later they were said to be well established. Other liberations followed, including at least one in West Gippsland. The deer remained in considerable numbers in these two areas until at least 1948, since when only occasional sightings have been reported in the Dandenongs, though rather more have survived in Gippsland. The reason for their comparative failure is uncertain; although the habitat in both areas is dissimilar to that of their native Java, and to that in which they have become more successfully naturalized near Sydney and in Papua New Guinea, they have established themselves in an equally unsuitable habitat in New Zealand. One probable reason is interspecific competition – and possible hybridization – with the more vigorous introduced sambar deer (*Cervus unicolor*).[2]

Shortly after 1885 (Frith (1979) says 1893) Timor deer from New Caledonia – to which they had been successfully introduced from the Dutch East Indies (probably Java) in about 1870 – were imported to New South Wales and released in the damp sclerophyll forests of the Royal National Park south of Sydney, where today herds of between 100 and 200 can be seen, especially around Bundeena in the east of the park. They have extended their range as far south as Nowra – a distance of some 110 km from their point of release.

In 1912 8 or 10 Timor deer of the race *C. t. moluccensis* were imported from Molucca (and possible also from the Celebes) by Mr Reginald A. C. Hockings to Thursday Island off the northern tip of the Cape York Peninsula, Queensland, from where they were later transferred to the neighbouring 520-ha Friday Island and released. They rapidly increased and before long had swum across the 1,300 m of crocodile- and shark-infested channel to the much larger (246 km^2) Prince of Wales Island, where by the late 1960s the population was estimated to number around 5,000. Both these islands are hilly and have an abundance of cover and forage in the form of trees, vines and shrubs; seaweed, lilies and so-called 'wild ginger' are also part of the deers' diet. On Prince of Wales in particular there is plenty of fresh water. In about 1960 the Japanese established a cultured pearl industry on Friday Island, and many of the deer were killed for food and sport.

In 1914 Mr H. N. Hockings moved some of the deer from Friday Island to Possession Island, which lies 25 km south-east of Thursday Island, between Prince of Wales Island and the mainland west of Cape York (not east of Cape York as stated and shown by Whitehead (1972)). From Possession Island Timor deer are believed to have swum the 3-km channel to the mainland of the Cape York Peninsula.

In 1952 an English missionary, Mr Fred Gray, shipped a buck and three does from Friday Island to his mission station in the 2,600-km^2 Aboriginal reserve on Groote Eylandt off the coast of Arnhem Land in the Gulf of Carpentaria. The deer were released on a 19-km^2 island off the northern coast of Groote Eylandt with an abundance of food, cover and fresh water. According to Bentley (1978), Timor deer, of unknown origin, are also reported to be established on Melville Island north of Darwin and along stretches of the Daly River to the south in the Northern Territory.

New Zealand

In November 1907 eight Timor deer (believed at the time to be sambar[3]) were imported (two by the New Zealand Tourist Department and six by Mr H. R. Benn of Lake Okareka near Rotorua) from Nouméa at the southeastern end of New Caledonia to North Island, where they were released at Galatea south-east of Rotorua in the region of Kaingaroa, where by the 1920s they were said to be thriving. For many years they were known locally as 'New Caledonian sambar' or 'little sambar', and it was not until 1955 that they were correctly identified as Timor deer. Since the deer introduced to New Caledonia are believed to have been of the Javan subspecies *C. t. russa* those imported to New Zealand are probably of the same race. In New Zealand the Timor deer rut takes place in the same months (July and August) as in New Caledonia, whereas the New Zealand sambar rut in March and April.

Several hundred Timor deer are now established in swamplands and dense manuka, fern and toetoe scrub on the fringe of native forest (which they seldom penetrate) near farmland around their point of release, from which there has been little spread. They lie up in cover during the day, emerging at night to forage on pastures and cultivated crops, where occasional damage is reported. Timor deer cause little if any harm to forestry interests but they have, by opening up the thick scrub, provided a favourable habitat for alien brush-tailed possums (*Trichosurus vulpecula*).[4]

Papua New Guinea

In about 1900 Mr Walter Grose, then general manager of Burns, Philp & Co. of Port Moresby, introduced some Timor deer of the Javan race *C. t. russa* from Ceylon (Sri Lanka) (one account says 'probably from Australia') to what is now Papua New Guinea, where they were released about 30 km north-west of Port Moresby in the Vaigaui swamps near the Brown and Loloki rivers. By the outbreak of the First World War they had spread south-east as far as Rigo (from where they later moved inland through the foothills of the Owen Stanley Mountains) and northwards into the foothills of Mount Brown near Wau, respectively 50 and 250 km from Port Moresby. They now occupy a range bounded by the Vanapa River and the Galley Reach in the north-west, by the fringe of the rainforest on the lower slopes of the Owen Stanley Range in the north and by the Kemp Welch River in the south-east. In this area, where in 1972 the population was believed to number between 700 and 800 divided into herds of between 40 and 50, the habitat consists of ridges (which they frequent mainly in the rainy season between November and April), lowlands, alluvial plains along the Brown River from Karema to the Galley Reach, swamps – such as the Waigani – evergreen forest and savannah, especially in the Astrolabe Range and on the Sogeri Plateau. One of their most favoured food plants is the young shoots of kunai grass. Although hunted in this area the deer seem to be maintaining their numbers satisfactorily.

A smaller population is established in the region of Madang on the north coast, where the deer range from Binnen Harbour to Yabob in the south-east, north-west to Sisiak and the Meiro plantation, and inland possibly as far south as the Gumm River beyond Uligen in the Adelbert Mountains.

In the province of West Irian Timor deer were first introduced near Merauke on the south coast by the Dutch in about 1920. This population, which Bentley (1978) describes as 'without a doubt, the biggest herd of deer in the Australia region (excluding New Zealand)', has been variously estimated to number between 15,000 and 'some hundreds of thousands'. Their main range extends from the extensive plains around Merauke eastwards past the Bensbach River to the delta of the Morehead River in the Trans-Fly (the region west of the Fly River). They have greatly increased in numbers and distribution during the past 30 years, and some are shot for their venison throughout the scrub and lightly forested country and along the coast and estuaries south of the Fly. They are occasionally reported from as far north as Boset near the intersection of the Fly River with the West Irian border, 200 km north of the coast; at such settlements as Suki Lagoon and Maderi plantation on the south bank of the Fly; from as far east as Daru Island; and on Uapumba, Dibiri, Wabuda, Umada, Kiwei and Mibu islands in the Fly delta.

In 1913 Timor deer of the Moluccan form *C. t. moluccensis* were imported from Ceram Island to the Onin Penisula (Bombarai) in Irian Jaya by Herr Raedt van Olderbarnevalt; today they are established on the north-west coast of Geelvink Bay (Teluk Irian) around Manokwari, Momi and along the Muturi River on the east coast of the Vogelkop (Berau) Peninsula (Tjendrawasih). Timor deer of the same race found on Rumperpon (or Roemberpon) Island on the west side of Geelvink Bay and around Hollandia (Djajapura) – now known as Sukarnapura – are descended from some imported by Lulofs from the island of Halmahera in the Moluccan Sea in 1920.[5]

Indonesia

Several islands in Indonesia have also been stocked with Timor deer. In about 1680 the subspecies *C. t russa* was imported by the Dutch from Java to southern Borneo, where it soon established itself on the grassy plains near Pula Lampej and in the Tareh Laut near Bandjarmasin. It is believed that Timor deer on Amboina (Ambon) in the Moluccas are descended from seventeenth century Dutch importations from Java (*C. t. russa*) and the Celebes (*C. t. macassaricus*). In 1855 Timor deer of the form *C. t. moluccensis* were imported by Cleerens to Aru Island in the Halmahera group from Ceram, Wammer and Wasior (Irian Jaya), where the present stock is said to be all descended from a single pair from Wasior. In about 1930 Diepenheim imported some of the race *C. t. moluccensis* from the Moluccas to the island of Belang-Belang in the Obi group.

In about 1908, while on a visit to Sydney, Australia, a German planter, Heinrich Rudolph Wahlen, met Mr Dudley Le Souef, then Director of the Melbourne Zoological Gardens in Royal Park. In the following year Le Souef despatched three 'Timor rusa' (a male and two females which had originally been imported from Timor) from Melbourne via Sydney on a Norddeutscher Lloyd vessel to Wahlen's plantation of 'Wahlenburg' on Maron Island in the Hermit Isles, north of Papua New Guinea. Shortly after landing one hind ran headlong into a coconut palm and broke her neck. The surviving pair flourished and some of their offspring later crossed a sandy bar to the nearby island of Arkeb (Luf) where there were said to be over 200 in 1954. Germans are also believed to have been responsible for the introduction of Timor deer to the Ninigo Islands north-west of the Hermit group.

The Annual Report of the German Protectorate of New Guinea for 1900–1 recorded that on Credner Island in St George's Channel off Kokopo in German New Guinea (between New Ireland and New Britain) 'there are also a few red deer, presented to the Government by the Governor'. (Whitehead (1972) incorrectly postdates this introduction to 1910.) These are believed to have been Timor deer of the form *C. t. moluccensis*, whose descendants, from photographic evidence, are now established in patches of kunai grass on New Britain around Ulatawa in the Rabaul/Kokopo district and between Cape Gazelle in the east and Varzin and Toma in the west, from where some may penetrate inland through wild rubber trees, *Ficus*, south of the Warangoi River. The development of land for cocoa plantations has resulted in a decline in the population in recent years.

In about 1970 Timor deer of the form *C. t. russa* were imported from the Dutch East Indies (probably Java) to New Caledonia, where by 1883 damage to crops was being reported. Other importations (possibly of another race) may have followed. In the late 1930s, when the population was said to number about 225,000, some 46,000 hides were annually exported: by the early 1950s the population had fallen slightly to around 200,000, and the number of hides exported had been reduced to 25,000–30,000.

Timor deer are also believed to have been imported to Horsburgh Island in the Cocos or Keeling group and to Sumatra. In 1972 Whitehead wrote of deer on Wakaya in the Fiji Islands that 'from the description and photograph of a stag shot about 1954, they would appear to be Rusa'; they are in fact fallow deer (*Cervus dama*).[6]

NOTES

1. For much of the information on Timor deer in Australasia I am indebted to Bentley & Downes (1969) and Bentley (1978).
2. See p. 224.
3. See p. 226.
4. See p. 10.
5. *Scientific Contributions to the New York Zoological Society* **41**(4) (1920).
6. See p. 186.

Sambar
(Cervus unicolor)

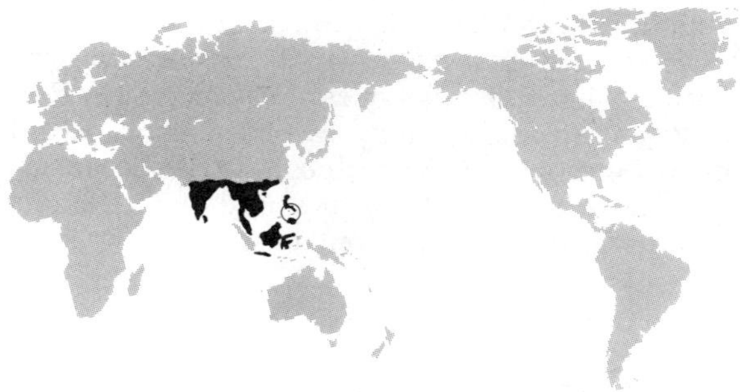

NATURAL DISTRIBUTION
India and Sri Lanka to southern China, Java, Borneo, the Celebes and the Philippines.

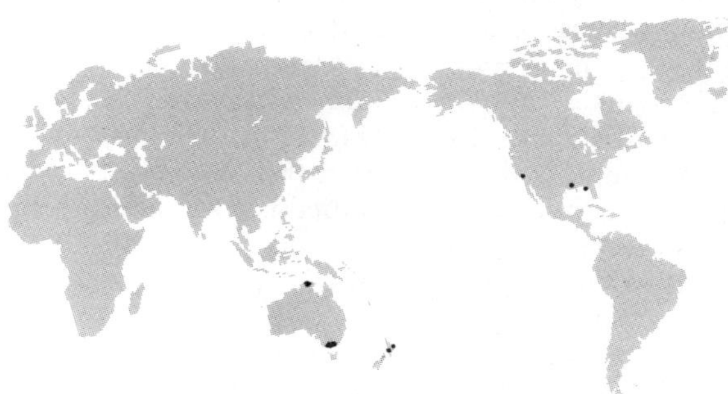

NATURALIZED DISTRIBUTION
North America: USA; Australasia: Australia, New Zealand.

NORTH AMERICA

USA

Small numbers of sambar are established in California, Texas and Florida. In California a herd of about a dozen near San Simeon in San Luis Obispo County is descended from animals which escaped from the neighbouring Hearst Ranch. In Texas sambar have been introduced to, among others, the Rickenbacker Ranch near San Antonio, where they are said to 'dislike the heat and spend the summer hiding in wooded canyons coming out at night'. In 1966 there were 60 sambar on three ranches in Texas; by 1971 the statewide population had risen to 85, but fell to 50 in 1974 and to only 22 in 1979. In northwestern Florida a herd of sambar established by Dr Roy V. Pierce on the privately owned offshore island of St Vincent near Apalachicola in 1908 had increased to around 50 by the mid-1950s, and a scattered population still remains.

AUSTRALASIA

Australia[1]

In 1861 Mr Edward Wilson, on behalf of the Victoria Acclimatization Society (of which in the same year he was the founder and first president), shipped on board the *Jeddo* from Galle in Ceylon (Sri Lanka) 'four splendid full-grown elk [the Singhalese name for sambar *C. u. unicolor*] – two stags and two hinds'. During the voyage a calf was born, but it and one hind died before their arrival in Melbourne. Other shipments, believed to have been of *C. u. equinus* from India, were received in subsequent years.

In 1863 five of these deer were released from the Royal Park on the slopes of Mount Sugarloaf near Kinglake, some 50 km north-east of Melbourne, and more were liberated in the vicinity of King Parrot Creek. Three years later others were freed on Snake Island (4,600 ha) off Port Welshpool east of Wilson's Promontory, and on Mr William Lyall's Harewood estate near Cranbourne which included two extensive swamps – the Koo-wee-rup (16,000 ha) and the Tobin Yallock of 2,800 ha – some 65 km south-east of Melbourne. In the same year seven sambar were liberated at Western Port, and others were freed on Mount Zero in the Grampians and on French Island, 65 km south of Melbourne. In 1871 sambar are believed to have been released on Sir Samuel Wilson's property at Ercildoune near Burrumbeet in western Victoria, in which case they were probably the ancestors of some of the present population in the Mount Cole State Forest Reserve (where a number were also freed in 1873) in the Grampians.

All the sambar turned down in Victoria appear to have flourished, multiplied and spread. Within a decade those on the Harewood property were extending their range through the Koo-wee-rup swamp, and by the turn of the century the species' southern range included Lang Lang, Heath Hill, Nyora and Glen Forbes. Ten years later damage to agricultural crops was being reported, and shooting licences were issued by the Fisheries and Game Department; in 1919 deer were removed from the list of protected species and an open hunting season was declared. For several years thereafter no decline in the overall population of sambar was apparent; the main reason for a localized decrease – and some dispersal in southwestern Gippsland near the south-east coast – was the drainage in the early 1920s of the Koo-wee-rup swamp coupled with denser human settlement, which forced sambar into the Baw Baw foothills from Drouin west to Pakenham, where they joined up with those released in the Grampians. From the late 1920s to the late 1930s the species' southern frontage extended from Gembrook through Tonimbuk, Labertouche, Jindivick and Neerin. Bentley (1978) considered that 'bushfires prepare samber habitat . . . the tragic fires of January 1939 when most of Victoria was virtually ablaze were, in my opinion, the most important single element or agency in the extension of the sambar's range', which by the late 1960s covered much of Gippsland. Two sambar shot at about the same time in the Rhyll swamp on Phillip Island are believed to have swum there either from French Island or from the mainland.

Today, sambar are established in Gippsland between Wilson's Promontory and Sale, 140 km to the north-east, and in the mountains west of Melbourne – particularly around Mount Cole in the Grampians north of Hamilton. The largest concentrations, however, are believed to occur in the eastern highlands from around Healesville north-east to Albury on the border with New South Wales – a distance of nearly 200 km. Bentley believed that sambar were continuing their colonization in that direction through the moist sclerophyll forests and ravines of the mountain ranges. In Victoria sambar spend most of the daylight hours lying up in dense cover, from which they emerge at night to feed; in autumn and winter agricultural crops – especially potatoes, turnips and oats – and orchards suffer some depredation; control is principally by hunting with hounds, which is said to account for around 200 per annum.

From 1863 to 1868 a number of Manila, Basilan or blackish sambar (*C. u. nigricans*) were kept for breeding by the Victoria Acclimatization Society in Royal Park; whether any were ever released is unknown.

In October 1912 three sambar from the Melbourne Zoological Gardens in Royal Park (part of a consignment – the largest ever made in Australia – which included 52 fallow deer (*Cervus dama*)[2] and five barasingha or swamp deer (*C. duvauceli*)) were shipped by the Gilruth Administration from Sydney to Darwin in the Northern Territory on board the SS *Mataram*, in the charge of the zoo's director, Mr Dudley Le Souef, and Mr H.B. Slaney. From Darwin the deer were trans-shipped on a 60-ton lugger under the command of one Captain Mugg to Port Essington on the Cobourg Peninsula, where they were landed and released near the camp of Jimmy Kafoo, a Macassar *bèche-de-mèr* Aboriginal fisherman. The fallow and barasingha soon disappeared, but the sambar survived and still persist over most of the peninsula on the Arnhem Land Aboriginal Reserve; although they are seldom seen their slots[3] and droppings are common along the western shore of Port Bremer, around Victoria and Port Essington and on parts of the south coast. Some have also been reported from the vicinity of Port Smith and Port Black. On the Cobourg Peninsula sambar frequent the dense mangrove and monsoonal-forest edges of saltwater inlets and creeks and freshwater swamps, from which they seem seldom to penetrate into the open woodlands.

In the early 1900s Mr T. H. Clarke of Campbell Town is said to have introduced sambar to Tasmania; if so they must have died out as there are none on the island now. Reports of sambar introduced from New Caledonia to the Royal National Park south of Sydney, New South Wales, almost certainly refer to Timor deer (*C. timorensis*).[4]

Sambar are not gregarious, usually occurring singly or in (possibly family) groups of three or four individuals. Although stags of the Indian race *C. u. equinus*, which is believed to have been introduced to Australia, may stand 150 cm high at the shoulder and weigh up to 320 kg, Australian sambar seldom exceed 130 cm and 270 kg. According to Frith (1979), 'the reason for the smaller size is not known. It could be due to inbreeding, the large population having developed from a small initial introduction. Inadequate nutrition due to poor habitat or even over-hunting may be included in the causes.'

Although Bentley's (1978) claim that 'the sambar's impact on native species, plant and animal alike, would appear not to have been detrimental' is supported in part by the presence of kangaroos and wallabies in the sambar's range, Frith (1979) considered that 'one of the most pressing needs in [Australian] wildlife conservation is to discover what is the place of the Sambar in the ecology of the south-east highlands. If the presence of the deer is adverse to native flora and fauna, then how can they be controlled? If they are not adverse, then how could they be managed for long-term productivity as game animals?'

New Zealand[5]

On 1 June 1875 Mr Falconer Larkworthy received from Ceylon (Sri Lanka) a pair of sambar of the race *C. u. unicolor* which were landed at Wellington from the *Teviotdale*, from where they were transported to his property at Carnarvon near the mouth of the Rangitikei River in North Island. Here the deer flourished, and before long had increased to over a hundred head. The Annual Report of the Wellington Acclimatization Society for 1894 recorded that 'The Ceylon Elk (Sambur [*sic*] Deer) imported into the Carnarvon district, Manawatu, by Mr. Larkworthy . . . now number about thirty'. Following this decline, brought about largely, it is believed, by poaching, the deer were given security by the government under the Animals' Protection Act. As a result, by 1898 the herd was reported to have increased to around 100 again, 'but there is good reason to think that they are really more numerous'. Donne (1924) reported that 'there is now a large number of sambar in the Upper Manawatu district'. Wodzicki (1950) estimated the size of this population in 1947 at between 400 and 500 head. By the mid-1950s the deer had spread north-west and had colonized a coastal strip of sandhill country, swamps and small plantations 100 km in length by 10 km wide between the Waitotara River in the north and the Hokio River in the south.

In 1876 the Auckland Acclimatization Society was presented by Larkworthy with a pair of sambar, of which the hind died shortly thereafter; the stag turned vicious and once attacked a man with a wheelbarrow who, with considerable presence of mind, turned it upside down and took refuge inside it until help arrived. Larkworthy obtained a replacement mate for this stag from Colombo, and the pair were released between Cambridge and Morrinsville, Waikato, where for a time they thrived and multiplied, but by the late 1940s their progeny had all been exterminated by shooting.

In 1915 a stag and six hinds were translocated by the Government Tourist Department from Manawatu to Lake Rerewhakaitu in the Hot Lakes district south-east of Rotorua. Four years later the Department removed eight sambar from the Himitangi district and in the following year liberated them near mounts Tarawera and Ruatoki east of Rotorua in the Kaingaroa State Forest north of Lake Rerewhakaitu, and at Taneatua near Whakatane. Once again the deer flourished and by the 1940s were reported to be spreading into new areas.

The eight sambar said to have been released in 1907 near Galatea were almost certainly Timor deer.[6]

Sambar in New Zealand are currently established on the west coast between Otaki and Wanganui, on the southern side of Lake Taupo, to the south-east of Rotorua, and in the watershed of the Whakatane River. Although in their native range mainly a woodland species, in New Zealand sambar exhibit a more catholic choice of habitat, occurring in raupo swamps with dense stands of manuka (*Leptospermum* sp.) scrub, bracken, fern, lupin, flax (*Phormium* sp.), and giant bunch-grass (toetoe) near rough pastureland; they shelter in both native forests and exotic pine plantations, into which, however, they seldom penetrate very far. Their diet includes flax, grass, swedes and the leaves and bark of some indigenous trees; their taste for the bark and young shoots of radiata pine saplings causes considerable silvicultural damage in some areas – though not to the same extent as in Sri Lanka where they are regarded as a serious economic pest. Although said to be unusually tolerant of human activities and disturbance, loss of habitat to the demands of agriculture is likely to curtail their future dispersal in New Zealand, which was formerly restricted by heavy shooting.

Sambar are also said at one time to have been introduced by the Spanish to Guam and Rota in Micronesia. References to the presence[7] of sambar on the island of Mauritius almost certainly concern the superficially somewhat similar Timor deer.[8]

Sambar were among the species introduced to Groote Schuur, near Cape Town, South Africa by Cecil Rhodes; they currently exist on several enclosed estates in the Western Cape.

NOTES

1. For much of the information on sambar in Australia I am indebted to Bentley (1978) and Frith (1979).
2. See p. 182.
3. The tracks or footprints of deer.
4. See p. 220.
5. For much of the information on sambar in New Zealand I am indebted to Wodzicki (1950).
6. See p. 221.
7. J. R. Maingard, *Journal of the Bombay Natural History Society* **50**: 648–9(1954); Chasen (1940); and *Country Life* (4 May):1381 (1951).
8. Owadally & Butzler (1972).

Reindeer or Caribou

(*Rangifer tarandus*)

Note: Natural distribution much altered by local extinction and replacement by domesticated stock.

NATURAL DISTRIBUTION

The tundra zone of the Palaearctic (where it is known as the reindeer) south in the taiga zone to between 52° and 54° N in the Altai Mountains and northern Mongolia. In the Nearctic (where it is known as the caribou) from Alaska westwards through northern Canada and the Arctic islands to western and eastern Greenland.

Note: Only naturalized populations outside the species' natural distribution are shown.

NATURALIZED DISTRIBUTION

Europe: Iceland (acclimatized in the British Isles); North America: USA*, Canada*, Greenland*; South America: ?Chile, ?Argentina; Oceania: South Georgia, Kerguelen. (*Reindeer introduced to caribou range.)

EUROPE

British Isles

Despite repeated assertions to the contrary (from Harting (1880) to Whitehead (1972) and others),

> there is no good evidence from subfossil finds for their [reindeers'] survival into the post-glacial period. A reference in the Orkney Saga to the hunting of reindeer in Caithness in the twelfth century is often interpreted as evidence of their survival in Scotland to that date. However in view of alternative explanations – confusion with red deer, introduction of reindeer by the Vikings – this cannot be considered conclusive.[1]

In 1738 Mr William Hamilton brought back from Sweden to his estate at Beltrim Castle in County Tyrone, Northern Ireland, a bull and cow reindeer which he released in the Sperrin Mountains, where according to his son, Claude, three were shot in 1770 'by persons unknown'.

In 1790 the Duke of Atholl introduced 14 reindeer to his estate at Blair Atholl in Perthshire, where, however, according to A. I. McConnochie,[2] they failed to 'find a permanent home in the forest'. The *Sporting Magazine* for August 1799 records that 'Mr Brooks . . . the celebrated collector of foreign animals brought to this town [London] from Lapland . . . twelve reindeer on their road to the Duke of Norfolk's estate near Penrith [Cumberland]'. What became of them is unknown. Early in the following century Sir Thomas Liddell (later Lord Ravensworth) failed in an attempt to establish five reindeer from Sweden at his seats, Ravensworth Castle and Eslington Park, in Northumberland. In 1816 three reindeer from Archangel were liberated near Kirkwall in the Orkneys, and some four years later the Earl of Fife introduced a number to the high tops of Mar Forest, Braemar, Aberdeenshire; like their predecessors, both these attempts were unsuccessful. Their failure has been attributed[3] at least partially to the fact that, whereas the original native Scottish reindeer were probably of the forest or woodland race *R. t. fennicus*, the attempted reintroductions were made with the tundra or mountain race *R. t. tarandus*.

In April 1952 eight reindeer from Jokkmokk in Arctic Sweden, under the care of Mr Mikel Utsi, were placed in a 120-ha fenced enclosure near Loch Morlich on Glen More (Queen's Forest), Aviemore, Inverness-shire, by the recently formed Reindeer Council of the United Kingdom. The object of the introduction was to try to establish reindeer in Scotland as an economic source (of venison and skins) rather than as an aesthetic amenity. In 1953 a further area on the northern slopes of Airgiod Meall (Silver Mount) was enclosed, and a year later 17 reindeer (the balance of the total of 25 authorized by the Secretary of State for Scotland in 1951) were placed therein. In later years more mountain reindeer were imported from Sweden, Russia and southern Norway, as well as some of the forest type. Their progeny – which in 1982 numbered some 60 head – roam over nearly 2,400 ha of the Cairngorms as an 'acclimatized' (but not 'naturalized') managed herd.

Iceland

The possibility of introducing reindeer from Scandinavia to Iceland was first put forward by the then Governor-General, Niels Fuhrmann, but he died in 1733 before any action could be taken. The earliest factual information available is to be found in the Annals of Kjósarsyslu (*Kjose Syssels Description Geographica*) where on 28 April 1746 J. Hjaltalín records that Count J. L. Holstein King Frederick V's Counsellor and President of the newly formed Danish Scientific Academy, had, in a letter dated 8 June 1743, instructed the Icelandic Governor-General, J. Lafrentz, to conduct a survey of the country to establish, *inter alia*, the feasibility of establishing reindeer on the island. Lafrentz's report was presumably satisfactory since on 19 January 1751 the king ordered that six reindeer – four bulls and two cows – be purchased in Norway at an estimated cost to the Danish Treasury of some 40 *rikisdalir* and consigned to Iceland in the care of Samish herdsmen to provide relief from a recent famine; it appears, however, that the required animals could not be caught (or the Danish government may have had second thoughts concerning the practicality of the scheme) and this shipment was never made.

Between 1771 and 1787 domesticated reindeer of the form *R. t. tarandus* were introduced on four occasions from Norway to Iceland where they were released. According to the Annals of Espólin (chapter 10, page 101 of the year book of 1771 –

quoted by Thorvaldsson (1960)) after a further famine, and in an attempt to introduce some variety to Icelandic husbandry, 'In the summer of 1771 13 animals arrived in the Westmann Islands [off the south coast] on the instructions of Governor-General Olaf Stephánsson. Ten died during the winter and three lived'. The survivors (one bull and two cows) of this initial stock, which came from the island of Sørø in Finmark in northern Norway, were later transferred to Landeyjar in Rangárvalla Sýsla on the mainland of southern Iceland, where 13 bull calves were born and where the herd died out by about 1786.

In 1777 23 reindeer (the survivors of a shipment of 6 bulls and 24 cows) from the same source were landed at Hafnarfjördur in Gullbringu Sýsla in southwestern Iceland, where by the middle of the nineteenth century they had increased to a maximum of 100 head on the Reykjanes Peninsula but spread no further east than Thingvallavatn and became extinct between 1920 and 1930; according to Thórisson (1983) the most likely causes of their demise were severe winters towards the end of the nineteenth century coupled with possible overgrazing of pastures.

In 1784 35 reindeer from Kautokeino in Finmark were landed at Eyjafjördur in Eyjafjardar Sýsla in northern Iceland; this population rapidly increased and by the turn of the century local farmers were complaining about the effect grazing by reindeer was having on the lichen *Cetraria islandica*, which was at that time used for human consumption: this herd did not, however, begin to expand its range until the mid-1820s when, as the result of a succession of severe winters and probable overgrazing, it spread eastwards, becoming established in Thingeyjar Sýsla between Axarfjördur in the north and Mývatn in the south, but especially south of Reykjaheioi. Between 1830 and 1840 this region was deserted in favour of an even more easterly and northerly one between Melrakkaslétta in the north, Hólsfjöll in the south and Thistilfjördur in the east. The population in this area reached a peak around 1850 but thereafter declined (probably for the same reasons as the preceding introduction) until by the turn of the century only a few survived, and this remnant finally disappeared in 1936.

In 1787 35 reindeer (five bulls and 30 cows) from Avjovarre in Finmark were landed at Vopnafjördur in Nordur Múla in northeastern Iceland (which lies at much the same latitude as Finmark), and from these animals are descended the reindeer established in the country today. Once again a rapid increase in the population brought complaints from local farmers of overgrazing of *Cetraria islandica* and some deterioration of winter pasturage. This herd is believed to have reached a peak in the mid-nineteenth century, when it was established on the highland plateaux of the east from Sandfell in the north to Vatnajökull in the south, and from Möorudalur in the west eastwards to the coastal fjords, where some were occasionally captured and redomesticated by local farmers; thereafter it declined, probably for the same reasons as before, coupled with the eruption on 28 March 1875 of the Askja volcano which deposited a 5–10-cm thick layer of ash on the animals' main winter pastures; by 1939 only some 100–300 (probably nearer the lower figure) remained in Kringilsárrani, Jökulsa á Dal and Vesturóraefi around Snaefell at the northeastern corner of the Vatnajōkull icecap. This herd, according to Thórisson (1983), showed no definite seasonal migrations, not returning to its former winter grazing on Fljótsdalsheidi for some 30 or more years.

Until 1790, when King Christian VII gave permission for 20 to be shot during the next two years, reindeer in Iceland were strictly protected. Further licences were issued in 1794 for the following three years and, as a result of an increase in the size of the population and the concomitant need for culling, the unlimited hunting of bulls under licence was allowed in 1798. This authorization was confirmed in 1849 and remained in force until 1882 when limited shooting was reintroduced. Between 1901 and 1943 reindeer in Iceland were totally protected, and for the next decade only bulls could be legally killed. From 1953 to 1965 the annual quota was fixed at 600 and between about 200 and 550 were actually killed; no shooting was allowed in 1965–7 and 1970–1, but from the latter year until 1978 the quota varied from over 800 to nearly 1,600, and between more than 400 and 1,000 were shot per annum. In 1983 1,100 licences were issued and 786 reindeer were killed, and in 1984 600 licences were granted. At these figures annual recruitment in most years seems more than to compensate for the numbers shot. At the time of writing (1984) the open season, determined by the Ministry of Education and Culture, lasts from 1 August to 15 September.

The popular belief that excessive hunting was

responsible for the decline in the Icelandic reindeer population that occurred in the nineteenth century is considered by Thorsteinsson et al. (1970) and Thórisson (1980) to be a myth. According to the former, 'Hunting pressure is thought to have been low at all times, and indeed the decline in the population continued despite long periods of total protection. The distribution of the reindeer was always limited to three areas with relatively light snow. The main factor causing the decline of reindeer in Iceland is thought to have been difficult snow condition limiting access to grazing in winter. . . . It seems unlikely that reindeer can maintain themselves for any length of time in those parts of Iceland where they are not presently found'.

In 1940 annual censuses of the reindeer population began; until 1956 these were conducted on horseback but thereafter from the air (mainly in late July)', and the results showed a steady increase in the national herd (with some fluctuations) to around 1,200 by 1950, 2,500 by 1960, 2,600 by 1970 and 3,600 by 1976; in 1984 some 1,200 were counted, and the total may have been in excess of 3,000.

Thórisson (1983) has traced the spread and seasonal migrations since 1940 of Iceland's reindeer; until the mid-1950s, when the population numbered about 1,800, their summer range consisted of Kringilsárrani and Vesturóraefi from where they dispersed in winter mainly to Fljótsdalsheidi. After about 1955, when the population increased, some animals began to winter in the eastern and southeastern fjords and others in Jökuldalsheidi to the west and north-west. Between 1955 and 1960 regular migrations seem to have begun to and from Kringilsárrani and Jökuldalsheidi. Reindeer wintering in the eastern fjords until 1955 were mainly bulls and yearlings and their numbers did not greatly increase; between 1965 and 1970 this area was permanently settled and, as a result of both breeding and emigration from the highland plateaux, the total multiplied rapidly. By the late 1970s 50 per cent of the population (about 1,700 animals) spent the entire year in the eastern fjords; about 600 summered on the high plateau of Vesturóraefi but migrated in autumn to spend the winters mainly in the southernmost eastern fjords; a similar number migrated in autumn from Vesturóraefi to Fljótsdalsheidi and a like figure from Kringilsárrani to Jökuldalsheidi. The total in the two last-named areas had increased slightly but declined in the two former ones, probably due to emigration as a result of overgrazing in the Fljótsdalsheidi.

Thórisson (1980) summed up the distribution of reindeer in Iceland as follows: in summer 80–90 per cent (Thorsteinsson et al., 1970) of the population occurs south of 65° N and north of the Vatnajökull icecap. In the west and east their range is bordered by the rivers Jökulsá á Brú and Jökulsá i Fljótsdal, respectively. A small number spend the summer west of the Jökulsá á Brú at Kringilsárrani. Local residents believe that more animals are beginning to summer outwith these areas, and small herds are now scattered in lowlying regions east and southeast of the rivers Jökulsá and Lagarfljót. In late summer or early autumn the main population migrates northeastwards between the Jökulsá á Brú and Jökulsá i Fljótsdal, and northwards to the west of the former to winter on the highland plateaux of Jökuldalsheidi, Fljótsdalsheidi and Fellaheidi and in the lowlands of Hróarstunga: some animals are believed to continue their migration across the Jökulsá i Fljótsdal to lowlands along the Lagarfljót and the eastern fjords and river valleys. According to Thorvaldsson (1960), stragglers have from time to time occurred on the south-east and even south-west coasts.

Thórisson (1980 and 1983) found that the habitat of reindeer in Iceland can be broadly divided into two; high ground plateaux (500–600 m) with some isolated ridges and mountains between the glacial rivers Jökulsá i Fljótsdal and Jökulsá á Dal (mainly Vesturóraefi, Fljótsdalsheidi and Fellaheidi) and between the latter and Jökulsá á Fjöllum (mainly Kringilsárrani, Saudafell, Jökuldalsheidi and Tunguheidi), and lowlying valleys (up to 200 m) of the eastern fjords.

West of Jökulsá á Brú plant growth is sparse, with continuous vegetative cover found mainly in depressions where plant communities receive protection from wind and soil erosion and moisture from melt-water and subterranean springs. Between the Jökulsá á Brú and Jökulsá i Fljótsdal plant growth is more luxuriant; wetlands comprise some 50 per cent of fertile land, with *Carex bigelowii, C. rostrata, C. rariflora, Eriophorum scheuchzeri, E. angustifolium* and *Salix callicarpaea* as dominants. Traces of tundra are visible in many boggy areas with occasional palsas or tundra-mounds; sedge heaths are plentiful in those places dominated by *C. bigelowii*, grasses (Gramineae) and dwarf shrubs – mainly *S. callicarpaea* in damp areas and *Kobresia myosuroides, Juncus trifidus* and *Empetrum nigrum* elsewhere. Snow-patch communities, where *Salix herbacea* and *Sibbaldia procumbens* are the dominant species, are abundant, and lichens are widely but sparsely distributed – possibly, according to Thorsteinsson et al., 1970, as a result of overgrazing. The lowlying valleys of the eastern fjords are mainly vegetated by chionofobus dwarf bush heaths such as *Betula pubescens* and grasses, communities dominated by *Empetrum nigrum, Vaccinium* spp., *B. nana* and grasses. Sedge heath and snow-patch communities, and to a lesser extent bogs, are mainly used for grazing; the bulk of the reindeers' diet consists of woody species such as *Salix herbacea, S. callicarpaea* and *Vaccinium uliginosum*, and grasses – principally *Festuca* spp. and *Agrostis* spp. Lesser quantities of sedges, horsetail, herbs and lichens are also eaten.

Icelandic reindeer start their spring migration in late April, with calves being born – mainly on Vesturöraefi, Kringilsárrani and Saudafell – in mid-May, although there is evidence in recent years of increasing calving outside these areas. The return autumn migration begins in late August or early September before the late September to mid-October rut. No accurate mortality figures are available for Iceland's reindeer; the only native carnivore is the Arctic fox (*Alopex lagopus*), which may take some calves and sick or injured adults. Parasites and biting insects are of little consequence in Iceland where warble (Oestrudae) and nostril (Tabanidae) flies and mosquitoes (Culicidae) are unknown. According to Thórisson (1980 : 768) 'the main causes of natural mortality are thought to be the animals falling over cliffs, and starvation'. A further hazard, at present being studied, is the proposed Austurlandsvirkjun hydroelectric power scheme, which would involve damming the Jökulsá i Fljótsdal and Jökulsá á Dal and could seriously affect the animals' highland habitat.

NORTH AMERICA

USA

Between 1891 (when 10 were brought in) and 1892 (171) to 1902 a total of 1,280 reindeer were imported to Alaska by the US Bureau of Education (at the instigation of Sheldon Jackson) from the Chukotsk Peninsula and the coastal regions of the Gulf of Anadyr in eastern Siberia. Most of the reindeer, which had been imported to provide a reliable source of meat and hides for the Eskimos, were turned out near Teller on the Seward Peninsula, facing the Bering Strait. By 1922 the population had increased to around 250,000; by 1930 – when 1.14 million kg of meat and 20,000 hides were shipped to the rest of the United States – it had doubled to around half a million; and by 1936 – when it stretched from Point Barrow on the Arctic Ocean in the north to Kodiak Island in the Gulf of Alaska in the south, and from the Seward Peninsula eastwards to Ophir – it had increased to between 600,000 and a million. In 1937 the US Government acquired those reindeer still in private hands and turned them over to the Eskimos. Thereafter, the population declined dramatically, until by 1950 it was doubtful if more than 25,000 remained. The reason is not far to seek. Unlike the Laplanders, who herd and manage the semi-domesticated reindeer in northern Scandinavia, following them on their seasonal migrations, the Eskimo, being primarily a sedentary hunter, proved reluctant to adopt the role of nomadic herdsman. Many thousands of reindeer were slaughtered, and the remainder permitted to destroy the lichens (which are slow in regenerating) that are essential to the animals' winter survival.[4] One of the largest remaining reindeer populations in Alaska is established on the Kotzebue Peninsula, where, as elsewhere in Alaska, hybridization has occurred with the local race of the native caribou (*R. t. granti*).

In the summer or autumn of 1911 the US government landed 40 reindeer on the Pribilof Islands[5] in the Bering Sea, south-west of Alaska, to provide the islanders with a sustained source of fresh meat. These animals were descended from those introduced to the Seward Peninsula on the Alaskan mainland between 1891 and 1902, from where they were subsequently transferred south to Unalakleet on the coast of Norton Sound, and thence, on the revenue cutter *Bear*, to the Pribilofs.[6] Four males – one of which was injured and later died – and 21 females were landed on the island of St Paul, and three males – one adult and two yearlings – and 12 females on St George, 55 km to the south, each accompanied by an Eskimo herdsman. Both introductions were an immediate success, 17 or 18 calves being born in the following year on St Paul and 11 on St George.

The reindeer in the Pribilofs were established in a succinct ecosystem with few if any alien pressures, no competition from native species and an absence of potential predators. As G. D. Hanna suggested, 'It would seem that here is the place to maintain model reindeer herds and to determine many of the needed facts for the propagation of these animals on a large scale. At no other place are conditions so favourable.'[7]

By 1922, when Hanna published his report, the herd on St George had reached its peak of 222 individuals, from which it was soon to decline to a smaller, but apparently stable, population of between 40 and 60. The herd on St Paul increased slowly but steadily until, in the early 1930s, there was a population explosion resulting in an increase by 1938 to 2,046 animals at a density of 1:5 ha – some 150 per cent greater than the carrying capacity of the range. Twelve years later the herd had collapsed dramatically to a total of eight. What were the reasons for the rapid increase, followed by a catastrophic decline, in the population on St Paul, and why do the development patterns differ so markedly on two adjoining islands?

According to L. J. Palmer, 'the lichens can not be said to be necessary for reindeer maintenance because of their nature or nutritive qualities, yet from the standpoint of a readily accessible winter food supply they are essential'.[8] Scheffer (1951) concurred: 'Observations on St Paul over the period 1940 to 1950 and examination of the records point to the inescapable conclusion that the lichen flora of the island is the key to the behaviour of the herd. Certain lichens, chiefly the taller, shrub-like forms of *Cladonia* and *Cetraria*, serve as emergency rations for reindeer.'

While the reindeer population on St Paul numbered only a few hundred, the speed of the lichens' regeneration (which Hanna (1922) considered was more rapid in the Pribilofs than on the mainland) made them self-sustaining. By about 1940, however, the lichens had all but disappeared; this, coupled with a sequence of exceptionally severe winters, possible poaching by the army during the Second World War, disease, fighting, accidental deaths, an unfavourable ratio between the sexes, and possibly inbreeding, may all have contributed to the decline. In the final analysis, however, the principal reason was overpopulation resulting in winter starvation.

Just why the reindeer population on St George, where it has never exceeded one per 40 ha, did not erupt like that on St Paul is something of a mystery. Although the islands are in such close proximity the mean southern limit of the Arctic drift ice lies approximately midway between them; it is possible that the warmer and damper climate of St George was less suitable than that of more northerly St Paul.

On August 20 1944 5 males and 24 females were landed on St Matthew Island (350 km north of St Paul) in the Bering Sea. The deer, which were transported on the US Coast Guard cutter *Clover* to establish an emergency war-time food source, were all two-year-olds from a herd near Nash Harbour on Nunivak Island. By 1957 the population had multiplied to around 1,350 – an average annual increment of 34 per cent – and by 1963 to some 6,000. By the following year it had plunged disastrously to barely 50, and by 1966 to only 42. (Roots, writing in 1976, claimed that there were still 6,000 reindeer on St Matthew.) Again, as on St Paul, the reason for the fall seems to have been winter starvation caused by the virtual elimination of slowly regenerating lichens, especially *Cladonia rangiferina*, through overbrowsing by an excessive population, coupled with exceptional falls of snow during the winter of 1963–4.

Roots (1976) claimed that 'Despite their low numbers they [reindeer] are still well established in many areas of western Alaska, on the Pribilof Islands and on Kodiak, Umnak and Nunivak islands'. In fact they occur almost exclusively in a few places on the Alaskan mainland and on Nunivak Island.

Reindeer have also been introduced to Atka Island in the Aleutians as a source of protein for the islanders where, as on the Alaskan mainland, they hybridized with the resident Grant's caribou. They were, however, only made use of so long as they remained in the immediate vicinity of settlements.

An unsuccessful attempt was at one time made to establish tundra reindeer in the state of Michigan.

Canada

On Christmas Day 1929 two Lapland herdsmen, Dan Crowley and Andrew Bahr, set out from the Selawik River in western Alaska with a herd of 3,400 reindeer destined for Kittigazuit east of the Mackenzie River delta near Kugmallit Bay on the Beaufort Sea in the Northwest Territories – a distance of some 1,200 km. After an epic journey lasting over five years, one of the most extraordinary animal-drives ever recorded in terms of duration, distance and the ruggedness of the territory covered (which it had been estimated would take some 18 months), Crowley and Bahr finally reached their destination in February 1935 with 2,370 animals. Only about 470 were members of the herd which had set out from the Selawik, the remainder having died or disappeared *en route*; the balance of around 1,900 were young animals born during the journey. Of those whieh arrived 1,500 were breeding females.

By 1952 the population in the Mackenzie Delta, which is established in a reserve of 15,000 km^2, numbered around 7,000, of which 1,000 were cropped annually for their meat and hides. The calves are dropped in April and May in the south of the range, near Reindeer Depot, the entire population moving north to the sea-coast in summer to escape from flies and mosquitoes. Hybridization with the Barren-ground caribou (*R. t. groenlandicus*) is known to have occurred.

Greenland

Domesticated reindeer from Norway have been introduced to (and since 1952 have become established in the wild in) parts of western Greenland.

SOUTH AMERICA

Argentina; Chile

In 1947 and 1948 reindeer were released on Isla Grande de Tierra del Fuego, on Isla de los Estados and on Georgia del Sud in southern Argentina, where they are said to have been exterminated by poachers. In 1971 a number were turned out on the Chilean side of Tierra del Fuego where their present status – if indeed they still exist – is unknown.

OCEANIA

South Georgia[9]

South Georgia lies in the South Atlantic Ocean 500 km south of the Antarctic Convergence. It has no native land mammals. Captain James Cook, who discovered the island in 1775, described it thus:

> The Wild rocks raised their lofty summits till they were lost in the Clouds and the Vallies laid buried in everlasting snow. Not a tree or shrub was to be seen, no not even big enough to make a toothpick.[10]

To this seemingly inhospitable land sealers – and later whalers – came in the nineteenth century, and in 1904 a whaling station was established by a Norwegian, Captain C. A. Larsen, at Grytviken. In November 1911 Captain Larsen and his brother, L. E. Larsen, landed some reindeer, for food and sport, from Valders in Numedal in central-southern Norway, at New Fortuna (since 1954 Ocean) Bay on the Barff Peninsula, north-east of Cumberland Bay. Three males and seven females were put ashore from the whaler *Ocean* of the Larsens' Compañia Argentina de Pesca. In the whaling season of 1911–12 another company, Christian Salvesen, landed two males and three females at Leith Harbour in Stromness Bay. 1925 a further consignment of three males and four females was put ashore on a peninsula between Busen Fiord and Cumberland Bay in Husvik Harbour, Stromness Bay.

The reindeer turned out on the Barff Peninsula increased rapidly – at a rate of about 40 per cent per annum in the early years. By 1916 they numbered 45 head, by 1920 about 120, almost 300 by 1921–2, between 400 and 500 by 1928 (when between 150 and 200 had been shot), 2,000 by 1953 and 4,000 by 1955–7 – an overall rate of expansion of some 14 per cent per annum. Although legally protected since their introduction (currently by the Falkland Islands Dependencies Conservation Ordinance of 1975), licences for shooting,

– mainly in the Barff herd – were issued to whalers once the population had become viable, and from the 1930s to 1950s up to a hundred were shot annually. Since 1964 (except for 1972–6 when 370 were removed from the Barff herd, 100 from that at Royal Bay and 120 from the Busen subpopulation for research purposes) less than 30 have been harvested annually. After 1958 they spread southeastwards across or along a narrow beach in front of the crevasses of the Cook and Heaney glaciers (which had hitherto been thought impassable) into Royal Bay: although reindeer were first observed in this area in 1959, it is thought that the main colonization took place between 1961 and 1965, when the population had increased to up to 5,000. This expansion of range is believed to have been due to increasing pressure on grazing following a growth in numbers after the closure of the whaling station at Grytviken – and with it the abandonment of shooting – in 1964. A considerable amount of vegetative degredation, especially of the previously abundant coastal tussock grass *Poa flabellata*, which is the reindeers' sole winter food, resulted in a decline in the population on the Barff Peninsula to around 1,300 and in Royal Bay to some 800 being reported in 1972 by the British Antarctic Survey team on HMS *Endurance*. In 1976 there were 1,000 yearlings and adults in the Barff subpopulation and 550 in the Royal Bay herd. A small number were shot by the invading Argentines during their illegal occupation of South Georgia in 1982, but this is unlikely to have affected the overall population.

The animals landed at Leith increased to around 20 before the entire herd was engulfed by an avalanche. Those at Busen – due to poaching during the Second World War – had multiplied to only about 40 by 1953; three years later this population was estimated at between 100 and 200, and since the cessation of whaling, and thus of shooting, after 1964 it increased still further to around 800 by 1972. In 1976 there were 450 adults and yearlings in the Busen herd. This subpopulation appears to be more sedentary than that on the Barff Peninsula and does not fully utilize all apparently suitable habitats; this, together with differences in morphology (the deer in the Busen herd are generally larger than those in the Barff and Royal Bay subpopulations) is probably due to their derivation from different stocks. It is therefore impossible to separate phenotypic from genotypic differences in body size.

The terrain of South Georgia is generally rugged and mountainous, rising in places to over 750 m. *Poa flabellata* is the only grass protruding through the snow in winter and is thus a vital source of food at that time; as a result, wherever reindeer are established it has become greatly reduced. In summer a wider choice of forage becomes available, but even so most lichens and the burnet *Acaena magellanica* have been severely grazed and trampled. Apart from the remaining patches of tussock grass the reindeer feed mainly on *Festuca erecta*, the burnets *Acaena tenera* and *A. adscenderis, Poa annua, Deschampsia antarctica* and the rush *Rostkovia magellanica*. Although lichens are slow to regenerate burnets, grasses and mosses do so quite rapidly, and the nutritious summer grazing maintains a high pregnancy rate and encourages early breeding. Many males, however, perhaps due to intensive rutting activity, die before the supplies of forage become reduced in early winter. The long-term future of reindeer in South Georgia must remain somewhat uncertain. They appear to have adapted well to their transference from the Northern to the Southern Hemisphere, where they calve in November – the southern spring. Although their only winter forage, *Poa flabellata*, has been severely reduced, it regenerates much more quickly than does *Cladonia rangiferina*, so a calamitous fall in the population, such as occurred on the islands of St Matthew and St Paul in the Pribilofs, seems improbable. A more likely pattern is a continuing gradual decline in numbers as increasing grazing pressure results in a reduction in the overall winter range carrying-capacity of the island.

Kerguelen[11]

In 1956 (some accounts say as early as 1953) two males and five females from Swedish Lapland were landed on the island of Grande Terre and a further male and two females on Île Haute. By 1965 those introduced to the latter had multiplied to 34 – an annual increase of 35 per cent – due mainly, it is believed, to the mild climate, an abundance of nutritious food, and freedom from predators and disease. In January 1968 the population on Île

Haute had risen to about 90, and it was judged that the island could support a maximum of 150 animals – a total which could have been reached by about 1970. Instead, by May 1969 it had fallen by a third to only 60.

There are two possible reasons for this sharp decline; between 1965 and 1967 10 were slaughtered for venison, and in 1968 a bull was observed swimming beyond the mudflats of seaweed *Macrocystis pyrifera* in Van Hoffen Sound, less than 200 m from Grande Terre, suggesting the probability of emigration. From 1970 to 1973 a total of 58 animals (at least 23 of which were females) were slaughtered: surveys in May 1971, May 1972, October 1972 (after the annual cull), March 1973 and December 1973 revealed totals of, respectively, 93, 99, 61, 50–60 and 70–80. Taking into account both the slaughtered animals and probable emigrations to Grande Terre the annual rate of increase is around 30 per cent. Lesel and Derenne (1975) concluded that as the result of the rise in the population and competition for grazing with introduced mouflon (*Ovis musimon*),[12] the majority of the herd had emigrated to Grande Terre and that they would eventually be joined by the remainder.

On Grande Terre – a much larger island than Île Haute with a more stable biotope – reindeer have adapted well: in 1974 the population was estimated at around 2,000, of which about 10 per cent were recruitments from Île Haute; there was still ample range for the future expansion of the population.

In 1924 10 reindeer were landed for experimental farming purposes by the Tokyo Fisheries Bureau on the island of Shinshiru in the Kurils, now part of the Soviet Union. By 1940 the population had increased to 224, of which 177 were adults and 47 immatures. Since then this herd has died out.

NOTES

1. Corbet & Southern (1977: 454).
2. *The deer and deer forest of Scotland* (H. F. G. Witherby, London, 1923.)
3. Fitter (1959: 61).
4. Palmer & Rouse (1945) and Leopold & Darling (1953).
5. See Scheffer (1951).
6. B. W. Evermann, *US Bureau of Fisheries Document* No. 780 (1913) and Lantis (1950).
7. *Scientific Monthly* **15**: 181 (1922).
8. Palmer (1934) and *The Alaska tundra and its use by reindeer* (US Department of Indian Affairs, 1945).
9. For much of the information on reindeer in South Georgia I am indebted to Leader-Williams (1978). See also *Report of the Inter-departmental Committee on Research and Development in the Dependencies of the Falkland Islands* (*HMSO Paper* 657) (HMSO, London, 1920), pp. 20–1; Olstad (1930); Tennant (1949); and Bonner (1958).
10. J. C. Beaglehole, ed., *The journals of Captain James Cook on his voyage of discovery. Vol. 2. The voyage of the Resolution and Adventure, 1772–1775* (Cambridge University Press for the Hakluyt Society, London, 1961), p. 622.
11. See *Polar Record* **7**: 518 (1955), Lesel (1967) and Lesel & Derenne (1975).
12. See p. 312.

Mule or Black-tailed Deer

(Odocoileus hemionus)

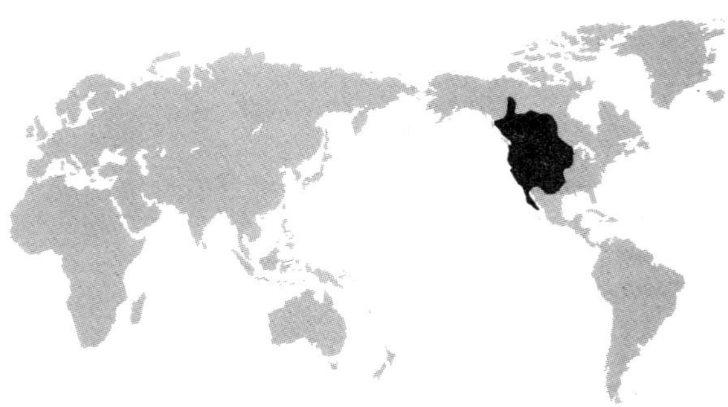

NATURAL DISTRIBUTION
Southern Alaska to the western and central United States, south to northern Mexico.

NATURALIZED DISTRIBUTION
Oceania: Hawaii.

OCEANIA

Hawaii[1]

Tomich (1969) records that in 'Telling the Kamaaina' in the *Honolulu Star Bulletin* of 13 April 1939 B. Clark claimed that Captain Alexander Adams, who voyaged between North America and Hawaii in the early nineteenth century, may have transported from the Pacific North-West to Hawaii in about 1816 on board the brig *Forester* two 'plain North American deer' as food for the ailing Billy Pitt, formerly adviser to King Kamehameha I. In *Voyages in the North Pacific* (1896) P. Corney wrote of the return voyage that 'Mr.

Biggot, the supercargo, took command, and got one Adams to navigate and some islanders to work the ship. He then sailed from Karakakooa [Kealakekua] Bay for the coast of California'. In February 1817 *Forester* (renamed *Kaahumanu*) was fitted out at Lahaina on the island of Maui for a voyage to Canton under the command of Alexander Adams. According to S. W. Tinker[2] a '200-pound buck' was shipped to the islands from Oregon in 1857 by Captain Collins on the brig *Advance*. It seems probable that these three animals were all mule deer.

On 10 June 1961 five pairs of wild mule deer fawns (presumably of the coastal subspecies *O. h. columbianus* which ranges from southwestern British Columbia – including Vancouver Island – south along the coast through Washington and Oregon into northwestern California), captured near the coast by Mr John McKean of the Oregon Department of Game and Fish, were imported to Honolulu by the Hawaiian Division of Game and Fish as a tourist attraction and sporting asset to replace feral goats, which it was planned to eliminate. From Honolulu the deer were transported by land to the Barber's Point Naval Air Station, from where they were flown by the US Navy on the following day to Bonham Air Base on Kauai, and thence by road to a release pen near Trail No. 4 on Polihale ridge in the Puu Ka Pele Game Management Area and Forest Reserve, where they were freed on 19 June. On 12 June 1962 a further 2 bucks and 8 does were released in the same locality, followed in November 1965 by 5 more does and on 13 April 1966 by an additional 2 bucks and 13 does – a grand total of 40 deer comprising 9 bucks and 31 does.

By June 1965 the population of mule deer on Kauai had reached 100; a year later – due to poaching – it had increased to only between 120 and 160, but two years later permits were issued for limited hunting. In 1981 the population, in the west of the island, numbered about 350.

The mule deer on Kauai – where they are known by the Hawaiian name of *kia* – range over more than 3,600 ha of land in the west of the island at an altitude of between 600 and 900 m. Apart from the sea cliffs, Puu Ka Pele is thickly vegetated with such native species as koa (*Acacia koa*), ohia (*Metrosideros collina*) and kukui (*Aleurites moluccana*), and aliens such as silk oak (*Grevillea robusta*) and eucalyptus. The understorey is also composed of mixed native and exotic species, including lantana (*Lantana camara*), molasses grass (*Melinus minutiflora*), aali (*Dodonaea eriocarpa*) and pukiawe (*Styphelia tameiameiae*). The deer prefer introduced plants and fruits to native vegetation, and their adverse effect on the environment (mainly by the distribution of the seeds of exotics through their droppings) is not great. As the result of heavy poaching the population is not currently increasing.

Although Harrison Matthews (1952) includes the black-tailed deer in his 'List of living British mammals',[3] this species has never become established in the wild in the British Isles.

In North America mule deer have been translocated to at least two localities outside their natural range. Between 1931 and 1935 50 were obtained from Salt Lake City, Utah, USA, by Mr LeRoy G. Denman of the San Antonio Loan Trust Company, who released them on the St Charles Ranch in the Blackjack Peninsula on the coast of southern Texas. All had died before 1939. They have been successfully transplanted on Santa Catalina in the Channel Islands off the coast of California.

Mule deer have also been successfully introduced to Sidney Island (713 ha) at the southern end of the Strait of Georgia in British Columbia, Canada, which they share with other imported birds and mammals, including fallow deer (*Cervus dama*).[4]

Concern has been expressed that mule deer stocked on several ranches in Neuquen province in the foothills of the Andes could escape into some of the adjacent national parks.

Mule deer have been introduced to Australia but have never succeeded in establishing wild populations.

In 1877 the Auckland Acclimatization Society unsuccessfully released 11 mule deer near Mercer and Piako in North Island, New Zealand where they were all shot by settlers and Maoris. In 1905 Mr T. E. Donne of the New Zealand Tourist Department obtained five mule deer (probably *O. h. hemionus*) from Santa Fé, New Mexico, which were released for hunting at Runuanga, Tarawera, in the Hawke's Bay district of North Island. A decade later they were reported by the Hawke's Bay Society to be 'increasing', but seem to have died out (or been killed) shortly thereafter. In the same year nine more are said to have been liberated by the Southland Acclimatization Society in the Rees Valley near Lake Wakatipu in South Island, but these too failed to become established.[5]

NOTES

1. For information on mule deer in Hawaii I am indebted to Tomich (1969) and Kramer (1971).
2. *Animals of Hawaii* (Nippu Jiji Co., Honolulu, 1938).
3. p. 388.
4. See p. 177.
5. Anon., *Acclimatization in Southland* (n.d.)

White-tailed or Virginia Deer

(*Odocoileus virginianus*)

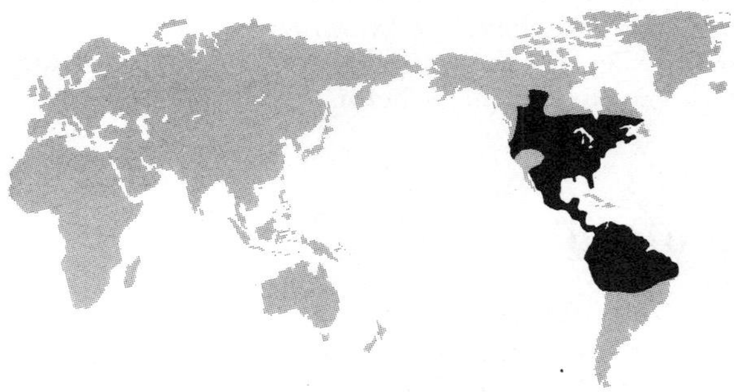

NATURAL DISTRIBUTION
The most widely distributed deer in North and Central America; ranges from southern Canada (south of approximately 52° N) southwards to Peru and northern Bolivia and Brazil.

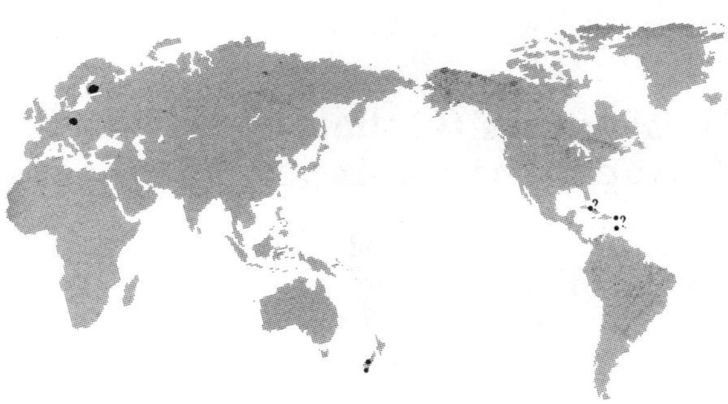

NATURALIZED DISTRIBUTION
Europe: Czechoslovakia, Finland; North America: West Indies (?Cuba, ?Curaçao, ?St Croix, ?St Thomas); Australasia: New Zealand.

EUROPE

Czechoslovakia

Sometime after 1890 Prince Colleredo-Mannsfeld, who in 1905 introduced muskrats (*Ondatra zibethicus*) to Czechoslovakia,[1] imported some white-tailed deer from southeastern Canada (possibly through the animal dealer Carl Hagenbeck of Hamburg) to his estate at Dobrisch, 40 km southwest of Prague. After the breakup of the estate the deer were allowed to run free, and today a population believed to number about 200 lives mainly in the submontane zone of Brdy and Hrebeny in central Bohemia.

Finland

In 1934 five white-tailed deer (a buck and four does) from Minnesota, USA, were released for hunting purposes 160 km north of Helsinki, where in 1947 the population was estimated to number more than 200; a further three pairs were imported to provide new blood in 1948. By the late 1960s the population in south-west Finland had increased to around 2,500.

NORTH AMERICA

West Indies (Cuba, Curaçao, St Croix, St Thomas)

White-tailed deer from Mexico or the southeastern United States (possibly the so-called 'Florida Key' subspecies *O. v. clavium*) were introduced to Cuba around the middle of the last century. The variety sometimes known as the 'bare-eared deer' (*O. gymnotis curassavicus*) is believed to have been introduced to the island of Curaçao off the coast of Venezuela from the Guajira Peninsula in neighbouring Colombia.[2] In the seventeenth century white-tailed deer were reputedly imported as game animals from Virginia to Antigua and Barbuda in the Leeward Islands; today only fallow deer (*Cervus dama*)[3] occur on Barbuda (and on the island of Guana, 100 m off the north-west coast of Antigua) – to both of which they were probably introduced in the early eighteenth century by the Codrington family of Gloucestershire, England[4] – so the earlier importation must be in doubt. As early as 1790 and as recently as 1945 white-tailed deer were introduced to St Croix and St Thomas in the US Virgin Islands. Deer of this genus are also said to have been imported to the Dominican Republic (Hispaniola), Jamaica, Dominica, Grenada and possibly some other islands in the Caribbean, on none of which, however, they are established in the wild today.

AUSTRALASIA

New Zealand[5]

White-tailed deer have twice been imported to New Zealand. In 1901 two pairs were released by Mr T. E. Donne of the Government Tourist Department for hunting in the Takaka Valley, Nelson, in the north of South Island. Four years later a buck from an original shipment of 22 (not 19, as stated by Whitehead (1972)), from New Hampshire, USA – presumably *O. v. borealis* – was freed at Motueka near the previous site; the remaining 18 (three having died on the voyage) were divided equally for liberation in the Rees Valley at the western end of Lake Wakatipu in Fiordland, and at Port Pegasus (2 bucks and 7 does) on Stewart Island.

The first of these plantings failed, and that at Lake Wakatipu was only moderately successful. For many years the deer remained in a small area, and did not begin to spread towards the Forbes mountains until the late 1950s; today they occur mainly in the lightly forested Rees, Dart and Routeburn valleys at the head of Lake Wakatipu. The rate of natural mortality in this area – where the habitat is not ideal – is high, and the population, which for many years has been steadily declining, may soon become extinct.

The introduction to Stewart Island, however, proved much more successful. By the early 1920s the deer were increasing and spreading rapidly throughout the island's more densely vegetated coastal areas, where by 1926–7 they were apparently widely distributed. A decade later it was stated that they 'inhabit the coastal strip only, and, although thinly scattered along almost all the coast-line, are not in any large concentrations in any locality'. In the late 1940s Wodzicki (1950) reported them 'from most of the coastal bush-covered parts of the island'.

Although introduced to New Zealand for hunting purposes, their establishment in relatively inaccessible areas has rendered white-tailed deer of doubtful sporting value. Writing of the population on Stewart Island in 1909, L. Cockayne asked: 'In this case, can the value of these animals for sport be held as a plea since the trivial amount of open ground, the dense character of the sub-alpine scrub and the close forest growth renders Stewart Island quite unsuitable as a hunting ground'.[6] Until 1920, when limited shooting was permitted, the white-tailed deer on Stewart Island were strictly protected; although full protection was abolished in 1925 serious hunting did not begin until 1937.

In *The naturalization of animals and plants in New Zealand* Thomson (1922) wrote of the white-tailed deer on Stewart Island that they 'threaten to destroy much of the vegetation besides opening up the country – which is a reserve for native birds – to deer-stalkers'. It was also feared that hunting the deer with hounds would pose a threat to ground-nesting birds such as kiwis (*Apteryx* spp.) and weka rails (*Gallirallus australis*).

By the early 1960s white-tailed deer were found over much of Stewart Island, where they mainly frequent bush below 300 m and gently sloping coastal valleys, seldom penetrating the subalpine belt of dense scrub to the mountain tops. They are, perhaps, more common in the south than in the north of the island. On dry, rocky and open spurs browsing by the deer led to the disappearance of ferns – except the hard fern (*Blechnum discolor*) which actually increased even where already common. Tree cover gradually thinned out until even such resilient species as rata (*Metrosideros lucida*) became scarce. The removal of dense growth from spurs and ridges gave the deer access to hitherto inaccessible damp gullies, where they found such succulent ferns and shrubs as *Schefflera* sp., *Nothopanax* sp. and *Asplenium* sp., some of which have been replaced by more woody plants. On cliffs and near the coast succulent herbs such as *Anistome* spp. and *Stilbocarpa lyallii* have in places been almost eaten out; the latter is now rare on the main island but remains common on deer-free outlying islets. In about 1930 *Pseudopanax crassifolium* – previously untouched – began to be barked.

Wodzicki believed that in opening up areas of dense scrub and forest on Stewart Island the more wide ranging white-tailed deer assisted in the establishment of the previously introduced red deer (*Cervus elaphus*),[7] who were reluctant to penetrate dense and tangled thickets which were far removed from their mainland *Nothofagus* forest haunts.

White-tailed deer are said to have been released in Germany, but there are none in the wild there today.

NOTES

1. See p. 334.
2. Westermann (1953).
3. See p. 177.
4. See p. 178.
5. For much of the information on white-tailed deer in New Zealand I am indebted to Wodzicki (1950).
6. *Report on a botanical survey of Stewart Island* (New Zealand Department of Lands, 1909).
7. See p. 203.

Moose
(*Alces alces*)

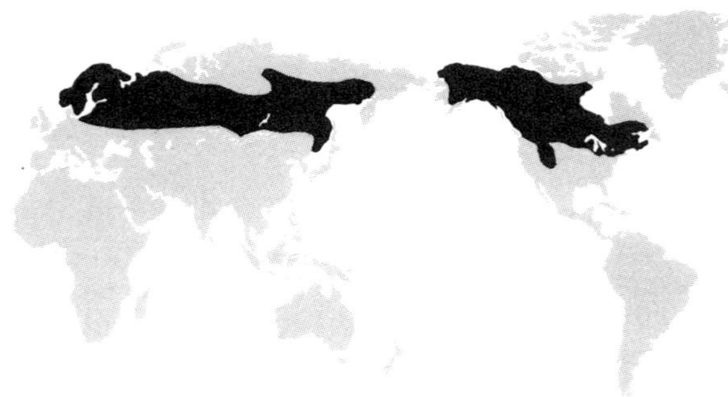

NATURAL DISTRIBUTION
From Scandinavia and eastern Poland to eastern Siberia in Eurasia (where it is known as the elk) south to the southern Urals, the Altai and lower Amur. In North America from northern USA and the Great Lakes to approximately 66° N in Alaska and Canada. Formerly extended further west in Europe.

NATURALIZED DISTRIBUTION
Australasia: ?New Zealand.

AUSTRALASIA

New Zealand

Following representations to the Canadian Government by Sir James Carroll, the Native Minister in New Zealand, 14 young moose (probably *A. a. americana* but possibly *A. a. andersoni*) were despatched from Vancouver by the Hudson Bay Company on 5 January 1900 (not in 1899 as stated by Whitehead (1972)) on board the SS *Aorangi*. Ten unfortunately died during a storm off Sydney, Australia, where the four survivors – two bulls and two nine-month old cows – were transferred to another vessel which berthed in Wellington on 8 February: two days later the deer were moved to Greymouth, and were later released in the Hokitika Valley in Westland. By 1903 only one cow remained in the valley and a second in the Hokitika River gorge, and these disappeared shortly thereafter.

In 1907 Mr T.E. Donne of the New Zealand Tourist Department recommended a second attempt; three years later 10 moose (believed to be *A. a. andersoni*) were captured in Saskatchewan and transported to Banff, Alberta, and thence to Vancouver for shipment to New Zealand. The entire consignment survived the rigours of the voyage and arrived in good condition in Wellington via Sydney, from where they were transferred by the steam yacht *Hinemoa* to Dusky Sound in the Fiordland National Park in Southland, where they were released at Supper Cove.

Although this second introduction was successful, little seems to have been recorded of the animals' early history. According to Thomson (1922), 'these deer were in considerable numbers round the Sound', and by 1924 they had spread to the Seaforth Valley. Wodzicki (1950) recorded that a line running south from Supper Cove in Dusky Sound to the estuary of the Richard Burn in Long Sound was their western boundary; the southern limit of their range ran from Long Sound to the shores of Lake Hauroko, then north to the junction of the Heath and Merrie mountains, and thence back to Supper Cove. Wodzicki believed that 'the precipitous slopes of the valleys, as well as the very slow increase of the herd are responsible for their present limited distribution', which was confined to the 'few valleys surrounding the place of their original liberation'.

Moose in New Zealand appear to eat mainly pokaka (*Eleaeocarpus hookerianus*), five finger (*Nothopanax arboreum*), koromiko (*Hebe salicifolia*), punga ferns (*Cyathea dealbata*), flax, mosses, beech leaves and other shrubs.

As Wodzicki pointed out, the main climatic differences between Saskatchewan (latitude 59° N) and Dusky Sound (45° S) are the latter's higher mean temperature and rainfall: this, together with the mooses' inability to climb the steep slopes from the dense rainforest to the mountain tops in warm weather, and interspecific competition from the infiltration of their range by introduced red deer (*Cervus elaphus*),[1] probably accounts for their dwindling range and numbers. Wodzicki estimated the population in the late 1940s at no more than 30–40 compared with an earlier peak of 60–70, and recent evidence suggests a continuing decline which may well end in extinction – if indeed this has not already occurred.

In 1878 and 1904 moose were successfully translocated from the North American mainland to insular Newfoundland, where they became established.

Unsuccessful attempts have been made to establish **barasingha** or **swamp deer** (*Cervus duvauceli*) – natives of central and northern India and southern Nepal – in Australia and on the Rickenbacker Ranch in Texas, USA, where they were said to be 'frail and weak, and not good breeders'.

In 1867–8 Mr A. Grote of Calcutta, India, presented a number of barasingha to the Victoria Acclimatization Society of Australia, by whom they were successfully bred in the Royal Park at Melbourne. There were seven in the Society's collection in 1871 but only one by 1885, and it seems probable that the others were released in the intervening years. The *Victorian Historical Review* (quoted by Bentley (1978)) records that 'the liberations took place in the mountainous parts of Gippsland' and that a few years later 'they had penetrated deeply and thrived to a limited degree'. Although Bentley considered that 'the possible survival of this deer in Victoria, in some remote area, cannot be discounted entirely' it must be deemed highly improbable.

In 1921 five barasingha from the Melbourne Zoo were among the large consignment of deer shipped on the SS *Mataram* from Sydney to Darwin in the Northern Territory. One of the barasingha is said to have escaped from its crate and jumped overboard off the Cape York Peninsula in northern Queensland, where it may have succeeded in swimming ashore. Bentley considered that if this animal was a hind in calf it could have been the origin of a number of unidentified deer reported on the east coast of Cape York in 1958. It is much more likely, however, that these were Timor deer (*C. timorensis*)[2] which in 1914 were translocated from Friday Island to Possession Island off Cape York, on which they subsequently landed. Bentley also believed that barasingha may be established on the Cobourg Peninsula in the Northern Territory, where the four survivors of the 1912 shipment are believed to have been released; the only deer known to be in this area, however, are sambar (*C. unicolor*).[3]

Although native to England, Scotland and Wales, and common in England in medieval times, **roe deer** (*Capreolus capreolus*) were extinct in England and Wales (except, perhaps, near the Scottish border) before about 1730. The main reasons for their decline appear to have been overhunting and extensive deforestation. Roe, probably from Scotland, were subsequently reintroduced to Petworth Park, Sussex, by the Earl of Egremont in the 1790s. In 1817 the *Sporting Magazine* reported that 'In the year 1800 the Earl of Dorchester turned into his woods near Milton Abbey in the County of Dorset four or five brace of roe bucks; some of which he procured from the menagerie of Mr Brook in London at considerable expence [*sic*] and some were presented to him by the Earl of Egremont'. These sources of Lord Dorchester's roe are confirmed in the 1834 edition of Gilpin's *Remarks on forest scenery*, although tradition has long held that the deer were brought to Dorset in 'padded horse boxes' from Perthshire, Scotland. In 1884 Sir William Dalziel Mackenzie obtained some roe from Württemberg, Germany, which he turned out in Great Queen's Wood on his estate (now part of Thetford Chase) at Santon Downham in Norfolk.[4] These, and other introductions later in the nineteenth century supplemented by translocations and escapes, are the ancestors of the present roe population in southern and eastern England. Roe have also been introduced to Ireland where they are not indigenous, but have failed to become established. They no longer occur in Wales.

At some unrecorded date Prince Albert or Queen Victoria is said to have presented six roe deer to the Victoria Acclimatization Society of Australia. Later, in 1875, a pair arrived in Melbourne from England on the *Shannon*, but died before they could be put ashore. Whether any others were ever released is unknown.

Small numbers of **Père David's deer** (*Elaphurus davidianus*) are said to be held on several ranches in the mountainous province of Neuquen in southwestern Argentina, where concern has been expressed that they could escape into some of the neighbouring national parks. Some 20 years ago a nucleus breeding stock, derived from the Duke of Bedford's captive herd at Woburn Abbey in England, was presented to the Peking Zoo in China (the species' native home) where the last animals had died in 1921.[5]

In 1861 Mr Edward Wilson imported to Victoria, Australia, from Ceylon (Sri Lanka) on board the *Jeddo* a shipment of deer which included 'five moose deer which fought with such ferocity that three of them died of their injuries, whilst the others died from want of company'. There must have been at least one further importation, as this species – known also as the **chevrotain, musk deer** or **mouse deer** (*Moschus moschiferus*) – was kept in captivity in the Royal Park for a number of years, but there is no record that any were ever released into the wild.

In 1966 a pair of deer (? sp.) were released on San Cristóbal in the Galápagos Islands, where within a year they had disappeared.

In 1870 the Auckland Acclimatization Society of New Zealand imported three **Chilean guemals** or **huemuls** (*Hippocamelus bisculus*); whether these were ever released is unrecorded, but if they had been the Southern Alps would probably have provided a near-ideal habitat.

NOTES

1. See pp. 200.
2. See p. 220.
3. See p. 225.
4. R. Prior, *Country Life* **172**: 250 (1982).
5. Fisher, Simon & Vincent (1969).

Pronghorn

(*Antilocapra americana*)

NATURAL DISTRIBUTION

Formerly western and central USA and southern Canada south to central Mexico: now confined mainly to Oregan, Montana, Nevada, Wyoming and Texas.

NATURALIZED DISTRIBUTION

Oceania: Hawaii.

OCEANIA

Hawaii

On 30 November 1959 56 pronghorn antelopes[1] were trapped in Montana by the state's Department of Fish and Game. Twelve died on the voyage from Seattle to Hawaii, where four were presented to the zoo in Honolulu. At the instigation of Mr Lyman Nichols, Jr, the resident wildlife biologist on Lanai, the remaining 24 were flown to that island on 10 December, two more dying (presumably from stress) during the flight.[2]

The six adult females which were among the survivors gave birth to eight young in the following June and July, when the total population, following the death of further adults after release, numbered 23. After a slow start due to injuries sustained in the bush and as a result of drinking salt-water, the pronghorns adapted well to life on Lanai's dry open grassland plateau which covers an area of around 90 km^2 at altitudes of between 250 and 550 m in the north-west of the island. By 1966 (when limited shooting began) the population (the first to become established in the wild outside North America) was estimated to number between 150 and 250. Pronghorns – known locally as '*Anekelopa* – are not presently (1981) open to public hunting on Lanai.

Pronghorn antelopes reintroduced to the more remote regions of the state of San Luis Potosi in central Mexico are reported to have spread as far north and west as the state of Durango at the southern end of the Golfo de California.

NOTES

1. Pronghorn antelopes – the sole survivors of a large North American artiodactyl (even-toed ungulate) family (the Antilocapridae) unique in shedding the branched horn-sheath from a blade-like bony core – are not true antelopes (Bovidae), which, unlike Cervidae (deer) who annually shed their bony antlers, normally retain a permanent horn-sheath on a bony core.
2. One account says that 38 were released on Lanai; another reports that 40 were shipped from Seattle and that only 18 were landed safely on Lanai.

Nilgai

(Boselaphus tragocamelus)

NATURAL DISTRIBUTION
India.

NATURALIZED DISTRIBUTION
North America: USA, ?Mexico.

NORTH AMERICA

USA, Mexico

Nilgai were first introduced to the United States in 1924[1] when two cows and a bull were released by the King Ranch Inc. on their Norias Division in southern Texas. Further liberations, mostly in groups of 3–5, continued until 1949 (when the total population was judged to number about 100), the largest individual releases comprising about a dozen in 1929–30 and eight cows and four bulls in 1941; most of the stock was obtained from San Diego Zoo, California. From the King Ranch nilgai were subsequently trans-located to ranches in nine other counties in

Texas and one, near Nuevo Laredo, in Mexico.

In Texas nilgai extended their range naturally over 2,600 km^2 of Kenedy and Willacy counties; the Norias Division of the King Ranch comprises about 29 per cent of this range and holds more than 50 per cent of the total Texan population, which by 1964 had multiplied to about 3,300 and by 1971 to around 4,000, but had decreased to 2,786 by 1974 and to 2,241 (2,057 of which were free-ranging) in 15 counties by 1979, thus following a classic sigmoid growth curve.

Today, nilgai still occur mainly in Kenedy and Willacy counties in the Rio Grande Plains and Gulf prairies of southern Texas, but also in Blanco, Kleberg and Zapata counties. Their largest population, including that on the Norias Division, extends from Baffin Bay south to the southern boundary of Willacy County and from Laguna Madre west to Highway 77. Although the population has somewhat declined in recent years they are breeding well and appear slowly to be extending their range. Their preferred habitats are improved rangeland where dominant grasses include bristlegrass (*Setaria* spp.), King Ranch bluestem (*Bothriochloa ischaemum*), bufflegrass (*Cenchrus ciliare*) and Gulf Coast cord grass (*Spartina spartina*); prickly pear cactus (*Opuntia lindheimeri*), live oak (*Quercus virginiana*) forest, and mesquite scrub (*Prosopis juliflora*) provide cover. In Texas nilgai feed mainly on the leaves of oak and mesquite, on acorns and on grass.

Various factors account for the success of nilgai in southern Texas. Since the original stock came from zoos rather than direct from India it was at least one generation removed from the wild, and was thus already acclimatized before its release. Limited hunting only, the prevention of poaching and the absence of major predators initially allowed a viable breeding stock to become established. Since nilgai are not subject to state game laws (being regarded as domestic stock) there is no regular shooting season and individual landowners may regulate hunting as they wish. The semi-arid climate, the woody scrub vegetation and the rainfall of the South Texas Plains approximate quite closely to those of India. Finally, an extended breeding period, with the birth of young all the year round, comes to a peak in Texas in October at the start of the drier, cooler weather, which provides optimum conditions for the calves' survival.

Nilgai have been introduced for sporting purposes to several large (85,000 ha or more) fenced private ranches in the Orange Free State, South Africa.

NOTE

1. Some accounts say 1929–30. Whichever date is correct this was the earliest introduction of an exotic artiodactyl to Texas.

Water Buffalo
(*Bubalus bubalis*)

Note: Natural distribution shown is original range of wild ancestor.

NATURAL DISTRIBUTION
The wild Indian water buffalo (*B. arnee*) – ancestor of the domestic *B. bubalis* – was formerly widespread in India and southern (and possibly western) Asia. It is now confirmed to parts of Nepal and southeastern Asia.

NATURALIZED DISTRIBUTION
Africa: Tunisia; South America: Argentina, ?Brazil, ?Venezuela; Australasia: Australia, Papua New Guinea, New Britain, New Ireland, ?Borneo, Java, ?Andaman Islands.

AFRICA

Tunisia

According to Haltenorth and Diller (1980), water buffalo were:

Very probably already established since Roman times as a wild-living herd and royal hunting quarry at Lake Ischkeul, at foot of Ischkeul mountains not far from Bizerta, N. Tunisia, with varying fortunes. In 1957 all but 3 calves were shot; from these a new herd has been built up which now has its own special reserve at Ischkeul. For 2,000 years has lived in the wild, but original descent unknown.

SOUTH AMERICA

Argentina; Brazil; Venezuela

According to Petrides (1975b):

A large number of additional species have been introduced [to Argentina] for specialized sport-hunting. This may possibly be true of the Water Buffalo (*Bubalus bubalis*), which has been established in Corrientes Province [south of Paraguay in the extreme north-east] since 1900 . . . As advertized in North America [in 1973] . . . opportunities to hunt Water Buffalo on Marejo Island [Brazil], in the mouth of the Amazon river, seem to indicate that this species is present there in some abundance, and may have escaped into the wild. . . . The Asiatic Water Buffalo, on several occasions over the last 35–40 years, was imported as domestic stock [to Venezuela], probably always from Trinidad. Between 1935 and 1936 some of the early imports escaped in the Rio Limon Valley, but are believed to have been exterminated by hunters.

AUSTRALASIA

Australia[1]

In September 1824 an expedition from England under the command of Captain (later Sir) James Bremer arrived at Port Essington on the Cobourg Peninsula in the Northern Territory of Australia; 'but fresh water not having been readily discovered', wrote Earl (1846) in *Enterprise in tropical Australia*, 'the ships went on to Melville Island', 240 km to the west, where a settlement named Fort Dundas under the command of Captain Maurice Barlow was established at King's Cove on the eastern side of the Apsley Strait. 'A vessel', Earl continues, 'was instantly despatched to Koepang (Dutch Timor) [800 km to the north-west] and soon returned with supplies of stock and other refreshments.' Although not specifically mentioned, the 'stock' almost certainly included water buffaloes. In 1826 16 water buffaloes arrived at Fort Dundas from Timor where within a fortnight several had expired; Mr J. Campbell, who had them in his charge, ordered some spacious sheds to be constructed, in which the animals sheltered from the scorching sun between 1000 and 1500 hours; this, together with the provision of twice daily wallows, seems considerably to have reduced the rate of mortality. Subsequent shipments brought the total number of buffaloes imported for working and for meat to about 50.

In 1827 a second settlement (Fort Wellington) was established at Raffles Bay on the Cobourg Peninsula, some 30 km east of Port Essington; two years later Fort Dundas was abandoned – some of the stock (including buffaloes) being released, while the remainder and the stores were transferred to Raffles Bay. Earl confirms that 'the settlements at Melville Island and Raffles Bay [were] supplied with stock from the Dutch town of Koepang, Timor'. In August of the same year (1829) the garrison at Fort Wellington removed to New South Wales, the stores and some of the stock being despatched to a new settlement on the Swan River in Western Australia; the rest of the stock (including some buffaloes) escaped or were released at Fort Wellington. Thus as early as 1829 a nucleus of feral water buffaloes had become established on Melville Island and on the Cobourg Peninsula in the Northern Territory.

In 1833 Port Essington was re-occupied by Bremer, and Victoria Settlement was established. In the same year Earl records that the schooner *Essington* 'obtained a full cargo of 19 buffaloes [one of which died *en route*], 60 pigs, more than 100 sheep and a quantity of poultry . . .' from the island of Kisar (Kissa) in the Moluccas. 'The buffaloes that had been left behind when the settlement at Raffles Bay was abandoned', Earl continues, 'have increased to a surprising degree and wander about the country near the neck of the Peninsula in herds of forty and fifty. Stragglers have often been encountered in the immediate neighbourhood of the settlement – large unwieldy bulls that have been driven out of the herd by others stronger than themselves.'

In 1843 Captain I. Everard Home of HMS *North Star* reported that 'wild ponies, pigs, buffaloes and red cattle, the greater part of which is the offspring of the stock left at Port Raffles when the settlement was broken up', roamed the woods around the Victoria Settlement.[2] In April of the same year Home reported the presence of 30 buffaloes at Port Essington, and that 'on the mainland opposite to Goulbourn Island two herds of wild buffaloes consisting of about 50 each have been observed feeding by the water side'.[3] In the same month the acting commandant of the Victoria Settlement, Captain James McArthur, wrote that he had 'received by the *Heroine* a supply of 49 buffaloes . . . [whose] meat is not only inferior, but absolutely injurious to some . . . we have a few men who can never eat it'.[4]

In 1844 the explorer Ludwig Leichhardt – the first man to reach the Top End overland from the south-east – saw buffaloes at Red Lily Lagoon on the Murgenella Plain near Oenpelli Mission on the east bank of the East Alligator River, some 200–250 km from Port Essington. Yet 18 years later John McDouall Stuart, following the Mary River northwards to the coast, saw none; nor, five years later in 1867, did Francis Cadell – while surveying the coast of Arnhem Land – find any east of the Liverpool River.

When the garrison at Port Essington was disbanded and transferred to Sydney in 1849 the buffaloes there were set free to fend for themselves; they were the largest single source of feral water buffaloes in the Northern Territory.

In 1857 the barque *Florence Street* carrying an unknown number of buffaloes from Malaysia and Burma to provide meat for the miners in the goldfields of New South Wales was wrecked off Cambridge Gulf near the East Kimberleys in Western Australia. Some buffaloes succeeded in swimming ashore near the mouth of the Ord River, but no large population developed.

Between 1864 and 1866 the schooner *Beatrice* made several voyages to Timor to supply buffaloes for a surveying party camping at Escape Cliffs near the mouth of the Adelaide River; they were described as 'a smaller variety than those we now see locally' and were 'turned adrift' when the camp was abandoned.

In 1869 a permanent settlement named Palmerston (changed in 1911 to Darwin) was established on Port Darwin Bay; a few buffaloes were introduced when the town was founded, but there is no record of any having been released.

J. A Gilruth records that in 1886 'Indian buffaloes (2 cows and one bull) were brought to Port Darwin . . . with the intention of establishing the *ghi* (or buffalo butter) industry. Some of the descendants of these buffaloes may still be seen being employed as beasts of burden'.[5]

It is thus apparent that most of the feral water buffaloes in Australia's Top End today are descended from a small original nucleus – those liberated when the settlement at Port Essington was abandoned for the second time in 1849 providing the principal source. The progeny of those liberated at Fort Dundas on Melville Island in 1829 were probably able to colonize a few outlying islets, but would have found the 24 km sea-crossing to the mainland an insuperable barrier.

From their original foothold on the Cobourg Peninsula feral water buffaloes began to spread slowly over the subcoastal plains and river basins of Arnhem Land bordering the Arafura Sea and Van Diemen Gulf. Since buffaloes are slow breeders, and as the original nucleus was small, the population increased and expanded only gradually; it was not long, however, before most of the suitable available habitat was occupied, and by the time the first pastoral lease was granted (to Francis Cadell in 1868) tens of thousands of feral buffaloes are estimated to have roamed over Melville Island and the Cobourg Peninsula. In 1874 John Lewis (who at the time held a lease from the government on the peninsula) travelled overland from Pine Creek, 150 km south-east of Darwin, to Port Essington, when he reported 'a large number of buffalo on the Peninsula as well as other animals . . . they were numerous on the Cobourg Peninsula, and just outside the Peninsula, on Scott's Plain, towards the Alligator'.[6] He estimated the buffalo population to number about 20,000. In 1885 the celebrated buffalo-hunter, Paddy Cahill, estimated the population around Oenpelli on the East Alligator River at around 60,000 and that on Melville Island at more than 6,000. Yet when W. J Sowden voyaged up the Adelaide River to Manton Hill in 1882 on the steam launch *Maggie* he found no trace of buffaloes on the newly established station at Marrakai.

By now, water buffaloes were established in most of the western region of Arnhem Land and in the main riverine floodlands as far west as the Adelaide River. Why the apparently suitable coastal plains and river valleys west of the Adelaide were not colonized is uncertain; possibly the population was not dense enough to force further expansion, and increasing human settlement around the Stuart Highway may have been an inhibiting factor, as the drier climate and more rugged terrain were in the south and east.

A professional buffalo hunter, E. O. Robinson, who killed several thousand for their hides on Melville Island in the 1880s, is believed to have been the first man to attempt to exploit feral buffaloes commercially. His example was soon followed by others, among them Paddy Cahill, who initiated the technique of shooting from horseback; using this method he was able to amass 2,000 hides a month during the dry season, which he sold for £1 each.

Between 1886 and the outbreak of the First World War a total of around 100,000 hides (averaging nearly 4,000 a year) were harvested in the Top End. From 1914 to 1939 (when the Buffalo Protection Ordinance limited shooting to licence-holders only) the annual take rose to around 7,000, and between 1945 and 1956 (when the international hide market collapsed) to about 12,000, with a record in 1937–8 of 16,549 and a grand total between 1886 and 1956 of around 390,000 – most of which were exported to England and Turkey. Exports slumped from 5,663 in 1955–6 to barely 100 in 1957–8, representing an annual loss in revenue of over £100,000.

Probably only about ten to a dozen hunters a season were employed in shooting buffaloes in the Top End, helped by a few dozen Aboriginal skinners. The hides were either overlanded to Darwin or were shipped by river on small steamers.

Between 1956 and 1962 some 400 buffaloes were exported alive to Hong Kong as draught animals or for slaughter; this enterprise was, however, eventually abandoned when a Chinese stockman was fatally gored.

In 1959 a trade began in buffalo meat for both domestic pets and human consumption; in the first year some 2,500 buffaloes (yielding 400 t of boned meat valued at $A100,000) were killed, and by 1964 some 7,600 had been processed for pet food and a further 11,900 slaughtered for human consumption. Between 1966 and 1968 an average of 7,000 buffaloes were killed annually for human consumption; between 1968 and 1971 the annual average rose to around 8,000, and between 1972 and 1974 it increased dramatically to an average of 21,000 head yielding, in 1974, $A1.6 million. In 1975 the mustering by helicopter of live buffaloes for subsequent slaughter was introduced, and in the same year an export trade in buffalo meat as 'game' (which was cancelled in 1977) was started to West Germany, and more recently to Sweden. In 1977–8, 26,718 buffaloes passed through the three abattoirs then in operation (Mudginberri, Jimmy's Creek (Port Stuart) and Wild Boar (Marrakai)) for human consumption, and a further 7,500 were slaughtered for pet food, yielding a total of around 5,000 t valued at $A3.8 million. Most of this meat (for the home market) came from feral field-shot buffaloes; the eviscerated carcasses were airfreighted daily from the abattoirs to Darwin; meat for export was obtained from live-caught animals which were subject to ante-mortem examination. The future of the industry largely depends on the availability of markets, operational costs, and the price obtained for buffalo meat.

The first serious attempt to redomesticate feral water buffaloes began in 1920 when a lease on 260 km^2 of land at Woolner was granted to Messrs. Gibson, Hollowood and Sullivan. For unknown reasons this enterprise proved unsuccessful. In 1958 Mr R. Willams and Mr K. Waldcock captured several hundred on Marrakai and Woolner, some of which were transported to a government experimental station at Beatrice Hill, some 55 km south-east of Darwin. More were subsequently caught and slaughtered at Marrakai, while the herd at Beatrice Hill was maintained for research.

Governmental and private interest in redomesticating buffaloes was stimulated by the successes at Marrakai and Beatrice Hill, and by 1970 captive herds – of up to 2,000–3,000 head each – had been established at a further eight stations: Annaburro, Ban Ban Springs, Goodparla, Mount Bundey, Mudginberri, Munmalary, Oenpelli Mission and Port Stuart. Longer leases, greater security of tenure and relaxation of the Brands Ordinance to allow the private ownership of buffaloes (all of which belonged to the Crown until 1970) gave encouragement to would-be buffalo-farmers.

By about 1973 several major obstacles to the successful redomestication of buffaloes had become apparent: feral animals shot in the field, or mustered and transported alive to an abattoir, fetched the same price per pound as animals bred in captivity, thus rendering the expense and time involved in maintaining domesticated herds hard to justify, especially as the 1970s fall in the price of beef was paralleled by that of buffalo meat; losses from heat stress and exhaustion during capture; the necessity of providing shelter, fresh water and adequate fencing; excessive stocking rates per hectare resulting in loss of condition and disease – all these factors have mitigated against the successful redomestication of feral water buffaloes. The future of viable buffalo husbandry depends largely on the development and maintenance of reliable outlets and the successful control of disease. (To date large-scale buffalo farming has been practised successfully mainly in northeastern Brazil.)

Trade in the export of live buffaloes from Australia was revived in 1974 when 100 heifers and 20 bulls were shipped to Papua New Guinea; by 1976 over 1,000 young (mainly female) buffaloes (to a value of around $A50,000 a year) had been air-freighted to Venezuela, Papua New Guinea, Hong Kong, Nigeria and Guyana. In 1976–7 small numbers of adult buffaloes for slaughter were shipped to Brunei. The future of the export trade of live buffaloes depends on the continued interest of importing countries and strict quality control of animals exported.

The contribution by buffaloes to Australia's economy has thus been largely in the form of hides and meat. By 1955–6 nearly 400,000 had been shot for their hides, and between 1960 and 1975 a further 140,000 were slaughtered for pet food and for human consumption. In recent years an increasing number of tourists from the Nourlangie Safari Camp (and others), between the South and East Alligator rivers, have photographed buffaloes in the Top End, and 'sportsmen' shoot easy trophies – the experience being aptly described by Frith (1979) as 'not unlike that of shooting an unusually large and extra wary dairy cow with a high velocity rifle in exotic surroundings'. These activities together produce an annual revenue amounting to some $A4 million and help to provide 80–100 jobs for both Europeans and Aborigines in an area of high unemployment.

Feral water buffaloes are today the dominant herbivore over some 66,000 km^2 of Australia's Top End. Around 90 per cent are concentrated in the hot and humid subcoastal plains and in the Marrakai and Koolpinyah land systems, encompassing the drainages of the Adelaide, Mary, Wildman, West Alligator, South Alligator, East Alligator and Liverpool rivers, and in the Cobourg Peninsula. Water buffaloes are well distributed throughout this region, especially in the plain of the Mary River and in the Murgenella Plain. They range over an area of about 100,000 km^2 between 11° and 14 °S and 130° and 134 °E, east, west and south of the Van Diemen Gulf, eastwards from the Stuart Highway and north of an east/west line through Pine Creek. Smaller numbers extend inland for up to 150 km or more in Arnhem Land, part of the Gulf country bordering the Barkly Tableland, all of the Daly River drainage and a small portion of the Victoria River region, including Newcastle Waters, Brunette Downs, Lawn Hill, west of Burketown, and east into Queensland and west to Kununurra in Western Australia. Terry (1965) reported sighting individual bulls in 1965 800 km south-west of the Cobourg Peninsula in the Victoria River area, 650 km to the south-east in the Barkly Tableland, and 1,450 km east into northern Queensland. McKnight's surveys in 1966 and 1971 found buffaloes as far east as the Mitchell River in the Cape York Peninsula of Queensland, and between Derby and Broome in the Dampier Peninsula of Western Australia. Buffaloes are currently increasing in the Waigait and Finniss river areas, and there is evidence of a further gradual extension of range into the Roper and Victoria river valleys; by 1979 individual bulls had been observed as far west as Broome in Western Australia, Yuendumu and Elkedra in the Alice Springs district to the south, and the coast of northern Queensland in the east. The Moyle River valley (between the Daly and Fitzmaurice rivers) and the Arafura Swamp (the floodplain of the Goyder) are possible areas of further expansion.

No accurate census has ever been made of Australia's feral water buffaloes; in 1961 Letts (1964) estimated the population at between 150,000 and 200,000; from aerial counts official government estimates for the late 1960s and 1970s were constant at around 150,000 head.

D. G Tulloch of the Northern Territory Animal Industry and Agriculture branch has made intensive studies of the biology of feral water buffaloes.

Their preferred habitat is the flat subcoastal plains along the lower reaches of the Top End's rivers, and the neighbouring rolling areas which rise above the floodplains. Here, the poorly drained river plains are criss-crossed by meandering tidal streams which, though often of some depth during the monsoonal period between September or October and March or April (when the rainfall averages 130–150 cm), dry out to become billabongs and brackish rivulets during the dry season. Most of this region is open eucalyptus forest interspersed with marshes and swamps which can be flooded to a depth of 4 m or more for from three to six months a year. Water buffaloes are much better adapted to life in this kind of habitat than are domestic cattle; they are also less prone to diseases and maintain better condition than their domesticated congeners.

When green forage is abundant, during and immediately after the five to seven months rainy season, buffaloes graze all over the higher ground; during the hot dry season, however, they are mainly found near watercourses where green fodder and water still remain. Tulloch reported that buffaloes frequent a limited home range area, containing drinking points, dunging areas, resting places, rubbing trees and wallows; in the last of these, in hot weather, buffaloes spend five to six hours when the sun is highest. They tend to congregate in small groups of between 10 and 20 which form part of a larger herd; the leader is usually an old cow, and most of the group are probably her offspring. Bulls are normally solitary or associate in all-male groups. Up to a dozen calves are sometimes guarded in a 'crèche' by a lone female, to whom the mothers return when danger threatens.

Although calves may be born at any time the peak period is in February towards the close of the wet season, when food and water are most abundant. Apart from man buffaloes have few predators, though estuarine crocodiles (*Crocodylus porosus*), dingoes (*Canis familiaris dingo*)[7] and feral pigs (*Sus scrofa*)[8] take some calves and aged or bogged adults. Tuberculosis, which occurs with some frequency, seldom proves fatal. The buffalo fly (*Siphona exigua*) is now regarded as endemic in the Top End, but rubbing and wallowing seems to provide adequate protection. The buffalo louse (*Haematopinus tuberculatus*) and the cattle tick (*Boophilus microplus*) seem to cause few problems. Beef nodules (*Onchocerca gibsoni*) are common. Starvation in times of drought, bogging (which is believed to have accounted for 20,000 in 1969) and, above all, predation by man are probably the buffaloes' main controlling factors.

In 1964 the Wildlife Conservation and Control Ordinance designated a protected area extending westwards from the Arnhem Land Aboriginal Reserve as far as the Adelaide River, which includes the greater part of the prime buffalo habitat. Here, buffaloes (which prior to 1975 were designated as 'wildlife' but are now officially 'stock') may only be killed by licensed commercial hunters or captured alive by licensed leaseholders. West and south of the protected area (where they cause damage to fencing, stock yards and water supplies) buffaloes are officially designated as pests, and their numbers are being gradually reduced as commercial and 'sport' hunting increases. East of the protected area, in the Aboriginal reserve, access to outsiders is strictly controlled and shooting almost entirely prohibited; here the buffalo population is likely to remain fairly static unless there should be any significant change in official government policy.

'The most important animal', wrote Tulloch (1975), 'on the black soil plains in terms of its ability to modify the environment is the water buffalo.' Yet there is doubt about the degree to which buffaloes have affected their ecosystem. Tulloch was inclined to believe that the damage caused by buffalo was exaggerated. Frith (1979), however, pointed out that in some places buffaloes have destroyed stands of spikerush (*Eleocharis* sp.) which is one of the main nest-building materials of the semi-palmated, magpie or pied goose (*Anseranas semipalmata*), which is confined to northern Australia and southern Papua New Guinea. These distinctive birds have been a considerable tourist attraction since they were first studied (as possible pests) when experimental rich-paddies were planted on Humpty Doo station near Darwin in 1953.

Large forests of paperbark (*Melaleuca* sp.) grow on the subcoastal plains, especially east of the Adelaide River. Erosion channels caused by buffalo grazing and trampling have broken through the levees and banks, allowing the entry of salt water which has killed large numbers of trees. Many of the paperbark swamps are impenetrable by man even during the dry season, and thus provide buffaloes with an effective sanctuary.

Monsoon forest in the Top End is floristically rich and provides a habitat for a distinctive fauna; much has been destroyed by buffaloes who have also heavily browsed freshwater mangroves and, by using them as rubbing trees, have felled sand palms (*Livistonia* sp.). Water lilies (lotus) in deep water and hydrophytes on the borders of billabongs and lagoons have been destroyed and previously vegetated areas, on which, during the dry season, magpie geese largely depend, have been converted to clear water. Buffaloes also occur in wildlife sanctuaries occupied by magpie geese, such as the Woolwonga Reserve, where the habitat is being seriously degraded.

In June 1978 the Northern Territory Legislative Assembly appointed a board of inquiry 'To enquire into, report and make recommendations to the Administrator concerning feral animals in the Northern Territory'. In March of the following year the board issued its report,[9] a major part of which was concerned with feral water buffalo. It recommended, *inter alia*, that 'a project should be undertaken . . . to assess more accurately the population and distribution of the buffalo in the Northern Territory as soon as possible and to monitor changes periodically'. The Board concluded that, so far as practicable, buffaloes should be eliminated from 'areas of conservation significance (e.g. parks and santuaries)', and that the risk of future re-encroachment should be prevented by the erection of fencing. The report recommended that immediate steps should be taken to counter the intrusion of salt water on the coastal plains, and advocated the attempted restoration of the environment in reserves such as Woolwonga and Sweet's Lookout.

More attention should be given as a matter of urgency, the Board suggested, to the protection of areas of special environmental significance such as rainforest and plant communities. New areas of conservation, for example in the Finniss River/Sweet's Lookout region, should be designated and given adequate protection. Three 'buffalo zones' should be defined consisting of a central zone encompassing properties on which buffaloes have been the most important source of revenue, a peripheral eradication zone, and a buffer zone between the two.

West of the Stuart Highway the report recommended that efforts should be intensified to restrict the further spread of buffaloes and to eliminate them wherever pastoralists are endeavouring to establish new forms of husbandry. Where feral buffaloes are maintained, as a tourist attraction or as a source of food, in the vicinity of conservation or elimination areas adequate fencing and some degree of management should be provided. Research should be undertaken to assess the effect of feral water buffaloes on the biota, and in particular on the connection between buffalo and weed (especially *Mimosa pigra*) infestation, and on more effective methods of controlling the further spread of such weeds.

The promotion of buffalo meat for human consumption, both at home and overseas should, the report suggested, be undertaken by the Australian Meat and Livestock Corporation and the Department of Trade; to help achieve this, Northern Territory government experimental farms and the Marrakai Reserve could be used for demonstrations of buffalo domestication, and research into breeding improvements could be initiated.

Finally, the Board recommended that the government should try to persuade pastoralists to bring buffalo herds under a higher level of control, and to this end should assist financially with the construction of fences and roads, and should provide gratis advisory and technical services.

As McKnight (1976) points out, the feral water buffalo 'is now a well established and dominant element of the biota of much of the Top End, and it is probably a stable element in the ecosystem': its future in Australia thus seems assured, though just how it will develop remains uncertain.

Papua New Guinea; New Britain; New Ireland; Borneo; Java; Andaman Islands

Water buffaloes from the Philippines were introduced into what is now Papua New Guinea between 1900 and 1903 during the period of German administration; further importations were made in 1906 from Indonesia and in 1913 again from the Philippines. Many became feral between the two world wars, most of which were slaughtered by the Japanese during the Second War. Since then there have been many importations from Australia and New Britain, and some feral animals have been recaptured to found re-domesticated herds on the north coast around Madang. Some feral animals still remain on the Gazelle Peninsula in New Britain and in New Ireland, and possibly also in Borneo. A population of some 200 feral water buffalo is established in dry forestland and extensive savannah around Baluran in eastern Java.

According to Roots (1976), 'Indian buffalo' are established and thriving in the Andaman Islands in the Indian Ocean.

In the Orange Free State, South Africa herds of water buffalo are maintained for sporting purposes on private fenced ranches of 85,000 or more hectares.

NOTES

1. For much of the history of the water buffalo in Australia I am indebted to Letts (1960, 1964), Tulloch (1979) and McKnight (1976).
2. New South Wales Governor's Despatches, 42.
3. Ibid.
4. Ibid.
5. *Proceedings of the Royal Geographical Society of Australia* **16**: 11 (1914).
6. *Bulletin of the Department of External Affairs, Melbourne* No. 1: 16 (1912).
7. See p. 47.
8. See p. 123.
9. Letts, Bassingthwaighte & de Vos (1979).

Banteng
(*Bos javanicus*)

Note: Present natural distribution is much reduced from the original range shown.

NATURAL DISTRIBUTION
Originally Borneo, Malaysia, Java and Burma, but now restricted to a few isolated herds on Java and in southeastern Asia. Ancestor of domestic Bali cattle bred on Bali, Timor, the Celebes, Java, Borneo and Sumatra.

NATURALIZED DISTRIBUTION
Australasia: Australia, Indonesia (Bali, Sangihe and Enggano).

AUSTRALASIA

Australia

According to Frith (1979), banteng were being domesticated as beasts of burden in Java certainly by 1709, and probably for several hundred years before that. 'There can be no doubt', Frith writes, 'that it was this domestic strain [of Bali cattle] that was transported to Australia.' Since the animals in Australia have reverted to the wild they are here referred to as banteng.

Exactly when banteng were first introduced to Australia is uncertain, since early records do not specify which breeds of cattle were imported. Between 1829 and 1840 'cattle' were shipped from Koepang on Timor and from Kissa (Kisar) Island to early settlements on Melville Island and at Raffles Bay and Port Essington.[1] It may well be that banteng were included in these shipments, but certain proof is lacking. G W. Earl, who was well acquainted with banteng in their native land, makes no mention of them in *Enterprises, discoveries and adventures in Australia* (1839). On 1 December 1845, however, the acting commandant of the Victoria Settlement in the Northern Territory, Captain James McArthur, wrote in a letter that 'the Bali cattle disappeared so suddenly and so decidedly, without showing any change of disposition that I do not expect to see them again. It is probable they are got away to the South-ward. We can discover no trace of them.' (In fact the cattle did subsequently reappear following a heavy fall of rain.) In a subsequent despatch dated 23 May 1849 McArthur reported that:

> A vessel was sent to us in January by Mr Langé, merchant at Balli Badong[2] [*sic*], and we received twenty Balli Sapis [*sapi* = Malay for cattle: the plural is *sapi sapi* not *sapis*], a peculiar breed of oxen . . . the oxen will be reserved now, as they appear to adapt themselves well to the soil; but they require much labor in looking after them. They must be constantly tailed, or in a few weeks would become perfectly wild.[3]

Apart from a brief reference to them by John Lewis[4] in 1875 as 'Sourabaya cattle', feral banteng on the Cobourg Peninsula (the result of releases when the settlements were abandoned) remained unstudied and largely forgotten for nearly a century. They were rediscovered during an expedition to Arnhem Land in 1948 by D. Johnson, who unfortunately failed to publish his findings until 1964.[5] He found banteng around the upper end of Port Essington, on the opposite side of the peninsula at James Creek, and eastward almost to the base of the peninsula. An Animal Industry Branch expedition to the Cobourg Peninsula, led by G. A. Letts in 1960, confirmed Johnson's identification.

Letts (1964) found feral banteng ranging over the whole of the 1,800 km^2 peninsula (especially in the northeastern corner), and reported that from time to time they moved south-east towards the black soil plains of the Murganella River. He estimated that there were some 370 between Port Essington and the isthmus and that the total population numbered around 1,500, giving an average density of one per 140 ha. Although banteng occur throughout the peninsula, they are most numerous in and around freshwater swamps and lagoons near the coast, especially in the Danger Point, Smith Point and Port Bremer areas. In August 1978 an estimate by the Primary Industry Division put the population at 1,070.

Banteng associate in family groups of cows and calves, usually accompanied by one or more bulls; bachelor groups of bulls also occur. They feed largely in the open, retiring when the sun is highest to shelter in patches of monsoonal forest. Banteng are extremely wary, and when alarmed take refuge in flight, if necessary crossing lagoons and even coastal estuaries.

According to a Commonwealth Scientific and Industrial Research Organization report of 1974:

> The effect of Bali cattle on the native fauna is not known, but the sandy plains appear overgrazed and trampled. This has caused some erosion. The common sedge (*Fimbristylis cymosa*) on these plains seems to be an especially favoured food. In areas where cattle congregate and shelter, ground vegetation is considerably damaged and there is a marked browse line on certain species of trees. The burrows of rodents on the sandy plains are often broken into by the feet of the cattle.[6]

In 1961–2 10 cows and 5 bulls (followed between 1962 and 1973 by a further 73) were transferred to the Government Experiment Farms at Berrimah and Beatrice Hill, where they are bred for research and have been crossed with Brahman cattle.

On the pretext of protecting the environment within the Cobourg Peninsula Fauna and Flora Reserve, the Wildlife Service culled over 300 banteng between 1975 and 1978. Pending a reappraisal of future policy an eight-year moratorium on shooting was then declared, which was strongly endorsed in the report of the Board of Inquiry into Feral Animals in the Northern Territory in 1979. The report also recommended the implementation of a suggestion, first put forward in 1846 by Captain John Lort Stokes, that a substantial 11-km long fence should be erected across the neck of the peninsula to prevent the movement of banteng and other large feral ungulates into or out of the reserve. The board considered it desirable that if possible banteng should eventually be removed from the peninsula entirely by translocating them to other areas where the genotype could be preserved, with the possible exception of a small securely fenced herd retained as a tourist attraction.

Banteng are able to maintain good condition remarkedly well with poor nutrition. Although carcase weights may be high in comparison with other bovids, it is improbable that they could prove economically viable in the wild: breeds based on interfertile banteng crosses have, however, been successfully developed on Madura and elsewhere; this, together with conception rates of 90–100 per cent, has encouraged the Primary Industry Division to attempt to develop a new cross containing banteng genes.

Indonesia (Bali, Sangihe, Enggano)

Banteng are currently established *ferae naturae* in the Bali Barat Game Reserve in western Bali; in the Sangihe Islands north-east of Sulawesi; and on Enggano Island off Sumatra, where a special hunting reserve for them is being formed.

NOTES

1. See p. 265.
2. Badung, now Denpasar, is on the southern side of Bali.
3. Quoted by Calaby & Keith (1974).
4. *Bulletin of the Department of External Affairs, Melbourne* No. 1: 16 (1912).
5. *Records of the American–Australian Scientific Expedition to Arnhem Land; Zoology* **4**: 427–515 (1964).
6. Quoted by Letts, Bassingthwaighte & de Vos (1979: 45).

Domestic Cattle

(*Bos taurus*)

NATURAL DISTRIBUTION

Domestic cattle are derived from the aurochs or wild ox (*B. primigenius*) which lived in Europe and western Asia until becoming extinct in the wild in the seventeenth century.[1] (For original distribution of extinct wild ancestor see text.)

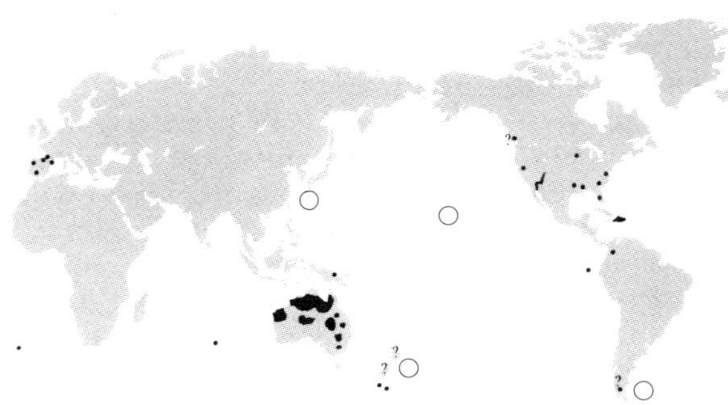

Note: Only principal feral populations mentioned in the text are shown.

NATURALIZED DISTRIBUTION

Europe: France, Spain; Asia: Japan; North America: USA, ?Canada, West Indies (Hispaniola); South America: Columbia, Argentina, Falkland Islands; Australasia: Australia, New Zealand, Papua New Guinea; Oceania: Hawaii, Galápagos, ?Tristan da Cunha, ?Nouvelle Amsterdam, ? other oceanic islands.

EUROPE

France; Spain

A small herd of feral Bétiso (plural, Bétisoak) cattle is established on land owned by M. Jean-Pierre Seiliez in the Haut Ariège region of the French Pyrenees. A second population formerly lived on the lower slopes of Mount Rhune in the Atlantic Pyrenees, where in 1924 it was destroyed in the interests of tourism to make way for a funicular railway.

Across the border in Spain (where they are known as Betizuak = Basque for 'wild cattle') small herds exist in the forests around Trucios and Goizueta in northern Navarra; in 1975 the latter numbered only 14. Other small groups occur at Urnieta in Guipuzcoa and on the Marquis de los Arcos' estate at Olague in Navarra.

In colour, bulls vary according to season from red to reddish-brown and cows from red to reddish-dun; the former weigh up to 400 kg and stand around 1.3 m at the shoulder, while the latter can weigh as little as 200 kg and stand up to 1.1 m high. Summer is spent, under the leadership of a master bull, either on the northern side of the high tops or in damp and wooded valleys. In autumn the cattle move further down the mountain and split into two smaller groups – one composed of cows and calves, the other of bulls and immature and barren cows – which are relatively sedentary and keep to the southern slopes. In spring the herd reforms and returns to the high tops where it ranges more widely.

How long Bétisoak cattle have roamed the Pyrenees is uncertain, but the German-Swiss naturalist Konrad Gessner, writing of the aurochs in his *Historia animalium* (1551–8), records that 'some say it is still to be found on the mountains between France and Spain'. Since Bétisoak cattle are not dissimilar, in both pelage and conformation, to the primeval wild ox, it is possible that the former are the descendants of a surviving remnant of the latter, modified by some interbreeding that is known to have occurred with domestic stock.

East of Andorra in the eastern French Pyrenees 300–400 small Massanaise cattle are established near the Massane River on the northern slopes of the Albères Mountains in the Pyrénées Orientales. Over the border in Spain (where they take their name from the mountain range) larger numbers occur on the southern slopes of the Albères.

About 170 (in 1975) long-horned light-coloured Cachena ('small') cattle are established near Entrino in the south of the Orense Province of Galicia in northwestern Spain. A small population of diminutive Pajuna cattle occurs in the Doñana National Park in Andalusia.

ASIA

Japan

For at least a hundred years a herd of feral cattle has been in existence on Kuchinoshima in the Tokara Islands south-west of Kyushu; in 1979 the population consisted of 5 adult bulls, 14 adult cows, 2 young bulls and 3 heifers. They are small in stature, the bulls standing only some 110 cm high and the cows 10 cm less. In colour, they are black, piebald, red, red and white, 'pink' or grey.

NORTH AMERICA

USA[2]

Feral cattle, although more numerous in the past than at the present time, have never been abundant or widespread in the United States, where they are currently established in small numbers, mainly in the rugged rimrock and canyon country of the west and in the marshlands and swamps of the east. Being descendants of animals that strayed from domestic herds and managed to evade subsequent roundups, they are, in comparison to other feral ungulates, both inconspicuous and relatively innocuous. Table 15 gives the distribution of feral cattle in the United States.

A few feral cattle are established in most western (especially southwestern) states, usually in small groups in remote areas. The largest populations – each totalling several hundred – are reported from Arizona, Colorado and Utah; smaller numbers occur in California, Nevada, New Mexico and Wyoming. Feral cattle are also found in small numbers in poorly drained river valleys and coastal forests in the south-east. Although on the mainland populations in general are probably not of long-standing, and in many cases may be short-lived, on Hawaii (see below) and the islands off Alaska feral cattle are the descendants of age-old stock.

Writing of feral cattle in the United States

Table 15 Distribution of feral cattle in the United States

State	*Distribution*
Alabama	Clark and Washington counties
Alaska	Shumagin and Chirikof islands
Arizona	Arizona Strip; Fort Apache and San Carlos Indian Reservation; Graham and Superstition mountains; Kofa Game Range; Navajo Indian Reservation
California	Granite and Santa Rosa mountains, Riverside County; Bontgunwale canyon region of Tehama County; Santa Catalina Island; Hunter Mountain district and Grapevine Range, Death Valley National Monument
Florida	Several localities on the west coast
Georgia	Fort Stewart Military Reservation
Hawaii	(see text, p. 268)
Louisiana	Ouachita River bottomland, Caldwell parish
Nevada	East of Pyramid Lake
New Mexico	San Mateo Mountains
Utah	Dark Canyon Plateau, Manti-La Sal National Forest; Henry Mountains; Book Cliffs region
Virginia	Dismal Swamp
Wyoming	Various areas

Source: After McKnight (1964), who obtained his data from a questionnaire/survey

McKnight (1964) said 'their numbers are few, their range is restricted, and their importance is negligible . . . they are incidental and/or temporary in occurrence, and will probably continue in that status, though with a continuing decrease in numbers'.

Canada

Cattle (together with other domestic stock from France) were first introduced to New France in 1541 by Jaques Cartier, whose crew, due to the length of the voyage, were, according to Richard Hakluyt in his *Principall navigations, voiages, and discoveries of the English nation* (1589), 'constrained to water [the animals] with Sider and other drinke'. Shortly after 1550 Portuguese sailors landed a large number of cattle (and pigs) on Sable Island, 150 km off the coast of Nova Scotia, probably as a source of food for shipwrecked mariners and fisherman on the Grand Banks. The English navigator Sir Humphrey Gilbert is believed to have revictualled his vessels on Sable Island during his voyage to colonize Newfoundland in 1583. When the Marquis de la Roche abandoned the colonists on Sable Island to their fate in 1598, they were able to survive during the ensuing five years solely due to the presence of these feral domestic stock. More were imported by Samuel de Champlain to New France in 1617, where a small but flourishing herd was established first at Quebec and, after 1626, at Cap Tourmente. The Jesuit missionaries in Quebec in 1634 maintained a small herd of two cows, two heifers and a young bull, as well as 10 pigs; due to inadequate fencing these animals eventually escaped into the wild where they caused havoc among the settlers' crops. It was not, however, until the following year, according to Père Le Jeune in the *Jesuit relations* of that year, that cattle and oxen were first used to plough the land for cultivation. The Jesuits subsequently transported cattle (and pigs and domestic fowl) up the Ottawa River by canoe to the Hurons of Sainte-Marie; the *Jesuit relations* of 1645–6 records how 'these good people now see in their country another kind of animal, of which they never had knowledge; there are little bulls and heifers which have been brought hither with great labours'.[3]

Domestic cattle abandoned in the Tofino area on the west coast of Vancouver Island, British Columbia, were hunted for sport until about 1942 when the last one was killed. Until at least the late 1950s feral cattle were established in the vicinity of Estevan Point and on Cape Scott. Herds of considerable size are said to have been established along the northern and eastern coasts of Graham Island in the Queen Charlottes after the abandonment of the early settlements before the First World War. Populations in the north disappeared in the early 1920s, but until at least August 1962 a herd of around 40 survived in the vicinity of the Oeanda River, and 18 were counted on Cape Ball in July of the following year.

West Indies (Hispaniola)

Cattle were probably first introduced to the West Indies – to Hispaniola – by the Spaniards soon after their discovery of the island in 1492. They were certainly there by 1526, when the French corsair brothers, Jean and Raoul Parmentier, saw large numbers near Santo Domingo in what is now the Dominican Republic. Nine years later the Spanish historian/naturalist Oviedo y Valdès wrote that 'the first cattle were brought from Spain to this island, [where] they are now so numerous that the ships return loaded with their hides'. In the late eighteenth century Moreau de Saint-Méry referred to the presence of feral cattle at Anse-à-Pitre, at Saltrou in the Massif de la Selle in the south-east, and in the 'cockpit country' of the Mornes de Plymouth north of Les Cayes. Early in the following century M. E. Descourtilz mentioned that feral cattle were widespread and caused considerable damage to crops. Today, some are still established on the island in arid limestone areas of sparse soil with xerophytic brush vegetation – especially in the peninsula of Baradères, in the Tapion de Papaye region of the northwestern peninsula, and on the south coast near Grand-Gozier and Anse-à-Pitre in Haiti.

SOUTH AMERICA

Colombia

The brown or tawny long-haired feral cattle of the *páramos* (high plateaux) between 3,300 and 4,500 m in the Sierra Nevada de Santa Marta in Colombia are probably descended from stock introduced by the Spanish conquistadors. Some of the calves are regularly caught and redomesticated by the local Kogi Indians. Most are believed to live in the Parque Naçional Natural Sierra Nevada de Santa Marta (designated in 1977), and possibly also in the Llanos Orientales (eastern lowlands).

Argentina

Feral cattle are currently established in Argentina as far south as Isla de los Estados, where a herd of about a dozen was reported at Bahia Crossley in 1971.

Falkland Islands

Strange (1972) has traced the history of feral cattle in the archipelago.

In 1764 the French navigator Louis Antoine de Bougainville landed seven heifers and two bulls at Port Louis on Berkeley Bay, which were the founder stock of one of the finest populations of cattle in the Southern Hemisphere. In 1837 Captain Grey of HMS *Cleopatra* reported the presence of 14 or 15 herds of feral cattle – each numbering around 40 head – in the vicinity of San Carlos, and wrote:

> The wild cattle are now calculated at some 20,000 head. I have no idea how it has been possible to form such an estimate, but even after hearing this I was surprised at the numbers I saw. They are evidently increasing from the number of calves in every flock . . . the wonderful increase of these animals clearly shows how well adapted the country is for grazing . . .

A year after Grey's report, in 1838, the first officer of HMS *Arrow*, while surveying East Falkland, reported that some 30,000 fine cattle roamed over that island; their pelage varied from dark in the north to almost pure white in the extreme south.

In 1839 Charles Melville, master of the schooner *Montgomery*, landed the first cattle on West Falkland when he transported 55 cows and 11 bulls from Port Pleasant to White Rock Bay. Eighteen years later, in 1857, Governor Moore found between 2,000 and 3,000 head of cattle in the valley of the Warrah River. Helped by feral horses and pigs these huge herds of feral cattle were responsible for the destruction by the 1840s, through overgrazing, of most of the islands' tussock grass (*Poa flabellata*); the resulting erosion removed much of the shelter and nesting-sites for seabirds such as the Magellan penguin (*Spheniscus magellanicus*), petrels, wildfowl and waders.

The potential value of the immense herds of feral cattle was recognized in 1823 when the United Provinces government offered the cattle (and fishing) rights to Louis Vernet and Jorge Pacheco. By 1831 Vernet had established a community of some 90 settlers at Port Soledad, whose main sources of income were fish, cattle hides and 'jacked beef' – a form of dried meat popular in South America.

When Britain permanently colonized the islands in 1833 the cattle were declared Crown property, and steps were taken to prevent visiting sealers and whalers from salting hides and carcases on their way south to the Antarctic hunting-grounds. Between 1833 and 1842 cattle were on occasion used in lieu of cash in the settlement at Port Louis: in 1834, for example, Antonina Roxa agreed to accept every other calf from each cow that she succeeded in taming for milking purposes, and employees of the British Naval Superintendent agreed 'That none of us can go out and catch cattle without orders from Lt Tyssen or his Capitaz, and also when we shall want hides for lassoos, sinchas or for balls to aquaint the Capitaz and to obtain his order to take them'. In 1840 Tyssen estimated that there were some 40,000 head of feral cattle on East Falkland, and in 1842 the islands' first governor, Richard C. Moody, confirmed this figure. By 1846 the total had risen to around 60,000, and by the following year to 80,000. The only accurate census ever made, by Surveyor-General Arthur Bailey in 1859, revealed a total of 6,611 on East Falkland.

In 1845 Mr S. Lafone, a wealthy cattle and hide merchant in Argentina, obtained from the British government a grant of 600,000 acres (243,000 ha) of land in East Falkland and possession of all the feral cattle it contained for a term of six years, for a payment of £10,000 down, and £20,000 in 10 years from 1852. In 1851, however, Lafone's interest in his land – Lafonia – was acquired for £30,000 by the newly incorporated Falkland Islands Company. These enterprises effectively sounded the knell for feral cattle in the Falklands.

Lafone's manager embarked on a policy of indiscriminate slaughter of the cattle, mainly for their hides and tallow. To prevent the cattle from escaping his gauchos constructed a turf wall across the isthmus which divided the north of East Falkland from the south, to the north of which the cattle were hunted by 1851 to the point of extinction.

In 1860 the Crown reassumed ownership of the few remaining feral cattle, and a fine of £20 was declared for each animal taken outside Lafonia without the governor's permission. This did not, however, prevent repeated rustling by sealers, whalers, settlers, and even by the Falkland Islands Company itself. In 1862 Moore informed the Colonial Secretary, the Duke of Newcastle, that the company had been caught salting the hides and carcass of 54 cattle freshly killed on Crown property: 'I have it on the best authority procurable that the Company's Camp Manager knowingly permits and indeed orders that the Company's servants at their chief establishment . . . shall be supplied with fresh beef only from the wild cattle on Crown Lands.' Eventually the Privy Council in London decided in favour of the company on the grounds that although the Crown had forbidden the taking of feral cattle it had failed to specify that it retained the right to slaughter cattle on all land grants. The fate of the feral herds was finally sealed.

No records exist to indicate what breed of cattle was introduced to the Falklands. They have been described as long-haired with wide spreading horns and the broad fore limbs and slender hind quarters of the Spanish Longhorn fighting breed. Although the descendants of the original stock were killed off many years ago, some of their genetic characteristics may still occur in a small group of about 40 animals at Volunteer Point near Johnson's Harbour; this herd is probably composed of escaped domestic stock with periodic recruitment from current domesticated herds. Most of these animals are black in colour, but red and red-and-white individuals also occur. Inbreeding and poor nutrition have been responsible for a reversion to the 'wild-type' conformation of heavy fore quarters and slim haunches. Escaped domestic cattle were formerly also established on the Wickham Heights, where they are believed to have been exterminated during the Anglo-Argentine war of 1982.

AUSTRALASIA

Australia[4]

The First Fleet in 1788 brought with it from England five cows and two bulls with which to start a breeding herd at Botany Bay. Unfortunately, within a few months all but one cow had escaped to found a herd 65 km south-west of Sydney which by 1806 had increased to over 3,000 head. In 1792 David Collins (secretary with Governor Arthur Phillip at Botany Bay and later Governor of Tasmania) reported the arrival of the *Atlantic* from Calcutta with 'two bulls and a cow of the Bengal breed' (i.e. zebus, '*B. indicus*'). This led R. B. Kelley[5] to suggest that feral cattle reported a few years later from the Cow Pastures area of Camden, Sydney, might have been zebus; they might equally, however, have been cattle brought out from England by the early settlers.

When cattle were first imported to Tasmania is uncertain, but the following editorial from the *Hobart Town Courier*, quoted by Rolls (1969), shows that by the mid-1820s feral herds of considerable size were clearly well established:

> The herds came like the plagues of Egypt and worse than a shower of locusts make a clean sweep of every green leaf. Who can protect himself against such enemies? We defy anything except a troop of mounted and expert horsemen to surround these ungovernable brutes and conduct them to the pound. We ourselves have had crops on crops destroyed in this way which were well and decently fenced: and what was worse our tame cattle and working beasts inveigled away to join them in the wild and inaccessible passes of the woods.

The Lieutenant-Governor, Colonel Arthur, complained that feral cattle 'provide an inexhaustible stock on which runaway convicts can draw'. Together with the chief police magistrate, the Crown Solicitor, the Attorney-General and the Colonial Secretary, Arthur was one of the defendants in a lawsuit brought by Sir John Owen and Mr Edward Lord (who was known regularly to muster large herds of feral cattle) when the government dismissed their claims to ownership of several unbranded ('cleanskin') herds grazing on the mountains south and west of Hobart. Litigation ended, unresolved, in June 1836 when few feral cattle remained at large.

Cattle were shipped from New South Wales and from Kisar (Kissa) Island in the Moluccas to Melville Island in 1824, to Raffles Bay in 1829 and to Port Essington in 1838 (all in the Northern Territory), and in 1840 to Port Essington from Koepang, Timor: between 1838 and 1849 over a hundred head of cattle arrived in Port Essington from Sydney.[6] Although Letts (1964) considered that these shipments were all of English cattle, there were most probably some zebus among them.

During the 1840s large numbers of cattle escaped or were released from the Cobourg Peninsula settlements when they were abandoned; more were freed when the surveying encampment at Escape Cliffs north of Darwin was disbanded in 1866. In 1873 John Lewis found 'quite a number of English cattle on the Peninsula but they did not thrive very well'.[7]

J. A. Gilruth[8] refers to a number of early overland translocations of cattle to Australia's Top End; in 1872 D'Arcy Wentworth herded about 400 from Queensland via the Roper River, while four years later Mr T. Nelson brought a herd from the MacDonnell Ranges in the Northern Territory over 1,000 km to a butcher in what is now Darwin; in 1879 2,000 cattle and 12,000 sheep were overlanded to the Katherine River. These introductions are believed to have included both zebus and English stock.

Zebus were included each year from 1863 in an inventory of species held by the Victoria Acclimatization Society. According to Rolls (1969), in 1910 the Curator of the Zoological Gardens, Colonel W. H. D. Le Souef, sold three zebu bulls to pastoralists in northern Queensland; today, more than 70 years later, the McDowell and Cove herds on Christmas Creek and the Blue Range west of Ingham, and the Fischer herd on the Daintree River, retain a significant percentage of zebu blood. Rolls believed that most of the feral cattle in Australia today contain an admixture of zebu genes.

Feral cattle have become established in Australia in three ways: some, as the result of careless herding, wander off into the bush; most of these remain fairly sedentary except where a preferred food-plant, such as parakeelya (*Calandrinia balonensis*), grows in desert regions; in 1950 cattle branded and ear-tagged at Billiluna Station on the Sturt Creek, 100 km south of Hall's Creek in Western Australia near the edge of the Great Sandy Desert, wandered off, later to reappear at Mount Doreen Station north-west of Alice Springs, 400 km to the south-east. Some, especially when early settlements in the Top End of the Northern Territory were abandoned in the nineteenth century, were deliberately released to fend for themselves; others escaped by stampeding from drovers during overland herding. Until as late as 1959 these drives continued – nearly 600 once being lost in a single stampede during a journey from the Kimberleys in Western Australia to Queensland.

Most of Australia's feral cattle today are concentrated in the northern quarter of the country north of the eighteenth parallel and around Pilbara in Western Australia: here, where ranches are large and fencing tends to be incomplete, water supplies are more numerous and unallocated Crown lands and Aboriginal reserves are more extensive. The larger herds are established in the

Hamersley Ranges; in the Ashburton and Fortescue drainages of the Pilbara region; in the King Leopold Ranges and other rough parts of the South Kimberleys near Meekatharra and Wiluna; in the Gulf country of the Northern Territory; and in extreme northwestern Queensland. Some feral cattle are also scattered throughout the Pilbara region, the Kimberleys, the Top End, the Gulf country and the Cape York Peninsula of Queensland. Long (1972) records that in Western Australia herds up to 30 strong were reported on two sheep stations along the Fitzroy River in the Kimberleys in 1956. Around 500 were estimated to be in the Meekatharra area in 1960, when 3,000 were said to roam the Carnegie district east of Wiluna. In the south-west feral cattle have reached Oldfield and Munglinup – 110 km west of Esperance – and coastal areas near Walpole. North of Perth they have been reported along the coast near Moora, between Hill River and Green Islet and north of Mullewa.

Smaller numbers of cattle occur in the southern part of the Northern Territory, especially north-west of Alice Springs; in the Channel country of southwestern Queensland; in hilly districts around the upper Burdekin drainage of northern Queensland; in several unimproved sheep ranges in central Queensland; in scattered localities of the eastern highlands of southeastern Queensland; in a few places in central New South Wales, particularly near the Willandra billabong; and in some of the montane valleys of eastern Victoria. McKnight (1976), who stressed the difficulty of achieving an accurate census when, as is so often the case, it is uncertain whether cattle are domesticated or feral, estimated the total population to be divided roughly as in Table 16. The report of the board of inquiry into feral animals in the Northern Territory, however (Letts, Bassingthwaighte & de Vos 1979), said that the Primary Industry Division had estimated that there were up to a quarter of a million feral cattle (15 per cent of the total herd) in that state alone: in terms of economic potential this represents a value of over $A10 million and could produce additional beef worth $A7.5 million annually.

Table 16 Distribution of feral cattle in Australia

State	*Estimated population*
Northern Territory	20,000–50,000
Queensland	20,000–40,000
Western Australia	15,000–30,000
New South Wales	2,000–5,000
Victoria	1,000–2,000
South Australia	0–1,000
Tasmania	0–1,000
Total	58,000–129,000

Source: From McKnight (1976)

Since most of Australia's domesticated cattle are either Herefords or Shorthorns it follows that most feral animals are derived from the same breeds, especially the latter. In some places black animals (presumably of Aberdeen Angus origin) occur, while for many years feral cattle – some spotted – with large heads and horns have been established in the headwaters of the drainage of the Dawson River near Injune in Queensland. Feral cattle on the subcoastal plains of the Top End contain a high percentage of the blood of zebus and of dairy breeds such as Friesian and Ayrshire. In recent years the Townsend family has successfully established a herd of some 18,000 Brahman cross cattle on their Stapleton Station on the Reynolds River, which could prove the source of additional feral herds in the future.

Domesticated cattle which escape into the wild seem to change their conformation comparatively little though, after several generations, some may develop longer horns, larger heads and a cleaner and more angular appearance. Where they are numerous they associate in herds of between 10 and 30; where there are few they range about in groups of from three to a dozen or so.

The trouble caused by feral cattle in Australia is economically insignificant: they compete to a limited extent with domestic stock for both grazing and water, and from time to time damage fences and water troughs; they sometimes disturb sheep, especially during mustering; on some ranches domestic cows are lured away from the herd by feral 'scrub' bulls. The reduction to a single remnant colony of the Queensland hairy-nosed wombat (*Lasiorhinus krefftii*) has been attributed largely to overgrazing and habitat degradation by feral cattle.

In 1959 it was found that 'cracker' (old and out of condition) cows yielded adequate meat for the American hamburger trade; exporters paid such high prices for these animals that Queensland pastoralists began redomesticating feral herds. 'Crackers' are also sometimes used for dingo bait, and 'mickeys' (younger animals) have been utilized for both home consumption and also as foundation stock for new domestic herds.

Because of the hot and humid climate of much of their Australian range feral cattle are generally at a disadvantage *vis-à-vis* water buffaloes.[9] In recent years their numbers have markedly declined as the result of slaughter for hides and meat. Their long-term future depends largely on the existence of a ready market for these products. In the absence of dung beetles the imported buffalo fly is able to breed undisturbed in the droppings of cattle and buffalo, which favours the growth of rank and unpalatable plants. To combat this problem the Commonwealth Scientific and Industrial Research Organization released large numbers of African dung bettles – especially the dung-burying beetle (*Onthophagus bubalus*) and the dung-ball rolling beetle (*Sisyphus rubripes*) – in northern Australia both to attack the buffalo fly larvae and to bury or break up the dung pats, thus providing the land with much-needed nutriment.

New Zealand

Robert McNab records that on 30 March 1833 John Bell sailed from Sydney for Mana Island with 10 head of cattle, and that 'with the exception of the domestic animals which accompanied the expeditions of Cook and Vancouver, this is the first record of any such having been taken to New Zealand'.[10] The Revd Samuel Marsden, however, claims that he had shipped cattle from New South Wales to the Bay of Islands 18 years previously.[11] However that may be, E. Jerningham Wakefield says that he saw feral cattle in 1839 on the hills surrounding the entrance to Pelorous Sound.[12] In the following year he found the descendants of some bartered with the Maori for flax to be abundant on Kapiti Island. Mr S. Thorne George told Thomson (1922) that feral cattle were widespread on Kawau when he was first there in 1869, and Mr A. C. Yarborough said that 'forty years ago' (i.e. in about 1880) they were common in Kohu Kohu. By Thomson's time they were well established in the high country between Lake Wakatipu and the west coast and in the valley of the Rockburn.

First introduced to Chatham Island in 1841, escaped cattle were being recaptured by the islanders by the early 1860s. Others, introduced to the Auckland Islands in 1849 by Captain Enderby, had been exterminated by sealers by 1894. In the following year more were landed on some small offlying islands at the northern tip of the archipelago, where they caused considerable damage to the vegetation through overgrazing, with consequent widespread erosion. According to Cockayne (1903), the descendants of 12 Shorthorns landed from the *Hinemoa* on 690-ha Enderby Island and on Rose Island in 1895–6 had barely maintained their numbers by 1903. Aston (1911) reported that by about 1910 overgrazing by feral cattle had exterminated tussocks of *Poa littorosa* on Enderby. In 1903 8 Ayrshire/Shorthorn crosses were landed on Campbell Island, where some of their descendants remain. Cattle released on the Antipodes Islands between 1886 and 1903 all died or were killed for food by castaways. Cattle have also been introduced on at least one occasion to Macquarie Island.

Aston (1915) wrote: 'Wild cattle are abundant in unfrequented valleys and gorges of the Tararua Range. They eat out many species of native plants, and have destroyed great numbers of *Ligusticum dissectum*, which is one of the most abundant and characteristic plants of the higher ground.' Aston lists tahoe or hina hina (*Melicytus ramiflorus*), karamu (*Coprosma grandifolia* and *C. tenuifolia*), broadleaf (*Griselinia littoralis*), mangrove (*Avicennia officinalis*), tawa (*Beilschmiedia tawa*) and karaka (*Corynocarpus laevigata*) as among the preferred food plants of cattle. To this list Thomson (1922) adds *Pittosporum tenuifolium, Hoheria populena, Coriaria ruscifolia, Panax (Nothopanax) arboreum, Schefflera digitata, Coprosma robusta, Myrsine (Rapanea) urvillei, Olea lanceolata, Geniostoma ligustrifolia, Solanum aviculare, Veronica salicifolia, Vitex lucens, Freycinetia banksii, Cyathea medullaris*, bracken fern (*Pteris aquilinia*), flax (*Phormium tenax*), cabbage tree (*Cordyline australis*) and occasionally ngaio (*Myoporum laetum*) and anise (*Angelica gingidium*).

After the Second World War feral cattle were still common in remote areas of forest and scrub on both main islands and on Stewart Island: they were widely scattered throughout the volcanic plateau of the Wanganui River region; in the valleys of the main North Island ranges (especially in the Kaimanawas, Ahimanawas, in the headwaters of the Mohaka River, the Urewera country and on Pirongia Mountain) from East Cape to the Ruahines; in the back-country of South Island from Farewell Spit to Fiordland; and in the Ruggedy Range on Stewart Island.

The spread of intensive farming, together with the desire to preserve the country's native biota, led to the gradual eradication of New Zealand's feral cattle. The Stewart Island population was exterminated by the late 1940s, and since then many herds on the two main islands have also been eliminated – one of the most recent, in 1975, being that on Farewell Spit. On Enderby Island, where the population has recently varied between around 40 and 60, most feral cattle are black or black and white, though some are reddish brown with white markings. Many of the cows and some of the bulls are naturally polled. On Campbell Island a small herd of between 12 and about 20 is confined to the 440-ha limestone corner of the 11,700-ha island. The majority are red-brown with white markings or white with red markings, but some are self-coloured red or sandy brown; most are horned. These two insular populations are the only ones which, since their introduction, have remained competely isolated from the possible effects of subsequent introductions. In 1976 a few feral cattle remained on the southern Tablelands and on Pitt Island in the Chatham group.

Papua New Guinea

Some feral cattle are established near the coast north of Madang in eastern Papua New Guinea.

OCEANIA

Hawaii[13]

Cattle were first introduced to Hawaii by Captain George Vancouver in 1793[14] during his second voyage of circumnavigation.[15] Vancouver originally intended to bring the cattle as a gift directly to King Kamehameha I at Kealakekua Bay, Hawaii, but dropped anchor first on 16 February at Kawaihae, where flat calms followed by contrary winds delayed his onward journey. One cow and one bull, in a weakened condition, were offloaded into a dugout canoe at Kawaihae on 19 February, but the cow died before she could be put ashore and the bull lived for only a few days. The five remaining cows, 'in a healthy state though in low condition', were safely landed in Kamehameha's canoes at Kealakekua Bay on 22 February.

On 15 January 1794, during his third voyage, Vancouver landed at Kealakekua 'a young bull calf nearly full grown, two fine cows, and two very fine bull calves, all in high condition'.[16] Those left behind in the previous year had thrived, one cow having given birth to a heifer; 'I had little doubt', Vancouver concluded, 'by this second importation of having at length effected the very desirable object of establishing in this island a breed of these valuable animals.'[17] At Vancouver's request King Kamehameha reluctantly agreed to place a 10-year *kapu* (taboo) on the slaughter of the cattle 'and other European animals', which, Vancouver considered, 'cannot fail to render the extirpation of these animals a task not easy to be accomplished'.[18]

Of the Waimea Plain where the stock was released Vancouver wrote: 'In this valley is a great tract of luxuriant, natural pasture, whither all the cattle and sheep imported by me were driven, there to roam unrestrained, to increase and multiply.'[19] The 1793 shipment came from Mission San Carlos near Monterey, California, while that of the following year originated in 'New Albion', possibly in the region of the lower Columbia River.

During an address to the Royal Hawaiian Agricultural Society in August 1850 R. C. Wyllie summed up Vancouver's introductions:

> On the 19th February, 1793, he landed a bull and a cow from California for Kamehameha I in the canoe of Krimamahoo, off the coast of Hawaii. On the 22nd of February, 1793, he landed five cows, two ewes and a ram, in the bay of Kealakekua for Kamehameha I; on the 15th of January, 1794, he landed a bull, two cows, two bull calves . . .

By 1801, according to Captain Delano, commander of the US merchantman *Perseverance* in 1817, the cattle had greatly increased and, following the relaxation of Kamehameha's *kapu*, were already being harvested for beef. One of the bulls imported in 1794 had been transferred to Maui

– the first introduction to an island other than Hawaii. In *Voyages in the North Pacific* (1896) P. Corney wrote that in Hawaii in 1818 'the cattle go about wild, and are not allowed to be shot without permission from King Kamehameha'.[20]

Andrew Bloxam, naturalist on the *Blonde* during her voyage from England to the Sandwich Islands in 1824–5, noted that on Oahu 'cattle, goats and horses were very plentiful' and that on Maui 'there are some cows on the southern part of the island kept for breeding, and also a few horses, but none in the neighbourhood of Lahaina'.

On Hawaii cattle were at an early stage put to a variety of uses, and Kamehameha's *kapu* was relaxed partly at least because of the great increase in the size of the herds. As the Hawaiian *Advertizer* of 11 August 1859, reported, 'here [on the upland slopes of Hualalai] they rapidly increased'.

Between 1811 and 1835 cattle were being regularly shipped from Hawaii for meat and hides to settlers in the Pacific Northwest. By the mid-1850s Kramer (1971) records that 'tame, semi-wild and wild cattle apparently ranged at will around the environs of Honolulu . . .', and that by the end of the century a large number of cattle ranches had becomes established.

W. L. Hall[21] recorded that as early as the 1850s escaped feral cattle were reported to be causing damage in forests by trampling down and browsing sapling trees and understorey vegetation; around the native *mamane* forests on Mauna Kea on Hawaii, farmers were forced to erect fences to confine cattle (of which A.W. Carter, Manager of the Parker Ranch, recorded there were some 10,000 head) to the higher slopes, where they were not finally eradicated until the 1930s. On Oahu, where, according to H. L. Lyon,[22] 'cattle have been the greatest factor in pushing the forests back to their present narrow limits, and at certain vital points cattle are still allowed to penetrate the remaining forests', they were heavily hunted until the mid-1940s but were not finally eliminated until 1962.

By the late 1960s feral cattle, known locally as *pipi*, of Hereford (and possibly Shorthorn) origin survived only in near-inaccessible forests of the Puu Ka Pele and Kokee regions of Kauai; on some of the lava fields of Hawaii (where Tomich[23] reported some 2,000 head of stray or feral cattle),[24] and in the eastern end of Molokai. Most appear to be concentrated on the Honaunau Forest Reserve above Kealakekua Bay and above 1,000 m in the Kapua-Honomalino region. In the 1970s visiting 'sportsmen' were being offered a 'Vancouver Wild Bull' trophy for $100.

Galápagos Islands

Cattle were probably first imported to the island of Floreana (Charles) from Ecuador by the early colonists led by General José Villamil in 1832. They were probably also introduced to San Cristóbal (Chatham) when that island was first permanently colonized in 1869 and to southern Isabela (Albemarle) in about 1893.

Dr Simeon Habel, who landed on the Galápagos in 1868, reported the presence of 'wild cattle' on Charles, from which they were doubtless transported to other islands. O. Salvin, who visited the archipelago in July 1875 on board HMS *Petrel*, wrote that 'of animals introduced, it was estimated that there existed 2500 head of cattle on Charles Island, 4000 on Chatham none on James [Santiago] and about 20 on Albemarle'.[26] It is possible that the cattle seen by Salvin on Isabela were descended from some which may have been left behind by Villamil when he visited that island in 1851.[26]

Rollo H. Beck, a member of the California Academy of Sciences' expedition to the Galápagos in 1905,[27] reported the presence on Isabela of 'thousands of cattle'. At that time the villagers of Santo Tomás (named after the Bishop of Panama who discovered the islands in 1535), shipped large numbers of cattle to the mainland both on the hoof and as hides and dried meat.

Cattle were first introduced to the island of Santa Cruz in 1923 by Captain Rafael A. Castro who irresponsibly released a bull and two cows on the south coast at Academy Bay. In about 1940 they began to spread into the higher altitude humid zone, partly because of a reduction in the annual rainfall and partly because of an increasing population, which eventually began to experience a shortage of fodder through overgrazing. In the severe drought of 1963 the cattle suffered further from lack of food and water, and also from a heavy infestation of ticks which has been introduced to the island with stock from Isabela.[28]

On Floreana, where the endemic giant tortoise (*Testudo* (*Geochelone*) *elephantopus*) died out between about 1850 and 1875, and on San Cristóbal, cattle have for many years been a staple diet of the islanders and a nuisance to growing crops.[29]

Although feral cattle trample on the nests of giant tortoises and on the nests and burrows of dark-rumped petrels (*Pterodroma phaeopygia*) and land iguanas (*Conolophus subcristatus*), and by their overgrazing deplete the vegetation and accelerate soil erosion, they are, apart from horses, the least-destructive introduced ungulate on the islands. Their main controlling agent, apart from man, is feral dogs (*Canis familiaris*),[30] which prey mainly on calves and old and sickly adults. In 1982 the total population was estimated at well over 2,000 head.

Other oceanic islands

Cattle were introduced to Tristan da Cunha in about 1817 – some six years after the island was first colonized by settlers from Massachusetts, USA – and were soon free-ranging on the southern lowlands around Cave Point and Stony Beach, where some still remain; by overgrazing the vegetation has been converted to a low sward (with consequent erosion) in which are incorporated many exotic species. Cattle, which later became feral but had died out by the 1960s, were also introduced to Inaccessible Island. On Nouvelle Amsterdam, where they were first imported in 1871, cattle have been largely responsible for the elimination of bush scrub and tussock grassland, and much of the eroded volcanic terrain now supports only a close-cropped sward of alien grasses. The population – at one time between 1,000 and 2,000 – has considerably declined.

In 1950 some cattle were landed on Kerguelen Island in the South Atlantic, but have since died out. Other oceanic island groups to which cattle have at one time or another been imported include Fatuhiva in the Marquesas; Raivavae, Tubuai and Rurutu in the Austral or Tubuai and Rapa islands; Tahiti; the Seychelles; Walpole in New Caledonia; and Easter Island. On Masatierra and Masafuera in the Juan Fernandez Islands off the coast of Chile feral cattle are established in some numbers.

During the English Wars of the Roses (1451–85), when all able-bodied men were put under arms, many domestic animals were abandoned to fend for themselves. In both Radnor and Colwyn forests feral cattle survived until at least the 1580s and 1590s, when their ownership was the subject of civil dispute. In Tudor times (1485–1603) sheep-farming became so profitable that more cattle were released into the wild, where their descendants survived into the seventeenth century. These animals were never rounded up, but from time to time were culled for meat.

Although there are no truly feral cattle in the British Isles today several semi-feral herds of ancient and primitive emparked white cattle are believed to have survived relatively unchanged since medieval times. The most pure-bred herd, and the one with the best provenance, is that at Chillingham in Northumberland, England, where the animals have reddish-brown ears, black-tipped horns and black muzzels and hooves. Another English herd exists at Woburn in Bedfordshire (to which it was removed from Chartley. Staffordshire in 1905), with Welsh herds at Vaynol in Carnaervonshire and at Dynevor in Carmarthenshire, and a single Scottish herd at Cadzow, Lanarkshire.

In Argentina concern has been expressed that **European bison** (*Bison bonasus*) could spread from ranches in the mountainous Nequen Province into the adjacent national parks.

The **American bison** (*Bison bison*) has been successfully transplanted on Santa Catalina in the Channel Islands off the coast of California.

NOTES

1. The aurochs was hunted by early Stone Age man, and domesticated as a beast of burden and for beef and milk in the late Stone Age. The earliest known domestic cattle date from the Halafian period (7,000 years or more ago) in northern Iraq. The last aurochsen died in a park in Poland in 1627.
2. For much of the information on feral cattle in the United States I am indebted to McKnight (1964).
3. Saunders (1935).
4. For much of the history of feral cattle in Australia I am indebted to McKnight (1976).
5. *Native and adapted cattle* (Angus & Robertson, Sydney, 1959).

6. See p. 258.
7. *Bulletin of the Department of External Affairs, Melbourne* No. 1: 16 (1912).
8. Ibid.
9. See p. 254.
10. *From Tasman to Marsden: a history of northern New Zealand from 1642 to 1818* (J. Wilkie, Dunedin, 1914).
11. Letter from Marsden to the Governor of New South Wales, dated 'Paramatta, 30th May, 1815'.
12. *Adventure in New Zealand from 1839 to 1844*, ed. Sir R. Stout (Whitcombe & Tombs, Christchurch, 1908).
13. For much of the history of feral cattle in Hawaii I am indebted to Tomich (1969) and Kramer (1971).
14. Brookshier, however, claims that a Spaniard, Don Marin, imported longhorn cattle from Mexico to Oahu in 1791.
15. *A voyage of discovery to the North Pacific Ocean, and round the world in 1790–1795. . .* (London, 1798), Vol. 2, pp. 120, 127, 147.
16. Ibid., Vol. 3, p. 11.
17. Ibid.
18. Ibid.
19. Ibid., Vol. 3, p. 64.
20. p. 109.
21. *Planters Monthly* (1904).
22. *Hawaiian Planters Record* (1919).
23. 1969: 103.
24. McKnight (1964), however, claimed that 'only a few dozen are left'.
25. Salvin (1877: 456). For a more detailed history of the settlement of the archipelago see p. 359.
26. Naveda (1950).
27. van Denburgh (1914).
28. Kastdalen (1982).
29. M. Wittmer, *Floreana* (1961).
30. See p. 50.

Blackbuck

(*Antilope cervicapra*)

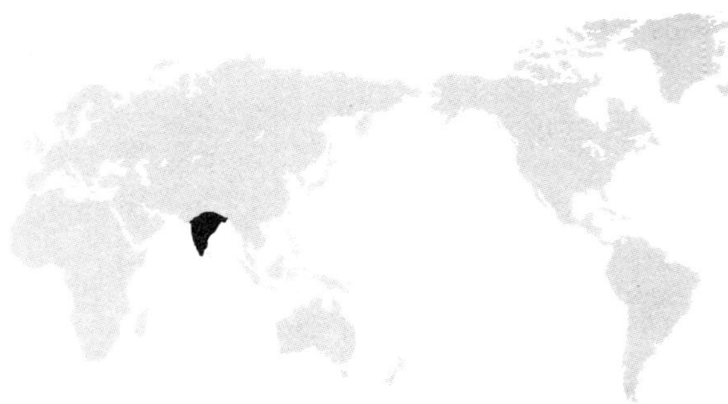

Note: Present natural distribution is much reduced from the original range shown.

NATURAL DISTRIBUTION

Originally from West Pakistan along the foot of the Himalayas from the Punjab through Uttar Pradesh and Nepal to west Bengal and East Pakistan and throughout peninsula India, where until the nineteenth century it was the most abundant ungulate with a total population of some four million. Blackbuck have never occurred in Sri Lanka, and have greatly declined in their earlier range on the Indian subcontinent.

NATURALIZED DISTRIBUTION

North America: USA; South America: Argentina; Australasia: ?Australia.

NORTH AMERICA

USA

The earliest known release of blackbuck in the United States took place in Kerr County, Texas, in 1932.[1] Five years later Mr Richard Friedrich obtained 18 blackbuck from Mr Fred Stark, director of the San Antonio Zoo, and turned them loose on his Bear Creek Ranch in Kerr County. After early setbacks the animals eventually began to thrive, and many of their progeny were translocated to other ranches. A few years after the original release a number of bucks were added to inject new blood to the herd, which by 1955 had increased to about 300.

In September 1938 10 pairs of blackbuck were released in live oak (*Quercus virginiana*) woodland on the Norias Division of the King Ranch. A number of young were born but are believed to have been killed by bobcats (*Felis rufus*) and coyotes (*Canis latrans*). Some of the survivors dispersed to Laguna Madre, where a further two males and three females were released in April 1939. By the end of the war the population numbered about 35.

In 1951 Bear Creek Ranch was acquired by Captain Eddie Rickenbacker, the celebrated aviator, whose son, David, began renting blackbuck-shooting and the sale of surplus animals to other ranches as breeding stock – the latter resulting in the dispersal of blackbuck to various parts of the state. In 1953 blackbuck were introduced to the Y-O Ranch at Mountain Home, which now claims to be the largest commercial game-ranching enterprise in the United States. They have also been released on the Bar-O Ranch.

The San Antonio and other zoos have continued to supply Texas ranches with blackbuck from their own surplus stock, and there has been a constant interchange of bucks between ranches to prevent inbreeding.

Blackbuck are sedentary animals and seem reluctant to colonize new land, only doing so in Texas when densities reach about one per 1.5 ha. Densities of one per 3 ha are the highest recorded in unhunted hill country populations, which seem to have become stabilized and from which the animals are free to disperse.

In 1955 it was estimated that the Texas population of blackbuck amounted to between 1,000 and 1,500: by 1963 the total had risen to about 3,700 on 35 ranches in 21 counties, and three years later to 4,100 on 56 ranches in 33 counties. By 1971 Texas supported some 5,500 blackbuck distributed over 42 counties, of which 84.5 per cent inhabited the Edwards Plateau. This region of southern central Texas is characterized by rough, stony hills intersected by several river systems. Elevations range from 400 to 1,000 m. The climax vegetation is savannah and open grassland interspersed with dry scrub and woody plants such as mesquite (*Prosopis glandulosa*), live oak, shinnery oaks (*Quercus* spp.) and ashe juniper (*Juniperus ashei*).

A 1974 census revealed a total of 7,339 blackbuck in 51 counties in Texas, with 80.7 per cent established on the Edwards Plateau. By 1979 the population had increased still further to 9,639 (of which 2,593 were free-ranging) in 55 counties, most of which were concentrated in Bandera, Bexar, Blanco, Edwards, Kendall, Kerr, La Salle, Live Oak, Llano, Menard and Real counties.

The most important factors in the successful establishment of blackbuck in Texas (which lies at about the same latitude as India) are the similarity of climate and habitat to those of their Indian homeland. Blackbuck make use of scrub for cover and seem more tolerant of overstorey canopy than the native pronghorn (*Antilocapra americana*): they cannot, on the other hand, withstand lengthy periods of freezing temperatures. Population trends in the future will depend largely on the continuing popularity of trophy-hunting and the economic viability of the game-ranching programme.

In 1971 10 blackbuck from Texas were repatriated to their native range in Sind's Cholistan Desert in Pakistan, where they were exterminated by hunting in the 1950s and 1960s. Although by 1980 they had increased to 23 it was felt that the injection of new blood might be beneficial. As a result a further male and five females were presented to Pakistan by the Copenhagen Zoo in Denmark. The population today of blackbuck in the United States and Argentina (see below) exceeds the whole of that in the Indian subcontinent.

SOUTH AMERICA

Argentina

In 1906 Señor Pedro Luro, a prosperous immigrant from the Basque country of northern Spain, released some blackbuck on his large *estancia* in La Pampa in southern-central Argentina, where today their descendants graze freely with domestic cattle. In 1912 further releases of blackbuck were made in Santa Fé, Cordoba and Buenos Aires provinces, and again in Buenos Aires in 1940 and more recently, and in Santa Fé also in the 1960s. Blackbuck are now established over wide areas of Argentina, being especially numerous in the east of the country, and several hundred are culled annually.

AUSTRALASIA

Australia

Blackbuck were introduced to Kojarena, Wiluna, Roelands near Bunbury and Numarracarra near Geraldton in Western Australia around the turn of the century. In 1912 a pair were released at Dinninup and a further three pairs were liberated at Numarracarra by Mr Mackenzie Grant. In the following year further consignments were despatched to Numarracarra, Roelands and Wiluna.

By 1920 blackbuck are said to have been well established in the Murchison country and around Wiluna. Nine years later blackbuck released some 500 km north of Perth had increased to pest proportions and required culling. A free-ranging herd of 100–150 was still in existence at Numarracarra in 1946. During the 1960s blackbuck were maintained and protected in fenced enclosures at Numarracarra and Coolya stations, where many are said to have died as the result of eating poisoned oats put out for rabbits. A remnant population of about 30 at Kojarena in the early 1970s may still survive.

The **saiga antelope** (*Saiga tatarica*), which ranges from the southern USSR east to Mongolia, and the **goitered gazelle** (*Gazella subgutturosa*) of Arabia to Mongolia, have been released and have become established on the islands of Bulla and Glinyany in the Caspian Sea. The former species, which is not a true antelope but belongs to a monospecific genus closely related only to the chiru or Tibetan antelope (*Pantholops hodgsoni*), was almost exterminated through hunting by the 1920s. Since then, as the result of strict protection, it has staged a dramatic recovery, and is now the most abundant wild ungulate in the Soviet Union yielding some 6,000 t of meat annually.[2]

By the late 1950s it had become apparent that the **Arabian oryx** (*Oryx leucoryx*)[3] was – due to excessive hunting – all but extinct in the wild. In April 1962 the Fauna Preservation Society (as it then was), supported by the World Wildlife Fund, succeeded in capturing three oryx in the plains northeast of Sanau in the Peoples' Democratic Republic of Yemen. From this small nucleus a herd – later augmented by others bred in captivity elsewhere – was built up in the Phoenix and San Diego zoos in the United States.

On 31 January 1982 four males, four females, a yearling, and a calf born in Oman under the White Oryx Project of Sultan Qaboos bin Said, were released into the wild at Yalooni in the Jiddat-al-Harasis – where the species had last been seen in the country – under the wardenship of Bedu rangers of the Harasis tribe. For the first three weeks after their liberation the oryx (which had been given total protection in Oman since 1976) remained close to their release site, and received supplementary feeding until torrential rain brought a flush of new grass. Thereafter they began to range further afield, and in March 1982 the first calf was born in the wild. By the end of that month the herd, which before its release had developed a stable social structure and five of whose members had been fitted with radio-collars, had established a home range of some 60 km^2. By 1984 the herd had increased to 13 and had expanded its range to 1,500 km^2 within 50–75 km of Yalooni. On 4 April 1984 a further 10 animals were released in the same area.

In October 1983 the Royal Jordanian Conservation Society released 31 oryx into the enclosed 22-km^2 Shaumari Reserve near Azraq. It seems unlikely that they could survive in the open desert since available sites are too close both to populated areas in Jordan and to the border with Saudi Arabia where they could not be protected.

At the time of writing (1984) this is the classic example of the successful reintroduction to the wild of a captive-bred mammalian species.

NOTES

1. For much of the information on blackbuck in Texas

I am indebted to Mungall (1977).

2. W. Linnard, *Oryx* **7**: 30–3 (1963).
3. A. S. Talhouk, W. Büttiker, J. G. & P. R. G. Gasperetti & D. Massey, *The wildlife of Oman and its neighbours* (London, Stacey International, 1981), p. 6; Anon, *Oryx* **16**: 305 (1982); R. S. R. Fitter, *Oryx* **16**: 406–10 (1982), **18**: 136–7 (1984).

Chamois

(*Rupicapra rupicapra*)

NATURAL DISTRIBUTION
The mountains of mainland Europe from northern Spain east to the Caucasus and north to the High Tatras (Tatrys) of Poland and Czechoslovakia, in a large number of discrete populations – especially in the Pyrenees, Alps, Jura, Abruzzi Apennines, Carpathians, Tatras, Balkans, Caucasus and Taurus Mountains.

NATURALIZED DISTRIBUTION
Australasia: New Zealand.

AUSTRALASIA

New Zealand

As early as 1880 Mr (later Sir) Julius von Haast (Surveyor-General of Canterbury) and Mr G. M. Thomson were making enquiries about the possibility of importing chamois from Europe for sporting purposes. Their enquiries were followed by others in 1888–9, but it was several years before they were to come to fruition. On 17 January 1907, after lengthy negotiations by Mr T. E. Donne, the General Manager of the New Zealand Department of Tourist and Health Resorts, eight chamois – a gift to New Zealand from Franz Josef, Emperor of Austria – were despatched to London and thence, on 23 January on board the SS *Turakina*, to Wellington (via Cape Town and Hobart), where they arrived, 'in splendid condition' according to Donne, on 15 March, having been insured for the sum of £600. During part of the voyage Donne (1924) reported that 'three of the chamois had been off their food . . . human passengers also are often like this at sea'. After a period of quarantine on Some's Island in Wellington Harbour the eight chamois (two males and six females, two of which were pregnant) were released on the slopes of Mount Cook, where by the following year (1908) they were reported to be breeding.

On 21 December 1913 three chamois from Franz Josef's private herd at Schonbrunn Palace (attempts to capture more wild ones having proved too hazardous to the animals) were despatched on the SS *Turul* from Austria to New Zealand via the Suez Canal; during the voyage one died of heat exhaustion between the Red Sea and Bombay, and the surviving pair became so debilitated that on arrival in Australia (the first chamois to reach that country) they were landed and placed in the care of the Melbourne Zoo; here they soon recovered and continued their voyage to New Zealand where, after the customary quarantine period, they were released on 28 May 1914 on the east bank of the Hooker River. The female soon disappeared, and was believed to have joined those already established in the wild; the male, who remained close to the point of release, became a danger to incautious tourists and eventually had to be killed.

By 1917 20 chamois were seen together in a single group, and two years later a herd of 70 was reported from near Mount Cook; by 1921 the total population had increased to around 100. Shortly thereafter chamois crossed the Hooker River south of the Hermitage and were soon spreading out towards Mounts Wakefield and Glenmary and the Franz Josef glacier. Thomson (1922) mentions that 'herds of 30, 40 and 70 were being noticed at one time', and chamois were reported from about 20 km north of Lake Ohau and south of the Hermitage at Mount Cook. By the autumn of 1934 they had reached Mount Brewster, near Haast Pass, in the south and Hope Pass in the north. Within a further six years chamois arrived in the Travers Valley at the head of Lake Rotoiti in Nelson and at the head of the D'Urville River near Lake Rotoroa. By the late 1940s chamois had, under pressure from an expanding population, colonized most suitable habitats along the Main Divide as far north as lakes Rotoroa and Rotoiti and south to lakes Wanaka and Hawea. Even at this stage the distribution showed a considerably greater northwards expansion than to the south. By the mid-1960s the population had extended along the Main Divide 350 km to the north but less than 200 km to the south of the original point of release.

Chamois are small goat-like animals with short, erect and sharply hooked horns; they stand about 80 cm high on long slender legs and weigh around 45 kg. They are gregarious, the females and juveniles forming small groups which in winter may join together to form herds of a hundred or more. Adult males tend to lead a solitary existence, only joining the females during the rut in May and June. In summer chamois occupy open tussock grassland and precipitous rocky country (over which they travel with remarkable nimbleness) as high as the snowfields around 2,500 m; in autumn and winter during heavy snowfalls they descend to as low as 500 m in subalpine forest and scrub. Their main foraging range, however, is between about 1,500 and 2,000 m where the herbaceous cover is more or less continuous.

In the area around Mount Cook chamois feed mainly on *Gaultheria rupestris*, puharetaiko (*Celmisia spectabilis*), koromiko (*Hebe salicifolia*), and snow- and silver-tussock (*Danthonia* spp.). Other plants eaten by chamois (and Himalayan tahr (*Hemitragus jemlahicus*[1])) include, in decreasing order of preference, lacebark (*Hoheria* sp.), Mount Cook lily (*Ranunculus lyallii*), speargrass (*Aciphylla colensoi*), *Carmichaelia* sp., alpine *Gentiana, Dracophyllum uniflorum* and *Celmisia linearis.* Like the tahr the chamois is thus a selective feeder, resulting in a chamois/tahr induced plant community.

As early as the 1940s the expanding populations of chamois and tahr were reported by Wodzicki (1950) to be reducing the sheep-carrying capacity of the land, and by overgrazing the steep alpine slopes and rocky screes were increasing erosion in a region already naturally prone to that problem. An absence of predators, a generally milder climate than in their European range (although some young kids fail to survive the winter), protection after their initial release until 1930, the fact that they were planted in a peculiarly suitable alpine habitat, and the difficulty of access to that habitat – except by the fit and athletic sportsman – have jointly helped to ensure the survival of chamois and tahr in New Zealand.

Today, breeding populations of chamois occur as far north as Lake Rotoiti and south to Lake Wakatipu, while the mountains of northern Fiordland are being gradually invaded. Although released a few years later than the first tahr with whom their range in New Zealand overlaps, chamois have been by far the more successful colonists; indeed, they have spread to a greater distance from their original point of release than any other introduced ungulate, and have travelled at an average rate of around 9.5 km a year. Official culling began in 1936, and by 1968 over 82,000 had been killed by government rangers. Large numbers are also shot annually for sport. The New Zealand population is believed to be the only one successfully naturalized outside their European range.

In about 1905 some chamois from the high Austrian Alps were translocated outside their natural range to the lower altitudes of the Jeseniky and Luzicke mountains of northwestern Czechoslovakia. In both ranges they became successfully established, the present populations numbering about 300 and 100, respectively.

Chamois have also been successfully translocated outside their natural range in Switzerland, France and Austria.

In Europe the **ibex** (*Capra ibex*) is now confined naturally solely to the Alps; it has been widely reintroduced elsewhere in Austria, Switzerland and France and, after reduction to a remnant herd in the Gran Paradiso National Park, in Italy. Ibex have also been planted in parts of Germany, Yugoslavia, Spain and the Soviet Union, as well as in the Carpathian Mountains on the Hungarian/Polish border.

In the High Tatra (Tatry) Mountains on the borders of Poland and Czechoslovakia ibex became extinct during the first half of the last century; in 1901 Prince Hohenlohe obtained some from Prince Pless in Austria which he reintroduced to the Tatras for hunting purposes. Some years later two Asiatic species – the **bezoar** or **Persian ibex** (*C. aegagrus*) and the **Sinaitic goat** (*C. nubiana*) – were also introduced to the Tatras, where they interbred with the ibex.[2]

Both chamois and ibex are currently stocked on several ranches in the mountainous Neuquen Province of west-central Argentina, where concern has been expressed that they may spread into adjacent national parks.

The **musk ox** (*Ovibos moschatus*), originally a native of the American arctic and Greenland, has been extinct in the Eurasian arctic since the end of the last glaciation. In 1929 17 musk oxen from eastern Greenland were introduced to Spitzbergen and in 1932 others were released in the Dovre Fjell in southern Norway, from where they have spread into adjacent parts of central-southern Sweden. In 1930 34 musk oxen were transported from Greenland to Alaska; five years later they and their progeny were transferred to Nunivak Island, where by 1966 the herd had increased to over 400. In 1978 the Alaskan population was estimated to number 1,500. In 1975 six cows and four bulls were flown from Banks Island in the Canadian arctic to the eastern shores of Lake Taymyr in northern Siberia, in an attempt to re-establish the species in the Soviet Union. In the same year a number were air-freighted from Nunivak to Wrangel Island in the East Siberian Sea, where by 1982 the herd numbered 27, including nine calves which had been born on the island. These plantings appear to have been successful, whereas importations to other parts of Sweden, and to Germany and Iceland have apparently failed.

NOTES

1. See p. 281.
2. Turcek (1951).

Himalayan Tahr

(*Hemitragus jemlahicus*)

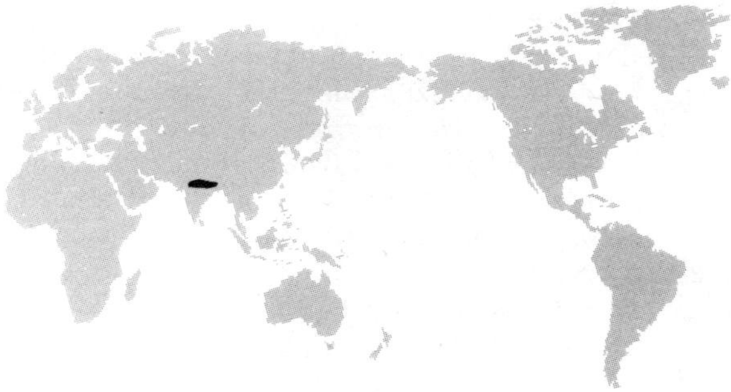

NATURAL DISTRIBUTION
Himalayas (Pir Panjal, Kashmir, Punjab, Kumaon, Nepal) up to an altitude of nearly 5,000 m.

NATURALIZED DISTRIBUTION
Africa: South Africa; Australasia: New Zealand.

AFRICA

South Africa

On 2 September 1935 a pair of Himalayan tahr were sent by Dr R. Bigalke from the National Zoological Gardens in Pretoria to the zoo at Groote Schuur; from here, two years later, they (or their progeny) escaped onto the nearby slopes of Table Mountain where, and on the adjoining peaks, they became established; fortunately they have not spread to the neighbouring Constantiaberg, from which they are separated by the shallow Constantia Nek.

By 1962 the population was judged to number

about 50, which by late 1974 had multiplied to around 600. As a result of the increasing damage inflicted by tahr on the montane vegetation the Cape Town municipal authorities began a culling programme in 1976. By the middle of 1981 around 600 tahr had been killed and the remaining population was claimed to be less than 100: since the present size of the individual groups has decreased and the tahr are more wary than they once were (thus making them more difficult to observe) this figure may be a considerable underestimate. [1]

AUSTRALASIA

New Zealand

In 1904, at the suggestion of Mr St George Littledale, the Duke of Bedford presented three pairs of tahr from his private menagerie at Woburn Abbey, England (to which their progenitors had been sent direct from India by the animal dealer Carl Jamrach between 1894 and 1897) to the New Zealand Tourist Board as potential game animals. On 6 April the tahr were shipped from Plymouth on board the SS *Corinthic* which berthed in Wellington in the following month, where one male succeeded in jumping overboard: the remaining 5, however, were successfully placed in quarantine on Some's Island in Wellington Harbour, from where they were subsequently released in highland forest, montane scrub and tussock grasslands near the Hermitage in the Mount Cook region of Westland – probably on the slopes of Mount Wakefield.

In 1909 the Duke despatched a further six males and two females (part of a consignment which included axis deer (*Cervus axis*),[2] llamas (*Lama glama*) and rheas (Rheidae) by the SS *Wakanui*, which in April were liberated in the same region. Later in the same year the Government Tourist Department purchased and unsuccessfully released a pair of adult tahr with a young female in the Hot Springs district of Rotorua on North Island. In 1931 a further three tahr were released on the Franz Josef glacier, north of Mount Cook, and six years later four more, from Wellington Zoo, were liberated in the Sealey Range.

By 1917 the earlier plantings had increased to at least 18–20 individuals and by the following year to about 50: by 1924 the total population was estimated to have reached 100 – all within 20 or so kilometres of the point of release. Wodzicki (1950) found that by the late 1940s tahr occupied a range in the Southern Alps measuring nearly 100 by 40 km.

The early spread of tahr in New Zealand was northwards into the Murchison Valley, and thence to the Cass and Godley valleys. The next divide to the north, Rangitata – occupied by three main herds at Havelock, Clyde and Lawrence – was probably first colonized between about 1940 and 1954. North of Rangitata three herds – centred on Rakaia, Mathias and Wilberforce – invaded the Rahaia catchment between around 1941 and 1955, presumably by way of the McCoy tributary of the Clyde.

Tahr first crossed the Mathias Pass in 1964, probably from the Hokitika catchment, when a herd of about 70 was located at the head of the Orari Valley.

To the south of the original point of release on the east of the divide lie the Dobson, Hopkins, Huxley, Ahuriri and Hunter valleys: tahr first appeared in the Dobson (via the Tasman) in the early years of colonization, in the Hopkins in 1947 and in the Ahuriri (via the Huxley branch of the Hopkins) in about 1958.

West of the divide the pattern of dispersal has been more complex. Travelling presumably from the Hopkins and/or Huxley via the Elcho or Broderick passes, tahr were established and breeding in the Landsborough region by 1947. The Douglas Valley was invaded in about 1950 and shortly thereafter the Karangarua – probably by dispersal south from the Douglas. Northwards, the next divide to be occupied may have been the Copland (probably via the Regina from the Douglas) in about 1955. The Makawhio or Jacobs Valley south of Kara was colonized in about 1960, probably via the Troyt.

Expansion of the population in the south-west has been principally northwards. The northwestern population probably originated in the Perth branch of the Whataroa via the Sealey Pass and Scone Creek from the Godley, or via the Denistoun Pass from the Havelock. The Perth Valley was probably occupied between 1951 and 1955, with tahr spreading south through the Wharatoa head and north via the Wanganui catchment to the Whitcombe Valley by about 1961. By

1970, their range extended from the Mathias and Whitcombe rivers in the north (and south into the Two Thumbs Salient in the north-east) south as far as the source of the Dingle Burn. This area includes parts of both sides of the Main Divide in the Southern Alps, and covers an area of over 3,600 km^2, which had been colonized at a maximum rate of 3.2–4.8 km a year – much slower than that of chamois (*Rupicapra rupicapra*).[3]

The development of the overall distribution and breeding range of tahr between the 1940s and 1960s shows that after an initial period of establishment the general rate of spread was approximately constant, though dispersal northwards was about twice as rapid as it was to the south, and was apparently largely a case of voluntary diffusion rather than as a result of population pressure. The age distribution of tahr on the perimeter of their breeding ranges suggests that dispersal is not confined solely to juveniles. Before the rut some males in search of females fortuitously move outside the previous breeding range into a new zone, where they encounter wandering females.

Mr Graeme Caughley, who conducted a survey of tahr in New Zealand in the late 1960s, considered that future expansion northwards could be expected at the present rate until Harper Pass was reached, after which it was likely to decline because the country north of the pass – neither as high nor as precipitous as that to the south – offers poor tahr habitat. North of Lewis Pass the rate of progress was likely to increase again as the favourable terrain of the Spenser Mountains was colonized. In time, parts of the high country of north-west Nelson and the Kaikouras could also become occupied.

Caughley believed that the Westland gap in the breeding range would eventually be closed through the expansion north-west of the south-western population and a corresponding extension of range south by the north-west population. The herds in the south-west, Caughley considered, should by degrees occupy the heads of the valleys draining the Hooker Range and then, with some members of the eastern population spreading into the Hunter and Wills, would cross the Haast Valley. They would then disperse down the hill country between lakes Wanaka, Wakatipu, Te Anau and the Tasman Sea. Dispersal west of the divide would be restricted by the lack of suitable habitat.

Tahr are small goat-like creatures standing about 100 cm high and weighing around 80–90 kg, with a dense and long pelage which in mature males includes a thick knee-length ruff or mane. A specialized hoof structure enables them to negotiate their precipitous terrain with ease. In New Zealand tahr occupy the subalpine zone of grassland and exposed crags far above the bush-line on both sides of the Main Divide of the Southern Alps, where the absence of covers renders them conspicuous; they are found mostly at altitudes of between 1,500 and 1,800 m – usually near a bluff system to which they retreat when disturbed – but seldom in valleys where the ridge crest is below 2,000 m. During periods of severe winter weather they may descend for shelter to the lower montane scrub belt and even, on occasion, to high altitude woodlands. Outside the rutting season (from late April to early June) tahr associate in separate all-male and female/immatures herds.

In the Mount Cook region tahr feed largely on puharetaiko (*Celmisia spectabilis*), snow- and silver-tussock (*Danthonia* spp.) and *Gaultheria rupestris*. Other plants browsed or grazed by tahr (and chamois) include, in decreasing order of preference, korimoko (*Hebe salicifolia*), lacebark (*Hoheria* sp.), Mount Cook lily (*Ranunculus lyallii*), speargrass (*Aciphylla colensoi*), *Carmichaelia* sp., alpine *Gentiana, Dracophyllum uniflorum* and *Celmisia linearis*. This shows that both animals are selective eaters resulting in a tahr/chamois induced vegetation.

Wodzicki (1950) reported that as early as the 1940s the expanding population of both species had reduced the sheep-carrying capacity of their habitat, especially east of the divide, and that in an area already subject to natural erosion the problem was being exacerbated by overgrazing on the rocky screes and precipitous alpine slopes. The lack of predators, a generally milder climate than in their native range (though many kids die after a prolonged spell of bad weather), protection after their initial liberation until 1930, and the difficulty of access, except by the most ardent of sportsmen, to their habitat have all combined to ensure the survival in New Zealand of both tahr and chamois.

By the late 1970s tahr were established along over 220 km of the Southern Alps, on both the Canterbury and Westland sides of the Main Divide, from the Hunter Valley at the head of Lake

Hawea in the south to the perimeter of the Waimakariri catchment in the north. Where they occur in high densities they have, by their selective feeding habits, considerably modified the native vegetation. Their marked preference for montane pastures of tussock grass and alpine flowers has resulted in the elimination in many areas of endemic mountain daisies, mountain lilies and mountain buttercups.

Official culling of tahr was commenced in 1937, and since then well over 30,000 have been shot by government rangers and many thousand more by (mainly tourist) sportsmen. Nevertheless, they continue slowly to expand their range, and individual males and even small groups are periodically reported from well outside their known breeding territory.

Himalayan tahr are also presently held on several ranches in the mountainous Neuquen Province of west-central Argentina, from which there is concern that they may escape into the adjacent national parks.

Some tahr from the Toronto Zoo, Canada, released near Peterborough, Ontario, survived in the wild for a time but did not spread. Similarly, some transferred by the Duke of Bedford from Woburn Abbey, Bedfordshire, England to Cairnsmore and Ketterick, Kirkcudbrightshire, Scotland before the First World War persisted for a while and even bred, but eventually died out after the disappearance of the males.

NOTES

1. Siegfried (1962), Bigalke (1977), Haltenorth & Diller (1980) and P.H. Lloyd (personal communication, 1 July 1981).
2. See p. 155.
3. See p. 278.

Domestic Goat
(*Capra hircus*)

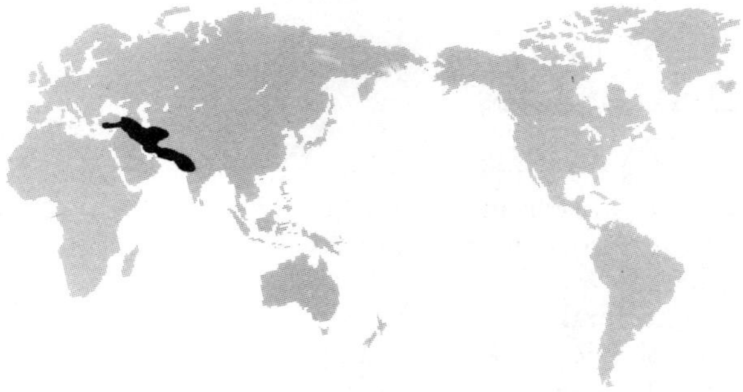

Note: Natural distribution shown is that of wild ancestor.

NATURAL DISTRIBUTION

The wild goat or bezoar (*C. aegagrus*), from which the domestic goat is descended, ranges from Asia Minor through the Caucasus and southern Turkmenia to Iran, Iraq and Baluchistan to western Sind, India.

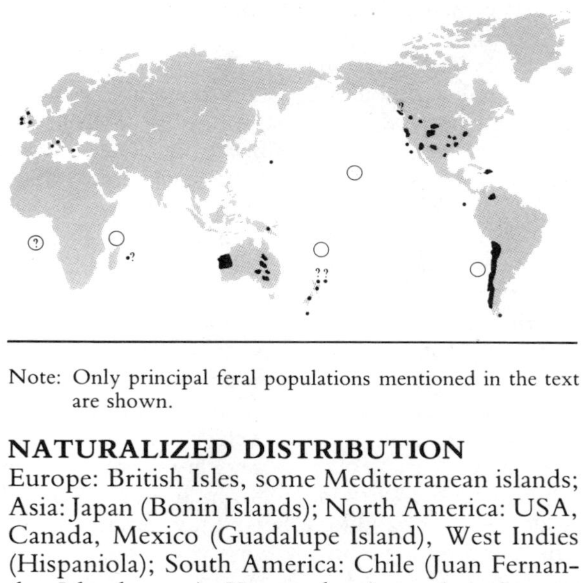

Note: Only principal feral populations mentioned in the text are shown.

NATURALIZED DISTRIBUTION

Europe: British Isles, some Mediterranean islands; Asia: Japan (Bonin Islands); North America: USA, Canada, Mexico (Guadalupe Island), West Indies (Hispaniola); South America: Chile (Juan Fernandez Islands, etc.), Venezeula, Argentina; Australasia: Australia, New Zealand; Oceania: Hawaii, Galápagos, Seychelles, ?Mauritius, Fiji, ?St Helena, various other oceanic islands.

EUROPE

British Isles

There is no evidence to show that any species of goat was ever indigenous to Britain, but domesticated goats were imported as early as Neolithic times.

The feral goats in Britain today are in many cases the descendants of ancestors which had been wild for centuries, which were themselves the offspring of feral domestic stock, some of which may have derived from those pastured in forestland as early as the fourteenth century. Others strayed from the large numbers which were driven from Ireland to England through Wales, and from the lowlands of Scotland to the Highlands, where depopulation in the nineteenth century led to the abandonment of many herds. In more recent times the feral population has been augmented by further escapes from domesticity.

At present, feral goats occur only locally in England, Ireland and Wales (although they are probably more prevalent in Wales than is sometimes believed) and are widely but thinly distributed in Scotland. There seems to be little if any extension of range or migration. The earliest specific reference to 'wild' (or more properly 'feral') goats in Wales is by Thomas Pennant who, in his *History of quadrupeds* (1781), refers to their presence on the Rhinogs in the early 1770s. Their main strongholds in Wales today are Merionethshire and the Snowdonia National Park, where the total population is said to be around 300.

In England there were goats on Lundy Island in the Bristol Channel in the eighteenth century, but these had disappeared before the outbreak of the First World War; more were introduced in 1926, and their descendants remain on the island today. There have long been a few feral goats on the Lynton Rocks in Devon, while those on the Isle of Man may be descended from some imported by lighthouse-keepers between 1818 and 1875. A number of flourishing herds, numbering several hundred individuals, exist on the Cheviot Hills of Northumberland, especially in the fells[1] of Kielder Forest.

In Ireland up to a thousand feral goats are established in small discrete herds in the mountains of counties Antrim, Armagh, Clare, Dublin, Galway, Kerry, Mayo, Sligo, Tipperary, Waterford and Wicklow, and on some islands off the coast of Fermanagh.

In Scotland – their principal present British stronghold – the goats at Inversnaid on Loch Lomond in Perthshire are reputed to possess pedigrees dating from before 1306, in the time of Robert the Bruce. Feral goats were part of the game driven to Mary Queen of Scots at the tinchel[2] organized in Glen Tilt on the Braes of Atholl in Perthshire in 1564, which suggests that their existence there was of long-standing. In 1590 'Mad' Colin Campbell and his men encountered a tribe of feral goats on Stuchd an Lochain in Glenlyon in the same county. A tribe of large white goats on Mamlorn, again in Perthshire, is mentioned in a report on the rebellion of Hew Murray from 1661–79, when it was claimed that 'he and his men would never be subdued while they could get a goat on Creag Mhor'.

Feral goats have for long been established on a number of Scottish island; there was a tribe of magnificent white goats (said to be descended from the survivors of a wreck of the Spanish Armada in 1588) on Ailsa Craig off the coast of Perthshire until as recently as 1925. Martin Martin, in *A description of the Western Islands of Scotland* (1703), says that goats were present on the Isle of Lewis at least by 1697. Pennant saw about 80 goats on the island of Jura and others on Arran in the early 1770s. The present tribe at Smaull near Kilchoman on Islay dates from a planting in 1787, and those surviving until recently on Canna Island off Rum and on Ardinaddy date from 1793 and 1812 respectively.

In Scotland several thousand feral goats are today thinly but widely distributed throughout the Highlands and Islands, as well as in a number of Lowland and Border counties. These include Argyllshire, Ayrshire, Dumfriesshire, Inverness-shire, Kirkudbrightshire, Morayshire, Nairnshire, Perthshire, Ross and Cromarty, Roxburghshire, Selkirkshire, Stirlingshire, Sutherlandshire, and the islands of Cara, Colonsay, Islay, Jura, Little Colonsay, Mull, Oronsay and Texa, and Holy Island to the east of Arran. In *The influence of man on animal life in Scotland* Ritchie (1920) wrote of goats that they were:

> exceedingly destructive; young seedlings and even sturdy trees up to there or four feet high are ruined by the demolition of their foliage and leading shoots and by the gnawing of the bark. So that when nature claims an aged and time worn member of the forest, there is

no fresh growth spontaneously to take its place; and the woodland, which was prevented from extending by those aliens, is also prevented by them from recuperating in its old age. Thus there has resulted a general decay of forests . . .

In the absence of other predators the future of feral goats in the British Isles depends entirely on the tolerance of man by whom, through shooting or rounding-up with sheep-dogs, they could at any time easily be exterminated. The present population is estimated to number over 4,500.

Continental Europe

Feral domestic goats are found over much of the Mediterranean region and on several offshore islands (e.g. off the Dalmatian coast of Yugoslavia; on Crete, Cyprus, Theodorou, Gioura, Erimomilos and Samothrace (Greece); on Monte Cristo (Italy); on Corsica (France); and on Ibiza and Majorca in the Balearics (Spain).

ASIA

Japan (Bonin Islands)

Long-horned goats of variable colours (among which cream with white spots is the most common) are established on the Ogasawara Gunto (Bonin) Islands in the Pacific Ocean, 900 km south-east of Tokyo. Adult males stand some 70 cm high. In 1968 the population was estimated to be about 1,000. They are said to be descended from some landed on the islands by the American Admiral Perry in 1853. In 1955 six primitive Tokara billys and four nannies were placed on Tomogasima Island, where they had increased to 20 by 1958.

NORTH AMERICA

USA[3]

Although nowhere abundant goats are the most widely distributed feral ungulate in the United States, where McKnight reported them from no fewer than 27 states. With the exception of California and Hawaii (see below) the various populations are thought to be of relatively recent origin, and individual tribes seldom number more than a couple of dozen or so: most are derived from escaped or abandoned animals owned by smallholders. They favour steeply sloping hill-land, canyon walls, riverine bluffs and rocky outcrops. Because they are few in number and in the main occupy marginal land they are of little economic value (apart from shooting for sport) and cause only limited damage; overbrowsing may accelerate soil erosion in their hilly habitat, and some competition for forage has been reported with deer.

Table 17 gives details of the distribution of feral goats in the United States. McKnight summed up the overall distribution of feral goats as follows:

1. They are found in all the Appalachian states with the possible exception of Maryland and New York.
2. They are reported from the central hill country of Arkansas, Oklahoma, Missouri, and Illinois.
3. They are known on islands off the coasts of Virginia and Alaska, and in Louisiana.
4. They occur in every western state with the possible exception of Washington, but including the Dakotas and Texas.

   In any one of the states mentioned above the total number present apparently does not exceed a few hundred, with the largest populations in Alabama, West Virginia, Arizona, Colorado, California, and Oregon.
5. In addition, there are outstanding occurrences of feral goats in two isolated island groups: Hawaii and the Channel Islands of California.[4]

The origin of the goats on Santa Catalina in the Channel Islands is uncertain. Until recently they were believed to have been landed either by early Spanish explorers such as Juan Cabrillo and Sebastian Vizcaino or by English buccaneers who preyed on their vessels: McKnight considered that 'these animals show relatively ancient Spanish ancestry in their appearance, which encompasses a long beard, extensive horns, and long body hair'. It is now, however, believed that goats were first landed on the island in the early years of the nineteenth century (before 1827) by traders who, to evade the payment of duty, left a portion of each cargo on Santa Catalina, and were then able to

Table 17 Distribution of feral goats in the United States

State	*Distribution*
Alabama	Morgan, Perry and Dallas counties
Alaska	Northern Kodiak Island
Arizona	Tucson, Santa Catalina, Cayateno (near Nogales), Granite Wash (Yuma County) and Cerbat (Mohave County) Mountains; south-east and south-west of Winkleman; Kofa Game Range; drainage of Bill Williams River, and South Mountain near Phoenix
Arkansas	Heavener area of Ouachita National Forest
California	North-west of Lava Beds National Monument, Siskiyou County; King's Peak area, Humboldt County; Cow Mountain, Medocino County; Mount Diablo region, Contra Costa County; New Idria and Pinnacles area, San Benito County; Panoche Hills; Groveland area of Tuolumne Canyon; between Groveland and Coulterville, Mariposa County; Black Mountain district, Fresno County; Delonegha Springs, Kern County; Coleville, Mono County; Paoha Island, Mono Lake; Hall and Pleasant canyons west of the Panamint Mountains; Martinez Canyon, Santa Rosa Mountains; Big Maria Mountains and Moreno Badlands, Riverside County; Santa Catalina, San Clemente and Santa Cruz in the Channel Islands
Colorado	Piceance Creek near Meeker; south of Grand Junction in the Unaweep Canyon; near Rifle; Glenwood Canyon, Colorado River; near Red Canyon and Basalt south of Glenwood Springs; between Gypsum and Dotsero; south-west of Kremmling around Big Piney Ridge; east of the Laramie River near the Wyoming border; west of Fort Collins in Poudre Canyon; Idaho Springs; between Turkey Creek and Georgetown; Arkansas River canyon near Royal Gorge; and Archuleta mesa in Southern Ute Indian Reservation
Georgia	An island in Lake Lanier near Gainsville, and Edgewood Arsenal
Hawaii	(See text, p. 296)
Idaho	Cuddy Mountain district north-east of Cambridge
Illinois	Alexander County
Kentucky	Fort Knox
Louisiana	Catahoula Wildlife Management Preserve, Grant parish
Missouri	Around Lake of the Ozarks, Camden and Benton counties
Montana	'Breaks' of Musselshell River; north-east of Winnet around Blood Creek; Black Butte, north-east of Lewiston; Madison River valley, near Norris; and Bridger Canyon near Bozeman
Nevada	An island in Pyramid Lake
New Mexico	Carlsbad Caverns National Park; southern Guadalupe Mountains; east of Organ Mountains; San Andres Mountains; Cimarron Canyon, Colfax County; and east of Belen
North Dakota	Bullion Butte, Golden Valley County, and Theodore Roosevelt National Memorial
Oregon	Heppner, Canyonville and La Grande districts; Jackson County and South Umpqua and Lower Rouge river drainages
Pennyslvania	Around Emporium
South Dakota	Black Hills
Texas	Central Hill country and Big Bend National Park
Utah	Below Moab in the Colorado River valley; around Hill Creek, near Rock House; west of Mount Nebo and in the Dinosaur National Monument
Virginia	Shenandoah National Park; around Big Stone Gap, Wise County; and Parramore, Chincoteague and Assateague islands
West Virginia	Blue Creek region, Kanawha County; Black Mountain area, near Marlinton; drainage of Blake Fork Creek, near Parsons; Great Cacapon Mountain, Morgan County, and Nathaniel Mountain and North River areas, Hamshire County
Wyoming	Southern Wind River Mountains; around Sweetwater Rocks, Natrona County; Ted Butte, west of Lander; Indian Butte, east of Riverton; Shirley Mountains, south of Casper; Powder River drainage, near Rozet; southeastern Converse County; and 'breaks' of Platte River, near Wendover

Source: After McKnight (1964)

smuggle the balance of the contraband onto the mainland. When J. F. Curtis visited the island in 1864 during the Civil War he found that goats were well established, and Charles Hillinger claimed in 1958[5] that they have numbered in the tens of thousands since the 1860s; several thousand still remain on Santa Catalina, with 300–400 only (in 1982) on San Clemente (where they may have been introduced in 1875) and where 16,000 were killed in 1975–7 and a further 8,000 in 1979–82.

It has never been clearly proved that goats were solely responsible for the original defoliation of Santa Catalina; overgrazing by sheep (22,000 in 1864) and cattle (4,000 in the early years of this century) must have been a contributary factor.

Where they persist, however, goats are, by their overgrazing, continuing the process of degradation of the island's flora and, by trampling the ground into trails with their sharp hooves, are accelerating the normal course of erosion, which adversely affects the littoral marine ecosystem. These activities have resulted in sharp declines in the populations of the island's endemic subspecies of the California quail (*Lophortyx californicus*) and three of the five endemic terrestrial mammals – the island grey fox (*Vulpes littoralis*), Slevin's deer mouse (*Peromyscus slevini*) and the western harvest mouse (*Reithrodontomys megalotis*). The goats' survival on Santa Catalina is due to their success in out-competing for food the introduced mule deer (*Odocoileus hemionus*) and feral pigs (*Sus domestica*)[6] and to the absence of predators.

On the neighbouring island of Santa Barbara, destruction of the native vegetation by goats (and rabbits) is believed to have been responsible for the extinction of the local race of the song sparrow (*Zonotrichia melodia graminea*).

Canada

Goats (together with cattle and pigs) were first imported to New France in 1541 by Jacques Cartier, whose crew, due to the unexpected prolongation of the voyage, was, according to Richard Hakluyt in his *Principall navigations, voiages, and discoveries of the English nation* (1589), 'constrained to water [them] with Sider and other drinke'. Whether any of Cartier's livestock escaped into the wild is unknown.

Goats are (or until quite recently were) established in the wild in Canada on Saturna, Prevost, Sidney, Galiano, Saltspring, Ramsay, Mayne (where a small tribe lived on Flaghill from 1942 to 1956), Lasqueti and Texada islands between Vancouver Island (where a number were established on Mount Tzuhalum) and the mainland of British Columbia, where at one time they also occurred on the mountain slopes around the head of Saanich Inlet, in the Sooke area and in the Highland District of Vancouver.

In the 1890s Mr J. Pain introduced a number of goats to Salturna Island; a second importation took place in about 1925, when Mr L. C. Harris transferred some (which he believed to be of the 'Bejornfelts' race) from Curlew Island. By the late 1950s a feral population of between 60 and 100 had become established on the west coast. These animals are morphologically diverse, with normally a long, coarse black, grey, mottled, white or red pelage which lacks, however, the dense undercoat of the Saltspring and Galiano islands' populations; they are believed to be derived from the English type with an admixture of Toggenburg blood. Both sexes may be horned or polled. On Saturna goats share their range with semi-feral domestic sheep.

In about 1920 Mr J. Cook released some goats on Galiano Island, where by the late 1950s a tribe of all-white animals was established on the western slope of Mount Sutall and on Mount Galiano.

In 1925 goats were introduced to Saltspring Island to establish a cheese-making industry; this flourished until the outbreak of the Second World War in 1939, when both the business and the goats were abandoned: the latter became established on Mount Bruce where they survived until 1957, when the last were shot by local hunters. In the late 1930s Mr J. Whimms released a number of goats on Saltspring whose descendants became established on a long, narrow, forested ridge west of the Walker Hook road, where by the late 1950s the tribe numbered about 15: their large body-size and horns and dense, coarse, white pelage led to the suggestion of Saanen and/or Schwarzhals ancestry. On all three islands feral goats share their range with native mule deer.

When McKnight (1964) carried out his survey of feral livestock in Anglo-America he mentioned only one colony of feral goats in Canada – on Hen Island in Lake Erie, Ontario.

Mexico

A population of black, grey and white goats – descended from a number introduced from Russia more than 150 years ago – is established on the island of Guadalupe off the coast of Baja California. The original importation is said to have been of the Orenburg type of the Kirgiz fleece goat which is still to be found in southeastern Bashkiria and in Chkalov Province. Adult males weigh 50–70 kg and females 20–35 kg. Under the coarse outer coat is a softer undercoat of down (*pashmina*). Kids are born between February and May.

At one time the population stood at 21,000, of which 41 per cent were adult males, 37 per cent adult females and 22 per cent juveniles. In 1971 98 per cent of the population was slaughtered, allegedly to further a programme of reafforestation. By 1978 the surviving 2 per cent had multiplied to over 2,000 head.

According to Roots (1976), the originally rich flora and fauna of Guadalupe has been devastated through overbrowsing by goats. The destruction of the vegetation reduced the supply of seeds and insects which has resulted in the loss of several species of endemic birds.

West Indies (Hispaniola); Curaçao; Bonaire

Although goats are believed to have been introduced to Hispaniola by the Spaniards shortly after their discovery of the island in 1492, those seen by the Spanish naturalist/historian Oviedo y Valdès in 1535 were apparently a West African (possibly French Guinean) breed. Feral goats are at present established on the island in dry limestone regions of poor soil with xerophytic scrub vegetation, especially in the Baradères Pensinula and in the Tapion de Papaye area of the northwest peninsula; on the south coast near Grand-Gozier; at Anse-à-Pitre (Haiti); on Morne Bienac near Gonaïves; and in the west of the Massif de la Hotte. Much of this range is shared with herds of feral cattle.[8]

Populations of at least 1,000 feral goats on each of Curaçao and Bonaire in the Netherlands Antilles are believed to be descended from domestic stock introduced in the eighteenth century.

SOUTH AMERICA

Chile;[9] Venezuela; Argentina

Between 4,100 and 4,400 feral goats are established on the Juan Fernandez Islands National Park off Santiago, Chile, of which 3,800–4,100 are on Alejandro Selkirk Island, 250–280 on Robinson Crusoe Island and 30–45 on Santa Clara Island: they are reputedly descended from a small number landed in 1572 by the Spanish navigator Juan Fernandez, who discovered the islands in 1563. Jacob le Maire and Willem Cornelis Schouten found goats (and pigs) in great abundance on the islands when they called to re-victual their vessels in 1616, and in 1680 a buccaneer named Sharp was able to kill and salt more than a hundred goats for his crew. The goats thrived on the luxuriant subtropical vegetation and became such a valuable source of meat to seventeenth century buccaneers that, six years later in 1686, in a partially successful attempt to control them, the Spanish introduced mastiff dogs to the islands. (It is this which may have given rise to the myth that dogs were released on Santiago in the Galápagos in 1685 to control goats introduced to the islands by seventeenth-century buccaneers.) When Captain Woodes Rogers, in command of two privateers fitted out by merchants of Bristol in England to cruise against the Spanish in the south seas during the War of Spanish Succession, rescued Alexander Selkirk – prototype of Robinson Crusoe – from Juan Fernandez in 1709, he found goats (and cats and rats) in considerable numbers, suggesting that the mastiffs were of only limited success as a controlling agent. The dogs, however, increased to such an extent and became so ferocious that they were exterminated by about 1830. Thereafter the goats rapidly increased again to some 3,000 by 1877. They have devastated the islands' native vegetation which, except in inaccessible ravines and on high ridges, has been largely replaced by alien plants, and soil erosion is a continuing problem.

The normal colour of goats in the Juan Fernandez Islands is reddish-yellow or reddish-grey with a black dorsal stripe running from head to tail and a transverse stripe across the shoulders – a pattern known as *chamoisé*. Partial or total melanism occurs in 35–45 per cent of the populations on Alejandro Selkirk (Masafuera) and Robinson Crusoe (Masatierra) and, as the result of importations of white domestic goats (e.g. Angoras) in the early 1900s, small percentages of the goats on these two islands also carry white markings. On Santa Clara the basic colour is paler and the dorsal stripe less well defined. Adult billies stand about 60–65 cm high and weigh about 50 kg. Between 300 and 1,200 are culled each year to prevent an increase in the population and in the percentage of unacceptably coloured animals; these are a valuable source of protein to the islanders and thus help to reduce the pressure to introduce cattle and sheep into the national park.

On the Chilean mainland feral goats are widespread. Elsewhere on the continent they occur in Argentina from Guayapa in Rioja in the north as far south as Isla de los Estados, where in 1971 around 60 were counted at Bahia Crossley, and north to Venezuela where Bates (1956) claimed that 'they might well be called the ecological dominant over . . . the Venezuelan Andes'.

AUSTRALASIA

Australia[10]

According to Frith (1979), 19 ordinary domestic goats were among the livestock landed by the First Fleet at Botany Bay in 1788. Thereafter, there were doubtless many more unrecorded importations, and before long goats came to be regarded as one of the most valuable animals – for their meat, milk, cheese and butter – owned by the early colonists. Although their grazing was supposed to be confined to certain designated common land, in practice they were allowed to roam freely in and around the colonists' settlements, and around mines, farms, fishing and pearling villages, boundary-fence riders' homes, overland telegraph stations, and railway gangers' and doggers' camps. They mostly occurred in small tribes of no more than a few dozen of so, but sometimes – as, for example, at Nutwood Downs in the Northern Territory – in herds of more than 300. In its heyday in the early 1900s the mining boom town of Canbelego in New South Wales supported three separate herds of over 200 each.

In Australia – as elsewhere – goats proved hardy and adaptable animals which required little herding. Many were allowed to range free, and later more were deliberately turned out to fend for themselves. When, for example, Mootwingee Hotel north-east of Broken Hill in New South Wales, which was for many years an important stagecoach halt, was closed down when the coming of the railway made horse-drawn coaches redundant, the hotel's herd of goats was abandoned. Similarly, when the Yudnamutna copper mines in the North Flinders Range of South Australia were worked out and deserted, the miners' goats were turned loose.

From the early nineteenth century deliberate attempts were made to establish herds of feral goats both on the mainland (e.g. in the Main Barrier Range in New South Wales) and, especially, on such offshore islands as the Whitsunday group off the coast of Queensland, the Recherche archipelago off the south coast of Western Australia, on a number of islands in the Bass Strait between Victoria and Tasmania, and on Bernier Island in Shark Bay off the west coast of Western Australia; these insular releases were intended to supply food for castaways and sealers.

After about 1861 some attempts were made to establish Angora goats from Asia Minor and Cashmere goats from Tibet in Australia as a valuable source of wool: the former were to be farmed on the drier lowland plains while the latter were to live in the high montane regions of the south-east. Although most of the Cashmeres failed to survive the voyage to Melbourne, the Angoras did – albeit briefly – become acclimatized. Difficulties in handling them, crossbreeding with ordinary domestic goats and the fall in the value of their wool in line with the rise of that of the more easily managed merino sheep all contributed to their eventual failure: many were turned loose into the wild where they doubtless joined up with the already established feral herds of domestic goats and pure white English Saanens.

Although most Australia's goats are white or pale-coloured, they also occur in all shades of brown and piebald. A. J. Fraser said that in Shark Bay, Western Australia, '. . . goats in the Islands are all pale coloured and are easily visible'.[11] Goats on the Queensland mainland and offshore islands and in New South Wales are almost invariably whitish or cream.

Feral goats in Australia are hardy and prolific. They are usually larger and heavier than their domesticated counterparts, except when stunted by severe inbreeding, and grow longer beards and more massive horns. They are seldom found on flat and open land, preferring rough, rocky, scrub country and a well-wooded terrain. They are primarily browsers rather than grazers and, like the East African gerenuk (*Litocranius walleri*), frequently stand on their hind legs to reach bushes which would otherwise remain out of reach; they have even been known to climb trees. Although solitary individuals sometimes occur, feral goats are basically gregarious, normally associating in single-sex groups of between 6 and 20, though on occasion in herds of between 100 and 500. They range far in search of forage and water, and are believed sometimes to follow the Outback rains. Although they are wild, timid and difficult to approach they are – because of their habit of clustering together – easy to round up.

The damage caused by feral goats is much the same as that produced by feral horses and donkeys.[12] Most serious is the erosion created by their overbrowsing and overgrazing, and their competition with domestic livestock for food and water. The harsh conditions of drought and heat in the arid Outback are goats' most important limiting factors, although some (especially kids) also fall prey to feral dogs (*Canis familiaris*),[13] dingoes (*C. f. dingo*),[14] red foxes (*Vulpes vulpes*)[15] and wedge-tailed eagles (*Aquila audax*). Goats are also controlled by man by rounding-up and shooting. Some carcasses are exported for pet food, and for human consumption in Japan, in underdeveloped countries in South East Asia (especially Singapore and Malaysia), in the Middle East, and in parts of southern Europe and the West Indies (especially Trinidad). Those close to sheep- and cattle-stations are sometimes shot for home consumption and for trophy heads.

Feral goats in most parts of Australia probably reached their zenith between the 1920s and early 1940s, when they were absent only from arid desert areas and in the rainforests and swamplands of the monsoonal north. Since the end of the Second World War their population and distribution have generally declined. In eastern Australia between 1966 and 1976 the population decreased still further, due largely to government-sponsored extermination campaigns. Goats are, none the less, more widely established in Australia than any other species of feral livestock. The densest concentrations are in the goldfields of Western Australia, the western division of New South Wales and adjoining portions of South Australia, and along the Great Dividing Range in New South Wales, Victoria and southern Queensland.

Table 18 gives total numbers for feral goats in Australia. Rolls (1969) records that in about 1870 Mr Arthur Blackwood of Shark Bay acquired 50 goats from the Victoria Acclimatization Society,

Table 18 Distribution of feral goats in Australia

State	*Estimated population*
Western Australia	100,000–200,000
New South Wales	40,000–80,000
Queensland	20,000–50,000
South Australia	15,000–40,000
Northern Territory	2,000–4,000
Victoria	1,000–2,000
Total	178,000–376,000

Source: From McKnight (1976)

and it seems probable that these were the first goats to be seen in Western Australia; within 25 years there were some 4,500 in the state. Long (1972) records that by about 1903 domestic goats were established in small numbers throughout the settled regions of Western Australia, and that large herds were abandoned when the market for mohair failed to achieve its early promise. By 1911 goats were established in every township and around most homesteads and telegraph stations north of Port Hedland: herds were also reported along the De Grey River, around Carnot and Beagle bays, and at Derby. In the upper Gascoyne district goats were officially declared vermin in 1928, at Marble Bar and Port Hedland in the following year, at Mullewa and Meekatharra in 1954, on Bernier and Dorre islands in Shark Bay in 1958, and in the Recherche archipelago in the following year.

By the early 1970s feral goats were established in the Kimberleys, North West, Pilbara and Murchison areas and in parts of the south-west of the state, and on Bernier Island and in the Recherche archipelago. In the East and West Kimberleys the combined population was estimated to be around 10,000 in herds up to 300 strong. In the North West, Pilbara and Murchison areas goats were widely distributed in suitably rocky terrain from the De Grey River in the north via Meekatharra and Wiluna to Menzies in the Eastern Goldfields, throughout the Gascoyne and Murchison areas and at least as far south as Geraldton and Lake Moore. The densest populations were to be found in the North West and Murchison areas where herds of several hundred were not uncommon. In the south-west small colonies were reported from Jarrahdale, Donnybrook and between 40 and 50 in the Recherche archipelago.

By the late 1970s the major concentrations

were around Shark Bay (especially on Bernier, Dirk Hartog and Faure islands and at Carnarvon) and in the Murchison, Gascoyne and Yalgoo districts. In Western Australia – in contrast to the rest of the country – goats are at present said to be increasing (especially in the north-west) in spite of a growing interest in shooting and mustering for the export market. On Bernier overbrowsing and overgrazing by goats has severely damaged the vegetation and thus the habitat of the endangered banded hare-wallaby (*Lagostrophus fasciatus*), the western hare-wallaby (*Lagorchestes hirsutus*), Lesueur's rat-kangaroo or boodie (*Bettongia lesueurii*), the little marl or barred bandicoot (*Perameles bougainvilleii*), the ashy-grey mouse (*Pseudomys* (*Geomys*) *albocinereus*) and the Shark Bay mouse (*P.* (*Thetamys*) *praeconis*).

In some parts of New South Wales the goat population has fallen by up to 90 per cent from that of the 1950s and 1960s, although considerable numbers remain in some places in the west of the state: in other parts of the country the pattern varies, but there appears to be a generally downward trend.

In spite of many opportunities goats have seldom succeeded in establishing feral colonies on the mainland of the Northern Territory.[16] By about 1912 many stations in the territory supported domestic herds up to 300 strong, from which some wandered off from time to time into the bush where they were mostly killed by dingoes. A herd of 50 deliberately released near the Mary River in 1956 had all disappeared by 1964: since then the same fate has befallen small herds at Port Stuart Station, near Alice Springs and near Hart's Range.

In the 1950s an English missionary, Mr F. Gray, transferred some goats from Umbakumba Mission on Groote Eylandt to the 4-km^2 North East Island to provide food for the Aboriginals. Following a population explosion in the early 1970s, in 1974 the Umbakumba Council and the Bureau of Meteorology sought assistance in culling the goats to reduce their devastation of the vegetation and the undermining of the automatic weather recording station: in that year wildlife officers shot 363 goats on North East Island, but the 20 which escaped had multiplied to 50 by 1979.

Feral goats (descended from a domestic herd) are also an increasing problem on Vanderlin Island in the Sir Edward Pellew group in the Gulf of Carpentaria; the 1967 population of 900 had increased to more than 1,000 by 1976 and to double that figure by 1979. On both North East and Vanderlin islands feral goats, in addition to their environmental depredations, are potential carriers of such exotic diseases as screw fly worm. On neither island do natural predators occur.

In the shire of Avoca in the west of Victoria in 1869 there were said to be some 60,000 feral goats; a century later a few still remained in this region, with others at Bacchus Marsh (50 km from Melbourne), Kulkyne, upstream of Mildura, where the Mallee fronts onto the Murray River. Large numbers still occur in parts of Queensland and South Australia. They are also found on the following islands: Woody, Queensland; Moreton, Queensland (where they were released by the Royal Navy in the 1860s and now number about 1,000); Kangaroo, Althorpe and Gambier (all off South Australia); Tasmania; and possibly on Lord Howe and Norfolk.

McKnight considered that the future of goats in Australia depends largely on the market for their meat and that shooting with the aim of at least local eradication will play an important role, especially in Western Australia. Due to the inaccessibility of much of their habitat and their high degree of replacement in good breeding seasons, general elimination seems unlikely. McKnight believed that, although an overall decline can be expected, there are likely to be minor population explosions in some districts, until numbers again reach commercially harvestable figures.

New Zealand[17]

Captain James Cook introduced the first goats, which he had obtained at the Cape of Good Hope, to New Zealand on his second voyage, as he records in his *Journal*:

> On 2nd June, 1773, I sent on shore, on the East side of the [Queen Charlotte] Sound, two goats, male and female. The former was something more than a year old, but the latter was much older. She had two fine kids some time before we arrived in Dusky Bay, which were killed by cold.[18]

George Forster, the naturalist on Cook's second expedition, says in his *Journal* that the goats were landed by Captain Tobias Furneaux of the *Ad-*

venture in an unfrequented part of East Bay, 'this place being fixed on in hopes that they would there remain unmolested by the natives, who indeed were the only enemies they had to fear'.[19]

On his third voyage, during which *Resolution* was in Queen Charlotte Sound from 12–25 February 1777, Cook records that:

> I gave Matahouah [probably near Bald Hill, East Bay, on Arapawa Island] two goats, a male and a female with kid (and to Tomatongeauooranuc two pigs, a boar and a sow). They made me a promise not to kill them; though I must own I put no great faith in this. The animals which Captain Furneaux sent on shore here, and which soon after fell into the hands of the natives, I was now told were all dead.[20]

Most of the feral goats in New Zealand today – especially in the north of South Island – are in all probability at least partially the descendants of Cook's introductions, augmented by further subsequent importations. Goats were among the livestock landed by the early settlers in Tautuku Bay east of Invercargill in 1843, and Ernest Dieffenbach in *Travels in New Zealand* (published in the same year) wrote that around Port Underwood 'goats thrive better than cattle, to the introduction of which the almost total want of grass is a most serious objection'.[21] J. C. Crawford also refers to the presence of goats at Port Underwood in *Recollections of travel in New Zealand and Australia* (1880). F. E. Gibbs told Thomson (1922) that tethered goats on the Goat Hills in the Maitai Valley escaped into the back country in the 1850s, and many more escaped from or were abandoned by prospectors, goldminers and road and railway constructors when their camps were deserted.

In 1867 three Cashmere goats and a pair of Angoras were imported from Australia to New Zealand by the Canterbury Acclimatization Society; in 1876 a local newspaper reported that 'a flock of 120 Angora goats on the Port Hills (Lyttleton), chiefly descended from two pair introduced into New Zealand by the Melbourne Acclimatization Society, has recently been dispersed and sold'. Also in 1867 the Otago Acclimatization Society imported four Angora goats (again presumably from Australia) which they released, although exactly where is unknown. Two years later the Auckland Acclimatization Society imported a number which they are believed to have liberated near Hawke's Bay.

The Department of Agriculture's annual report for 1903, quoted by Thomson (1922), said that 'The original flock [of Angoras] imported from Victoria and South Australia has now assumed considerable proportions, partly though the natural increase and the purchase of nineteen grade nannies from Mr Taylor White of Wimbledon, Hawke's Bay . . . '. Many more Angora goats were subsequently imported by private individuals from whom, inevitably, some escaped, which led Aston (1911) to say that 'I hear the Angora is hybridizing with the common goat in some parts of Marlborough'.

Mr L. Hotop of Queenstown told Thomson that in the Lake Wakatipu region in 1916:

> There is an immense number spread all over the Lakes district – a moderate estimate gives them as many as 30,000 . . . they are principally at Moonlight, Skippers, Sandhills, and at the lower end of the Lake, seriously interfering with the pasturage in these localities. One runholder has paid year after year for as many as a thousand during the season.

By the early 1920s Thomson found that feral goats were common in many parts of Marlborough, in rocky and precipitous country around Palliser Bay east of Wellington, on Kapiti Island and on Mount Egmont, 'where they are doing much damage'. They were less abundant in the sparse scrub on the cliffs of the west coast, north and south of Hokianga Harbour in Northland. In South Island Thomson found goats still surviving in reduced numbers on Galloway Station in central Otago, and larger numbers near Kaikoura in the extreme south-west.

In *The game animals of New Zealand* Donne (1924) wrote that: 'Goat shooting is now obtainable in many localities including Cape Kidnappers, Orongorongo, the Tararua Ranges, Picton, West Coast and in some of the high mountains of the Cold Lakes districts, in fact their distribution is fairly general throughout the Dominion.'

According to Wodzicki (1950) many goats were deliberately released to check the spread of blackberry (*Rubus* spp.) and other alien weeds, and to help in clearing scrub, bush, bracken, fern, gorse and manuka for cattle and sheep; he considered that the worst infestations, such as at Taranaki, Marlborough and Hawke's Bay, were derived from this source. Goats were also often released on mutton bird islets to destroy punui (*Stilbocarpa lyallii*) and thus improve the habitat for mutton birds which were an important source of food. Their relatively sedentary nature and high fecundity resulted in the creation of big herds, little large-scale expansion of range and the early exhaustion of forage. As an example of the rapidity with which a population of goats can increase Wodzicki cites the case of 80–100 freed by the Native Department in 1943 on native lands near Mauriceville in the Wairarapa to control the encroaching blackberry. In the short space of a few weeks they had invaded the neighbouring State Forest Reserve where they speedily multiplied and spread; three years later Wodzicki found the entire forest infested with herds of between 30 and 60, in spite of the fact that many had already been shot.

By the late 1940s Wodzicki reported that, although feral goats ranged from the extreme north of Northland on North Island southwards to Otago on South Island and Stewart Island, they were mainly confined to a number of relatively localized regions in some of which, however, such as Marlborough and Otago, the population had expanded over a sizeable area. Wodzicki's enquiries revealed that most populations remained close to their original source; that wide rivers prove effective barriers to spread; that there is little evidence of soil preference; and that although in general goats favour drier mountainous areas such as in Otago and Marlborough and on the east coast of North Island, they are sufficiently adaptable to survive in the damp of the west coast, around Mount Egmont, in southern Westland, in the Tararua ranges and in Taranaki.

The havoc caused by feral goats to the New Zealand ecosystem has been considerable. As early as 1895 Kirk reported the extermination of tainui (*Pomaderris apetala*) at Kawhia. Aston (1911), who wrote that 'in the Marlborough Sounds goats are fast killing out the species of *Panax*, which are completely barked up to 4 feet or 5 feet from the ground', also reported that the shrub *Senecio greyii* was virtually confined to Palliser Bay. In 1935 G. F. Yerex reported[22] that in the Egmont National Park goats had killed, by barking, *Nothopanax colensoi*, *Coprosma grandifolia*, *C. tenuifolia*, and *Pseudopanax crassifolium*; goats also ate, Yerex continued, 'fern fronds, grasses, supple-jack, vine tips, the leaves and shoots of a very wide variety of trees and shrubs – seedlings of such major species as rimu, miro, kamahi, totara, etc.', some of which were also frayed[23] by the animals' horns. Goats also overgrazed snow tussock and other alpine vegetation. Zotov, who examined the flora on Mount Matthews, 1,000 m up in the Tararuas, in 1933 and 1947, reported that goats:

> appear to eat everything within reach, which is in general any shrub or plant up to four feet high. The windward places, which are not necessarily 'sunny' faces are most thoroughly denuded. The only indication that any undergrowth ever grew is the presence of fairly numerous seedlings an inch or so high or at most young plants cropped to that height. The consolidation of soil is everywhere in evidence. There is a wholesale destruction of bryophytes (mosses and liverworts) as well as all other herbaceous cover. The growth of the numerous extensive slips on the mountain is probably connected with the increased run-off associated with this phenomenon.[24]

Among the preferred food plants of goats in New Zealand are *Nothopanax colensoi*, wineberry (*Aristotelia serrata*), kamahi (*Weinmannia racemosa*), *Olearia arborescens*, putaputaweta (*Carpodetus serratus*) and *Melicytus lanceolatus*; also favoured are the leaves, fruit or seeds of haumakoroa (*Nothopanax simplex*), karamu (*Coprosma foetidissima*), raurekau (*C. grandifolia*), *C. tenuifolia*, and broadleaf (*Griselinia littoralis*), and to a lesser extent native broom (*Carmichhaelia* sp.), a lichen (*Sticta* sp.), *Asplenium flaccidum*, and the fern *Phormium colensoi*.

Goats have also been liberated on a number of New Zealand offshore islands. Table 19 gives brief details of the various recorded introductions. On the Auckland Islands goats were released, as potential food for sealers, whalers and shipwrecked sailors, on at least 10 occasions between 1850 and about 1895. In the former year Captain Greig of the SS *Southland* put a number ashore, and more were landed by the Victoria Acclimatization Society in 1865; in the same year Captain Norman released four at Terror Cove which soon, however, disappeared, and more were landed by the captain of the *Hinemoa* before 1890. According to Cockayne (1903): 'Two or three were landed on Ewing Island in 1895, but none have been seen recently. On Ocean Island, a very small island in the Auckland group, goats are numerous at the present time, but I have no details as to how they got there.' Thomson was informed by Captain Bollons that:

Table 19 Introductions of goats to New Zealand offshore islands

Island	*Date*	*History and current status*
Arapawa	1777 (by Capt. Cook); also subsequent introductions during the 1830s	Between about 100 and 200 in the north (see also text)
Kermadecs		
Macauley Island	1836	3,200 shot out in 1966–7 (see also text)
Raoul Island	Before 1837; 1842 (from Samoa)	Periodically culled; serious control began in 1972 (see also text)
Auckland (and Ewing and Ocean islands)	1850 to *c.*1895 (from Australia and the New Zealand mainland); 1915 (from Raoul)	Only a small population remains
Adams	1885	Died out by 1888
Enderby	1850 (by Capt. Enderby); 1865 (by Capt. Norman)	Died out by 1889
Kapiti	Before 1854; some later transferred to Cape Palliser	Eliminated by 1928
Campbell	1868 (5 from SS *Southland*); 1888; 1890 (by Capt. Fairchild)	Died out
Antipodes	1886; 1888 (3 from the *Stella*); more until 1903	Died out
Snares	1889 (or 1900) (by Capt. Bollons); 1890 (by Capt. Norman)	Died out
Three Kings Group (Great Island)	*c.*1889–90	Eradicated by Department of Internal Affairs in 1946 (see also text)
Cuvier	Late 1890s	Exterminated between 1951 and 1961 (see also text)
Mokohinau (Burgess Island)	?	Eradicated by Department of Internal Affairs in 1973
Great Barrier	?	Eradication by NZ Forest Service began in 1976
Cavalli	?	Exterminated in 1971
Great Mercury	?	Present in 1976
Whale	?	?
East	?	Exterminated by Maoris in 1968
D'Urville	?	Abundant in 1976
Forsyth	?	?
Herekopare	By mutton-birders	Present in 1976
Mangere	?	Extinct
South-east	?	Present between 1900 and 1915; all killed 1914–16
Chatham	?	Still present

Source: After Whitaker and Rudge (1976)

Goats have been sent down from time to time to the Auckland Islands since 1890, most of which have either died or been killed off for food by castaways. At the Snares they were liberated about 1889 but soon died off. At Campbell Island some were landed in 1888 and 1890, and several were alive when the main island was taken up for a sheep run in 1896. At the Antipodes several were liberated between 1886 and 1900, but were either used for food by castaways or died off.

Great Island in the Three Kings group was occupied and cultivated by the Maoris before their departure in about 1840. Forest growth soon reclaimed the cultivated land and forest succession progressed towards a climax type. When T. F. Cheeseman[25] visited Great Island in 1887 (before goats were introduced) he found 143 species of plants on the island. In 1934 – 35 years after goats had been released – W. R. B. Oliver[26] could find only 70 remaining. E. G. Turbott[27] in 1946 found that the earlier manuka (*Leptospermum scoparium*) dominant association had been replaced by an unpalatable kanuka (*L. ericoides*), ngaio (*Myoporum laetum*) and *Carex* community. Although mature trees such as pohutukawa (*Metrosideros tormentosa*), wharangi (*Melicope ternata*), mahoe (*Melicytus ramiflorus*), mangeao (*Litsaea calicaris*), the Three Kings' milk tree (*Paratrophis smithii*), puriri (*Vitex lucens*), tawapou (*Sideroxylon costatum*) were all blossoming and fruiting, browsing by goats had prevented the establishment of seedlings. Species such as wharangi and the bird-catching tree (*Pisonia brunoniana*) were frequently ring-barked.

In 1946–7 the entire population of 393 goats on Great Island was exterminated, and a repetition of the earlier natural succession of a luxuriant and floristically rich plant community developed, with green shoots of the cabbage tree (*Cordyline* sp.), low branches of the Three Kings' milk tree and seedlings of tawapou, wharangi and mangeao appearing. Turbott considered that:

> had the considerable goat population been allowed to remain on the island, it would seem inevitable that all vegetation except *Leptospermum ericoides/Carex* shrubland would have disappeared. Although most of Great Island is still clothed in the goat-induced *Leptospermum* community, there is an unmistakable barren aspect on all areas to which goats can penetrate, contrasting forcibly with the deep green on South-West Island.

By 1968 a mixed forest had been re-established, with endemic spcies re-invading from adjacent islands and a new fauna exploiting the secondary forest.[28]

Before their extermination on Cuvier Island in the Hauraki Gulf, feral goats had, in little over half a century, eradicated several species of flora, altered the forest succession, and destroyed the entire coastal scrub community.

On the Aucklands goats have been accused in the past of endangering several species of rare endemic plants, but the relict population on the northwestern side of Port Ross on the main island (near where the first release took place and which is one of the driest and mildest parts of the island) is no longer regarded as a threat. Most are uniformly white or cream, some with greyish patches on the neck and shoulders; the remainder are various combinations of white, brown and black, or are entirely black. Their coat is short, and both sexes are horned and bearded. Due to the peaty substrata their hooves tend to become overgrown and split, causing them to move awkwardly.[29]

On Raoul and Macauley islands in the Kermadecs Captain W. B. Rhodes, who visited the archipelago in 1837, was told by a settler, a Mr Reid, that feral goats (and pigs) were well established; the former were later reinforced by Reid's partner, a Mr Baker, with importations from Samoa in 1842. Since the early years of the century Raoul had been an important base for whalers, who most probably introduced goats as a source of meat and milk.

The goats on 7,300-ha Arapawa Island in Marlborough Sounds are in many respects the most interesting in New Zealand. There is a conflict of opinion as to whether they are entirely or partially descended from Cook's introduction in 1777, or whether they originate from those imported from New South Wales by whalers as a source of food in the 1830s. What is generally conceded is that in appearance they resemble the extinct 'Old English' goat from which they may well be descended; if this is so, they are probably the last representatives of the breed in the wild today, and are a valuable gene bank for future breeding purposes. Most are predominantly brown, with lesser numbers black or white or piebald, and most have dorsal-stripes and/or face-stripes. Their ears are erect and there are no throat tassels. The horns of the billies, which weigh about 40 kg, are long, divergent and twisted. Unfortunately the goats on Arapawa live in a designated forest reserve, and many have been culled by the New Zealand Forest Service.

Feral goats in New Zealand are in general a mixture of Angoras, 'improved' dairy breeds such as Alpine, Toggenberg and Saanen, and 'unimproved' or 'scrub' animals; they are widely distributed, but are mainly confined to the lower parts of forested ranges and rocky outcrops, where they browse on the ecotone dividing the open high tops from the bush line, where the ecological balance is most precarious and the soil most prone to erosion. In South Island they are principally confined to the area north of Lake Wakatipu; between Westport and Reefton; and from north-west of the Clarence River to Nelson and the Queen Charlotte and Pelorous Sounds area. In North Island they occur mainly north of Wellington and Cape Palliser, in Taranaki and around Hawke's Bay, west of the Motu River, and on the Coromandel Peninsula.

Since official control operations were commenced in 1937 well over half a million goats have been shot by government rangers and a further 750,000 by sportsmen and private landowners.

OCEANIA

Hawaii[30]

The earliest goats in Hawaii were introduced during Captain James Cook's third voyage when, on 1 or 2 February 1778, he landed three on the island of Niihau.

> On the return of the boat, I went myself with the pinnace and launch up to the point . . . taking with me a ram-goat and two ewes . . . being very desirous of benefiting these poor people, by furnishing them with some additional articles of food.[31]

Although within a year the two 'ewes' had given birth to three kids, all were shortly thereafter killed in a dispute over their ownership between the respective chiefs of Niihau and Kauai.[32] Tomich (1969) believes that Cook probably left more goats in Hawaii during the winter of 1778–9 shortly before his assassination in Kealakekua Bay on 14 February.

One of Cook's midshipmen, George Vancouver, later returned to Hawaii with his own command, where in 1793 he presented chief Kahomotoo of Kawaihae on Hawaii with a pair of goats (and also landed another pair on Kauai) and noted that another chief, Tianna of Kealakekua, already possessed several.[33] A year later Vancouver landed a number of goats on Maui and noted that Kahomotoo 'had taken the greatest care of the goats I had presented him with on a former occasion . . .',[34] having placed them under the same royal taboo which applied to sheep and cattle.[35]

It thus seems reasonable to assume that by the mid-1790s goats were well established as domestic livestock in Hawaii; Andrew Bloxam, naturalist on the *Blonde* during her voyage to the Sandwich Islands in 1824–5, noted the presence of goats on Maui and remarked that on Oahu 'cattle, goats and horses were very plentiful'. By the middle of the century goats were widely distributed throughout the archipelago; between 1844 and 1854 a total of 245,862 skins were exported to North America at an average price of 15 cents each.

Exactly when goats first escaped into the wild in Hawaii is uncertain: they are known to have been feral on Kahoolawe before 1863, and some four years later King Umi's *heiau* (shrine) on the saddle between Hualalai and Mauna Loa was being used as a trap and holding-yard.

Goats (and sheep) were introduced to Lanai after 1861, where the former severely browsed the native trees which by 1870 were said to have been reduced to 'a motheaten remnant'.

In the 1890s Angora goats were introduced to the islands for the production of mohair for making lariats and saddle-girths for Mexican cowhands imported to round up feral cattle.

Between 1844 and 1900 no fewer than 1,581,000 goat skins were exported from Hawaii – most to North America. By 1905 it was estimated that only some 500 of the 10,000 goats on Oahu were domesticated, and that on Molokai (where a few years previously there had been 10,000) there were now only about 2,000. Three years later there were judged to be some 10,000 on Lanai and ten times that number on Hawaii: by the end of the decade feral goats were becoming an increasing problem to foresters and farmers, especially on Hawaii and Maui, where they were controlled mainly by shooting. By 1930 only 75,000 remained in Hawaii where, except on Big Island, they were said to be 'well under control'.[36] There were believed to be few on Lanai and none on Niihau and Kahoolawe.[37]

By the early 1980s feral goats (most of which are black, black and brown or brown, some with white flank patches and are known locally as *kao* or *kunana*) were present on the islands of Hawaii, Maui, Molokai, Lanai (especially in the east), Oahu, Kauai (especially on the north-west coast) and Kahoolawe. About 100 were established in the more rugged parts of the Koolau and Waianae ranges on Oahu; on Kauai there were 2,800, most of which were in the Waimea canyon or on the Na Pali coast; the majority of the 800 on Molokai occurred on the cliffs between Halawa and Wailau Valley; in 1963 it was estimated that there were at least 600 in or around the Haleakala National Park on Maui, where 20 years later an undetermined number remained; and there were between 200 and 300 on Kahoolawe.

Where feral goats exist in coastal lowland scrub, transition forests and mountain parkland formations, much damage has occurred. In some places the endemic mamani trees (*Sophora chrysophylla*) have long since disappeared. As recently as 1979 goats on the slopes of Mauna Kea volcano on Hawaii were found to be destroying the mamani and naio (*Myoporum sandwichensis*) trees which constituted the last known habitat of the Hawaiian honeycreeper, the palila (*Loxioides bailleui*); after a petition from the Sierra Club and the National Audubon Society, who accused the Hawaiian Department of Land and Natural Resources of maintaining the goats for sporting purposes, a federal district judge ordered their removal: nevertheless a few years later – due to their high replacement rate – more than 5,000 remained in the Hawaii Volcanoes National Park, where, however, there had been an estimated 14,000 in 1971.

Apart from man himself feral goats have been the most destructive animal in the ecosystem of Hawaii. They range from sea-level up to 3,000 m and occur most frequently in semi-arid regions on sea cliffs, in deep gulches and on steep mountain slopes. By overgrazing, overbrowsing and trampling with their sharp hooves goats have in many places eliminated a large number of plants, and thus ground cover for such endangered birds as the Hawaiian goose or nene (*Branta sandvicensis*), and have also caused accelerated soil erosion and landslides. Through the dissemination of undigested seeds in their faeces goats have helped in the spread of such exotic plants as blackberry.

At the present time goats in Hawaii – though still a serious pest – are not out of control. Although on Molokai, where in recent years they have caused a severe decline in much of the native vegetation, they are strictly controlled by shooting, on Kahoolawe they are left virtually undisturbed. On Maui, except perhaps in the Haleakala National Park where they are controlled, they will probably continue to survive. In the Volcanoes National Park on Hawaii it appears likely that goats will exist at least in the foreseeable future.

Galápagos Islands

Several authors have claimed that goats were first released on Isla Santiago (James) in the seventeenth century by English buccaneers who careened and reprovisioned their ships at a semi-permanent base which they set up at James Bay, and that in 1685 the Viceroy of Peru introduced dogs to the island in an attempt to control them.[38] There is, however, no firm evidence for introductions in the seventeenth (or indeed eighteenth) century of either animal, and in view of the abundance of easily obtainable food in the form of tortoises, turtles and ground-nesting birds and their eggs, the introduction of a new food species seems highly unlikely. When Captain Woodes Rogers, commanding a small British squadron during the War of Spanish Succession, visited the Galápagos in 1709, he made no mention of goats – or indeed of any other mammals – in his zoological observations.

The first goats in the Galápagos – on Santiago – came from the USS *Essex* under the command of Captain David Porter who, while harrying the British whaling fleet, spent several months in the islands in 1813. During this time he sent out several shore-parties in search of fresh water, and from his own and their observations compiled copious notes on the wildlife of the archipelago; in these, however, he made no mention of introduced animals other than black rats which infested his ship. Slevin (1959) recounts that:

> While here, [James Bay on Santiago] Porter landed [on 4 August 1813] four goats [a billy and three nannies] and some sheep from the *Essex*, and, as he states, they being so tame, left them without a keeper, carrying water ashore for them each morning. One morning, however, they disappeared and a searching party failed to locate them, so he concluded they had found some fresh water and would remain inland. These animals were afterwards seen by the crew of H.M.S. *Tagus* on July 30, 1814.[39]

Bewailing the loss of his goats, Porter noted sadly that 'future navigators may perhaps obtain an abundant supply of goats' meat'.

The earliest deliberate importation of goats to the Galápagos was probably made by the first settlers from Ecuador who, under the command of General José Villamil, established a colony on Floreana (Charles) in 1832. Temporary bases were later set up on various islands – the first permanent ones being founded on San Cristóbal (Chatham) by Señor Mañuel Cobos in 1869 and on southern Isabela (Albemarle) in about 1893 by Señor Antonio Gil,[40] to both of which goats were most probably imported.

O. Salvin, who visited the archipelago in July 1875 on board HMS *Petrel*, wrote (1877):

> Pigs and goats do not seem to have been turned out in the Galápagos so soon after their discovery as was usually the case with other islands. . . . Mr Darwin says that on Charles Island there were many wild pigs and goats. This was in 1835. In 1813 Captain Porter relates how some goats belonging to his ship, which had been tethered on shore in James Island, got loose and were not recovered. . . . of animals introduced, it was estimated that [in 1875] there existed . . . of goats, 300 or 400 on Charles, a few on Chatham, but none elsewhere.

This century goats were introduced to Santiago from Baltra (South Seymour) by Rollo H. Beck of the California Academy of Sciences' expedition in 1906; to Santa Cruz (Indefatigable) from Santa Fé (Barrington), Baltra and Santiago by early settlers for milking sometime after 1925; and to Pinta (Abingdon) – a billy and two nannies – by local fishermen as a future source of food in 1957. Since Beck went to the trouble of shipping goats from Baltra to Santiago in 1906 and as Darwin apparently saw them only on Floreana in 1835, it seems probable that the four which escaped from Porter on Santiago in 1813 did not succeed in establishing a permanent population.

Table 20[41] lists which islands have supported populations of feral goats from 1813 to 1982. Feral goats have been established on two other islands in the Galápagos – Rábida (Jervis) south of Santiago and on Plaza Sur off the east coast of Santa Cruz – on both of which they were exterminated before 1974.

Two sharply conflicting views have been expressed about the feral goats of the Galápagos Islands: 'they could prove of considerable economic value for meat, because they thrive in the arid zone that otherwise produces no commercial products';[42] and 'the goat is probably the greatest single threat today to the Galápagos flora and fauna.[43] In the light of what follows the fallacy of the former opinion and the truth of the latter will become readily apparent.

On Santa Fé overbrowsing by goats virtually eliminated the shrub layer and the vegetation was reduced to an open quasi-parkland. Some species, such as *Scalesia helleri*, survived only on the steepest cliff-sides which were inaccessible even to goats,[44] and the endemic Galápagos tomato and wild cotton completely disappeared.

After an intensive shooting campaign by the National Park Service in the late 1960s goats were finally eliminated on Santa Fé towards the end of 1971. Thereafter the regeneration of such woody plants as *S. helleri*, *Waltheria ovata*, *Cordia lutea* and *Lantana pendicularis* notably increased, though some (e.g. *Ipomoea habeliana*) had apparently been eradicated. Loss of cover due to overgrazing and browsing by goats is said to have been respon-

Table 20 Islands in the Galápagos group that have supported populations of feral goats from 1813 to 1982

	1813	*1835*	*1875*	*1905–6*	*1957*	*1966*	*1971*	*1974*	*1977*	*1982*
Floreana	–	×	×	×	×	×	×	×	×	×
San Cristóbal	–	–	×	×	×	×	×	×	×	×
Santa Cruz	–	–	–	–	×	×	×	×	×	×
Isabela	–	–	–	–	×	×	×	×	×	×
Española	–	–	–	×	×	×	×	×	×	–
Santa Fé	–	–	–	×	×	×	–	–	–	–
Baltra	–	–	–	×	×	×	×	×	×	–
Santiago	×	?	?	×	×	×	×	×	×	×
Marchena	–	–	–	–	–	–	×	×	×	–
Pinta	–	–	–	–	×	×	×	×	×	×

sible for the decline on Santa Fé of the medium ground-finch (*Geospiza fortis*), the warbler-finch (*Certhidea olivacea*), the large tree-finch (*Camarhynchus psittacula*) and the small tree-finch (*C. parvulus*), and to have exposed the hatchlings of the Santa Fé land iguana (*Conolophus pallidus*), which occurs on no other island, to predation by Galápagos hawks (*Buteo galapagoensis*): now that the goats have been removed, the land iguana – with which goats competed for food such as fallen *Opuntia* cactus pads – is making a satisfactory recovery.

On Pinta, where three goats were put ashore in 1957, the population had multiplied within a decade to between 3,000 and 5,000, and on the lower slopes four endemic shrubs – *Alternanthera flavicoma* and *A. nudicaulis, Gossypium barbadense* and *Ipomoea habeliana* – had been exterminated, and five species – one an endemic – had been greatly reduced in numbers. On the densely populated southern shores six trees and shrubs – four of which are endemic – could no longer be found, though they still occurred in the centre and north of the island, where the number of goats was significantly less.[45] At least 44 of the archipelago's 250 or so endemic plants grow on Pinta; many have been seriously depleted by the grazing and browsing of goats – the only feral mammals on the island – which are also believed to have been responsible for the extinction on the island in the early 1960s of the endemic giant tortoise (*Testudo (Geochelone) elephantopus*). By 1970–1, when goats had spread all over the island (even into the humid zone where they eliminated much of the dense understorey) and had destroyed most of the low vegetation in the arid and transitional zones, the population on Pinta was variously estimated to be between 5,000 and 10,000[46] or not less than 20,000.[47] Between 1971 and 1980 a total of 40,000 were destroyed by the National Park Service and by the latter year only around 200 remained, most of which were concentrated on the island's eastern slopes and where the west coast's steep *barrancos* plunge up to 500 m to the sea. As a result the arid lowlands began to regenerate and the density and number of woody plants increased.

Feral goats were apparently well established on Española at the time of the California Academy of Sciences' visit in 1905–6, when seven plants recorded by Snodgrass and Heller towards the end of the nineteenth century could no longer be found.[48] Before their eradication on Española goats were responsible for the elimination of at least 10 plant species – *Alternanthera radicata, A. echinocephala, Croton scouleri, Gossypium barbadense, Acanthospermum lecocarpoides, Amaranthus sclerantthoides, Oxalis cornelli, Kallstroemia adscendens, Pectis hookeri* and *Euphorbia punctata* and 11 more have been severely depleted;[49] the plants thus affected represent 77 per cent of the island's species, many of which, however, are still to be found on the goat-free offlying Gardner Islet in Gardner Bay. This progressive impoverishment by goats of the vegetation on Española is believed to have been the principal cause of the decline of the island's endemic giant tortoise.

On Santa Cruz, where they have spread dramatically since their introduction in the late 1920s and are especially common at Academy Bay between the Darwin Station and Puerto Nuñez, feral goats are said to compete with giant tortoises for the fallen pads of *Opuntia* tree cacti and, as on Santa Fé, to have largely restricted *Scalesia helleri* to inaccessible cliffside sites.

On Isabela it was discovered in 1979 that goats had managed to cross the hitherto seemingly impassable lava of the Perry Isthmus, which divides the island in two; this could pose a threat to the fauna and flora of the previously goat-free northern part of the island.

A very rare succulent, *Calandrinia galapagosa*, which occurs only in the Sappho Cove region of San Cristóbal, was collected by the California Academy of Sciences' expedition in 1906 and seen by U. Eliasson in 1967.[50] The 20 or so surviving specimens seen by Eliasson were all sterile and severely damaged by goats.

A second endemic succulent, *Portulaca howellii*, occurs on the islets of Bartolomé, Rábida, Daphne Major, Eden, North Seymour, Plaza, Brattle, Onslow, Champion, Enderby, Caldwell and Gardner-by-Charles; it was not found by Eliasson on any of the goat-infested larger islands – Santiago, Santa Cruz, Isabela and Floreana – off which these islets lie. Two stunted specimens only were observed near the coast on Pinta, on which goats had been landed in 1957 and which at the time of Eliasson's visit a decade later had multiplied to between 3,000 and 5,000 individuals. For most of the year *howelli* exists only in the form of bare succulent stems, in which much of the plant's nutriment is stored, and which goats are believed to find palatable. If, as seems likely, *howelli* is now extinct on Pinta, it may only survive on some small islets which are uninfested by goats.

Since feral dogs were eliminated from Floreana goats have caused increasing damage to the vegetation and have been proving a nuisance – mainly by trampling on their burrows – to nesting dark-rumped petrels (*Pterodroma phaeopygia*).

Isla Santiago has the largest population of feral goats – estimated in 1975 to number between 80,000 and 100,000 individuals – in the Galápagos. The goats associate in tribes of between 2 and 10, containing either or both sexes. Kids are born throughout the year, the rate of twining being around 20 per cent. Most of the population is black, and 80 per cent is black, light brown (*café*), red, piebald or light brown and white; the remaining 20 per cent are grey, mid-brown (*marrón*), dark brown (*castano*) or any combination thereof. Billies have long, and nannies shorter, spirally spreading horns.[51]

Studies of vegetation on Santiago both inside and outside exclosures have shown that all canopy species and most shrubs are not regenerating successfully due to the browsing of seedlings by goats. Because of the goats' selective feeding a number of endemic species have virtually disappeared, surviving only in protected areas. Some annual herbs which are favoured by grazing have taken over and supplanted perennial species, thus changing the composition and dominance of plant communities. Of the 585 km^2 of Santiago some 25 per cent is lava with no vegetative cover, while a further previously vegetated 15 per cent has been reduced by goats to grassland or barren soil. Studies of exclosures on Santiago (and of permanent plots on Santa Cruz, Pinta and Santa Fé), however, have shown that several trees and perennial herbs are able to regenerate successfully when protected from browsing by goats.[52]

Eliasson (1968) reported that the stems of some *Opuntia* tree cacti on Santiago nearly 60 cm in diameter had been completely gnawed through by goats. Near the coast he observed that goats had browsed the branches of *Cryptocarpus pyriformis* as high up the bushes as they could reach, and were even seen clambering 2 m above the ground to browse on the higher branches of *Clerodendron*. Eliasson believed that the distribution of the endemic *Scalesia* was gradually contracting, due mainly to pressure from browsing goats: three *Scalesia* species are endemic to Santiago, and the survival of two is gravely threatened. Eliasson was unable to find *S. darwinii* in its previous location near James Bay, and saw only three small examples of *S. atractyloides* growing on a steep cliffside which was inaccessible even to goats: the third species, *S. stewartii*, Eliasson found on the goat-free islet of Bartolomé off the east coast of Santiago, and on a new lava field on the main island opposite the islet. A fourth species, *S. pedunculata* – a tree-like variety growing to a height of over 10 m and common on Santiago at the turn of the century at altitudes of around 650 m – Eliasson found only as scattered full-grown specimens, with no signs of regeneration. Other lowland species which have suffered include *Bursera graveolens, Eruthrina velutina, Cordia lutea, Croton scouleri* and *Acacia* spp. The destruction by goats (and pigs) of the principal highland climax tree, *Zanthoxylum fagara*, which supports epiphytic mosses, provides shade, and gathers the damp air of the *garua* (wet) season, thereby creating favourable conditions for a varied understorey of *Tournefortia* spp. and *Psychotria rufipes*, is forming dust bowls which accelerate soil erosion. Elsewhere, *Alternanthera echinocephala*, *Gossypium barbadense* and *Waltheria reticulata* have apparently disappeared, and seven more species – of which five are endemic – are seriously threatened.

Clearance of the highland forest on Santiago, followed by the invasion of weedy species normally confined to lowland areas, has affected the diet and distribution of seed-eating ground-finches (*Geospiza* spp.) and Galápagos doves (*Zenaida galapagoensis*). The destruction of leaf-litter and increased aridity of the soil as the result of decreasing protection by plants have adversely influenced those birds, such as the sharp-billed ground-finch (*G. difficilis*), which feed mainly on leaf-litter invertebrates and seeds. On the credit side, the clearance by goats of dense highland vegetation has made prey more readily available to the endangered and non-territorial endemic Galápagos hawk, which on Santiago and other islands also scavenges on goat carcasses.

In an attempt to eliminate feral goats in the Galápagos, the National Park Service, with advice from the Charles Darwin Research Station on Santa Cruz, has in recent years been conducting an intensive shooting campaign. At first, mainly nannies were killed, both to decrease numbers and also to inhibit juvenile recruitment, and by 1982 some 60,000 goats had been killed and the populations on several small and medium-sized islands had been eradicated. The removal of goats from the larger islands is likely to prove more difficult, but must be achieved if the long-term survival of the Galápagos endemic fauna and flora is to be secured.

Seychelles[53]

Between 450 and 600 goats are established on two of the four islands of the Aldabra group (politically part of the Seychelles) north of Madagascar (the Malagasy Republic). Their origin is uncertain; one theory is that they may have been landed by Arab corsairs in the seventeenth or eighteenth centuries, while another suggests that they were introduced by Europeans in about 1890.

On Île Malabar there are 200–250 mainly black goats, with reddish-brown stripes on their face, legs and belly. The 250–350 animals on Grande Terre show a much more variable coloration, among which is red-brown, black, red and white, black and white, black and white and red, and innumerable intermediate patterns and shades. The goats on Grande Terre are larger than those on Île Malabar; on the former billies weigh on average 36 kg and stand 78 cm high, while nannies weigh 32 kg and measure 71 cm; on the latter the figures are respectively 30 kg /72 cm and 26 kg /64 cm. Although the pelage of goats on Grande Terre (as on Ile Malabar) is basically short, some individuals grow a shaggy or wavy coat, particularly on the shoulders, back and hindquarters. In both populations all animals are horned. Kids are born mainly between June and August: on Grande Terre at least 20 per cent are twins; on Île Malabar all are singletons. Feral goats are also present on St Pierre in the nearby Providence group.

Mauritius

As early as 1512 the Portuguese navigator Pedro Mascarenhas released 'hogs, goats and fowls' on the island of Mauritius in the Mascarenes, where in the late seventeenth century the two mammals were a major factor in the extinction of the dodo. The introduction in 1844 of goats, which have now been shot out, (and rabbits) to Round Island, Mauritius, is frequently cited as a classic example of the ecological problems which alien species can cause to the biota of a small island.[54] The extermination of the goats is believed to be responsible for the increase in thickets and newly mature lantan palms, the establishment of some young bottle palms and increased screwpine regeneration observed by Bullock and North in 1982.

Fiji

Free-ranging goats are currently established on many of the smaller (especially uninhabited) islands in the Fijian archipelago, including Yaqaga, Devilau, Yadua, Yaduataba and Monuriki, and perhaps also on Aiwa, Wakaya and Makodroga.

According to Gibbons (1984), the indirect effect goats have on iguanas in Fiji can be as devastating as the direct one exerted by feral cats. By browsing the forest understorey (especially in times of drought and when they reach a high density) goats inhibit regeneration, and the removal of ground cover renders juvenile and hatchling iguanas susceptible to predation by kingfishers and hawks and increases soil erosion, while the burning of scrub to provide new grazing has destroyed valuable iguana forest habitat. In 1979 a species new to science, the Fijian crested iguana (*Brachylophus vitiensis*), was discovered by Dr Gibbons on the uninhabited 80-ha island of Yaduataba which, with some islands in the Mamanuca and Yasawa groups, appears to be its only home. Unfortunately, some years previously the villagers of Nakorolevu on a neighbouring island had released some goats on Yaduataba where they were rapidly destroying the vegetation where the crested iguana lives and feeds. On 7 August 1980 an agreement was signed creating Fiji's first nature reserve on the island, from which it is intended to remove all the goats. If this can be accomplished a harmless and interesting species will have been saved from untimely extinction.

St Helena

The island of St Helena in the South Atlantic provides a striking example of the devastation of an insular flora by a single grazing mammal. A. R. Wallace in *Island life* (1895) wrote:

When first discovered, in the year 1501, St Helena was densely covered with a luxuriant forest vegetation, the trees overhanging the seaward precipices and covering every part of the surface with an evergreen mantle. This indigenous vegetation has been almost wholly destroyed [and] the general aspect of the island is now barren and forbidding . . . when [the vegetation] was destroyed the heavy tropical rains soon washed away the soil and left a vast expanse of bare rock and sterile clay. This irreparable destruction was caused in the first place by goats, which were introduced by the Portuguese [Fernando Lopez] in 1513 and increased so rapidly that in 1588 they existed in thousands.

In 1709 the Governor of the island complained that the timber was rapidly disappearing, and asked that the goats be destroyed, to preserve the woods and because the island was suffering from droughts. The reply was 'the goats are not to be destroyed, being more valuable than ebony'. In 1809 the Governor reported the total destruction of the forests . . .

Today, small relict patches of native vegetation (which has been largely replaced by imported exotics) occur only on the highest ridges and in inaccessible ravines, and soil erosion continues to be a problem. Of the island's 33 endemic plants no fewer than 18 are officially classified as 'endangered'.

Other oceanic islands

At one time or another goats have become established on Île St Paul; on Île Amsterdam; on Île de la Possession; on Macquarie Island; on Pitcairn Island; on Fatuhiva Island in the Marquesas; on Rapa, Raivavae, Tubuai, Rurutu and possibly Rimatara in the Austral or Tubuai and Rapa islands; on Meetia (Mehetia) and Tahiti in the Society Islands; on Malden atoll in the Equatorial or Line Island; on Aguijan (south of Tinian) in the Marianas; on Walpole Island in New Caledonia; and possibly on Maron and Luf islands near Papua New Guinea, as well as on the mainland near Garaina in Morobe Province. On Ascension Island they died out by 1944.

Until their removal, excessive grazing and browsing by feral goats had reduced Nonsuch Island in Bermuda – the home of nesting seabirds – in the Castle Harbour Island National Park (designated in the 1960s) to a semi-desert.

On Tristan da Cunha goats were probably first landed by the Portuguese shortly after their discovery of the islands in 1506; in 1790 they were said to be numerous,[55] and although they were still present in 1829[56] they were shortly thereafter exterminated. Others which were subsequently introduced died out by 1951. Surprisingly, they seem to have caused little damage to the native vegetation which has, however, been devastated by herds of feral cattle.[57] On Inaccessible and Gough islands (where they were introduced respectively in 1882–3 and as recently as 1958) goats died out in the 1960s.

Goats were first introduced to the Falkland Islands in 1764 by the French explorer Louis Antoine de Bougainville, and later on a number of offlying islands by nineteenth-century sealers and whalers. By 1846 feral goats were established in the mountains of East Falkland, and in 1870 several hundred were exported from Grand Jason Island. There are no feral goats in the Falklands today.

NOTES

1. Stretches of moorland in northern England.
2. Scottish deer-drive.
3. For information on goats in the United States I am indebted to McKnight (1964).
4. McKnight (1964: 36–7).
5. *The California islands* (Academy Publishers, Los Angeles, 1958).
6. See pp. 237 and 117.
7. For information on goats on Guadalupe I am indebted to Mason (1979).
8. See p. 262.
9. Mason (1979).
10. For information on goats in Australia I am indebted to Rolls (1969), McKnight (1976) and Evans (1980).
11. *Fauna Bulletin, Western Australia Fisheries Department* No. 2 (1962).
12. See pp. 97 and 109.
13. See p. 50.
14. See p. 47.
15. See p. 40.
16. Letts, Bassingthwaighte & De Vos (1979).
17. For the history of goats in New Zealand I am indebted to Thomson (1922) and Wodzicki (1950).
18. *A voyage towards the South Pole and round the world, performed in H.M.S. 'Resolution' and 'Adventure' in the years 1772–5* (London, 1777).
19. *A voyage round the world in H.B.M.S. Resolution during the years 1772–5* (London, 1777).

20. *A voyage to the Pacific Ocean . . . in H.M.S. 'Resolution' and 'Discovery' in the years 1776–80* (London, 1784).
21. Vol. 1, p. 41.
22. Quoted by Wodzicki (1950).
23. Damage to tree bark by rubbing with antlers or horns.
24. Zotov, Elder, Beddie et al. (1939).
25. *Transactions of the New Zealand Institute* **20**: 141 (1888) and **23**: 408 (1891).
26. *Records of the Auckland Institute and Museum* **3**: 211–38 (1948).
27. Ibid., **3**: 253–72 (1948).
28. See also E. G. Turbott in Gressitt (1963: 485–98).
29. Rudge & Campbell (1977).
30. For the history of goats in Hawaii I am indebted to Tomich (1969) and Kramer (1971).
31. Cook, *A voyage to the Pacific Ocean* . . . Vol. 2, p. 217.
32. Ibid., Vol. 3, p. 95.
33. *A voyage of discovery to the North Pacific Ocean and round the world in 1790–5 under Captain George Vancouver* (London, 1798), Vol. 1, p. 156.
34. Ibid., Vol. 2, p. 115.
35. L. W. Bryan, *Paradise of the Pacific* **42**: 77 (1930); and see pp. 268 and 319.
36. Bryan, ibid.
37. Anon., *The Friend*, 193 (1930).
38. e.g. V. W. von Hagen, *Ecuador and the Galápagos Islands* (1949); T. Heyerdahl & A. Skjölsvold, *American Antiquarian* **22**: 1–71 (1956); Thornton (1971); and Roots (1976).
39. Koford (1966) misquotes Slevin by claiming that 'in 1814 Captain Porter even released goats on James Island in order to stock it'.
40. For a more detailed history of the settlement of the Galápagos see p. 359.
41. Sources are: Slevin (1931, 1959); Salvin (1877); van Denburgh (1914); R. I. Bowman, 'Report on a biological reconnaissance of the Galápagos Islands during 1957' (mimeo for Unesco, Paris, 1960); Koford (1966); Thornton (1971); Werner (1977); and personal information.
42. Koford (1966).
43. Thornton (1971).
44. Eliasson (1968) and Hamann (1975).
45. *Charles Darwin Research Station Scientific and Conservation Report* No. 14 (1968) and Eckhardt (1972: 588).
46. Eckhardt, ibid., per Mr Rolf Sievers, manager of the Charles Darwin Research Station.
47. *Noticias de Galápagos* **21**(1973).
48. B. L. Robinson, *Proceedings of the American Academy of Arts and Sciences* **38**: 77–279 (1902).
49. See note 45.
50. Eliasson (1968), Hamann (1975) and Thornton (1971).
51. Mason (1979).
52. L. Calvopiña, Abstract of paper presented at the 18th World Conference of the International Council for Bird Preservation at Cambridge, England, August 1982; see Moors (in press).
53. For information on goats in the Seychelles I am indebted to Mason (1979).
54. For full details see under rabbit (p. 409).
55. P. A. Munch, *Results of the Norwegian Scientific Expedition to Tristan da Cunha, 1937–38*, No. 13 (1945).
56. B. Morrell, *A narrative of four voyages* . . . (1832).
57. See p. 270.

Barbary Sheep or Aoudad

(*Ammotragus lervia*)

Note: Natural distribution in the Sahara Desert is discontinuous.

NATURAL DISTRIBUTION

Originally distributed throughout North Africa from the Atlantic coast eastwards to the Red Sea and south to the Sudan and the northern bend of the Niger River, Barbary sheep or aoudads[1] in Africa are now virtually confined to parts of the Saharan region. They are decreasing rapidly in the north of their range.

NATURALIZED DISTRIBUTION

North America: USA, Mexico.

NORTH AMERICA

USA

Barbary sheep were first imported to the United States by the New York and Washington, DC, zoos around the turn of the century. In 1924 the newspaper proprietor Mr William Randolph

Hearst established a private menagerie at his ranch on Lone Tree Hill near San Simeon in San Luis Obispo County, California, with the acquisition of a small number of Barbary sheep, probably from the Fleishacker Zoo in San Francisco.[2]

By 1937 the flock had increased to between 20 and 30 on the 100,000-ha Hearst Ranch, of which only about 810 ha were fenced. Censuses on the ranch revealed a total of 98 Barbary sheep in 1949, 58 in 1950, 55 in 1951 and 67 in 1952; in the following year Hearst died and the zoo/ranch was abandoned, allowing a total of 85[3] Barbary sheep to escape into the wild. (In 1950–1 81 sheep were sold to an animal dealer, Mr Louis Goebel of Thousand Oaks, California, some of which were to form the nucleus stock of the present flocks in the Canadian River Gorge, New Mexico, and in the Palo Duro Canyon, Texas.)

In the year following their escape from the Hearst collection the first Barbary sheep (believed to be rams) were seen 14 km to the south on Vulture Rock, where within a decade a breeding flock of around 100 had become established. By this time it was estimated that there were at least 172 Barbary sheep between Hearst Castle and Red Rock (1.5 km to the east), within the confines of the present 30,800-ha Hearst Ranch (Rancho Piedra Blanca.)

In late August 1960 the Weferling burn scorched 22,300 ha of thick chaparral (dense thicket of thorny shrubs) from Red Rock east over the Santa Lucia Range to the edge of the Nacimiento Reservoir; aerial re-seeding in the fall (autumn) improved the habitat for Barbary sheep considerably, and allowed the formation of two new flocks centred on large rocky outcrops – one on Glazier Ridge to the west and one on Cline Peak to the east of the Santa Lucia Crest. Thus, by 1964 four ewe/lamb flocks of Barbary sheep totalling around 500 animals were established west and south-west of San Simeon, and individual rams were being reported from up to 30 km to the north and 65 km to the south.

In 1964, when the shooting of rams for sport commenced, the combined population of the Red Rock and Glazier Ridge flocks was estimated at about 350. The Red Rock flock numbered at least 258 in the following year and 249 in 1966: counts made in 1976 and 1977 revealed totals of 160 and 154 respectively, with an accompanying decline in the ram:ewe ratio.

In 1965 the Glazier Ridge flock numbered 130 and in the following year at least 143; since then regeneration of the chaparral, resulting in the exclusion of much of the herbaceous vegetation, together with predation by cougars or pumas (*Felis concolor*), bobcats (*Felis rufus*), coyotes (*Canis latrans*) and human poaching, had caused a decline to no more than about 50 sheep by 1979.

Eighty-six Barbary sheep were counted in the Cline Peak flock in 1965, which by the following year had increased to around 100; by 1979, however, for much the same reasons as for the decline in the Glazier Ridge flock, the number had fallen to around 50.

The flock on Vulture Rock, which competes with beef cattle for the sparse forage, was estimated to number about 100 in 1964. Largely as a result of shooting the population now tends to fluctuate between 40 and 60 head. Indeed, it is generally considered that shooting rams for trophies has been largely responsible for preventing the establishment of further flocks of wild Barbary sheep (for which there are no bag limits or closed season) in California, where the total 1978 population was estimated at around 400.

Barbary sheep were first imported to the state of New Mexico[4] in 1940 when Mr Joe McKnight, who ranched at Picacho near Riverside on the Hondo River west of Roswell, purchased three ewes and two rams (including one lamb of either sex) from the zoo at St Louis, Missouri. Shortly after their release into the wild the mature ram died, so McKnight obtained a further adult ram and two lambs (one of each sex) from the San Diego Zoo in California in 1941: so far as is known these eight animals comprise the only deliberate introduction of Barbary sheep into south-eastern New Mexico.

Due to inadequate fencing and incompetent management the sheep began escaping from their 810-ha enclosure as early as 1943; between that year and 1979 an average of 10–20 animals were permitted to escape annually, with mass exoduses of about 100 in 1965 and 50 in 1977. These figures indicate that some 500 Barbary sheep were allowed to escape at Picacho into the wild during the 36-year period.

By the late 1940s Barbary sheep were established in parts of the Hondo Valley and Rio Felix drainages west and south of the McKnight Ranch. In 1951 two mature rams were reported in the northern Guadalupe Mountains, and in 1955 another adult male was observed in the Brokeoff Mountains some 15 km north of the New Mexico/Texas border.

By 1960 Barbary sheep had spread rapidly westward up the Hondo Valley to near Ruidoso, east to the Boarder Hill region and south beyond the Rio Penasco. Little further expansion of range was apparent until around 1966, when mixed-sex flocks were reported from the Brokeoff Mountains of New Mexico and the Guadalupe Mountains in Texas: in the same year breeding groups were also observed north of the Hondo River and up the Rio Bonito tributary nearly as far as Lincoln. In 1967 three small flocks were reported occasionally to frequent Dry and Fresnal canyons north-east of Alamogordo.

Between 1970 and 1975 the Barbary sheeps' expansion of range was almost entirely confined to north of the Hondo River and up the Rio Bonito, and to the Guadalupe Mountains area – in the latter principally in the 3 Forks, Dark Canyon and Rim districts of the Lincoln National Forest.

By 1979 the breeding population had spread east and west along the Rio Felix and Rio Penasco drainages, and mixed-sex flocks were observed in the Carlsbad Caverns National Park. Isolated rams were reported south-west of Vaughn and in the San Andres National Wildlife Refuge in 1976, and from west of Eunice in 1977.

Dickinson and Simpson[5] consider that the dispersal of Barbary sheep in southeastern New Mexico is the result of an innate tendency to extend their range into unoccupied and even inhospitable habitats, independent of population size and density. Although initial dispersal from the Hondo Valley appears to have been random in direction, the successful establishment of new populations, which is usually heralded by the advance appearance of single mature males, has been mainly south of the Hondo River, and largely between the 2,300-m and 1,200-m contours. The vegetation varies widely from creosote bush (*Larrea tridentata*)/tarbush (*Flourensia cernua*) associations at lower elevations to ponderosa pine (*Pinus ponderosa*)/white pine (*P. strobiformis*) forests at higher altitudes. Between these extremes occur juniper (*Juniperus* spp.), mountain mahogany (*Cercocarpus montanus*), broom snake-weed (*Xanthocephalum sarothrae*), sacahuista (*Nolina micrantha*), sotol (*Dasylirian leiophyllum*), wavy-leaf oak (*Quercus undulata*), skunkbush (*Rhus aromatica*) and yucca (*Yucca* spp.). The shooting of Barbary sheep began in the Hondo Valley in 1967 and in the Guadalupe Mountains in 1975.

In January 1950 the New Mexico Department of Game and Fish purchased eight ewes and four rams from McKnight and planted them in Old Mills Canyon near Roy on the Canadian River in northeastern New Mexico. A further 26 ewes (including 6 lambs) and 19 rams (5 of which were lambs), obtained from Hearst's Ranch in California, were released on a state game refuge near the northern end of the Canadian River gorge in the following November.[6] In 1955 or 1956 the San Juan County Wildlife Federation obtained 21 Barbary sheep from Louis Goebel, and (unknown to the State Wildlife Agency) released them in Canyon Largo near Farmington in northwestern New Mexico. Today, a flock of about 250 ranges over some 48,500 ha. About 180 km to the south some Barbary sheep are believed to have escaped in 1975 from a private collection near Grants in the Mount Taylor region of the San Mateo Mountains in the Cibola National Forest, where three years later they were estimated to number between 35 and 50.

The Canadian River population increased so rapidly that by 1955 (when they were estimated to number 200 and ranged over 70 km of the gorge) Barbary sheep were officially declared game animals in New Mexico. By 1958 the Canadian River colony had increased still further to around 1,000 and by 1962 to some 1,500, and had spread northwards as far as Colorado – a distance of some 130 km – and 80 km south to beyond the Conchas Reservoir. Occasional stragglers have even been reported eastward to Oklahoma. The extension of range occurred mostly within the rocky brush-covered rims of the canyon, and the sheep have not spread into the rolling grass plains to the east and west. This population, which is currently the largest in North America, is principally concentrated in the main gorge and in some of its branching canyons from near Taylor Springs south to around Sabinso. A few sheep also occur in the smaller La Cinta Canyon which runs parallel to and east of the Canadian River gorge. Most of the land, which consists primarily of grassy benches (terraces), brush-covered slopes and stands of ponderosa pines, is privately owned, and the lower elevations are used for the rearing of beef cattle; in the vicinity of Old Mills a 25-km strip on both sides of the main canyon is owned by the US Forest Service.

The Canadian River canyon population has, since its inception, been the subject of intensive study by the New Mexico Department of Game and Fish. Their investigations have revealed no adverse affects caused by Barbary sheep on either domestic livestock or on native mule or black-tailed deer (*Odocoileus hemionus*); the former have difficulty in gaining access to the Barbary sheep's rocky territory and the population of the latter has noticeably increased in recent years. The latest (1976) survey suggested a total of 1,340 Barbary sheep in New Mexico.

In February and December 1957 and again in December 1958 a total of 44 Barbary sheep (12 rams, 22 ewes and 10 unsexed lambs) were released by the Texas Game and Fish Commission in the Palo Duro Canyon south of Amarillo in northern Texas; this was at the time the only state-sponsored stocking of exotic game animals in Texas. Of these, 31 were freed in Armstrong County south-west of Claude in 1957 and 13 in the following year near Quitaque in Briscoe County. Further plantings, by private individuals, were made in the Canadian River canyon north of Amarillo in 1959, on Mr Mark A. Moss's ranch near Llano on the Edwards Plateau, and on the Bar-O Ranch in southwestern Texas.

By 1967 the Palo Duro flock had increased to around 600 and had spread throughout the eroded cliffs and mesas[7] of the escarpment within the canyon and the adjacent cap rock areas of the Texas panhandle. A census in 1964 revealed a total of around 1,600 Barbary sheep in the whole of Texas; the population had declined to around 1,300 by 1966, but had increased again to 3,200 by 1971, to 3,531 by 1974 (of which about 650 were in the Palo Duro Canyon and a total of 1,050 were free-ranging) and to 8,451 distributed over 50 counties by 1979, of which, however, only 3,709 in 17 counties were considered to be free-ranging, the remainder being confined within deer-proof fences on 72 ranches in 40 counties. The population is concentrated mainly in the Texas panhandle and in the south-west, principally in Armstrong, Bexar, Brewster, Briscoe, Floyd, Motley, Palo Pinto, Randall, Real, Robertson and Stephens counties.

At present, only California, New Mexico, Texas (and possibly Oklahoma and Colorado) are known to harbour free-ranging populations of Barbary sheep in the United States. In the 1960s, however, the Nevada State Wildlife Agency proposed their introduction to public lands in the Seven Troughs Range north-west of Lovelock: in the face of public opposition this scheme was eventually abandoned.

Mexico[8]

Barbary sheep have been released by private landowners at three sites in Mexico not, as in the United States, for shooting but either as additions to already existing collections of exotics or as a potential source of protein.

The earliest liberation of Barbary sheep in Mexico was probably that on the 2,800-ha Sierra Morena Ranch of Señor Octaviano Longorias near Nuevo Laredo in the state of Nuevo Leon: this long-established major cattle ranch supported a number of free-ranging exotic ungulates in addition to Barbary sheep. After several changes of ownership the ranch was eventually subdivided into a number of smaller units (*ejidos*) to comply with land reform legislation, and a group of Barbary sheep spread into the eastern foothills of the Sierra Madre Range, where they became established; this flock has increased to more than 100 animals and has dispersed along the range, yielding a number of trophy heads to local sportsmen.

A second release was made in the Sierra Pajaros Azules along the western boundary of Nuevo Leon and Coahuila, where a population (of unknown size) has become established on the eastern slopes of the Sierra Madre Oriental.

In 1975–6 12 Barbary sheep (8 ewes, 2 rams and 1 lamb of either sex) were released in the northwestern part of the state of San Luis Potosi; in spite of heavy predation by coyotes the flock has increased but had not, by 1979, dispersed far from its release site.

If the behaviour of free-ranging Barbary sheep in the United States is typical then their progressive dispersal in Mexico from the Sierra Madre Oriental can be confidently expected – probably westward into the more arid hinterland of the Bolson de Mapimi (a southern extension of the Chihuahuan Desert) which offers a similar habitat to that occupied by Barbary sheep in New Mexico and Texas, and from where further expansion westwards into the Sierra Madre Occidental (the sole remaining habitat in Mexico of the desert bighorn sheep (*Ovis canadensis mexicana*)) would be relatively unrestricted. In the Bolson de Mapimi the principal vegetation consists of creosote bush, tarbush, coldenia (*Coldenia* spp.), mesquite (*Prosopis* spp.), whitethorn (*Acacia constricta*), dalea (*Dalea* spp.), prickly pear (*Opuntia* spp.), lecheguilla (*Agave lecheguilla*), yucca, sotol, bear grass, (*Ephedra* spp.), and candelilla (*Euphorbia* spp.). Grasses include tobosagrass (*Hilaria mutica*), burro grass (*Scleropogon brevifolius*), sacaton (*Sporobolus airoides*) and lovegrasses (*Eragrostis* spp.). Trees are mostly oaks, juniper and pinyon pine (*Pinus cembroides*). Game animals, including pronghorn antelope (*Antilocapra americana*), white-tailed deer (*Odocoileus virginianus*) and mule deer, occur only in small numbers.

The principal reason for the successful naturalization of Barbary sheep in the United States and Mexico is the suitability of the climate and terrain of the release sites, coupled with their known adaptability to a variety of habitats. As previously mentioned, Barbary sheep in the Canadian River canyon of New Mexico appear not to compete for food with either native deer or domestic cattle; they frequent the steepest canyon walls or benches under the cap rock, whereas deer prefer an intermediate position between them and cattle which feed primarily on the flat canyon-floor or on lower benches. In Largo Canyon and in the Guadalupe Mountains of New Mexico and in Palo Duro Canyon, Texas, however, Barbary sheep and native mule deer show a significant dietary overlap. Since Barbary sheep are exceptionally well adapted to near-desert and rocky terrains they are likely to dominate natives in interspecific competition. In general, Barbary sheep appear able to co-exist satisfactorily with predators such as bobcats, coyotes, cougars, black bears (*Ursus americanus*), bald and golden eagles (*Haliaeetus leucocephalus* and *Aquila chrysaetos*), although coyotes and cougars especially may take a few lambs and even an occasional adult. Concern has been expressed about the possible impact Barbary sheep might have, through interspecific competition, on the already rare native bighorn or mountain sheep, were the two ever to come into contact, and this clearly requires close monitoring.

Barbary sheep have been introduced for sporting purposes to several large (85,000 ha or more) fenced private ranches in the Orange Free State, South Africa.

NOTES

1. R. Valdez, in *Wild Sheep International* **2**(1): 2(1980), suggests that the name 'aoudad' should be substituted for that of 'Barbary sheep' as the animals are not taxonomically true sheep and anatomically most closely resemble goats (*Capra*). Some authorities (e.g. Corbet (1978) in *The mammals of the Palaearctic Region*) now include the species in *Capra*. Here, however, the more usual term 'Barbary sheep' has been retained.
2. For the history of Barbary sheep in California I am indebted to R. H. Barrett, *Proceedings of the Symposium on Ecology and Management of Barbary sheep, Texas Tech University, November 1978* (1980), pp. 46–50.
3. J. Yoakum, ibid. p. 10, says 83.
4. For information on Barbary sheep in New Mexico I am indebted to B. L Morrison, *Proceedings of the Symposium on Ecology and Management of Barbary Sheep, Texas Tech University, November 1979* (1980), pp. 15–17; T. G. Dickinson & C. D. Simpson, ibid. pp. 33–45; and Ogren (1965).
5. See under note 4.
6. Some accounts say that 4 rams and 4 ewes were obtained from McKnight, and 27 ewes and 17 rams from Hearst. Roots (1976) gives the date incorrectly as 1949.
7. Flat-topped rocky hills with steeply sloping sides.
8. For information on Barbary sheep in Mexico I am indebted to E. R Woodyard & C. D Simpson, *Proceedings of the Symposium on Ecology and Management of Barbary Sheep, Texas Tech University, November 1978* (1980), pp. 30–2.

Mouflon

(*Ovis musimon*)

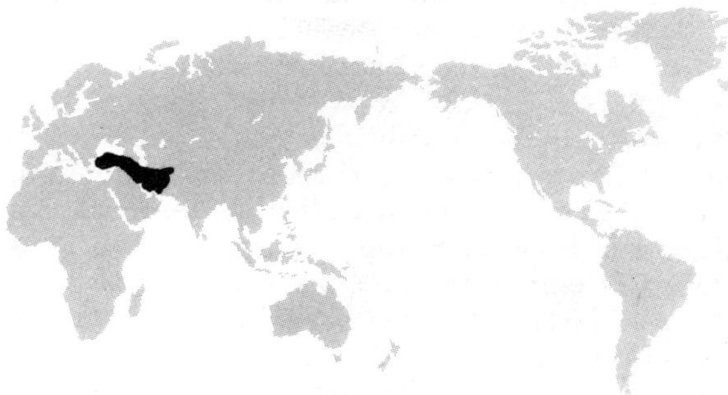

Note: Natural distribution shown is that of probable wild ancestor. Status on Corsica, Sardinia and Cyprus (unmarked) is uncertain.

NATURAL DISTRIBUTION
The probable ancestor of the European mouflon,[1] the Asiatic mouflon (*O. orientalis*), ranges from Iran to Asia Minor.

NATURALIZED DISTRIBUTION
Europe: British Isles, Germany, France, Switzerland, Austria, Hungary, Czechoslovakia, Italy, Spain (including Tenerife, Canary Islands), Poland, Yugoslavia, USSR, Romania, ? Belgium, ? Sardinia, ? Corsica, ? Cyprus; North America: USA; South America: Chile; Oceania: Hawaii, Kerguelen.

EUROPE

British Isles

No wild sheep have occurred in Britain in post-glacial times, but the domestic type has been present since the Neolithic. The Soay sheep of the St Kilda group of islands, which lie some 80 km west of Harris in the Outer Hebrides of Scotland, are the only truly feral breed (apart from Borerays[2]) in the British Isles today, and are very similar in appearance to the European mouflon; they are the most primitive form of domesticated sheep surviving in Europe, and probably bear as close a resemblance to the peat-sheep or turbary-sheep of the late Neolithic period as any existing in the world at the present time.

Sheep may originally have been introduced to St Kilda in prehistoric times. The bones of present-day Soays closely resemble those of sheep excavated from late Neolithic deposits on Orkney and in late Bronze Age/early Iron Age (*c.* sixth to first centuries BC) sites in the Shetlands, as well as in Romano-British settlements in southern Scotland and the south of England.

It is known that the island of North Rona – situated about mid-way between the Shetlands and St Kilda – was inhabited between about AD 300 and 400, in pre-Viking times. This, therefore, is another possible time for the introduction of sheep to the Western Isles. The known domination of the Vikings over the Outer Isles, however, together with direct evidence of their occupation of St Kilda coupled with the close resemblance between the bones of Soay sheep and those of a breed found in Viking settlements in Greenland, all point to a probable Norse origin. If this is so, they may well have been brought first to Orkney and the Shetlands from Denmark and Scandinavia around AD 800, and thence to St Kilda. It has been shown that the long-tailed field mouse or wood mouse (*Apodemus sylvaticus*) and (now extinct) house mouse (*Mus musculus muralis*) of St Kilda were probably introduced accidentally from Denmark and Scandinavia in hay and other fodder which the Norsemen imported to feed their domestic stock; among these would most probably have been sheep, which subsequently escaped from domestication and became feral. Soay sheep are thus probably the direct descendants of this early domesticated breed, and have remained intact on the island of Soay for over a thousand years. Here they were seen in about 1520 by the Scottish historian Hector Boece (or Boethius) and in 1571 by the Bishop of Ross, John Leslie (or Lesley) during a pilgrimage of St Kilda.

In medieval times a four-horned breed, known as the 'St Kilda' or 'Hebridean' sheep, was imported from Scandinavia to Hirta (the main island in the group) and to Boreray, but not to Soay, to provide new blood. In the 1870s Scottish blackface sheep (which probably originated in the Pennine Hills[3]) were imported to the islands of Hirta and Boreray, where they remained the principal breed until the islands were evacuated in 1930 by the human inhabitants, who took with them the blackface from Hirta but left behind the blackface on Boreray and the Soays on Soay. Two years later 107 Soay sheep were transferred from the 97-ha island of Soay to the 600-ha island of Hirta, where by the outbreak of the Second World War the population had risen to about 500; since then it has fluctuated considerably, the most recent count by the Nature Conservancy Council (of 1981) giving a total of more than 1,400. The low ratio of mature rams to ewes is believed to be due to the drain on the energy resources of the former during autumn tupping, which reduces their ability to survive the long St Kilda winter.

Soay sheep are small goat-like animals standing some 60 cm high; adult rams weigh around 35 kg and ewes about 10 kg less. There are two main colour forms – dark chocolate-brown and pale oatmeal – both of which have a coat pattern which resembles that of the wild mouflon, with a white belly and paler hair around the eyes and on the chin. 'Self-coloured' individuals, lacking these points, and some with considerable degrees of white on both head and body are not uncommon. The rams grow heavy, curling horns, while ewes may have short spiky horns or be polled. Tupping takes place in November, and the lambs are born in the following April. Preferred grazing is on *Holcus/Agrostis/Festuca* maritime meadows. Late winter and early spring storms, which strike when the resistance of adults is lowest and during lambing, are the sheeps' most important controlling factor. The ruined buildings of the islanders' abandoned village on Hirta and its surrounding *cleitans* (small stone cells built as drying chambers) provide invaluable winter shelter.

The Soay sheep of St Kilda exist in a stable ecosystem completely isolated from outside influences, where there have been no removals or introductions of new blood for many years; they face no competition from other herbivores and, in the absence of ground predators, have only to face light predation by hooded crows (*Corvus corone cornix*), ravens (*C. corax*), and greater and lesser black-backed gulls (*Larus marinus* and *L. fuscus*), which take some lambs and sick or injured adults. Soay sheep have been aptly described by K. Williamson and J.M. Boyd as 'one of the faunal treasures of Britain'.[4]

Feral flocks of Soay sheep have been established on several other British offshore islands, to which they have been introduced from St Kilda via the mainland: these include Lundy Island in the Bristol Channel, England; Ailsa Craig off the Ayrshire coast in Scotland; and the islands of Skokholm, Cardigan, Middleholm, St Margaret's and Skomer off the coast of Pembrokeshire in south Wales. Only those on Lundy, Cardigan and possibly Ailsa Craig remain today.

Austria; Hungary; Czechoslovakia; USSR

Mouflon have been imported as a game animal to central and southern Europe, to the Iberian peninsula and to the Crimea (Krym) in the Soviet Union from Corsica and Sardinia[5] since the middle of the nineteenth century. They are first known to have been introduced to the Hluboká Reserve in Bohemia (Czechoslovakia) from the Leinz Reserve near Vienna, Austria in 1858, though there may well have been much earlier introductions to Bohemia. In 1977 there were 11,674 mouflon in the wild and in reserves in Czechoslovakia. In the Soviet Union they are established and hunted in at least one region of the Crimea, where they were originally introduced in 1913–14: the population is controlled by the scarcity of shelter and heavy snowfalls in winter.

On Corsica the horns of ewes are normally small, while on Sardinia and in most of the introduced populations of mainland Europe they are usually absent altogether. On the Mediterranean islands mouflon mainly frequent open mountains near the tree-line and rough scrub, whereas in mainland Europe (where they are perhaps most abundant in Austria, Hungary and Czechoslovakia) they prefer lowland woodlands with open clearings or adjacent grassland.

NORTH AMERICA

USA

Although mouflon are said to have been introduced to desert ranges by ranchers in the state of New Mexico and to California, they are currently believed to be established in the wild in the United States only in Texas.

They were first introduced to the King Ranch (then the largest in the world) in southern Texas in the late 1930s and early 1940s, and subsequently in 1946 by Mr Mark Moss of Llano, and by the Bar-O Ranch. By 1966 the population was estimated to number around 10,000; 16,169 were counted in 1971, 15,254 (1,250 of which were free-ranging) in 1974 and 9,536 in 1979, of which 2,538 were free-ranging, when they were present in 52 counties, the majority being in Bandera, Bell, Bexar, Blanco, Bosque, Brooks, Brown, Burnet, Kendall, Kerr, Kinney, Llano, Mason, Medina, Palo Pinto, Presidio, Real, Stephens, Sutton, Uvalde, Williamson and Wild counties – mostly on the Edwards Plateau, and in the Trans-Pecos, the Rolling Plains and the Cross Timbers and Prairies. Extensive crossbreeding among mouflons, domestic barbados sheep from West Africa and other domestic sheep has resulted in the virtual elimination of recognizably pure-bred animals; only those considered to be relatively pure were included in the 1979 survey.

SOUTH AMERICA

Chile; Argentina

In southern Chile mouflon have for some time been ranched for commercial hunting in Osorno Province, where they are also established in the wild on an island in Lake Rupanco. Concern has been expressed that mouflon on ranches in the mountainous Neuquen Province of Argentina could spread into the adjacent national parks.

OCEANIA

Hawaii

Mouflon were first introduced to the island of Lanai in the Hawaiian group on 30 July 1954 for sporting purposes by the Division of Fish and Game. The original planting, on a ridge between Keone and Naupaka (Anapuka) gulches on the island's western slope, consisted of two mature rams, three adult ewes and five ewe lambs, donated by the Honolulu Zoo. During the next eight years a further 35 were released, which by 1964 (when shooting was commenced) had increased to over 250.

In 1958 two adult pairs of mouflon were turned down in the Na Pali region on Kauai, but they soon disappeared, having probably fallen victim to poachers.

In 1957 Dr William Graf of San Jose State University suggested that by cross-breeding mouflon with feral domestic sheep which are established on Mauna Kea on Hawaii,[6] a hybrid might result which would prove less destructive to the endemic vegetation. Between June 1962 and February 1966 46 pure mouflon rams and 48 pure-bred ewes were planted around Pohakula on Mauna Kea, while 33 hybrid rams, 66 hybrid ewes and 16 pure mouflons were released in the Puu Laau area, and 78 pure mouflons in the Kahinahina section.

Today, mouflons (known locally as *hipa*) are established in the low hundreds on Lanai – mainly at the western end of the island where they are believed to be spreading. On Hawaii, where both hybrids and pure-bred animals tend to remain within a few kilometres of their 1960s release sites, their status is less certain; their future is probably closely linked to that of feral domestic sheep, which some hunters prefer for their approachability and more tender meat. If feral domestic rams are not eliminated they will, by back-crossing, adversely affect the hybridizing programme by tending to obliterate the positive characteristics of mouflon offspring.

Future releases of mouflon on Kauai and possibly on Maui have been suggested.

Kerguelen

In 1957 two pairs of mouflon from the Parc Zoologique de Paris were released on Île Blakeney and a further two pairs from the same source on Île Haute, Kerguelen. Only the latter introduction was successful, and the mouflon had multiplied to around 300 by 1974 when it was considered that, like the reindeer (*Rangifer tarandus*),[7] they might eventually swim to Île Grande Terre if Île Haute became overpopulated.

NOTES

1. Several authorities now believe that the European mouflon is a feral sheep. To avoid possible confusion it is here treated separately from feral domestic sheep (*O. aries*), but with feral Soay sheep which closely resemble it.
2. See p. 314.
3. See p. 314.
4. *St Kilda summer* (Hutchinson, London, 1960).
5. It is probable that the relict populations on Corsica and Sardinia (and on Cyprus) are descended from animals introduced by man, which were domesticated but were still very similar to their wild ancestors.
6. See p. 319.
7. See p. 234.

Domestic Sheep
(*Ovis aries*)

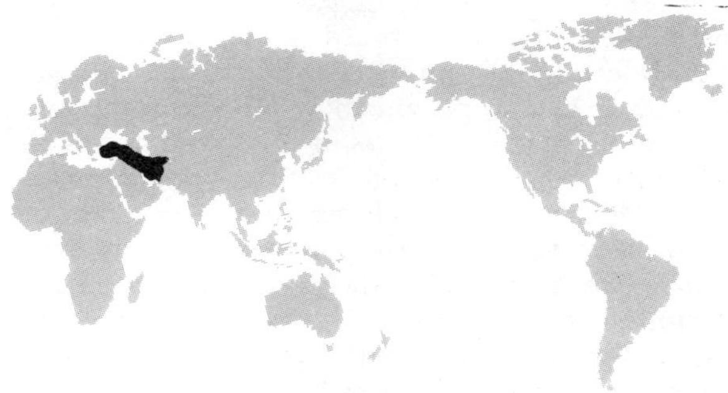

Note: Natural distribution shown is that of wild ancestor.

NATURAL DISTRIBUTION
Domestic sheep[1] are probably mainly descended, like the European mouflon (*O. musimon*), from the Asiatic mouflon (*O. orientalis*) of Iran and Asia Minor, and possibly also from the argali (*O. ammon*) which ranges from the Altai to the Himalayas.

Note: Only principal feral populations mentioned in the text are shown.

NATURALIZED DISTRIBUTION
Europe: British Isles, Norway, Sweden; North America: USA; Australasia: ?Australia, New Zealand; Oceania: Hawaii, Kerguelen and various other oceanic islands.

EUROPE

British Isles

Sheep may have been established on the St Kilda group of islands in the Outer Hebrides of Scotland since prehistoric times.[2] In the medieval period a four-horned breed – the 'St Kilda' or 'Hebridean' – was imported to the islands of Boreray and Hirta to provide new blood; on the former both Martin Martin in 1697 and Kenneth Macaulay in 1758 estimated the population at about 400, the latter adding that 'every one of these sheep has two horns and many of them have four'. Since then the sheep on Boreray have doubtless been changed by introductions from Hirta (the main island in the group), unlike those on the island of Soay which have remained unaltered since the Viking period. On none of the islands of St Kilda are four-horned sheep found today.

In the 1870s Scottish blackface sheep (which arrived in Scotland from northern England in the eighteenth century) were introduced to Hirta and Boreray; according to Boyd (1981) the sheep kept on Hirta for centuries previously had been the 'Old Scottish Shortwool' or 'Dunface', which seems to have been the predominant breed in Europe between 500 BC and AD 1500 and whose characteristics are represented today by the Shetland and North Ronaldsay (Orkney) breeds. Although the present Boreray sheep resemble a cross between a Soay[3] and a modern Scottish blackface (and are similar in appearance to blackfaces depicted in mid-nineteenth-century paintings), they are in fact the descendants of a cross between the introduced nineteenth-century Scottish blackface and the 'Old Scottish Shortwool', and have survived for over half a century on the 77-ha island, undisturbed and uncontaminated by further introductions since the last human inhabitants departed in 1930.

About 66 per cent of the population, which in 1980 Bullock (1983) estimated to number about 700, have a cream-coloured or off-white fleece, 20 per cent are grey or grey-brown and the remainder are tan or blackish. Some 50 per cent have off-white faces with some grey, grey-brown or black markings, and as Boyd had noted, there is sometimes a conspicuous wide collar in both rams and ewes that extends from the nape of the neck to the front legs. The tail is short and sometimes dark with a dark anal patch. Both sexes are horned – those of the rams being massive, long and curving. Rams stand around 60 cm high at the withers and ewes about 5 cm less; their autumn pre-tupping weights are some 45 and 25–30 kg respectively.

Boreray sheep live at a high density of nearly 14 per hectare on around 50 ha of rich martime pasture heavily fertilized by guano; apart from the tupping season the rams and ewes run in separate flocks – the latter with lambs and yearling rams. As with the Soay sheep on Hirta the ratio of 1 ram to 10 ewes is unusually low, probably due – as on Hirta – to the drainage of energy resources of the former during autumn tupping, which reduces their chances of winter survival.

Boreray sheep have adapted well to a harsh environment, and although their presence may exert an inhibiting effect on nesting seabirds, their close grazing appears to prevent soil erosion by creating a dense sward which may also provide a greater diversity of species than might exist in their absence. They are the result of over half a century of natural selection with no outside interference from man, and as the descendants of an early domesticated breed are of exceptional scientific and agricultural interest, and should be preserved genetically intact.

Norway; Sweden

A semi-feral population of sheep (which is strictly protected) exists in southwestern Norway. Another flock of about 100 ewes and 6 rams of the original Swedish Landrace breed (now known as Goths) lives on Lilla Karlsö off Gotland in the Baltic.

NORTH AMERICA

USA[4]

Feral sheep – usually in small flocks of less than a dozen or so – have been reported on the mainland of the United States in recent years only from the hill country of Morgan County in northern Alabama; the Picnic Flats area of the Southern Ute Indian Reservation in south-west Colorado; the Split Mountain Gorge in northeastern Utah; and the Trout and Hay Creek drainages of Jefferson County, Oregon.

On the Channel Islands off the southern coast of California sheep-ranching on Santa Cruz began in about 1835 and by 1890 the population numbered at least 50,000. Ranching was abandoned in the 1930s, and the 10,000–20,000 animals that remain today (which are smaller than their merino ancestors and partially shed their wool) have, by overgrazing and trampling, severely affected the islands' vegetation and the associated bird communities.

Few shepherds would agree with McKnight's (1964) criticism of their charges that 'Their lack of intelligence in caring for themselves makes them susceptible to inclement weather . . .'; indeed, of all livestock sheep are perhaps the best adapted to survival in a harsh environment, and are more likely to suffer in a hot climate (e.g. from attacks by ticks and maggots and by the matting of unshorn wool) than from the cold. They provide, on the other hand, an easy target for carnivores, from whom the protection offered by predator-free island groups gives welcome immunity. Feral sheep are relatively easy to control and their survival is largely dependent on the tolerance of man.

AUSTRALASIA

Australia

There appear to be few early records of the importation of sheep to Australia; one of the first introductions (of merinos from Spain via England) took place in about 1795, the direct descendants of which still survive (in captivity) in Camden Park, Sydney (see also below under New Zealand). Sheep are believed to have first become feral in Australia on Big Green Island in the Furneaux group in the Bass Strait in about 1830, where they were at one time thought to have caused a decline in the short-tailed shearwater (*Puffinus tenuirostris*); research carried out between 1965 and 1968, however, revealed that trampling by sheep has not prevented the excavation of nesting burrows, and that grazing activities have not resulted in a shortage of cover.

On the Australian mainland few flocks of feral sheep have ever become established for any length of time; some occasionally escape into the wild from sheep stations, but few exist for long. In the late-1920s over 300 sheep were released in the Kimberleys area of Western Australia, where their descendants survived in a small remnant flock some 40 years later; they were said to be of large size with sparse wool, and were severely inbred.[5]

New Zealand[6]

Captain James Cook, on his second voyage of discovery, obtained two rams and four ewes when, in the *Resolution*, he called at the Cape of Good Hope in November 1772; of these, however, only one of either sex – both 'suffering from an inveterate sea-scurvy' – survived when *Resolution* dropped anchor in Dusky Sound in March 1773;[7] the expedition's naturalist, George Forster, wrote in his *Journal* that 'they were in so wretched a condition, that their further preservation was very doubtful'.[8] Cook considered that the rough, forested country surrounding Dusky Sound would be unsuitable for sheep, and took them on with him to Queen Charlotte Sound, where he landed them on 18 May. In his *Journal* he wrote:

> On the 22nd in the morning, the ewe and ram, I had with so much care and trouble brought to this place, were both found dead, occasioned, as was supposed, by eating some poisonous plant. Thus my hopes of stocking this country with a breed of sheep were blasted in a moment.[9]

In a letter from Paramatta dated 30 May 1815 to the governor of New South Wales, the Revd Samuel Marsden said that in the previous year he had imported from Sydney to the Bay of Islands one merino ram and four ewes, and claimed that these animals formed the nucleus of the earliest flock of sheep in New Zealand. R. McNab tells how an early settler on Mana Island, John Bell, imported 102 merinos from New South Wales, and that four years later he was able to export four bales of wool to Sydney.[10]

When regular settlement began in the 1840s sheep were increasingly imported from New South Wales – indeed the *New Zealand Gazette and Wellington Spectator* reported that in 1841 no fewer than five ships carrying sheep arrived within a fortnight. Most of the sheep introduced during this period came from Australia, and were predominantly merinos. Many were allowed to range free, being gathered only for shearing or tailing; some, inevitably, escaped mustering, and were soon widely established in the wild, especially in the mountains of South Island, in the Hawkes Bay area and doubtless elsewhere.

In the second half of the nineteenth century the sheep population increased enormously, both by domestic breeding and massive importations from Australia. Numbers rose from 750,000 in 1855 to 2.8 million in 1861, 4.9 million in 1864 and 8.4 million in 1867: by 1880 there were estimated to be over 13 million (mainly merino) sheep in the country, a figure which by 1890 had risen to 17.5 million of which, however, only 34 per cent were merinos.

Aston (1911) wrote:

> On the North-West side of Isolated Hill [Marlborough] . . . are hundreds of wild sheep in small flocks of about half-a-dozen each. All . . . are . . . in exellent condition. Some were curiously marked and coloured . . . the rams had large horns, and all were tamer than ordinary domestic sheep. Their food appears to consist of the Silver Tussock, (*Poa caespitosa*); a Poa-like *P. colensoi*; the spear-grass *Aciphylla colensoi* . . . these sheep destroy the Gaya trees (*Gaya lyallii*) by eating the bark, which we watched one stripping off in large sheets.

As late as the early 1920s Thomson (1922) was able to say that feral sheep were 'still abundant in some of the wilder parts of the country, and are especially numerous in the high limestone country of Marlborough'.

As sheep farming grew increasingly intensive and better organized in the early years of the present century feral flocks became gradually confined to the more remote and inaccessible regions. Since the inception of wildlife control policies feral sheep have been regularly culled together with deer and goats, and since the Second World War determined efforts have been made to eliminate sheep from certain areas, such as Marlborough; Wodzicki (1950) records that in 1946 over 5,000 were killed. Between 1951 and 1961 an average of more than 2,000 sheep a year were removed from the wild without any apparent drop in the overall population; during the 1960s, however, several feral flocks were eradicated, and in the 1970s the decline in the number of flocks and total population continued.

At the present time flocks of feral sheep survive on the mainland in about a dozen localities from Raglan Harbour to Southland, on the distant offshore islands of Campbell and Chatham, and on Arapawa Island in Marlborough Sound. The following particulars are largely derived from Whitaker and Rudge (1976)

On a peninsula in Raglan harbour, west of Hamilton, a feral flock of some 300 sheep has been established since the early 1930s; they are typical 'Romney' sheep although very open-faced and 'leggy'. In the Hawkes Bay region there are two discrete flocks – one in the northeastern Ruahines and the other along the Mohaka River. The former (known as the 'Omahaki' flock) ranges over some 7,000 hectares of scrub, scree and mountain bluffs around the confluence of the Ngaruroro and Taruarau rivers in the Eastern Ruahine State Forest and on Big Hill and Omahaki stations. This flock has been established in the wild for more than 50 years, and is probably the remnant of a much larger one which originated in free-ranging merinos of the late nineteenth century. The current population numbers several hundred.

The Mohaka flock consists of several groups on or near the banks of the Mohaka River between the Te Hoe and Waipunga rivers and behind Tutira. Rather less than 1,000 sheep currently range over an area of more than 30,000 ha, and numbers are currently declining due to loss of habitat to farming. H. Guthrie-Smith refers to the existence of feral merino sheep in this region in the 1880s.[11] Since the change from merinos to longwools was virtually complete at Tutira by the turn of the century it would seem that the Mohaka flock has been established for about the same length of time as that at Omahaki.

The Marlborough region – long a stronghold of feral sheep – at present supports two flocks in the Wairau and Clarence respectively. The former, which is of uncertain age and numbers only a few hundred distributed over the headwaters of the Branch, Leatham and Waihopai, is believed to be augmented from time to time by merino and cross-bred sheep which range free nearby. The latter consists of two groups – each of between 200 and 300 head – respectively in the vicinity of the Boundary and Ouse rivers and along the Mead River. They are believed to be relicts of feral flocks of the 1880s and to contain a larger proportion of pure-bred merinos than the Wairau flock. Both flocks are rigorously culled.

The feral sheep on Arapawa Island[12] are in many respects the most interesting in New Zealand. The present population numbers around 120 and occupies some 700 ha of woodland and cliff-tops in a Marlborough Maritime Park Board reserve on the seaward slopes of the island.

The earliest record of domestic sheep on Arapawa is in 1867 when some were imported from the Wairau Valley; these were almost certainly Australian merinos descended from Flaxborne stock shipped from Sydney in the early 1840s. The first mention of feral sheep on the island is in 1901 when they were described as 'extremely abundant'. Although the present stock has the horns, wool and general conformation of merinos their follicle density count and secondary/primary ratio are both considerably below the lowest figures previously recorded for any merinos – those of Camden Park, Sydney, reputedly the oldest Australian line and the descendants of the first merinos imported from Spain (via England) in about 1795 (see above). This places them in the range of 'down' sheep, although apart from the follicle figures they have no apparent 'down' characteristics. Assuming that feral sheep on Arapawa became established before about 1885 (allowing them 15 years to increase to 'extremely abundant') they would have been in existence when the very few 'down' flocks in New Zealand were well documented. Researchers are currently trying to discover if a change in follicle density is a concomitant of artificial selection for heavier fleeces; if this is so it is possible that the sheep of Arapawa may bear a close resemblance to early New Zealand merinos. About three-quarters of the flock is pigmented, brown or black with white heads (or blazes) and half-white tails, while the remaining quarter is all white; the rams grow long, curling horns. The biological, historical and conservation aspects of this unique flock are clearly considerable.

South of Oamaru a 50–60-year old flock of about 1,000 merino sheep exists in the Waianakarua catchment at the eastern end of the Kakanui Range; it has been suggested that they interfere with domestic stock through interbreeding and competition for grazing.

The Hokonui Hills in Southland support a flock of feral sheep, variously estimated to number between 'a few hundred' and 'over a thousand', which are reputedly the descendants of 400 Saxon merinos imported from Tasmania in 1858. Attempts are currently being made to eradicate this flock, which is distributed in bush ecotone and clearings particularly in the south and east of the range, on the unsubstantiated grounds that they are vectors of bovine tuberculosis.

Saxon merinos were translocated from the Hutt Valley to South East Island in the Chatham group in 1841 by Baron von Alsdorf, who transferred their progeny to Pitt Island a few years later. By 1855 the Chatham Islands population numbered only around 200; intensified farming raised the total tenfold by 1866 and, according to Thomson (1922), to some 60,000 by the turn of the century, when, Thomson wrote, 'they have profoundly altered the native vegetation by eating out many species, such as *Myosotidium nobile*, *Aciphylla traversii*, *Veronica diefenbachii*, and allied species, all of which they eat greedily'. When long-wool breeds supplanted merinos around the turn of the century these sheep were abandoned by their owners and left to fend for themselves. In 1976 a total of between 2,000 and 3,000 ranged over some 2,500 ha of the southern half of Pitt Island. At the same time a very small flock of merinos was established near the south-west corner of the main Chatham Island.

The principal distinguishing feature of the feral sheep on Pitt Island is the colour of their fleece, which in 95 per cent of the population is black with white patches on face, throat and tail. The population 50 and 30 years ago was 80 per cent and 50 per cent all-white respectively, and a few white individuals remain today. Accompanying this colour-change has been the development of a higher proportion of coarse to fine wool fibres and the ability to self-shed the fleece. Some animals show a tendency towards the brown 'moorit' colour of Hebridean sheep, which may indicate the re-emergence of a primitive gene combination. As in their merino ancestors both rams and ewes are horned.

In 1981 the Department of Lands and Survey created a 200-ha experimental reserve on Pitt Island, which was stocked with some 300 sheep. The effect of the animals on the island's flora will be closely studied (feral sheep tend to browse more than domesticated breeds which are primarily grazers), and attempts will be made to assess their potential commercial value, if any.

In November 1894[13] James Gordon was given a grazing lease (an act rightly described by Thomson (1922) – though with the benefit of hindsight – as 'a piece of vandalism on the part of the man who did it, and the government which granted it') on over 11,300 ha of New Zealand's subantarctic Campbell Island, on which in the following year he landed a flock of 350 sheep of unknown breed. In 1901 2,000 two-toothed merinos were introduced, followed in the next year by a further 1,000, also of unknown breed. In 1903 the total population was said to be 4,500. In August 1931, in face of the approaching economic depression, the island was evacuated and its sheep abandoned.

The maximum population of about 8,500 was achieved in 1906 or 1916 (accounts differ), when, according to Thomson, the 'transformation and destruction of the native flora was going on at a great rate': this figure had halved by 1931. By 1950, when the sheep population had fallen to 1,500, the dominant tussock *Poa foliosa* and *Danthonia flavescens* grass had been greatly reduced, and *P. litorosa* was coming under pressure. Bare eroding peat covered many slopes and erosion increased as tussock grass was replaced by the alien *P. annua* and the unpalatable native *Chrysobactron rossii*; the characteristic *Pleurophyllum hookeri* was, through excessive grazing by sheep, virtually confined to a few inaccessible cliff faces. The number of sheep continued to decline until in the early 1960s only 900–1,000 remained; by 1969, however, a dramatic and unexplained increase had restored the population to around 3,000. In 1970, as part of a study into the effects of sheep on subantarctic vegetation and seabirds, a fence was constructed across the island and all the sheep (about 1,500) to the north were shot; since then there has been a spectacular regeneration of the vegetation which includes several endemic plants.

The present population is quarter- to half-bred merino/longwool cross, of which the longwool element has been variously described as Lincoln, Leicester or Romney. The sheeps' grazing activities appear to have had some slight detrimental affect on nesting grey-headed and black-browed albatrosses (*Diomedea chrysostoma* and *D. melanophris*), but have, on the other hand, been instrumental in providing more nesting-sites for royal albatrosses (*D. epomophora*) whose continued survival on the island may well be linked to that of the sheep.[14]

The Campbell Island flock is the only one which is known for certain to contain other than pure-bred merino blood. Other feral flocks, however, exhibit interesting differences between them which may be attributable to the type originally introduced, to the degree and duration of isolation, and/or to selective pressures exerted by their environment. The flocks on Arapawa and Pitt islands both have a high percentage of pigmented animals, and the proportion of black sheep in the Omahaki flock is said to be increasing. Guthrie-Smith[15] reported a high ratio of blacks at Tutira in the 1890s – a situation parallelled among feral sheep on Kapiti Island in the 1920s.

Little appears to be known about feral populations now extinct. Sheep released on the Antipodes islands between 1886 and 1900, on Enderby in 1850 and 1897, on Adams and on Auckland Island at various dates between about 1840 and 1900 had all died or been eaten by castaways by at least the early 1920s.

Of the now extinct feral sheep on Kapiti Island H. B. Kirk and W. E. Bendall wrote in 1919:

> Several hundred . . . are running on the island which is now a plant and animal sanctuary . . . Not only do they prevent to a very large extent the growth of young trees, but they open up the forest to the sweep of the wind. They prepare it for invasion by grass, tauhinu (*Pomaderris phylicaefolia*), manuka (*Leptospermum scoparium*), and other hardy plants . . . where manuka has replaced the forest, much soil has been removed, and in no great time the manuka itself will be unable to retain its footing. In such cases the manuka marks a phase in the passage to utter barrenness.[16]

Feral flocks which have recently become extinct include one in the Takitimu mountains of Southland, which may have been of considerable interest as it was situated in a region where merinos no longer run (a relict population may possibly survive) and others in the Glens of Tekoa and Oxford State Forests in Canterbury, and Romney-cross flocks on South-East and Mangere islands in the Chatham group, which had, however, been feral for only two or three years before their removal in 1961 and 1968 respectively. The former island remains the only known breeding place of the Chatham Island petrel (*Pterodroma axillaris*), and the latter of the Chatham Island robin-flycatcher (*Petroica traversi*) whose entire population was translocated in the late 1970s from the deteriorating habitat of Little Mangere to the rapidly regenerating scrub forest on Mangere.

OCEANIA

Hawaii[17]

Captain James Colnett of the merchantman *Argonaut* is said to have landed the first sheep – one ram and two ewes from the Pacific coast of North America – on the island of Kauai in April 1791.[18] On 14 February 1793[19] Captain George Vancouver landed one ram, two ewes and a lamb born *en route* from California (believed to have been cross-bred merinos) at Kawaihae on Hawaii, in the care of Kahomotoo (to whom he had entrusted some goats in the previous year). From Kawaihae Vancouver sailed on to Kealakekua Bay, where on 22 February[20] he put ashore his remaining stock of a further ram and two ewes as a gift for King Kamehameha I.

On his third visit to Hawaii in 1794 Vancouver presented a further two rams and two ewes (one account says five of each) to King Kamehameha at Kealakekua on 15 (or 19) January,[21] where they were released under the protection of a royal *kapu* (taboo). Coasting northwards from the island of Hawaii Vancouver noted on 5 March that he 'had reserved some sheep for establishing the breed in each of the islands'.[22] As he lay hove to off the coast of Molokai, however, an expected visit from a local chief failed to materialize, the chosen day being unfortunately *kapu*. He was therefore forced to continue northwards with his remaining stock of six sheep; whether more were landed on any of the other Hawaiian islands at this stage is uncertain, but if not such introductions must have taken place shortly thereafter.[23]

When the first governmental body was formed the ownership of feral sheep in Hawaii became vested jointly in King Kamehameha and the government, from whom permission had to be obtained before any could be taken. There is a gap in the official records between 1822 and 1845, but by at least the latter year (when the first pure-bred merinos were introduced) sheep are known to have been established in the wild on the island of Kauai. By 1851 it was estimated that there were some 3,000 feral sheep on Hawaii, and, according to Kramer (1971), 'uncounted numbers' on Maui, Molokai and Kauai.

While ascending Mauna Kea on Hawaii in 1822 the Revd Joseph Goodrich noted that: 'Very near to the summit, upon one of the peaks I found eight or ten dead sheep; they probably fled up there to seek a refuge from the wild dogs; I have heard that there are many wild dogs, sheep and goats'.

By 1853 Saxon and Southdown sheep had been imported, and in 1862 the Emperor of France, Napoleon III, presented 'four picked rams from the Royal flock at Rambouillet to the King of Hawaii'.[24] Subsequent introductions included Shropshires and possibly De Laines.

In 1859 a sheep ranch was started on Kahoolawe but was abandoned within a year as the result of sheep scab, the remaining flock of 2,075 being presumably left to fend for themselves. After 1861 a ranch was opened on Lanai, and on 4 April 1864 permission was granted to transfer 3,000 sheep from Molokai to found a ranch on Niihau.

From 1879 to 1884 between 101,000 and 121,000 sheep were raised in Hawaii annually, but the industry subsequently declined until by 1928 only some 25,000 remained. Before 1910, when the Lanai Company Ltd switched from sheep- to cattle-raising, that island supported up to 50,000 sheep – a figure which by 1920 had declined to 860. On Molokai in 1900 (when Tunis sheep were first imported) there were nearly 16,000 head; by 1928 the total had fallen to barely 200. The abandonment of sheep-farming in favour of cattle is believed to have been largely due to overgrazing and the proclivity to lameness caused by the thorns of dry mesquite kiawe (*Prosopis chilensis*).

In 1937 it was estimated that some 40,000 feral sheep roamed the 32,000 ha of the Mauna Kea Forest Reserve – where they had severely damaged the endemic mamane trees (*Sophora chrysophylla*) – and the lava fields of Mount Hualalai and Mauna Loa. An 80-km fence was erected around the lower boundary of the park in an attempt to keep the sheep, several thousand goats and an 'enormous number' of feral pigs off the ranchlands below.[25] Between 1922 and 1947 nearly 47,000 sheep were removed from the park.[26] Official culling was abandoned in 1946 and was replaced by unrestricted sport-hunting. By 1950 the population had dwindled to around 200, due partially, it has been suggested, to increased predation by feral dogs and pigs, reduced fecundity, and possibly some epizootic disease. At this point the Hawaiian Division of Fish and Game assumed responsibility for management of the reserve, and a moratorium on hunting was declared until 1953 when the population had risen to about 1,000; by

1975 it had increased to 1,700. By cross-breeding feral sheep with mouflon (*O. musimon*),[27] it is hoped to produce a hybrid which is less damaging to the vegetation (through associating in smaller flocks), is wilder (and therefore a more sporting target) and possesses a superior trophy head. It is said that only when mouflon rams are released singly and in the absence of mouflon ewes will they join feral domestic sheep. Cross-breeds, however, tend to associate with feral domestic flocks.

It has been estimated that some 90 per cent of the topsoil on Mauna Kea has been eroded through overgrazing by sheep, goats and pigs, and the process continues. Nearly half of the mamane trees are dead and a further quarter are dying, thereby gravely endangering the future of an endemic honeycreeper, the palila (*Loxioides bailleui*), which depends largely on the survival of the trees, on whose seeds it feeds. According to Kramer (1971), 'feral and domestic sheep have probably wrought more changes on the original Hawaiian ecosystem than any other animal, at least on certain areas of Molokai, Lanai, Kahoolawe and Hawaii'. Feral sheep, known locally as *hipa*, were eliminated some years ago on Molokai and Lanai and more recently on Hawaii, but still survive on the uninhabited island of Kahoolawe where they number between 300 and 400.

Kerguelen

Sheep have been farmed on Kerguelen from 1909 to 1932 and since 1952. Semi-feral flocks of mixed race (among which the large black-and-white fleeced Bizet is dominant) on two or three islands in the Golfe Morbihan provide fresh mutton for the staff at Port-aux-Français. In 1973 about 70 sheep were established in the wild on the Péninsule Courbet on Grande Terre, to which they had escaped while being transferred for slaughter from one island to another. The future of this flock depends entirely on the ewes of the year since, after one season, the growth of caudal fleece prevents further copulation.[28]

Other oceanic islands

Commercial sheep farming has also been conducted on Tristan da Cunha (since about 1817); on Gough Island, where 10 sheep (the first grazing animals) were imported in 1956 to provide mutton for the South African weather station,[29] they have been confined to part of the Glen and above Capsize Sands, where they have severely reduced *Poa flabellata* tussocks; on Inaccessible (from 1938 until the 1960s) and on Nightingale Island; on the Falklands (since their colonization by the French and British in 1764 and 1765); on Nouvelle Amsterdam; on St Paul (until 1958); on Marion (1947–67); on Macquarie (for a few years after 1947); on South Georgia; on Île de la Possession (between 1969 and 1981); on Eiao and Montane (Mohotani) in the Marquesas; on Robinson Crusoe (Masatierra) and Santa Clara (about 70) in the Juan Fernandez group; and on Easter Island where their overbrowsing is believed to have been responsible for the extermination of the island's only native tree *Sophora toromiro*. In 1962 sheep were introduced to the Sierra Negra region of southeastern Isabela in the Galápagos Islands from where they have since disappeared, possibly due to predation by feral dogs.[30] They also occur on Ascension Island.

Marc Lescarbot[31] claims to have introduced the first pair of sheep to Canada in 1607, one of which was lost at Port Mouton. For several years after about the middle of the century an average of around 50 sheep a year were despatched from the mother country to New France, many of which died *en route*. When in 1661 Jean Talon, the first intendant of the new crown colony, asked for larger numbers to be imported, he was rebuked by Louis XIV's financial minister, Jean Colbert, who pointed out that the Spaniards had taken but few sheep to their colonies in the south, where careful husbandry had enabled them to increase dramatically.[32] Some of the early colonists' sheep may well have escaped from captivity, but so far as is known none are presently established in the wild in Canada. A few semi-feral animals occur on Lasqueti, Saltspring, Chatham, De Courcey and Saturna islands, British Columbia.

NOTES

1. Domestic sheep differ from their wild relatives in having a predominantly woolly pelage which lacks the guard hairs that usually conceal the winter wool of the wild forms.

2. See p. 310.
3. See p. 310.
4. For information on feral sheep in the United States I am indebted to McKnight (1964).
5. *The Bulletin* **86:** 73 (11 July 1963).
6. For much of the history of sheep in New Zealand I am indebted to Thomson (1922) and Whitaker & Rudge (1976).
7. *A voyage towards the South Pole and around the world, performed in H.M.S. 'Resolution' and 'Adventure' in the years 1772–5* (London, 1777).
8. *A voyage round the world in H.B.M.S. Resolution during the years 1772–5* (London, 1777).
9. See note 7.
10. *The old whaling days: a history of southern New Zealand, from 1830 to 1840* (Whitcombe & Tombs, Christchurch, 1913).
11. *Tutira – the story of a New Zealand sheep station* (Blackwood, Edinburgh and London, 1921).
12. A. H. Whitaker, *Ark* **2**: 171–2 (1975).
13. An unsubstantiated account claims that sheep, which later died out, were released on Campbell Island as early as 1865, and Thomson (1922) says they were liberated there between 1880 and 1890.
14. Thomson (1922) made the interesting suggestion that 'the remarkable development of the carnivorous habit of the kea or mountain parrot (*Nestor*), a bird which originally fed, chiefly, if not exclusively, on a vegetable diet . . .' could be attributed to the introduction to New Zealand of sheep, on whose carcasses and offspring keas then fed.
15. See note 11.
16. *Transactions of the New Zealand Institute* **51**: 468–9 (1919).
17. For the history of feral sheep in Hawaii I am indebted to Mcknight (1964), Tomich (1969) and Kramer (1971).
18. R. C. Wyllie, *Transactions of the Royal Hawaiian Agricultural Society* **1**: 36–49 (1850).
19. *A voyage of discovery to the North Pacific Ocean and around the world in 1790–5 under Captain George Vancouver* (London, 1798), Vol. 2, p. 115.
20. Ibid., Vol. 2, p. 127.
21. Ibid., Vol. 3, p. 11.
22. Ibid., Vol. 3, p. 70.
23. L. W. Bryan, *Paradise of the Pacific* **49**: 19 (1937).
24. L. A. Henke, *University of Hawaii Research Publication* No. 5 (1929).
25. R. E. Warner, *Pacific Discovery* **13**: 9 (1960).
26. L. W. Bryan, *Paradise of the Pacific* **62**: 113 (1950).
27. See p. 312.
28. Lesel & Derenne (1975).
29. M. W. Holdgate, *Mountains in the sea: the story of the Gough Islands expedition* (Macmillan, London, 1958).
30. D. C. Duffy, *Noticias de Galápagos* **33**: 21 (1981).
31. *The history of New France*, ed. W.L. Grant (Champlain Society, Toronto, 1907–14), Vol. 2, pp. 229 and 266.
32. Talon à Colbert, *Compte Rendu*, 103–6 (1669) and Colbert à Talon, 5 January and 5 April (1666) (P. G. Roy, ed. *Rapport de l'archiviste de la Province de Quebec, 1930–31*, pp. 44–5), all quoted by Saunders (1935).

Grey Squirrel

(*Sciurus carolinensis*)

NATURAL DISTRIBUTION
Eastern North America from the Gulf of Mexico north to the Great Lakes and the St Lawrence, west to the Prairies.

NATURALIZED DISTRIBUTION
Europe: British Isles; Africa: South Africa.

EUROPE

British Isles

The first recorded occurrence of the grey squirrel in Britain was at Llandisilio Hall, Denbighshire, north Wales, in October 1828; others are said to have been established at Llanfair Caereinion, Llan Eurvyl and Cwm Llwynog in Montgomeryshire prior to 1830: when and by whom they were introduced is unknown.

The earliest recorded release of the grey squirrel in Britain was in 1876, when Mr T. V. Brocklehurst liberated a pair, imported from America, at Henbury Park near Macclesfield, Cheshire. In 1889 Mr G. S. Page of New Jersey released five grey squirrels from America in Bushy Park, Middlesex, where they apparently failed to become established. Ten more squirrels, imported by Mr Page in 1890, were released by the Duke of Bedford at Woburn Abbey, Bedfordshire, where they rapidly increased. Two years later a pair were liberated at Finnart on Loch Long, on the borders of Dunbarton and Argyll, Scotland, from where, within a period of 25 years, they had succeeded in colonizing an area of nearly 800 km^2.

Between 1902 and 1929 a veritable wave of introductions took place throughout the British Isles, some of the stock coming direct from America and some from Woburn; at least one consignment (from which an individual was apparently released in Buckinghamshire in 1909) is said to have come from South Africa, where the species had been introduced (almost certainly from Britain) by Cecil Rhodes a decade or so earlier.

By 1930 grey squirrels were firmly established throughout south-east England as far north as Northamptonshire and Warwickshire over an area of around 25,000 km^2. By the outbreak of war in 1939 the species covered an area of about 50,000 km^2 roughly extending from Sussex and Kent in the south and east as far as Cheshire and Wrexham in north Wales in the north and west; by the end of hostilities in 1945 the animals had further extended their range into northern and eastern Yorkshire.

Between 1945 (when approximately 90,000 km^2 had been colonized) and 1956 grey squirrels moved further east into southern Essex and west into south Wales, southwestern Dorset, Devon and Somerset. They appeared in Westmorland for the first time in 1944, in Suffolk (1947), Cardiganshire and south Lancashire (1950), Merionethshire (1951) and Cornwall (1955), and spread further into Brecknockshire, Derbyshire, Durham, Montgomeryshire, Radnorshire, Shropshire, Staffordshire, Worcestershire, Lincolnshire and eastern Cambridgeshire. In Scotland, grey squirrels spread east through Fife and northwards into Perthshire.

By the early 1960s the species had colonized hitherto vacant areas of Devon, Wiltshire, eastern Cornwall, south-west Yorkshire and parts of Norfolk, Essex, eastern Suffolk, Westmorland, Cumberland and Northumberland.

Today, grey squirrels occur in suitable habitats throughout England and Wales, except for the Isle of Wight and some other offshore islands, northern Norfolk, and parts of the Lake District, Northumberland and north Durham. The Scottish population is confined mainly to the central Lowlands, but appears to be spreading slowly south-west into Peebleshire and up the valley of the Tay.

In Ireland by 1938 the descendants of stock liberated at Castle Forbes, County Longford by the Earl of Granard in 1913 had spread into Leitrim, Roscommon and Westmeath; by 1956 they had reached Armagh, Down, Fermanagh and Tyrone and are continuing to extend their range which is, however, still confined principally to some of the central and eastern counties.

Wherever the grey squirrel is not to be found, its absence is probably due to the existence of impassable barriers, unsuitable moorland or marshland habitats, and the presence of coniferous rather than deciduous forests.

In the early years of the life of the grey squirrel in Britain, the native red squirrel (*S. vulgaris*) was already decreasing in many areas, almost certainly due to disease between about 1904 and 1914, at a time when the grey had yet to obtain a firm foothold. By the time that the reds had recovered from their epidemic in the 1920s the greys were sufficiently well established to be able to resist their return, and in some places may now even be driving the reds out of their remaining coniferous strongholds.

The grey squirrel, as well as causing much damage to orchard, garden and farm crops, is both a predator of eggs and nestling birds, and a serious woodland pest: in Britain it has few if any predators, and its numbers are unlikely ever permanently to decrease.

AFRICA

South Africa

A. Haagner recorded that 'soon after 1900 the grey squirrel was introduced into Cape Town . . . by the late Cecil Rhodes [Prime Minister of the Cape 1890–6], and has multiplied to such an extent that it has now over-run the Cape Peninsula and has become a pest in gardens and orchards'.[1] Subsequent authorities have suggested that Rhodes introduced the animals during the 1890s.

The progeny of the few pairs released by Rhodes (which almost certainly came from Britain) on his Groote Schuur estate on the eastern slopes of Table Mountain spread within a decade into the nearby suburbs of Cape Town, and by about 1920 extended from the peninsula across the Cape Flats, and were established in the Paarl and Stellenbosch districts at Suider Paarl, Simondium, Groot Drakenstein, Pniel and Jonkershoek.

By the 1930s they had reached the foothills of the Franschhoek and Hottentots Holland Mountains in the Somerset West area, establishing colonies *en route* at Helderberg, Vergelegen and elsewhere. They spread east through the mountains – some being shot in Oak Valley near Elgin as early as 1933 – and by 1939 the whole of the Steenbras Catchment area near Sir Lowry Pass was infested.

Table 21 Earliest records (by Davis 1950) of grey squirrels in South Africa

Date	*Locality*
Cape Peninsula	
1901–2	Groote Schuur
1908	Tokai
1910–11	Wynberg
1914	Plumstead
1914	Hout Bay
*c.*1914	Simonstown
1919	Retreat
1932	Eerste River
Bellville District	
1930	Kraaifontein
Stellenbosch District	
Before 1914	Jonkershoek
1920	Stellenbosch
1932	Eerste River
Paarl District	
1919	Groot Drakenstein
1942–5	La Motte (Franschhoek)
Caledon District	
1933	Oak Valley, Elgin
1939	Helderfontein, Glen Elgin, Kromme River and Steenbras Catchment area
1943	Elgin
1948	Lebanon Plantation, near Groenland Mountains

Table 21 lists Davis's (1950) earliest records of grey squirrels in South Africa. Although grey squirrels were established in large numbers at Groot Drakenstein, at the foot of the Franschhoek Valley, by 1942, they did not penetrate the valley itself until the early 1950s, when they also spread into Somerset West and over the Sir Lowry Pass into Elgin. By this time they occupied an area of several hundred square kilometres over the Cape, Wynberg, Simonstown, Bellville, Stellenbosch, Paarl, Somerset West and Caledon (Elgin) districts, with a 65-km radius extending from Cape Town to Paarl and Klein Drakenstein and Franschhoek in the north-east and to Grabouw in the south-east. The animals were continuing to extend their range in the Franschhoek Valley and in and around Elgin as afforestation increased and pine plantations matured. Additional dates for the first recorded arrival of grey squirrels in South Africa are provided by Millar (1980) (Table 22).

A comparison of the range of the grey squirrel in South Africa as recorded by Davis in 1950 and the present distribution indicated by Millar in 1980 suggests that in most areas the animal has significantly extended its range; the sole exception appears to be in the Cape Peninsula where increasing urbanization and, since 1969, the removal of exotic vegetation such as pines, has continued to reduce the amount of habitat favoured by squirrels and has led to a small decrease in the population. In the Stellenbosch and Paarl divisions a consolidation and limited expansion has taken place, while in Caledon the extension of range has been considerable. After two recent misguided introductions new small colonies have become established at Swellendam and Ceres. Where distributional changes have occurred they are directly connected to alterations in the quantity and/or quality of habitat, to an expansion into previously uncolonized suitable areas or into places where agricultural or silvicultural activities have created favourable habitats.

In the Cape Division grey squirrels are still well established in the remaining stands of *Pinus pinea* above Kloofnek Road on Signal Hill, between Kloofnek and the Glen at Camps Bay, in Higgovale, in upper Gardens, in Oranjezicht, below Tafelberg Road, in the Cecilia Forest Plan-

Table 22 First recorded arrivals (by Millar 1980) of grey squirrels in South Africa

Date	*Locality*
Cape Division	
1910	Diep River
Since 1923	Rosebank
Since mid-1930s	Pinelands
1940	Durbanville: Kenridge area (Altydgedacht)
Since at least 1947	Camps Bay
1964	Durbanville (Welbeloon) (disappeared by 1970)
1965	Durbanville (Groot Fisantekraal) (disappeared by 1970)
1965	Tygerberg area (De Grendel Estate)
Stellenbosch Division	
1924	Schoogezight and Rustenberg
1920s	Banghoek
1924–8	Vlottenberg
1925	Koelenhof (Muratie Farm)
1926	Lynedoch
1930	Firgrove and Foure
1941	Muldersvlei (East Hill Farm)
1943	Kuils River (Dormershire)
Paarl Division	
1914	Paarl (Pontac Farm)
1915	Simondium
1922	Helderberg (Nature Reserve)
1928	Pniel
1929	Suider Paarl, Paarl Mountain and Franschhoek (Middagrans Farm)
1930	Bovlei, Wellington
1930	Hottentotsvallei (escaped pets from Cape Town)
1937	La Motte (Franschhoek) (from Groot Drakenstein)
1940	Klein Drakenstein and Gordon's Bay (Kariba Farm)
Paarl Division (*cont.*)	
1942–68	Wellington (Alphenhorst (1942), Welvanpas (1960), Olifantskop (1961), Welbedacht (1966), Diemiesfontein (1968) and Vrugbaar (1968) farms)
1952	Sir Lowry Pass (Beaconsfield Farm)
1970	Klapmuts (Uit Komst Farm)
Caledon Division	
1925	Grabouw (probably from the Cape Peninsula)
1930	Oak Valley, Elgin
1934	Eikenhof, Elgin (escaped pets from Grabouw and natural spread from Somerset West)
1934	Glen Erskine, Elgin (escaped pets from the Cape Peninsula)
1946	Houwhoek (Korteshoven and Oakdale)
1955	Vyeboom (Rus-en-Vrede)
1957	Vyeboom (Patryskop) (from Grabouw)
Probably after 1960	Eerstehoop (Twaalffontein) (from Vyeboom)
1963	Villiersdorp
Before 1967	Nuweberg Plantation
1967	Villiersdorp (Bo Radyn)
1969	Bot River (Goedvertrou)
1970	Villiersdorp (High Noon and Elandskloof Valley)
Swellendam Division	
1957	Forest Plantation and along the Koringlands River in northern Swellendam (introduced from Groot Drakenstein)
Ceres Division	
1968	Belmont Hotel and Municipal Camping Area (introduced from Stellenbosch)

tation on Devil's Peak and especially on the Groote Schuur Estate from Devil's Peak to Newlands. They are also still abundant in Pinelands, upper Claremont, Bishopscourt, Wynberg, Constantia Diep River, Tokai, Westlake, Noordhoek, on the Silvermine Plateau and in Hout Bay, and occur in smaller numbers in *P. pinaster* plantations on the summit of Table Mountain, on Rondebosch Common and on the Rondebosch, Royal Cape, Mowbray and Westlake golf courses. Two small discrete colonies of squirrels survive in *P. pinea* stands at Glencairn and Kommetjie, but there are no longer any around Simonstown. In the northern parts of the Cape Division Millar recorded apparently isolated colonies at Tygerberg and Kenridge and on pastoral land over the surrounding valley east of Durbanville and at Fisantekraal.

From Millar's survey it would appear that in the Stellenbosch Division grey squirrels were considerably more widespread by 1950 than indicated by Davis. As well as having reached Koelenhof, Muldersvlei, Kuils River, Firgrove and Faure (see above), they were also presumably present along the banks of the Eerste River. Since 1950 their main direction of expansion seems to have been northwards from Koelenhof and Bottelaryberg into previously tree-less veld now converted to vineyards and orchards with windbreaks of *P. radiata*. If further land is developed in this area there is likely to be an accompanying extension of range by grey squirrels.

In the Paarl Division squirrels had colonized most of the area they now occupy – from Bovlei and Wellington in the north to Pneil in the south and the head of the Franschhoek Valley in the east – by the time of Davis's survey in 1950. Millar reported expansions of range to Klapmuts in the west and along the Klapmuts River; around and over Paarl Mountain as far west as Windmill; from Wellington north to the Berg River valley in the west; and to northern Groenberg in the east. Expansion is only likely to continue in this region if fruit-farming and forestry spread further northwards.

In the Ceres Division two small colonies – totalling perhaps 100 individuals – have recently become established in a grove of oaks (*Quercus robur*) near the Belmont Hotel in the town of Ceres and in a plantation of *P. pinaster* in the municipal camping area; these colonies derive from a single pair foolishly translocated from Stellenbosch in about 1968 by the present owner of the hotel.

The most significant extensions of range and, especially of numbers, by grey squirrels in South Africa have occurred in the Caledon Division, thus confirming Davis's 1950 prediction that this was likely to occur as afforestation increased and existing exotic plantations matured. Millar records that between 1950 and 1969 afforested land in Caledon increased by 77 per cent from 7,740 ha to 13,671 ha. The spread of intensive fruit-farming, and of *P. pinaster* into protea-heath *fynbos* and pastureland, also added considerably to the available habitat suitable for squirrels. Since 1950 the animals have consolidated their base in the Elgin basin and have spread up the Steenbras River catchment to the west, into the plantations of *P. pinaster* on the slopes of the Hottentots Holland and Groenland Mountains to the north, from the Kromme River and Helderfontein and Glen Elgin eastwards, and from the Lebanon Forest Plantation to Houwhoek.

The main new areas of colonization in the Caledon Division, however, are over and north-east of the Groenland Mountains into and throughout Villiersdorp and the upper Sondereind River basin to Franschhoek Pass in the north and Villiersdorp in the east.

In 1957 Muller and Alphons Conradie, the two sons of the foreman of the Swellendam Forestry Plantation, Mr G. D. Conradie, irresponsibly released seven grey squirrels from Paarl in the plantation, situated about 200 km from Cape Town and 100 km from Villiersdorp; here they are at present confined to a section of 16 ha of *P. pinaster*, and to groves of oaks (*Q. robur*) along the Koringlands River to the north of Swellendam.

The grey squirrel in South Africa is restricted to areas of mixed timber plantations with higher than average rainfall, and to orchards and vineyards; it is generally absent from pastoral land, exotic *Eucalyptus* spp. plantations, bush and veld, and only occurs in urban surroundings with an adequate supply of tree cover and watercourses. The existence of at least one of the three staple seed-bearing food species – *Pinus pinaster*, *P. pinea* and *Quercus robur* – seems essential to its survival.

If afforestation with exotic species, fruit-farming and viniculture continue to increase in South Africa, Millar concluded that the most likely areas of further colonization by grey squirrels would be:

1. Forest plantations of *P. pinaster* and/or *P. canariensis* in, for example, Kluitjieskraal and Waterval near Wolseley, close to existing populations at Ceres and Wellington.
2. Deciduous fruit-farming areas near pine and/or oak plantations of the Klein Berg and Breede River valley near Tulbagh, Wolseley, Goudini and Rawsonville.
3. The southern and eastern slopes of the Paardeberg near Malmesbury.
4. From the Villiersdorp basin eastwards down the Sondereind River valley towards Greyton, Genadental and Swellendam.
5. Isolated exotic pine plantations at, for instance, Garcia near Riversdale, the Southern Cape Forest Region and the Tsitsikamma Forest Region. Grey squirrels are unlikely to reach these areas through a natural extension of

range, but if misguidedly translocated could develop into one of the largest and most extensive populations in South Africa.

In South Africa the grey squirrel's main foods include acorns from *Q. robur* and pine seeds and cones, especially from *P. pinea*, *P. pinaster* and *P. canariensis*, fruits (especially pears, peaches and grapes), bulbs (mainly daffodil), green shoots (principally those of cannas and vines), green gum capsules (especially of *Eucalyptus calophylla* and *E. ficifolia*), fungi, almonds, green mealies (maize), lizards, insect larvae and eggs and nestling birds. The animal's principal predators include feral and wild cats and boomslangs (*Dispholidus typus*); Argentine ants may kill some young. Although there is, surprisingly, no proof of damage by grey squirrels to South African forestry plantations such as occurs in the British Isles, complaints are made from time to time about depredations on fruit-farms and in gardens, and to the country's native birdlife. As early as 1918 grey squirrels were declared vermin in South Africa, and retained this status until the bounty system was abolished in 1958.

Barrett (1934) recorded that 'many years ago' grey squirrels were liberated by the late Sir Frederick Sargood on his estate, Ripponlea, between Elsternwick and Balaclava, near Melbourne. In about 1940 a single pair of grey squirrels was released in the botanic gardens at Ballarat, Victoria. Although Millar (1980, quoting Barrett (1934) and J. C. Taylor (personal communication, 1973)) states that 'In Australia . . . the squirrel is confined to gardens in the city of Melbourne', the animals disappeared from there in the 1930s and from Ballarat in the late 1960s.

At some time before 1914 three or four pairs of grey squirrels (probably from Ontario) were released in the 400-ha Stanley Park, Vancouver, British Columbia, some 2,000 km west of their nearest natural range. By about 1920 the population, which is prevented from expanding by the Pacific on three sides and by the city on the fourth, had reached the density at which it has remained more or less unchanged ever since.

In the autumn of 1966 the Vandermeer Game Farm in Metchosin, southern Vancouver Island, acquired two females and one male from Ontario. These soon escaped, and their descendants, according to Guiguet (1975), are currently established in the areas around Weir's Beach, East Sooke, Esquimalt and Saanich.

Grey squirrels have also been translocated outside their natural range in North America to Saskatoon, Saskatchewan, southern Quebec, New Brunswick and Nova Scotia in Canada, and to Oregon, California, Washington and Montana in the United States.

It is a curious fact that, whereas in Britain the grey squirrel is a serious pest species, little if any damage is reported from either South Africa or its North American homeland; in the latter, indeed, it is regarded as a valued sporting asset (over 130 million are said to be shot every year) and is esteemed as a culinary delicacy.

In about 1898 Lieutenant-Colonel E. A. Le Souef released some **northern Indian palm squirrels** (also known as five-striped or five-lined palm squirrels) – (*Funambulus pennanti*) – in the grounds of the South Perth Zoological Gardens, Western Australia, of which he was at the time director. Here they settled down well, multiplied and were soon firmly established.

For some years palm squirrels have also occurred – though in lesser numbers – outside the zoo grounds. In 1967 one was seen in Raphael Park, some 5 km from the zoo, and by the following year they were being regularly reported from residential districts of south Perth and Como. They have spread only slowly, however, and now occupy an area bounded by the Swan River foreshore from the Causeway to Canning Bridge, but have crossed the Canning Highway to its southern side.

Palm squirrels seem not to favour native eucalypts, but instead prefer gardens with exotic trees, especially pines and palms, with an abundant supply of water. It seems likely that the sand plain, scrub and farmland surrounding Perth have prevented their expansion in the past. Recent development, however, resulting in the extension of residential suburbs and gardens, may allow them eventually to spread throughout the Perth metropolitan area, outside which they are unlikely to penetrate.

In about 1942 some palm squirrels were translocated to Taronga Park Zoo, Sydney, New South Wales, where a few escaped and established a colony in a park nearby.

Although in the 1940s consideration was given to exterminating palm squirrels, they appear to cause little if any damage in Australia.

Between 1966 and 1970 two or three pairs (later followed by at least two more pairs) of **Barbary ground squirrels** (*Atlantoxerus getulus*), which are natives of Morocco and Algeria and not unlike palm squirrels in appearance, were released on Fuerteventura in the Canary Islands. By February 1978 they had spread over much of the arid montane area of the island. Little is yet known about their diet on Fuerteventura, but they could well become a pest to crops and reduce still further the sparse vegetation of the island's arid zone.[2]

The **Formosan golden-backed** or **grey-bellied tree squirrel** (*Callosciurus caniceps thaiwanensis*) was introduced from southern Formosa to zoological gardens on Oshima Island, south of Tokyo, Japan, some time after 1940. During the war the animals escaped, and by 1950 had increased to an estimated population of 20,000. They strip the bark from trees and shrubs (mizuki (*Cornus controversa*) and yashabushi (*Alnus firma*)) and eat commercially grown nuts. The principal harm they cause, however, is by preventing the insectivorous and nectar-seeking white-eyes (*Zosterops palpebrosa* or *Z. japonica*) from pollinating camellia bushes, whose oil was once the island's main product, and by eating the camellia seeds. Their principal predators are crows, hawks and snakes (*Elaphe climacophora*).

One or two pairs of the **red-bellied** or **Guatemalan grey squirrel** (*Sciurus aureogaster*) – a native of central Mexico and Guatemala – were in 1938 or 1939 released by a plantation owner on Elliott Key, Dade County, in Biscayne Bay, Florida, USA. Although by the 1950s they are said to have increased to several hundred, by 1969 only eight were counted in 'dense, subtropical, hammock-forest' in the Elliott Key County Park, where they were observed to feed largely on the fleshy yellow fruit of mastic (*Mastichodendron foetidissimum*) and on seeds of the West Indian mahogany tree (*Swietenia mahogoni*). In the same year the US Congress designated Biscayne Bay National Monument (which includes Elliott Key), so the future of this unique colony may have been safeguarded.[3]

The **Altai squirrel** (*Sciurus vulgaris altaicus*) was in 1937 introduced to the Teberda State Reserve in the Soviet Union, where it has become established in river basin forests in Kuban, Teberda, Rion and elsewhere in Transcaucasia. The **large-toothed souslik** (*Spermophilus fulvus*) from Turkestan has also been successfully transplanted in the Soviet Union.

Although the **Eurasian red squirrel** (*Sciurus vulgaris*) is indigenous on the mainland of the British Isles it was extinct in Ireland and the Lowlands of Scotland by the early eighteenth century, and nearly so in the Scottish Highlands in the late eighteenth/early nineteenth centuries. Its present status in Scotland is due to reintroductions (mainly from England) to at least 10 localities between 1772 (at Dalkeith, and 1793 at Dunkeld) and 1872, and in Ireland to reintroductions from England to about 10 sites between 1815 and 1856.[4]

The **alpine marmot** (*Marmota marmota*) of central Europe has been translocated to Germany (1887), Austria, Switzerland (1883), northern Yugoslavia (1953), the Carpathians, the Pyrenees,[5] the northern Apennines, Italy, and the Soviet Union.

Colonies of the **Siberian chipmunk** (*Tamias sibiricus*), whose natural distribution is northern Russia and Siberia, have become established – as the result of escapes from captivity – in parts of France, West Germany, the Netherlands and Austria. *T. asiaticus uthensis* (now *T. sibiricus*) was said to be established in the 1930s in the Tiba Prefecture of Japan. In about 1906 Mr P. R. Sargood unsuccessfully released a pair of **eastern chipmunks** (*T. striatus*) and a pair of 'brown Californian squirrels' (?*Spermophilus beecheyi*) at Dunedin, New Zealand. Recently some *S. beecheyi* arrived on the island of Kauai in Hawaii in a consignment of containerized freight from

California; fortunately they were eradicated before they could become established.

The **American red squirrel** (*Tamiasciurus hudsonicus*) has been successfully transplanted from the Canadian mainland to insular Newfoundland.

Transplanted populations of the **black-tailed prairie dog** (*Cynomys ludovicianus*) – a native of the central United States – exist, or formerly existed, in Sac County, Iowa; Monroe, Louisiana; Seneca, South Carolina; and on Nantucket Island, Massachusetts. In 1965 a small colony of four males and one female was trapped and destroyed near O'Neill Park, 16 km east of El Toro in Orange County, California. A few years ago some black-tailed prairie dogs were illegally landed on Kauai in Hawaii. Contrary to one report (Linn 1979), there are no free-ranging prairie dogs in the British Isles.

NOTES

1. *South African mammals: short manual for the use of field naturalists, sportsmen and travellers* (H. F. & G. Witherby, London, 1920).
2. *Oryx* **15**: 7(1979).
3. Brown & McGuire (1969).
4. Shorten (1954).
5. M. A. J. Couturier, *Säugetierkundliche Mitteilungen* **3**: 105–8 (1955).

North American or Canadian Beaver

(*Castor canadensis*)

Note: Present natural distribution much reduced from original range shown.

NATURAL DISTRIBUTION
Formerly all forested parts of North America south to northern Mexico. Now, as a result of deforestation and persecution by man, absent from many of its former haunts, but extensive reintroduction programmes have taken place in the USA and Canada.

NATURALIZED DISTRIBUTION
Europe: Finland, ?France; Eurasia: USSR; South America: ?Argentina.

EUROPE

Finland[1]

In 1937 seven American beavers were imported to Finland from New York State, to which they had been translocated from the Canadian Rocky Mountains and the Yellowstone National Park in

Wyoming, USA. They were released in three regions in southern Finland, to two of which 19 European beavers had been reintroduced from Norway in 1935–6. In Sääminki in southeastern Finland the North American beavers did exceptionally well and by 1945, when a number were transferred to other parts of the country (especially in the north and east), the population in Sääminki alone was estimated to number between 130 and 140. A decade later it had increased in the country as a whole to between 450 and 500 and was expanding its range especially in the south-east. By 1960 the combined national population of American and European beavers had risen to 500–600, by 1965 to 1,800–3,000, by 1970 to an estimated maximum of 3,200 and by 1975 to some 6,000 – most of which were believed to be of North American origin.

Today, over half the Finnish beaver population is situated in the east of the country, which is the only area where numbers have noticeably increased since about 1965; the lakelands surrounding Lake Saimaa in the south-east also support a sizeable number. In both these areas the population consists entirely of North American animals.

The construction by beavers of dams and lodges causes a considerable amount of timber damage in Finland by flooding; in 1970 approximately 900 ha of plantations were killed, and the growth-rate of a further 1,600 ha was seriously impaired by lengthy inundation of the trees' root-systems. Between 1954 and 1974 beaver trapping in Finland yielded about 1,600 pelts.

American beavers are at present more widely distributed and exist in larger numbers in Finland than the native European beaver (*Castor fiber*), which is generally less prolific; although in some areas the two species live sympatrically, interbreeding is apparently unknown.

France

In 1975 two pairs of American beavers presented by the Quebec government were released on a small lake in the grounds of the Parc Naturel Saint Herbert de Boutissant near St Fargeau, Loiret: they soon escaped and constructed a dam, and subsequently a lodge, on the neighbouring Reservoir du Bourdon.

Apart from the possible existence of a population of allegedly American beavers in Poland (the identity of which has been disputed) the Finnish/USSR populations (and possibly that in France) are believed to be the only ones currently extant in Europe.

EURASIA

USSR

In 1933–4 seven American beavers released in the Rovno district of the Ukraine succeeded in establishing a small population which had, however, died out by 1957.

In the early 1950s (one account says not until about 1960) American beavers crossed the frontier from Finland into western Karelia and the Leningrad Province of the Soviet Union. By 1977 the population was estimated at around 1,000, and was continuing to expand.

In 1969 and 1971 54 American beavers from Leningrad Province were released on the Obor and Kur rivers in the Khabarovsk Territory. In 1975–6 a further 100 or more were liberated on the Selgon River in Khabarovsk and in the Amur area. More recently American beavers have been introduced to Kamchatka in the northeastern USSR, where they are now breeding. Plans have also been announced for experimental introductions to the Amur-Ussri Territory and to Sakhalin Island.

Russian scientists consider that *C. canadensis* is more suited to the Soviet Far East than *C. fiber* because it adapts better to life in lotic montane streams and to a harsher climate, and because its tendency to feed, even in summer, on twigs and leaves of trees (primarily birch (*Betula* sp.)) favours its success in the northeastern catchment areas which are poor in herbaceous food-plants but rich in the number and variety of trees and shrubs. Although it is almost certain that the two species cannot hybridize, there is the additional advantage that in the Far East (unlike in western Europe) their ranges do not overlap.

SOUTH AMERICA

Argentina

Canadian beavers introduced in 1956 to Lago Fagnano on Isla Grande de Tierra del Fuego were still present in 1971, when damage was observed at Bahia Buen Sucesco.

The **European** (or **Eurasian**) **beaver** (*Castor fiber*) formerly ranged throughout the forested areas of the Palaearctic region apart from the Mediterranean zone and Japan. In Spain it died out in the sixth century, in Britain in about the twelfth century, and in Italy in the sixteenth century, having been hunted to extinction by man for its flesh, its fur and for its *castoreum* – a glandular secretion reputedly possessing valuable medicinal properties.

By the 1920s the European beaver survived only in parts of Russia (in the west; in north-west Siberia and in the Altai Mountains); in Poland; along a stretch of the middle Elbe in Germany; in the Camargue delta of the Petit Rhône in southern France; and on the River Nid in south-east Norway. Since then it has been successfully reintroduced to Sweden, Finland, Switzerland and Austria, and the German, French and Western Siberian and European Russian populations have been reinforced. The possible reintroduction of beavers to Belgium is under consideration, but a planned liberation in the Netherlands had to be abandoned due to the destruction of the proposed release site in the Delta Project. In several countries damage to crops and timber (by flooding and debarking) has been reported.

In 1874 the Marquess of Bute attempted to acclimatize a pair of North American or Canadian beavers (and a pair of European beavers) in the woods near Kilchattan Bay on the Isle of Bute, Scotland, but they survived only until 1890. In that year Sir Edmund Loder, Bt, introduced some American beavers to his estate at Leonardslee near Horsham in Sussex, where the last one died in 1948. Both these colonies were within enclosures. In about 1870 a number of American beavers were released into the wild at Sotterley Park in Suffolk where they subsequently bred, two of their progeny eventually reaching Benacre Broad, some 6–8 km away: within a few years this colony also had died out or had been exterminated.

A misguided proposal in 1977 to reintroduce the European beaver to Britain fortunately found little support and much opposition.

NOTE

1. Lahti & Helminen (1974).

Muskrat
(*Ondatra zibethicus*)

NATURAL DISTRIBUTION
North America from the Barren Grounds of the Arctic south to the Rio Grande in central Mexico, and from the Pacific to the Atlantic coast.

NATURALIZED DISTRIBUTION
Europe: British Isles (1922–35), Czechoslovakia, Germany, Austria, Hungary, Poland, Bulgaria, Albania, Romania, Yugoslavia, France, Belgium, Greece, Italy, Switzerland, the Netherlands, Luxembourg, Finland, Sweden, Norway; Eurasia: USSR; Asia: Mongolia, China, Japan, Korea; South America: Argentina, Chile.

EUROPE

British Isles

Muskrats, whose fur is known to the trade as musquash, were first introduced to Britain in about 1920, when a number turned down near Oban in Argyllshire, Scotland, failed to become established. However, of six pairs placed in a fenced enclosure at Feddal, near Braco in Perthshire, several soon escaped, and within two years they and their progeny had built no fewer than 16 lodges on a marsh over 3 km distant. Before long, a total of nearly 90 km of the rivers Forth, Teith and (especially) the Earn and of the Allan Water had become infested.

A pair of muskrats imported from Malachi, Ontario, Canada to Annabeg near Nenagh in County Tipperary, Ireland in July 1927 promptly escaped, and within five years their offspring had colonized an area of nearly 400 km^2.

In England at least 76 musquash farms are known to have been established by 1930; between 1929 and 1931 muskrats had escaped or been deliberately released from no fewer than 14 English and 5 Scottish farms, and within two or three years had been reported from 14 English, 9 Scottish, 1 Welsh and 2 Irish counties. The main area of infestation in England was centred on Shrawardine, 10 km west of Shrewsbury in Shropshire, where 120 muskrats were introduced from North Rice Lake, Ontario. In 1930, following an escape, some made their way to the Severn and by late the following year had overrun 30 km of the river (whose banks in places they seriously undermined) above Shrewsbury, eventually spreading over a distance of almost 60 km, with outliers being reported from as far distant as Welshpool upriver and downstream to Bewdley.

In March 1932 the Destructive Imported Animals Act made it illegal in future to keep muskrats in Britain without a permit, and in the following year their farming was completely banned. Similar legislation was also passed in Ireland.

In the five main regions where muskrats were most firmly entrenched – in the upper Severn Valley (mainly in Shropshire), on the River Wey near Farnham in Surrey, in the marshes and dykes bordering the River Arun in west Sussex, and around Feddal and Nenagh – intensive trapping was started in June 1932 which resulted by the end of 1943 in the capture of no fewer than 2,672 muskrats in the Severn district, 910 in Scotland, 159 in Sussex and 52 in Surrey – a total of 3,793. Between September 1933 and May 1934 487 muskrats were trapped in Ireland, where the creatures were then regarded as exterminated. By the end of 1934 muskrats must also have been nearly eradicated in Britain, as no young were trapped in 1935, and before the end of the decade natural causes had accounted for any survivors.

Today, nearly half a century later, the muskrat remains the only established alien vertebrate to have been successfully eliminated from the British Isles.

Czechoslovakia; Germany; Austria; Hungary; Poland; Romania; Bulgaria; Albania; Italy; Greece; Yugoslavia[1]

In the spring of 1905 Prince Colloredo (or Colleredo) Mannsfeld brought back to Bohemia[2] from a hunting trip to Alaska three pairs of muskrats of which one, a male, died on the journey. The survivors were released into two ponds on the Prince's estate near Dobřiš (Dobrisch) on the River Kocaba, 40 km south-west of Prague, by the Forest and Domain Director, Mr K. Waschatko. Within two years they were breeding prolifically, and damage to watercourse embankments was being reported. A few years later they were augmented by a shipment of seven pairs from the animal dealer Carl Hagenbeck, who is believed to have derived his stock either from Quebec, Canada or from Ohio, USA.

Between 1905 and 1913 the rate of the muskrats' expansion in Bohemia varied from 4 to 30 km a year, and within a decade of their release the whole of the region had been overrun; having crossed the border near the River Regen, about 140 km from Dobřiš, the first muskrat in Germany was killed in Bavaria in 1914, where between 1923 and 1924 the animals are said to have spread at between 50 and 70 km a year.

In 1914, nine years after the initial liberation near Dobřiš, the Bohemian muskrat population was estimated to amount to two million. In the same year the Upper Palatinate and Lower Bavaria in Germany and Lower Austria were infested; Upper Austria and Moravia were overrun by the following year; Saxony (Germany) by 1917 (possibly by 1913–14); Thuringia (Germany) by 1919; Prussia (Germany) by 1922; Bundesland, Vienna, Austria by 1924 (with a few isolated records from 1922); Hungary by 1924 (a single report in 1922); Prussian Silesia (Germany) by 1924 and probably also Czechoslovakian Silesia; and Salzburg and the Burgenland (both Austria) from 1925, although isolated cases had been reported in the last in 1915 and 1923. A single appearance occurred in Styria in Austria in 1924, and in 1927 parts of Middle and Lower Franconia in Bavaria, and in the following year Brandenburg in Prussia, were colonized.

By 1929 muskrats had travelled 420 km south-east down the Danube into Hungary, but only around 200 km north-east as far as the Glatzer Mountains in southwestern Poland, where they were first seen during the winter of 1929–30 in Oberschlesien, and subsequently in the Kreissen of Rybnik, Teschen, Pless, and by 1933 in Wojewodschaff Kielee. From 1932 the deliberate translocation of muskrats in Poland was punishable by a fine of 2,000 zloty or six weeks imprisonment.

By 1930 the drainage-system of the River Elbe had been occupied from its source to Magdeburg; that of the Danube from Ingolstadt over Gran into Hungary; the Weser's, especially in its source river, the Werra (in 1921), and the Rhine's in the Main; there was also a single colony of muskrats on the Oder. The occupied territory stretched from around 47° to 52° N and from 10° to 18° E and embraced some 200,000 km^2.

It was soon found how destructive – by burrowing and by their consumption of vegetable crops – muskrats could be: intensive trapping – initiated in Bavaria in 1917 – had little effect on either the increase in population or on its expansion of range, which by the late 1960s in Germany had spread further downstream on the Rhine to beyond the confluence of the Lippe, to the west of which a small discrete colony was established at Xanten. In northwestern Germany muskrats continued to spread towards the Netherlands along the Ems/Vechta Canal, while a number were trapped as far downstream as Aschendorf, north of the Hümmling Hills. By the early 1970s, with the exception of a few areas in Schleswig-Holstein, in northern Oldenburg and along the east bank of the Rhine, most of Germany had been overrun.

From Germany muskrats spread south-east along the Danube via Austria into Yugoslavia, and via Czechoslovakia and Hungary into Romania; by 1960 they had spread over much of Banat and Crisana in the west and in the Danube delta in the east, and were becoming established in parts of Maramures, the lowlands and plateau of Transylvania, in swamplands along the Danube, in western Dobruja and in the eastern and southern half of Moldavia in Romania. From the last muskrats have extended their range into northern Bulgaria, Albania, northern Greece and northern Italy. In Bulgaria, Czechoslovakia, Hungary, Poland, Romania, Yugoslavia and Finland (see below) between 50,000 and 200,000 pelts are harvested annually.

France; Switzerland

Musquash-farming is believed to have been first tried experimentally in France in about 1919, but was not widely undertaken until 1927–8. Muskrats were first successfully bred in captivity in France by Madame L. Pereyrol, who described her experiences in *Élevage et Fourrure* in September 1928. As with so many other captive animals escapes soon took place (some were apparently also deliberately released 'to satisfy the biological requirements of this ungovernable rodent'),[3] the first capture in the wild in France, at Grenoble in 1930, being described in the March issue of *Procès verbaux de la Société dauphinoise d'Études biologiques*. The earliest centres of infestation appear to have been near Amiens, in the Seine-Maritime and near Sélestat in the Bas-Rhin. The April 1933 edition of the *Bulletin de la Société des Amis des Sciences naturelles de Rouen* reported captures in Eure and Lower Seine, and in the March issue of the *Bulletin de la Société nationale d'Acclimatation* of the following year A. Chappellier, Director of the Service of Injurious Vertebrates of the National Centre of Agronomical Research at Versailles, summed up the current available information about muskrats in the wild in France.

Chappellier recorded a considerable number of areas of infestation; in the agriculturally important basins of the Seine and Somme north and west of Paris muskrats were established in the *départements* of Eure, Somme, Seine, Oise, Pas-de Calais and Lower Seine – in the last of which they were especially abundant; in eastern France they had settled in Ardennes; near the Belgian frontier in Meuse; in Meurthe et Moselle; in Alsace; in Doubs; and in the Territoire de Belfort near the border with Switzerland (where they were first seen near Vierwaldstätter See in about 1934 and where they are at present confined to the cantons of Bern and Basle); the central and southeastern *départements* of Allier, Loire and Isère had also been colonized, via the Rhône. The most serious damage – mostly from burrowing – was reported from Eure, Belfort and Doubs.

By 1953 23 *départements* north of a line through Avranches, Laval, Le Mans, Blois, Orléans, Melun, Laon, Nancy and Belfort had been overrun: between 1953 and 1960 muskrats further expanded their range to the east and west, while from 1960–7 they dispersed southwards – their rate of spread being slowed but not prevented by trapping activities. Thus, by the end of 1967 the southern boundary of the French area of infestation ran from the mouth of the Charente river through the *départements* of Charente-Maritime, Deux-Sèvres, Vienne, Indre, Creuse, Allier, Saône-et-Loire, Ain, Jura and Doubs to where the Rhône/Rhine canal crosses the Franco/German border at Le Russey. A further isolated pocket of infestation occurred along the Rhône between the confluence of the Drôme and Ardèche.[4]

Belgium[5]

Musquash-farming in Belgium began in 1928 at Begijnedijk near Aarschot and at Huissignies and Manderfeld. When, two years later, further farming was prohibited the last two farms were closed down, but some of the stock at Aarschot managed to escape and settled in swampland between the rivers Nèthe and Démer: from here muskrats spread steadily northwards, and by 1941 had crossed the Dutch border into northern Limburg; four years later the entire province of Antwerp and the north of Limburg and Brabant had been overrun; by 1950 muskrats had spread westward to parts of East Flanders, and in Brabant as far south as the River Dyle near Louvain.

Despite an official trapping campaign begun in 1953 muskrats continued their spread around Ghent in East Flanders, south of Louvain in Brabant, in southern Limburg and in northern Liège. After 1965 the position rapidly deteriorated still further; the rate of expansion westwards into East Flanders increased apace, and colonization was intensified in northern and eastern Hainaut and the provinces of Antwerp and Brabant.

Extreme southern Belgium remained uninfested until about 1959, when muskrats spread westwards from Germany via the rivers Sure, Attert, Eisch and Pal (tributaries of the Moselle) and across the hills between the Moselle and Semois basins, and from between Mézières and Belfort in northeastern France (animals from these last two centres are believed to have met and interbred somewhere south of the Grand Duchy of Luxembourg, giving rise to a hardier and more prolific strain with a characteristic black, red or chocolate-brown pelage): by 1961 the whole of the Semois basin had been overrun. In the previous year the River Our in Liège, and later the rivers Warche and Amblève, had been colonized, and muskrats began spreading northwards into Germany; by 1961 the rivers Houille, Lesse and Meuse (as far as Dinant) had become infested, and outliers appeared 25 km further north near Namur. In the province of Liège muskrats became established along stretches of the smaller Roer, and in 1963 along the Ourthe (downstream to within 6 km of the city of Liège), and later in the basins of the Virton and Sambre.

Between 1964 and 1966 (when a small discrete colony of melanistic mutants was discovered 35 km to the east near Hasselt) the original colony at Begijnendijk merged with muskrats moving up from the south – especially in the Gette and Dyle basins, with migrants re-infesting the River Geer. Within the space of two years the populations in southern Belgium had occupied a larger area than that colonized from Begijnendijk in 35 years, and the success of a trapping campaign which between 1953 and 1963 had eliminated muskrats from some 15,000 km^2 of the province of Limburg had been nullified.

In 1955 muskrats from Eure-Seine in northern France crossed the Belgian border via the Yser basin in West Flanders; seven years later this population began spreading north and east, and by the end of the decade virtually the whole of the country had been overrun.

Netherlands[6]

The first muskrat in the Netherlands was caught in 1941 – allegedly one of a number smuggled over the border from Belgium and sold, to obtain a government bounty, to Dutch trappers. By 1951 (when 569 were trapped) muskrats had colonized an area of 750 km^2. The number captured (and their range) gradually increased during the next decade to an annual total of around 700. The hard winter of 1962–3 was responsible for a population explosion which resulted in an increase in the annual take to more than 3,000 in 1963, over 5,000 in 1964, and 11,700 in 1967. At the same time the area of infestation increased from some 1,600 km^2 in 1962 to 2,300 km^2 in 1967; this expansion of range was achieved largely by the invasion of the tidal saltwater marshes of the province of Zeeland and those of the River Scheldt in North Brabant (where large populations became established as far as the Maas), which are more vulnerable to muskrat-damage than are eastern North Brabant and Limburg, and where difficulty of access hindered effective control. Previously it had been possible to intercept invading muskrats from Belgium, and few young were born in the Netherlands; the lengthening of the muskrats' irregular front made efficient control increasingly difficult, and before long animals were crossing the German border into the provinces of Guelderland and Overyssel.

In recent years land development – including the industrialization of the west bank of the Scheldt in Belgium and the construction of the Scheldt/Rhine canal and the concomitant reclamation of the 'Schorren' (salt-marshes) of Bath – have influenced the expected pattern of expansion which has proceeded in various directions.

The most important migratory route into Zeeland has been along the Scheldt/Rhine canal. The main infiltration of Limburg has been via Germany, and the lower reaches of the Maas and parts of the Peel have become infested. Since 1968–9 there is evidence of an intensifying influx from Germany into the eastern border regions, while north of the Zuiderzee the fact that muskrats have been seen on German islands in the Waddenzee and in or near sea-dykes along the Netherlands' Waddenzee coast suggests that they are migrating outside dykes along the Waddenzee coast from East Friesland in Germany.

Luxembourg[7]

Muskrats first appeared in the Grand Duchy of Luxembourg – probably from France – in 1955, and until 1960 were confined to the south of the country. Thereafter they began to extend their range progressively northwards and westwards, and by the end of the decade had occupied all suitable habitats and were rapidly approaching the Belgian, Dutch and northern German borders. In the decade after the start of official trapping in 1958 a total of 16,788 muskrats were caught in the Grand Duchy.

Finland[8]

The earliest (unsuccessful) release of muskrats in Finland was in Lake Ruuhijärvi near Kajaani in 1919 or 1920. This was followed by a second and more successful liberation in Kruunupyy in 1922, where the animals became established and began to spread. Another introduction is said to have been made in Kronoby north of Vaasa on the Gulf of Bothnia in 1919 or 1922: others followed in 1923 in Ylivieska (Uleaborg *län*), 80 km north-east of Kronoby, and in Tunlos (Tavastehus *län*), 120 km north of Helsinki. From the latter, situated in a lakeland region, muskrats spread – both naturally and with the assistance of trappers – at a rate of 30–40 km a year.

During the 1920s and 1930s a total of around 2,300 muskrats were released at no fewer than 293 localities in Finland from Hankoniemi in the south to well inside the Arctic Circle at Inari. During this period the wild population gradually built up, and after 1937 muskrats were only turned down in the more remote northern and central regions. The highest number of releases (61) took place in the Vuoksi drainage basin, whereas there was only one in the Ahvenanmaa Islands and in the basin of the River Suojoki. The largest individual release (200 pairs) was in Lake Kirkkosurnunen in Pieksämäki; other sizeable liberations included 50 pairs in Lake Kuusjärvi, Virrat, 29 pairs at Häggesböle, Kirkkonummi, 20 pairs in Lake Kirakkajärvi, Inari, in Lake Portimojärvi, Ylitornio and in Lake Kiurujärvi, Savonranta. The greatest number of muskrats (579) were turned down in the Kymijoki drainage basin. Most of the stock was imported direct from Ontario, Canada – a small proportion (about 103) coming from Czechoslovakia; the releases were highly successful.

Some 300 so-called 'Virginian' muskrats (*O. z. macrodon*) – the dark 'blue' or 'black' phase – imported via Maryland and Pennsylvania and released in 46 localities fared less well than the nominate race. Introduced to all the country's main water-systems except in Lapland – in the Tornion/Muonionjoki River basin, to the waters of the Kuusamo, in the drainage areas of northern Pohjanmaa, to the watercourses of the Uusimaa, on the Ahvenanmaa Islands and in the Suojoki River basin – 'Virginian' muskrats today remain only in small numbers in the Saimaa area, where they comprise only 2–3 per cent of the annual catch. The probable reason for their comparative failure is the milder climate of their native range.

By 1927 muskrats had been released into most Finnish water-systems. Their distribution, however, remained mainly local, except for Pohjanmaa and the Kokemäenjoki drainage basin where they were slightly more widespread.

By 1932, partially as a result of a further 105 releases, large areas of northern and southern Pohjanmaa and the Kokemäenjoki basin were colonized. In Uusimaa the muskrats' range increased dramatically – the only area from which the animals were absent being Kuusamo, where none had ever been turned down.

By 1937, following 137 more releases, all discrete populations in western Finland had merged: muskrats were by now more or less continuously distributed over the major part of the south-west and centre of the country, and in the north the range of isolated colonies had greatly increased. From the Tornionjoki River basin, however, muskrats had unaccountably disappeared.

By 1942 muskrats were only absent in central Finland from the Vuoksi and Oulujoki drainage basins: in those of the Simojoki and Kiiminkijoki they had multiplied and increased their range. A further liberation had been made in the Kuusamo River system, two more in the drainage basin of the Tornionjoki, and an additional five in that of the Kemijoki, which were universally successful: in the last of these areas discrete populations had merged. In Finnish Lapland, on the other hand, the population had died out, and an attempt to revive it with the release of 20 pairs from the Canadian tundra proved unsuccessful.

By 1947 there had been a slight expansion of range in the Oulujoki drainage basin (except in the east where there was some contraction), whereas mukrats had disappeared entirely from the Simojoki River system. Two new introductions had been made in the Kuusamo region, and the population in the middle reaches of the Kemijoki drainage basin was now continuous.

Between 1952 and 1957 muskrats extended their range in northern Finland – principally in the water systems of the Tornionjoki and Kemijoki, where in the former they were continuously distributed up to Pello, whence they spread to adjacent waters in neighbouring Sweden. During this period gaps in the occupancy of the Oulujoki drainage basin and in eastern parts of Kuusamo were infilled, and Hailuoto was invaded. By 1960, when nearly all the suitable water systems of Finland had been colonized, a further expansion of range in the north had populated Muonio, and in the upper reaches of the Kemijoki muskrats had extended even further north to the River Kaptugas at 68° 30′ N.

Muskrats became successfully established in Finland largely because of the similarity of the climate to that of northern Canada, from whence most of the original stock had been imported. In addition they found congenial habitats and an abundant supply of food for which they compete only with water voles (*Arvicola terrestris*) and brown rats (*Rattus norvegicus*). Most of their preferred food plants – *Typha latifolia*, *Sparganium erectum* and *S. simplex*, *Scirpus caespitosus*, *Carex* spp., *Equisetum limosum*, *Isoëtes lacustris* and *Phragmites communis* – have suffered from the depredations of muskrats, whose only important Finnish predators are American and European mink (*Mustela vison* and *M. lutreola*), polecats (*M. putorius*), foxes (*Vulpes vulpes*) and man. As might be expected, muskrats in Finland have tended to spread at a slower rate in the north than in the south. Their main limiting factor appears to be the length of the period of iced-up waters in winter.

Sweden

Although the importation of muskrats into Sweden was banned as early as 1916, some from neighbouring Finland were illegally released in the early 1940s at two points along tributaries of the River Torne, where they were probably established by the end of the decade. They were reinforced in about 1950 by natural immigration from Finland via the Torne and the Haparanda archipelago at the head of the Gulf of Bothnia.

By the middle 1960s muskrats occupied most suitable territory between the Finnish border in the east and the River Kalix in the west, northward to just south of Pajala, and south to the mouth of the Kalix (40 km west of the Haparanda archipelago, which was also colonized), and were extending south and west towards the Lule and Pite rivers. The most densely populated parts of Norrbotten were Sangis, Haparanda and the lower Torne Valley.

Muskrats in Sweden spread at a rate of around 10 km a year (slightly faster in the south) mainly in spring and autumn, although it was often many years before all suitable waters in a new territory were occupied. The animals seem equally at home in fresh or brackish water or in marshland and swamps. In their present Swedish range the water is frozen over for between 200 and 240 days a year.

Norway

The first recorded muskrat in Norway was shot in June 1970 in Jarfjord, in the south Varanger district of Finnmark, bordering the Barents Sea.

EURASIA

USSR[9]

The desirability of naturalizing muskrats as an economic resource in Russia was discussed as early as 1915. Because of opposition from various quarters no action was taken until 1927, when 11 were imported from Finland to the K. A. Timiryazev Moscow Young Naturalists' Biological Station in the Sokol'niki; the only female in the consignment, however, soon died, the experiment thus ending in failure.

Later in the same year a further 20 muskrats were imported from Finland, which it was intended to release on Great Solovetskiy Island and on Karaginskiy Island off the north-east coast of Kamchatka; because of the onset of winter, however, the animals were kept in captivity in Kiev until the following spring.

In March 1928 40 muskrats (of which all but one died on the voyage) were acquired in North America by the Kamchatka Joint-Stock Company and were shipped to Vladivostok. Two months later the Company entrusted Mr Walter J. Eyerdam of the Wilson Fisheries vessel *Apex* with the task of escorting a shipment of a further 40 to Karaginskiy Island. Three days before sailing 36 muskrats arrived in Seattle, Washington, USA by express train from Ontario, Canada. In his subsequent account of the voyage Eyerdam reported that: 'One of the males, which was a particularly vicious animal, dashed at me one morning, as was his custom, and fell overboard.' After making much of the difficulties of looking after his charges and giving graphic accounts of being attacked by savage male muskrats, Eyerdam rather smugly concluded: 'On arrival at Petropavlovsk, thirty-two days after leaving Seattle, thirty-five rats were turned over in perfect condition to the Russian keeper who was instructed in their proper care.' These animals were subsequently liberated on Karaginskiy Island.

In September and October 1928 four more small shipments, totalling 110 animals, arrived in Russia from Finland, of which 99 were released in the lakes of Great Solovetskiy Island. Ten muskrats imported from the United States in the following year were despatched to the Pushkin *zverosovkhoz* (wild animal state farm) near Moscow. In May 1929 about 1,000 muskrats were shipped from Montreal, Canada, of which 705 arrived safely in Leningrad, 157 of the survivors being sent to Great Solovetskiy Island and the remainder to the *zverosovkhoz* at Pushkin: from the last some were translocated in the following autumn to the Krasnoyarsk Territory and to Archangel and Tyumensk provinces. In 1932 1,495 muskrats were received from Finland and 167 from Britain.

Between 1928 and 1932 more than 1,646 muskrats from Finland, North America and Britain were released in nine Russian provinces, territories and autonomous republics as follows: Karaginskiy Island, Kamchatka Province, 35; Archangel (including Great Solovetskiy Island), 669; Buryat-Mongol ASSR, 100; Vologodsk, 312; Krasnoyarsk Territory, 234; Murmansk, 86; Tyumensk, 90; and Yakutsk, 120. An unknown number were also turned down in Sverdlovsk Province. Nearly all were of the nominate species *O. z. zibethicus*, only 10 being of the more valuable 'black' subspecies *O. z. macrodon*.

Lavrov (1960) divides the gradual naturalization of muskrats in the Soviet Union into six stages.

Between 1928 and 1930 attempts were made to breed the animals in semi-captivity, while at the same time small numbers (totalling for the period about 670) were released in seven provinces in the taiga zone and on offshore islands as far south as 58° N. By the end of 1930 muskrats were confined to Great Solovetskiy and Karaginskiy islands; to Lake Slobodskoye with its inflowing and outflowing rivers in Archangel Province; to Lake Katromskoye in Vologodsk Province and to a few lakes in the valleys of the Tavda (Sverdlovsk Province); Dem'yanka (Tyumensk Province); Yelogui (Krasnoyarsk Territory); and to Tokko in Yakutsk ASSR.

In 1932 the importation of muskrats from abroad ceased, future introductions being translocations within the Soviet Union. Between 1931 and 1935 a total of 7,120 muskrats were released in 10 new provinces and those mentioned above received a further 4,210.

By the end of 1935, when releases were being made in some southern provinces (about 45° N) and experimental trapping had begun, muskrats were fairly widely distributed in the basins of the Umba, Teriberka, Ponoy, Chapoma, Strel'na, Iokan'ga and Varzuga rivers, and in Lake Imandra. Their range in Archangel Province had also considerably expanded to include the central catchment area of the Onega (adjoining the Karelian ASSR); Onezhskaya Bay; the Zolotitze; the central reaches and lower basin of the Northern Dvina; the basins of the Kuloy and Vashka; and the left bank of the Pinega (Nel'da).

In Vologodsk Province muskrats had colonized only two small areas (Lake Katromskoye and its outflow the Katroma, and the headwaters of the Vozehga) and in Komi ASSR only three areas consisting of the upper catchment area of the Vychegda and two tributaries of the Pechora, the Vuktyl and Tsil'ma. During this period muskrats were translocated into the Karelian ASSR, to waterways adjoining Lake Onega, and to the lakeland of the Idel'.

In Transuralia in Sverdlovsk Province muskrats were to be found in lakes in the lower basins of the Southern Sos'va and Loz'va and on the Tavda and its tributary the Chornaya. In Tyumensk they were generally distributed along the Little Sos'va; on the middle reaches of the Kazym and its tributary the Amka; along the Sabun and the Vakh near Korlika; along the Dem'yanka and downstream of Alymsk on the Irtysh; and along the Noska, Layma and Alymka.

In Tomsk Province distribution was confined to the middle basin of the Tym, and in Novosibirsk Province to a few lakes around Vengerovsk and Chanovsk. In Krasnoyarsk Territory muskrats occurred on the left bank of the Yenisei and along the Lower Baikha (a tributary of the Turukhan); near Dudinka; in the Yelogui catchment area; and in several lakes of the Sym River system.

The introduction of muskrats into Irkutsk Province began in 1935, with releases being made into lakes near Nepa in the upper basin of the Lower Tunguska, and in waters in the valleys of two tributaries of the Lena, the Mansurka and Anga in Kachugsk. Muskrats colonized the basin of a tributary of the Irkut (the Tunka), the central valley of the Selenga and the Greater Angara in the Buryat-Mongol ASSR. In Chitinsk Province they occurred in lakes near Nelyat and Kokarevsk on the Vitim. In the Yakutsk ASSR their range was restricted to part of the floodland of the Tokko River and some lakes in Kobyaysk, and to the Berga (a tributary of the Lena) and near Kokuysk.

As mentioned above, stocking in some southern provinces began in 1935, muskrats being released in the Teli-Kul' Lake system (into which the River Sary-Su flows) in Kzyl-Ordinsk Province, and in the upper regions of the Ilya Delta in Alma-Atinsk Province.

Between 1936 (when the first state-owned farm was established in the Ili Delta, Balkhash) and 1940 some 5,460 muskrats were translocated into 13 new provinces, and those already stocked received a further 25,270. As a result of natural and artificial expansion muskrats increased their range considerably, establishing new populations in Yefimovsk district in southeastern Leningrad Province – possibly as the result of invasion from the Karelian isthmus by Finnish muskrats; in the lower reaches of the Pesha and Pechora rivers in Archangel Province; along the Sar-Yu and the Pok-Yu in the Komi ASSR; in the upper basin of the Vym; and on the Gremyachaya, a tributary of the Vyatka, in Kirovsk Province. In 1937 muskrats were released in Lake Chany west of Novosibirsk.

In the forest-steppes of Western Siberia muskrats were established over a vast area. In the

Altai Territory they settled in many land-locked lakes and in others running into or out of the rivers Barnaulka, Kosmol, Kulunda, Karasuk and Bogan; on the Ainka, a tributary of the Lebd', in the Altai Mountains; in the catchments of the Tomi and Ini in Kemerovsk Province; in the central basin of the Lower Tunguska; and on the Chuna, the Podkamennaya Tunguska, the Serezha and the Susha (in the Khakassk autonomous Province) in Krasnoyarsk Territory, where those established in the basins of the Kas, Taseyevaya and Ketya amalgamated with those in the Ona, Uda, Iya, Oka and Kitoy lake-systems in south-western Irkutsk.

In the Buryat-Mongol ASSR muskrats settled in lakes in the central catchment areas of the Barguzin, Khilk, Chikoy, Ona and Tsipa rivers; in lakes Dukhovoye and Kotokel'skoye near the eastern shores of Lake Baikal; in the Selenga Delta; in lakes Yeravnoye, Isinginskoye, Busanya and Baunt; and in the basins of the Ivolga, Sukhara and lower Dzhida. In Chitinsk Province muskrats were released in the Chikoy Valley lakes; in Amur Province in the basin of the Urkan, a tributary of the Zeya; in Khabarovsk Territory in Lake Mukhtel'; and in the lower Tugur and Uda. In the Yakutsk ASSR they were released at the mouth of the Undyulyun' in the Zhigansk district; in lakes in the lower basins of the Markha, Vilyui, Namana, Aldan and Nyuya; in lakes near Yakutsk in the central and lower catchment area of the Amga; and on the Churapcha, Cherendeya and Kempedyay – tributaries of the Tatta, Lena and Vilyui.

In the Kazakh SSR muskrats were turned down near Kzyl-Orda, from where they dispersed to the lower Syr-Darya. Some which escaped from a farm became established on the Aksayka and Little Almaatinka in Alma-Atinsk Province.

In the first few years of the last war, between 1941 and 1943, less than 2,000 muskrats were translocated, of which only 76 were released in new provinces – Taldy-Kurgansk, Arzamassk and Semipalatinsk.

After 1943 work on stocking muskrats was resumed, and in 1944–5 around 6,300 were introduced into 21 unoccupied provinces, especially in the south.

Muskrats were first turned down in the Kuban floodplains in Krasnodar Territory in 1944, and others later colonized the reed-thickets of the lower reaches of the Terek, Kuban, Eya and Sulak rivers. Here they became established in about 400,000 ha of floodlands, and in about 75,000 ha of steppe streams which provided a near-ideal habitat of an almost continuous chain of overgrown ponds; in spite of being preyed on by raccoon-dogs (*Nyctereutes procyonoides*), wild boars (*Sus scrofa*), brown rats (*Rattus norvegicus*), mink (*Mustela lutreola*), marsh harriers (*Circus aeruginosus*), snakes, pike (*Esox lucius*) and catfish (*Silurus glanis*), they managed to maintain a mainly high, though fluctuating, population.

Also in 1944 120 muskrats from the Makushino district and Kurgan region were introduced for the first time to Zhovganova in the Dnieper marshes in the Zaporozha region of the Ukraine, and others crossed the Czechoslovakian border into Transcarpathia and the Polish border into the Brest district of the Lvov and Volyn regions in Belorussia. In 1945–6 1,750 were transferred from the Priozernyi area in the Archangel region to nine steppe and forest regions in the Ukraine.

In the mid-1940s muskrats were also introduced to the Astrakhan region, to the Stavropol Territory and to the Kalmyk ASSR. Those (355) released on the lower reaches of the Amu Darya in the Kara-Kalpak ASSR in 1944 multiplied so rapidly that 2,200 were trapped two years later. In the same year 336 were liberated in the Amu Darya delta in Uzbekistan, where they became established on the floodlands and marshes, and 295 from Kazakhstan were introduced to the Issyk-Kul (and subsequently to other waters) in Kirgizia.

The years between 1946 and 1950 were marked by intensification of muskrat-stocking, and by a rapid increase in population and in the harvest of pelts. During the period in excess of 36,000 animals were stocked into 31 uncolonized provinces, of which 27 were in the European part of the Soviet Union.

In 1946 96 muskrats were introduced – for the first time – to the Ivanov region, from where they dispersed to the adjacent Yaroslavl region. In the following year 372 from Buryat-Mongol were introduced to the Maritime Territory; these were later augmented by translocations of their locally bred progeny to other localities. In 1946–7 846 muskrats – reinforced by more a decade later – were turned down in the Bashkiria, Tatar, Mari, Chuvash and Mordvinian ASSRs, and in the Ul'yanovsk and Knibyshev regions in the Volga-

Kama Territory. Muskrats introduced for the first time in the same years to the Kuchurgan Estuary in Moldavia soon spread to the lower Dniester, where they quickly became established. Others released in Estoniya in 1947 are now abundant in the lakes of the south and east; a survey in 1979 showed that their destruction of the larger species of the hydrosere has resulted in a sharp decline in the numbers of fish, crayfish and molluscs contained in the lakes. In 1950 muskrats were released in several lakes in the Luzhsk and Gatchinsk districts of Leningrad Province.

From 1951 to 1955 muskrat-stocking and pelt production grew apace, and by the latter year the animals occupied an area stretching from about 39° 15′ N to 71° 30′ N – further north than in their North American homeland.

In 1954 1,195 muskrats were introduced to the Ikryanoe, Liman, Travin and Marfino regions of Astrakhan. In a high spring flood the following year many were drowned, but numbers in Ikryanoe and Liman soon recovered; their main limiting factors were found to be high spring and autumn floods, a fall in the water level in winter, the freezing-over of shallow waters, reed-burning and fishing.

In 1952 77 muskrats trapped in the lower reaches of the Amur River were translocated to wetlands bordering the Tym' on Sakhalin Island, where by 1966 they had become established in all suitable habitats. In the same year 44 were transferred from the Gor'ki University Biological Station to the River Dobritsa in the Ivanovo region, and in the following year others were introduced to the Khoresm and Tashkent regions.

In Lithuania naturalization of muskrats began in 1954 when 82 from Archangel were introduced to the Anikshchyai and Trakai districts. Two years later 204 from Kazakhstan were turned down on lakes in the Zarasai and Lazoiyai districts, where they quickly became established. From 1953 to 1956 2,080 muskrats were liberated in six lakes in the Minsk, Gomel and Vitebsk regions.

In 1957 96 were transferred from Gor'ki University to the River Lyulekh in the Ivanovo region, and in 1958 a like number to the Parska River. In 1956–7 animals from the nearby Yaroslavl and Gor'ki regions spread naturally into other parts of Ivanovo, where they settled down in an ideal habitat of marshes, lakes, ancient peat-bogs, and large and small rivers with abundant aquatic and marginal vegetation.

Between 1928 and 1958 (when a total of more than three million animals were trapped) over 160,000 muskrats were stocked in the Soviet Union – the vast majority of which were translocated within the country – and by the middle 1960s, when the annual harvest of pelts exceeded that of the whole of Canada, muskrats covered a huge area; they ranged from Kaliningrad Bay on the Baltic and the Danube in the west to the Karaginskiy and Sakhalin islands in the east, and from the Kola Peninsula, the mouth of the Yenisei and the lower Kolyma in the north to the Black Sea, the Kopet-Dagh Range and the shores of Pos'yet Bay in the south. They were established in all the main water-systems of the Kola Peninsula; on the Solovetskiy islands; in almost the whole of the Karelo-Finnish region south of the Kem area; in the Archangel and Volodga regions; in the basins of the Vychegda and middle Pechora rivers; in several localities along tributaries of the Ob and Irtysh rivers and on the left bank of the Yenisei; in the Urals; in western afforested parts (e.g. Omsk, Krasnoyarsk) of Siberia; in a number of water-systems in the Irkutsk and Chita regions; in the Buryat-Mongol and Yakutsk republics, especially in the basins of the Lena, Vilyui, Kolyma, Aldan and Indigirka rivers; on Karaginskiy Island in the Bering Sea; on Lake Mukhetel in the Lower Amur region; in the steppe lakes in the Omsk, Chelyabinsk and Novosibirsk regions and in the Altai Territory of western Siberia; in Kazakhstan in the Aktyubinsk region; along the lower reaches of the Sary-Su and Ili rivers and in the lakelands of the lower Syr Darya River; in the Uzbeck and Kirgiz republics; in the Khabarovsk Territory; around Magadan on the northern coast of the Sea of Okhotsk; on Sakhalin Island; and on Kamchatka. In the Issyk region muskrats had penetrated into the Kentei region via the Selenga and Chikoi rivers in northern Mongolia. In the Maritime Territory they ranged from the northern tundra to the southern mixed forest and steppe being especially common along both banks of the Amur; on Lake Khanka; on the Zeya-Bureya Plain; in the valleys of the Nyukzha, Bira and Tunguska rivers; on the plains of the Evoron and Chukchagirskoe lakes and in the eastern part of the valley of the Amgum River; in the floodplains of the lower Bikin, Khor, Kiya and Ussuri rivers; and in the valleys of the Maya, Tugar and Uda rivers.

The optimum conditions and highest density of muskrat populations in the Soviet Union seem to be in the deltas of the Ili, Amu-Darya and Selenga rivers, in forest steppe and steppe lakes in western Siberia, and in areas of neighbouring Kazakhstan, where the numerous lakes provide an abundant supply of food and ideal breeding conditions, and the vast areas of dense reed-bed thickets furnish plenty of shelter. Muskrats were slow to colonize the Volga Delta until the waters were regulated in 1961–2 and flooding decreased. The taiga supports a smaller population than most areas because of the meagre supply of food and the frequent flooding of burrows. Floodlands of such rivers as the Ob, however, and areas of taiga such as Vasyugan, support considerable populations due to the numerous floodplains of the watershed lakes which are rich in food plants and provide good breeding sites.

In the Soviet Union, as elsewhere, muskrats cause a certain amount of damage by burrowing into the banks of rivers, lakes, dykes, dams, drainage ditches and irrigation canals (which results in the flooding of farmland), and railway, road and bridge embankments; they consume valuable farm crops – especially corn and sugar-beet – and damage to fishermens' nets is also sometimes reported. Muskrats are additionally believed to be vectors of *Leptospira icterohaemorrhagiae*, which causes Weil's disease in man.

In some parts of the Soviet Union (and in areas of western Europe) muskrats have declined somewhat in recent years as a result of their annihilation of reed-beds and other water plants and a reduction of aquatic macrophyte thickets. They nevertheless remain in many millions in much of western and central Europe north of the Alps, and in Soviet Europe and Asia, where they are an important economic resource, yielding around five million pelts annually.

The success of the muskrat in becoming widely established in much of western Europe has been attributed to a number of factors: the relatively mild climate; an abundant supply of both natural plants and cultivated crops for food; a variety of suitable habitats; a shortage of endo- and ectoparasites, and an absence of epizootics and predators; an adaptability, where and when necessary, to climatic extremes of cold and drought and to polluted and/or brackish water; and, perhaps most important of all, the availability of an empty and suitable ecological niche devoid of interspecific competitors.

ASIA

Mongolia

Four discrete muskrat populations are established in the Mongolian People's Republic: in the late 1940s and early 1950s natural immigrants from the Soviet Union via the Ili, Ussuri, Irtysch, Selenga and Amur rivers, settled in the Selenga area and in the Onon River system of eastern Mongolia; muskrats released in 1967 at lakes Char-usnur and Has in western Mongolia (where they were being cropped by the mid-1970s) are reported to be extending their range, and in the early 1970s muskrats dispersing from Kasakhstan via the People's Republic of China spread into Bulgan-gol in southwestern Mongolia. The annual harvest of pelts is about 100,000.

China; Korea

The first muskrats in the Chinese People's Republic were collected in the Amur region of Hehlung Kiang in northern Manchuria, having probably crossed the border from eastern Mongolia, by Mr Tsong of the Animal Products Company in the early summer of 1954. In the following spring Mr C. Chang visited the Hu Ma district and collected 21 muskrats. Mr T. H. Lo found more in the same area in 1956, since when they have become established. They are said to have first arrived in Hu Ma in the late 1940s.[10] Muskrats also occur on the Kunges River in Sinkiang and at Changwu in Liao-ning, again possibly as immigrants from Mongolia, and in parts of Korea. In China about half a million pelts are taken per annum.

Japan

Muskrats from the United States were imported into central Honshu for musquash-farming probably in the late 1920s (Kuroda (1939, 1955) says 'perhaps before 1945'); they soon escaped and began breeding in the wild in Tokyo, Chiba, Kanagawa and Saitama prefectures. They are at present confined to Tokyo and its environs, where, although they appear to have no detrimental affect on native species, their burrows cause a certain amount of flooding.

SOUTH AMERICA

Argentina; Chile

Muskrats introduced to Lago Fagnano on Isla Grande de Tierra del Fuego in Argentina between 1940 and 1956 had spread by 1971 west into Chile and south onto Isla Navarino (also in Chile) south of the Beagle Channel. In both countries they are abundant enough to be harvested annually for fur.

No other mammal deliberately spread by man has subsequently expanded its range as widely and rapidly as has muskrat: its Eurasian distribution is now greater than that of its North American homeland. As Elton (1958) points out, a purely Nearctic species has, within the space of half a century and with the assistance of man, become Holarctic.

NOTES

1. For much of the early history of the spread of the muskrat in Europe I am indebted to Mohr (1933).
2. Not also to Germany as stated by Roots (1976).
3. Bourdelle (1939).
4. E. Popelin, Report of the International Conference on the Muskrat, *EPPO Publications*, Ser. A No. 47: 16–17 (1969).
5. For much of the information on the spread of the muskrat in Belgium I am indebted to Engels (1969).
6. Oldeneel Tot Oldenzeel (1969).
7. Steffes (1969).
8. For much of the history of the muskrat in Finland I am indebted to Artimo (1960).
9. For much of the information on the spread of the muskrat in the Soviet Union I am indebted to Lavrov (1960) and Yanushevich (1966).
10. T. H. Shaw & C. Chu, *Shen Wu Hseuh Tung Pao (Biologica)* **19** (1955); T. H. Shaw & C. Chang, *Chinese Journal of Animal Husbandry* **2**: 57 (1957); and T. H. Shaw, *Nature* **180**: 1215 (1957).

Polynesian Rat

(*Rattus exulans*)

NATURAL DISTRIBUTION
Eastern Bangladesh, Burma, Thailand, Khmer, Laos, Vietnam, Malaysia, Sumatra, the Philippines, Indonesia and Papua New Guinea.

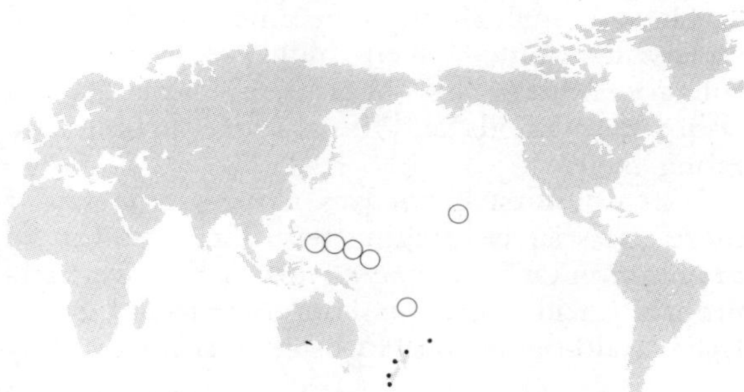

Note: Naturalized distribution includes most atolls and high islands of the Central and South Pacific, especially those in Micronesia.

NATURALIZED DISTRIBUTION
Australasia: New Zealand; Oceania: Hawaii, Fiji and many other Pacific islands, especially those in Micronesia.

AUSTRALASIA

New Zealand

The Polynesian (or Pacific) rat was reputedly introduced to New Zealand (where it has traditionally been known as the kiore or Maori rat) from Hawaiki (believed to be the island of Savaii in western Samoa) by the Maoris in about AD 1350.

The first rats may have got ashore from the war-canoe, *Horouta*, of the great chief, Te Kupe, or from one of the canoes – *Aotea*, *Aotearoa* or *Mataharua* – of the Great Fleet which followed. According to E. Best, 'The variety of sweet potato known as *Kakau* was placed on the vessel, as also were the *Kiore* rat, the swamp hen, and seeds of the *Karaka* tree, hence the famed saying, "*Aotea utanga nui*", or "*Aotea* of the important freight".[1] This is confirmed by F. von Hochstetter who in *New Zealand* (1867) quotes a Maori tradition that 'the Kumara, or sweet potato, the taro, the calabash-plant, the Karaki tree, the rat *Kiore*, the Pukeko and green parrot are said to have been imported from Hawaiki'.

White (1894), on the other hand, records that the kiore was present on Chatham Island with the pre-Maori immigrants, the Moriori, so the date of its arrival may have been considerably earlier.

The rats, which soon became abundant and widely distributed, were highly regarded as a culinary delicacy by the Maoris, who devised elaborate methods and rituals for catching and eating them.

When the first Europeans arrived some 400 or more years later, bringing with them the brown or common rat (*Rattus norvegicus*), kiores were still probably widespread in both North and South Islands, although von Hochstetter claimed that:

> this indigenous [*sic*] rat was so scarce already at the time of the arrival of the first Europeans, that a chief, on observing the large European rats on board one of the vessels, entreated the captain to let these rats run ashore, and thus enable the raising of some new and larger game.

R. A. Cruise, in his *Journal of ten months residence in New Zealand* (1843), told of speaking to a chief who 'expressed a wish for an importation of some rats from England to improve the breed, and thereby give him a more bountiful meal'.

When the Revd W. Colenso arrived in New Zealand in 1834 he could find little evidence of kiores remaining in North Island. Charles Darwin, who visited the Bay of Islands in the following year during the voyage of the *Beagle*, wrote: 'It is said that the common Norway rat, in the short space of two years, annihilated in this north end of the island the New Zealand species.'[2] Ernest Dieffenbach, in *Travels in New Zealand* (1843), wrote of North Island in 1839 that:

> There exists a frugivorous native rat, called Kiore maori (indigenous rat) by the natives on which they fed very largely in former times; but it has now become so scarce, owing to the extermination carried on against it by the European rat, that I could never obtain one. A few, however, are still found in the interior, viz. at Rotorua.

White (1894) and Kirk (1895) claimed that kiores survived only on a few islets off the north-east coast of North Island and in some northern districts of South Island.

Although it is clear that kiores were much reduced in numbers following the introduction of *norvegicus* by the early European voyagers, there is evidence that they managed to survive in some parts of North Island until the middle of the nineteenth century. J. S. Polak in *New Zealand* (1838), for example, implies that kiores were still being eaten by the Maoris at that time; in 1839 kiores were said still to be found at Rotorua, and between 1820 and 1840 in the Rauhine district.

The early settlers at New Plymouth in 1841–2 were plagued by rats which, because they were agile climbers, were presumably *exulans* rather than *norvegicus*. Between 1830 and 1845 kiores were undoubtedly present in the Urewera district, although by the early 1850s they had apparently disappeared.

Sir George Grey (Governor of New Zealand 1845–53 and 1861–7) sent a specimen of a kiore from Auckland to the British Museum in 1851; H. Guthrie-Smith reported[3] that three kiores had been captured at Tutira Station between 1879 and 1906. The general impression, therefore, is that kiores persisted, albeit in reduced numbers, in at least some parts of North Island until 1850–60. This, as Atkinson (1973) points out, is consistent with his hypothesis that *exulans* could not resist the invasion and agonistic behaviour of the black or ship rat (*Rattus rattus*) at about this time.

In parts of South Island kiores managed to survive in large numbers – sometimes, indeed, reaching plague proportions – until the arrival, according to Atkinson, of *rattus* in about the late 1880s. The earliest of these irruptions (erroneously believed by Thomson (1922) to be *rattus*) took place in Queen Charlotte Sound during Cook's second visit in 1773; others occurred between 1855 and 1860 at Collingwood, Nelson; at Shotover, Otago in 1863, and at Picton and in various parts of the Nelson district in 1863, 1872, 1884 and 1888. The greatest spread took place in 1884, when huge numbers were reported from

Cape Farewell in the north-west through Nelson to Blenheim in the east. According to Rutland (1890), 'they follow after heavy crops of fruit throughout our beech forests . . . they gradually increase in numbers until November, when all disappear again gradually, as they came'. Thereafter, the population in South Island appears to have declined dramatically, so that Thomson (1922) believed that the kiore in New Zealand was extinct.

Stead (1937), however, revealed that during visits to Taranga (Hen Island) in 1910 and 1933 he had observed rats which he considered to be of this species. In 1949 kiores were found also still to exist on White Island, and a survey in 1956 revealed that the species was far from extinct. It was discovered on (from north to south) the Cavalli Isles; Fanel Island in the Moko Hinau group; Taranga; Little Barrier Island; Cuvier Island (where it is believed to have exterminated the pied tit (*Petroica macrocephala toitoi*)) and Rurima Rocks – all off the northeastern coast of North Island. Since *rattus* and *norvegicus* did not occur on other islands off the northeastern coastline, it was considered probable that rats on Mayor, Mercury, Waiheke and Cape Colville islands were also *exulans*.

In South Island's Fiordland kiores were found still to survive in small numbers in the Notornis/Takahe Valley, Te Anau, and around Doubtful Sound, Manapouri. In 1950 one was trapped on the southern slopes of Mount Rakiahua in central Stewart Island (where they still occur) – the only place in New Zealand where all three species appear to have coexisted this century. Kiores have subsequently been discovered on some other islands such as Kapiti, which they share with *norvegicus*, Inner Chetwode (where they have apparently eliminated the yellow-breasted tit (*P. m. macrocephala*)), Codfish, Tiritiri Matangi, Mercury and Motuara. In the mid 1960s a rat believed to be *exulans* was trapped near Jackson's Bay, Westland.

The virtual disappearance of the kiore from the New Zealand mainland has been attributed by Atkinson (1973) largely to interspecific competition with the larger and more aggressive black rat, and by R. H. Taylor[4] to competition with the house mouse (*Mus musculus*). In the tropics *exulans* coexists satisfactorily on many islands with other rats – in Malaysia, for example, with *rattus*, and on Guam with the closely allied *mindanensis*; on Hawaii *R. e. hawaiiensis* lives side by side with both *rattus* and *norvegicus*.

In New Zealand, where it is near the southern limit of its range, climatic conditions may hinder *exulans* in its competition with *rattus* and *M. musculus*; it would be strange if *norvegicus*, which is even larger and more aggressive than *rattus*, had not also played a contributory role in the kiore's decline.

Kiores are virtually ominivorous, in winter feeding mainly on berries (especially hinau (*Elaeocarpus dentatus*)) and beech mast: they favour scrub and forest areas of high country and the summits of ranges well away from agricultural regions. Loss of habitat through the clearing of forest and scrub for agriculture (together with the introduction of mustelid predators towards the end of the nineteenth century) may have been contributory factors in the decline of the kiore in New Zealand.

OCEANIA

Hawaii

Rattus exulans hawaiiensis first arrived in Hawaii with the early Polynesian colonists from the central Pacific some 1,000–1,500 or more years ago.

It is first mentioned by Captain James Cook who, on visiting Kauai in 1778, noted that 'There were some rats resembling those seen at every island at which we had, as yet, touch'.[5] R. Titian Peale, naturalist with the American exploring expedition of 1838–42 led by Charles Wilkes, wrote: 'Nearly all the coral islands visited were found to be inhabited by rats, so like the common Brown or Norway Rat that they might be pronounced a diminutive variety.'[6]

Stokes (1917) recounts how the early Hawaiians indulged in the sport of *Pana iole* – shooting rats with bow and arrow. On the island of Raratonga in the mid-Pacific W. W. Gill, in *Historical sketches of savage life in Polynesia* (1876), tells how 'In those days – ere the cat had been introduced – rats were very plentiful. Rat hunting was the grave employment of bearded men, the flesh being regarded as delicious'. On Tonga, where the sport was known as *fanna gooma*, the king and his entourage used to shoot rats along the forest trials with bows and arrows, both 2 m long. J. Martin, in *Natives of the Tonga Islands* (1817), notes that 'Whichever party kills ten rats first, wins the game. If there be plenty of rats, they generally play three or four games'. Elton (1958) develops this theme: 'There were elaborate rules, as we should now have for football or hunting: precedence, offside, and – a wonderful dispensation – if you shot a bird you could count it as a rat!'

For many years, before it was discovered that it had hitherto been confused with the young of the grey form, *alexandrinus*, of the black rat, *exulans* was believed to be extinct in Hawaii. Today it occurs on Hawaii, Kauai, Oahu, Molokai, Lanai, Mauai, Kahoolawe, Kure Atoll, Popoia Islet and Mokumanu, and perhaps on Niihau, Kaula, Lehua and Mokolii.

On Kure Atoll (where it was first found in large numbers by the Tanager Expedition in 1923) kiores have killed numbers of incubating Laysan albatrosses (*Diomedia immutabilis*), sooty terns (*Sterna fuscata*), brown noddies (*Anous stolidus*), Bonin Island petrels (*Pterodroma hypoleuca*) and red-tailed tropic birds (*Phaethon rubricauda*).

Other oceanic islands

Many other Pacific islands are inhabited by Polynesian rats introduced by early Melanesian or Polynesian travellers during their great voyages of discovery between 1500 BC (or earlier) and around AD 800. They are, for example, widespread on Raoul (Sunday) Island in the Kermadecs (where they were first recorded in 1877), where they live sympatrically with brown rats introduced during the Second World War and prey on the Kermadec petrel (*Pterodroma neglecta*).

Of his visit to Easter Island in the Pacific in 1774, Cook notes: 'They also have rats, which it seems they eat; for I saw a man with some dead ones in his hand, and he seemed unwilling to part with them, giving me to understand they were for food.'[7]

On Fiji, where they were regarded by the early islanders as a valuable source of protein, Polynesian rats are abundant in all habitats on all the islands.

On Howland Island, in the Line group in the Pacific, Polynesian rats were reported in 1854 to be a serious threat to the eggs and young of nesting seabirds.[8]

'Throughout Micronesia . . . both geographically and ecologically the roof [black] rat and the Polynesian rat are the most widespread, being reported from nearly all sizeable land areas.'[9]

Although in South-East Asia Polynesian rats are largely commensals of man, in the Pacific islands (where their wide dispersal is believed to be associated with the pattern of early human settlement) they live independently of man in a variety of habitats, favouring especially low vegetation, where they nest in holes or other cover close to the ground, seeking food here as well as in trees which they climb with ease. They are a vector of diseases, a pest to crops and prey on both seabirds and landbirds, although in general they affect fewer species than black and brown rats. On many Pacific islands Polynesian rats have been exterminated by the arrival of domestic cats (*Felis catus*) and by the aggressiveness of the larger black and brown rats, but they are still the fourth most widely distributed rodent after *Mus musculus, R. rattus* and *norvegicus*.

NOTES

1. *Dominion Museum Bulletin, Wellington* **14**: 423 (1942).
2. *The voyage of the Beagle* (London, 1845).
3. *Tutira – the story of a New Zealand sheep station* (Blackwood, Edinburgh and London, 1921).
4. *New Zealand Department of Lands and Survey Information Series* No. 4: 137–8 (1978).
5. *A voyage to the Pacific Ocean . . . in H.M.S. 'Resolution' and 'Discovery' in the years 1776–80* (London, 1784), Vol. 2, p. 228.
6. *Mammals: U.S. Exploration Expedition*, 2nd edn (1858), p. 38.
7. *A voyage towards the South Pole and round the world, performed in H.M.S. 'Resolution' and 'Adventure' in the years 1772–5* (London, 1777).
8. I.A.E. Atkinson, *Pacific Science* **9**: 104–6 (1955).
9. Storer (1962).

Brown Rat

(*Rattus norvegicus*)

NATURAL DISTRIBUTION
Originally southeastern Asia south of the Himalayas.

NATURALIZED DISTRIBUTION
Widely distributed as the result of introductions by man.

EUROPE

British Isles

According to Thomas Pennant, in *British zoology* (1776), the first brown or common rats (also known as Norway rats – hence the creature's trivial name – from the erroneous belief that they first arrived in Britain in Norwegian timber-vessels) reached England around 1728 or 1729, probably in ships sailing from Russian ports.

Thereafter, they spread with amazing rapidity, usurping the place of the smaller and less aggressive black rat (*Rattus rattus*) wherever the two came into contact. In his *Mammalia Scotica* (compiled between 1764 and 1774) John Walker recounts how:

> the Norway Rat, First brought, as they say, to Scotland in ships from Norway; wheresoever it pitches its abode, it pitches out the Black ratten utterly: the Black Ratten and the Norway Rat were previously entirely unknown in Annan; however about twenty years agone the Norway Rat was cast on shore from ships driven to the mouth of the River Annan and now is scattered through almost the whole region of Annan.

The new statistical account of Scotland for the parish of Newlands in Peebles relates that:

> The brown rat a good many years ago invaded Tweedale, to the total extirmination [*sic*] of the former black rat inhabitants. Their first appearance was at Selkirk, about the year 1776 or 1777, and passing from Selkirk, they were next heard of in the mill at Traquair; from thence they appeared in the mill at Peebles . . . they arrived at Flemington Mill, in this parish, about the year 1791 or 1792.

Brown rats were first recorded in the Scottish Highlands in 1814, when the Revd G. Gordon saw some in Morayshire.

According to John Rutty, in *An essay towards the natural history of the County of Dublin* (1772), brown rats 'first began to infest these parts about the year 1722'. On the island of Anglesey in Wales brown rats were so numerous in August 1762 that they are said to have eaten ears of standing corn while it was being reaped.

A number of offshore islands have at one time or another been overrun by brown rats, almost invariably to the detriment of the existing fauna. On Priestholm, off the coast of north Wales, for example, rats which swam ashore from a wrecked Prussian ship in about 1816 almost wiped out the resident population of rabbits (*Oryctolagus cuniculus*) and puffins (*Fratercula arctica*). The fauna of Ailsa Craig and Little Cumbrae in the Firth of Clyde, and Manx shearwaters (*Puffinus puffinus*) on Lundy Island in the Bristol Channel, have similarly suffered.

Today, brown rats occur in almost every type of habitat (apart from the few remaining dockland areas and quaysides still occupied by black rats) throughout the British mainland, as well as on many of the offshore islands. The two species almost invariably occupy different habitats in any one locality.

Continental Europe

The brown rat appears to have reached mainland Europe via Russia in the early eighteenth century. The German naturalist, Peter Simon Pallas, records that vast hordes migrated to the west from Asia following a severe earthquake in 1727: they infested houses in Astrakhan, swam the Volga and spread further west to the remainder of European Russia and the Baltic. They arrived in Denmark (at Copenhagen) from Asia when the Russian fleet visited the town in 1716, and by 1750 had reached Paris and parts of eastern Prussia. They were first recorded on the Norwegian mainland in 1762; on the Faeroe Islands (which they are said to have reached on wreckage from the *King of Prussia* which foundered off the Isle of Lewis, Scotland) in 1768; in Brunswick and Greenland around 1780, and in Sweden in 1790. They arrived in Spain and Italy in the mid- to late eighteenth century and in Switzerland in about 1809.

AFRICA

Morocco

According to Roots (1976), brown rats did not cross the Straits of Gibraltar until 1930, reaching Marrakech in the south four years later.

South Africa

Firmly established in all suitable habitats, brown (and black) rats together were, as long ago as 1945, estimated to destroy at least R1 million worth of grain alone annually.

NORTH AMERICA

USA

Brown rats first appeared in the United States, having crossed the Atlantic presumably in English ships, in about the first year of the Revolution, 1775. According to John James Audubon, they had not penetrated as far west as the Pacific coast by the time of his death (1851), though they must have arrived there shortly afterwards, having presumably travelled overland with the early settlers. By 1855 they were reported to abound in several Pacific coast localities, such as San Francisco, California, Astoria, Oregon, and Steilacoom, Washington. They first appeared in Colorado, at Denver, in 1886, and by 1907 most of the towns in the state were infested. In 1888 they were unknown in Utah but had occurred in Arizona and were abundant in Albuquerque, New Mexico. Wyoming was the penultimate state to be invaded (in about 1919) followed by Montana in 1923.

Canada

According to the explorer St John Richardson, brown rats in Canada had, by 1825, spread no further west than Kingston, Ontario. By 1887, however, they were said to be well established in Vancouver, New Westminster and Victoria, and seven years later they had penetrated to Chilliwack, some 100 km up the Fraser River. They crossed the border from North Dakota and Minnesota into Manitoba around the turn of the century, where by the outbreak of war in 1914 they were established as far north as the Assiniboine River and had crossed into Saskatchewan. During the two decades between the wars almost all the urban areas of both provinces were overrun. Large rivers and mountain ranges slowed their rate of spread – Alberta, for example, remaining uninfested until as late as 1948.

The present North American range of the brown rat consists of the whole of the United States and Mexico and southern-central Canada. On the Canadian east coast the species ranges north into southern Quebec, while on the milder west coast it extends along the coast as far north as Alaska, USA.

Bermuda

In Bermuda both brown rats (which are believed to have arrived in the mid-eighteenth century) and black rats are significant predators of nesting seabirds and landbirds. Until their extinction in the Castle Harbour Island National Park brown rats especially preyed on such nesting seabirds as the Bermuda petrel or cahow (*Pterodroma cahow*) (rediscovered in 1951 after being considered extinct for over 300 years), Audubon's shearwater (*Puffinus l'herminieri*), the common tern (*Sterna hirundo*) and white-tailed tropic birds (*Phaethon lepturus*). Brown rats are still common in Bermuda – mainly in urban areas – where they frequent rubbish dumps and similar sites.

SOUTH AMERICA

Argentina

In South America brown rats occur in suitable habitats as far south as Bahia Crossley (a penal settlement between 1890 and 1902) on Isla de los Estados off Tierra del Fuego.

South Georgia

Although on South Georgia, which they are believed to have first colonized in about 1870 (though possibly as early as around 1800), brown rats consume mostly vegetable matter, invertebrates and carrion, there is some evidence for avian predation. Pye and Bonner (1980) found a negative correlation between the distribution of rats and breeding subantarctic pipits (*Anthus antarctica*) and South Georgia or yellow-billed pintails (*Anas georgica*). The nestlings of white-chinned petrels (*Procellaria aequinoctialis*) are also sometimes taken as, possibly, are burrow-nesting Antarctic or dove prions (*Pachyptila desolata*), South Georgia diving petrels (*Pelecanoides georgicus*) and blue petrels (*Halobaena caerulea*).

Falkland Islands

Brown rats probably landed on the Falklands from the ships of the first French and English colonists in 1764 and 1765. In 1842 Governor Richard C. Moody found them only around the settlement of Port Louis on Berkeley Sound. Today, they are fairly well distributed, and locally common around some settlements and on a number of tussock grass offlying islands.

AUSTRALASIA

Australia

Brown rats probably first arrived soon after European settlement began in 1788. By 1830 they were said to be plentiful on Swan Island in Banks Strait. In Australia the brown rat is closely associated with human settlement, being abundant only in wharves, drains and sewers in coastal cities, in heavily developed rural seaboard areas, and in some towns in the interior close to large watercourses.

New Zealand

According to the Revd William Yate, in his *Account of New Zealand* (1835), 'The natives tell us that rats were introduced in the first ship, by Tasman', the Dutch navigator who landed in Cook Strait in 1642. It is, however, generally believed that brown rats actually first reached New Zealand either with Captain James Cook in 1769, 1773 or 1776, or with the early whalers and sealers who established shore-stations there towards the end of the century; by about 1800 they had apparently become generally distributed throughout both main islands.

The Frenchman, Crozet, who visited the Bay of Islands in North Island in 1772, wrote in *Voyage to Tasmania, New Zealand . . . 1771–1772* that 'The rats are the same species as those we have in our fields and forests'. At that time *norvegicus* was the common rat of France. R. Wakefield, in *The handbook for New Zealand* (1848), wrote of 'the large Norway rat, which exterminates the other race [the kiore] in its extensive migrations all over the country'.

Charles Darwin, who visited the Bay of Islands in 1835 during the voyage of the *Beagle*, reported that 'It is said that the common Norway rat, in the short space of two years, annihilated in this north end of the island the New Zealand species'.[1] Alfred Russel Wallace agreed with Darwin: 'This invading rat has now been carried by commerce all over the world, and in New Zealand has completely extirpated a native rat, which the Maoris allege they brought with them from their home in the Pacific.'[2] A. Saunders, in *New Zealand* (1868), and White (1894) complain that in South Island brown rats were a considerable nuisance to the early settlers in Canterbury and Otago. Thomson (1922) tells how: 'During visits to Stewart Island and the West Coast Sounds between 1874 and 1880, I was struck by the abundance of these animals in regions uninhabited and almost unvisited by man. One day . . . in Paterson Inlet . . . a wide stretch of beach nearly a mile long was alive with rats.'

By 1868 brown rats were said to be common on Chatham Island (where they still remain), and in 1884 they were reported by Reischek (1888) to be abundant in Dusky, Chalky and other west coast sounds. H. F. von Haast, in his biography (1948) of his father, Sir Julius von Haast, quotes from the latter's journal an account of an expedition in 1862 to Tasman Glacier in the Southern Alps, where 'the rats all belong to the large grey Norwegian species'.

Before the introduction to New Zealand of mustelid predators in the 1880s, brown rats were virtually ubiquitous in both main islands from sea-level to the mountain tops. Today, they are patchily distributed in North and South islands (though widespread on Stewart Island), living largely as commensals of man at ground floor level in farms and towns, and along the banks of watercourses. Their principal predator and limiting factor in New Zealand is the introduced stoat (*Mustela erminea*).[3] Brown rats do not dominate black rats in New Zealand as they do in Britain and the United States, though the two species seldom coexist.

In the New Zealand subantarctic brown rats (only) occur on Campbell Island where they have been established since before 1883. Recently, the Campbell Island flightless teal (*Anas aucklandica nesiotis*), having for some time been thought to be extinct as the result of predation by brown rats, was rediscovered on the tiny offlying Dent Island. Only on Stewart Island have *R. norvegicus, rattus* and *exulans* lived sympatrically this century.

OCEANIA

Hawaii

Atkinson (1977) refers to the fact that prior to the construction of the first wharf in Hawaii (at Honolulu) in 1825, all ships had been compelled to anchor off-shore, thus reducing the likelihood of rats reaching land. Within a decade, however, a new and larger rat (presumably *norvegicus*) had joined the long-established Polynesian or kiore rat in Hawaii. In 1903 D. Malo in *Moolelo Hawaii* wrote of animals which had arrived since the time (around 1835) of King Kamehameha I: 'the rabbit, or *iole-lapaki*, which makes excellent food, the rat or iole-nui, and the mouse, or *iole-liilii*'. R. Titian Peale, naturalist with Charles Wilkes's US Exploring Expedition of 1838–42, was, perhaps, the first European to write of brown rats in Hawaii, which the expedition visited in 1840: 'The common brown or Norway rat was observed at the Hawaiian, and some other islands in the Pacific Ocean, where it has been introduced by commercial intercourse with Europe and America. We did not find it on islands uninhabited by mankind.'[4]

Today, brown rats occur on Hawaii, Kauai, Oahu, Molokai, Lanai, Maui and on Ford Island in Pearl Harbor, and possibly on some other small inhabited islands. With the black rat they prey on, among other species, the Hawaiian duck (*Anas platyrhynchos wyvilliana*), the Hawaiian race of the dark-rumped petrel (*Pterodroma phaeopygia sandwichensis*), and the Hawaiian goose or nene (*Branta sandvicensis*).

Fiji; Juan Fernandez

Brown rats, which arrived in the nineteenth century, are common in agricultural, urban suburban and coastal areas on all the main islands of the Fijian group. They also occur on Robinson Crusoe (Masatierra) and Alejandro Selkirk (Masafuera) in the Juan Fernandez group.

At the International Council for Bird Preservation's Eighteenth World Conference at Cambridge, England, in August 1982,[5] Dr Ian Atkinson of New Zealand traced the spread of brown (and other) rats to the world's oceanic islands.

Between about 1710 and 1830 the previously almost ubiquitous black rat (*R. rattus*) was well-nigh completely displaced throughout western Europe by the spread or increase of the brown rat, which thus became the common rat aboard European ships, and introductions of rats to oceanic islands during this period involved almost exclusively *norvegicus*.

After about 1830 – particularly from the middle of the century – the black rat (for some unknown reasons) reappeared as the common shipboard rat, and for the last century and a half both brown and (more frequently) black rats have been spread by man to further islands throughout the world, and few major oceanic islands are now uninfested by one or other species. Within many archipelagoes, however, small offlying islands, including a number of great biological value, have so far remained uncolonized.

In general, brown rats appear to prefer temperate to tropical climates. In Micronesia, for example, where black and Polynesian or kiore rats are virtually ubiquitous, the brown rat occurs only on the islands of Saipan, Koror (Palau) and Ponape.[6] Brown rats burrow and swim well and are almost entirely terrestrial. Ground- or burrow-nesting (especially colonial) seabirds and landbirds are thus most vulnerable to predation by brown rats, which are themselves an important prey for a number of introduced carnivores (e.g. mongooses (*Herpestes auropunctatus*) and feral cats (*Felis catus*)[7]), whose increased populations then exert an increasing pressure on avian species.

Brown and black rats are without doubt the world's worst introduced pests – as vectors of disease (e.g. plague and rabies), as pillagers of stored and growing crops, and as predators of other (mainly avian) species. It has, for example, been estimated that since about 1600 introduced rats have exterminated at least 18 species or subspecies of birds, and that they have decimated a further 40 or more. One of the most recent to suffer, the Puerto Rican yellow-shouldered blackbird (*Agelaius xanthomus*), was officially declared endangered in 1977 by the US Fish and Wildlife Service, largely through predation by rats in its coastal mangrove habitat.

The **ricefield rat** (*Rattus argentiventer*), which ranges from Indochina to Java, Borneo and the Philippines, and the **Himalayan rat** (*R. nitidus*), which is distributed from the Himalayas to Indochina, and in the Philippines, have both been introduced to the Celebes and to Papua New Guinea.

NOTES

1. *The voyage of the Beagle* (London, 1845).
2. *Island life* (Macmillan, London, 1902).
3. See p. 62.
4. *Mammals: U.S. Exploration Expedition* (1848), p. 47.
5. See Moors (in press).
6. Storer (1962).
7. See pp. 75 and 84.

Black Rat

(Rattus rattus)

NATURAL DISTRIBUTION
Originally probably confined to southeastern Asia.

NATURALIZED DISTRIBUTION
Widely distributed as the result of introductions by man.

EUROPE

British Isles

Until as recently as 1979 it was traditionally believed that the black, ship or roof rat was first introduced to Britain in the baggage of returning twelfth-century Crusaders. In that year, however, and in 1984, it was announced that remains, subsequently identified as those of *R. rattus*, had been discovered in Roman deposits in York and London; radiocarbon determinations suggest a third or fourth century date for these specimens, thus antedating the Crusades by at least 800 years.[1]

The earliest reference to black rats in the British Isles (in Ireland) seems to be that made by Giraldus Cambrensis (Gerald de Barri) in his *Topographia Hibernica* (compiled between 1183 and 1186)[2]: some years later in Wales the same author speaks of *murium majorum, qui vulgariter Rati dicuntur* ('the larger species of mice, commonly called rats') which were said to have attacked a man, and to have eaten a woman and her child at Merrion, in Pembroke.

By the late thirteenth century rats were clearly becoming a nuisance in many parts of England; in 1297 some were sold for ¼*d.* each in Weston, and more were trapped in Oxford in 1335 and 1363. The fourteenth-century chronicler Ranulf Higden, in his *Polychronicon*, describes the creatures as *nocentissimos* ('most harmful').

The contemporaneous Black Death (bubonic plague) – transmitted by the rat flea – was prevalent over a large area of England (and in continental Europe) during the fourteenth century, which suggests that the animal was at the time rapidly extending its range.

In the *Pardonere's tale* (*c.* 1390), Geoffrey Chaucer tells of a harrassed householder who:

> Forth he goeth, no lenger would he tary,
> Into the toun unto a Pothecary,
> and praied him that he him wolde sell
> Som poison, that he might his ratounes quell.

In about 1377 damage caused by rats or mice is mentioned in the Register of Archbishop Sweteman of County Armagh, Northern Ireland, and a rat-trap is listed as an item of expenditure in the accounts for 1469 of the churchwardens of St Michael's, Cornhill, in the City of London.

The earliest positive mention of rats in Scotland refers to their presence in 'Catteynes' (Caithness) in about 1630; Martin Martin, in his *Description of the Western Isles of Scotland* (1703), tells how he saw 'a great many Rats in the Village Rowdil [on the island of Harris], which have become very troublesome to the Natives'.

The fortunes of the black rat in Britain declined dramatically following the introduction of the larger and more aggressive brown rat (*Rattus norvegicus*)[3] in 1728–9. Today, the black rat is confined solely to a number of offshore islands and to the docks and warehouses of some major seaports, and is continuing to decline.

ASIA

Malaysia

Rats, which eat the eggs of brown boobies (*Sula leucogaster*) and other seabirds on the island of Pulau Perak (130 km west-north-west of Penang), were in 1973 found to be indistinguishable from *R. r. diardii*, the common house-rat of the Malayan mainland. 'This population is probably, therefore, of recent origin on Pulau Perak – introduced and reinforced from visiting fishing boats.'[4]

AFRICA

Mauritius

On the island of Mauritius rats, with the assistance of cats and the introduced crab-eating macaque (*Macaca fascicularis*),[5] have been responsible for drastically reducing the populations of the Mauritius cuckoo-shrike (*Coracinia typica*), the Mauritius fody (*Foudia rubra*), a flycatcher known locally as the *coq-du-bois* and a weaver. They may also have been partly responsible (with man) for the extermination by around 1681 of the dodo.

On the African mainland black rats are widespread in areas of human settlement, especially along trade routes such as the Mombasa (Kenya) to Kampala (Uganda) railway, but are absent in unsettled regions. In Zambia there are records of black rats from archaeological sites dating from the sixteenth century, but the animals still do not occur in the west of the country.[6] In South Africa black rats are firmly established in all the larger coastal settlements and in inhabited parts of the interior. As long ago as 1945 black and brown rats were together estimated to destroy at least R1 million worth of grain annually.

NORTH AMERICA

USA; Canada

The first positive mention of rats in North America refers to their presence at Jamestown in 1609. John Smith, in *The proceedings of the English colony in Virginia* (1612), tells that: 'In searching our Casked Corne, we found it halfe rotten, the rest so consumed with the many thousand Rattes (increased first from the ships) that wee knew not how to keepe that little wee had.'[7]

By the early eighteenth century black rats of the nominal subspecies *R. r. rattus* were generally distributed along the coasts and in the seaports of much of eastern North America, and by the middle of the nineteenth century they had penetrated north to Montreal and Halifax, Nova Scotia; on Sable Island, off the coast of the latter, the lighthouse-keeper was said to be faced with starvation through the consumption of his stores by rats.

At about the same time black rats arrived on the Pacific coast, where they were first established at San Diego and Humbolt Bay, California. As in the British Isles, the arrival of the brown rat (in about 1775) signalled the decline of the black rat, which is now confined largely to coastal areas on both seaboards, extending inland eastwards to Illinois, but mainly in some of the southeastern states.

The grey-bellied Alexandrine rat (*R. r. alexandrinus*) of the Mediterranean region – said to have arrived in North America from Spain and Italy – is common in parts of Mexico north to the Dismal Swamp, Virginia.

Bermuda

Black rats first arrived in Bermuda in 1613 on board a captured Spanish grain vessel which was towed into St George's. Thereafter they multiplied explosively, and within the space of a year the population had reached plague proportions. In *Memorials of the discovery and early settlement of the Bermudas or Somers Islands, 1515–1685* (1877–9), Sir John Henry Le Froy (Governor of Bermuda 1871–9) quotes a reference in *A general narrative of the Government of Captain Daniel Tucker* (taken from John Smith's *Proceedings of the English colony in Virginia* of 1612) to a famine caused by black rats in 1615:

> an increase of silly [ordinary] rats, in the space of two years [i.e. by 1615] so abounded that they filled not only those places where they first landed, but swimming from place to place, spread themselves into all parts of the Countrey, insomuch that there was no Iland but it was pestered with them. They spared not the fruits of the plants, or trees, nor the very plants themselves, but ate them up. When they had set their corne, the rats would come by troupes in the night and scratch it out of the ground.

Le Froy also quotes Richard Norwood who in 1662–3 wrote that in 1616 'the whole Country [of Bermuda] was then greivously afflicted with ratts', and refers to a complaint made at an Assize Court in October 1617 that 'the Rattes doe so much hurt' and that the Brother islands were 'full of ratts'. The near-extinction of the endemic cahow or Bermuda petrel (*Pterodroma cahow*) can indirectly be attributed to rats, since the famine they caused forced the colonists to slaughter vast numbers of the birds in order to survive. Thereafter the rat population became stabilized, and black rats are today still a common pest in Bermuda – mainly in rural areas, where they are significant predators of nesting landbirds and seabirds.

West Indies

In the West Indies R. P. J. B. Du Tertre, in his *Histoire générale des îles de S. Christophe, de la Guadeloupe, de la Martinique, et des autres dans l'Amérique* (1654), mentions the numbers and voracity of rats in the French islands, and complains that they destroyed crops of fruit and sugar-cane.

C. C. Rochefort, in his *Histoire naturelle et morale des îles de l'Amerique* (1658),[8] confirms the arrival of black rats (in the Leeward Islands) with the early colonists from Europe: 'Rats are creatures heretofore unknown to the Caribbians; but now since the coming in of so many Ships to those Islands they have got to land, and are so multiply'd that in some places they do abundance of mischief among the Potatoes, Pease, Beans and Wheat [maize].'[9]

G. Hughes, in *The natural history of Barbados* (1750), laments that rats 'are so very numerous, and so very destructive to Sugar-canes, that the yearly loss to The Inhabitants of the Parishes of St Joseph's and St Andrew's alone, is computed to be no less than Two or Three Thousand Pounds'.

Patrick Browne, in the *Civil and natural history of Jamaica* (1789), claimed that rats often destroyed more than a quarter of the island's sugar-cane crop: 'There are great numbers of them in every plantation, though they take great pains to get rid of them. I have been informed by men of character, who have tasted of them, that they are very delicate meat.'

SOUTH AMERICA

Peru

According to Thomas Pennant, in his *History of quadrupeds* (1781), the first black rats in the New World arrived in Peru in the ships of early explorers in 1544, during the time (1544–6) of the Viceroy Blasco Nuñez. They are today widely distributed in suitable habitats throughout the continent.

AUSTRALASIA

Australia

Exactly when black rats first reached Australia is uncertain, but it may well have been in the mid-nineteenth century when the species also first arrived in New Zealand. Rolls (1969) quotes from an early Tasmanian newspaper: 'The number of rats leaving the convict ship now tied up in the Bay [presumably Storm Bay, Hobart] has to be seen to be believed'; if the date of publication was before about 1830[10] these, of course, may well have been *norvegicus*. By 1867 the animals had penetrated to the Cape York Peninsula in northern Queensland, and 20 years later the first irruption was recorded, at Ivanhoe in western New South Wales. Today, black rats are widely distributed throughout Australia wherever there is human habitation, and on farmland and in other habitats altered by man.

New Zealand

It has usually been assumed that black rats first reached New Zealand with Captain James Cook in 1769, 1773 or 1776, or with the early sealers and whalers who arrived shortly afterwards, and that by about the turn of the century they were generally distributed throughout both main islands. Atkinson (1973),[11] however, argues persuasively that, although black rats were probably present in isolated localities earlier, it is more likely that they did not begin to spread in North Island until after 1860 and in South Island until the late 1880s or early '90s.

Most of the vessels visiting New Zealand between 1769 and about 1850 sailed from England, France, America or Australia; during this period brown rats were rapidly replacing black rats in the first three countries, and had probably yet to reach Australia. The likelihood of the latter arriving in any numbers in New Zealand at this time is, therefore, comparatively small.

During the second half of the nineteenth century the species of rat normally carried by ships appears to have changed from *norvegicus* to *rattus*: R. H. Taylor suggested to Atkinson that the alteration may have been associated with the replacement of sailing vessels by steamships, whose engine-rooms would have provided a more favourable microclimate for black rats. The first steamship to visit New Zealand arrived at Auckland in 1846. After 1842 there was an increase of sea-trade between Britain and Egypt, and the opening of the Suez Canal in 1869 provided a shorter route between New Zealand and the Middle East and Europe; in both Egypt and the Middle East the common urban rat was the black rat.

Only two authorities before 1860 refer to rats other than Polynesian rats or kiores and brown rats as being present in New Zealand: Ernest Dieffenbach, in *Travels in New Zealand* (1843), makes a confusing statement which could be taken to mean that he believed he had seen black rats in the country, and the Revd W. Colenso[12] may have confused the kiore, *exulans*, with *rattus*.

The earliest observations of black rats in New Zealand may have been made in Northland in the 1860s by D. C. Wilson.[13] Definite evidence of the presence of *rattus* in North Island is given by Buller (1871), who misidentified as *exulans* a specimen captured at Wangaehu in 1870. Other early dates for black rats identified in North Island include 1870 (Big Omaha and Raglan), 1871 (Wellington) and 1876 at Napier. From this time onwards there is increasing evidence of *rattus* in North Island, although it may still not have been widespread by the turn of the century.

What may be the earliest record of a black rat in South Island is one caught in a house in Greymouth in November 1882.[14] In South Westland C. E. Douglas wrote of his expedition to the upper Waiho River and Franz Josef Glacier in 1893 that 'The Norwegian rat, which swarmed in the country at one time, is now becoming extinct, and the native and black rats are taking its place'[15] – a claim with which other contemporary authors agree. Other early dates for the appearance of *rattus* in South Island include 1893–4 at Collingwood, Motueka, Riwaka, Awaroa (near Totaranui) and Nelson, and 1895 at Opawa (Christchurch).

The spread of black rats through much of South Island was probably slow, and infestation of the southern regions appears to have taken place only after 1894, when they had not yet reached Otago or Southland. The earliest records for Stewart Island (1911) are described by H. Guthrie-Smith.[16]

In New Zealand podocarp forests black rats, being largely arboreal, can be a major factor in the destruction of the seeds of rimu (*Dacrydium cupressinum*), kahikatea (*Podocarpus dacrydiodes*), miro (*P. ferrugineus*), matai (*P. spicatus*) and hinau (*Elaeocarpus dentatus*). In mixed podocarp/dicotylous forest black rats also eat the seeds and inflorescence of such hardwoods as the kiekie (*Freycinetia banksii*), the nikau palm (*Rhopalostylis sapida*) and the supplejack (*Ripogonum scandens*). They also consume the roots of umbellifers such as *Ligusticum*, *Angelica* and *Aciphylla*, the tubers of *Gastrodia cunninghamii*, and the fruit of the passion flower (*Passiflora tetrandra*).

On Big South Cape Island southwest of Stewart Island (where they were first recorded in 1955) an irruption of black rats between 1962 and 1965 resulted in the eradication of Stead's bush wren (*Xenicus longipes variabilis*), the South Island saddleback (*Creadion carunculatus carunculatus*), the Stewart Island robin (*Petroica australis rakiura*), the Stewart Island fernbird (*Bowdleria punctata stewartiana*), and the subantarctic snipe (*Coenocorypha aucklandica iredalei*). The only known local colony of the Stewart Island bat (*Mystacina tuberculata robusta*) was also exterminated, and populations of the red-fronted parakeet (*Cyanoramphus novaeseelandiae novaeseelandiae*), the yellow-fronted parakeet (*C. auriceps auriceps*) and bellbirds (*Anthornis melanura*) all greatly declined. Tuis (*Prosthemadera novaeseelandiae*), grey-breasted silvereyes (*Zosterops lateralis*), grey gerygone warblers (*Gerygone igata*) and hedge sparrows or dunnocks (*Prunella modularis*), on the other hand, had all noticeably increased. This havoc on Big South Cape Island suggests a no less catastrophic effect on the country's avifauna during the black rat's expansion on the mainland. As Atkinson (1973) points out, the probable period of the black rat's spread in North Island largely coincides with the decline there of the bellbird, robin, stitchbird (*Notiomystis cincta*), saddleback and New Zealand thrush or piopio (*Turnagra capensis*). Atkinson also considers that the spread of black rats may have been responsible for the virtual extinction on the mainland of the Polynesian rat or kiore.

In New Zealand black rats are widespread on both main islands inside buildings (where, being agile climbers, they tend to occupy upper floors) and outside, and throughout the bush and forest areas from near sea-level to exposed mountain tops. They also occur on Chatham Island. In New Zealand and Australia (unlike in Britain and the United States) the numbers and distribution of black rats are not seriously affected by competition from brown rats, though the two species usually live allopatrically.

OCEANIA

Hawaii

Although in the past it has been assumed that black rats reached Hawaii in the first European ships towards the end of the eighteenth century, Atkinson (1977) cites circumstantial but convincing evidence which suggests that the animal did not in fact arrive until after 1840 – probably between about 1870 and 1880. Today, black rats occur on all main islands of the group – Hawaii, Maui, Kahoolawe, Lanai, Molokai, Oahu, Niihau and Kauai – and on Kaohikaipu, Mokuoloe, Mokolii, Mokuauia, Sand and Eastern islands, as well as on Ford Island in Pearl Harbor and possibly on some other small uninhabited islets. On Midway Atoll (where they escaped from ships in 1943–4) black rats prey on the white tern (*Gygis alba*) and on nesting Laysan albatrosses (*Diomedea immutabilis*), and have exterminated the Laysan rail (*Porzanula palmeri*) and the Laysan finchbill (*Psittirostra cantans cantans*), which had nested on the island after translocation from Laysan Island.[17] On Maui they also eat the eggs and young of the Hawaiian race of the dark-rumped petrel (*Pterodroma phaeopygia sandwichensis*) and of the Hawaiian goose or nene (*Branta sandvicensis*), and on some of the main islands are believed to prey on the eggs and nestlings of endemic honeycreepers, such as Captain King's '*perroquet*', the ou (*Psittirostra psittacea*).

Aleutian Islands

Black rats, accidentally introduced to the Aleutian group in the Bering Sea during the Second World War, have (together with Arctic foxes (*Alopex lagopus*),[18] and feral cats and dogs) been responsible for confining the breeding grounds of the distinctive Aleutian Islands Canada goose (*Branta canadensis leucopareia*) to the 1,720-ha island of Buldir.

Galápagos Islands

An interesting and plausible paper has traced the hypothetical origins of black rats in the Galápagos archipelago.[19]

Discovered by Fray Tomás de Berlanga, Bishop of Panama, in 1535 the islands remained uninhabited for a century and a half. Between 1684 and the early eighteenth century English privateers established a semi-permanent base near James Bay on the west coast of the island of Santiago (San Salvador or James) where their vessels could be reprovisioned and careened while plundering the Spanish Main. It seems probable that this quasi-settlement was the origin of the first black rats in the Galápagos. The earliest recorded specimens in the islands were caught only at Sullivan Bay on Santiago during Charles Darwin's historic visit in 1835; it is noteworthy that Darwin does not refer in his journal to the presence of the species on any of the other islands in the group on which he landed, namely San Cristóbal (Chatham), Floreana (Charles or Santa Maria) and Isabela (Albemarle).

The earliest permanent inhabitant of the archipelago seems to have been a British sailor, one Patrick Watkins, who was either marooned or left at his own request on Floreana where he cultivated crops for a number of years but did not, so far as is known, import any livestock.

The second main period of human settlement began in 1832 when General José Villamil, a native of Louisiana, USA, who had fought for Ecuador in her war of liberation against the Spanish, annexed the islands on behalf of Ecuador, and founded a colony some kilometres inland from Black's Beach on Floreana – so-called in honour of Juan José Flores, President of Ecuador. Further temporary settlements were subsequently established, followed by more permanent ones at El Progreso in the highlands of San Cristóbal by Señor Manuel Cobos in 1869 and on southern Isabela by Señor Antonio Gil in about 1893. During this period the islands were visited annually by between 60 and 70 (mostly American) whalers, and there was a considerable amount of intercourse between the individual island settlements. There were thus many opportunities for the importation of black rats from North America (where they first arrived on the Pacific coast in the mid-nineteenth century) and for their dissemination between the three inhabited islands, on which the earliest recorded specimens were caught between 1891 and 1899 (on San Cristóbal in 1898).

The third period of human settlement – on Baltra (South Seymour) and Santa Cruz (Indefatigable) – took place in the 1930s and 1940s. Although the first colonists arrived on Santa Cruz in about 1905 Patton, Yang and Myers (1975) conclude that on two counts the introduction of black rats is likely to have occurred more recently: first, the melanistic colour morph (which is generally characteristic of recent introductions of shipborne black rats throughout the Pacific)[20] is the only one present on Santa Cruz and is nearly fixed on Baltra; second, when the Crane Pacific Expedition visited Academy and Conway bays on Santa Cruz in 1929 they collected only specimens of the endemic rice rats *Nesoryzomys darwini* and *N. indefessus*. Although Patton, Yang and Myers state that 'This date represents the last sighting of either species', D. A. Clark (1984) records that an early settler on Santa Cruz, Anders Rambeck, stated that rice rats were abundant and black rats were unknown on the island from the time of his arrival in 1926 until his departure in early 1930. In April of the latter year J. P. Chapin collected 14 specimens of *N. darwini* in the highlands of Santa Cruz. When in November 1934 Rambeck returned to Santa Cruz black rats were abundant and widespread, and the last known specimen of a Santa Cruz rice rat – *N. indefessus* – was collected in the following month. When Chapin returned to Santa Cruz in March 1935 during the Astor and Templeton Crocker expeditions he caught only black rats at Conway Bay (on the uninhabited north-west coast) where both *Nesoryzomys* species had been taken in 1929. Both Brosset (1963) and Niethammer (1964) point to the strong circumstantial evidence linking the arrival in the Galápagos of black rats with the extinction of rice rats, which on Santa Cruz are thus both likely to have occurred between 1930 and 1934. (It is significant that only on Fernandina (Narborough) and Santa Fé (Barrington), which have never been invaded by black rats, do native rice rats – respectively *Nesoryzomys narboroughi* and *Oryzomys bauri* – remain today.)

Baltra Island north of Santa Cruz remained uninhabited until the construction of a US Army Air Corps base on the island during the Second World War; it seems likely that the arrival of black rats on Baltra dates from this period.

The only island infested with black rats which has never been settled by man is Pinzón (Duncan) between Isabela and Santa Cruz. Since the island was frequently visited by whalers during the first half of the nineteenth century colonization by black rats probably occurred during that time. Specimens were first collected on Pinzón in 1891.

Patton, Yang and Myers (1975) sum up the probable history of black rat arrivals in the Galápagos as follows:

1. Initial introduction at Isla Santiago in the late 1600s, possibly from a direct source in England. The Isleta Bartolomé population undoubtedly was derived directly by across-the-water transport from adjacent Sullivan Bay on Santiago.
2. Introduction through Floreana with spread to San Cristóbal and Isabela, augmented by new immigrants, during the 1800s and perhaps from a source in west coast South America.
3. Introduction to Baltra and Santa Cruz, primarily as a result of U.S. armed forces occupation during World War II with rats possibly originating from California, Hawaii or Panama . . . The Isla Pinzón population . . . may represent a fourth point for outside rat introduction. Alternatively, it may also have been derived directly from the Floreana–Isabela–San Cristóbal group . . .

The three population groups, each with distinctive characteristics, of black rats in the Galápagos (Santiago; Baltra and Santa Cruz; Isabela, San Cristóbal and Pinzón; with Floreana as part of either of the last two groups) is plausibly accounted for by Patton, Yang and Myers through this theory of multiple invasions.

At the present time black rats occur in a wide spectrum of vegetated habitats from desert scrub to lush montane forest on these seven islands, as well as on a number of smaller ones. They are suspected, on strong circumstantial evidence, of being responsible for the extermination this century of four of the endemic species of rice rats in the Galápagos (*Nesoryzomys swarthi* on Santiago, *N. darwini* and *N. indefessus* on Santa Cruz, and *Oryzomys galapagoensis* on San Cristóbal) and, probably in the nineteenth century, of a muskrat-sized rodent (*Megaoryzomys curioi),* which was only discovered, from bones found in the highlands of Santa Cruz, as recently as 1962.

J. L. Patton and M. S. Hafner[21] have concluded that the Galápagos *Oryzomys* originate in a relatively recent colonization of the archipelago – perhaps through an introduction of *O. xantheolus* from Peru by aboriginal voyagers within the last few hundred years. The same authors suggest that the ancestral stock of *Nesoryzomys* was also related to *O. xantheolus*: their evidence points to a much earlier immigration to the Galápagos for *Nesoryzomys* than for *Oryzomys* – possibly by 'rafting' the 1,000 km from the Guayas River on the mainland of Ecuador. D. W. Steadman and C. E. Ray[22] consider that the closest relation to *Megaoryzomys* is the South American cricetine genus *Thomasomys*, which belongs to a different tribe from that of other Galápagos rodents, and must presumably be derived from a third invasion of the archipelago by a mainland rodent.

Brosset (1963) has suggested three ways in which black rats could have brought about the demise of the native rice rats: they could have out-competed the indigines for food and shelter; they could have attacked and killed the (generally smaller) native rodents; or they could have brought with them some parasite of disease to which the endemic species had never been exposed. Clark (1984) believed that the third explanation would most reasonably account for the rapid disappearance of *Nesoryzomys* on Santa Cruz, and Brosset mentions that, within a week of their arrival from Santa Fé at the Charles Darwin Research Station on Santa Cruz (where black rats are abundant), 14 captive *O. bauri* had died from some unidentified disease. On Santiago, on the other hand, both *Nesoryzomys* and *Rattus* survived sympatrically for some 300 years before the former eventually died out in 1906. It therefore seems likely that all three factors played a part in the extinction of the various insular species of rice rats in the archipelago.

Clark (1981) has succinctly summed up the impact of black rats on the Galápagos ecosystem. Several factors combine to affect the island's flora: in the Galápagos black rats are primarily herbivorous (vegetation comprises an average of 80 per cent of stomach contents) and they seem to prefer the reproductive parts of plants; they destroy rather than disperse (through their faeces) nearly all ingested seeds, and often concentrate on less common species. Their selectivity in feeding may also considerably affect the composition of plant communities. Although animals form a small proportion of their diet, a large majority of the population does consume such food. As well as having been almost certainly responsible for the extinction of the five native rats referred to above, predation by black rats of hatchling tortoises on Pinzón prevents natural recruitment in that population,[23] and their destruction of eggs and chicks is believed to be responsible for the continuing decline of the local race of the endangered dark-rumped petrel (*Pterodroma phaeopygia phaeopygia*) on Santa Cruz, Isabela, San Cristóbal and Floreana. Black rats in the Galápagos may also prey to a lesser extent on a number of endemic reptiles such as the gecko *Phyllodactylus galapagoenisis*, larva lizards (*Tropidurus* spp.) and snakes (*Dromicus* sp.). Invertebrates, too, form an important part of their diet, with a wide variety of (in some cases endemic) species – especially bugs and the larvae of Lepidoptera – being eaten in all habitats.

In addition to their direct impact on the Galápagos' fauna, black rats may be having important indirect effects on species whose food they eat. On Pinzón, for example, they may be competing with granivorous doves and finches for the rich Croton seeds, while in the *Scalesia* forest they may be reducing the biomass and diversity of fruits available to frugivorous tree finches (*Platyspiza crassirostris*). Their main controlling agent in the Galápagos – apart from man – is feral cats.[24]

The impact of black rats in the Galápagos is likely to be greater on those islands which never supported populations of native rice rats, whose ecological role and, especially, diet, were probably similar to those of the introduced species. This supposition is borne out by the fact that only on Pinzón (where native rats never occurred) are black rats known to prevent tortoise recruitment. 'With their great diet breadth and their highly preferential feeding behaviour', Clark (1981) concludes, 'black rats may be causing many severe impacts on the flora and fauna of the Galápagos Islands.'

Fiji

Black (and brown) rats arrived in Fiji in the nineteenth century, and are today locally abundant in agricultural, urban, suburban and coastal areas and in plantations on all the main islands.

Tristan da Cunha

Black rats swam shore on Tristan da Cunha in the South Atlantic from a wrecked ship in 1882. In spite of the pastor, who urged that they be killed, they were permitted to become established, and were soon infesting the fields of potatoes – the islanders' principal crop. In the wilder and more mountainous hinterland they preyed on ground-nesting birds and, together with feral cats, are believed to have exterminated the Tristan rail (*Porphyriornis nesiotis*), and to have decimated the Atlantic frigate-petrel (*Pelagodroma marina hypoleuca*) and the Antarctic or dove prion (*Pachyptila desolata*).

Île Amsterdam; Île St Paul

Île St Paul in the Indian Ocean has been infested with black rats, especially around Port Jeanne d'Arc, since their introduction by nineteenth-century whalers. In the first half century after their introduction to St Paul and to Île Amsterdam (possibly with the help of brown rats) they are believed to have caused a heavy reduction in the breeding populations of broad-billed prions (*Pachyptila vittata*) and perhaps also of white-bellied storm petrels (*Fregetta grallaria*).[25]

Seychelles

In the Seychelles group black rats occur on Providence Island, on Desroches in the Amirantes (and probably on several other Amirantes islands) and on most of the Seychelles islands. They are, however, absent from the smaller islands of the Farquhar Atoll, Cerf Island in the Providence group, and Desnoeufs, Marie Louise, Boudeuse, Etoile and African Banks in the Amirantes.

Predation by black rats (and in the past by feral cats) has been responsible for confining the Seychelles magpie robin (*Copsychus sechellarum*) – known locally as the *pie chanteuse* – (which formerly nested on Marianne, La Digue, Félicité, Mahé, Aride, Praslin, Silhouette and Frégate) to the last named island, from which cats (but not rats) have recently been eradicated. On Bird Island, however, to which black rats were accidentally introduced in 1967 from Praslin in a cargo of leaves for thatching, and where by 1972 they covered the entire island, they appear to have had little or no effect on nesting sooty terns (*Sterna fuscata*), white terns (*Gygis alba*), wedge-tailed shearwaters (*Puffinus pacificus*) and brown noddies (*Anous stolidus*).[26]

Kerguelen

The first (dead) black rat was found on this group in the southern Indian Ocean in 1956; since then the species has become established throughout South Grande Terre.

Kermadec

Recent observation suggests that black rats (and cats) are endangering a colony of sooty terns at Denham Bay on Raoul Island in the Kermadecs.[27]

Lord Howe

Black rats are believed to have first appeared on the island (600 km off the east coast of Australia) in 1918 after the grounding of the SS *Makambo*: by 1938 they had succeeded in exterminating five members of the island's avifauna – the vinous-tinted thrush (*Turdus xanthopus vinitinctus*), the Lord Howe Island warbler (*Gerygone insularis*), the Lord Howe Island fantail (*Rhipidura cervina*), the robust silvereye (*Zosterops strenua*) and the Lord Howe Island starling (*Aplonis fuscus hullianus*). A large insect – *Dryococelus australis* – and two large land snails, *Placostylus bivaricosus* and *Gudeococcha sophiae*, were also eradicated, and the population of the Lord Howe Island woodhen (*Tricholimnas sylvestris*) was seriously depleted. Owls, introduced to control the rats, preyed instead on the white tern.[28]

South Georgia; Macquarie

On South Georgia in the South Atlantic black rats are said to have been introduced to Prince Olav Harbour – once known as Rat Harbour – at least by 1800, and to Macquarie Island off New Zealand a few years later by early sealers and whalers. On the former they no longer occur, but on the latter (assisted by feral cats and weka rails (*Gallirallus australis scotti*) imported from New Zealand for food), they had by 1890 exterminated the last of the island's endemic parakeets (*Cyanoramphus novaeseelandiae erythrotis*) and, four years later, the endemic banded rail (*Rallus philippensis macquariensis*). There is, however, little evidence that black rats today are significant predators of birds on Macquarie Island. They are largely confined to tussock-grassland, where their diet is primarily vegetarian, with some invertebrate intake.[29] Although direct evidence is lacking the destruction of some nestling sooty shearwaters (*Puffinus griseus*) and white-headed petrels (*Pterodroma lessoni*) has, however, been attributed to rat predation.

Ian Atkinson of New Zealand has traced the dispersal of black (and other) rats to the world's oceanic islands.[30]

Between about AD 800 and 1500 most oceanic islands outside the Pacific remained undiscovered and thus unaffected by commensal rats. Exploration and trading by Arab and Asian peoples, however, probably dispersed black rats to a few continental islands.

As a result of increasing long-distance voyaging by European traders and explorers between 1500 and about 1710 black rats gained access to many islands in the Atlantic and Indian oceans. Few islands in the Pacific, however, received introductions of black rats during this period – a notable exception being Santiago in the Galápagos.[31]

From about 1710 to around 1830 black rats were almost completely displaced in western Europe by the spread or increase of brown rats, which thus became the common rat aboard European ships. After about 1830 – especially from mid-century – black rats staged a comeback as the common shipboard rat: the reasons for this change are not yet fully understood. During the last 150 years both black and (less frequently) brown rats have been spread by man to further islands throughout the world, and few major oceanic groups are now without one or other species. 'With the exception of man, the rat is the mammal that has undoubtedly had the greatest impact on the environment and economy of most Pacific islands . . . throughout Micronesia . . . both geographically and ecologically the roof [black] rat and the Polynesian rat are the most widespread, being reported from nearly all sizeable land areas.'[32] Within many archipelagos, however, small islets, including some of considerable biological interest, remain uninfested.

In general, black rats tend to thrive in tropical rather than in temperate regions, and in areas without an extensive fauna of small indigenous mammals. They are extremely agile climbers, often nesting in trees but seldom in burrows, and are thus a particularly serious threat to (especially small hole-nesting) passerines. They are themselves a major prey of some alien carnivores (e.g. feral cats (*Felis catus*) and mongooses (*Herpestes auropunctatus*)[33]) whose increased populations then exert a greater pressure on avian species. Atkinson[34] has put forward an intriguing (but as yet untested) suggestion that on small islands the presence of land crabs may possibly affect the vulnerability of an island avifauna to predation by rats.

Black and brown rats are undoubtedly the world's worst introduced pests – as vectors of disease (e.g. bubonic plague or Black Death, caused by the organism *Pasteurella (Yersinia) pestis* and transmitted by the flea *Xenopsylla cheopis*, which has ravaged mankind since the third century BC), as plunderers of growing and stored crops and as predators of other (principally avian) species. According to Clutton-Brock (1981), the black rat 'as a carrier of diseases is believed to have caused a greater number of deaths in the human species than any natural catastrophe or war'; it has also been estimated that since about 1600 introduced rats have been responsible for the extermination of at least 18 species or races of birds, and that they have put at risk a further 40 or more. One of the most recent to suffer, the Puerto Rican yellow-shouldered blackbird (*Agelaius xanthomus*), was in 1977 officially listed by the US Fish and Wildlife Service as endangered, largely through predation by rats in its coastal mangrove habitat. As D. B. Clark (1980) has rightly said, '*Rattus rattus* is one of the world's most successful mammalian 'tramp' species. With humans as its dispersal agent, it has colonized six continents and thousands of islands.'

NOTES

1. Rackham (1979) and Armitage, West & Steedman (1984).
2. Cadwaladr Vendigaid (i.e. the Blessed), the semi-mythical king of the Britons, is said to have died of the plague which was ravaging Ireland in AD 664. If this is true then rats must have been present by at least the early seventh century (see Twigg 1978).
3. See p. 349.
4. Langham, Wells & Charles (1974).
5. See p. 30.
6. W. F. H. Ansell, *The mammals of Zambia* (National Parks and Wildlife Service, Chilanga, Zambia, 1978), p. 82.
7. In S. Purchas, *Narratives of early Virginia* (1613), p. 185.
8. Transl. by J. Davies, as *The history of the Caribby-Islands* (1666).
9. p. 154.
10. See p. 352.
11. I am indebted to this paper for much of the history of the black rat in New Zealand.
12. *Tasmanian Journal of Natural Science* **2**: 81–107(1846); *Transactions of the New Zealand Institute* **25**: 495–503 (1893).
13. *Transactions of the New Zealand Institute* **10**: 239–42 (1878).
14. Helms (1883).
15. *Appendices to the Journal of the New Zealand House of Representatives* C.1: 71–5 (1894).
16. *Mutton birds and other birds* (Whitcombe & Tombs, Christchurch and London, 1914).
17. P. H. Baldwin, *Audubon* **47**: 343–8 (1945).
18. See p. 41.
19. Patton, Yang & Myers (1975). (I am grateful to Dr Deborah A. Clark for drawing my attention to this paper.)
20. Storer (1962).
21. 'Biosystematics of the native rodents of the Galápagos Archipelago, Ecuador', in R. I. Bowman & A.E. Leviton, eds, *Patterns of evolution in Galápagos organisms* (American Association for the Advancement of Science, San Francisco, 1982).

22. *Smithsonian Contributions to Paleobiology* No. 51 (1982).
23. C. G. MacFarland, J. Villa & B. Toro, *Biological Conservation* **6**: 118–33 (1974).
24. See p. 88.
25. Segonzac (1972).
26. C. J. Feare, *Atoll Research Bulletin* No. 226: 10 (1979); A. W. Diamond & C. J. Feare, *Proceedings of the 4th Pan-African Ornithological Congress*, 89–98 (1980)
27. R. H. Taylor, *Notornis* **26**: 199–202 (1979).
28. K. A. Hindwood, *Emu* **40**: 1–86 (1941); H. F. Recher & S. S. Clark, *Biological Conservation* **6**: 263, 273 (1974).
29. Jenkin, Johnstone & Copson (1982).
30. Paper presented at the 18th World Conference of the International Council for Bird Preservation at Cambridge, England, August 1982; see Moors (in press).
31. See p. 359.
32. Storer (1962).
33. See pp. 84 and 75.
34. See note 30; and personal communication (1983).

House Mouse

(*Mus musculus*)

NATURAL DISTRIBUTION
Originally the dry steppe zone of the southern Palaearctic, and possibly the Mediterranean region.

NATURALIZED DISTRIBUTION
Virtually worldwide.[1]

EUROPE

British Isles

According to G. B. Corbet, 'The house mouse was probably the first mammal to be introduced to Britain by human agency . . . one recent record seems to confirm its presence in the pre-Roman Iron Age. This is based on the rostrum of two animals and one mandible, from a site at Gussage All Saints, Dorset.'[2]

The tendency of isolated offshore islands to allow the development of different races of one species displaying new genetic characteristics is typified in Britain by the now extinct St Kilda house mouse (*M. m. muralis*) of the Outer Hebrides (see below) which more closely resembled the mice of the Shetland Islands and of the island of Mykines in the Faeroes (see below) than those of mainland Britain. It thus seems likely that the Norsemen brought the species from Scandinavia – possibly in cargoes of hay imported to feed their domestic stock – more than a thousand years ago.

Faeroe Islands[3]

In 1981, S. Easteal wrote:

> The colonization of a new area by a species may be a major event in the evolution of that species and can result in the formation of new species. This can occur if the colonizing event itself causes isolation between different populations, which then diverge genetically as the result of micro-evolutionary processes, or if the colonizing event itself, in cases where it involves very few individuals, brings about a radical genetic change in the founding population.[4]

The house mice of the Faeroe Islands (lying approximately midway between Iceland and Norway) are frequently cited as spectacular examples of this microevolutionary divergence.

W. Eagle Clarke of the Royal Scottish Museum, Edinburgh, was the first to recognize the distinctiveness of house mice on the Faeroes, when he described specimens from the island of Nolsøy as 'remarkable for their great size, indeed they are veritable giants, being considerably larger than the type and of any of its numerous geographical races'.[5] On the sole basis of their 'immense size' Clarke placed them in a new subspecies, *faeroensis*. A few years later G. S. Miller promoted the Faeroe Islands mice to full specific rank since they 'differ so conspicuously from other known members of the group'.[6] In 1942 M. Degerbøl[7] created another subspecies, *mykinessiensis*, for individuals on the island of Mykines, in which the mesopterygoid fossa on the base of the skull tapers to a point (unlike that of most house mice) but like that of mice in the Shetland Islands and of the extinct St Kilda house mouse (see above). Degerbøl regarded another island in the Faeroe group, Fugløy, as inhabited by a 'special stock of mice', but did not accord them taxonomic rank. He considered specimens on Streymøy as probably 'fairly pure representatives of the original Faroe mouse, originating through isolation for a long space of time, while the big Nolsøy mice have been developed by adaptation to the leaping life on the bird cliffs'.

In the same year (1942) that Degerbøl published his findings, J. S. Huxley pointed out that although the formation of new subspecies normally takes some 5,000 years, 'the facts concerning rats and mice show that sub-speciation may sometimes occur much more quickly. In particular, the Faroe house mouse, which was introduced into the islands not much more than 250 years ago, is now so distinct that certain modern authorities have assigned full specific status to it'.[8] A decade later Matthews (1952) commented that 'these island races of mouse are particularly interesting in showing the rate at which isolated wild populations can develop new genetic characters, for it is definitely known that their forerunners could not have been introduced into the Faeroes less than two hundred and fifty or more than about a thousand years ago'.[9] As Berry (1981) points out, however, the microevolutionary divergence of house mice in the Faeroe Islands is in fact a much more recent phenomenon.

Of the six Faroe islands where mice occur, Hestur seems to have acquired them since 1939 while both Fugløy and Mykines were recorded as being free from mice in 1800. [According to A. Reinert[10] house mice colonized the island of Bordøy soon after the opening of a shop there in 1838, but were exterminated between 1915 and 1920 by rats which colonized the island after the establishment of a whaling station.] Notwithstanding all the island races are very distinct from each other and from other North European populations for both morphometric and biochemical traits . . . Mykines is the most distinct population, although its separation from others is less on biochemical than on metrical data . . . the genetically and morphologically most variable population came from Sandøy, one of the southern Faroe islands. The southern islands are more fertile and intensively farmed than the northern islands, and it seems plausible that the original establishment of mice in Faroe was on Sandøy, and that they spread from there to the other islands, although both Hestur and Nolsøy were probably colonized from the large neighbouring island of Streymøy. There is much that is speculation about the Faroe mouse story, but there can be little doubt that the present differentiation has resulted from the colonization of different islands by small numbers of animals in fairly recent times. Chance segregation of available variation has been much more important than progressive adaptation. Almost certainly there is a recurring theme in house mouse evolution; many populations probably have a short existence in terms of numbers of generations, and the species survives and succeeds by opportunistic establishment and rapid growth of new colonies.

Since, as previously mentioned, narrowing of the mesopterygoid fossa is found in mice on Mykines in the Faeroes, on the Shetland Islands and on St Kilda, it seems reasonable to assume that the populations of all three groups originated in a single source – the marauding Vikings from Scandinavia.

ASIA

USSR

One of the few recorded introductions of house mice in Asia took place in 1870, when several mice were offloaded from the schooner *Justus* on Bering Island in the Commander Group off Kamchatka in a cargo of flour from San Francisco.

NORTH AMERICA

USA; Canada; Mexico; West Indies; Bermuda

The Arctic explorer and naturalist Sir John Richardson records in his *Journal* (1851) that house mice were first introduced to North America by Stephen Harriman Long's expedition of 1819–20 at Engineer Cantonment near Council Bluffs on the Missouri River, Iowa. Richardson himself was shown a dead mouse among goods shipped from England in the Hudson's Bay Company's store at York Factory, Manitoba, Canada, in 1829, and by the following year the animals were said to be common in Ontario. By 1855 house mice were to be found in many other inland localities, such as Prairie Mer Rouge, Louisiana; Fort Riley, Kansas; Fort Pierre, South Dakota; Fort Redding, California; and Parras, Coahuila, Mexico. In 1891 house mice were introduced in a wagonload of seedgrain into the Huachuca Mountains, Arizona. R. P. J. B. Du Tertre, in his *Histoire générale des îles de S. Christophe, de la Guadeloupe, de la Martinique, et des autres dans l'Amérique* (1654), records that house mice were at that time established in the French islands in the West Indies, but that they were still uncommon there. Exactly when house mice first reached Bermuda – where they are today a common pest – is uncertain, but they probably arrived soon after British settlement began in 1612.

SOUTH AMERICA

Chile; Argentina

On the mainland of South America house mice are established as far south as Punta Arenas, Chile, and Patagonia, Argentina.

Falkland Islands

House mice probably landed on the Falklands from the ships of the first French colonists in 1764, although they appear not to have been recorded in the English settlement at Port Egmont on Saunders Island until 1774, nine years after it was established. In 1842 Governor Richard C. Moody found them only around the settlement of Port Louis on Berkeley Sound. Today, mice are widely distributed in areas of human habitation on both East and West Falkland, and even on some remote offlying islands such as Steeple Jason.

AUSTRALASIA

Australia

House mice almost certainly first arrived in Australia with the early colonists who landed at Botany Bay in January 1788.

From time to time in eastern Australia the population reaches plague proportions, and vast areas of crops are destroyed. The first such was in the early 1860s in the Cooper Creek area of Queensland and South Australia. A major outbreak took place in 1917 in the wheat-growing areas of Victoria and New South Wales; at Sheep Hills, in the Mallee, for example, no less than 7 tons of mice were poisoned in a single night. The most recent outbreak occurred in 1980 along the west coast of South Australia and in north-west Victoria. In the former state the area affected stretched from Cactus in the west through Penong, Thevenard and Munipa to about 100 km east of Ceduna, and south to Streaky Bay; in Victoria over 250,000 km^2 of valuable farmland were infested, and losses and damage in both states were estimated to have cost millions of pounds.

New Zealand

Although the exact date of the arrival of the house mouse in New Zealand is uncertain, Thomson (1922) was of the opinion that it was originally introduced early in the nineteenth century. Ernest Dieffenbach, in his *Travels in New Zealand* (1843), was the first author to record its presence in North Island, and H. Guthrie-Smith[11] states that it was to be seen around the Bay of

Islands Mission Station in Northland in about 1830, coincident with the arrival from England of the first ships carrying cargoes of grain.

Of South Island, Robert Gilles wrote:

It is quite certain that there were no mice in Otago in 1852, but a year or perhaps two years after they were noticed in Dunedin first. As soon as the mice appeared the Rats disappeared . . . the Molyneux River stopped their southern migration for a time, and it was considerably later before Molyneux Island (Inchclutha) was touched by them.[12]

White (1890b) records that mice first made their appearance in the Canterbury Plains in about 1855. Pastor J. F. H. Wohlers, a German missionary on Ruapuke in the Foveaux Straits, in his autobiography (1895) writes that mice first arrived on the island when the *Elizabeth Henrietta* was wrecked off the coast in 1824, and that as late as 1873 they were still being referred to by the islanders as 'Henriettas'.

As in Australia and elsewhere, periodic population explosions occur in New Zealand; one such, in January 1918, was thus described:

There is a plague of mice in the district west of the Waiau. From Bluecliff to the Knife and Steel near the Big River, and beyond, . . . was overrun with them . . . on the beach and in the dense bush, wherever we went, they were plentiful. At the Hump, near Lake Hauroto, they were as numerous as elsewhere. This prevalence of mice is certainly not usual; I have been on the Hump four or five times since 1911, and last year tramped along the Knife and Steel, and apart from an odd one or two, no mice were in evidence on former trips.[13]

Today, house mice are widely distributed throughout North and South Islands (as well as on many of the offshore islands including the Antipodes, Chatham and the Aucklands – where they have been on Enderby since about 1843 and on the main island since about 1820 – but excluding Stewart Island), living both as commensals of man and in the open countryside. In the latter they occur in a wide variety of habitats, ranging from sand-dunes with wild lupins and marram grass to the open tussock land of Marlborough and south Canterbury, the gorse and heathland of Otago, and the scree of the high mountain tops. They appear sometimes to have difficulty in becoming established where large populations of brown rats (*Rattus norvegicus*) exist. As elsewhere throughout the world, mice in New Zealand are a serious economic pest and potential vectors of a variety of diseases. They are preyed on by introduced mustelids, cats, harriers, moreporks, small brown owls and possibly kingfishers.

OCEANIA

Hawaii

At one time or another mice have been introduced to a large number of the world's oceanic islands. They are first referred to in Hawaii by O. von Kotzebue who commanded a Russian naval expedition which visited the islands in 1816. In 1903 D. Malo in *Moolelo Hawaii*, writing of animals which had been imported since the time (around 1835) of King Kamehameha I, described 'the mouse, or *iole-liilii*' as a 'late importation'. Today, they are common on Hawaii, Maui, Molokai, Lanai, Oahu, Kauai, Mokuoloe, Manana (Rabbit Island), Kekepa, Kepapa, and probably on Kaula, Kahoolawe and Niihau.

Other oceanic islands

On Fiji, where they are today locally abundant in both urban and agricultural areas on Viti Levu and Vanua Levu, mice were first introduced by the American Exploration Expedition of 1838–42, led by Charles Wilkes, which visited the islands in 1840. In Micronesia house mice occur on some but not all hilly islands, but are virtually absent from low-lying atolls.[14] During the Second World War they were introduced to the Galápagos Islands, where their arrival resulted in an increase in both barn owls (*Tyto alba punctatissima*) and short-eared owls (*Asio flammeus galapagoensis*).

Among the more remote oceanic islands infested with house mice are Tristan da Cunha (where they have been found up to high altitudes in montane feldmark and heath since about 1830) and Gough Island (since before 1887) in the South Atlantic; South Georgia, east of Cape Horn (where they have been established for around 150 years); Juan Fernandez off the coast of Chile; Macquarie, south-west of New Zealand; and Amsterdam, St Paul, Île aux Cochons, Île de la Possession (Crozet), Kerguelen (since before 1874), the Seychelles, and Marion in the southern Indian Ocean. They also occur on Ascension Island.

Mice were probably introduced to Macquarie Island in the late nineteenth century by sealers and whalers, who also imported weka rails (*Gallirallus australis*) from New Zealand as a source of food. The earliest written evidence of mice on Macquarie was made by Edward Wilson, who visited the island during Scott's polar expedition in 1901, when they were said to have been present for over a decade. The mice of Macquarie are quite distinctive, being large and long-bodied – somewhat similar to those on Tristan da Cunha and Gough – but with a relatively short tail. A favourite food is the Macquarie cabbage (*Stilbocarpa polaris*). Marion mice are short-bodied but heavy and those of South Georgia are also large, though none are as big as those found on islands such as St Kilda, the Faeroes and Foula in the North Atlantic.[15] The mice of South Georgia, which exist on a remote headland surrounded by glaciers where they may be snowed in at any time of the year and where for 20 weeks the average temperature is below freezing, possess a great deal of brown fat (an indicator of physiological adaptation to low temperatures, and double or treble that of mice on Marion and Macquarie) which suggests a greater physiological stress on mice inhabiting South Georgia than on those infesting islands outside the Antarctic Convergence. On South Georgia, where mice shelter in the peaty bases of tussock grass, the breeding season extends from about October to April, whereas on Marion and Macquarie mice breed at all seasons. On all subantarctic islands the temperature is nearly always so low that elsewhere reproduction would cease, suggesting that these ecads must be living at or near their physiological limits.[16]

Summing up the achievements of the house mouse as a colonist R. J. Berry wrote:

> the house mouse is astonishing in its flexibility. In Hawaii it lives in sugar-cane plantations where the rainfall is 135 in. [3,430 mm] a year, and 10 miles [16 km] away it flourishes in near desert with 7 in. [179 mm] a year; and the population extends without a break up to 9,000 ft. [2,750 m] on the volcanoes of the island, where the temperature averages 40 deg F [5 °C] lower than at the coast. In Peru house mice out-compete the native *Phyllotis* rodents [leaf-eared mice] at the desert's edge, and occur also at the upper limits of human occupation around 15,000 ft [4,570 m] up the Andes . . . the mouse achieves greater prodigies of adaptation than any other mammal except possibly ourselves.[17]

Although the **long-tailed field mouse** or **wood mouse** (*Apodemus sylvaticus*) is a native of mainland Britain, it probably owes its presence on many offshore islands to human intervention. Those in the Outer Hebrides, the Shetlands, and St Kilda, Scotland, and in Ireland (and Iceland) more closely resemble the field mice of Norway than they do those of mainland Britain, and it appears probable that they were introduced by Norsemen from Scandinavia more than a thousand years ago, possibly in the hay which they imported to feed their domestic stock.[18] Field mice in the Orkneys were probably introduced by Neolithic people.[19]

The **Cairo spiny mouse** (*Acomys cahirinus*), which ranges from Pakistan through Arabia to northern and eastern Africa, has become established as a commensal of man on the island of Crete (where it is also known as the **Cretan spiny mouse** (*A. minous*)), and possibly also on Cyprus.

NOTES

1. Berry (1981) points out that although Linnaeus was not strictly accurate in writing that house mice 'inhabit all parts of the world except the Arctic' (they are, for example, still relatively uncommon in central Africa and were unknown in Uganda and Zambia until the mid 1960s), 'they have colonized an enormous range of habitats, from coral atolls in the Pacific to near Antarctic conditions on South Georgia, from coastal tussock grass in temperature regions to 4,000 metres above sea-level in the Andes, from central heating ducts to refrigerated stores'.
2. In 'The distribution of mammals in historic times', in D. L. Hawksworth, ed., *The changing flora and fauna of Britain* (Systematics Association Special Vol. No. 6) (Academic Press, London and New York, 1974), pp. 187–8.
3. For information on the house mouse in the Faeroe

Islands I am indebted to Berry, Jakobson & Peters (1978) and Berry (1981).

4. S. Easteal, *Biological Journal of the Linnean Society* **16**: 93–113 (1981).
5. *Proceedings of the Royal Physical Society of Edinburgh* **15**: 160–7 (1904).
6. *Catalogue of the mammals of Western Europe in the collection of the British Museum* (BM (NH), London, 1912).
7. 'Mammalia', in *The zoology of the Faeroes* **65**: 1–133 (1942).
8. *Evolution, the modern synthesis* (Allen & Unwin, London, 1942).
9. See also F. C. Evans & H. G. Vevers, *Journal of Animal Ecology* **7**: 290–7 (1938).
10. *Danmarks Natur* **10**: 537–8 (1971).
11. H. Guthrie-Smith, *Tutira – the story of a New Zealand sheep station* (Blackwood, Edinburgh and London, 1921).
12. *Transactions of the New Zealand Institute* **10**: 306–24 (1877).
13. Quoted by Thomson (1922).
14. Storer (1962).
15. Thus conforming to Bergmann's Rule that the body size of homoiothermic species and subspecies tends to increase with the decreasing mean temperature of their habitat (i.e. going north in the Northern Hemisphere and south in the Southern Hemisphere).
16. Berry, Peters & van Aarde (1978) and Berry, Bonner & Peters (1979).
17. *Country Life* **170**: 1994–6 (1981).
18. R. J. Berry, *Journal of Zoology* **159**: 311–28 (1969); M. Degerbøl, *Appendix to Zoology of Iceland* **76**: 39–52 (1939); G. B. Corbet (1966: 216) and personal communication (1984).
19. G. B. Corbet, 'Appendix E', in A. C. Renfrew, ed., *Investigations in Orkney* (Reports of the Research Commission of the Society of Antiquaries No. 38) (Society of Antiquaries, London, 1979), pp. 135–7.

Edible Dormouse
(*Glis glis*)

NATURAL DISTRIBUTION
Widespread in southern and eastern Europe east to Iran. Absent from most of Spain and Portugal, northern France, the Low Countries, Denmark and Scandinavia.

NATURALIZED DISTRIBUTION
Europe: British Isles (England).

EUROPE

British Isles (England)

Major Stanley Flower records[1] that 'On or about February 4th, 1902,[2] some individuals of *Glis glis* [also known as the squirrel-tailed or fat dormouse] from the Continent of Europe (probably Germany or Switzerland) were turned loose in Tring Park [Hertfordshire] by Walter (later Lord) Rothschild'.

The dormice, which were highly esteemed by Roman gourmets who fattened them for the table in special earthenware vessels called *gliraria*, multiplied rapidly and caused considerable damage to corn and vegetable crops, and also to thatch, as a result of which a campaign was conducted which was thought nearly to have exterminated them. In 1925, however, they were once again observed in Tring Park, and in the following year the neighbouring estate of Pendley Manor was overrun.

By the autumn of 1931 edible dormice had spread to Whipsnade in Bedfordshire, and by 1935 to Aldbury. Since 1941 they have been observed sporadically at Great Pednor, Buckinghamshire, and since 1945 and 1946 in small numbers in Amersham and Berkhamsted respectively.

In 1951 a census was made of occupied sites, which resulted in the discovery of at least two dozen fresh colonies, some in entirely new districts such as Ashley Green, Cholesbury, Hyde Heath, Great Missenden and Pitstone in Buckinghamshire, and Ashridge Park, Little Gaddesden, Ringshall and Rossway in Hertfordshire.

In the first three decades following their introduction, edible dormice succeeded in colonizing only a limited number of places, the spread to their present limits taking place almost entirely during the war. They are currently confined mainly to a triangle of around 260 km^2 in the Chiltern Hills formed by Beaconsfield, Aylesbury and Luton; records from outside this area include Ibstone and Bledlow Ridge on the Buckinghamshire/Oxfordshire border in the west, and Potters Bar, Middlesex in the east.

In continental Europe edible dormice at times cause considerable damage to corn, fruit and vegetable crops, and also to coniferous timber plantations, but in England only the last of these have suffered depredations in recent years. Edible dormice are preyed on by rats, cats, weasels, stoats and tawny owls, and are easily poisoned and trapped, so control of their numbers should not prove difficult, especially as for some unexplained reason they have so far shown no propensity to expand their present range.

Two forms (*orcadensis* and *sarnius*) of the **common vole** (*Microtus arvalis*) – which, although unknown elsewhere in Britain, is one of the commonest voles in continental Europe – occur on, respectively, the Orkney Islands (mainland, Westray, Sanday, South Ronaldsay and Rousay) Scotland, and on Guernsey in the Channel Islands, England. They are believed to have been introduced, no doubt accidentally, from continental Europe (possibly Iberia) by Neolithic or early Bronze Age people.[3] The presence of *M. arvalis* on Spitzbergen – far from the rest of its natural range – is almost certainly also due to human intervention.[4]

In 1870 the **northern red-backed vole** (*Clethrionomys rutilus*) was translocated from Kamchatka in the Soviet Union to Bering Island in the Commander group, where within a decade it was reported to be omnipresent.

In 1964 a population of the **bank vole** (*Clethrionomys glareolus*) – almost certainly the result of a recent introduction (possibly about 1950) – was discovered in south-west Ireland; by 1971 its presence was confirmed in counties Limerick, Cork, Kerry, southern Clare and Tipperary.[5] On the British mainland the bank vole is indigenous, but its presence on the islands of Raasay, Skomer, Ramsey and Jersey is also probably as the result of (earlier) introductions.[6]

In 1973, at the conclusion of filming a natural history programme for television near Fishbourne, 4 km west of Ryde on the Isle of Wight, England, a number of rodents – including several **Mongolian gerbils** (*Meriones unguiculatus*) – were either deliberately released or permitted to escape. The gerbils settled down on some waste-ground and in a nearby timberyard, and within three years had increased in number to about 100.

In 1971 **gerbils** (? sp.) were found in the wild on Thorne and Goole moors in Yorkshire, northern England. Others were reported in Armthorpe in 1972–3 and 1975–7, in the Bradford area (1975), Harrogate (1977) and Doncaster (1980) – all in the same county. Those observed on Thorne and Goole moors, which are far from human habitation, were almost certainly deliberately released pets, while the animals in Armthorpe originated from a number that escaped from the laboratory of a local school.[7]

The **golden hamster** (*Mesocricetus auratus*) from Asia Minor has become established in parts of the Netherlands and Germany, where it is a local threat to growing and stored crops. In England six golden hamsters escaped into the basement of a pet shop in Bath, Somerset, in 1957, where in the following year more than 50 were caught. Several other colonies have been recorded, including one of more than 70 in Bury St Edmunds, Suffolk in 1964. In late 1981 a number became established in the suburb of Burnt Oak in Barnet on the northwestern outskirts of London. By the following August some 180 had been trapped, and the total population was estimated at several hundred. Hamsters are best controlled by trapping or gassing or by flooding their burrows. Poisoning seems ineffective for two reasons: first, because the animals hoard much more food than they require and are thus unlikely to ingest lethal concentrations; and second, because their habit of storing food in their cheek-pouches gives the poison time to react and be detected and rejected before being swallowed. Since they successfully survived the severe winter of 1981–2 in hibernation, it seems probable that, unless controlled, they could become established in southern England, where they might well become a serious agricultural pest.[8]

Ord's kangaroo-rat (*Dipodomys ordi*) of the western United States and northeastern Mexico has been translocated to the shores of Lake Erie, Ohio, where for a time it became established in sand dunes near Fairport.

The **Brazilian agouti** (*Dasyprocta aguti*) was introduced in pre-Columbian times from South America to St Croix, St Kitts, Antigua, Montserrat, Dominica, St Vincent, Grenada, Martinique, St Lucia, Guadeloupe and possibly Barbados in the West Indies. On Grenada it had been nearly exterminated before about 1910 by the introduced small Indian mongoose (*Herpestes auropunctatus*).[9] It was not imported to St Thomas until after European settlement in 1657 (but before 1852), since although it was a favourite food of the Caribs no remains have been found in their middens on that island. The **red agouti** (*D. punctata*) from Central America and northern South America and the **Mexican agouti** (*D. mexicana*) were introduced in the late nineteenth century to the Sierra de los Organos, Pinar del Río, and the Sierra Cristal, Oriente, in Cuba. At around the same time the former was also introduced to the Cayman Islands in the Caribbean. Because of their extensive burrowing both species have made some inland regions impossible to cultivate.[10] The closely related **paca** (*Cuniculus paca*) of southern Mexico and northern South America has been introduced into Cuba.

In 1952 and 1954 **chinchillas** (*Chinchilla* sp.) were introduced from South America into California, USA, where they failed to become established. In 1964 9 males and 10 females were released in the Darvaz Range, Tajikistan, USSR, apparently without success.[11]

Guinea-pigs (*Cavia porcellus*) are said to have been unsuccessfully introduced to New Zealand by the Auckland Acclimatization Society in 1869. A century later a number escaped or were released on Santa Cruz in the Galápagos Islands, where by the early 1980s the population had increased to between 300 and 500. Being largely granivorous they could assist in the dissemination in the highland regions of introduced plants such as avocado, balsa, bamboo, guava and hibiscus. The dispersal of Chinchona, which has become a problem in the *Miconia* zone of Santa Cruz, could also be aggravated by the presence on the island of feral guinea-pigs.

A report published in the correspondence columns of *The Times* newspaper in 1983 (13 May) that guinea-pigs 'roamed wild on Bodmin Moor', Cornwall, England, proved on investigation to be a hoax.

In 1969 a pair of **Himalayan** or **Hodgson's porcupines** (*Hystrix hodgsoni*) escaped from the Pine Valley Wildlife Park near Okehampton in Devon, England, where a small population became established in the wooded valleys of the rivers Okement and Torridge. Following reports of extensive damage to conifers by debarking, a trapping campaign, instituted by the Ministry of Agriculture, resulted in the capture of one male and three females; one other animal was killed by a dog and another was found dead in the entrance to a badger (*Meles meles*) sett. The number remaining at large (if any) is unknown. In the summer of 1972 a pair of **crested porcupines** (*H. cristata*), natives of northern Africa, escaped from the botanical gardens at Alton Towers near Stoke-on-Trent in Staffordshire, England, where they caused extensive damage to larch (*Larix europaeus*) plantations. Crested porcupines occur on the mainland of Italy and in Sicily, possibly as a result of introductions by man. More recently they are believed to have been introduced into Albania and thence to Yugoslavia.

NOTES

1. *Proceedings of the Zoological Society of London*, 769 (1929).
2. Not, as De Vos, Manville & van Gelder (1956) and Roots (1976) say, 1886.
3. Corbet (1961).
4. E. Alendal, *Fauna, Oslo* **30**: 8–11 (1977).
5. Fairley (1971) and Smal & Fairley (1978).
6. Corbet (1961).
7. C. Howes, '"Free-range" gerbils', *Mammal Society Newsletter* No. 56 (1983).
8. Lever (1983a).
9. See p. 78.
10. C. Swabey & C. B. Lewis, *Development and welfare in the West Indies*, *Bulletin* No. 23: 11 (1946).
11. M. Pavlov & A. Senelnikov, *IUCN Bulletin* No. 17 (1965).

Coypu
(*Myocastor coypus*)

NATURAL DISTRIBUTION
Argentina, Uruguay, Paraguay, southern Brazil, southern Bolivia, southern Peru, Chile, the Chonos Archipelago, Chilöe Island, south to the Straits of Magellan.

NATURALIZED DISTRIBUTION[1]
Europe: British Isles (England), France, the Netherlands, Germany, Italy, Belgium, ?Denmark; Eurasia: USSR; Asia: Turkey, Israel, Japan; Africa: Kenya, ?Zambia, ?Tanzania; North America: USA, ?Canada.

EUROPE

British Isles (England)

Coypus, which produce a fur known in the trade as nutria, were first introduced to England in 1929 to stock fur-farms, mainly in Surrey, Sussex, Hampshire, Norfolk and Devon. About 50 farms were established, but a fall in the value of fur and the outbreak of war resulted in the closure of all by 1939. The first escapes were reported from Horsham in Sussex in 1932, and by the 1940s colonies in the wild were established in Norfolk and Suffolk, and at Dorney near Slough in Buckinghamshire where they remained until about 1954.[2] Isolated instances of escaped coypus were also reported from Bedfordshire, Cheshire, Essex, Gloucestershire, Somerset and Staffordshire between 1936 and 1944. Coypus were also bred between 1929 and 1939 on two farms in Yorkshire and Cumberland from which, strangely enough, there were no recorded escapes.

In Scotland coypus were reared at Auchinroath near Rothes on the River Spey in Morayshire, from which there were some escapes in 1934, and at Perth. In Wales a farm was established near the confluence of the rivers Vyrnwy and Severn in Montgomeryshire, from which escapes were reported in 1936.

From about 1940 wild coypu colonies largely disappeared in the south, those in Norfolk and Suffolk alone surviving. In the former county the first coypus escaped from farms in the valley of the River Yare in 1937; during the war they spread up and down the Yare and its tributaries and along the rivers Wensum and Tass, colonizing some 65 km of those waterways by 1944. In the same year coypus were becoming increasingly numerous in the marshland around Surlingham, Wheatfen and Rockland Broads,[3] where they found an ideal habitat. By the following year they had spread east to Cantley, Reedham and the Langley marshes, and south-west to Northwold on the River Wissey and to Wroxham.

Coypus were first seen on Hickling Broad in 1948 and on Hornsey in the following year. In the early 1950s they appeared in many inland localities away from the main river-system, where they formed new and discrete colonies on isolated ponds and flooded gravel-pits, having followed little streams from the rivers Waveney, Yare, Wensum and Bure. By the mid-1950s coypus had arrived on the River Glaven in north Norfolk, and colonization of Suffolk was well under way. By the late 1950s coypus had reached Oulton Broad and the rivers Ant and Thurne; east and west Suffolk and Cambridgeshire as far south and east as Holbrook near Ipswich; south and west to Mildenhall on the River Lark; Wisbech and King's Lynn; and as far north as Melton Constable. Isolated individuals were reported from as far west as Thrapston in Northamptonshire and south to Mountnessing in Essex.

Coypus in England are a serious, albeit local, economic pest. They raid gardens for brassicas and farmland for cereals and root crops – swedes, mangolds, potatoes and especially sugar-beet. Their burrows undermine the banks of rivers and dykes which then collapse, resulting in the flooding of valuable farmland. They trample down and crush marsh vegetation and eat the young shoots of reeds (*Phragmites australis*), cutting up the rhizomes and often completely clearing entire beds, thus converting them into open water. They eat reed-mace (*Typha latifolia*) and lesser reed-mace (*T. angustifolia*), pulling them up by the roots and eating the submerged parts – mainly the rhizomes – leaving the remainder floating. The reed meadow grass (*Glyceria maxima*) is also extensively eaten, as is saw-sedge (*Cladium mariscus*), tufted sedge (*Carex elata*), great pond sedge (*C. riparia*), burr-reed (*Sparganium erectum*), yellow water-lily (*Nuphar lutea*), reed-grass (*Phalaris arundinacea*), the great water dock (*Rumex hydrolapathum*) and, where they occur, water-parsnip (*Sium erectum*) and cowbane or water-hemlock (*Cicuta virosa*). Damage to ozier beds and larch saplings has also been reported. Although the removal of large areas of reed-beds did help in some measure to keep the waterways clear and prevent silting, it also threatened the habitat and survival of such rare marshbirds as bitterns (*Botaurus stellaris*), bearded tits (*Panurus biarmicus*) and marsh harriers (*Circus aeruginosus*), and led in 1962 to the coypu's addition to the Destructive Imported Animals Act.

In England coypus have few enemies apart from man. Indeed, other than human interference, exceptionally cold weather is their main controlling factor. Between 1945 and 1962 – apart from the severe winter of 1946–7 – coypus steadily increased in range and numbers in the east of the country, reaching a peak population of around 200,000 by 1962. A government-sponsored trapping scheme resulted in the capture of some 97,000 coypus in the two years up to August 1962 without, however, any apparent reduction in either distribution or numbers. Indeed, the animals' range was actually expanding at this time, and coypus began increasingly to be reported from Lincolnshire, Cambridgeshire, the Isle of Ely, Huntingdonshire, north-east Bedfordshire, south-west Hertfordshire and in various parts of Essex and Sussex; isolated sightings came from as far afield as Sheffield, Leicestershire, Northamptonshire and Derbyshire.

Between 1962 and 1965 an intensified trapping campaign resulted in the capture of 40,461 coypus. During the exceptionally severe winter of 1962–3 an estimated 80–90 per cent of the coypu population succumbed, and by the middle of the latter year the survivors were confined to 6,000 km^2 of Norfolk and eastern Suffolk. Colder than average winters and continued trapping kept the population low and declining to 1969. From 1970 to 1975 the winters were mild or average, and the population rose sharply from about 1,600 in 1970 to around 8,500 in 1973 and some 12,000 by 1975; following the cold winter of 1975–6 the population has fluctuated between about 6,000 and 14,000, with an annual kill of around 6,000 to 12,000.

In 1980, when the population had been reduced again to the 1975 level, the Ministry of Agriculture, Fisheries and Food accepted the findings of an independent body, the Coypu Strategy Group, which recommended an even more intensive campaign (which started officially in April 1981) designed to eradicate coypus in eastern England – where they are today confined – by around 1990, at an estimated cost of £1.7 million.

Unlike other parts of the world, coypus in the wild in Britain have seldom been of significant economic value either for meat or fur.

France

As early as 1882 Pays-Mellier described in the *Bulletin de la Société national d'Acclimatation de France* how he was currently breeding coypus in captivity on his estate, Patandière, at Champigny-sur-Veude, Indre et Loire in Touraine. Six years later the *Bulletin* published an account by Edgard Roger of the rearing of coypus in the park of the Château de Naudy, Seine et Marne, and the *Revue d'Histoire Naturelle appliqué de la Société d'Acclimatation* for the years 1890–2 mentioned that H. Sharland was breeding coypus at Fontaine St Cyr near Tours in Indre et Loire.

Most of these enterprises closed down during the First World War, and were not revived until between about 1925 and 1928 when the price for nutria was again high. Most farms were established in the centre or south-west of the country, but there were also several in the east and north and in Brittany. The world economic crisis of the 1930s again forced the closure of most nutria-farms, whose liberated stock soon became established in the wild. The principal populations today, which are both expanding their range, are established in the Camargue and in the valley of the Loire.

Netherlands[4]

Exactly when nutria-farming began in the Netherlands is uncertain, but the industry was flourishing by around 1930, and before the end of the decade escaped coypus were being reported from farms near Bunde, Diepenveen (where young are said to have first been born in the wild) and in East-Gelderland. After the Second World War the breeding of coypus increased considerably, as did the number of escaped animals reported in the wild. Due to the relatively poor quality of fur of the subspecies farmed (*M. c. bonariensis*) and large importations of superior-quality pelts from eastern Europe, followed by the fall from favour of nutria as a fashionable fur, commercial breeding ceased in the early 1950s and the stock of many farms was released: by 1955 wild populations were established in most provinces and widespread agricultural and horticultural damage was reported. Subsequent attempts to re-open nutria farms (e.g. at Nederweert in the mid-1970s) and to breed coypus for their meat, were unsuccessful.

During the hard winter of 1955–6 the greater part of the free-living population perished – principally from starvation: small numbers survived only in places where the surface water did not freeze over (coypus are very susceptible to frost-bite in their toes and tail), such as in the basin of the River Roer in Mid-Limburg where the population increased between 1957 and 1962, and in Vriezenveen where a new colony was formed in 1961.

The severe winter of 1962–3 seems to have virtually wiped out the entire population of coypus in the Netherlands. By the late autumn of the latter year, however, small numbers – members of the relict-population which survived in the neighbouring German part of the river basin – began to re-appear along the banks of the Roer in Mid-Limburg. A succession of mild winters between 1970 and 1975 then resulted in a slow but steady growth in this population which gradually spread out over the basins of the Roer and Maas and from the latter via its tributaries into the surrounding marshes, creeks and clay-pits and gravel-pits. Colonies were formed as far north and south as Obbicht and Sanbeek; the greater northward spread was due to winter floods from which the coypus escaped downstream. Reports from Appeltern and Geffen in 1970, along the Niers (1972) – where a small population died out three years later – Lithoyen (1973), Maasbommel and Woudrichem (1975) and Wellseind and Willemstad (1976) suggest that these dispersals resulted in a discontinuous distribution from the original population near the German border.

Reports of sightings and catches in the Brabant-Limburg Peel region and near Asten in 1974, Bakel, Griendsveen, Helmond and Horst in 1977 and Oostrum in 1978 are probably referable to those distributed along the Maas, while others (e.g. at Uden and Zeeland in 1978) are more likely to be of animals which had recently escaped or had been released from zoos and private collections, as are individuals seen at Doorenburg in 1970, Ootmarsum (1976), Engelbert and Meerkerk (1977) and at Vlenten in 1979. Colonies which became established in gravel-workings south of Maastricht in Zuid-Limburg and in the Geul Valley were probably descended from a number of animals which escaped from a nutria-farm in Lanaeken across the border in Belgium. The Geul Valley population died out in about 1974, while that south of Maastricht – and another discrete colony in the former coal-mining region of northern Zuid-Limburg – survived only until January 1979 when, as the result of intensive trapping, heavy snowfalls and freezing temperatures accompanied by a high wind-chill factor, many other colonies perished: the survival of a small remnant population along the Roer and Maas in Mid-Limburg has been attributed to the raising of the water temperature by thermal pollution.

Reliable data on the number of free-ranging coypus in the Netherlands is unobtainable. As elsewhere in Europe hard winters such as those of 1955–6, 1962–3, 1978–9 and 1981–2 decimate coypu stocks which, in intervening years, slowly recover: thus by late 1962 the population in Mid-Limburg was estimated to number between 200 and 300, while a year later it had fallen to less than 10. A series of relatively mild winters after 1963 allowed the population and its distribution slowly to expand, until within a decade there were believed to be several thousand in Mid-Limburg and serious damage to farm and garden crops was being reported. The principal harm caused by coypus in the Netherlands is by undermining river banks through tunnelling; the consumption of crops (grass and cereals in early spring, young corn and sugar beet in summer, and mainly sugar beet in autumn and winter) and riverside and aquatic plants (which causes respectively erosion and silting); and trampling of riverine vegetation. As a result of these depredations a trapping campaign was organized in the winter of 1974–5 which continued until the severe winter of 1978–9; the number of coypu trapped or found dead fell from 673 in 1974 to 602 (1975), 177 (1976), 127 (1977), 117 (1978) and 84 in 1979 – a total of 1,780, of which only 14 were taken from outside Limburg.

In the summer of 1977 the total number of coypus in the Netherlands was estimated to be no more than 100, and following the severe winter of 1978–9 had fallen still lower. In January of the latter year traces of coypus were found only along unfrozen stretches of the rivers Maas, Roer and Swalm, and in heated water drainages as for example, at Buggenum and Maasbracht: two months later evidence for the survival of a small relict population was found only at Buggenum, Swalmen, Roermond, Paarlo, St Odilienberg, Ool and Maasbracht. This indicates that although severely depleted by trapping and the winter of 1978–9 coypus in the Netherlands were not virtually exterminated as during the 1962–3 winter. Their future depends largely on the intensity of the trapping campaign, which at present is mainly concentrated on the periphery of the infested area to prevent further spread, and on winter weather conditions.

Germany

In Germany large numbers of nutria-farms were established in the late 1920s and 1930s: although the difficulty in obtaining fodder and in maintaining enclosures during and immediately after the Second World War resulted in the escape or release of considerable numbers of animals into the wild, by 1949 no fewer than 2,744 farms holding a total of 33,299 coypus remained in what is now East Germany, of which 1,246 farms (45 per cent) containing some 16,000 individuals (48 per cent) were situated in the Sachsen-Anhalt region – mainly along the Elbe and Saale between Magdeburg and Leipzig.

In the early years of their establishment in the wild much of the damage caused by coypus was mistakenly attributed to muskrats; cabbages, potatoes, sugar-beet and mangolds were the agricultural crops most seriously affected, and in some places water plants and sapling trees suffered severe depredation; there were also reports of flooding caused by the collapse of river banks undermined by burrowing.[5]

By 1960 some 5,000 coypus were estimated to be established in the wild in what is now West Germany. The principal areas of infestation were in the Rhenish-Palatinate, especially along the River Glan and in the oxbow[6] of the Rhine between Wörth and Speyer south of Mannheim. Other large populations occurred in Northrhenish-Westphalia, in Hessen, and in the Saarland.[7] Although the hard winter of 1962–3 resulted in the death, mainly through starvation, of the greater part of the population, a sequence of mild winters after 1963 – especially between 1970 and 1975 – enabled the animals to stage a rapid recovery and, as mentioned above, coypus subsequently crossed the border into Limburg province in the southern Netherlands.

Italy

Coypus were first introduced to Italy for fur-farming in 1928. Escapes soon took place, and the animals are now well established in southern Tuscany, where they are spreading throughout the Ombrone river system, including the Maremma National Park. Elsewhere coypus occur in Umbria and in Latium; along the River Tiber (especially at Lago di Nazzano, some 50 km from Rome) and its tributaries; at Lago di Bracciano, 30 km north-west of Rome, and along the Volturno, Sele and Tusciano rivers in Campania.

In Italy, in addition to crop and vegetation damage reported from other countries, coypus appear to be having a detrimental affect, through habitat degradation and disturbance, on the already declining otter (*Lutra lutra*).

Coypus are also presently established in the wild in parts of Belgium, and possibly in Denmark. In Austria, Norway, Czechoslovakia, Poland, Yugoslavia, Greece and Spain coypus have also been bred in captivity but, so far as is known, none have ever become settled in the wild (except for a short period in Austria in 1935).

EURASIA

USSR

In 1930 113 (one account says 123) coypus from Argentina were imported to the Soviet Union to stock nutria-farms and to try to acclimatize populations in the wild. In the following two years a further 2,500 were received, mostly from farms in England and Germany, some of which were released – with mixed success – in Kazakhstan, Turkmenia, Dagestan, the Krasnodar region and in Georgia in the Caucasus; others freed in the Karasu stream network in the Kura River basin in Azerbaijan soon began to breed, but were preyed on by jackals, jungle cats (*Felis chaus*), foxes and wolves.

In 1940 50 coypus from the Karayazy farm in Azerbaijan were released at the source of the Sev-Dzhur River in Lake Aigerlich, Armenia; they quickly multiplied and spread to the Araxes River and to Arazdayan. By 1947–8 they had arrived in the lower reaches of the Arpachai River, Nakhichevan, and gradually colonized the estuaries of the Akara and Okhichichai.

In 1949 60 coypus from Azerbaijan were introduced to the *plavni* (floodland lakes) of the Vakhsh River valley, where they were soon established and breeding.

An attempt to establish coypus in the wild in Uzbekistan in 1931 failed. A second attempt in 1950, when 170 were released on a farm in the Syr Darya *plavni*, was more successful, and coypus became acclimatized on several farms in the Termez region.

In Kirgizia six coypus introduced to the Tokmak Reserve in 1954, where there are numerous streams and swamps which do not freeze over in winter, had multiplied to around 100 by 1958. They were subsequently reinforced in 1962 by the addition of 50 more from the Dzhilikul *promkhoz* (hunting reserve) in Tadzhikistan, followed by a further 78 which were freed near Chernyi.

As recently as 1971 coypus were experimentally released in the Korugi hunting reserve of the Sagaredzhoisk region, Georgia.

Attempts to establish coypus in the wild in the Soviet Union have shown that without assistance from man breeding results are usually poor, due partly to predation but mainly because of the severity of the winters and the consequent freezing of lakes and rivers. Coypus have done best on spring-fed waters in Azerbaijan (Lenkoran and the lower Kura basin) in Transcaucasia, in Georgia (Poti and the lower reaches of the Rion), in Armenia (the Araks and its tributaries) and in the lower reaches of the Vakhsh in Tadzhikistan. Russian scientists believe that successive generations of wild coypus may acquire an increased tolerance to harsh conditions. Their pelts are a valuable economic asset to the Soviet Union.

ASIA

Turkey

Following a natural extension of range from Georgia and Armenia in the Soviet Union, coypus have been established since 1973 in the clear, deep lotic waters of the Kara Su near Kars, Arahk, in eastern Turkey.

Israel

Because they are one of the few animals to produce good-quality fur in a hot climate, coypus from Chile were in about 1953 introduced to Kafr Masaryk in northern Palestine (Israel). In the following year eight escaped, and settled in some fishponds and nearby swamps around Afeq; by 1956 they had removed all reed-growth from the upper reaches of the Naaman River, and by 1960 were regarded as established throughout the Naaman region.

In April 1957 181 coypus were introduced from Chile to two ponds and one of the drainage canals of Kafr Rupin, where by the end of the year the population had increased to between 430 and 530. A year later a number escaped, and soon colonized the nearby swamps and marshland.

In Israel the coypus' main predators are golden jackals (*Canis aureus*), red foxes (*Vulpes vulpes*), jungle cats (*Felis chaus*), Egyptian mongooses (*Herpestes ichneumon*), otters (*Lutra lutra*) and beech or stone martens (*Martes foina*).

Japan

In 1931 seven or eight coypus were imported to Tokyo from Europe by Prince Takamatsu, and in the same year a further eight were introduced from France to Oji, Tokyo, soon becoming established in parts of central Tyôsen.

By about 1940 coypus were being raised in captivity for both meat and fur in several places in Japan; two years later the population was said to be 1,427, and by 1944 around 40,000 coypus were being farmed in Tokyo and westwards to Shikoku and Kyushu.

Between 1938 and 1955 at least 48 nutria-farms, from a number of which coypus escaped, were established in the Okayama Prefecture. By 1949 an estimated 500 coypus were settled in the wild along the coast of Kojima Bay, south of Okayama City. When the demand for nutria fell many farms closed down, and their liberated stock became established in small numbers in the network of the Yoshii and Asahi rivers. Here they have recently been increasing in numbers, and are extending their range north along the Yoshii and Asahi rivers and westward along the Oda.

Some damage by trampling and eating rice in paddyfields has been reported, mainly in October.

AFRICA

Kenya[8]

Coypus were first introduced to Kenya by Mr R. Hook to his farm near Nanyuki (140 km north of Nairobi) in about 1940. Whether the animals escaped or, having failed to reach a commercially acceptable standard, were deliberately released, is uncertain, but they were established in a moat on the farm by 1950 and in a nearby dam two years later.

During the 1950s and 1960s the population increased and dispersed, presumably into the headwaters of the Uaso Nyiro River and thence up the Narok River into Lake Ol Bolossat in the Aberdare Mountains via Thomson's Falls; it was possibly augmented by other introductions around Ol Bolossat or on the Kinangop Plateau to the south.

Coypus were first seen on the western shore of Lake Ol Bolossat in 1970, but it is probable that they had already been there for some time. From Ol Bolossat they are believed to have moved into the headwaters of the Malewa River and then on south by about 1968 to Lake Naivasha. At some stage they must also have climbed the escarpment and debouched onto the Kinangop Plateau, probably via the Turasha River and its tributaries, and also moved east through the Aberdares towards Nyeri by way of the Maringate River.

In Lake Naivasha, west of the Aberdares, where the current population is considerable, coypus are suspected of being responsible for the near extinction of the blue water-lily *Nymphaea caerula*. East of the Aberdares coypus have been seen with increasing frequency since about 1970 in dams around the Mount Kenya Safari Club near Nanyuki. On the Kinangop Plateau to the south and elsewhere the retaining walls of at least nine dams have been severely affected by burrowing, and one large dam, at Seminis, is said to have been damaged beyond repair.

Tanzania; Zambia

According to Haltenorth and Diller (1980), 'from 1960 the coypu has gone feral in the coastal swamps in Tanzania'. Dr Antoon de Vos says that 'copyu have been established in the wild in Zambia'.[9] They have also been farmed in Zimbabwe where, so far as is known, none have ever escaped. Others were introduced in the 1950s and 1960s to the Republic of South Africa, where by the mid-1960s several escapes were reported from farms in the eastern part of Cape Province; none, fortunately, ever became established.

NORTH AMERICA

USA

As long ago as 1899 an unsuccessful attempt was made to establish a nutria-farm at Elizabeth Lake in California. In the 1930s coypus were released in Louisiana marshlands near New Orleans, but were soon trapped out. Unconfirmed reports of coypus in the wild in the late 1930s came from Colorado, Louisiana and Ohio.

In 1937 a nutria-farm on the Pecos River, Chaves County, New Mexico, lost around 500 coypus during a flood: in the same year another flood was responsible for the escape of coypus in Tillamook County on the coast of Oregon, where they quickly became established on Garrison Lake and on the north coast of Lincoln County, spreading in the following year along the Nestucca River.

In March 1938 a Louisiana industrialist, Mr E. A. McIlhenny, imported six pairs (one account says 10 pairs) from Argentina to the marshland of Avery Island (a small mainland township) in Iberia parish. Here they did well, and some are believed to have escaped into the nearby coastal marshland. In August 1940 a hurricane flooded the marshes, and 150 more swam over the top of the fences of their enclosures to freedom. With an abundance of aquatic food plants and few predators they rapidly increased, and were soon extending their range via the innumerable bayous[10] intersecting the marshland; within a month of their escape one was observed over 100 km away on the Lacassine National Wildlife Refuge.

Coypus were first introduced into Utah[11] in 1939 by Mr Clarence Holmstead, a fur-farmer of Lehi near the source of the Jordan River, who imported 50 pairs from a commercial breeder in Roswell, New Mexico; by 1941 the entire stock had escaped into the surrounding countryside, and by 1950 coypus were being reported up to 25 km from their point of escape.

In 1940 Mr Dean Willis of Lehi purchased several pairs of coypus from Holmstead, most of which had escaped within a year. Also in 1940 Mr Bruce A. Hartman, proprietor of the Salt Lake Fur Farm at West Jordan, acquired a few pairs from Holmstead, the majority of which again escaped in less than a year.

In 1941 Mr Jim Smyth, a Salt Lake City hatter, imported a number of coypus from a fur-farm in Colorado and also purchased some from Hartman, to provide a ready source of supply for the manufacture of 'beaver' hats. The animals were kept in Murray along the Jordan River, from where by the autumn of 1942 some 15 had managed to escape. In the spring of the following year Smyth deliberately released the balance of his stock – nine pairs – at the New State Gun Club west of Woods Cross, all of which were in poor condition. Five were found dead in the marshes during the following winter, but the remaining 13 produced 17 young in 1944.

In 1942 Mr A. M. Creamer of Logan bought the balance of Hartman's stock and installed them on his farm on the Valley View Highway, from where four subsequently escaped. By 1950 coypus were said to be 'thriving in Utah marshes'.

By 1946 coypus were found from the west bank of the Mississippi as far as south-east Texas. From here they spread into most suitable habitats along the entire western Gulf Coast from western Florida to southwestern Texas, inland in Texas, Louisiana (especially in St Mary, Iberia, Vermilion and east Cameron counties) and Mississippi and up the Mississippi River to northwestern Louisiana, southwestern Arkansas and southern Missouri. Some of this extension of range was augmented by deliberate releases, and coastal colonies of independent origin became established in the Mobile Delta area of Alabama (since 1948), in the Hillsborough River drainage area of west-central Florida (since the 1950s), in southern Georgia, and, since 1941, from Cape Hatteras in North Carolina to southern Virginia.

In the 1940s coypus were deliberately released in eastern Maryland, where they soon spread over much of the marshland as far as the Blackwater National Wildlife Refuge at Cambridge.

On the west coast a fox-fur farmer introduced 30 coypus from Ramsey, New Jersey, to his farm near Oakdale, California, in 1942. Two years later he sold his stock, then numbering 33, to a neighbour, who also acquired a further eight from Avery Island. In or before 1945 several escaped, and by 1948 damage (originally attributed to muskrats (*Ondatra zibethicus*)) was being reported to irrigation canals. By the end of 1951 over 100 coypus had been trapped, and records showed that the population in Stanislaus County extended at least as far as the Stanislaus and San Joaquin rivers, and occupied much of the marshland and waterways bordering the Tuolumn River.

In Washington nutria-farms were established in the mid-1930s at Seattle, Bothell, Bremerton, Bellingham and Maple Valley; escaped or released coypus from Seattle had settled in Puget Sound by about 1937, and in Maple Valley by 1939–40.

In 1944–5 a trapper in Montana caught three coypus near Corvallis in Bitterroot Valley, Ravalli County: they were almost certainly escapes from a fur-farm belonging to a Mr Erdman 8 km upstream on a branch of the Gird Creek, where in 1944 a broken irrigation ditch caused a flood which allowed his stock (imported direct from South America) to escape.

Away from the coast other small populations were established by the mid-1950s along the Grande Ronde River in Union County, eastern Oregon; on the St Maries River watershed in Benewah County and in Gallia County, Idaho; several in southern Michigan and the southern part of the 'Thumb'; and possibly some near Fort Knox in Kentucky. A decade later coypus were established in the wild in at least 20 states (including all those in the south-east except Tennessee), mainly in Washington, Oregon and Idaho; in southern Texas, Oklahoma, and Arkansas; in Louisiana; and in parts of Alabama, Florida, Maryland, North and South Carolina, Georgia and Virginia. In 1957 the total number of coypus in the wild in the United States was estimated at well over 2 million.

The spread of nutria-farming in the United States was accompanied by unscrupulous advertising claiming huge potential profits, which persuaded some gullible farmers in the late 1950s to pay up to $1,000 for a standard breeding pair. 'New fur-bearing animal now being bred in U.S.

offers high profits to breeders. This fur is next to mink *now* in price and will soon surpass it', is a typical example. Others suggested that by their consumption of aquatic vegetation coypus ('four-legged weed-cutters') would greatly improve fisheries, and many were released unsuccessfully in Louisiana and eastern Texas for this purpose. As one disgruntled South Carolina resident put it, 'we were led to believe this animal would do everything but lay hard-boiled eggs'.

In 1958 it was estimated that about 1,000 nutria-farms in the United States were raising a total of between 25,000 and 30,000 coypus, and rare colour mutations were fetching an exorbitant $1,750 a breeding pair.

The extravagant claims made for nutria and for the ease and cheapness with which coypus could be reared, coupled with the excessive prices for breeding stock, attracted the attentions of the National Better Business Bureau, Inc. After thorough investigation the bureau was able to disprove most of these claims, adding that 'The demand for nutria is so limited in the United States that about 65–70% is exported to Europe'.

Partly as a result of the bureau's investigation the demand for nutria fell sharply, and many disillusioned farmers all over the country went out of business, allowing their redundant stock to escape into the countryside, where they augmented the already large wild population.

Before long, damage by coypus was being reported from many parts of the United States. Louisiana ricefields were heavily grazed and their water-pumping stations and the levees[12] used when irrigating the fields were damaged by burrowing, sometimes causing entire ricefields to wash out. Elsewhere damage to corn (maize), cabbage, lettuce, peas, alfalfa (lucerne), ryegrass and sugar-cane was soon being reported, and drainage-ditches and dykes were undermined. Through their consumption and destruction of hydrophytes coypus also tend to compete for food with the more economically valuable muskrat and with wildfowl.

In the mid-1960s the demand for nutria recovered and trapping became once more an economic enterprise yielding, in Louisiana alone, an annual income from fur and meat of several million dollars.

Canada

In about 1938 coypus were introduced to British Columbia to stock a nutria-farm at Ladner, Vancouver, from which they escaped and became established in small numbers in marshland on the lower delta of the Fraser River near Vancouver and New Westminster. Coypus have also been established in Canada in the Whitefish River drainage area of Thunder Bay, Ontario, and in the Ottawa River drainage of western Quebec and eastern Ontario.

NOTES

1. Where recorded (e.g. in the British Isles, the Netherlands, Germany, Italy, the Soviet Union and North America) the subspecies imported was *M. c. bonariensis* from subtropical northern Uruguay and Argentina.
2. Roots (1976) wrote 'they are also established on the sewage farm at Slough' some 22 years after they had disappeared. His map, on the other hand, omits established populations in Italy, Turkey, Israel and the Soviet Union.
3. Flooded medieval peat workings, which now form shallow lakes and waterways extending out of the rivers of north-east Norfolk.
4. For much of the information on copyus in the Netherlands I am indebted to Litjens (1980b).
5. Hoffman (1952b).
6. A loop in a river which is cut off from the main stream by silt banks and eventually becomes a pond or lake.
7. Müller-Using (1965).
8. For much of the information on copyus in Kenya I am indebted to Gosling (1976).
9. Letter to author (1980).
10. Small creeks and minor watercourses.
11. *Game, fur animals and fish introductions into Utah* (US Department of Fish and Game Publication No. 4, 1951).
12. Embankments used to prevent unwanted inundation and to contain areas of land to be flooded.

Brown Hare

(*Lepus capensis*)

NATURAL DISTRIBUTION

If the Eurasian form (*europaeus*) is correctly identified with the African *capensis* (which is still debatable) the brown hare is one of the most eurytopic of living mammals. It ranges from the savannah, steppe and semi-desert regions of Africa, and all Europe south of (and below) the coniferous forest limit, through similar habitats across Asia as far east as central China.

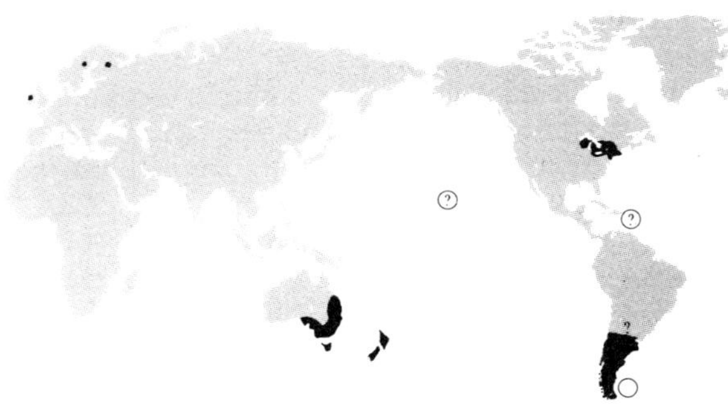

NATURALIZED DISTRIBUTION

Europe: British Isles (Ireland), Finland, Sweden; North America: USA, Canada, West Indies (Barbados, ?Guadeloupe, ?Grenadines (Balliceaux)); South America: Argentina, Brazil, Chile, Uruguay, Falkland Islands; Australasia: Australia, New Zealand; Oceania: ? Hawaii.

EUROPE

British Isles (Ireland)

Although the brown hare is a native of the British mainland it is not indigenous in Ireland where it

Table 23 Earliest records of the introduction of the brown hare to Ireland

Date	*County*	*Locality*	*Introduced by*
*c.*1852	Cork	Castlemartyr	Lord Shannon
*c.*1858	Cork	Castle Hyde, Fermoy	Mr Hyde
*c.*1865–6	Wicklow	Powerscourt	Lord Powerscourt (from Mr Edward Heneage, Lincolnshire, England)
*c.*1868	Armagh	Brownlow House	Lord Lurgan
*c.*1868	Down	Copeland Island	Mr David Kerr
1876–7	Tyrone	Strabane	Mr John Herdman
?1880s	Galway	Salrock, Killary Bay	General Thompson et al.
?1880s	Fermanagh	Cleenish Island, Lough Erne	Capt. Collum
?1880s	Londonderry	?	Mr Stevenson
*c.*1892	Donegal	Ballyconnell	Mr Olpert
?	Cork	Trabulgan Coursing Ground	Lord Fermoy

Source: From Barrett-Hamilton (1898)

was introduced in the nineteenth century (see Table 23). Despite Barrett-Hamilton's (1898) claim that brown hares had 'succeeded in establishing themselves wherever they were given a fair chance', and that near Strabane, County Tyrone, they were 'even said to be the only kind of hares found', they are today confined, in a small numbers only, principally to parts of County Donegal.

Brown hares were reputedly first introduced to the Isle of Man, England, by Sir John Stanley when he was granted the Lordship of the island by Henry IV in 1405.[1] They have also been introduced to several other smaller British offshore islands.

Finland; Sweden

Brown hares have been introduced to several North Sea islands and to parts of central Finland and Sweden, where to some extent their range overlaps with that of the Arctic or mountain hare (*Lepus timidus*).

ASIA

USSR

Outside their natural range in the Soviet Union brown hares have been translocated to several districts in Siberia. In 1933 a number were imported from central Europe to the upper Yenisei River where, in spite of the severe winters, they had soon colonized a considerable area. Two years later 200 from Bashkiria were released in Salsir on the River Berd, in Moslyanino District, Novosibirsk Province. In 1938–9 223 hares were released in Zalari, Irkutsk Oblast, where a 1962 census established a population density of only one per 9 km^2, due, it was believed, to late spring and early autumn frosts which kill large numbers of leverets. In 1939 119 hares were turned out in the Minusinsk District, Krasnoyarsk region, 56 in the Emelyanovskii District, 115 in the Uyar District, and 182 40 km from Büsk. Up to the mid-1960s hares had reached only as far as the Tatarsk District of Novosibirsk Oblast, although some had been released in the Omsk region. Brown hares are also currently established on Ostrov Barsa-Kel'mes in the Aral Sea.

NORTH AMERICA

USA

According to Dean and de Vos (1965), the earliest introduction of brown hares – from England – was in 1888 (Presnall 1958 gives 1886) at Jobstown, New Jersey. Others followed at Bethlehem, Pennsylvania, at White Plains, New York, and in 1890 on the Dietrich Estate,

Dutchess County, New York. The most successful seems to have been the release of several hundred from Hungary at Millbrook, Dutchess County in 1893, which was augmented by further importations – of up to 500 animals each at a cost of some $10 a head – every four or five years until 1910–11. Subsequent liberations were made in Connecticut (shortly before 1920), on the islands of Nantucket and Martha's Vineyard off the coast of Massachusetts (in the late 1920s), and in Maryland in about 1940.

From these centres hares spread slowly but steadily east into western Litchfield and Fairfield counties, Connecticut, to southern Berkshire County, Massachusetts, and by 1924 to Skelton, Middlesex County, New Jersey, southern Vermont and, to a limited extent, westwards across the Hudson and Delaware rivers into extreme eastern Pennsylvania. By 1950 they were present in Michigan, northern Wisconsin and possibly Minnesota, from the St Lawrence across northeastern New York (in northeastern Dutchess, southeastern Columbia, southern Montgomery and west-central Washington counties), in northwestern Connecticut, in Kent County, Maryland, in New Jersey, in southwestern Vermont, and in eastern Pennsylvania.

Importations of brown hares – for sporting purposes – were succeeded by a sigmoid pattern of rapid increase (accompanied by considerable damage in the 1920s to orchard and farm crops, amounting one year to more than $100,000 worth in Dutchess County alone), followed by a decline and stabilization at a level where damage – now reported to be negligible – is outweighed by sporting potential. Brown hares are today locally common in the Hudson River valley of eastern New York State, and in western Connecticut, where they are a highly regarded game animal.[2]

Canada

Only one importation of hares to Canada is on record.[3] Mr Otto Herold records how:

> In February, 1912 when I was Manager of the Bow Park Farm, near Brantford, [Ontario], I imported from Germany seven females and two males. They belong to the species *Lepus timidus*.[4] When they arrived I kept them fenced in order to see how they would do in the climate. However, they found a loophole and escaped. My idea was to keep some on the Bow Park Farm which is nearly surrounded by water of the Grand River, but in winter when the ice came they escaped.

The first leverets in the wild were born in the summer of 1912, and thereafter the animals steadily expanded their range. In the south they reached Aylmer in 1919, Ingersoll (1921), Simcoe (1923), and Port Rowan on Lake Erie by about 1925. By 1928 they had spread west to Highgate, and by the following year they had colonized most of southwestern Ontario. Waterloo and Wellington in the north were reached by 1921, Walkerton, Flesherton, Meaford, Tottenham, Collingwood, Penetanguishene and Georgian Bay by 1928, and Burk's Falls, Parry Sound by 1948. In the east the Niagara Falls Peninsula and Wentworth were colonized by 1921; from here, hares progressed eastwards through the country north of Lake Ontario, reaching Woodbridge, Erindale, Toronto and Maple by 1925. They arrived in Markham and Uxbridge by 1928, and Darlington Township by 1930. Between 1930 and 1936 hares crossed Northumberland County, and in the latter year were first reported from Prince Edward County. By 1948 they had reached Kingston and Frontenac County.

Between 1912 and 1928 – in the absence of important predators and competitors – the hares' extensions of range absorbed most of their annual population increment. By 1928 they had colonized most of southwestern Ontario, comprising the most suitable habitats in the province. Some crossed into Michigan, but the Great Lakes and their connecting waterways proved effective barriers to further large-scale expansion, and the Ontario population south and west of Lake Simcoe soon built up.

From 1928 to 1941 many thousands of hares were killed by sportsmen without any apparent affect on the population: in the early 1940s, however (curiously at a time when the war had reduced shooting pressure), a gradual decline in numbers, though not in distribution, became apparent. During the 1948–9 breeding season the population was clearly recovering, and by the mid-1950s hares were established in most of Ontario south of the Precambrian Shield (in Lambton, Middlesex, Elgin, Norfolk, Haldimand, Brant, Lincoln, Welland, Huron, Wellington, Oxford, Perth, Waterloo and Dufferin counties, and especially in the 'apple-belt' of Grey and southern Simcoe counties), but not on islands off the Bruce Peninsula in Lake Huron, at the western end of Lake Erie, or at the eastern end of Lake Ontario.

In the late 1950s and early 1960s hares slowly extended their range northwards, especially in eastern Frontenac County, and became established around Sharbot Lake. During this period they also continued their more rapid eastward spread, reaching the Gananoque area around Leeds and Lansdowne and western Grenville County. Isolated records came from north Burgess Township, Lanark County, and from north of Maitland, Grenville County, in the vicinity of Smiths Falls in Lanark and Leeds counties. It is possible that the northern distribution was augmented by releases at about this time jointly by the Bancroft Fish and Game Association and the Frankford Fish and Game Club.

At the eastern end of Lake Ontario, Amherst and Wolfe islands were colonized in the late 1950s and Main Duck Island before 1965 – probably when the lake froze over in winter. By this time the brown hare in Canada had spread some 480 km in an easterly direction and around 240 km northwards. Today, hares are well established locally in southern Ontario and possibly in parts of southwestern Quebec; they cause only limited economic damage, principally to autumn wheat, fruit-trees and market-garden crops, and are a valued sporting asset.

Several attempts by Colonel L. S. Dear of Port Arthur to introduce hares for sporting purposes to Lakehead, near Fort William, on Thunder Bay in Lake Superior between 1941 and 1945 failed to succeed, as did another attempt at Cochrane in eastern Ontario after 1940.

West Indies

Brown hares (and rabbits) from England were introduced in 1842 by Thomas Trotman to the Bulkeley Estate in St George's Parish, Barbados, from which they escaped during a cloudburst which destroyed their enclosure. During the 1870s some 200–300 were shot annually, but both species were decimated in the late 1880s by the introduced small Indian mongoose (*Herpestes auropunctatus*).[5] According to R. P. J. B. Du Tertre, in his *Histoire générale des îles de S. Christophe, de la Guadeloupe, de la Martinique et des autres dans l'Amerique* (1654), both hares and rabbits had been introduced before that date to Guadeloupe. Both species have also been released on the island of Balliceaux in the Grenadines.

SOUTH AMERICA

Argentina

> The european hare would have been introduced into the country [Argentina] for the first time in the year 1888, by the then German Consul in Rosario, Santa Fe, Senor Woeltje Tietjen, who had four pairs sent from his homeland which were later released on the estancia La Hansa, between Canada de Gomez and Las Rosas, for sporting purposes. From there the hares spread through all the zone and neighbouring provinces. Similarly there have existed other importations since the magazine 'La Agriculture' No. 205, makes reference that in 1896 Senor Sulpicio Gomez introduced from Austria a batch of this same species which was let loose in the fields of Tandil [Buenos Aires Province] . . . the owners of the estancia 'Las Isletas' in San Luis, acquired examples of this species to breed on the said establishment, from whence they also spread rapidly.[6]

Hares have also been introduced to Tierra del Fuego, where they are found as far south as Isla Observatorio and Islas Ano Nuevo. Eleswhere, they have spread over the greater part of central and southern Argentina.

Chile

From Argentina (where they are said to compete for grazing with the Patagonian cavy or mara (*Dolichotis patagona*)) hares spread before 1928 to Malleco Province in Chile. Five years later they were reported to have reached plague proportions in the Fundo El Vergel, where 300 rauli trees (*Nothofagus alpina*) were destroyed.

Possibly due to a dearth of foxes and other predators as a result of trapping, the hare population increased markedly in 1959–60; the animals first appeared in Central Valley, Malleco Province in 1961, when there were also indications that they had reached the coast at Paicavi in Arauco Province.

By 1966, when they were first seen at the Laguna Malleco (at an altitude of 980 m), hares were said to be more abundant in the Cordillera de Nahuelbuta than in either Central Valley or the Cordillera de los Andes. They now occur in large numbers in many parts of central and southern Chile, at least as far north as Illapel, and at altitudes of up to 2,600 m. Although a certain amount of damage to crops is reported, their pelt is regarded as a valuable economic asset. They have also occurred around Punta Arenas and Ultima Esperanza, especially near Lake Sarmiento.

Uruguay; Brazil

Brown hares have also been introduced to Uruguay, and in the early 1960s the first hare was reported from Brazil. The species now occurs over much of South America south of 28 °S.

Falkland Islands

On the Falkland Islands (where they occur only on East Falkland – mainly at Goose Green and North Arm) hares may have been introduced as a source of food in the early 1800s by sealers and whalers. The population at Cape Pembroke near Stanley is believed to have declined considerably due to increased use of the airport during and after the Anglo-Argentine war of 1982, although nearby minefields may provide refuges for the survivors.

AUSTRALASIA

Australia[7]

The earliest recorded attempt to establish hares in Australia took place in 1837, when a pair were sent to Tasmania, but failed to breed. An unspecified number despatched from England on the *Courier* in 1840 appear never to have arrived at their destination in South Australia: they may have failed to survive the rigours of the voyage, like a party of nine *en route* to Tasmania in 1868 which expired, it is said, due to disturbance by the crew of the ship who were celebrating New Year's Eve.

Mr William Lyall of Western Port Bay, Victoria was probably the first to introduce hares successfully to the Australian mainland in 1859[8]: by March 1862 he was able to report to the Acclimatization Society of Victoria that the animals were thriving and were breeding successfully in the wild.

Only a few months after Lyall's introduction, Mr Thomas Austin landed two pairs of hares (and two dozen rabbits) on Christmas Day, 1859, at Hobson's Bay, Melbourne, but unfortunately the two does died shortly afterwards. Of another pair despatched to Austin on the *Great Britain* in 1862, the buck died on the voyage out from England.

In the following year Austin – presumably despairing of otherwise receiving a viable stock – sailed to England, returning in February 1864 on the *Yorkshire* with a group of 11 hares, of which 9 arrived safely in Australia. These were turned out into a 1.6-ha enclosure on his estate at Barwon Park, Winchelsea near Geelong.

In December 1862 the *Adelaide Observer* objected to the shooting of hares at the Reedbeds near Adelaide:

> This destruction threatens to put an end to an important and apparently successful experiment in acclimatization. From a letter forwarded by the last mail from England to the Victoria Acclimatization Society by Mr Edward Wilson [1814–78; Vice-President of the British Acclimatization Society and Founder of the Victoria Society] we infer that the difficulty of keeping hares alive on a long sea-voyage is very great indeed.

The Annual Reports of the Victoria Acclimatization Society of the time carry several references to importations of hares (and other animals): 'Measures are now in progress for the speedy introduction and acclimatization of . . . Hares, Sparrows, and Songbirds from England' (1862); 'quadrupeds that have been introduced, either by the direct action of the Council or by the kindly aid of friends' included '5 English Hares' (1863); the council refer with particular satisfaction to the introduction of the hares, four of those received (from the Zoological Society of London) being despatched to Phillip Island, where they were soon breeding. 'During the past year there have been liberated . . . at Phillip Island 6 hares . . . at the Royal Park 4 hares' (1864); 'The Council are glad to be able to assure the members that the English hare may be looked upon as fairly established. At the Royal Park the two brace that were turned loose have bred repeatedly, and the animals are seen every day; whilst at Phillip Island they have increased so rapidly that Mr [John] M'Haffie estimates their number as exceeding 200' (1865).

From Phillip Island the Society distributed hares to various parts of the mainland – to Castlemaine, Mount Emu, Eaglehawk, Ovens and Warranook, for example (and to Tasmania and New Zealand) – and by 1870 they were established and spreading throughout Victoria and South Australia. Releases continued until 1873, when the animals ceased to be protected by the Australian Game Act.

In Victoria hares thrived and rapidly increased; before the relaxation of the Game Act they were established in considerable numbers around Melbourne (where in the market they fetched around 8*s*. (40p) each), and there were several complaints of agricultural damage – in 1871, for example, a farmer near Colac lost 300 sapling trees through ring-barking. Hares were introduced before the First World War to West Sister Island in the Bass Strait, where they still survive.

In South Australia hares spread gradually to the northern border of the wheat belt (around Wonoka by the 1880s) and were well established in the Flinders Range before the turn of the century; they began to decline with the gradual abandonment of wheat-growing, and are seldom found there today. They still occur, however, on the flats between Quorn and Mount Arden, wherever wheat continues to be grown.

In the early 1870s the only hares in New South Wales (the descendants of a pair received from the state's acclimatization society) were on the estate of Mr Edward Cox of Mulgoa, from which they spread by the end of the decade through much of the south of the state.

From 1870 to 1873 hares were coursed (illegally) and from 1873 to 1953 legally in Australia with greyhounds. They are currently established in a crescent extending from near Ceduna in South Australia, through Victoria and most of New South Wales to as far north as Cairns in north-eastern Queensland. They favour arable and pastoral open country interspersed with tussock-grass and rough grazing. Contrary to Roots's (1976) claim that in Australia hares have 'never become an economic pest', they graze vegetable, cereal and lucerne crops and new-sown leys, consuming much valuable pasturage, and ring-bark sapling fruit-trees and vines, in some districts causing serious depredation.

New Zealand

The earliest mention of the introduction of hares to New Zealand appears to refer to a shipment to Akaroa, near Christchurch, in 1850, all of which, however, died on the voyage from England. The first hares to land in New Zealand arrived in Lyttlton Harbour, Christchurch, in 1851 on board the *Eagle*, from which they escaped through an open porthole and swam ashore to the Banks Peninsula, where they became established in the Port Hills.

Table 24 gives particulars of importations of hares to New Zealand between 1863 and 1905, where they rapidly increased in both numbers and distribution. In 1872 the *Christchurch Illustrated Press* pontificated: 'We understand there are a brace of hares in the neighbourhood of Prebbleton and that several persons have amused themselves by trying to catch them on the last two Sundays. Such a proceeding is a disgrace to the district and we hope active measures will be taken to suppress it.' Yet only four years later the Christchurch Acclimatization Society was able to announce in its Annual Report that 'hares have increased in such numbers that they might be said to be successfully established'.

Hares spread quickly over the flatter parts of South Island, especially in cultivated districts in south Canterbury free from rabbits; Thomson (1922) reported them to be 'common from Cook Strait to Foveaux Strait'. In North Island hares spread south from Auckland, Thomson describing them as increasing around Pahiatua, common about Marton and in the Eketahuna district, ranging above the bush line to altitudes of over 2,000 m on Mount Egmont, and well distributed over the volcanic plateau from Rotorua to Waiouru.

By the late 1940s Wodzicki (1950) found hares established in all suitable habitats on both main

Table 24 Importations of brown hares to New Zealand between 1863 and 1905

Date	*Number*	*Importer*	*Liberated*	*Source*	*Remarks*
1863	1	Otago AS	?	?	
1865	?	Mr Michael Studholme	South Canterbury	England	Most died in hot weather; the survivors and their progeny were liberated at Waimate in the early 1870s and by 1880 'were quite numerous in the district'
1867	1 buck; 2 does	Mr John Ewen for Otago AS	Lake Waihola	Geelong, Victoria, Australia	Common by 1869
1867 or 1868	1	Canterbury AS	?	Dr J. Macdonald, surgeon of the *Blue Jacket*	
1868	1	Canterbury AS	?	Capt. Ross (or Rose) of the *Mermaid*	
1868	1 buck	Canterbury AS	?	Melbourne AS, Australia	
1868 or 1869	2	Auckland AS	?	Victoria AS, Australia	
1869	1	Otago AS	?	?	
1869	?	Southland AS	?	Victoria AS, Australia	Coursing began in 1878
1871	3	Canterbury AS	?	Capt. Fox of the *Charlotte Gladstone* (? from England)	
1871	5	Auckland AS	?	?	
1871	2	Southland AS	Half Way Bush	?	
1872	?	Nelson AS	?	Victoria AS, Australia	Soon increased to pest proportions
1872	4	Mr Gordon Holmes of Pigeon Bay, Canterbury	?	Melbourne, Australia	
1873	4 leverets	Canterbury AS	?	Wood Bros. flourmerchants	
1873	5 does	Canterbury AS	?	Australia	
1874	2	Southland AS	Half Way Bush	?	
1874	2	Wellington AS	?	?	By 1885 reported to be abundant around Wellington and at the lower end of the Wairarapa Valley
1875	3	Otago AS	?	?	
1875	14	Wellington AS	?	?	
1876	4	Wellington AS	?	?	
1876	?	Taranaki AS	?	?	Reported to be thriving by 1884
1876	44	Wellington AS	?	?	
1887	40	Southland AS	?	?	
1905	?	Waimarino AS	?	?	So numerous by 1907 that shooting was allowed

Source: After Thomson (1922) and Wodzicki (1950)
AS, Acclimatization Society

islands: open tussock country, farmland, swamps, marshes and riverbeds are the most favoured places, followed by the fringes of native bush, secondary growth and young plantations. Climate too, is an important factor governing hares' distribution in New Zealand, the drier and cooler parts of both islands being preferred to wetter areas.

Hares were introduced to New Zealand for coursing and shooting. Donne (1924) wrote: 'Coursing clubs were established in North and South Islands, and the sport of coursing was maintained with considerable vitality for many years.' Coursing was finally declared illegal in 1960. Hares cause a certain amount of damage to such crops as turnips, swedes, oats, barley, wheat, maize, cabbage, carrots and sweet blue lupins, and in some districts compete with sheep for winter grazing; some damage to young fruit, coniferous and eucalyptus trees also occurs. In alpine meadows, where they are sometimes numerous, overgrazing by hares causes some erosion. Leverets are preyed on by introduced mustelids, cats and dogs, and by native raptors; breeding – and therefore replacement – is slow.

Flux (1981), who estimated the current population in New Zealand to be about one million, states that hares occur today throughout the two main islands on suitable grasslands from sea-level to the snowline at around 2,000 m except for two small areas: Fiordland, west of Lake Te Anau in South Island, where hares are believed never to have penetrated; and a small region of Northland from Auckland to about level with Kawau Island, from which hares disappeared some 50 years ago.

OCEANIA

Hawaii

Brown hares, believed to be of Russian origin, are said to occur in small numbers on two tiny uninhabited islets off Oahu.

Arctic or **mountain hares** (*Lepus timidus*) were in 1930–1 introduced to Spitzbergen (Svalbard). In 1854 or 1855 a pair of hares (*L. t. timidus* and *L. t. canescens*) were introduced from Norway to the Faeroes, where their progeny gave rise to a new race, *seclusus*: after their introduction, all the hares turned white in winter, but they now retain their normal coloration throughout the year. In the British Isles the mountain hare is indigenous in the Scottish Highlands and in Ireland. It has been introduced (mostly in the last century) to the Shetland and Orkney islands, the Outer Hebrides, and the islands of Skye, Raasay, Scalpay, Eigg (now extinct), Mull, Islay (now extinct) and Jura. The Lowlands (Southern Uplands) of Scotland were colonized around the middle of the nineteenth century, and the Pennine regions of south Yorkshire and Derbyshire from introductions of about 1880. In north Wales a few remain from plantings near Bangor in around 1885.[9]

Black-naped hares (*Lepus nigricollis*) from India and Sri Lanka have been introduced to the island of Java in Indonesia (where they now occur around Djakarta, Bogor and Bandung), and also to Cousin in the Seychelles, to Mauritius and to Gunner's Quoin, a small islet off Round Island in the Mascarenes.

Hares (? sp.) have also been transplanted outside their natural range in the Far East and on the Sino-Russian border.

In the 1930s **black-tailed jack-rabbits** (*Lepus californicus*) were translocated for greyhound-training purposes to land now occupied by the Miami International Airport in Florida, USA, where they still occur; they could cause considerable economic damage were they ever to spread from the airport to the state's improved cattle-ranching pastureland.

Hares (? sp.), introduced to the Agaléga Islands in the Indian Ocean after their colonization by the French in 1818, were described in 1981 as 'rare'.[10]

NOTES

1. Fargher (1977).
2. Rolls (1976) describes them erroneously as 'abundant' also in New Jersey, Massachusetts, Vermont and Wisconsin.

3. For much of the history of brown hares in Canada I am indebted to Reynolds (1955).
4. Although Herold states that the species he imported was *timidus*, the hares of North America are in fact *capensis*. Until at least 1874 *capensis* (previously *europaeus*) was known as *timidus*, the name now given to the mountain or Arctic hare.
5. See p. 78.
6. C. Godoy, *Fauna silvestre*, Vol. 1, p. 202 in *Evaluacion de los reursos naturales de la Argentina*, Vol. 3 (Conseje Federal de Inversiones, Buenos Aires, 1963). I am indebted to Dr John E. C. Flux of the Department of Scientific and Industrial Research, New Zealand, for drawing to my attention this reference sent to him by Dr John Jackson.
7. For much of the history of hares in Australia I am indebted to Rolls (1969).
8. Roots (1976) incorrectly gives 1870.
9. R. Hewson in Corbet & Southern (1977: 147).
10. J. Guého & F. Staub, *Proceedings of the Royal Society of Arts and Sciences of Mauritius* **4**: 17 (1983).

Rabbit

(*Oryctolagus cuniculus*)

Note: Original presence in north-west Africa is uncertain.

NATURAL DISTRIBUTION

Originally the Iberian Peninsula north to the French Pyrenees, and possibly northwestern Africa.

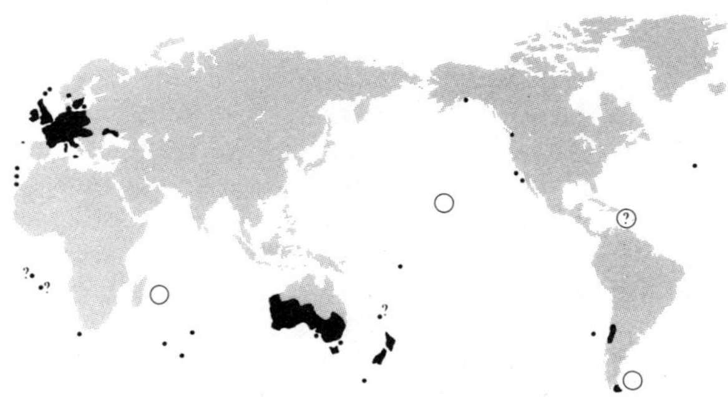

Note: Also occurs on many other oceanic islands.

NATURALIZED DISTRIBUTION

Europe: most of western Europe including some Mediterranean islands (but only in southern Scandinavia and absent from most of Italy and the Balkans) east to eastern Poland and Hungary; North Atlantic islands: Madeira, Azores, Canaries, Selvagens; Asia: ?Sri Lanka; Africa: South Africa; North America: USA, Canada, West Indies (?Grenadines (Balliceaux), ?Guadeloupe, Barbados); South America: Chile, Falkland Islands; Australasia: Australia, New Zealand; Oceania: Hawaii, ?St Helena, ?Ascension, and many other oceanic islands.

EUROPE

British Isles

Fitter (1959) advanced an interesting theory about the origin of the wild rabbit, which was unknown in Italy (even in captivity) until the third century BC, and did not appear in northern and central Europe until the Middle Ages:

> Why did an animal so eminently suited to life in the British Isles and other parts of northern and western Europe not spread there by natural means long before? . . . The answer must be that the wild rabbit with which European man has plagued much of the earth's surface is not a true wild rabbit at all, but a feral domesticated rabbit . . . It seems . . . likely that the Romans, who are known to have domesticated rabbits, turned the Spanish stock into a hardier and more adaptable strain. It is this strain, it would seem, that was imported into northern Europe in the Middle Ages.

In his *Rerum rusticarum*, compiled in 54 BC, Marcus Terrentius Varro records that rabbits were first introduced to Britain from Spain (where Stone Age cave-paintings depicting the animals have been found) by the Romans. No evidence, however, exists to show that the present British population is derived from Roman stock; as Fitter points out, there is no Anglo-Saxon or Celtic word for the rabbit, and the lack of mention of rabbit-warrens in the Domesday Book (the great survey of England completed in 1086) suggests that rabbits were not established in Britain by the eleventh century.

England

Veale (1957) is responsible for having discovered the earliest references to date of rabbits in the British Isles:

> In 1176 there were rabbits in the Scilly Isles, where Richard de Wyka granted to the abbey of Tavistock his title *de cuniculis*, 'which for some time I had unlawfully withheld, believing that tithes were not payable on things of this sort'.[1] At some time between 1183 and 1219 the tenant of Lundy Island was entitled to take fifty rabbits a year from certain *chovis* (coves?) on the island.[2]
>
> Evidence also survives as to the existence of rabbits in the early thirteenth century on the Isle of Wight, where in 1225 there was a *custod' cuniculorum* in the manor of Bowcombe, Carisbrook.[3]

The earliest rabbits' remains so far discovered on the English mainland – in a kitchen-midden at Rayleigh Castle in Essex – appear to date from the late twelfth to early thirteenth centuries.[4] Sheail (1971) refers to rabbit bones from the Buttermarket in Ipswich, Suffolk, which also appear to date from about the twelfth century.

These dates tend to suggest that rabbits were introduced to Britain either during the Norman period (1066–1154) or in early Plantagenet times.

The first written reference to rabbits on the English mainland occurs in 1235, when Henry III made a gift of *decem cuninos vivos* from the royal park at Guildford in Surrey. During the next few years the number of coneygarths and warrens appears to have multiplied, and rabbits began to feature prominently on the menus at feasts and banquets. They were also increasingly transported alive from one part of the country to another, and during these movements it seems likely that many escaped into the countryside where, as early as 1254–7, the burgesses of Dunster in Somerset were complaining of their destructiveness. From the time of their introduction until the late fourteenth century rabbit skin and flesh must, from the prices obtained for them, have been regarded as luxuries. These high prices suggest that rabbits were still at this time only locally distributed in England.

Scotland

The earliest mention of rabbits on the Scottish mainland occurs in a Statute protecting the royal warrens from poachers during the reign (1214–49) of Alexander II. On Scottish offshore islands rabbits are known to have been present on Little Cumbrae as early as 1453, and on Orkney, Lambholme, Sanday, Mull, Inchkenneth and several islands in the Firth of Forth by 1529. The Bass Rock, off North Berwick, was colonized by 1584, Great Cumbrae and Ailsa Craig by 1612, and by 1677 rabbits had spread as far as the Outer Hebrides. Seven years later warrens were established on Burray Island in the Orkneys and on Sanda Island off the Mull of Kintyre.

Rabbits were introduced to several districts in the Highlands of Scotland during the eighteenth century. They first appeared in Morayshire, on the golf-links at Lossiemouth, in 1750, having apparently swum ashore from a ship moored in Branderburgh harbour. They first appeared in

Caithness between 1743 (when they were unknown) and 1793 (when they were said to be abundant). Elsewhere in Scotland they were introduced to the island of Gigha in 1763, and first appeared in Dumfriesshire in 1815, in Kintyre in 1845 and in Wester Ross in 1850. Between 1840 and 1890 a number of introductions were made to Raasay and to North and South Uist in the Outer Hebrides. In the 1870s some pet rabbits were released on Foula in the Shetland Islands, where they were also introduced (to Hascosay) as recently as 1900.

Wales

Matheson (1941) has traced the early history of the rabbit in Wales. The first written evidence dates from 1282, when Richard le Forester was paid 3*s*. 6*d*. (about 17½p) for catching rabbits for the king at Rhuddlan Castle in Flintshire.[5] In a Pipe Roll of 1325–6 rabbits are referred to on the islands of Schalmey, Skokholm and Middleholm. By 1376 a warren had been established at Cardiff Castle, and by 1492 another on the island of Flat Holm. The Elizabethan historian, Rice Merrick,[6] refers to a number of rabbit-warrens established in Glamorgan by 1578. Writing in about 1540 the antiquary John Leland refers to numerous rabbits on St Tudwall's Island, St Dwynen (Llanddwyn) on the Isle of Anglesey, and on Hilbre Island at the mouth of the River Dee. During his tour in Wales in about 1773 Thomas Pennant observed rabbits on Priestholm and on the Skerries, both off the coast of Anglesey.

Ireland

Veale (1957) discovered the earliest reference in the British Isles to rights *in warrensis cunigariis* in a charter granting lands in Connaught (the mid-west of Ireland) to Hugh de Lacy in 1204. During the fourteenth and fifteenth centuries a not inconsiderable export trade in rabbit skins developed between Ireland and England. By the late 1770s Arthur Young, in his *Tour of Ireland* (1776–9), reported that rabbits were then common in many parts of the country. As recently as 1906 rabbits were deliberately introduced to Clare Island, County Mayo, and to Rathlin Island, County Antrim.

In their early days in the British Isles rabbits were largely confined on the mainland to the sandy links-land between the sea and the hinterland, and to a number of small offshore islands – probably because burrowing was easier in light sand, and because of the absence of predators.

Rabbits probably began to escape from captivity during the thirteenth century; by 1551 the Swiss naturalist, Conrad Gesner, was writing of *copia ingens cuniculorum* in lowland England, and remarked that although 'there are few countries wherein coneys do not breed, the most plenty of all is in England'. Even here, however, there were many districts where rabbits were scarce or unknown until as late as the early 1800s.

The second half of the nineteenth century witnessed a phenomenal explosion of the rabbit population in Britain, which has been attributed by Ritchie (1920) to the advances of agriculture which provided additional sources of food – especially in winter when they were most needed – and to the fashion for game-rearing with its concomitant destruction of predators.

By the late 1930s the wild rabbit population in Britain was estimated to number almost 30 million – a figure which, before the advent of myxomatosis in 1953, had staggeringly doubled or even trebled.

Myxomatosis – a virus first discovered in Uruguay in 1896 – is transmitted between individuals in Britain by the rabbit flea *Spilopsyllus cuniculi*, and in other parts of Europe and in Australia also by mosquitoes. In Europe, where it spread with astonishing rapidity, it was deliberately introduced in France in 1952 by a retired physician, Armand Delille. In England myxomatosis, within a few years of its arrival, destroyed a well-nigh unbelievable 99 per cent of the wild rabbit population without, however, noticeably decreasing the animal's distribution.

By the late 1960s and early 1970s rabbits were beginning to show signs of staging a recovery. Myxomatosis must, however, now be regarded as enzootic in Britain, although its virulence varies from year to year and from one part of the country to another. The behaviour of the virus is not fully understood, but in some districts, when rabbits become numerous, it seems capable of killing the entire annual increment.

In the British Isles – as elsewhere – rabbits cause immense economic damage to agricultural, horticultural and silvicultural crops. On the other hand, the extinction in 1979 of the large blue butterfly (*Maculinea arion*) is believed to have been largely caused by the decline in the number of rabbits, whose close-cropping of grassland allowed the wild thyme (*Thymus drucei* and *T. pulegioides*), on which the butterfly lays its eggs, to flourish; smaller numbers of rabbits have also resulted in an increase in the growth of scrub – especially hawthorn (*Crataegus monogyna*) – on open chalk and limestone grassland.

The principal predators of rabbits in Britain – apart from man – are badgers (*Meles meles*), foxes (*Vulpes vulpes*), stoats (*Mustela erminea*), weasels (*M. nivalis*), mink (*M. vison*) and birds of prey, none of which, however, appear to have any noticeable affect on the overall population or distribution.

Continental Europe and Mediterranean and North Atlantic islands

About 2,300 years ago rabbits were confined to the Iberian Peninsula – where the Phoenicians found them to be abundant – and perhaps to north-west Africa. Phoenician, and later Roman, traders spread them around the Mediterranean, where they gradually became established on the mainland and on several offshore islands. Domesticated forms probably originated in French monasteries between about AD 600 and 1000, although coloured varieties are not recorded before about 1550. Over much of western Europe domestic breeds probably arrived before wild stock spread naturally as a result of agricultural development. As previously mentioned, wild rabbits first appeared in northern and central continental Europe (e.g. Germany and the Netherlands) during the Middle Ages. They are now distributed through much of western Europe (apart from northern Scandinavia, most of Italy and the Balkans) east to Poland and Hungary, and have been introduced to the Ukrainian SSR on the north coast of the Black Sea and to some offshore islands in the Caspian Sea, to Denmark, to the south-east coast of Sweden, and to islands off the south-west coast of Norway. The earliest reference to the presence of rabbits in Hungary (near the eastern limit of their European range) is by J. B. Grossinger[7] who mentions that rabbits in the wild had greatly increased around Moson by 1779, although elsewhere in the country they were still known only in domestication. By 1860 rabbits were abundant in the wild in Hungary, and by the mid-1880s they had reached plague proportions.

In the Mediterranean rabbits are found on the Balearic Islands, on Corsica, Sardinia, Sicily and Crete, and possibly on Cyprus, and in the Atlantic on the Azores, the Canaries, and the Selvagens (or Salvages – between Madeira and the Canaries). In 1418 they were introduced to Porto Santo in the Madeira Islands.

ASIA

Sri Lanka

Niethammer's (1963) map suggests that rabbits have been introduced to Ceylon (Sri Lanka).

AFRICA

South Africa[8]

Domestic rabbits were first introduced from Holland to the 507-ha Robben Island (formerly a leper colony and now the site of a notorious prison) in Table Bay off Cape Town in April 1654 by Governor Jan van Riebeeck as a source of victuals for vessels *en route* to the East. This introduction failed, but rabbits were subsequently successfully reintroduced in 1656–8; by 1659 there were said to be 50 on the island, and two years later they were reported to be abundant. Efforts in the past to eradicate them failed, but their numbers are currently declining. In spite of the probable introduction over the years of domestic stock of variable colour they long ago reverted to the wild-type brown coloration and, having virtually exterminated all but inedible plants, have taken to eating seaweed.

Rabbits – probably from Robben Island – were translocated to Dassen Island (222 ha) at some time between 1662 and 1668. A number of white rabbits were introduced in the first half of the present century, but the coloration is now predominantly wild-type brown; a few individuals, however, showing black, gold, chestnut and white occur, especially near human habitation. A large grey and white domestic buck translocated from the mainland in 1971 had a conspicuous effect on the progeny of the next few years. The normally large population tends to fluctuate according to hunting pressure by man (which, being usually confined to brown animals, helps to preserve the coloured varieties) and by feral cats (*Felis catus*), and also to the amount of vegetation surviving the summer drought.

F. Levaillant, in *Travels from the Cape of Good Hope into the interior parts of Africa* (1790), makes the earliest reference to (probably domestic) rabbits on 41-ha Schaapen Island in 1781, where they were apparently introduced early in the eighteenth century. This population died out before about 1953 and was replaced by some smaller and slower albinos from the mainland which still survive.

By about 1850 most of the accumulated guano had been removed from many of South Africa's offshore islands: only then could soils form and wind-blown seeds germinate to provide forage and cover which made the introduction of rabbits a practicable proposition. During the second half of the nineteenth century rabbits – mostly probably of the domestic variety – were introduced to a number of other South African offshore islands, both as translocations from previously colonized islands and also as importations from captive stocks on the mainland.

On Possession Island (90 ha) off the coast of Namibia most rabbits are wild-type light brown, though chocolate-brown individuals also occur. Since the island is only sparsely vegetated with a few perennial bushes the rabbits' survival is probably dependent on supplemental feeding by man.

The small population of rabbits on 46-ha Jutten Island are mainly light brown in colour with a few showing traces of dark brown and white: they do not appear to depend for food on man. In late summer (as on Schaapen Island) when the terrestrial vegetation has dried out they enter the intertidal zone to consume marine algae not, apparently, for its nutrient value, but for its moisture content.

Vondeling Island (21 ha) supports a stock of small black rabbits with white markings and some white ones with brown markings which, although the island was evacuated by man in 1963, survive in small numbers, probably feeding on seaweed in late summer.

Bird Island (19 ha) in Algoa Bay was stocked with rabbits some time after 1755 – probably after the lighthouse was commissioned in 1852. The large population is mainly natural brown, with a few white and black individuals. The *Eastern Province Herald* for 23 June 1868 mentioned that a Mr Pilkington had recently shot 40 rabbits on the island. Seal Island, measuring only 250 by 200 m, also in Algoa Bay, was colonized with stock perhaps translocated from Bird Island: although rabbits certainly occurred on Seal Island in 1971 they are now extinct.

Evidence for the existence of rabbits on 12-ha St Croix Island in Algoa Bay rests on a report in the *Eastern Province Herald* for July 1920 which claimed that 'rabbits formerly flourished but died out when the vegetation was destroyed by a drought some five years ago'. Since St Croix has little vegetation this may have been a mistake for Bird Island. On Malagas Island (9 ha), where they were probably introduced after 1850, the last rabbits were killed in 1977; on Marcus Island (11 ha), to which they were also probably introduced after 1850, they had disappeared by December 1972 after the departure of the guano labourers during the 1960s. On 7-ha Meeuw or Meeuwen Island albino rabbits introduced before about 1953 died out shortly before February 1977 when droppings and scrapes but no animals were observed. In 1859 'a few pair' of rabbits were transferred from Robben Island to a 2.5-ha islet in the Keurbooms River estuary, where they died out probably before 1865.

The main limiting factor controlling rabbit populations on South African offshore islands is the short (June to October) growing season for vegetation, and during the long dry period many die from malnutrition or starvation. On Jutten, Schaapen, Dassen and probably other islands, rabbits invade the intertidal area in search of water and marine algae on halophytes.

M. E. Gillham[9] considered that browsing by rabbits on Meeuw and Schaapen islands had reduced the number of bushes in which crowned cormorants (*Phalacrocorax coronatus*) nested, and had also eliminated much of the vegetative nesting cover of kelp gulls (*Larus dominicanus*) and Hartlaub's gulls (*L. hartlaubii*). On Meeuw, where rabbits are now extinct, the vegetation is markedly superior to that on Schaapen where they still exist. Little is known of the original vegetation on Robben and Dassen islands, so the impact of rabbits is difficult to assess. Probably due to the flat and arid terrain rabbits have not caused the same erosion problems on South Africa's offshore islands as they have done on subantarctic ones.

Only on Dassen Island do mammalian predators (feral domestic cats) occur, and there rabbits are their principal prey. On Robben Island, where attempts were made in the seventeenth century to eliminate them to preserve the rabbits, mole snakes (*Pseudaspis cana*) may kill a few young ones.

Feral rabbits have existed on Robben and Dassen islands for well over 300 years and, as Brooke and Cooper (1980) and Cooper and Brooke (1982) point out, it is unlikely that their removal would result in a return of the original vegetation – whose nature is in any case unknown. Robben Island, which has been much altered by human occupation, is no longer an important breeding site for seabirds: Dassen Island does remain a major site for breeding species such as the jackass penguin (*Spheniscus demersus*); although rabbits make use of penguin burrows the reverse also occurs, and it seems likely that the decline in the penguin population has taken place independently of rabbits. As Brooke and Cooper (and others) point out, isolated island populations of apparently harmless rabbits which have been in existence for more than 300 years are worthy of study and perhaps even of preservation.

A case does, however, exist for eliminating the population on Schaapen Island where rabbits have degraded the vegetation and may well have adversely affected tree-nesting birds. The large rabbit population on Bird Island appears to have had no deleterious affect on breeding seabirds (principally Cape gannets (*Sula capensis*)), nor have the smaller colonies on Possession, Jutten and Vondeling.

Jutten, Schaapen and Dassen islands are expected shortly to become part of a West Coast National Park, when it may be considered desirable that exotic species should be removed; before any attempt is made to eradicate rabbits on Dassen Island, Brooke and Cooper recommend that they should first be eliminated on Schaapen Island to try to assess what effects, if any, this would have on the vegetation and nesting seabirds, and that a study should be made of the Dassen population.

Feral rabbits on South African islands are currently confined to those of more than 20 ha, probably because they are the only ones to support a complex flora; they are established on only one island with little vegetation – Possession – which is both large (90 ha) and also inhabited by man. Their continued existence on South African offshore islands is therefore probably governed by the nature and extent of the vegetation and the size of individual islands.

As C. J. Skead[10] points out, Jan van Riebeeck, when he introduced rabbits to Robben Island in the 1650s, specifically prohibited their release on the South African mainland, where they still do not occur. Although southern Africa possesses many mammalian, avian and reptilian predators capable of killing rabbits, it is more than likely that they could become established in plague proportions in pastoral regions of the southwestern cape, as they have done elsewhere throughout the world.

NORTH AMERICA

USA

In 1929 J. M. Edson of Bellingham in the State of Washington wrote:

> It is my impression that rabbits have been common on the [San Juan] islands [between Washington and Vancouver Island] since I came to this region. I visited Skipjack Island as early as June, 1895, and while I have no written record as to the rabbits, it is my recollection that a number of them of different colors were seen on that trip. I suspect that if their history can be traced it will run back to the days when the Hudson's Bay Company occupied this region.[11]

Around the turn of the century, when San Juan, Smith, Long, Lopez, Colville, Whidbey, Wasp, Jones, Spieden, Flat Top, Johns and Mateo Islands were also infested, the lighthouse-keeper on Smith (at the mouth of Puget Sound) imported domesticated varieties – Belgian Hares and Black Flemish – to augment his income by the sale of meat and pelts to Seattle. His successor gave these rabbits their freedom, their progeny reaching a density of about 70 per hectare by 1924; this in turn led to the undermining and collapse of buildings, and to a vegetation dominated by nonsucculent rabbit-resistant bracken (*Pteridium aquilium pubescens*), tarweed (*Media exigua*) and cheat (*Bromus* sp.). Douglas firs (*Pseudotsuga menziesii*), stunted by gales, were ring-barked, and overgrazing caused serious erosion. After visiting the islands in 1924 on behalf of the US Biological Survey, Couch (1929) wrote: 'We saw evidence of rabbit depredation everywhere, and animals . . . running in all directions . . . the destruction of forage and shrubbery no doubt accounted for the lack of nesting birds.' Couch and his companion, Theodore Scheffer, found only two robins' nests, some non-nesting tufted puffins (*Lunda cirrhata*) and a single 'emaciated' California murre (*Uria aalge californica*).

The success of rabbits on the San Juan Islands is probably due to the temperate climate, with cool dry summers and mild moist winters, and an absence of any serious predators.

In 1954 a newspaper in Columbus, Ohio reported: 'The European rabbit, which has ravaged Australia and New Zealand, has been recently introduced into Ohio and Pennsylvania by sportsmen.' According to Laycock (1966), some 7,000 were turned down in Pennsylvania, several thousand in Ohio, 500 in Indiana, and others in Illinois, New Jersey, Wisconsin and Maryland. Since the animals came from San Juan it was considered that the Lacey Act of 1900, prohibiting the importation and release of animals from abroad, did not apply. Fortunately, none of these misguided introductions was successful.

Apart from San Juan, the only other stocks of wild rabbits in the United States have been established on South Farallon Island, 50 km west of San Francisco, on Santa Barbara and Anacapa islands in the Channel Islands National Monument off Los Angeles (where they have eradicated several species of endemic plants, such as the giant coreopsis, *Coreopsis gigantea*) which has resulted in the probable extinction on the former of the local race of the song sparrow (*Zonotrichia melodia graminea*), and on Middleton Island in the Gulf of Alaska. On Santa Barbara they are much larger than wild European rabbits.

In the autumn of 1954 a buck and three does – all white – were released on Middleton, which lies 130 km south of Cordova and covers an area of around 13 km^2. A year later the population had increased to about 50 but, due to heavy winter mortality, had fallen to seven (two bucks and three does) by the following spring, only to multiply explosively to some 200 by the end of the same summer. When O'Farrell (1965) visited the island in late February 1961 he estimated the population to number between 3,600 and 3,700. He reported that at least five colour phases were represented, these being in diminishing order of dominance, black, white, grey, agouti and red, and was told of 'Himalayan-marked' individuals – white-pelaged albinos with dark ears, nose, tail, feet and scut. Mortality from disease, as well as from the results of severe winter weather, appeared to be high.

Canada[12]

Rabbits – presumably from the San Juan Islands – were at some unknown date released on Bare (Mandarte) Island (near Sidney on Vancouver Island) and on Strongtide Island in the Chatham-Discovery group, and in 1954 on Chatham Island, just over the border from San Juan. On Vancouver Island, feral domestic rabbits have from time to time become established locally, principally in the Sooke Road area and in the Highland District. Rabbits turned down in the McGee swamp region near Shawnigan Lake did not long survive. Domestic rabbits released on South Pender Island in about 1909–10 by Mr A. E. Stanford initially increased considerably, but had apparently died out by around 1916. Rabbits have also been introduced near Tlell on the east coast of Queen Charlotte Island and, prior to 1921, on Triangle Island (Scott Islands), but with apparently only limited success.

West Indies

R. B. J. B. Du Tertre, in *Histoire générale des îles de S. Christophe, de la Guadeloupe, de la Martinique, et des autres dans l'Amerique* (1654), reports the presence of European rabbits (and hares) on the island of Balliceaux in the Grenadines and on Guadeloupe, where they were said to be abundant. Rabbits and hares from England were first introduced, by Mr Thomas Trotman, to the Bulkeley Estate in St George's Parish, Barbados in 1842, from which they escaped when a rainstorm demolished their enclosure. Both species multiplied rapidly in the wild, and in the 1870s some 200–300 were shot annually. In about 1880 rabbits were imported from the Canaries to the island of Cuba. Both animals were much reduced in numbers in the late 1880s following the introduction of the small Indian mongoose (*Herpestes auropunctatus*).[13]

SOUTH AMERICA

Chile

Rabbits from Europe are believed to have been first introduced to the mainland of Chile in the mid-eighteenth century, where they became locally abundant.

In about 1900 more were released by European sportsmen, but they did not spread to central Chile until 1943. By 1960, however, they were present throughout – and in some places abundant in – Malleco Province, being found as high as 1,035 m at Fundo Bella Vista. Shortly afterwards they became established at an altitude of 1,500 m in the Cordillera de los Andes. The reclamation of scrub and forest for agricultural purposes increases rabbits' available habitat, and the trapping of the Colpeo fox (*Dusicyon culpaeus*) and the Argentine grey fox (*D. griseus*) removes potential predators. Some agricultural damage by rabbits occurs, especially in the Cordillera de Nahuelbuta west of Angol.

E. Lucas Bridges records the introduction in about 1880 of rabbits to the Tierra del Fuego region of Chile:

> In the Beagle Channel, and others further south, there are innumerable islands, mostly rocky, but with a good deal of bush, grass and wild celery growing on them. Father realized that these islands, if stocked with rabbits, would yield welcome food for the natives and for any shipwrecked crews who might be stranded there. Accordingly, he brought some rabbits from the Falklands [to which they had been introduced in the previous century – see below]. He took great care that they did not escape on the main island, nor did he release them on the larger islands in the Channel, lest they should become a pest to future farmers. But on any small island that he considered suitable he would land two or three pairs. Where there was good sandy soil and bush they thrived and increased exceedingly. Some years later H.M.S. *Sirius* anchored off one of these islands, and the whole ship's company landed in two batches on consecutive days. Hunting the descendants of the couple of pairs set free there by my father provided them with plenty of exercise, and over six hundred rabbits were caught – one for every man on the ship.[14]

In about 1910 small numbers of domesticated rabbits were released at Puerto Porvenir in the Straits of Magellan. Twenty years later, perhaps reinforced by further liberations, they were reported to be spreading inland where, by 1947, they had reached plague proportions and were damaging pastures and causing erosion by burrowing on two ranches of over 400,000 ha leased to the Sociedad Explotadora de Tierra del Fuego.

In 1953 myxomatosis was deliberately introduced by the Chilean government, and in the following year (when the wool clip and mutton production fell by 22 per cent and 24 per cent respectively because of malnutrition and reduced stocking capacity through competition for grazing by rabbits) more rabbits were inoculated with the virus by the Sociedad Explotadora: vast numbers of rabbits died and the disease is now regarded as epizootic. Since there are said to be no mosquitoes, rabbit fleas or other ectoparasites in Tierra del Fuego, the means by which myxomatosis spreads there is uncertain.

Today, rabbits are common in parts of central Chile and in Tierra del Fuego – in the latter especially around Porvenir and on the mainland near Punta Arenas, as well as along the Beagle Channel, and among the islands of Cape Horn, being particularly abundant on Lennox Island.

Falkland Islands

Rabbits were probably first introduced to the Falklands by the French navigator Louis Antoine de Bougainville in 1764. In the following year a shipment was landed at Port Egmont on Saunders Island by Commodore John Byron, the leader of an expedition despatched from England. Others may have been imported as a source of food in the early 1800s by sealers and whalers. Some forty years later Richard C. Moody, appointed first governor of the islands in 1841, spoke of the presence in the valleys near the settlement of Port Louis on Berkeley Sound (from which in former times the islanders had exported their pelts) of grey – and especially black – rabbits 'literally in myriads'. Today, they are established around a number of settlements on East Falkland – especially in the north at Rincon Grande and at Fannings Head near Port Carlos – and at Port Stephens (only) on West Falkland. Rabbits also occur on several smaller islands such as Bense off Roy Cove, on many in the Sound (especially on Flat Tyssen), and on New and Saunders islands. The animals on small islands are subject to periodic population explosions (followed by mass starvation and death) which, for some unexplained reasons, do not occur on the two main islands, and are frequently accompanied by the development of unusual colour phases. At one time rabbits were also found on Keppel Island, where they were exterminated by feral cats which themselves subsequently died out due to disease.

In 1872 rabbits, possibly from the Falklands, were introduced to neighbouring South Georgia, where they have since died out.

AUSTRALASIA

Australia[15]

The first rabbits – probably domestic Silver-Greys – to reach Australia arrived at Botany Bay, New South Wales in January 1788 with the First Fleet bringing out the early colonists from England. The commissary, Andrew Grey, listed five (among a quantity of other farm animals) in *An account of livestock in the settlement* on 1 May 1788. A second shipment – also probably Silver-Greys – arrived at Sydney (as the new settlement was called) on the *Gorgon* via South Africa in 1791.

In *Two years in New South Wales* (1827) a naval surgeon, Peter Cunningham, wrote: 'Rabbits are bred about houses, but we have yet no wild ones in enclosures, although there is a good scope of sandy country on the sea-coast between Port Jackson and Botany Bay fit for little else than goat pasture and rabbit warrens.'

According to George William Evans, Surveyor-General of the Colony, in his *History and description of the present State of Van Diemen's Land* (1824), domestic rabbits first reached Tasmania in the early 1820s. In 1827 Dr William Crowther of Hobart confirmed this: 'In the year 1824 I imported the wild grey rabbit into this colony, from which a considerable stock has increased.'[16] In the same year the *Colonial Times and Tasmania Advertiser* informed its readers that 'The common rabbit is becoming so numerous throughout the colony that they [*sic*] are running about on some large estates by hundreds.

In the 1820s a lease was granted of land on Betsy Island in the Derwent River, Tasmania, on which to breed Silver-Greys for the Chinese felt-market. A large number were shipped out from England on the *Adrian* in 1826, but the venture was not a success, and six years later rabbits were fetching only 2*s*.6*d*. to 3*s*. (about 12½ to 15p) a pair in the streets of Hobart.

In 1869 the Surveyor-General of Tasmania, James Calder, wrote home to England that 'The English rabbit has multiplied prodigiously . . . some of our great landowners employ men all the year round whose sole business it is to shoot them down . . . they have very generally acquired the rusty grey colour of the wild rabbit of England.' This colour change may, as Rolls suggests, have been the result of interbreeding – as the only other early change recorded in Australia was to sandy brown in desert areas – or it may simply be the natural reversion to wild type of a feral domesticated species.

During the debate on the Rabbit Nuisance Bill in 1883, Mr W. J. Lyne said in the New South Wales Parliament: 'I resided for some years in Tasmania . . . the rabbits appeared so suddenly and increased so rapidly that although strenuous measures were adopted to destroy them, it was totally impossible to cope.'

From the 1830s domesticated rabbits – both captive and feral – were established on the eastern and southern coast of Australia and on a number of offshore islands. John Lort Stokes, assistant surveyor of the *Beagle*, who visited Australia with Charles Darwin in 1836, released some rabbits (and pigs) on Deal Island in the Kent group in the Bass Strait, as victuals for shipwrecked mariners. A few days later the *Beagle* lay hove-to off a small islet off Wilson's Promontory:

> From the abundance of rabbits, we called it Rabbit Isle.[17] I have since learned that the animals had multiplied from a single pair turned loose by a praiseworthy sealer six years before and the sight of their number did not a little to encourage me to expect a similar result from the gift I had bestowed on Kent group.

In *The Yeoman and Australian Acclimatiser* in 1861 'An Old Bushman' recounted his success at breeding rabbits at Mount Alexander north of Castlemaine in Victoria in 1845 – the earliest extant evidence of rabbit-breeding in the interior. In 1880 Mr Albrecht Feez, then a Queensland politician, informed Parliament that in 1854 he had turned out rabbits in the hinterland of New England, New South Wales. By 1860–1 the animals had reached plague proportions in some areas, though it was not until the turn of the century that rabbits became a pest generally in New England.

The *Port Phillip Patriot* for 26 June 1846 noted with apparent satisfaction that: 'Numerous rabbits, which have from time immemorial escaped from the hutches and clutches of the vendors in the market place, have taken refuge under the police office and have increased rapidly. They appear to thrive well in the locality they have chosen.'[18]

All the rabbits so far introduced to Australia were domesticated varieties descended from – but very different in appearance to – the wild *Oryctolagus cuniculus*.

The first successful introduction of wild rabbits was made, according to H. M. Suttor in *Australian Milestones* (1925), by Mr Thomas Austin of Barwon Park, Winchelsea, near Geelong in Victoria. His brother, James, loaded 24 wild rabbits (and 72 partridges and 5 hares) on board the Black Ball clipper *Lightning* (Captain Clarke) in Liverpool. The *Lightning* cleared the Mersey on 6 October 1859, arriving off Gellisbrand Point, Melbourne on Christmas night. Only 6 partridges and 1 hare failed to survive the voyage.

The rabbits were placed in a fenced enclosure at Barwon, though if the Melbourne *Australasian* for 7 July 1865 is to be believed, 13 were at once released on the banks of the Barwon River. By 1862 the *Geelong Chronicle* was able to report that the rabbits (and partridges) were all thriving in Barwon Park – the former in thousands – 'but unfortunately they prove highly destructive to the young trees planted in the various plantations intended for pheasants and partridges'. Exactly when the fenced rabbits escaped or were released is unknown.

Austin's success prompted a flood of subsequent introductions – in Victoria, New South Wales and South Australia. On 3 December 1860 the *Sydney Morning Herald* announced that 'Those who recollect an English rabbit warren . . . will be glad to hear that an attempt is now being made to introduce the wild rabbit on the estate [the Warren] of Mr [Thomas] Holt at Cook's River [now the Sydney suburb of Marrickville]'. Holt's rabbits soon spread to Liverpool, and by 1883, they had extended their range west to Camden and south to Bulli.

In December 1861 *The Yeoman and Australian Acclimatiser* thus described an early rabbit shoot near Melbourne: 'Two of us commenced about half-past five and in little more than an hour we killed seven couple . . . Our friend established his warren only three or four years ago and now the rabbits are all over the place and positively require thinning.' Rabbit shooting for sport was soon followed by the introduction of coursing, as described by *Our Daily News* of Castlemaine on 28 April 1863:

> This favourite field sport has been introduced into this district with much spirit. The first meeting took place yesterday, and a posse of men of the right stamp were in attendance to give a jest to the game. The rabbits supplied by Mr R. Gravenor of the Rocky Water Holes were young, strong and wild . . . there were any number of the furay kind and their enimies had much difficulty in bringing them to grief.

Rabbit shooting at Barwon Park was on the grand scale; by as early as 1865 Austin reckoned he had killed over 20,000, and estimated he had half as many again remaining. Two years later, when the British *Field* magazine reported that 14,253 rabbits had been killed at Barwon in the previous season, Austin entertained Alfred, Duke of Edinburgh, at a shoot when over 600 were killed. In the following year the Duke was himself shot but nothing daunted he returned in 1869, and in one day shot 325 rabbits to his own gun out of a total bag of 1,200. By this time rabbits had spread almost 60 km from Barwon along the south bank of the river past Geelong to the coast.

It is probable that the early spread of rabbits in Australia was assisted by man. Fifty pairs, for example, released at Rocky Water Holes by Mr Gravenor in 1861, multiplied so quickly that two years later the local fishmonger changed his advertisements from 'Shell & Other Fish – Murray R. Cod' to 'Rabbits, Rabbits, Rabbits Always on Sale'.

In 1868 came the first reports of serious damage caused by rabbits. In the Warrion Hills, north of Colac, rabbits increased so rapidly that many hectares of valuable pasturage were destroyed, and trappers were soon killing 4,500 a week. The influential London *Times* in the same year was among the earliest to sound a warning note:

The irrepressible rabbit in Australia promises to be as great an enemy to rural contentment as he is in this country. Here he has been the cause of more ill-blood than church rates. He has set tenants against landlords and inspired the leaders of each local opposition with their most bitter and sarcastic invectives. He appears at the hustings at every general election, his name adorns the party squibs and his skin is hurled at the unpopular candidate. Our correspondent from Melbourne tells us that rabbits threaten to starve the sheep out of their runs. One large landholder and squatter William Robertson has been put to an expense of £4,000 or £5,000 and he estimates that it will cost him £10,000 in wages to trappers and killers before he has achieved any marked success.[19]

Within a year Robertson had already spent £9,500 of this sum and reckoned he had killed well over two million rabbits.

In the mid- to late 1860s rabbits began to become established also in South Australia (around Kapunda) and in New South Wales at Wentworth (at the confluence of the Murray and Darling rivers) and by the early 1870s at Balranald (on the Murrumbidgee) in the Riverina. Runs near Wentworth and 260 km^2 of land north of Balranald were the first seriously affected areas in New South Wales.

Rabbits in settled areas such as Adelaide, Geelong, Guildford, Cape Otway and Sydney were confined by topography and human habitation. In lightly settled country spectacular population explosions and extensions of range occurred; within a few short months of their arrival in 1872 at Julia Creek in South Australia, for example, rabbits had travelled 80 km north to Mount Bryan, and south-west as far as the Sheoak Log plain between Roseworthy and Daveyston.

Rabbits in the Mallee[20] (Victoria) now began also rapidly to extend their range; the progeny of four pairs transplanted from South Australia in 1866 to Morton Plains, 25 km north of Donald, had by 1878 travelled west to the Wimmera River and east to the Avoca (a spread of over 110 km), south to Donald and 160 km north to the Murray River, covering in all nearly 2 million hectares. Many sheep runs, overrun with rabbits, were abandoned: on innumerable farms wheat was eaten as soon as it appeared through the ground, and many farmers did not bother even to sow. As a contemporary reported:

No one would believe the mischief the rabbits are doing unless they could see it. The sheep farmers have not tried to clear them and now many of the runs are so bare that there is not grass enough to feed the rabbits, let alone the sheep. The consequence is that the rabbits are travelling in thousands in search of food. Last Friday morning I met a swarm coming from the hills. I never saw such a thing before. The ground was scarcely to be seen for about a mile in length. Five weeks since I could not find a rabbit on my land, but since then we have killed thousands. . . . the paddocks stink with the dead ones.[21]

Rabbits were by now abundant on some 6,500 km^2 of country from Eden Valley to Peterborough (Victoria), and beyond to the Murray River, which they crossed in about 1877 into New South Wales.

Although the Riverina seems hospitable in comparison to the Mallee, rabbits there only increased slowly; by 1879 they had spread up the Lachlan from Balranald only as far as Hillston, but were increasingly seen between the Murray and the Murrumbidgee west of Hay. Only near Balranald, where they had been turned out in 1870, were they at all abundant.

Rabbits now began crossing the border from South Australia along the Murray to Wentworth on the lower Darling in New South Wales. Those which eventually overran New South Wales and southern Queensland thus originated in the Victorian Mallee in the south and South Australia in the west. The Mallee stock, as we have seen, came from South Australia, and the latter's were either the progeny of Austin's 1859 importation or of subsequent introductions from England.

In 1883 rabbits in New South Wales began a northward move, and by the following year were being reported from three new areas – Cobar, Mudgee and Curranyalpa on the Darling, 150 km south of the Queensland border. Two years later rabbits began colonizing the Eyre Peninsula in South Australia (from Colton), and crossed the Victorian border from the Glenelg River region to overrun Naracoorte. They spread up the Paroo River and greatly increased along the Darling and Lachlan, where they took over the burrows of marsupial bilbies or rabbit-bandicoots (*Macrotis lagotis*); by the end of the year they had travelled from the Barrier Range along the Coko Range and on to the Grey Range as far as Wompah in Queensland. Those coming up the Darling had either followed the Paroo to Hungerford, Currawinya and Caiwarro, or had continued up the Culgoa. Apart from movements up the Barrier Range and from South Australia the general trend of progression was across the Murray and up the Darling: those established on the New South Wales side of the former were joined by others from the Mallee.

In the first eight months of 1887 a staggering 10 million rabbits were destroyed in New South Wales, and by the end of the year northeastern South Australia resembled a desert.

From Dubbo in east-central New South Wales rabbits travelled north along the Macquarie and Castleragh rivers, reaching Coonamble, 250 km east of Cobar, by 1898. They followed Cox's Creek northwards to the Namoi River, arriving at Boggabri by 1905. In areas of unsuitable habitat, such as heavily timbered country like the great Pilliga Scrub, they often followed in the wake of land clearance by railway engineers (whose tracks provided one of the main avenues of expansion) and pastoralists. Eastwards they travelled along the Liverpool Range and the Manning River, and north along the Great Dividing Range, keeping, wherever possible, to the red basalt country, where thorny brush – blackberry, lantana and (in the north) prickly pear – and the holes of wombats provided cover and security.

Rabbits clearly disliked the black soil in the Mullaley region of northern New South Wales: in 1904–5 they traversed the black plains east of the Barwon River but remained for not more than a couple of years. By 1925 rabbits had arrived, from the west, in the Tweed River district on the far northern coast of New South Wales. North-east from Tenterfield and Glen Innes along the Rocky River, and eastwards through Newton Boyd to Grafton, they easily became acclimatized to the subtropical climate from that of temperate New England. By now all suitable habitats in New South Wales had been overrun.

Most of Queensland escaped infestation. Large numbers in the 1870s were turned out around Helidon and further west on the Darling Downs, but apparently did not thrive, in part at least due to predation by dasyures or native cats (*Dasyurus* spp.). Until 1889 the only stations to suffer serious infestation were some on the Bulloo, Paroo and Balonne rivers, with small numbers reported from near Cunnamulla on the Warrego; by 1892 numbers had increased around Cunnamulla, but three years later – when another invasion from New South Wales halted west of Hungerford – there were still comparatively few rabbits in Queensland. These few had, however, eaten out Bulloo Downs and Thargomindah, and had forced the partial abandonment of Yuleema and Boorara west of the Paroo River; small numbers were also reported from the districts of Leichardt, Mitchell and Gregory North.

After a population explosion at Collymongle, rabbits are believed to have crossed into southern Queensland via the Culgoa River and across the Barwon at Mungindi, as a result of which the Australian Pastoral Company's stations on the Balonne (Cubbie, Narine, Bullamon, Gnoolama, Noondoo and Doondi) – and Yeranbah in New South Wales – were overrun.

In Western Australia[22] rabbits fared rather better. They were first reported in 1887 at Cheynes Beach and on Bald Island, and in 1893 from Northampton, Carnac Island and the Abrolhos Islands, having apparently been introduced at Lynton on the mainland in about 1888. By 1894 rabbits had been carried by man across the Nullarbor Plain from Fowler's Bay in South Australia (which they reached in about 1891) to Eucla in West Australia; two years later they had travelled another 300 km west to Eyre, and by 1898 had moved a further 250 km west to Israelite Bay.

By 1901 rabbits were reported in the interior at Leonora (450 km from the south coast), and in the following year from Edjudina, Coolgardie, Bulong, Kanowna, Southern Cross, Esperance and Nannine, and were established throughout the Coolgardie goldfields south to west of Esperance. Between 1906 and 1908 rabbits were reported inside the No. 2 rabbit-proof fence at Mullewa, Coorow, north of Cunderdin and at Lynton. By 1911–12 they had reached Mingenew and had spread from Northampton to Geraldton and Greenough. They also occurred at Wollal, and by 1912–13 had spread, probably from the Geraldton district, as far south as Moora.

Under their initial momentum rabbits ranged over most of the pastoral regions of Western Australia as far north as Port Hedland. Infestation of the more heavily forested southern regions of the state proceeded more slowly.

By the outbreak of the First World War rabbits

were inside the No.2 fence at Bremer Bay, and also occurred from 65 km north of Fremantle (which they reached in 1916) to the mouth of the Murchison River. In the extreme south-west rabbits were reported with increasing frequency in the early 1920s, apparently first reaching Manjimup (where they were well established by 1928) in 1925–6.

By the early 1930s rabbits appear to have colonized all suitable habitats in Western Australia. They are now established over most of the state north to the Kimberleys, but are not a serious economic pest north of Shark Bay or in inland pastoral areas where they are generally thinly distributed. On offshore islands they currently occur on Goose Island in the Recherche Archipelago, and perhaps on Sandy Island south of Pemberton.

One of the greatest threats to rabbits in Australia, from which, however, they invariably stage a recovery, is recurrent drought. This highlights the difference in the relative ability between sheep and rabbits to survive on arid and wet land; in areas of high precipitation rabbits eat grass so close to the ground that sheep can no longer graze, and the roots rabbits dig up support them for some time after top-growth has been consumed. Sheep, on the other hand, providing they have access to water, thrive in arid conditions on dry fodder which rabbits digest poorly.

It is sometimes difficult to distinguish between the damage to land due to overstocking of sheep or cattle and that caused by rabbits. By eating out palatable ephemerals and grasses in the Mallee, rabbits force livestock (and themselves) to browse on scrub. In the Riverina pithy kurrajong saplings, sandalwood, pine, belah, wild lemon and leopard trees are ring-barked by rabbits, and in many places saltbush, cottonbush, bluebush and lignum have been replaced by white everlasting *Helipterum* sp., which provides adequate fodder for livestock and kangaroos but not for rabbits, and by the introduced cockspur (*Centaurea melitensis*).

Because trees and shrubs are seldom barked in areas of high rainfall the effect of rabbits there is noticeably less severe; most nutritious plants, however, through repeated grazing, are seriously stunted, and are replaced by inedible thistles and nettles, and by unpalatable weeds.

Rabbits are undoubtedly the most pestilential animals ever introduced to Australia: at their zenith they yielded (through meat, 'felt', glue and size) an estimated annual revenue of some £3 million pounds, which was exceeded in damage they caused by tenfold.

Between 1883 and 1892 a series of so-called 'rabbit-proof' fences were erected in a vain attempt to control the rabbits' spread. One such stretched for more than 1,100 km from Corowa on the Murray River in New South Wales via Wagga, Orange, Dubbo and Bourke to east of Barringun on the Queensland border. In the late 1880s Victoria constructed a fence from Tyntynder below Swan Hill westwards through the Mallee. Western Australia intended to build a fence (No.1) 1,822 km in length from Starvation Boat Harbour, west of Esperance, to Eighty Mile Beach between Port Hedland and Broome; before it had been completed, however, rabbits had been transported inside it by gold-prospectors and overlanders, necessitating the construction of another fence (No. 2) over 1,100 km long north from Mary Ann Haven on the Bight, 120 km west of the No.1. Subsequently No.3 fence was built running east from north of Northampton to join up with the No.2. Between 1888 and 1890 the New South Wales/South Australian and the Queensland/South Australian borders were also fenced, the latter being some 800 km long. The fencing, which cost around £40 per kilometre and on which Western Australia alone expended no less than £413,531, had a mesh of about 32 mm, stood over a metre high, was buried 15 cm in the ground, and was constantly being broken down by wombats, kangaroos, emus, floods and tempest.

Until the introduction of myxomatosis in 1950 the main methods of rabbit control in Australia – apart from the largely ineffectual fencing – were trapping, poisoning, digging, gassing and shooting. Unfortunately, the first two (those most commonly used) also resulted in the inadvertent destruction of large numbers of harmless endemic species, such as bilbies or rabbit bandicoots, hare wallabies (*Lagorchestes* spp.), kangaroo-rats (*Dipodomys* spp.), wombats and even the duckbilled platypus (*Ornithorhynchus anatinus*), which, because of their slow rates of reproduction, were less successful than the fecund rabbit in population replacement. Natural predators, such as carnivorous dasyures or native cats, tiger cats, dingoes, introduced foxes, raptors and pythons take only a light toll: indeed, absence of serious predators – coupled with suitable habitats, climate and food, and absence of competition from native species – are the main reasons for the rabbit's phenomenal success in Australia.

In 1950, when the Australian rabbit population was estimated at around 750 million, myxomatosis was introduced experimentally at Gunbower in Victoria. By early 1953 90 per cent of Australia's rabbits had been annihilated, with a resulting rise in the annual wool-clip of some 40 million kilograms valued at around $A60 million. By 1955, however, as rabbits in Australia (as in Britain) acquired a degree of resistance to the virus, the population began to stage a recovery, though not to pre-myxomatosis levels. Today, the disease is reinforced as a controlling agent by the use of compound 1080 (sodium monofluoroacetate).

Within 20 years of Thomas Austin's importation of two dozen rabbits to Barwon Park, Victoria, in 1859, their descendants – augmented by subsequent introductions in Victoria and South Australia – were infesting thousands of hectares in the former state, had crossed the Murray River into New South Wales, had reached the Yanko Creek and the Murrumbidgee River in the Riverina, and were rapidly approaching the border with South Australia to join forces with others liberated there. By the turn of the century rabbits were thriving in the sandhills of central Australia south of Birdsville, which has an annual rainfall of less than 13 cm, and by 1925 were established in the subtropical Tweed River district with an annual precipitation of 190 cm. They are found today in suitable habitats throughout mainland Australia south of the Tropic of Capricorn, as well as an several offshore islands such as French, Phillip, Churchill, Elizabeth, Doughboy, Sunday and St Margaret's off the coast of Gippsland.

New Zealand[23]

The earliest reference to (domesticated) rabbits in New Zealand appears to be that of Abel du Petit-Thouars in his *Voyage autour du Monde sur la Frégate La Venus . . . 1836–9* (1841), who mentions that during his visit in 1838 there were 'still to be found some rabbits imported from New South Wales'.

Baron Ortsdorff is said to have bred and sold rabbits at the Hutt, near Wellington, as early as 1842. Referring to the country between the Clutha and Mataura rivers, T. Tuckett noted, on 10 May 1844, in his journal, during his expedition to South Island (in Thomas Morland Hocken's *Contributions to the early history of New Zealand* (1898)) that 'returning from Tapuke [Taukupu] we landed on the island, and with the assistance of a capital beagle caught six rabbits alive and uninjured'.

Rabbits were introduced to Otago and Southland by the early settlers, two of whom acquired stock from Sydney when they settled at Port Molyneux in the early 1840s. Others (black and grey) were released in about 1848–9 by the Revd A. Fenton of Dunedin in the sandhills between Invercargill and Riverton and at Queenstown, and the Julius brothers turned out more on their run at Rugged Ridge on the upper Waitaki River. Mr C. R. Carter – the founder of Carterton – brought rabbits with him when he arrived as an immigrant in 1857, and in the following year is believed to have released them in Nelson and Marlborough. T. Lindsay Buick, in *Old Marlborough* (1900), records that Silver-Greys (the earliest mention of a specific variety) were introduced to Kaikoura, Marlborough, by Captain Ruck-Keene, RN in the same year (1858) and to Nelson in about 1865. So-called 'French' rabbits – possibly Argentes – were at about the same time turned out at Taradale, on Hawke Bay, where they apparently prospered.

In 1857 long-eared black and white rabbits (the origin of which is unknown) were reported between the Tokomairiro and Witahuna rivers. Mr H. B. Flett told G. M. Thomson (1922) that he attributed their extermination to the weka rail or 'Maori hen' (*Gallirallus australis*) 'which was very plentiful at that time and for some years subsequent. They used to go into the holes and eat the young rabbits. I have seen a weka killing a half-grown rabbit.'

According to J. C. Anderson,[24] wild rabbits from England were first introduced to New Zealand in 1862 by Dr Menzies of Mataura who, according to Thomson (1922), 'is usually credited with having been the successful introducer of them in the south, an achievement the credit of which has not been very eagerly sought after'. Menzies' rabbits were turned out on sandhills between the estuaries of the Oreti (New) and Aparima rivers, near the Bluff, where they soon became established. Two years later others were released on the sand-dunes around Invercargill, and about 50 more were turned down by a Mr Telford on the banks of the Molyneux; these all soon disappeared, as did others liberated by a Mr Clapcott on Popotunoa Station.

During the middle 1860s various acclimatization societies began to show the same interest in the introduction of rabbits as they did with other species. The Annual Report for 1866 of the Canterbury Acclimatization Society informed members that: 'An enclosure has been set apart for the Silver-Grey Rabbits presented by Sir G. Grey [Governor of New Zealand 1845–53 and 1861–7; Prime Minister 1877–9], which have thriven well and increased to a great extent, and have been distributed to members far and near.' In the same year the Otago Acclimatization Society released 60 rabbits, followed by 23 in 1867 and 18 in 1868. In about 1870 some of the Lop-Eared variety were turned out at Palmerston, but soon died out.

The many early failures of rabbits to become established in New Zealand can be attributed to the fact that at least some were of so-called 'fancy' breeds. This comparative lack of success continued until an increasing preponderance of wild rabbits were imported between about 1862 (when a number were landed on Enderby Island in the Auckland group where their descendants remain) and 1867: it was, however, a further decade or more before rabbits began noticeably to increase in numbers and to extend their range, thus becoming a significant economic problem. The build up in the population is illustrated by the fact that in 1873 only 33,000 skins were exported; this had risen by 1877 to nearly one million, and by 1882 to over nine million. In spite of the revenue resulting from the sale of skins, it was stated in 1883 that the net annual loss to the colony amounted to some £1.7 million, and that the cumulative figure over the past decade totalled £10 million. The bulk of this figure was attributed to the reduced sheep-carrying capacity of overgrazed pasturage, and the resulting surrender to the government of some 700,000 ha of leasehold land.

By 1878 rabbits from the sand-dunes of Invercargill had 'reached the shores of Lake Wakatipu and swarmed into Tapanui. Then having crossed the Clutha river or (more probably) having been put across by prospectors, they went north into the Dunstan district and on again until they were stopped by the Waitaki River'.[25]

Before long much of northern and central Otago and the whole of Southland had been overrun.

Late in 1878 rabbits reached Orari Gorge, and by 1881 were established in the country between the south branch of the Ashburton and Rangitata rivers. In 1887 they invaded Canterbury, passing over the mountainous country between lakes Hawea and Ohau; here they split into two forces, one of which debouched onto the arid Waitaki Plains and travelled eastward along the Ahuriri River, while the other circled round the northern shores of Oahu, crossed the Hopkins and Dobson rivers and threatened to cross the Tasman. Further south they swam the Waitaki and invaded stations on the Waitaki Plains and Tiakatarawa. It was soon apparent that unless they were stopped all the hill country between the Waitaki and Rangitata rivers would be overrun. By 1890 they were entrenched in Rabbit Pass which divides Westland from Otago and lies between the headwaters of the Waiatoto and Wilkin rivers.

In Marlborough rabbits became established particularly quickly. Captain Ruck-Keene liberated some of his Silver-Greys at Kaikoura in 1867, as a result of which he is subsequently said to have lost £70,000 worth of stock. As A. D. McIntosh graphically describes it: 'They increased at an alarming rate after 1869 . . . the whole of the open land in the province was overrun. Hillsides were alive, pastures devastated, flocks starved, sheep-farmers ruined, and hundreds of thousands of acres had to be abandoned.'[26]

Records for North Island are less complete, but the general pattern is likely to have been much the same as in South Island. One of the earliest references mentions rabbits at Wairarapa in about 1863 where by 1890 nearly 5,000 ha of land had been overrun.

H. Guthrie-Smith[27] says that rabbits first arrived in his region between 1880 and 1890, but initially failed to become established. The main invasion of North Island began around the turn of the century, and by 1906 the country surrounding Wairoa had been overrun; five years later rabbits had crossed the Mohaka River and were within 15 km of the town of Wairoa. By 1945 they were established at Te Kao, North Auckland, and by the late 1940s had spread into

Taranaki in the west and to Poverty Bay in the east.

The earliest liberation on a distant New Zealand offshore island was made (with the Dutch domesticated variety) in about 1845 on Rose Island in the Auckland group. Enderby Island, also in the Aucklands, was colonized (with the French Argente de Champagne variety) in about 1862, Philip Island, off Norfolk Island, before 1865, and Mangere in the Chatham Group (where grazing rabbits have endangered the endemic giant deep blue forget-me-not) around the turn of the century. Rabbits were introduced unsuccessfully on Stewart Island in the 1860s and again in 1942.

Domestic 'French' rabbits were imported to Macquarie Island southwest of New Zealand in 1880 (or possibly a decade earlier) from Dunedin as a source of food for sealers and shipwrecked sailors; until about 1930 they were not very abundant, but were even so reported to be eradicating the Macquarie Island cabbage (*Stilbocarpa polaris*). Since then they have invaded all except the northern quarter of the island, and have virtually eliminated tussock grass (*Poa foliosa*) from many of the steep coastal cliffs, thereby causing severe erosion and the loss of nesting-sites for sooty shearwaters (*Puffinus griseus*) and white-headed petrels (*Pterodroma lessoni*). By providing an additional food source for great skuas (*Stercorarius skua lonnbergi*) rabbits have been responsible for an increase in the population of these predators, which has greatly intensified the pressure on small and medium-sized petrels. Cats too, which also multiplied considerably after the arrival of rabbits, began to eat seabirds, whose eggs were a staple part of the human population's diet; domestic dogs, turned loose to eradicate the cats, preferred instead to prey on more sedentary seals, which were also an important source of the islanders' food supply. This kind of chain reaction to the attempted biological control of an existing pest by another species has been repeated on many other island groups throughout the world.

In 1969 the National Parks and Wildlife Service of Tasmania instituted a programme of rabbit control on Macquarie Island. As shooting and poisoning had been shown to be ineffective, it was considered that the introduction of a virus (as for cats on Marion Island[28]) such as myxomatosis offered the best chance of success. Since, however, the rabbit population lacked a suitable blood-sucking parasite the first decade was devoted to the introduction of the rabbit flea *Spilopsyllus cuniculi*. By 1978 the flea was sufficiently widespread for the first importation of the myxoma virus to be made. Considerable localized deaths resulted and fresh introductions of the virus, which is able to survive the subantarctic winter, have since been made each summer. Although the availability of dead and dying rabbits may have resulted in a temporary increase in the feral cat population, it is confidently expected that the latter will begin to fall as the supply of the former decreases.[29] Whether the use of such methods to redress the balance of nature destroyed by the actions of man is ethical must, at the very least, be a matter for debate.

By the mid-1940s rabbits had colonized all suitable habitats in both main islands – only dense forest, heavy secondary growth and exposed montane regions being uninfested. Most settled arable areas were only lightly populated, the heavier infestations being concentrated in places with friable and well-drained sandy or loam soils. Although rabbits prosper in dry and mild as well as in cold and wet conditions, they appear to prefer the more temperate climate of South Island.

In 1947, when a total of 17 million skins were exported, the population of rabbits in New Zealand was estimated at some 50 million (reaching a density in parts of South Island of 100 per hectare), representing a loss of revenue – through reduced sheep-carrying capacity of the land, erosion and soil degredation – of nearly £4 million.

Early efforts at rabbit control in New Zealand included poisoning, trapping, shooting, gassing, the artificial spreading of diseases, and the introduction, between 1883 and 1888, of mustelid predators, none of which had any significant lasting effect. Between 1882 and 1885 a 'rabbit-proof' fence 65 km long was erected in North Island between Hawke Bay and North Wairarapa; in South Island, after rabbits had crossed the Waitaki River, a 130 km fence was built in South Canterbury, followed by another along the Boyle and Waian rivers in the Amuri district, none of which proved successful.

The notable success in rabbit control in New Zealand in the 1950s is attributed by Gibb and Flux (1973) to the determination of those responsible; the policy of giving responsibility to specially constituted rabbit boards; the dropping by aeroplane on agricultural land of poisoned baits; and the passing of legislation – to prevent them from being 'farmed' – prohibiting the sale of rabbits or their pelts.

The present policy of control is more realistic and effective than the former one of attempted extermination, and rabbits today are under at least some degree of control throughout New Zealand. Once control has removed the bulk of the population (and natural predators), isolated pockets of survivors can, however, live and breed more successfully than hitherto in inaccessible areas; as in other species, survival and reproductive rates vary inversely with the density of the population.

OCEANIA

Hawaii

Exactly when rabbits were first introduced to Hawaii is uncertain, but it must have been shortly after European settlement; Andrew Bloxam, naturalist on the *Blonde* which visited Hawaii on a voyage from England in 1824, observed that on a visit to the 'Pearl River' on Oahu there was 'in one spot an island of nearly one mile in length, on this a quantity of rabbits have been turned up and are now become wild and numerous; they are of a black and white colour, and the island is named for them [Rabbit Island, now known as Ford Island]'.[30] On 22 October 1853 *The Polynesian* reported that 'Mr Tanner has recently introduced from Australia . . . a pair of white rabbits', and 20 years later the Government issued a permit to allow the breeding of rabbits on Coconut Island in Hilo Bay, Hawaii.

In 1903 the German-born manager of a guano-mining firm, Max Schlemmer, introduced rabbits to Laysan Island to supplement the diet of his employees. Eight years later an expedition sponsored by the US Department of Agriculture and the University of Iowa reported that:

> Rabbits now literally swarm over the island by thousands. The amount of damage done by them can better be imagined than told. They are exterminating first one species of plant then another. Several species that were common eight years ago have entirely gone, others are already doomed.[31]

By overbrowsing shrubs, such as *Chenopodium*, *Achyranthes* and *Santalum* – resulting in a loss of shelter, nesting material and nectar-seeking insects – rabbits were instrumental in the extermination on the island (before they were themselves eradicated by 1923) of the honey-eater or apapane (*Himatione sanguinea freethi*), the Laysan millerbird (*Acrocephalus familiaris familiaris*) and the rail (*Porzanula palmeri*). (Some rails had previously been transplanted to Midway Atoll where they survived until the introduction of rats in 1943–4.) Assisted by plumage hunters, rabbits also nearly succeeded in exterminating the Laysan duck (*Anas platyrhynchos laysanensis*).

Shortly after introducing rabbits to Laysan, Schlemmer translocated a number to Lisianski Island, where, within 10 years they had defoliated the landscape and within a further decade had starved to death.

Today, rabbits occur in the Hawaiian Islands on Manana off Oahu, on Lehua (since 1930) and probably (since before 1954) on Molokini.

Galápagos Islands

Rabbits in the wild on San Cristóbal (Chatham) Island at least by 1965 soon disappeared, possibly because the climate was too damp and/or because the population was too small.

Phoenix Islands

Rabbits were introduced to the Phoenix group (between Hawaii and New Zealand) by an American guano-mining company in the late 1860s.

Mauritius (Round Island)

The introduction by Mr Thomas Corby in 1844 of rabbits – believed to have been of the Argente de Champagne variety – and goats[32] to Round Island, which lies 22 km north-north-east of Mauritius in the Indian Ocean and extends to 151 ha, is a classic example of the disasters that can follow the arrival of potential pests (originally introduced to provide a reserve food supply for Mauritius) in a small island ecosystem. Their destruction of the vegatation is almost invariably followed by a diminution of wild life, which on Round Island includes four endemic reptiles (two snakes, one gecko and one skink), as well as several more which are peculiar to the Mauritius group, and a number of endemic plants. Today, as a result of overgrazing and browsing by rabbits and goats, Round Island has the unenviable distinction of possessing more threatened animals and plants per hectare than anywhere else in the world.

Perhaps the worst loss as a result of overgrazing and browsing has been the dicotyledonous forest, which by 1900 had almost disappeared; the remaining indigenous fan palms (*Latania loddigesii*) and the few surviving native screwpines (*Pandanus vandermeerschi*) provide cover for, and are essential to the survival of, one of the endemic snakes *Casarea dussumieri* and the endemic gecko *Phelsuma guentherii*. The extreme rarity (or extinction) of the other endemic snake *Bolyeria multicarinata* may be due to the destruction of the hardwood forest which is believed to have been its primary habitat. The two snakes are of especial interest as the sole remaining survivors of the subfamily Bolyerinae, a group of primitive boas possibly linked to the cobras. Two endemic palms – the bottle palm (*Mascarena revaugnhnii*) and a hurricane palm (*Dictyosperma album*), both seriously affected by erosion caused by rabbits – are almost, if not quite, extinct.

Rabbits on Round Island live and breed entirely above ground. Their diet consists predominantly of *Passiflora*, especially the berries and leaves, and also *Portulaca oleracea*, *Solanum nigrum*, grasses, and the seedlings of *Latania*, *Pandanus*, *Mascarena* and *Dictyosperma*. Their pelage has not reverted to the 'wild type' coloration, and ranges from black to sandy, often with some white on the paws or head.

A concerted attempt to shoot out the rabbits on Round Island in 1976 was unsuccessful, and they remain very abundant. The population, which in 1975 Bullock estimated at between 650 and 1,500, was found by Bullock and North (1982) to have multiplied to between 2,450 and 2,900. Increased grazing pressure had noticeably reduced the number of seedling lantan palms, had severely restricted or arrested the regeneration of palm savannah over wide areas, had allowed plants such as tobacco (*Nicotiana tabacum*) to spread, and had made many areas susceptible to erosion.

Kerguelen

Introduced in 1874, rabbits now cover most of Grande Terre and several other islands in the group, and are believed to be responsible for the disappearance of *Pringlea antiscorbutica* and for the flourishing population of cats which were reintroduced in 1952.[33] The advent of myxomatosis in 1955 apparently had little effect.

Crozet Archipelago

Rabbits introduced to Île aux Cochons (in about 1850), to Île de la Possession and to Île de l'Est (where they are still established) are said on the first to reach a maximum population in summer of around 2,000. Mortality, caused chiefly by predation by cats in winter, is low.

Île St Paul

According to Gill, St Paul Island in the Indian Ocean is 'infested with both rats and hares'[34]: the latter are in fact rabbits, which were originally introduced some time after 1874.

Juan Fernandez

Rabbits were introduced with devastating effect to Alejandro Selkirk (Masafuera) and Robinson Crusoe (Masatierra) islands and to two accessible outlying islets, Santa Clara and Morro Vinilla, after the Juan Fernandez Archipelago off the coast of Chile was declared a national park in 1935. They have since disappeared from Masafuera but are still present on the other islands.[35]

Other oceanic islands

From time to time rabbits have been introduced to innumerable other oceanic (and to many continental) island groups. Flux and Fullagar (1983), who have records of their occurrence on some

600 individual islands from around 54°S to 62°N, are preparing an annotated list of occupied island sites.

In the 1960s the American **eastern cottontail rabbit** (*Sylvilagus floridanus*), which is not susceptible to myxomatosis, was introduced for sporting purposes to Italy (especially in the Po Valley) and subsequently to many places in southern France (especially near Mt Pellier) and in northern Spain. In the early 1980s **New England cottontails** (*S. transitionalis*), presumably because of their ability to survive in subalpine conditions, were released in parts of Germany. Recent research has shown that the imported cottontails, which could well prove harmful to agriculture, silviculture and viniculture, carry a flea (*Euhoplapsyllus glacialis*), which may affect the European rabbit and the brown hare (*Lepus capensis*). Since 1980–2 the Institut für Wildbiologie und Jagdkunde (Göttlangen), the Netherlands Research Institute for Nature Management, the Société française pour l'Étude et la Protection des Mammifères and the IUCN Lagomorph Specialist Group have all officially declared their opposition to introductions of *Sylvilagus*.

The **black-naped rabbit** (*Sylvilagus floridanus continentis*) is believed to have been introduced before 1894 from northeastern Venezuela to the islands of Curaçao and Aruba in the Lesser Antilles.[36]

NOTES

1. H. P. R. Finberg, *English Historical Review* **62**: 365(1947).
2. Exeter City Archives, Miscellaneous Deeds, D. 614.
3. Public Records Office, Exchequer, Foreign Rolls 8 Henry III, E.364.
4. E. B. Francis, *Transactions of the Essex Archaeological Society* **12**: 184(1913); M. A. C. Hinton, *Essex Naturalist* **17**: 16–21 (1912–13).
5. *The Antiquary* (August 1911).
6. J. A. Corbet, ed., *A booke of Glamorganshire's antiquities* (1825).
7. *Universa historia phisica Hungariae* (1793), Vol. 1, p. 214 (quoted by Szunyoghy 1958).
8. For information on rabbits in South Africa I am indebted to Brooke & Cooper (1980) and Cooper & Brooke (1982).
9. *Journal of Ecology* **51**: 275–94 (1963).
10. *Historical mammal incidence in the Cape Province, 1. The Western and Northern Cape* (Department of Nature and Environmental Conservation, Cape Town, 1980).
11. Couch (1929).
12. Carl & Guiget (1972).
13. See p. 78.
14. *Uttermost part of the Earth* (Hodder & Stoughton, London, 1948), pp. 100-1.
15. For much of the history of the rabbit in Australia I am indebted to Rolls (1969).
16. Despite Dr Crowther's statement, most authorities appear to have assumed that the importation was not of wild rabbits but of the domesticated Silver-Grey.
17. Since the comparatively recent extinction of rabbits on Rabbit Island many hitherto eroded areas have become restabilized and have been colonized by short-tailed shearwaters (*Puffinus tenuirostris*).
18. Quoted by Rolls (1969).
19. Quoted by Rolls (1969).
20. Mallee are low-growing eucalypts (especially *Eucalyptus dumosa* and *E. oleosa*), which give their name to the featureless, nearly waterless and densely overgrown country where they occur.
21. Quoted by Rolls (1969).
22. For much of the history of the rabbit in Western Australia I am indebted to Long (1972: 12–14).
23. For much of the history of the rabbit in New Zealand I am indebted to Wodzicki (1950).
24. *Jubilee history of South Canterbury* (1916).
25. R. M. Burdon, *High country: the evolution of a New Zealand sheep station* (Whitcombe & Tombs, Melbourne and London, 1938).
26. *Marlborough: a provincial history* (Marlborough Provincial Historical Committee, Blenheim, New Zealand, 1940).
27. *Tutira – the story of a New Zealand sheep station* (Blackwood, Edinburgh and London, 1921).
28. See p. 90.
29. G. W. Johnstone, 'Threats to birds on Subantarctic islands', paper presented at the 18th World Conference of the International Council for Bird Preservation, Cambridge, England, August 1982; see Moors (in press).
30. Quoted by Tomich (1969).
31. Quoted by Kramer (1971).
32. See p. 301.
33. See p. 89.
34. *Proceedings of the US National Museum* **123**: 1–33 (1967).
35. Dr W. R. P. Bourne, letter to author (1983).
36. E. Harteret, *Novitates Zoologicae* **1**: 40 (1894).

Appendix

Several exotic ungulates (see Table 25) have been acclimatized in fenced enclosures in the states of New Mexico and Texas, USA, where some may become established as free-ranging naturalized animals. The principal ranches in Texas on which exotic ungulates have been introduced include King, Bear Creek,[1] Y-O,[2] Powderhorn, Bar-O, Camp Cooley and Greenwood Valley. A survey carried out in 1979–80 found 72,147 exotic game animals of 51 different species in Texas (of which, however, 66,805 or 92.6 per cent were axis (*Cervus axis*),[3] fallow (*C. dama*)[4] and sika (*C. nippon*)[5] deer, mouflon × barbados and Barbary (*Ammotragus lervia*)[6] sheep, blackbuck (*Antilope cervicapra*)[7] and nilgai (*Boselaphus tragocamelus*)[8]), and revealed that increasing numbers of ranches are stocking exotics which are generally increasing their populations and distribution throughout the state, especially as free-ranging herds on the Edwards Plateau, where their impact on native species is currently being studied on the Kerr Wildlife Management Area near Hunt.

In July 1961 the New Mexico State Game Commission instructed the Department of Game and Fish to enquire into the feasibility of introducing further game animals (in addition to the Barbary sheep already established) into the arid regions of southeastern New Mexico. By November of the following year it was apparent that Siberian ibex, gemsbok or beisa oryx and greater kudu were among the species most likely to acclimatize successfully.

In the United States federal legislation[9] bans the direct introduction from abroad and release into the wild of exotic game species. Such animals may be imported by zoos and similar institutions only for holding as breeding stock; their offspring, however, may – subject to state regulations – be legally released.

The first importation (of four Siberian ibex from Mongolia, via Moscow, East and West Berlin, Hamburg and the US quarantine station at Clifton, New Jersey) was placed in specially constructed enclosures in the Rio Grande Zoo at Albuquerque in September 1962. They were followed in 1963 by a further pair, and by six female and two male gemsbok from the Kalahari Desert in Botswana as well as by eight greater kudu (two males and six females), supplemented by additional ibex and kudu in subsequent years.

Table 25 Exotic ungulates acclimatized in fenced enclosures in the states of New Mexico and Texas, USA

English name	*Scientific name*	*1980 Texas population*	*Natural distribution*
Cervidae			
Red deer*	*Cervus elaphus*	434	Western Europe, north-west Africa, western China
Barasingha (or swamp deer)†	*Cervus duvauceli*	79	India
Sambar†	*Cervus unicolor*	22	India to southern China and Java, Borneo, the Celebes, the Philippines, Sri Lanka
Indian muntjac	*Muntiacus muntjak*	10	India to southern China, Java, Borneo, Sri Lanka and Taiwan
Brocket (? sp.)	*Mazama* sp.	?	Central Mexico to northern South America
Bovidae			
Manx loghtan sheep	*Ovis* sp.	126	Isle of Man, England (probably introduced by the Vikings from Scandinavia *c.* AD 800)
Argali	*Ovis ammon*	104	Altai Mountains to the Himalayas
Hawaiian black sheep	*Ovis* sp.	?	Hawaii
Armenian (or Elburz) red sheep (or Asiatic mouflon)	*Ovis orientalis*	?	Iran to Asia Minor
Pygmy goat	*Capra* sp.	?	?
Bezoar (or Persian ibex)	*Capra aegagrus*	485 (Bezoar and Siberian ibex combined)	Pakistan to Asia Minor
Siberian ibex	*Capra ibex siberica*		Central Asia
Markhor	*Capra falconeri*	55	Western Himalayas and Afghanistan
Japanese serow	*Capricornis crispus*	?	Japan and Taiwan
Himalayan tahr‡	*Hemitragus jemlahicus*	?	Himalayas
Eland§	*Tragelaphus oryx*	150	Southern and eastern Africa
Sitatunga	*Tragelaphus spekei*	10	Africa south of the Sahara
Lesser kudu	*Tragelaphus imberbis*	8	Ethiopia to Tanzania
Greater kudu	*Tragelaphus strepsiceros*	12	Africa south of the Congo
Nyala (or mountain nyala)	*Tragelaphus angasi* (or *buxtoni*)	4	Natal to Malawi; or Ethiopia
Gemsbok (or beisa oryx¶)	*Oryx gazella*	118 (Gemsbok and Scimitar-horned oryx combined)	Eastern and southwestern Africa
Scimitar-horned oryx	*Oryx dammah*		Southern fringe of the Sahara
Mountain gazelle (idmi)	*Gazella gazella*	52	Arabia to Israel
Grant's gazelle	*Gazella granti*	1	Ethiopia to Tanzania
Thomson's gazelle	*Gazella thomsoni*	?	Sudan to Tanzania
Persian gazelle (or goitered gazelle)	*Gazella subgutturosa*	?	Arabia to Pakistan and Mongolia
Addax	*Addax nasomaculatus*	172	Sahara
Roan antelope	*Hippotragus equinus*	12	Africa south of the Sahara
Sable antelope	*Hippotragus niger*	28	Kenya to South Africa
Impala	*Aepyceros melampus*	16	Eastern and southern Africa

Continued

Table 25 *(continued)*

English name	*Scientific name*	*1980 Texas population*	*Natural distribution*
Bovidae *(cont.)*			
Blesbok (bontebok)	*Damaliscus dorcas*	11	Southern Africa
Topi (tssessebi)	*Damaliscus lunatus*	1	Africa south of the Sahara
Hartebeest (or Lichtenstein's hartebeest)	*Alcelaphus buselaphus* (or *lichtensteini*)	?	Eastern, western and south-west Africa; (or Angola to Tanzania)
Black wildebeest (or white-tailed gnu)	*Connochaetes gnou*	19 (both species)	Southern Africa
Blue wildebeest (or brindled gnu)	*Connochaetes taurinus*		Southern and eastern Africa
Water buffalo	*Bubalus bubalis*	1	India (ancestor of domestic form)
Banteng	*Bos javanicus*	3	Burma to Java and Borneo (ancestor of domestic Bali cattle)
Springbok	*Antidorcas marsupialis*	53	Southern Africa
Kob	*Kobus kob*	?	Upper Volta to Kenya
Lechwe	*Kobus leche*	1	Zambia
Waterbuck	*Kobus defassa*	24	Africa south of the Sahara
Suni	*Neotragus moschatus*	?	Kenya to Natal
Camelidae			
Dromedary (or Arabian camel)	*Camelus dromedarius*	23	? Arabic (known only in domestication)
Guanaco	*Lama guanicoe*	5	Peru to Pategonia (possible ancestor of domestic llama)
Llama	*Lama glama*	107	—
Vicugna	*Vicugna vicugna*	?	Peru to northern Chile
Equidae			
Common zebra (or Burchell's zebra)	*Equus burchelli*	76	Southern and eastern Africa
Giraffidae			
Giraffe	*Giraffa camelopardalis*	17	Africa south of the Sahara

Source: Adapted from Harmel (1980) and Mecke (1977a)

* First imported to the Bear Creek Ranch in Kerr County, Texas, in the late 1930s by Mr Richard Friedrich, and subsequently to the Camp Cooley Ranch, Robertson County, Texas, in 1958 by Mr Bert Wheeler. They occur principally on the Edwards Plateau and in the Rio Grande Plains. They interbreed freely with the native wapiti (*Cervus canadensis*)

† First introduced to the Bear Creek Ranch, Texas, in 1951 by Mr David Rickenbacker. Small herds of sambar have been established in the Gulf Prairies and Marshes and on the Edwards Plateau; they are also said to exist on the Hearst Ranch near San Simeon, California

‡ Tahr are believed to have spread from their original introduction to the Hearst Ranch in California into the Santa Lucia Mountains

§ First introduced in 1951 to the Bear Creek Ranch, Texas, by Mr David Rickenbacker. Eland have also been successfully bred in Florida

¶ Mr Bert Wheeler introduced some gemsbok or oryx to the Camp Cooley Ranch in Texas in 1960. It is hoped that the endangered Arabian oryx (*Oryx leucoryx*) may be saved from extinction by the small breeding nuclei maintained in the Phoenix and San Diego zoos in Arizona and California from where 10 were returned to the wild in Oman in 1982 (see p. 274). ('Gemsbok' = 'pointed-horned buck (Africaans); 'oryx' = 'pickaxe' (Greek))

In 1964 a 210-ha holding enclosure was constructed near Red Rock in southern Grant County near where the Gila River breaks out of its box canyon, some 700 km south of Santa Fe. This region of rugged, open, rolling desert-like plains, interspersed with dense thickets and bush intersected by watercourses at an altitude of 1,700 m, provides an approximately similar climate and terrain to those of the African savannah and parts of Central Asia.

Siberian ibex (offspring of the 1962 importation) were the first animals to be transferred to Red Rock, followed shortly by gemsbok and kudu. In April 1965 the Governor of the Province of Kermanshah in southern Iran presented one male and two female Persian ibex or bezoar to Mr Jack Campbell, then Governor of New Mexico. Two years later these were supplemented by an additional three females presented to the Department of Game and Fish by the Shikar-Safari Club. In 1966 and 1967 the two females imported in 1965 gave birth to seven kids (six females and one male) which were later transferred to Red Rock.

In April 1965 (at the time of the original importation of Persian ibex) a pair of Armenian (or Elburz) red sheep or Asiatic mouflon arrived from Iran; due to the possibility of interbreeding with the native bighorn or mountain sheep (*Ovis canadensis*) the plan of releasing red sheep (and kudu, which failed to acclimatize) into the wild was subsequently abandoned.

In May 1967 two male and three female Persian or goitered gazelles were imported by the Shikar-Safari Club on behalf of the Department of Game and Fish, followed by a shipment of markhor.

In 1969 offspring of the importation of gemsbok were released into the fenced enclosure of the White Sands Missile Range in the Tularosa Basin. Between 1970 and 1976 the New Mexico Bureau of Land Management and the Department of Game and Fish released a total of 77 Persian ibex in the Florida Mountains near the southwestern township of Deming. By 1979 the herd had increased to over 300, of which 121 were kids of the year; they currrently range over 8,500 ha of public land and an additional 120,000 ha of privately owned and state-controlled land. Recently a small number of Siberian ibex are said to have been released in the Canadian River gorge near the town of Roy in northwestern New Mexico.

As a result of public and state opposition the future release (if any) of exotic game animals in New Mexico is likely to be on privately owned rather than on public land.

According to the Silversteins (1974), Israel is currently trying to reintroduce to the country as many animals as possible which are referred to in the Bible. So far, gazelles, ibex, addax, Asiatic wild asses or onagers (*Equus hemionus*) (the ass on which Christ rode into Jerusalem), African asses (*E. africanus*), and the endangered Arabian oryx have been placed in no fewer than 160 reserves throughout the country.

NOTES

1. Sold to the Patio Foods Co. Inc. of San Antonio, Texas, this ranch – now known as the Patio-Rickenbacker Ranch – maintains some 17 species of exotic ungulates for shooting and/or selling as breeding stock.
2. The Y-O Ranch at Mountain Home claims to be the largest commercial game ranching enterprise in the United States, with some 10,000 exotic game animals of around 15 different species roaming over more than 325 km^2 of land.
3. See p. 151.
4. See p. 176.
5. See p. 213.
6. See p. 307.
7. See p. 273.
8. See p. 247.
9. 21 US C. 101–5; 111–35b; 612–14.

Table 26 Countries and major island groups with principal naturalized species*

Country	*Species*	*Date of earliest successful introduction*	*Introduced by*	*Reason*	*Natural range*
Europe					
Albania	Muskrat (*Ondatra zibethicus*)	? (Natural extension from Czechoslovakia)	—	—	North America
Austria	Raccoon-dog (*Nyctereutes procyonoides*)	1962 (natural extension from USSR)	—	—	Eastern Asia
	Fallow deer (*Cervus dama*)	1932	?	? Aesthetic and/or sport	Asia Minor and southern Europe
	Sika (*Cervus nippon*)	?	?	?	Southeastern Asia
	Mouflon (*Ovis musimon*)	Since before 1858	?	Sport	Originally parts of Asia
	Muskrat (*Ondatra zibethicus*)	1914–15 (natural extension from Czechoslovakia)	—	—	North America
Belgium	Fallow deer (*Cervus dama*)	1850	?	? Aesthetic and/or sport	Asia Minor and southern Europe
	? Mouflon (*Ovis musimon*)	Since mid-19th century	?	Sport	Originally parts of Asia
	Muskrat (*Ondatra zibethicus*)	1928	Fur-farmers	Fur	North America
	Coypu (*Myocastor coypus*)	? (? Natural extension from the Netherlands)	—	—	Much of South America
British Isles	Red-necked wallaby (*Macropus rufogriseus*)	1908	Sir Edmund Loder	Aesthetic	East and southeastern Australia and Tasmania
		1936	Capt. Henry Brocklehurst	Aesthetic	
	Domestic cat (*Felis catus*)	Before Middle Ages	?	Domestic	Originally North Africa
	Polecat-ferret (*Mustela putorius*)	Late 13th century (ferret)	? Plantagenets	Rabbit-hunting	Europe
	North American mink (*Mustela vison*)	1929	Fur-farmers	Fur	Most of North America
	Chinese water deer (*Hydropotes inermis*)	*c.*1900	Duke of Bedford	Aesthetic	Central and northeastern China and Korea
	Chinese muntjac (*Muntiacus reevesi*)	*c.*1900	Duke of Bedford	Aesthetic	Southern and eastern China and Taiwan
	Fallow deer (*Cervus dama*)	? After AD 43	? Romans	—	Asia Minor and southern Europe

* The rabbit is naturalized throughout most of western Europe, including some Mediterranean islands. The black rat, brown rat and house mouse are naturalized over much of the world. Such domesticates as the cat and dog have been carried by humans nearly everywhere that they have travelled and have become, to a greater and lesser extent respectively, widely feral.

Continued

Table 26 *(continued)*

Country	*Species*	*Date of earliest successful introduction*	*Introduced by*	*Reason*	*Natural range*
Europe					
British Isles (*cont.*)	Sika (*Cervus nippon*)	1860 (in captivity)	Zoological Society of London and Viscount Powerscourt, Ireland	—	Southeastern Asia
	Domestic goat (*Capra hircus*)	Late Stone Age/Early Bronze Age	Neolithic/Early Bronze Age man	Food	Eastern Mediterranean and Middle East
	Soay sheep (*Ovis musimon*)	? Prehistoric (or *c.* AD 800)	(Vikings)	Food and wool	Scandinavia
	Grey squirrel (*Sciurus carolinensis*)	1876 (first recorded 1828)	Mr T. V. Brocklehurst	Curiosity	Eastern North America
	Brown rat (*Rattus norvegicus*)	1728–9	Russian ships	Accidental	Originally southeastern Asia
	Black rat (*Rattus rattus*)	4th or 5th century AD	Romans	Accidental	Originally southeastern Asia
	House mouse (*Mus musculus*)	*c.*1000 BC	Iron Age Man	Accidental	Originally southern Palaearctic
	Edible dormouse (*Glis glis*)	1902	Lord Rothschild	Curiosity	Much of southern and eastern Europe
	Coypu (*Myocastor coypus*)	1929	Fur-farmers	Fur	Much of South America
	Brown hare (*Lepus capensis*)	*c.*1852 (Ireland) (native England, Scotland, Wales)	Lord Shannon	Sport	Much of Europe, Asia and Africa
	Rabbit (*Oryctolagus cuniculus*)	Before 1176 (Scilly Isles)	? Plantagenets	Food and pelt	Originally northwestern Africa and Iberian peninsula
Bulgaria	Raccoon-dog (*Nyctereutes procyonoides*)	1967 (natural extension from USSR)	—	—	Eastern Asia
	Fallow deer (*Cervus dama*)	1904	?	? Aesthetic and/or Sport	Asia Minor and southern Europe
	Muskrat (*Ondatra zibethicus*)	? (Natural extension from Czechoslovakia)	—	—	North America
Czechoslovakia	Raccoon-dog (*Nyctereutes procyonoides*)	1959 (natural extension from USSR)	—	—	Eastern Asia
	Fallow deer (*Cervus dama*)	1465 (in captivity) 19th century (in the wild)	?	Sport	Asia Minor and southern Europe
	Sika (*Cervus nippon*)	?	?	Sport	Southeastern Asia

Continued

Table 26 *(continued)*

Country	*Species*	*Date of earliest successful introduction*	*Introduced by*	*Reason*	*Natural range*
Europe					
Czechoslovakia (*cont.*)	White-tailed deer (*Odocoileus virginianus*)	After 1890	Prince Colleredo-Mannsfeld	Aesthetic and/or sport	North and central America
	Mouflon (*Ovis musimon*)	1858 (? much earlier)	?	Sport	Originally parts of Asia
	Muskrat (*Ondatra zibethicus*)	1905	Prince Colleredo-Mannsfeld	Fur	North America
Denmark	North American mink (*Mustela vison*)	? 1940s	Fur-farmers	Fur	Most of North America
	Fallow deer (*Cervus dama*)	13th century (reintroduction)	(Reign of) King Valdemar II	? Sport and/or Food	Asia Minor and southern Europe
	Sika (*Cervus nippon*)	1900	?	Aesthetic	Southeastern Asia
	Brown rat (*Rattus norvegicus*)	1716 (mainland)	Russian ships from Asia	Accidental	Originally southeastern Asia
		1768 (Faeroe Islands)	Wreckage of *King of Prussia*	Accidental	
	Coypu (*Myocastor coypus*)	? (? Extension of range)	—	—	Much of South America
Finland	Raccoon-dog (*Nyctereutes procyonoides*)	1935 or early 1950s (natural extension from USSR)	—	—	Eastern Asia
	North American mink (*Mustela vison*)	Late 1920s	Fur-farmers	Fur	Most of North America
	Fallow deer (*Cervus dama*)	1938	?	Aesthetic and/or Sport	Asia Minor and southern Europe
	White-tailed deer (*Odocoileus virginianus*)	1934	?	Sport	North and central America
	North American beaver (*Castor canadensis*)	1937	?	? Pelt	Originally much of North America
	Muskrat (*Ondatra zibethicus*)	1919 or 1922	?	Fur	North America
	Brown hare (*Lepus capensis*)	?	?	Sport	Much of Europe, Asia and Africa
France	Raccoon (*Procyon lotor*)	After 1934 (natural extension from West Germany)	—	—	Most of North and Central America
	North American mink (*Mustela vison*)	?	Fur-farmers	Fur	Most of North America

Continued

Table 26 *(continued)*

Country	*Species*	*Date of earliest successful introduction*	*Introduced by*	*Reason*	*Natural range*
Europe					
France (*cont.*)	Domestic horse (*Equus caballus*)	?	?	?	Originally eastern Europe to Mongolia
	? Chinese water deer (*Hydropotes inermis*)	*c.*1960–5	?	Sport	Central and northeastern China and Korea
	Fallow deer (*Cervus dama*)	16th century or earlier	?	? Food and/or sport	Asia Minor and southern Europe
	Sika (*Cervus nippon*)	1890	President François Sadi via the Emperor of Japan	Aesthetic	Southeastern Asia
	Mouflon (*Ovis musimon*)	Since mid-19th century (?Including –much earlier–Island of Corsica)	?	Sport	Originally parts of Asia
	Domestic cattle (*Bos taurus*)	?	?	? Food	Originally Europe and western Asia
	Muskrat (*Ondatra zibethicus*)	*c.*1919 (or 1927–8)	(Madame L. Pereyrol)	Fur	North America
	Brown rat (*Rattus norvegicus*)	1750	?	Accidental	Originally southeastern Asia
	Coypu (*Myocastor coypus*)	(Originally 1882) 1925–8	(Pays-Mellier) Fur-farmers	Fur	Much of South America
Germany (East and West)	Raccoon-dog (*Nyctereutes procyonoides*)	1962 (natural extension from USSR)	—	—	Eastern Asia
	Raccoon (*Procyon lotor*)	1934	?	? Fur .	Most of North and Central America
	Fallow Deer (*Cervus dama*)	16th century (in captivity); 1713 and 1883 (in the wild)	?	? Aesthetic and/or sport	Asia Minor and southern Europe
	Sika (*Cervus nippon*)	?	?	?	Southeastern Asia
	Mouflon (*Ovis musimon*)	Since mid-19th century	?	Sport	Originally parts of Asia
	Muskrat (*Ondatra zibethicus*)	1914 (natural extension from Czechoslovakia)	—	—	North America
	Brown Rat (*Rattus norvegicus*)	1750 (eastern Prussia) *c.* 1780 (Brunswick)	?	Accidental	Originally northeastern Asia
	Coypu (*Myocastor coypus*)	Late 1920s	Fur-farmers	Fur	Much of South America
	Rabbit (*Oryctolagus cuniculus*)	Middle Ages	?	Food and pelt	Originally northwestern Africa and Iberian peninsula

Continued

Table 26 (*continued*)

Country	Species	Date of earliest successful introduction	Introduced by	Reason	Natural range
Europe					
Gibraltar	Barbary Ape (*Macaca sylvanus*)	Before 1740	? Moors	?	Morocco and northern Algeria
Greece	Domestic horse (*Equus caballus*)	? (Island of Skyros)	?	?	Originally eastern Europe to Mongolia
	Fallow deer (*Cervus dama*)	Island of Rhodes, after 1309 (exterminated by Turks 1522–1912)	Knights Hospitallers of St John of Jerusalem	(Traditionally) to eradicate snakes	Asia Minor and southern Europe
		Reintroduced 1912–45	Italians	? Aesthetic	
	Mouflon (*Ovis musimon*)	? (Island of Cyprus)	?	Sport	Originally parts of Asia
	Domestic goat (*Capra hircus*)	? (Islands in the Aegean, Cyprus, etc.)	?	Food	Eastern Mediterranean and Middle East
	Muskrat (*Ondatra zibethicus*)	? (Natural extension from Czechoslovakia)	—	—	North America
Greenland	Brown rat (*Rattus norvegicus*)	*c.*1780	?	Accidental	Originally southeastern Asia
	Reindeer (*Rangifer tarandus*) (to caribou range)	1952	?	? Food, hide and draught	Northern Holarctic
Hungary	Raccoon-dog (*Nyctereutes procyonoides*)	1961 (natural extension from USSR)	—	—	Eastern Asia
	Fallow deer (*Cervus dama*)	Late 19th century	?	? Aesthetic and/or sport	Asia Minor and southern Europe
	Mouflon (*Ovis musimon*)	Since mid-19th century	?	Sport	Originally parts of Asia
	Muskrat (*Ondatra zibethicus*)	1924 (natural extension from Czechoslovakia)	—	—	North America
	Rabbit (*Oryctolagus cuniculus*)	Before 1779	?	Food and pelt	Originally northwestern Africa and Iberian peninsula
Iceland	North American mink (*Mustela vison*)	1930	Fur-farmers	Fur	Most of North America
	Reindeer (*Rangifer tarandus*)	1771–87	?	Draught, food and/or hide	Northern Holarctic

Continued

Table 26 *(continued)*

Country	*Species*	*Date of earliest successful introduction*	*Introduced by*	*Reason*	*Natural range*
Europe					
Italy	Fallow deer (*Cervus dama*)	? (On Sardinia – where now extinct – before 1774; near Trieste – by natural extension from Yugoslavia – since 1971)	?	? Aesthetic and/or sport	Asia Minor and southern Europe
	Domestic goat (*Capra hircus*)	? (Island of Monte Cristo)	?	Food	Eastern Mediterranean and Middle East
	Mouflon (*Ovis musimon*)	Since mid-19th century. (?Including – much earlier – Island of Sardinia)	?	Sport	Originally parts of Asia
	Muskrat (*Ondatra zibethicus*)	? (Natural extension from Czechoslovakia)	—	—	North America
	Brown rat (*Rattus norvegicus*)	Mid–late 18th century	?	Accidental	Originally southeastern Asia
	Coypu (*Myocastor coypus*)	1928	Fur-farmers	Fur	Much of South America
Luxembourg	Muskrat (*Ondatra zibethicus*)	1955 (? natural extension from France)	—	—	North America
Netherlands	Raccoon (*Procyon lotor*)	After 1934 (natural extension from West Germany)	—	—	Most of North and Central America
	Stoat (*Mustela erminea*)	1931 (to Terschelling Island)	Foresters	To kill water voles	Eurasia and Nearctic Region
	Fallow deer (*Cervus dama*)	Early 17th century (? in captivity earlier)	Prince Maurice of Nassau	?	Asia Minor and southern Europe
	Muskrat (*Ondatra zibethicus*)	1941 (allegedly smuggled over the border from Belgium)	Fur-farmers	Fur	North America
	Coypu (*Myocastor coypus*)	1949 (in the wild)	Fur-farmers	Fur	Much of South America
	Rabbit (*Oryctolagus cuniculus*)	Middle Ages	?	Food and pelt	Originally northwestern Africa and Iberian peninsula
Norway	North American mink (*Mustela vison*)	Late 1920s	Fur-farmers	Fur	Most of North America

Continued

Table 26 *(continued)*

Country	*Species*	*Date of earliest successful introduction*	*Introduced by*	*Reason*	*Natural range*
Europe					
Norway *(cont.)*	Wild boar (*Sus scrofa*)?	?	?	?	Europe, north Africa and Asia
	Fallow deer (*Cervus dama*)	? Early 12th century; 1645 (in captivity); before 1700 (in the wild)	?	? Hides and/or food	Asia Minor and southern Europe
	Domestic sheep (*Ovis aries*)	?	?	Food and/or wool	Originally Asia Minor and Iran
	Muskrat (*Ondatra zibethicus*)	First recorded 1970 (natural extension from Finland)	—	—	North America
	Brown rat (*Rattus norvegicus*)	1762	?	Accidental	Originally southeastern Asia
Poland	Raccoon-dog (*Nyctereutes procyonoides*)	1955 (natural extension from USSR)	—	—	Eastern Asia
	Fallow deer (*Cervus dama*)	17th century	?	? Aesthetic and/or sport	Asia Minor and southern Europe
	Sika (*Cervus nippon*)	?	?	?	Southeastern Asia
	Mouflon (*Ovis musimon*)	Since mid-19th century	?	Sport	Originally parts of Asia
	Muskrat (*Ondatra zibethicus*)	1929–30 (natural extension from Czechoslovakia)	—	—	North America
Portugal	Domestic horse (*Equus caballus*)	?	?	?	Originally eastern Europe to Mongolia
	? Fallow deer (*Cervus dama*)	? (Natural extension from Spain)	—	—	Asia Minor and southern Europe
Romania	Raccoon-dog (*Nyctereutes procyonoides*)	1951 (natural extension from USSR)	—	—	Eastern Asia
	Fallow deer (*Cervus dama*)	?	?	?	Asia Minor and southern Europe
	Mouflon (*Ovis musimon*)	Since mid-19th century	?	Sport	Originally parts of Asia
	Muskrat (*Ondatra zibethicus*)	? (Natural extension from Czechoslovakia)	—	—	North America
Spain	Domestic horse (*Equus caballus*)	?	?	?	Originally eastern Europe to Mongolia
	Fallow deer (*Cervus dama*)	Before 1893	? Royal estates	? Aesthetic and/or sport	Asia Minor and southern Europe

Continued

Table 26 *(continued)*

Country	*Species*	*Date of earliest successful introduction*	*Introduced by*	*Reason*	*Natural range*
Europe					
Spain *(cont.)*	Red deer (*Cervus elaphus*)	1954 (island of Fernando Poo, Gulf of Guinea)	?	?	Palaearctic region
	Domestic cattle (*Bos taurus*)	?	?	? Food	Originally Europe and western Asia
	Mouflon (*Ovis musimon*)	Since mid-19th century (mainland and Tenerife, Canary Islands)	?	Sport	Originally parts of Asia
	Brown rat (*Rattus norvegicus*)	Mid-late 18th century	?	Accidental	Originally southeastern Asia
Sweden	Raccoon-dog (*Nyctereutes procyonoides*)	1945 (natural extension from USSR)	—	—	Eastern Asia
	North American mink (*Mustela vison*)	1925 or 1928	Fur-farmers	Fur	Most of North America
	Wild boar (*Sus scrofa*)	?	?	?	Europe, north Africa and Asia
	Fallow deer (*Cervus dama*)	Before 1579	?	? Sport and/or food	Asia Minor and southern Europe
	Muskrat (*Ondatra zibethicus*)	Early 1940s	Fur-farmers	Fur	North America
	Brown rat (*Rattus norvegicus*)	1790	?	Accidental	Originally southeastern Asia
	Brown hare (*Lepus capensis*)	?	?	Sport	Much of Europe, Asia and Africa
Switzerland	Raccoon-dog (*Nyctereutes procyonoides*)	After 1977 (natural extension from USSR)	—	—	Eastern Asia
	Mouflon (*Ovis musimon*)	Since mid-19th century	?	Sport	Originally parts of Asia
	Muskrat (*Ondatra zibethicus*)	? (Natural extension from France, augmented by domestic escapes)	—	—	North America
	Brown rat (*Rattus norvegicus*)	*c.*1809	?	Accidental	Originally southeastern Asia
Yugoslavia	Axis deer (*Cervus axis*)	?	?	?	India, Nepal and Sri Lanka
	Fallow deer (*Cervus dama*)	?	?	? Sport	Asia Minor and southern Europe
	Domestic goat (*Capra hircus*)	? (Islands off Dalmation coast)	?	Food	Eastern Mediterranean and Middle East

Continued

Table 26 *(continued)*

Country	*Species*	*Date of earliest successful introduction*	*Introduced by*	*Reason*	*Natural range*
Europe					
Yugoslavia (*cont.*)	Mouflon (*Ovis musimon*)	Since mid-19th century	?	Sport	Originally parts of Asia
	Muskrat (*Ondatra zibethicus*)	? (Natural extension from Czechoslovakia)	—	—	North America
Asia					
China	Muskrat (*Ondatra zibethicus*)	1954 (natural extension from Mongolia)	—	—	North America
Iran	Domestic horse (*Equus caballus*)	? (Caspian Sea)	?	?	Originally eastern Europe to Mongolia
Israel	Coypu (*Myocastor coypus*)	*c.*1953	Fur-farmers	Fur	Much of South America
Japan	Domestic cattle (*Bos taurus*)	Before 1880	?	? Food	Originally Europe and western Asia
	Domestic goat (*Capra hircus*)	? 1853 (Bonin Islands)	? Admiral Perry (USA)	Food	Eastern Mediterranean and Middle East
	Muskrat (*Ondatra zibethicus*)	? Late 1920s or before 1945	Fur-farmers	Fur	North America
	Coypu (*Myocastor coypus*)	1931	Prince Takamatsu	Fur	Much of South America
Korea	Muskrat (*Ondatra zibethicus*)	? (Natural extension from China)	—	—	North America
Mongolia	Raccoon-dog (*Nyctereutes procyonoides*)	Early 1960s	?	Fur	Eastern Asia
	Muskrat (*Ondatra zibethicus*)	Late 1940s/early 1950s (natural extension from USSR)	—	—	North America
		Released 1967	?		
Saudi Arabia	Domestic ass (*Equus asinus*)	?	?	? Draught	Northeastern Africa
Sri Lanka	Domestic ass (*Equus asinus*)	?	?	? Draught	Northeastern Africa
	Hog deer (*Cervus porcinus*)	16th century (or 17th or early 18th century)	Portuguese settlers (or Dutch settlers)	Sport	Northwestern India to Indochina
Turkey	Coypu (*Myocastor coypus*)	1973 (natural extension from USSR)	—	—	Much of South America

Continued

Table 26 *(continued)*

Country	*Species*	*Date of earliest successful introduction*	*Introduced by*	*Reason*	*Natural range*
Asia					
Soviet Union (including European Russia)	Raccoon-dog (*Nyctereutes procyonoides*)	1928 or 1929	State fur-farmers	Fur	Eastern Asia
	Raccoon (*Procyon lotor*)	1936	State fur-farmers	Fur	Most of North and Central America
	North American mink (*Mustela vison*)	1928	State fur-farmers	Fur	Most of North America
	Axis deer (*Cervus axis*)	1939	State	Antlers	India, Nepal and Sri Lanka
	Fallow deer (*Cervus dama*)	16th century	?	? Sport	Asia Minor and southern Europe
	Sika (*Cervus nippon*)	?	?	?	Southeastern Asia
	Mouflon (*Ovis musimon*)	1913–14 (Crimea)	?	Sport	Originally parts of Asia
	North American beaver (*Castor canadensis*)	Europe: 1933–4 (died out by 1957); early 1950s (or *c.*1960) by natural extension from Finland. Asia: 1969 and 1971; 1975–6 and later	?	? Pelt	Originally much of North America
	Muskrat (*Ondatra zibethicus*)	1928	Kamchatka Joint-Stock Company	Fur	North America
	Brown rat (*Rattus norvegicus*)	*c.*1727 (natural extension of range)	—	—	Originally southeastern Asia
	House mouse (*Mus musculus*)	1870 (Bering Island, Asia)	Schooner *Justus*	Accidental	Originally southern Palaearctic
	Coypu (*Myocastor coypus*)	1930	State fur-farmers	Fur	Much of South America
	Rabbit (*Oryctolagus cuniculus*)	? (Ukraine and islands in Caspian Sea)	?	Food and pelt	Originally northwestern Africa and Iberian peninsula
Yemen	Domestic ass (*Equus asinus*)	? (Island of Socotra)	? Ancient Egyptians	? Draught	Northeastern Africa
Africa					
Algeria	Domestic ass (*Equus asinus*)	?	?	? Draught	Northeastern Africa
Chad	Domestic ass (*Equus asinus*)	?	?	? Draught	Northeastern Africa

Continued

Table 26 *(continued)*

Country	*Species*	*Date of earliest successful introduction*	*Introduced by*	*Reason*	*Natural range*
Africa					
Kenya	Coypu (*Myocastor coypus*)	*c.*1940	Mr R. Hook	Fur	Much of South America
Morocco	Red deer (*Cervus elaphus*)	1952	?	?	Palaearctic region
	Brown rat (*Rattus norvegicus*)	?1930	? Via Gibraltar	Accidental	Originally southeastern Asia
South Africa	Domestic cat (*Felis catus*)	Second half 19th century	?	?	Originally North Africa
	Domestic pig (*Sus domestica*)	1925	Foresters	To control moth larvae	Europe, north Africa and Asia
	Wild boar (*Sus scrofa*)	1930s	Department of Forestry	To control moth larvae	Europe, north Africa and Asia
	Fallow deer (*Cervus dama*)	1869 (in captivity)	?	Aesthetic	Asia Minor and southern Europe
	? Red deer (*Cervus elaphus*)	1870 (in captivity)	Mr Ohlsson	Aesthetic	Palaearctic region
	Himalayan tahr (*Hemitragus jemlahicus*)	1935 (escaped from zoo 1937)	Dr R. Bigalke	? Aesthetic	Himalayan Mountains
	Grey squirrel (*Sciurus carolinensis*)	Shortly after 1900 (or in the 1890s)	Mr Cecil Rhodes, Prime Minister	Curiosity	Eastern North America
	Brown rat (*Rattus norvegicus*)	?	?	Accidental	Originally southeastern Asia
	Black rat (*Rattus rattus*)	?	?	Accidental	Originally southeastern Asia
	Rabbit (*Oryctolagus cuniculus*)	1656–8 (Robben Island)	Governor Jan van Riebeeck	Food and pelt	Originally northwestern Africa and Iberian peninsula
Sudan	Domestic ass (*Equus asinus*)	?	?	? Draught	Northeastern Africa
Tanzania	?Coypu (*Myocastor coypus*)	'From 1960'	? Fur-farmers	? Fur	Much of South America
	Small Indian mongoose (*Herpestes auropunctatus*)	(Mafia Island)	?	Rat control	Northern Arabia to Malaysia
Tunisia	Water buffalo (*Bubalus bubalis*)	*c.*1957 (originally since Roman period)	?	Sport	India
North and Central America					
Bermuda	Black rat (*Rattus rattus*)	Before 1616	Early settlers	Accidental	Originally southeastern Asia

Continued

Table 26 *(continued)*

Country	*Species*	*Date of earliest successful introduction*	*Introduced by*	*Reason*	*Natural range*
North and Central America					
Bermuda *(cont.)*	Brown rat (*Rattus norvegicus*)	?	?	Accidental	Originally southeastern Asia
Canada	Domestic cat (*Felis catus*)	?	? Early settlers	Domestic	Originally North Africa
	Domestic horse (*Equus caballus*)	1610	Poutrincourt	Riding	Originally eastern Europe to Mongolia
	Fallow deer (*Cervus dama*)	1895	Mr Irving H. Wheatcroft	? Aesthetic and/or sport	Asia Minor and southern Europe
	Reindeer (*Rangifer tarandus*) (to caribou range)	1935	Mr Dan Crowley and Mr Andrew Bahr	Food and hides	Northern Holarctic
	Domestic cattle (*Bos taurus*)	?	?	Food	Originally Europe and Western Asia
	Domestic goat (*Capra hircus*)	First in 1541	?	Food	Eastern Mediterranean
	Brown rat (*Rattus norvegicus*)	Before 1825	?	Accidental	Originally southeastern Asia
	Black rat (*Rattus rattus*)	Mid-19th century (natural extension of range from USA)	—	—	Originally southeastern Asia
	House mouse (*Mus musculus*)	By 1829	? Hudson's Bay Co. from England	Accidental	Originally southern Palaearctic
	? Coypu (*Myocastor coypus*)	*c.*1938	Fur-farmers	Fur	Much of South America
	Brown hare (*Lepus capensis*)	1912	Mr Otto Herold, Bow Park Farm, Ontario	? Pelt	Much of Europe, Asia and Africa
	Rabbit (*Oryctolagus cuniculus*)	Before 1954 (islands off Vancouver)	(From San Juan islands, Washington, USA)	?	Originally northwestern Africa and Iberian peninsula
Mexico	Domestic cat (*Felis catus*)	?	?	?	Originally North Africa
	Domestic horse (*Equus caballus*)	1519	Hernando Cortes	Riding	Eastern Europe to Mongolia
	Nilgai (*Boselaphus tragocamelus*)	?	Ranch near Nuevo Laredo	? Sport	India
	Domestic goat (*Capra hircus*)	Before *c.*1830 (Island of Guadalupe)	?	Fleece	Eastern Mediterranean and Middle East

Continued

Table 26 *(continued)*

Country	Species	Date of earliest successful introduction	Introduced by	Reason	Natural range
North and Central America					
Mexico (*cont.*)	Barbary sheep (*Ammotragus lervia*)	?	Señor Octaviano Longorias	Food	Originally much of North Africa
	Brown rat (*Rattus norvegicus*)	? (Natural extension from USA)	—	—	Originally southeastern Asia
	House mouse (*Mus musculus*)	By 1855 (natural extension from USA)	—	—	Originally southern Palaearctic
United States	Nine-banded armadillo (*Dasypus novemcinctus*)	*c.*1924	Cocoa Zoo	?	Central South America to the southeastern USA
	Crab-eating macaque (*Macaca fascicularis*)	?	?	Tourism and research	Southeastern Asia
	Rhesus macaque (*Macaca mulatta*)	1933	?	?	Eastern Afghanistan to southern China
	Domestic cat (*Felis catus*)	?	?	?	Originally North Africa
	Domestic ass (*Equus asinus*)	16th century	Spanish conquistadors	Draught	Northeastern Africa
	Domestic horse (*Equus caballus*)	1528–36	Pánfilo de Narváez or Cabeza de Vaca	Riding	Originally eastern Europe to Mongolia
	Domestic pig (*Sus domestica*)	1493	Christopher Columbus	Food	Europe, North Africa and Asia
	Wild boar (*Sus scrofa*)	1912	Mr George Moore	Sport	Europe, North Africa and Asia
	Axis deer (*Cervus axis*)	1930s	King and Bear Creek Ranches (Mr Richard Friedrich)	Sport	India, Nepal and Sri Lanka
	Fallow deer (*Cervus dama*)	1878 (most of present stock are home-bred)	?	? Aesthetic and/or sport	Asia Minor and southern Europe
	Red deer (*Cervus elaphus*)	Before 1950	Woodlands National Wildlife Refuge	Aesthetic	Palaearctic region
	Sika (*Cervus nippon*)	1930s and early 1940s	Mr Richard Friedrich, Bear Creek Ranch	Sport	Southeastern Asia
	Sambar (*Cervus unicolor*)	Before 1900	?	Sport	India and Sri Lanka to southern China
	Reindeer (*Rangifer tarandus*) (to caribou range)	1891–2	US Bureau of Education	Food and hide	Northern Holarctic

Continued

Table 26 *(continued)*

Country	*Species*	*Date of earliest successful introduction*	*Introduced by*	*Reason*	*Natural range*
North and Central America					
United States (*cont.*)	Nilgai (*Boselaphus tragocamelus*)	1924 (or 1929–30)	King Ranch Inc.	Sport	India
	Domestic cattle (*Bos taurus*)	?	?	Food	Originally Europe and Western Asia
	Blackbuck (*Antelope cervicapra*)	1932 (and 1937)	(Mr Richard Friedrich, Bear Creek Ranch)	Sport	India
	Domestic goat (*Capra hircus*)	?	?	Food	Eastern Mediterranean and Middle East
	Barbary sheep (*Ammotragus lervia*)	*c.*1900 (and 1924)	(Mr William Randolph Hearst)	Sport	Originally much of North Africa
	Mouflon (*Ovis musimon*)	Late 1930s and early 1940s	King Ranch, Texas	Sport	Originally parts of Asia
	Domestic sheep (*Ovis aries*)	?	?	Food and/or wool	Originally Asia Minor and Iran
	Brown rat (*Rattus norvegicus*)	*c.*1775	English ships	Accidental	Originally southeastern Asia
	Black rat (*Rattus rattus*)	By 1609 (possibly considerably earlier)	John Smith	Accidental	Originally southeastern Asia
	House mouse (*Mus musculus*)	1819–20	Stephen Harriman Long's expedition	Accidental	Originally southern Palaearctic
	Coypu (*Myocastor coypus*)	1937	Fur-farmers	Fur	Much of South America
	Brown hare (*Lepus capensis*)	1886 or 1888	?	Sport	Much of Europe, Asia and Africa
	Rabbit (*Oryctolagus cuniculus*)	Before 1895 (San Juan Islands, Washington)	?	?	Originally northwestern Africa and Iberian peninsula
West Indies	Rhesus macaque (*Macaca mulatta*)	? (Puerto Rico)	Caribbean Primate Research Center	Research	Eastern Afghanistan to southern China
	Green monkey (*Cercopithecus aethiops*)	Second half 17th century (St Kitts and Nevis)	The French	?	Senegal and Somalia to South Africa
		Before 1750 (Barbados)	?	?	
	Raccoon (*Procyon lotor*)	Before 1750 (Barbados); (? Guadeloupe)	?	?	Most of North and Central America

Continued

Table 26 *(continued)*

Country	*Species*	*Date of earliest successful introduction*	*Introduced by*	*Reason*	*Natural range*
North and Central America					
West Indies (*cont.*)	Raccoon (*Procyon lotor*)	1932 or 1933 (Grand Bahama)	Mr Jack Morris	?	
	Domestic dog (*Canis familiaris*)	? (Cayman Brac, Little Cayman, Great Inagua, Cuba, etc.)	?	?	Originally Holarctic
	Small Indian mongoose (*Herpestes auropunctatus*)	1872 (Jamaica; subsequently to many other islands)	Mr W. Bancroft Espeut	Rat control	Northern Arabia to Malaysia
	Domestic cat (*Felis catus*)	Various islands	?	Domestic	Originally North Africa
	Domestic horse (*Equus caballus*)	1493 (Hispaniola)	Christopher Columbus	Riding	Originally eastern Europe to Mongolia
	Domestic pig (*Sus domestica*)	1492 (Hispaniola)	Christopher Columbus	Food	Europe, North Africa and Asia
		? (Puerto Rico, Mona Island and Virgin Islands)	?	? Food	
	Fallow deer (*Cervus dama*)	Early 18th century (Barbuda and Guana)	The Codrington family, England	? Aesthetic and/or food	Asia Minor and southern Europe
	? White-tailed deer (*Odocoileus virginianus*)	Mid-19th century (Cuba)	?	? Sport	North and Central America
		? (Curaçao)	?	? Sport	
		1790 and 1945 (St Croix and St Thomas)	?	? Sport	
	Domestic cattle (*Bos taurus*)	Shortly after 1492 (Hispaniola)	Spanish settlers	Food (and ? draught)	Originally Europe and western Asia
	Domestic goat (*Capra hircus*)	Shortly after 1492 (Hispaniola)	Spanish settlers	Food	Eastern Mediterranean and Middle East
	Black rat (*Rattus rattus*)	Before 1654 (St Christopher, Guadeloupe, Martinique, etc.); before 1750 (Barbados); before 1789 (Jamaica)	Early settlers	Accidental	Originally southeastern Asia
	House mouse (*Mus musculus*)	Before 1654 (St Christopher, Guadeloupe, Martinique, etc.)	Early settlers	Accidental	Originally southern Palaearctic

Continued

Table 26 *(continued)*

Country	*Species*	*Date of earliest successful introduction*	*Introduced by*	*Reason*	*Natural range*
North and Central America					
West Indies (*cont.*)	Brown hare (*Lepus capensis*)	Before 1654 (Guadeloupe and Balliceaux); 1842 (Barbados)	Mr Thomas Trotman (Barbados)	? Sport	Much of Europe, Asia and Africa
	Rabbit (*Oryctolagus cuniculus*)	Before 1654 (Balliceaux and Guadeloupe); 1842 (Barbados)	Mr Thomas Trotman	Sport	Originally northwestern Africa and Iberian peninsula
South America					
Argentina	Domestic pig (*Sus domestica*)	? *c.*1750	?	? Food	Europe, North Africa and Asia
	Axis deer (*Cervus axis*)	1906	Señor Pedro Luro	? Aesthetic and/or sport	India, Nepal and Sri Lanka
	Fallow deer (*Cervus dama*)	Shortly after 1900 (and 1905)	? (Señor Ernesto Tornquist)	? Aesthetic and/or sport	Asia Minor and southern Europe
	Red deer (*Cervus elaphus*)	1902 (and 1906)	(Señor Pedro Luro)	? Aesthetic and/or sport	Palaearctic region
	? Reindeer (*Rangifer tarandus*)	1947–8	?	? Food and/or draught	Northern Holarctic
	? Water buffalo (*Bubalus bubalis*)	Before 1900	?	Sport	India
	Domestic cattle (*Bos taurus*)	?	?	Food	Originally Europe and western Asia
	Blackbuck (*Antelope cervicapra*)	1906	Señor Pedro Luro	Aesthetic and/or sport	India
	Domestic goat (*Capra hircus*)	?	?	?	Eastern Mediterranean and Middle East
	? North American beaver (*Castor canadensis*)	1965	?	? Pelt	Originally much of North America
	Muskrat (*Ondatra zibethicus*)	1940–56	?	Fur	North America
	Brown rat (*Rattus norvegicus*)	?	?	Accidental	Originally southeastern Asia
	House mouse (*Mus musculus*)	?	?	Accidental	Originally southeastern Asia
	Brown hare (*Lepus capensis*)	1888	Woeltje Tietjen, German Consul in Rosario	Sport	Much of Europe, Asia and Africa
Brazil	Rhesus macaque (*Macaca mulatta*)	1940s	?	Medical research	Eastern Afghanistan to southern China

Continued

Table 26 *(continued)*

Country	*Species*	*Date of earliest successful introduction*	*Introduced by*	*Reason*	*Natural range*
South America					
Brazil *(cont.)*	Wild boar (*Sus scrofa*)	?	?	? Sport	Europe, North Africa and Asia
	Axis deer (*Cervus axis*)	? *c.* 1906 (or ?1930s)	? via Señor Pedro Luro	? Sport	India, Nepal and Sri Lanka
	? Water buffalo (*Bubalus bubalis*)	? (Marejo Island)	?	? Sport	India
	Brown hare (*Lepus capensis*)	Early 1960s (natural extension from Uruguay)	—	—	Much of Europe, Asia and Africa
Chile	Domestic pig (*Sus domestica*)	? *c.*1750 (mainland); 1616 (Juan Fernandez Islands)	?	Food	Europe, North Africa and Asia
	Fallow deer (*Cervus dama*)	1887	?	? Aesthetic	Asia Minor and southern Europe
	Red deer (*Cervus elaphus*)	1920s	?	? Aesthetic and/or sport	Palaearctic region
	? Reindeer (*Rangifer tarandus*)	1971	?	? Food and hide	Northern Holarctic
	Domestic goat (*Capra hircus*)	1574–80 (Juan Fernandez Islands); also on mainland	(Juan Fernandez)	Food	Eastern Mediterranean and Middle East
	Mouflon (*Ovis musimon*)	? (Island in Lake Rupanco)	?	Sport	Originally parts of Asia
	Muskrat (*Ondatra zibethicus*)	1971 (natural extension from Argentina)	—	—	North America
	House mouse (*Mus musculus*)	?	?	Accidental	Originally southern Palaearctic
	Brown hare (*Lepus capensis*)	Before 1928 (natural extension from Argentina)	—	—	Much of Europe, Asia and Africa
	Rabbit (*Oryctolagus cuniculus*)	Mid-18th century	?	Sport	Originally northwestern Africa and Iberian peninsula
	? Reindeer (*Rangifer tarandus*)	1971	?	? Food and/or draught	Northern Holarctic
Colombia	Small Indian mongoose (*Herpestes auropunctatus*)	? 1951	?	? Rat control	Northern Arabia to Malaysia
	Domestic horse (*Equus caballus*)	?	? Spanish conquistadors	? Riding	Originally eastern Europe to Mongolia
	Domestic pig (*Sus domestica*)	? *c.*1750	?	? Food	Europe, North Africa and Asia

Continued

Table 26 *(continued)*

Country	*Species*	*Date of earliest successful introduction*	*Introduced by*	*Reason*	*Natural range*
South America					
Colombia *(cont.)*	Domestic cattle *(Bos taurus)*	? 16th century	? Spanish settlers	Food(and ? draught)	Originally Europe and western Asia
French Guiana	Small Indian mongoose *(Herpestes auropunctatus)*	After 1900	?	? Rat control	Northern Arabia to Malaysia
Guyana	Small Indian mongoose *(Herpestes auropunctatus)*	After 1900	?	? Rat control	Northern Arabia to Malaysia
Peru	Domestic horse *(Equus caballus)*	Early 16th century	?	? Riding	Originally eastern Europe to Mongolia
	Fallow deer *(Cervus dama)*	1948	Huacraruco Ranch	Aesthetic and sport	Asia Minor and southern Europe
	Red deer *(Cervus elaphus)*	1948	Huacraruco Ranch	Aesthetic and sport	Palaearctic region
	Black rat *(Rattus rattus)*	1544	Early explorers	Accidental	Originally southeastern Asia
Surinam	Small Indian mongoose *(Herpestes auropunctatus)*	1900	?	? Rat control	Northern Arabia to Malaysia
Uruguay	Axis deer *(Cervus axis)*	1930s	Señor Aaron de Anchorena	? Aesthetic and/or sport	India, Nepal and Sri Lanka
	Fallow deer *(Cervus dama)*	1930s	Señor Aaron de Anchorena	? Aesthetic and/or sport	Asia Minor and southern Europe
	Brown hare *(Lepus capensis)*	?	?	? Sport	Much of Europe, Asia and Africa
Venezuela	? Water Buffalo *(Bubalus bubalis)*	1935–40	?	Food and/or draught	India
	Domestic goat *(Capra hircus)*	?	?	? Food and skin	Eastern Mediterranean and Middle East
Australasia					
Australia	Red fox *(Vulpes vulpes)*	1845	Mr T. H. Pyke	Hunting	Most of Palaearctic and Nearctic regions
	Dingo *(Canis familiaris dingo)*	3000–7000 years ago	Aboriginal voyagers	Domestic	Southeastern Asia
	Domestic dog *(Canis familiaris)*	? 1788	? First colonists	Domestic	Originally Holarctic
	Domestic cat *(Felis catus)*	1788	First colonists	Domestic	Originally North Africa

Continued

Table 26 *(continued)*

Country	*Species*	*Date of earliest successful introduction*	*Introduced by*	*Reason*	*Natural range*
Australasia					
Australia *(cont.)*	Domestic ass (*Equus asinus*)	1866 or 1886	Sir Thomas Elder	Draught	Northeastern Africa
	Domestic horse (*Equus caballus*)	Late 18th century	Early settlers	Riding	Originally eastern Europe to Mongolia
	Domestic pig (*Sus domestica*)	1788	First colonists	Food	Europe, north Africa and Asia
	Arabian camel (*Camelus dromedarius*)	1840	Per governor of South Australia, Mr George Gawler	Draught	Arabia
	Axis deer (*Cervus axis*)	1803	Mr John Harris	Sport	India, Nepal and Sri Lanka
	Fallow deer (*Cervus dama*)	1829, 1834 or 1836 (Tasmania)	Private landowners and acclimatization societies	Aesthetic and sport	Asia Minor and southern Europe
	Red deer (*Cervus elaphus*)	1860	Mr Thomas Chirnside via Queen Victoria and Prince Albert	Aesthetic and sport	Palaearctic region
	Hog deer (*Cervus porcinus*)	1858	Acclimatization Society of Victoria via Rajah Rajendro Mullick, Mr A. Grote of Calcutta and Mr C. P. Layard of Ceylon	Aesthetic and sport	Northwestern India to Indochina
	Timor deer (*Cervus timorensis*)	1868 (in captivity); 1890 (in the wild)	Victoria Acclimatization Society	Aesthetic and sport	Java, Celebes, Lesser Sunda Islands
	Sambar (*Cervus unicolor*)	1861	Mr Edward Wilson for the Victoria Acclimatization Society	Aesthetic and sport	India and Sri Lanka to southern China
	Water buffalo (*Bubalus bubalis*)	1824 or 1826	Capt. James Bremer	Food (and ? draught)	India
	Banteng (*Bos javanicus*)	1829–40 (or 1845)	? Capt. James Bremer (or Capt. James McArthur)	Food (and ? draught)	Southeastern Asia
	Domestic cattle (*Bos taurus*)	1788	First colonists	Food (and ? draught)	Originally Europe and western Asia
	? Blackbuck (*Antelope cervicapra*)	*c.*1900 (and 1912)	(Mr Mackenzie Grant)	Aesthetic and/or sport	India

Continued

Table 26 *(continued)*

Country	*Species*	*Date of earliest successful introduction*	*Introduced by*	*Reason*	*Natural range*
Australasia					
Australia (*cont.*)	Domestic goat (*Capra hircus*)	1788	First colonists	Food	Eastern Mediterranean and Middle East
	Domestic sheep (*Ovis aries*)	*c.*1795 or (? 1788)	(First colonists)	Food and/or wool	Originally Asia Minor and Iran
	Brown rat (*Rattus norvegicus*)	Shortly after 1788	Early colonists	Accidental	Originally southeastern Asia
	Black rat (*Rattus rattus*)	? Mid-19th century	Settlers	Accidental	Originally southeastern Asia
	House mouse (*Mus musculus*)	1788	First colonists	Accidental	Originally southern Palaearctic
	Brown hare (*Lepus capensis*)	1859	Mr William Lyall	Sport	Much of Europe, Asia and Africa
	Rabbit (*Oryctolagus cuniculus*)	1788 (domestic)	First colonists	Food and pelt	Originally northwestern Africa and Iberian peninsula
		1859 (wild)	Mr Thomas Austin	Sport	
New Zealand	Brush-tailed possum (*Trichosurus vulpecula*)	1837–40 (or 1858)	Capt. John Howell (or Mr Christopher Basstian)	Fur	Australia and Tasmania
	Tammar wallaby (*Macropus eugenii*)	*c.*1870	Sir George Grey	Aesthetic and/or sport	Southwestern and southern Australia
	Parma wallaby (*Macropus parma*)	*c.*1870	Sir George Grey	Aesthetic and/or sport	Southeastern Australia
	Red-necked wallaby (*Macropus rufogriseus*)	1867	Mr A. M. Johnson	Aesthetic and/or sport	Eastern and southern Australia and Tasmania
	Brush-tailed rock wallaby (*Petrogale penicillata*)	*c.*1870	Sir George Grey	Aesthetic and/or sport	Australia
	Hedgehog (*Erinaceus europaeus*)	1885	Otago Acclimatization Society	Insect control	Western Europe
	Stoat (*Mustela erminea*)	1885	NZ Government	Rabbit control	Eurasia and Nearctic region
	Weasel (*Mustela nivalis*)	1885	NZ Government	Rabbit control	Most of Holarctic region
	Ferret (*Mustela furo*)	1867	Canterbury Acclimatization Society	Rabbit control	—

Continued

Table 26 (*continued*)

Country	*Species*	*Date of earliest successful introduction*	*Introduced by*	*Reason*	*Natural range*
Australasia					
New Zealand (*cont.*)	Domestic cat (*Felis catus*)	Late 18th century	Sealers/whalers	Rat and mouse control	Originally North Africa
	Domestic horse (*Equus caballus*)	Early 19th century	Early settlers	Riding	Originally eastern Europe to Mongolia
	Domestic pig (*Sus domestica*)	1769	de Surville	Food	Europe, North Africa and Asia
	Fallow deer (*Cervus dama*)	1864	Nelson Acclimatization Society	Aesthetic and sport	Asia Minor and southern Europe
	Red deer (*Cervus elaphus*)	1862 or 1863	via Queen Victoria	Aesthetic and sport	Palaearctic region
	Wapiti (*Cervus canadensis*)	1870s	Sir George Grey	Aesthetic and sport	North America
	Sika (*Cervus nippon*)	1904	NZ Government via Duke of Bedford, England	Aesthetic and sport	Southeastern Asia
	Timor deer (*Cervus timorensis*)	1907	NZ Tourist Department and Mr H. R. Benn	Aesthetic and/or sport	Java, Celebes, Lesser Sunda Islands
	Sambar (*Cervus unicolor*)	1875	Mr Falconer Larkworthy	Aesthetic and/or sport	India and Sri Lanka to southern China
	White-tailed deer (*Odocoileus virginianus*)	1901 and 1905	Mr T. E. Donne for NZ Tourist Department	Tourist attraction and sport	North and Central America
	Moose (*Alces alces*)	1900 and 1907	Mr T. E. Donne for NZ Tourist Department	Tourist attraction and sport	Northern Holarctic
	Domestic cattle (*Bos taurus*)	1833 (or 1815)	Mr John Bell (or Revd Samuel Marsden)	Food (and ? draught)	Originally Europe and western Asia
	Chamois (*Rupicapra rupicapra*)	1907 and 1913	Mr T. E. Donne for NZ Tourist Department via Franz Josef, Emperor of Austria	Tourist attraction and sport	Mountains of mainland Europe
	Himalayan tahr (*Hemitragus jemlahicus*)	1904	NZ Tourist Department via Duke of Bedford, England	Sport	Himalayan Mountains
	Domestic goat (*Capra hircus*)	1773 and 1777	Capt. James Cook	Food	Eastern Mediterranean and Middle East
	Domestic sheep (*Ovis aries*)	1814	Revd Samuel Marsden	Food and/or wool	Originally Asia Minor and Iran

Continued

Table 26 *(continued)*

Country	*Species*	*Date of earliest successful introduction*	*Introduced by*	*Reason*	*Natural range*
Australasia					
New Zealand (*cont.*)	Polynesian rat or kiore (*Rattus exulans*)	*c.*1350 (or earlier)	Maoris	Food	Southeastern Asia
	Brown rat (*Rattus norvegicus*)	? 1642 (more probably 1769, 1773 or 1776 or later 18th century)	Tasman (Capt. James Cook or early sealers and whalers)	Accidental	Originally southeastern Asia
	Black rat (*Rattus rattus*)	Before 1860 (possibly second half of 18th century)	(Capt. James Cook or early sealers and whalers)	Accidental	Originally southeastern Asia
	House mouse (*Mus musculus*)	Early 19th century	Early grain-carrying ships	Accidental	Originally southern Palaearctic
	Brown hare (*Lepus capensis*)	1851	Ship *Eagle*	Sport	Much of Europe, Asia and Africa
	Rabbit (*Oryctolagus cuniculus*)	Before 1838 (domestic)	Early settlers	Food and pelt	Originally northwestern Africa and Iberian peninsula
		1862 (wild)	Dr Menzies	Sport	
Papua New Guinea	Domestic dog (*Canis familiaris*)	? Prehistoric	?	? Domestic	Originally Holarctic
	Domestic pig (*Sus domestica*)	?	?	? Food	Europe, North Africa and Asia
	Timor deer (*Cervus timorensis*)	*c.*1900	Mr Walter Grose	Sport and food	Java, Celebes, Lesser Sunda Islands
	Water buffalo (*Bubalus bubalis*)	1900–3	German settlers	Food and/or draught	India
	Domestic cattle (*Bos taurus*)	?	?	Food (and ? draught)	Originally Europe and western Asia
	Domestic goat (*Capra hircus*)	?	?	Food	Eastern Mediterranean and Middle East
Oceania					
Galápagos	Domestic dog (*Canis familiaris*)	1832	Early settlers from Ecuador	Domestic	Originally Holarctic
	Domestic cat (*Felis catus*)	18th century	Visiting ships	To dispose of unwanted kittens	Originally North Africa
	Domestic ass (*Equus asinus*)	1832	Early settlers from Ecuador	To transport turtle oil and sulphur	Northeastern Africa
	Domestic horse (*Equus caballus*)	1832	Early settlers from Ecuador	? Riding and draught	Originally eastern Europe to Mongolia

Continued

Table 26 *(continued)*

Country	*Species*	*Date of earliest successful introduction*	*Introduced by*	*Reason*	*Natural range*
Oceania					
Galápagos *(cont.)*	Domestic pig (*Sus domestica*)	1832	Early settlers from Ecuador	Food	Europe, North Africa and Asia
	Domestic cattle (*Bos taurus*)	1832	Early settlers from Ecuador	Food	Originally Europe and Western Asia
	Domestic goat (*Capra hircus*)	1813 (and 1832)	Capt. David Porter (and early settlers from Ecuador)	Food	Eastern Mediterranean and Middle East
	Black rat (*Rattus rattus*)	1684–early 1700s	English privateers	Accidental	Originally southeastern Asia
	House mouse (*Mus musculus*)	Second World War (1939–45)	? Military	Accidental	Originally southern Palaearctic
Hawaii	Brush-tailed rock wallaby (*Petrogale penicillata*)	1916	Mr Richard H. Trent	?	Australia
	Domestic dog (*Canis familiaris*)	16th century	Polynesian voyagers	Domestic	Originally Holarctic
	Small Indian mongoose (*Herpestes auropunctatus*)	1883	Mr J. Tucker, for Hilo Planters' Association	Rat control	Northern Arabia to Malaysia
	Domestic cat (*Felis catus*)	Early 19th century	Early European settlers	? Rat and mouse control	Originally North Africa
	Domestic ass (*Equus asinus*)	1825	Capt. Richard Charlton	? Draught	Northeastern Africa
	Domestic horse (*Equus caballus*)	1803	Capt. William Shaler	? Riding	Originally eastern Europe to Mongolia
	Domestic pig (*Sus domestica*)	*c.* AD 1000 (and 1778)	Polynesians (Capt. James Cook)	Food	Europe, North Africa and Asia
	Axis deer (*Cervus axis*)	1867	Dr William Hillebrand	Gift to King Kamehameha V	India, Nepal and Sri Lanka
	Mule deer (*Odocoileus hemionus*)	1961	Hawaiian Division of Game and Fish	Tourist attraction and sport	Western and Central North America
	Pronghorn (*Antilocapra americana*)	1959	Per Mr Lyman Nichols, Jr	Aesthetic and sport	Mainly Oregon, Montana, Navada, Wyoming and Texas, USA
	Domestic cattle (*Bos taurus*)	1793	Capt. James Vancouver	Gift to King Kamehameha I	Originally Europe and western Asia
	Domestic goat (*Capra hircus*)	1778	Capt. James Cook	Food	Eastern Mediterranean and Middle East
	Mouflon (*Ovis musimon*)	1954	Hawaiian Division of Fish and Game	Sport	Originally parts of Asia

Continued

Table 26 *(continued)*

Country	Species	Date of earliest successful introduction	Introduced by	Reason	Natural range
Oceania					
Hawaii (*cont.*)	Domestic sheep (*Ovis aries*)	1791 (1793)	Capt. James Colnett (Capt. George Vancouver)	? (Gift to King Kamehameha I)	Originally Asia Minor and Iran
	Polynesian rat or kiore (*Rattus exulans*)	*c.* AD 500–900	Polynesians	Food and sport	Southeastern Asia
	Brown rat (*Rattus norvegicus*)	*c.*1835	Visiting ships	Accidental	Originally southeastern Asia
	Black rat (*Rattus rattus*)	After 1840 (probably 1870–80)	Visiting ships	Accidental	Originally southeastern Asia
	House mouse (*Mus musculus*)	Before 1816	Visiting ships	Accidental	Originally southern Palaearctic
	? Brown hare (*Lepus capensis*)	? (Two islets off Oahu)	? From Russia	?	Much of Europe, Asia and Africa
	Rabbit (*Oryctolagus cuniculus*)	Shortly after European settlement: abundant before 1824	Early settlers	? Food and pelt	Originally northwestern Africa and Iberian peninsula

Bibliography

The published material on naturalized mammals is extensive. Only the more important references are given here, the principal ones being marked with an asterisk★.

A

Abbot, W. E. 1913. *The rabbit-pest and the balance of nature.* (Privately published.)

Abdulali, H. 1962. The wild pigs in the Andamans. *Journal of the Bombay Natural History Society* **59**: 280–3.

Ables, E. D. ed. 1977. *The axis deer in Texas.* Caesar Kleberg Research Program in Wildlife Ecology, Texas A & M University System, 86 pp.

Ables, E. D. & **Ramsey, C. W.** 1975. Indian mammals on Texas rangelands. *Journal of the Bombay Natural History Society* **71**: 18–25.

Adams, W. H. 1956. The nutria in coastal Louisiana. *Proceedings of the Louisiana Academy of Sciences* **19**: 28–41.

———. 1957. Nutria: for better or for worse. *Alabama Conservation* (June/July).

Ahrens, T. G. 1921. Muskrats in central Europe. *Journal of Mammalogy* **2**: 236–7.

Akkermann, R. 1975. Untersuchungen zur Ökologie und Populationsdynamik des Bisams [muskrat]. *Zeitschrift für angewandte Zoologie* **62**: 173–218.

Alex-Hansen, B. 1970. Danmarks klovbaerende vildt [fallow deer]. *Dansk Jagttidende* **87**: 166–71. (See also: **62**: 37–8, 48–9, 63–4, 74–7, 88–91, 103–5, 118–21; **67**: 118–23; **72**: 156–61; **77**: 150–6; **82**: 139–43.)

Aliev, F. F. 1956. Theoretical and practical foundations of coypu raising in Azerbaijan. *Trudȳ Instituta Zoologii, Akademiya Nauk Azerbaĭdzhanskoĭ SSR, Baku* **19**.

———. 1965a. [Coypus in the USSR.] *Journal of Mammalogy* **46**: 101–2.

———. 1965b. Extent and causes of nutria mortality in the water bodies of the southern USSR. *Mammalia* **29**: 435–7.

———. 1967. Numerical changes and the population structure of the coypu in different countries. *Säugetierkundliche Mitteilungen* **15**: 238–42.

Aliev, F. F. & **Sanderson, G. C.** 1966. Distribution and status of the raccoon in the Soviet Union. *Journal of Wildlife Management* **30**: 497–502.

———. 1970. The American mink in the USSR. *Säugetierkundliche Mitteilungen* **18**: 122–7.

Aliev, F. F. & **Schöps, P.** 1964. Foundations of animal breeding in the Caucasus. *Das Pelzgewerbe, Leipzig,* n.s. **15**: 68–71.

Allan, R. J. P. 1984. Scottish wild sheep. *The Ark* **6**: 162–6.

Allen, D. L. 1954. *Our wildlife legacy.* Funk & Wagnalls: New York.

Allen, G. M. 1911. Mammals of the West Indies. *Bulletin of the Museum of Comparative Zoology* **54**: 175–263.

Allen, R. & **Neill, W. T.** 1954. The Florida deer. *Florida Wildlife* **7**(9): 21, 37.

Allin, A. E. 1950. European hare introduced into the district of Thunder bay, Ontario. *Canadian Field Naturalist* **64**: 122–4.

Allison, C. & **Coombes, I.** 1969. *The Australian hunter.* Cassell: Australia.

Almăşan, H. 1971. Capre negre pe locuri vechi şi nov. *Vinatorul si Pescarul Sportif* **11**: 20–2.

Almăşan, H., Ilie, E. & **Nesterov, V.** 1972. [Mouflon in Romania.] *Vinatorul si Pescarul Sportif* **12**: 11–12.

Altshul, M. P. 1957. Acclimatization of muskrats in Leningrad Province. *Translations of the Russian Game Reports* **2**: 144–55. Canadian Wildlife Service.

Alvarez, F., Brazza, F. & **Norzagasay, A.** 1976. Etograma Cuantificado del Gama en libertad. Unpublished report, Estación Biológica de Doñana, Paraguay.

Anderson, G. D. & **Condy, P. R.** 1974. A note on the feral house cat and house mouse on Marion Island. *South African Journal of Antarctic Research* No. 4: 58–61.

***Anderson, J. A.** & **Henderson, J. B.** 1961. Himalayan tahr in New Zealand. *New Zealand Deerstalkers Association Special Publication* **2**: 1–37.

Anderson, R. M. 1923. Further notes on the European hare in Ontario. *Canadian Field Naturalist* **37**: 75–6.

———. 1934. Effect of the introduction of exotic animal forms. *Proceedings of the 5th Pacific Science Congress, June 1933* **1**: 769–78. University of Toronto Press: Toronto.

Anon. n.d. The rock apes of Gibraltar. Unpublished mimeo of the Gibraltar Regiment, issued by Gibraltar Tourist Office.

———. 1883 & 1885. The mungoose. *Planters Monthly* **1**: 307–8, 249–51; **4**: 65; **7**: 196; **16**: 437–8; **17**: 244.

———. 1901. The mischievous mongoose. *Paradise of the Pacific* **14**(6): 9.

———. 1902. The mongoose in Jamaica. *Paradise of the Pacific* **16**(1): 7–8.

———. 1904. The obnoxious mongoose. *Paradise of the Pacific* **17**(4): 8.

———. 1908. Camel breeding in Australia. *Western Australian Journal of Agriculture* **16**: 424–30.

———. 1914. The Afghan camel invasion. *Australian Life & Character* (11 August): 16–17.

———. 1918. Rats and mongooses in the West Indies. *Agricultural News*, in *Bulletin of the Department of Agriculture, Trinidad* **17**: 187–90.

———. 1919–33. British wild goats. *Monthly Circular of the British Goat Society* **12**(8): 2–3; **13**(9): 3; **13**(11): 5–7; **15**(11): 210–12. *Yearbook* 1923: 52–5. *Monthly Journal*, **24**(1): 11; **26**(11): 275.

———. 1919. Camels in the Australian hinterland. *Land Transport & Sea Trade* **2**: 8–10.

———. 1921. Über ein angebliches seuchenhaftes Aussterben der Bisamratte [muskrat] in der Tschechoslowakei. *Nachrichtenblatt für den Deutschen Pflanzenschutzdienst*, 50.

———. 1927. Bringing plants and animals to Hawaii. *Paradise of the Pacific* **40**(8): 22–3.

———. 1933. Wild pigs declared a pest. *Queensland Agricultural Journal* **39**: 140.

———. 1943. Wild pigs [in Australia]. *Walkabout* **9**: 29.

———. 1950. Furbearing animals. *23rd Biennial Report of the Idaho Department of Fish & Game*, 42–4.

———. 1952. Wild burros in California. *Outdoor California* **13**(25): 4–5.

———. 1953a. Iceland's mink are a menace to wildlife. *Modern Game Breeders & Hunters Club News* **23**(8): 13.

———. 1953b. California's wild burros given legal protection. *Outdoor California* **14**(20): 2.

———. 1954a. A new threat: European rabbits in the United States. *Nature Conservation News* **4**(5): 1.

———. 1954b. Die Vermehrung wertvoller Pelztiere in der Sowjetunion. *Umschau* **54**: 456.

———. 1954c. Sumpfbiber [coypus] im Altrhein bei Speyer. *DD. Jäger* **24**: 14.

———. 1955. Wild pigs now 'noxious' throughout the state. *Agricultural Gazette of New South Wales* **66**: 211.

———. 1960. Sumpfbiber [coypus] und Bisamratten [muskrats] in Hessen. *Die Pirsch* **12**(1): 28.

———. 1962. Many goldfields pioneers relied on camel trains. *Countryman* **77**: 11.

———. 1965a. Britain's wild wallabies. *Animals* **7**: 524.

———. 1965b. Buffalo in Northern Territory [of Australia]. *The Territorian* **2**: 36.

———. 1965c. Wild pigs [in Australia]. *The Territorian* **2**: 47.

*———. 1967. *Towards a new relationship of man and nature in temperate lands. Part III. Changes due to introduced species. Proceedings of the 10th IUCN Technical Meeting, Lucerne, June 1966, IUCN Publications* n.s., No. 9, 259 pp.

———. 1968a. *Symposium on introduced exotic animals: ecologic and socioeconomic considerations*, 25 pp. Caesar Kleberg Research Program in Wildlife Ecology, Texas A & M University: College Station.

———. 1968b. Problems in species' introduction. *IUCN Bulletin* **2**(7): 70–1.

———. 1969a. Deer species established in Victoria [Australia]. *Victoria's Resources* **11**(2): 26.

———. 1969b. Report of the International Conference on the Muskrat. *EPPO Publications* Ser. A, No. 47, 88 pp.

———. 1969c. Water buffaloes – asset or liability. *Australian & New Zealand Bank Quarterly Survey*: 13–15.

———. 1970. Policy on introductions to nature reserves. *SPNR Technical Publication* No. 2. Society for the Promotion of Nature Reserves: Lincoln.

———. 1971a. [Mink in Russia.] *Oryx* **11**: 208.

———. 1971b. Re-establishment of musk-oxen in Northern Alaska. *Polar Record* **15**: 747.

———. 1974. Starkes damwild [fallow deer] in Schweden. *Wild und Hund* **77**: 18.

———. 1975a. Predators in Estonia [raccoon-dog]. *Oryx* **13**: 136.

———. 1975b. Introduction of reindeer into Isles de Kerguelen [summarized from *Dagens Nyheter* 9 January 1955]. *Polar Record* **7**: 518.

———. 1976a. *Exclusion and detection manual.* State of California Department of Food & Agriculture, 37 pp.

———. 1976b. Tortoises, iguanas and the menace of feral dogs. *Noticias Galápagos* **25**: 1–2.

———. 1978. *Coypu.* Report of the Coypu Strategy Group, Ministry of Agriculture, Fisheries & Food. HMSO: London.

———. 1981. *The ecology and control of feral cats [in the UK]. Proceedings of a symposium, September 1980,* 1–99. Universities Federation of Animal Welfare.

Apps, P. 1984. Cats on Dassen Island. *Proceedings of the 3rd International Theriological Congress, Helsinki, August 1982. Acta zoologica fennica* (in press).

Armitage, P., West, B. & **Steedman, K.** 1984. New evidence of black rat in Roman London. *The London Archaeologist* **4**: 375–83.

Armstrong, W. E. 1976. Food habits of exotics. Unpublished report, Texas Parks & Wildlife Department Ref. Job 17, Project W-76-R-19.

Armstrong, W. E. & **Harmel, D. E.** 1981. Competing with the natives. *Texas Parks & Wildlife* **39**(2): 6–7.

Artimo, A. 1949. [Muskrats in Finland.] *Suomen Riista* **4**: 7–61 [in Finnish].

★———. 1960. The dispersal and acclimatization of the muskrat in Finland. *Papers on Game Research,* No. 21: 1–101.

Ashbrook, F. G. 1948. Nutrias grow in the United States. *Journal of Wildlife Management* **12**: 87–95.

———. 1953. Muskrat-nutria. *Louisiana Conservationist* **5**(6): 16–17.

Aspisoff, D. I. 1957. Acclimatization of muskrats in the Volga-Kama region. *Translations of the Russian Game Reports* **2**: 156–202. Canadian Wildlife Service.

Aston, B. C. 1911. Some effects of imported animals on the indigenous vegetation. *Transactions of the New Zealand Institute* **44**, *Proceedings*: 19–24.

———. 1915. The vegetation of the Tarawere Mountains, New Zealand. *Transactions of the New Zealand Institute* **48**: 304–14.

Atkinson, I. A. E. 1964. Relations between feral goats and vegetation in New Zealand. *Proceedings of the New Zealand Ecological Society* **11**: 39–44.

★———. 1973. Spread of the ship rat (*Rattus r. rattus* L.) in New Zealand. *Journal of the Royal Society of New Zealand* **3**: 457–72.

———. 1977. A reassessment of factors, particularly *Rattus rattus* L., that influenced the decline of endemic forest birds in the Hawaiian islands. *Pacific Science* **31**: 109–33.

Atkinson, I. A. E. & **Bell, B. D.** 1973. Offshore and outlying islands. In Williams, G. R. ed., *The natural history of New Zealand,* 372–92. A. H. & A. W. Reed: Wellington.

Atwood, E. L. 1950. Life history studies of nutria, or coypu in coastal Louisiana. *Journal of Wildlife Management* **14**: 249–65.

Australian acclimatization societies. *Annual reports.*

B

Bailey, A. M. & **Hendee, R. W.** 1926. Notes on the mammals of northwestern Alaska [reindeer from Siberia]. *Journal of Mammalogy* **7**: 9–28.

Bailey, A. M. & **Sorensen, J. H.** 1962. *Subantarctic Campbell Island.* Denver Museum of Natural History: Denver, Colorado.

Bailey, H. H. 1974. The armadillo in Florida and how it reached there. *Journal of Mammalogy* **5**: 264–5.

Bailey, R. C., Baker, R. S., Brown, D. S., von Hildebrand, P., Mittermeier, R. A., Sponsel, L. E. & **Wolf, K. E.** 1974. Progress of a breeding project for non-human primates in Colombia. *Nature, London* **248**: 453–5.

Baillie, J. L. 1929. The European hare, a new mammal for Toronto. *Canadian Field Naturalist* **43**(3): 64.

Baker, C. M. A. & **Manwell, C.** 1981. Fiercely feral: on the survival of domesticates without care from man. *Sonderdruck ans Zeitshrift für Tierzüch. und Züch* **98**: 241–57.

Baker, H. G. & **Stebbins, G. L.** eds. 1965. *The genetics of colonizing species.* Academic Press: New York.

Baker, J. K. 1979. The feral pig in Hawaii Volcanoes National Park. *Transactions of the US National Parks Service,* Proceedings Ser. No. 51: 365–7.

Baker, K. 1972. The history of the Blue Mountain fallow deer herd. *New Zealand Wildlife* **38**: 27–31.

Baldwin, P. H., Schwartz, C. W. & **Schwartz, E. R.** 1952. Life history and economic status of the mongoose in Hawaii. *Journal of Mammalogy* **33**: 335–56.

Balis M. 1971. [The up-to-date results of acclimatization and introduction of animals in Slovakia.] *Ochrana Fauny* **5**(1): 34–6 [English summary 36–7].

Banfield, A. W. F. 1974. *The mammals of Canada.* University of Toronto Press: Toronto.

Bankalow, F. S. & **Shorten, M.** 1973. *The world of the grey squirrel.* Lippincott: Philadelphia.

Banwell, D. B. 1966. *Wapiti in New Zealand.* A. H. & A. W. Reed: Wellington.

Barbour, T. 1930. Some faunistic changes in the

Lesser Antilles. *Proceedings of the New England Zoological Club* **11**: 73–85.

Barbu, P. 1972. Beiträge zum Studium des Marderhundes [raccoon-dog] aus dem Donaudelta. *Säugetierkundliche Mitteilungen* **20**: 375–405.

Barclay, A. 1938. With the buffalo hunters [of Australia]. *Walkabout* **4**: 12–20.

Barker, H. M. 1964. *Camels and the outback*. Angus & Robertson: Melbourne and London.

Barnes, W. B. 1955. Status of the San Juan rabbit in the mid-west. *Transactions of the 17th Midwest Wildlife Conference*, 1–7 (mimeo).

Barnes, W. C. 1930. Wild burros. *American Forests & Forest Life* **36**: 640–2.

Barnett, S. A. 1976. *The rat: a study in behaviour*. Australian National University Press: Canberra.

Barnum, C. C. 1930. Rat control [by mongooses] in Hawaii. *Hawaiian Planters Record* **34**: 421–3.

Barrau, J. & **Devamkez, L.**1957. [Some unexpected results of introductions in New Caledonia.] *Terre et Vie* **104**: 324–34 [in French].

Barrett, C. 1934. The grey squirrel in Melbourne. *Victorian Naturalist* **51**: 108–10.

Barrett, R. H. 1967. History and status of introduced ungulates on Rancho Piedra Blanca, California. Unpublished MS thesis, University of Michigan, 141 pp.

———. 1978. The feral hog on the Dye Creek Ranch, California. *Hilgardia* **46**: 283–355.

Barrett-Hamilton, G. E. H. 1898. Notes on the introduction of the brown hare into Ireland. *Irish Naturalist* **7**(3): 69–76.

Batchler, C. L. 1973. Estimation of population density changes. In Orwin, J., *Assessment and management of introduced animals in New Zealand forests*. New Zealand Forest Service (Forest Research Institute Symposia **14**).

Bates, M. 1956. Man as an agent in the spread of organisms. In Thomas, W. L. et al. eds, *Man's role in changing the face of the earth (Proceedings of an international symposium, Princeton, 1955)*, 788–804. Cambridge University Press.

Bathgate, A. 1897. Notes on acclimatization in New Zealand. *Transactions of the New Zealand Institute* **30**: 266–79.

Bavin, C. R. 1977. Wildlife importations. *Proceedings of the Annual Conference of the Western Association of State Game & Fish Commissioners* **57**: 241–6.

Beaucournu, J. C. 1980. Les ectoparasites du lapin de garenne; apports á son histoire. *Bulletin mensuel de l'Office national de la Chasse*, No. Spec. Scien. Tech. 23–35.

Becker, K. 1967. Populationsstudien an Bisamratten [muskrat]. *Zoologische Beiträge* **13**: 369–96.

———. 1972. Muskrats in central Europe. *Proceedings of the 5th Vertebrate Pest Conference, Fresno, California*, 18–21. University of California: Davis (for Vertebrate Pest Conference).

Bednarik, K. 1955. The San Juan rabbit. *Ohio Conservation Bulletin*, 8–9.

Belden, R. C. & **Frankenberger, W. G.** 1977. Management of feral hogs in Florida – past, present and future. In Wood, G. W. ed., *Research and management of wild hog populations*, 5–10. Belle W. Baruch Forest Science Institute, Clemson University: Clemson, South Carolina.

Belden, R. C. & **Pelton, M. R.** 1976. European wild hog rooting in the mountains of east Tennessee. *Proceedings of the Annual Conference of the Southeastern Association of Game & Fish Commissioners* **29**: 665–71.

Bell, B. D. 1974. Opossums (2). In Knox, R. ed., *New Zealand's nature heritage* **3**: 922–8. Hamlyn: Hong Kong.

———. 1975. Chatham Islands. In Knox, R. ed., *New Zealand's nature heritage*. Hamlyn: Hong Kong.

———. ed. 1981. *Proceedings of the 1st symposium on marsupials in New Zealand*. Victoria University: Wellington (Zoology Publications No. 74, 280 pp.).

Bell, B. D. & **Taylor, R. H.** 1970. The wild sheep of Campbell island. *Forest & Bird* **178**: 3–10.

Bell, B. D., Brockie, R. E. Cowman, P. E., Efford, M. G. & **White, A. J.** 1979. Demography of *Trichosurus vulpecula* in Australia and New Zealand. *Proceedings of the 49th ANZAAS Congress, Abstracts* **1**(2), Zoology: 319–22.

Bennetsen, E. 1977. Some facts about sika deer in Denmark. *Deer* **4**: 153–7.

Bentley, A. 1957. A brief account of the deer in Australia. *Journal of Wildlife Management* **21**: 221–5.

*———. 1978. *An introduction to the deer of Australia, with special reference to Victoria*. John Gartner: Melbourne (first publ. 1967).

Bentley, A. & **Downes, M. C.** 1968–9. Deer in New Guinea. *Papua & New Guinea Agricultural Journal* **20** (1–4), 1–14, 95–9.

Bentley, E. W. 1959. The distribution and status of *Rattus rattus* in the United Kingdom in 1951 and 1956. *Journal of Animal Ecology* **28**: 299–308.

———. 1964. A further loss of ground by *Rattus rattus* in the United Kingdom during 1956–61. *Journal of Animal Ecology* **33**: 371–3.

Berger, N. M. 1944. Acclimatization of fur animals in west Siberia. *Zoologicheskiĭ Zhurnal* **23**: 267–74 (*Biological Abstracts* **20**: 5080) [in Russian with English summary].

———. 1962. Prisposoblenie amerikanskoi norki k novym usloviyam obitaniya v Sibiri i ee prakticheskoe ispol' zovanie. *Vaprosy Ekologii* **6**: 22–3.

Berry, R. J. 1964. The evolution of an island population of the house mouse. *Evolution* **18**: 468–83.

———. 1968. The ecology of an island population of the house mouse. *Journal of Animal Ecology* **37**: 445–70.

———. 1970. The natural history of the house mouse. *Field Study* **3**: 212–62.

———. 1981. Town mouse, country mouse: adaptation and adaptability in *Mus domesticus* (*M. musculus domesticus*). *Mammal Review* **11**: 91–136.

Berry, R. J. & **Peters, J.** 1975. Macquarie Island house mice: a genetical isolate on a sub-antarctic island. *Journal of Zoology, London* **176**: 375–89.

Berry, R. J., Bonner, W. M. & **Peters, J.** 1979. Natural selection in house mice (*Mus musculus*) from South Georgia (south Atlantic Ocean). *Journal of Zoology, London* **189**: 385–98.

Berry, R. J., Jakobson, M. E. & **Peters, J.** 1978. The house mice of the Faroe Islands: a study in microdifferentiation. *Journal of Zoology, London* **185**: 73–92.

Berry, R. J., Peters. J. & **Van Aarde, R. J.** 1978. Sub-antarctic house mice: colonization, survival and selection. *Journal of Zoology, London* **184**: 127–41.

Bigalke, R. 1937a. The adulteration of the fauna and flora of our national parks. *South African Journal of Science* **33**: 221–5.

*———. 1937b. The naturalization of animals, with special reference to South Africa. *South African Journal of Science* **33**: 45–63.

———. 1977. The Himalayan tahr on Table Mountain. *Zoologica Africana* **12**: 504.

Bisdee, J. 1871. Rabbits. *Proceedings of the Royal Society of Tasmania* (May).

Blaser, E. 1963. Die Nutria [coypu] in der Wildbahn. *Die Pirsch* **15**(1): 12–13.

Blouch, R. I. 1954. Introductions all around. *Michigan Conservation* **23**(3): 6–9.

***Boitani, L.** ed. 1976. *Reintroductions: techniques and ethics*. World Wildlife Fund: Rome.

***Bonner, W. N.** 1958. The introduced reindeer of South Georgia. *Falkland Island Dependencies Survey Scientific Report*, No. 22: 1–8.

———. 1980. British biological research in the Antarctic. *Biological Journal of the Linnean Society* **14**: 1–10.

Borbidge, R. T. 1960. The problem of pigs run wild in Victoria. *Proceedings of the Vermin Control Conference, Hobart, Tasmania.*

Bostad, H. A. et al. 1962. *Minkinkasvatus*. Yhteispohjismainen minkinkasvattajain käsikirja: Helsinki.

Bourdelle, E. 1939. American mammals introduced into France in the contemporary period, especially *Myocastor* and *Ondatra*. *Journal of Mammalogy* **20**: 287–91.

Bourlière, F. 1955. *The natural history of mammals* (transl. by H. M. Parshley). Harrap: London.

Bourne, W. R. P. 1983a. Rats as avian predators: discussion. *Atoll Research Bulletin* **255**: 69–71.

———. 1983b. New Zealand ornithological survey of Juan Fernandez, South Pacific. *Ibis* **4**: 595.

Box, T. W. 1968. Introduced animals and their implications in range vegetation management. In *Proceedings of a symposium on introduced exotic animals: ecological and socioeconomic condiserations*, 17–20. Caesar Kleberg Research Program in Wildlife Ecology, Texas A & M University: College Station.

Boyd, J. M. 1953–5. The sheep population of Hirta, St Kilda. *Scottish Naturalist* **65**: 15–28; **68**: 10–13.

———. 1981. The Boreray sheep of St Kilda, Outer Hebrides, Scotland: the natural history of a feral population. *Biological Conservation* **20**: 215–27.

Boyd, J. M., Doney, J. M., Gunn, R. G. & **Jewell, P. A.** 1964. The Soay sheep on the Island of Hirta, St Kilda. *Proceedings of the Zoological Society of London*. **142**: 129–63.

Brockie, R. E. 1957. The hedgehog population and the invertebrate fauna of the west coast sand dunes. *Proceedings of the New Zealand Ecological Society* **5**: 27–9.

———. 1958. The ecology of the hedgehog in Wellington, New Zealand. Unpublished MSc thesis, Victoria University, Wellington.

———. 1959. Observations on the food of the hedgehog in New Zealand. *New Zealand Journal of Science* **2**: 121–36.

Broekhuizen, S. 1977. De beverrat [coypu] in Nederland. Unpublished report, Rijks Instituut voor Natuurbehear, Arnhem, 9 pp.

Brooke, C. H. 1980. Pacific ocean islands: feral populations of goats, pigs, cattle, donkeys, horses, sheep, and water buffalo. Unpublished mimeo, annotated 'incomplete and provisional', 3 pp.

———. 1981. Feral farm animals, endangered species and island ecosystems. *AAG Abstracts, Los Angeles, 1981*. Association of American Geographers: Washington DC.

———. 1984. An historical-geographic perspective of the feral Caprinae problem. In *Feral mammals – problems and potential* (*Proceedings of a SSC Caprinae Specialist Group Workshop, Helsinki, August 1982*), 17–30. IUCN: Morges.

Brooke R. K. & **Cooper, J.** 1980. *Oryctolagus cuniculus* as a South African species. Unpublished mimeo, 6 pp.

Brookshier, F. A. 1945. On the burro trail in Hawaii. *Paradise of the Pacific* **57**(12): 17–20.

Brosset, A. 1963. Statut actuel des mammiferes des îles Galápagos. *Mammalia* **27**: 323–38.

Brown, B. A. 1976. The population dynamics of nilgai in south Texas, with some management considerations. Unpublished PhD dissertation, Texas A & M University.

Brown, L. N. & **McGuire, R. M.** 1969. Status of the red-bellied squirrel in the Florida keys. *American Midland Naturalist* **82**: 629–30.

Brownlee, C. E. 1968. Habitat preference of exotics. Unpublished report, Texas Parks & Wildlife Ref. Job 18, Project W-76-11.

Bryan, E. H. 1938. The much maligned mongoose. *Paradise of the Pacific* **50**(4): 32–4.

Bryan, L. W. 1930. Hawaii's wild goat problem. *Para-*

dise of the Pacific **43**(12): 77–8.

———. 1937. Wild sheep, cattle and pigs in Hawaii. *Paradise of the Pacific* **49**(3): 19, 30; **49**(9): 30–2.

Buchanan, G. D. 1955. Nine-banded armadillo – invader from the south. *Animal Kingdom* **58**(3): 82–8.

———. 1958. The current range of the armadillo in the United States. *Texas Journal of Science* **10**: 349–51.

Buchanan, G. D. & **Talmage, R. V.** 1954. The geographical distribution of the armadillo in the United States. *Texas Journal of Science* **2**: 142–50.

Buchanan Smith, A. D. 1932. Wild goats in Scotland. *British Goat Society Yearbook*, 120–5.

Bucksath, C. S. 1956. Red deer in Argentina. *Transactions of the 2nd Congress of the International Union of Game Biologists*, 104–6.

Bull, P. C. 1953. Hedgehogs and ground-nesting birds. *Annual Reports of the Ornithological Society of New Zealand* **1**: 32–3.

———. 1962. Population dynamics of the wild rabbit. *New Zealand Science Review* **20**: 23–6.

Buller, W. I. 1871. On the New Zealand rat. *Transactions of the New Zealand Institute* **3**: 1–4.

Bulliet, R. W. 1975. *The camel and the wheel*. Harvard University Press: Cambridge, Mass.

Bullock, D. 1977. Round Island [Mauritius], a tale of destruction [by rabbits and goats]. *Oryx* **14**: 51–8.

———. 1983. Borerays, the other rare breed on St Kilda. *The Ark* **10**: 274–8.

Bullock, D. & **North, S.** 1984. Round Island in 1982. *Oryx* **18**: 36–41.

Bump, G. 1940. The introduction and transplantation of game birds and mammals into the state of New York. *Proceedings of the 5th North American Wildlife Conference*, 409–20.

———. 1951. Game introductions – when, where and how. *Transactions of the 16th North American Wildlife Conference*, 316–25.

———. 1952. How shall foreign species be introduced? *Atlantic Monthly* **7**: 112–17.

———. 1957. Foreign game introductions into the southeast. *Proceedings of the 11th Annual Conference of the Southeastern Association of Game & Fish Commissioners*, 17–20.

———. 1963. Status of the foreign game introduction program. *Transactions of the 28th North American Wildlife Conference*, 240–7.

———. 1968. Exotics and the role of the State–Federal foreign game investigation program. In *Proceedings of a symposium on introduced exotic animals: ecologic and socioeconomic considerations*, 5–8. Caesar Kleberg Research Program in Wildlife Ecology, Texas A & M University: College Station.

———. 1970. Acclimatization of game animals in the United States. *Transactions of the 9th International Congress of Game Biologists, Moscow, September 1969*, 136–40 [in English, with Russian summary].

Burckhardt, D. 1975. Reintroduction of animals, yes or no? *Schweizer Naturschutz* **41**.

Burt, T. & **Lee, L.** 1967. Exotic animals. In *New Mexico wildlife management*, 200–9.

Burton, F. D. 1972. The integration of biology and behaviour in the socialization of *Macaca sylvana* of Gibraltar. In Poirier, F. E. ed., *Primate socialization*, 29–62. Random House: New York.

Burton, F. D. & **Sawchuk, L.** 1974. Demography of *Macaca sylvanus* of Gibraltar. *Primates* **15**: 271–8.

Burton, M. 1969. *The hedgehog*. André Deutsch: London.

Buskirk, S. W. & **Gipson, P. S.** 1981. Zoogeography of Arctic foxes on Aleutian Islands. In Chapman, A. & Pursley, D. eds, *Proceedings of the 1st World Furbearers Conference, Maryland, 1980*. University of Maryland Press: Adelphi.

Butts, G. L. 1975. Food habits of exotics. Unpublished report, Texas Parks & Wildlife Department Job 17, Project W-76-R-18.

———. 1979. The status of exotic big game in Texas. *Rangelands* **1**: 52–3.

C

Cabrera, A. 1914. *Fauna Ibérica mamiféros*. Museo Nacional de Ciencias Naturales: Madrid.

———. 1945. *Caballos de América*. Editorial Sudamericana: Buenos Aires.

Cabrera, A. & **Yepes, J.** 1940. *Mamiféros Sud-Americanos*. Argentina de Editores: Buenos Aires.

Cahalane, V. H. 1955. Some effects of exotics on nature. *Proceedings of the International Technical Conference on the Protection of Nature*, 396–405. Unesco: Paris.

Calaby, J. H. 1977. Changing the face of a continent. *Australian Natural History* **19**(2): 62–5.

Calaby, J. H. & **Keith, K.** 1974. Fauna survey of the Port Essington district, Cobourg peninsula, Northern Territory of Australia (ed. Frith, H. J. & Calaby, J. H.). *CSIRO Division of Wildlife Research Technical Paper* No. 28: 179–208.

Calvopiña, L. H. 1977. Fluctuaciones de la poblacion de chivos salvajes en Bahia Bucanero, Isla San Salvador, Galápagos. *Revista de la Universided Catolica, Año 3*, No. 16: 131–43.

———. 1984. Effects and present status of the feral mammals on the biota of several Galápagos islands. *Proceedings of the 3rd International Theriological Congress, Helsinki, August 1982. Acta zoologica fennica* (in press).

Calvopiña, L. H. & **de Vries, T.** 1975. Estructura de la poblacion de cabras salvajes y los daños causados en la vegetacion de la Isla San Salvador, Galápagos. *Revista de la Universidad Catolica, Año 3*, No. 8: 219–41.

'Camelus'. 1966. With the camels in the north-west.

Hoofs & Horns, 6–9 (Jan.); 8–12 (Feb.); 10–13 (March).

Cameron, K. T. 1956. Queensland's struggle against rabbits, 1880–1930. *Quarterly Journal of the Royal Historical Society* **5**: 1201–17.

Camoens, J. K. 1976. The buffalo in Malaysia. *Bulletin of the Malaysian Ministry of Agriculture* **145**: 1–184.

Cansdale, G. S. 1953. *Animals and man*. F. A. Praeger: New York.

Carl, G. C. & **Guiguet, C. J.** 1972. *Alien animals in British Columbia* (British Columbia Provincial Museum Handbook No. 14, 94 pp.; first published 1958). Victoria.

Carlquist, S. 1974. *Island biology*. Columbia University Press: New York.

Carrick, R. 1957. The wildlife of Macquarie island. *Australian Museum Magazine* **12**: 255–60.

Carter, T. D., Hill, J. E. & **Tate, G. H. H.**1945. *Mammals of the Pacific World*. Macmillan: New York.

Caseby, J. A. 1936. The Welsh goat. *British Goat Society Yearbook*, 120–3.

Caughley, G. 1963. Dispersal rates of several ungulates introduced into New Zealand. *Nature, London* **200**: 280–1.

———. 1970a. Eruption of ungulate populations with emphasis on Himalayan tahr in New Zealand. *Ecology* **51**: 53–72.

*———. 1970b. Liberation, dispersal and distribution of Himalayan tahr in New Zealand. *New Zealand Journal of Science* **13**: 220–39.

Challies, C. N. 1975. Feral pigs on Auckland Island. *New Zealand Journal of Zoology* **2**: 479–90.

Chang, Cheih, Yien, Chih-Tang & **Hsu, Ping-Yu.** 1974. Age of muskrat populations. *Acta zoologica sinica* **20**: 89–104.

Chanin, P. & **Linn, I.** 1980. The diet of the feral mink (*Mustela vison*) in southwest Britain. *Journal of Zoology, London* **192**: 205–23.

***Chapman, D. I.** & **Chapman, N. G.** 1975. *Fallow deer: their history, distribution and biology*. Terence Dalton: Lavenham, Suffolk.

Chapman, J. A. & **Sandt, J. L.** 1977. The blacktailed jack-rabbit – new record in Maryland, U.S.A. *Chesapeake Science* **18**: 318–19.

***Chapman, N. G.** & **Chapman, D. I.** 1980. The distribution of fallow deer: a worldwide review. *Mammal Review* **10**: 61–138.

Chappellier, A. 1948. Le rat musqué en France. *Bulletin français de Pisciculture* **20**(149): 137–58.

Charles Darwin Research Station, Galápagos. 1982. *Annual Report*.

Chasen, F. N. 1940. A handlist of Malaysian mammals: a systematic list of the mammals of the Malay peninsula, Sumatra, Borneo and Java, including the adjacent small islands. *Bulletin of the Raffles Museum* **15**: 1–209.

Chashchukhin, V. A. 1975. Effect of muskrats on aquatic vegetation. *Byulletin' Moskovskogo Obshchestva Ispўtateleĭ Prirodў*, Otdel Biol. **80**(6): 21–9.

Cheel, E. 1934. Effect of the introduction of exotic plants and animals on Australian plant life. *Proceedings of the 5th Pacific Science Congress, June, 1933* **1**: 791–6. University of Toronto Press: Toronto.

Chesnokov, N. I. 1976. Pattern of muskrat acclimatization. *Soviet Journal of Ecology* **7**: 535–40 [transl. of *Ekologiya* **6**: 63–70].

Christian, C. S. & **Stewart, G. A.** 1953. *Survey of Katherine-Darwin region 1946*, Vol. 1. CSIRO Land Resources Service, Australia.

Christie, A. H. C. 1964. A note on the chamois in New Zealand. *Proceedings of the New Zealand Ecological Society* **11**: 32–6.

Christie, A. H. C. & **Andrews, J. R. H.** 1965. Introduced ungulates in New Zealand: Virginia deer. *Tuatara* **13**(1): 1–8.

Churcher, C. S. 1959. The specific status of the New World red fox. *Journal of Mammalogy* **40**: 513–20.

***Clark, A. H.** 1949. *The invasion of New Zealand by people, animals and plants*. Rutgers University Press: New Brunswick.

Clark, D. A. 1978. Black rat feeding ecology in the Galápagos Islands, Ecuador. PhD thesis, University of Wisconsin.

———. 1981. Foraging patterns of black rats across a desert-montane forest gradient in the Galápagos Islands. *Biotropica* **13**: 182–94.

———. 1984. Native land mammals. In Perry, P. ed., *Galápagos*. Pergamon Press: Oxford.

Clark, D. B. 1980. Population ecology of *Rattus rattus* across a desert-montane forest gradient in the Galápagos Islands. *Ecology* **61**: 1422–33.

Clarke, C. M. H. 1971. . . . red deer in northern South Island. *New Zealand Forest Science* **1**: 194–207.

———. 1976. Fallow deer in the north of South Island. *New Zealand Wildlife* **52**: 5–13.

Cleland, J. B. 1910. The Australian fauna and flora and exotic invasions. *Journal of the Natural History & Scientific Society of Western Australia* **3**(1).

———. 1911. The first introduction of camels into South Australia and the west. *Report of the 13th Meeting of the Australian Association for the Advancement of Science*, 314–16.

Clifford, C. G. 1952. Alien animals on Vancouver Island. *Victorian Naturalist* **9**(1): 7–10.

Clifford, C. G., Guiguet, C. J. & **Hardy, G. A.** 1951. Biology of the Scott Island group, British Columbia [rabbits]. *Report of the British Columbia Provincial Museum of Natural History & Anthropology 1950*, B21–B63.

Clutton-Brock, J. 1981. *Domesticated animals from early times*. Heinemann and British Museum (Natural History): London.

Coblentz, B. E. 1976. Wild goats of Santa Catalina.

Natural History **85**: 70–7.

———. 1978. Effects of feral goats on the Santa Catalina Island ecosystem. *Biological Conservation* **13**: 279–86.

Cockayne, L. 1903. A botanical excursion during midwinter to the southern islands of New Zealand. *Transactions of the New Zealand Institute* **36**: 225–333.

Combe, A. 1935. Sumpfbiber [coypus] in Württemberg. *Deutsche Jagd* **2**: 1032.

Conigrave, C. P. 1932. The camel in Australia. *New Nation Magazine* **72** (1 March): 25–6.

Conley, R. H. 1977. European wild hog in Tennessee. In Wood, G. W. ed., *Research and management of wild hog populations*. Belle W. Baruch Forest Science Institute, Clemson University: Clemson, South Carolina.

Connelly, J. 1948. Minnesota whitetails flown to Finland. *The Conservation Volunteer* **11**: 29–30.

Cook, R. L. 1973. Habitat preference of exotics. Unpublished report, Texas Parks & Wildlife Department Job 18, Project W-76-R-16.

———. 1974–5. Food habits of exotics. Unpublished report, Texas Parks & Wildlife Department Job 17, Project W-76-R-16.

Cooper, H. M. 1954. Kangaroo Island's wild pigs: their possible origin. *South Australian Naturalist* **28**: 57–61.

***Cooper, J.** & **Brooke, R. K.** 1982. Past and present distribution of the feral European rabbit on southern African offshore islands. *South African Journal of Wildlife Research* **12**(2): 71–5.

Copson, G. R., Brothers, N. P. & **Skira, I. J.** 1981. Distribution and abundance of the rabbit at subantarctic Macquarie Island. *Australian Wildlife Research* **8**: 597–611.

Corbet, G. B. 1961. Origin of the British insular races of small mammals and the 'Lusitanian' fauna. *Nature, London* **191**: 1037–40.

———. 1966. *The terrestrial mammals of western Europe*. Foulis: London.

———. 1974. *The distribution of mammals in historic times* (Systematics Association Special Vol. No. 6). Academic Press: London and New York.

———. 1978. *The mammals of the Palaearctic: a taxonomic review*. British Museum (Natural History): London; Cornell University Press: Ithaca.

Corbet, G. B. & **Hill, J. E.** 1980. *A world list of mammalian species*. British Museum (Natural History): London; Cornell University Press: Ithaca.

Corbet, G. B. & **Southern, N. N.** eds. 1977. *The handbook of British mammals*, 2nd edn. Blackwell: Oxford.

Corbett. L. K. 1974. Dingoes, feral dogs and crossbreeds in Victoria. *Australian Mammalogy* **1**: 303–4.

Cottam, C. 1950. The effect of uncontrolled introductions of plants and animals. *Proceedings & Papers of an International Technical Conference on the Protection of Nature, Lake Success, New York, August 1949*, 408–13. Unesco: Paris.

———. 1956. The problem of wildlife introductions – its successes and failures. *Proceedings of the 46th Convention of the International Association of Game, Fish & Conservation Commissioners*, 1–37, 94–111.

Cotton, K. E. 1963. The coypu. *River Boards Association Yearbook* **31.**

Couch, L. K. 1929. Introduced European rabbits in the San Juan islands, Washington. *Journal of Mammalogy* **10**: 334–6.

Cowan, D. 1980. *The wild rabbit*. Blandford Press: Poole, Dorset.

Cowling, S. J. 1975. A review of deer conservation in Australia. *Fish & Wildlife Paper, Victoria* **8**: 1–10.

Craig, J. C. 1970. The ecology of feral goats in Scotland. Unpublished MSc thesis, University of Edinburgh.

***Craighead, F. C.** & **Dasmann, R. F.** 1966. Exotic big game on public lands. Unpublished report, U.S. Department of the Interior Bureau of Land Management, 26 pp..

Crawford, A. 1908–16. *Annual reports of the Chief Inspector of Rabbits, 1907–15*. Agriculture Protection Board of Western Australia Research Library.

Creed, J. M. 1917. The hare in Australia. *The Field*, 740.

Crenshaw, J. 1976. Exotics and the future. *New Mexico Wildlife*, 16–18.

Crook, I. G. 1969. Feral goats in north Wales. *Animals* **12**(1): 13–15.

Crowcroft, P. 1966. *Mice all over*. Foulis: London.

Crowell, K. L. 1973. Experimental zoogeography: introductions of mice to small islands. *American Naturalist* **107**: 535–58.

Cuff, H. C. 1970. Buffaloes – a need for domestication. *Turnoff* **2**: 2–7.

Cuthbert, J. H. 1973. The origin and distribution of feral mink in Scotland. *Mammal Review* **3**: 97–103.

D

Dalrymple, B. W. 1965. The new immigrants: exotic game animals. *Elks Magazine*: 10–11, 44–6.

Dammerman, K. W. 1929. Fauna Buruana. *Mammalia* **7**: 149–63.

Danell, K. 1977. Dispersal and distribution of the muskrat in Sweden. *Viltrevy* **10**(1): 1–26.

———. 1978a. *Ecology of the muskrat in northern Sweden*. Swedish National Environment Protection Board.

———. 1978b. Muskrats in a shallow Swedish lake. *Journal of Animal Ecology* **47**: 697–709.

Daniel, M. J. 1962. Control of introduced deer in New Zealand. *Nature, London* **194**: 527–8.

———. 1969. A survey of rats on Kapiti Island, New Zealand. *New Zealand Journal of Science* **12**: 363–72.

———. 1972. Bionomics of the ship rat (*R.r. rattus*) in a New Zealand indigenous forest. *New Zealand Journal of Science* **15**: 313–41.

———. 1976. An imported fauna. *Australian Natural History* **18**: 370–3.

Danilov, P. I. 1972. [Acclimatization of Canadian beavers in Karelia.] *Ekologia* **3**: 102–14.

Danilov, P. I., Andreev, V. A. & **Markovsky, V. A.** 1974. [The status of beaver populations in the Karelian ASSR.] In *Problems of animal ecology*. Petrozavodsk.

Danilov, P. I., Ivanov, P. D., Novikov, G. A. & **Timofeeva, E. K.** 1974. Current distribution of some species of mammals in the northwest of the European U.S.S.R. [hare, mink, beaver and raccoon-dog]. *Byulletin' Moskovskogo Obshchestva Ispȳtateleĭ Prirodȳ*, Otdel Biol. **78**(5): 5–21.

Dansie, O. 1971. The spread of the remarkable muntjac. *Deer* **2**(4): 616–20.

Darlington, P. J. 1957. *Zoogeography: the geographical distribution of animals*. Wiley: New York.

Davidson, M. M. 1973. Characterics, liberation and dispersal of sika deer in New Zealand. *New Zealand Forest Science* **3**: 153–80.

Davidson, M. M. & **Kean, R. I.** 1960. Establishment of red deer range in the Tararua mountains. *New Zealand Journal of Forestry Science* **8**: 293–324.

***Davis, D. H. S.** 1950. Notes on the status of the American grey squirrel in the south-western Cape, South Africa. *Proceedings of the Zoological Society of London* **120**: 265–8.

Davis, G. S. 1953. Respite for the burro. *Nature Magazine* **46**: 370–4.

Davis, J. H. 1953. Biotic change in New Zealand. *Scientific Monthly* **76**: 109–11.

Davis, R. A. 1955. The coypu. *Discovery* (June).

———.1956. The coypu. *Journal of Agriculture, Ministry of Agriculture* **63**: 127–9.

———.1963. Feral coypus in Britain. *Annals of Applied Biology* **5**: 345–8.

———.1964. The coypu on the continent and in Britain. *Bulletin of the Mammal Society of the British Isles* **21**: 627.

———.1969. A note on feral coypu in the United Kingdom, 1930–1967. *EPPO Publications* Ser. A, No. 47: 63–6.

Davis, R. A. & **Jenson, O. G.** 1960. A note on the distribution of the coypu in Great Britain. *Journal of Animal Ecology* **29**: 397.

Day, A. M. 1948. Introduction of exotic species. Unpublished report, US Department of the Interior Fish & Wildlife Service, 13 pp.

———.1948. Introduction of exotic species. *Conventions of the International Association of Game & Fish Conservation Commissioners* **38**: 138–44.

de la Tour, G. D. 1949. La introducción de animales exóticos y su repercusion sobre la conservación de la fauna autóctona. *Diana, Buenos Aires* **114**: 1–4; **116**; **118**: 1–20.

———. 1950. Préjudices portés á la conservation de la fauna autochone par l'introduction d'espèces exotiques. *Proceedings and Papers of an International Technical Conference on the Protection of Nature, Lake Success, New York, August 1949*, 416–18. Unesco: Paris.

De Vos, A. & **Petrides, G. A.** 1967. Biological effects caused by terrestrial vertebrates introduced into non-native environments. *Proceedings of the 10th IUCN Technical Meeting, Lucerne, June 1966. IUCN Publications*, n.s. No. 9: 113–19.

***De Vos, A., Manville, R. H.** & **Van Gelder, R. G.** 1956. Introduced mammals and their influence on native biota. *Zoologica* **41**: 163–94.

De Vries, T. & **Calvopiña, L. H.** 1977. Papel de los chivos en los cambios de la vegetacion de la Isla San Salvador, Galápagos. *Revista de la Universidad Catolic Año 5*, No. 16: 145–69.

Deag, J. M. & **Crook, J. H.** 1971. Social behaviour and agonistic buffering in the wild Barbary macaque, *Macaca sylvana* L. *Folia Primatologia* **15:** 183–200.

Dean, P. B. & **De Vos, A.** 1965. The spread and present status of the European hare, *Lepus europaeus hybridus*, in North America. *Canadian Field Naturalist* **79**: 38–48.

Deane, C. D. 1972. Deer in Ireland. *Deer* **2**: 920–6.

Deane, C. D. & **O'Gorman, F.** 1969. The spread of feral mink in Ireland. *Irish Naturalists' Journal* **16**: 198–202.

Debach, P. 1974. *Biological control by natural enemies*. Cambridge University Press: Cambridge.

Dehnel, A. 1956. *Nyctereutes procyonoides* – Nowy ssak dla fauny polskiej. *Chrońmy Przyrodę Ojczystą, Kraków* **12**: 17–21.

———. 1957. Ein Neus Säugetier [raccoon-dog] der polnischen Fauna, *Nyctereutes procyonoides*. *Annales Universitatis Mariae Curie-Skłodowska, Lublin*, Sect. C **10**: 269–74.

Delattre, P. & **Louarn, H.** 1980. Réproduction des rats. *Mammalia* **44**: 233–43.

Dell, J. 1957. The European hare in New York. *New York State Conservationist* **11**(4): 20–1, 28–9.

Denburgh, J. van. 1914. The gigantic land tortoises of the Galápagos Islands. *Proceedings of the California Academy of Sciences*, Ser. 4 **2**: 203–374.

Denhardt, R. M. 1947. *The horse of the Americas*. University of Oklahoma Press: Norman.

Dent, A. A. 1983. Wild cattle of the Basque country. *Country Life* **173**: 1775.

Derenne, P. 1976. Notes sur la biologie du Chat heret de Kerguelen. *Mammalia* **40**: 531–95.

Derenne, P. & **Mougin, J. C.** 1976. Donneés écologiques sur le mammifères introduits de l'île aux Cochons, archipel Crozet. *Mammalia* **40**: 21–52.

Dickinson, T. G. & **Simpson, D. D.** 1979. Dispersal and establishment of Barbary sheep in the Southern Great Plains. *US Department of Agriculture Forest Service, Rocky Mountain Forest & Range Experimental Station Report*, 52 pp.

Dilks, P. J. 1979. Observations on the food of feral cats on Campbell Island. *New Zealand Journal of Ecology* **2**: 64–6.

Dilks, P. J. & **Wilson, P. R.** 1979. Feral sheep and cattle and Royal Albatrosses on Campbell Island. *New Zealand Journal of Zoology* **6**: 127–39.

Dingwall, P. R., Atkinson, I. A. E. & **Hay, C.** eds. 1978. *The ecology and control of rodents in New Zealand nature reserves. Proceedings of a symposium, Wellington 1976*. Department of Lands & Survey: Wellington (Information Series No. 4).

Dixon, J. S. 1929. Artificial distribution of fur-bearing mammals. *Journal of Mammalogy* **10**: 358–9.

Dollman, G. 1936. The rabbit menace. *Natural History Magazine* **5**: 297–311.

Donaldson, H. H. 1924. *The rat: data and reference tables for the albino rat (Mus norvegicus albinus) and the Norway rat (Mus norvegicus)*, 2nd edn. Philadelphia (Memoirs of the Wistar Institute of Anatomy and Biology, No. 6).

***Donne, T. E.** 1924. *The game animals of New Zealand*. John Murray: London.

Dorst, J. 1964. L'introduction d'espèces animales et leur impact sur l'environnement tropical. In *The ecology of man in the tropical environment. Proceedings of the 9th IUCN Technical Meeting, IUCN Publications* n.s. No. 4: 245–53.

Dorst, J. & **Giban, J.** 1954. Les mammifères acclimatés en France depuis un siècle. *Terre et Vie* **101**: 217–29.

Doty, R. E. 1945. Rat control [by mongooses] on Hawaiian sugar cane plantations. *Hawaiian Planters Record* **49**(2): 71–239.

Doude van Troostwijk, W. J. 1976. *The muskrat in the Netherlands: its ecological aspects and their consequences for man*, Verhandeling 7. Rijks Instituut voor Natuurbeheer: Arnhem, 136 pp.

———. 1978. Muskrat control in the Netherlands. *Proceedings of the 8th Vertebrate Pest Conference, Sacramento, California, March 1978*, 115–17. University of California: Davis (for Vertebrate Pest Conference).

Doughty, R. & **Choban, A.** 1977. Texas exotics. *Wildlife* **19**: 204–8.

Douglas, C. L. & **Norment, C.** 1979. Ecological studies of feral burros in Death Valley. *Transactions of the US National Parks Service*, Proceedings Ser. No. 5, **1**: 373–83.

Douglas, G. 1969. Draft checklist of Pacific islands. *Micronesia* **5**: 327–463.

Dozier, H. L. 1951. The present status and future of nutria in the southeastern states. *Proceedings of the 5th Annual Conference of the Southeastern Association of Game & Fish Commissioners*.

Dragoev, P. 1978. Introduction of game animals in Bulgaria. *Proceedings of the 2nd International Theriological Congress, Brno, June 1978, 32* (Abstract). Czechoslovak Academy of Sciences.

Dudzinski, W., Haber, A. & **Matuszewski, G.** 1963. Die Verbreitung des Marderhundes [raccoon-dog] in Polen. *Zeitschrift für Jagdwissenschaft, Hamburg-Berlin* **9**: 98–105.

———. 1965. Jumat *Nyctereutes procyonoides* w Polsce. *Chrońmy Przyrodę Ojczystą, Kraków* **21**: 21–30.

Dunmore, R. 1967. Analysis of study materials and field data relative to the life history and status of the European hare in New York. Unpublished report, New York Federal Aid Project W-84-R-13.

Duvigneaud, P. 1950. The introduction of exotic species. *Proceedings and papers of an International Technical Conference on the Protection of Nature, Lake Success, New York, August 1949*. Unesco: Paris.

Dymond, J. R. 1922. The European hare in Ontario. *Canadian Field Naturalist* **36**: 142–3.

———. 1928. Present range of the European hare in Ontario. *Canadian Field Naturalist* **42**: 95.

———. 1929. The spread of the European hare in Ontario. *Royal Ontario Museum Zoology Bulletin* **2**: 9–11.

Dyuozhkin, V. V. 1960. [The problem of the existence of Canadian beavers (*Castor canadensis*) in the Rovensky region.] *Proceedings of the Voronezhsky State Reserve*, issue G [in Russian].

E

Earl, G. W. 1846. *Enterprise in tropical Australia*. Madden & Malcolm: London.

Ecke, D. H. 1954. An invasion of Norway rats in south-west Georgia. *Journal of Mammalogy* **35**: 521–5.

Eckhardt, R. C. 1972. Introduced plants and animals in the Galápagos Islands. *BioScience* **22**: 585–90.

Edlin, H. L. 1952. *Changing wildlife of Britain*. Batsford: London.

Edmonds, J. W., Nolan, I. F., Shepherd, R. C. H., Backholer, J. R. & **Jackson, R.** 1976. Rabbits on Gippsland islands [Australia]. *Victorian Naturalist* **93**: 110–12.

Egilsson, K. 1983. *The food and pastures of reindeer in Iceland*. Orkustofnun: Reykjavík (OS-83073/VOD-07), pp. 235 [in Icelandic with English summary pp. 217–29].

Egilsson, K. & Thórisson, S. 1983. *Icelandic reindeer and the proposed Austurlandsvirkjun hydro-electric power scheme*. Orkustofnun: Reykjavík (OS-83074/VOD-08), pp. 65 [in Icelandic with English summary pp. 59–65].

Eliasson, U. 1968. On the influence of introduced animals on the natural vegetation of the Galápagos Islands. *Noticias Galápagos* **11**: 19–21.

Elkins, W. A. & **Nelson, U. C.** 1957. Wildlife introductions and transplants in Alaska. *Science in Alaska*, 29–30.

Elliott, H. W. 1977. Fallow and axis deer census at Point Reyes National seashore, spring 1977. Unpublished Report, National Parks Service, California.

*__Elton, C.__ 1927. *Animal ecology*. Sidgwick & Jackson: London.

*———. 1958. *The ecology of invasions by animals and plants*. Methuen: London.

Engels, L. 1969. Evolution de l'extension du rat musqué en Belgique. *EPPO Publications*, Ser. A, No. 47: 29–34.

England, D. 1955. Barbary apes of Gibraltar. *Zoo Life* **10**(3): 66.

Eric, A. 1896. The mongoose in Jamaica. *Planters Monthly* **15**: 106–8.

Errlich, S. 1956. Zjawiaka Zachdesace Populacji nutria a redukcla ich liesebnesci. *Ecologia Polska, Warsaw* **58**: 171–212.

Errington, P. L. 1963. *Muskrat populations*. Iowa State University Press: Ames.

Espeut, W. B. 1882. On the acclimatization of the Indian mungoos in Jamaica. *Proceedings of the Zoological Society of London*, 712–14.

Evans, J. V. 1980. Goats. In *Refresher course for veterinarians* (Proceedings), 52. University of Sydney: Sydney.

Evans, P. 1957. The Aoudad sheep, an exotic introduced in the Palo Duro canyon of Texas. *Proceedings of the 21st Annual Conference of the Southeastern Association of Game & Fish Commissioners*, 183–8.

Evers, A. et al. 1973. [Azores.] *Boletin do Museu Municipal do Funchal* **27**: 5–17.

Eyerdam, W. J. 1932. A shipment of muskrats to Kamchatka. *Journal of Mammalogy*, **13**: 281–2.

F

*__Fa, J. E.__ 1981. Apes on the Rock [Gibraltar]. *Oryx* **16**: 73–6.

———. 1983a. The Barbary macaque – the future. *Oryx* **17**: 62–7.

———. 1983b. Structure and dynamics of the Barbary macaque population in Gibraltar. In Fa, J. E. ed., *The Barbary macaque – a case study in conservation*. Plenum Press: New York.

Fa, J. E. & **Pankhurst, I.** 1982. Gibraltar Monkey Park – development proposals. Mimeo report to Gibraltar Government, 20 pp.

Fairley, J. S. 1971. The present distribution of the bank vole in Ireland. *Proceedings of the Royal Irish Academy* **71**(B): 183–9.

Falla, R. A. 1965. Birds and mammals of the subantarctic islands. *Proceedings of the New Zealand Ecological Society* No. 12: 63–8.

Fargher, S. E. 1977. The distribution of the brown hare and the mountain hare in the Isle of Man. *Journal of Zoology, London* **182**: 164–7.

Farwell, G. 1948. Camel town [Australia]. *Walkabout* **14**: 16–20.

Feldhamer, G. A. & **Chapman, J. A.** 1978. Sika deer and white-tailed deer on Maryland's eastern shore. *Wildlife Society Bulletin* **6**: 155–7.

Fenner, F. 1954. The rabbit plague. *Scientific American* **190**(2): 30–5.

Fennessy, B. V. 1958. The control of the European rabbit in New Zealand. *Technical Paper, CSIRO Wildlife Survey Section*, 1–40.

———. 1966. The impact of wildlife species on sheep production in Australia. *Australian Society of Animal Production Bulletin* **6**: 148–56.

Fisher, J. 1960. The black rat in London. *London Naturalist* **29**: 136.

Fisher, J., **Simon, N.** & **Vincent, J.** 1969. *The red book: wildlife in danger*. Collins: London.

Fitch, H. S., **Goodrum, P.** & **Newman, C.** 1952. The armadillo in the southeastern United States. *Journal of Mammalogy* **33**: 21–37.

Fitter, R. S. R. 1950. Man's additions to the British fauna. *Discovery* **11**: 58–62.

*———. 1959. *The ark in our midst*. Collins: London.

———. 1967. Animal introductions and their ecological effects in Europe. *Proceedings of the 10th IUCN Technical Meeting, Lucerne, June 1966, IUCN Publication* n.s. No. 9: 177–80.

Fitzgerald, B. M. 1964. The ecology of mustelids in New Zealand. Unpublished MSc thesis, University of Canterbury, 164 pp.

Fitzpatrick, J. 1950. The brumby muster. *South-West Pacific* **23**: 32–9.

Fjeldså, J. 1975. Recent changes in the water-fowl situation in the lakes (Myvatn & Vikingavatn). *Dansk Ornithologisk Forenings Tidsskrift* **69**: 89–101.

Fletcher, M. R. & **Waner, R. H.** 1976. Feral burro management at Bandelier National Monument, New Mexico. *Transactions of the Desert Bighorn Council* **20**: 54–5.

Flux, J. E. C. 1981. Prospects for hare farming in New Zealand. *New Zealand Agricultural Science* **15**: 24–9.

Flux, J. E. C. & **Fullagar, P. J.** 1983. World distribution of the rabbit. In Pulliainen, E. & Sulkawa, S. eds, *Proceedings of the 3rd International Theriological Congress, Helsinki, August 1982. VI. Acta zoologica fennica* No. 175: 75–7.

Flyger, V. F. 1959. Maryland's new deer citizens. *Maryland Conservationist* **36**: 23–5.

———. 1960. Sika deer on islands in Maryland and Virginia. *Journal of Mammalogy* **41**: 140.

Flynn, D. M. 1980. *Report on vertebrate pests and exotic animal diseases*. Australian Bureau of Animal Health: Canberra.

★**Ford, B. D.** & **Tulloch, D. G.** 1977. The Australian buffalo. *Northern Territory Department of Animal Industry & Agriculture Branch, Technical Bulletin* No. 18: 1–90.

Fosberg, F. R. 1963. Disturbance in island ecosystems. In Gressitt, J. L. ed., *Biogeography of the Pacific Basin*. Bishop Museum: Honolulu, Hawaii.

———. 1972. The axis deer problem [in Hawaii]. *Elepaio* **32**(9): 86–8.

Fowler, L. 1979. Population ecology and impact of the feral burros of Galápagos. *Annual Report of the Charles Darwin Research Station, Galápagos*, 111–13.

———. 1980. Feeding and population ecologies of the burros and tortoises of Volcan Alcedo, Isabela Island. *Annual Report of the Charles Darwin Research Station, Galápagos*, 112–14.

———. 1983. Feeding and population ecologies of the burros and tortoises of Volcan Alcedo, Galápagos: a study of the interactions between an introduced feral and an island endemic. Unpublished PhD dissertation, University of Florida.

Fraguglione, D. 1966. [European hare: a world review.] *Bulletin Special du Conseil Superiéur de la Chasse: Lièvre Communications*, No. 9: 1–94. Ministère de l'Agriculture: Paris.

Francis, G. W. 1862. Acclimatization of plants and animals. *Papers of the Philosophical Society of Adelaide* (13 May).

Frank, F. & Häerle, A. 1967. Derzeitiger Stand und vorauissichtliche Entwicklung des Bisambefalls [muskrat] in der Bundesrepublik Deutschland. *Nachrichtenblatt für den Deutschen Pflanzenschutzdienst, Braunschweig* **19**: 123–5.

Frankenberger, W. G. & **Belden, R. C.** 1978. Feral hogs in Florida. *Proceedings of the Annual Conference of the Southeastern Association of Fish & Wildlife Agencies* **30**: 641–4.

Freeman, W. G. 1918. The mongoose. *Bulletin of the Department of Agriculture, Trinidad* **17**: 166.

Freye, H. A. 1978. [Beavers.] In *Handbuch der Saugetiere Europas*, Vol. 1, 182–200. Akademische Verlagsgesellschaft: Wiesbaden.

Friedrich, R. 1940. *The Friedrich Bear Creek Ranch*. Hunt: Texas.

Friörikisson, S. 1960. [Icelandic reindeer.] *Náttúrufraedingurinn* **30**: 1–7 [in Icelandic].

★**Frith, H. J.** 1979. Acclimatization. In Frith, H. J., *Wildlife conservation*, 137–95. Angus & Robertson: Sydney (first published 1973).

Frith, H. J. & **Calaby, J. H.** Fauna survey of the Port Essington district, Cobourg peninsula, Northern Territory, Australia. *CSIRO Division of Wildlife Research Technical Paper* No. 28.

Furon, R. 1958. The gentle little goat: archdespoiler of the earth. *UNESCO Courier* **1**: 30–2.

G

Gagne, A. 1972. [Axis deer in Hawaii.] *Elepaio* **32**(9): 83–6.

Ganzak, J. 1964. [Mammals in Czechoslovakia.] *Okhota i Okhotnich'e Khozyaĭstvo* **5**: 52–3.

Gauthier-Pilters, H. & Dagg, A. I. 1981. *The camel: its evolution, ecology, behaviour, and relationship to man.* University of Chicago Press: Chicago.

Geist, V. 1960. Feral goats in British Columbia. *Murrelet.* **41**(3): 34–40.

Gerell, R. 1967 & 1968. Dispersal and acclimatization of the mink in Sweden. *Viltrevy* **5**: 1–38, 119–94.

———. 1970. Mink in north-western Europe. *Transactions of the 9th International Congress of Game Biologists, Moscow, September 1969*, 327–36. Moscow.

Gersdorf, E. 1969. Verbreitung und Bekämpfung des Bisams [muskrat] in Niedersachsen. *Neues Archiv für Niedersachsen* **18**: 352–4.

———. 1970. Zur Bisamabwehr [muskrat]. *Wasser und Boden* **2**: 41–2.

———. 1971. Beobachtungen an in Gefangenschaft gehaltenen Bisam [muskrat]. *Zeitschrift für angewandte Zoologie* **58**: 29–42.

Giban, J. & **Aubry, J.** 1956. Extension of range of the muskrat in France. *Mammalia* **20**: 34–45.

★**Gibb, J. A.** & **Flux, J. E. C.** 1973. Mammals. In Williams, G. R. ed., *The natural history of New Zealand*, 334–71. A. H. & W. Reed: Wellington.

Gibbons, J. 1984. Iguanas of the south Pacific. [Pigs, cats, goats, mongooses.] *Oryx* **18**: 82–91.

Gilmore, D. P. 1965. New Zealand's possum pests. *Animals* **7**(16 November): 498–501.

———. 1966. Studies of the biology of *Trichosurus vulpecula*. Unpublished PhD thesis, University of Canterbury.

Gilmore, R. M. 1946. [Mammals living in the region today but not represented in the material.] In, Mammals in archaeological collections from Southwestern Pennsylvania. *Journal of Mammalogy* **27**: 232–4.

Gilruth, J. A. 1912. Report on domestic animals in the Northern Territory. *Bulletin of the Northern Territory, Australia* **1**: 16.

Glass, B. P. & **Halloran, A. F.** 1961. The small mammals of the Wichita mountains wildlife refuge, Oklahoma [armadillo]. *Journal of Mammalogy* **42**: 234–9.

Glauert, L. 1956. Fauna conservation in Western Australia: problems of conservation. (c) Introduction of exotics. *Bulletin of the Western Australian Department of Fish & Fauna* No. 1.

Glazener, W. C. & **Cottam, C.** 1958. Exotics in Texas. *Texas Game & Fish* **16**: 4–7.

Glazner. M. 1958. First report of the growth of nutria in fish ponds [in Israel] and their influence on destruction of coarse vegetation. *Bulletin of Fish Culture, Israel* **10**: 32–5.

Gleeson, J. P. 1984. Feeding ecology of the house mouse on Marion Island. *Proceedings of the 3rd International Theriological Congress, Helsinki, August 1982. Acta zoologica fennica* (in press).

Gliwicx, J. 1984. Rodents on islands: living in a crowd. *Proceedings of the 3rd International Theriological Congress, Helsinki, August 1982. Acta zoologica fennica* (in press).

Godfrey, E. B. 1980. The economics of wild horse and burro management. *Proceedings of the Society of Range Management Annual Meeting* **33**(19).

Goodall, C. P. 1956. Buffalo hunting . . . .*Northern Australian Monthly* **3**: 43–4.

Gore, R. 1978. Florida, Noah's ark for exotic newcomers. *National Geographic* **150**: 538–58.

Gorion, A. 1936. Nutriajagd auf dem Laacher See. *Deutsch Jagd* **5**: 226.

Gorman, M. L. 1975. The diet of feral *Herpestes auropunctatus* in the Fijian islands. *Journal of the Zoological Society of London* **175**: 273–8.

Gosling, L. M. 1973–5. Porcupines at large in Devon and Staffordshire. Internal report, Ministry of Agriculture, Fisheries & Food, London in *Current Topics* (21 October 1974). (With Wright, M.) Feral porcupines in Devon (unpublished mimeo, 3 July 1973); Feral porcupines in Staffordshire (unpublished mimeo, 11 June 1974); Wiltshire porcupines (unpublished mimeo, 19 March 1975).

———. 1974a. The coypu in East Anglia. *Transactions of the Norfolk & Norwich Naturalists' Society (Norfolk Bird & Mammal Report)* **23**(1): 49–59.

———. 1974b. A preliminary account of the population ecology of the coypu population in East Anglia, England. *Second Report of the Working Party on Field Rodents*, 24–7. Ministry of Agriculture, Fisheries & Food, London.

———. 1975. A preliminary account of the population ecology of the coypu population in East Anglia, England. *EPPO Publications* Ser. C, No. 37: 24–7.

———. 1976a. Coypus in Kenya; a preliminary survey. Unpublished report, Ministry of Agriculture, Fisheries & Food, London.

———. 1976b. Research on the control of coypus – 1973–76. *Transaction of the Norfolk & Norwich Naturalists' Society* **24**(1): 36–8.

Gossart, H. 1929. Die Ausbreitung der Bisamratte [muskrat] in Deutschland. *Mitteilungen der Fischereivereine für die Provinz Brandenburg* **1**: 487–9.

Gottschalk, J. S. 1967. The introduction of exotic animals into the United States. *Proceedings of the 10th IUCN Technical Meeting Lucerne, June 1966. IUCN Publications* n.s. No. 9: 1–29.

Goudie, R. I. 1978. Red squirrels in the Salmonier river valley, Newfoundland. *Canadian Field Naturalist* **92**: 193–4.

Gould, M. S. 1979. The behavioural ecology of the feral goats of Aldabra Island. Unpublished PhD dissertation, Duke University, Durham, North Carolina.

Graf, W. 1958. Observations and comments on deer control and wildlife management in New Zealand. Report on Agriculture and Forestry for the Territory of Hawaii, Division of Fish & Game, Honolulu.

———. 1959a. Axis deer in the rainzone of Molokai Island [Hawaii]. *Yearbook of the American Philosophical Society*, 236–8.

———. 1959b. The deer problem in Hawaii. Report on Agriculture and Forestry for the Territory of Hawaii, Division of Fish & Game, Honolulu.

Graf, W. & Nichols, L. 1967. The axis deer in Hawaii. *Journal of the Bombay Natural History Society* **63**: 629–734.

Green, D. 1979. The effect of feral pigs on the nesting success of the East Pacific green turtle in the Galápagos Islands. Preliminary (mimeo) report for the Charles Darwin Research Station, Galápagos, 10 pp.

Greer, J. K. 1966. Mammals of Malleco Province, Chile. *Publications of Michigan State University Museum,* Biological Series **3**(2): 49–152.

Gressitt, J. L. ed. 1963. *Biogeography of the Pacific Basin*. Bishop Museum: Honolulu, Hawaii.

Groves, C. 1980. Notes on the systematics of *Babyrousa. Zoölogische Mededeelingen* **55**(3): 29–46.

———. 1981. Ancestors for the pigs: taxonomy and phylogeny of the genus *Sus. Technical Bulletin, Department of Prehistory, Research School of Pacific Studies, Australian National University* No. 3: 1–96.

Grummt, W. 1965. Das Vorkommen de Waschbären [raccoon] in Berliner Raum Beiträge zur Jagd-und Wildforschung IV.11. *Tagung der Arbeitsgemeinschaft für Jagd-und Wildforschung am 8 und 9.1. in Gatersleben,* 179–80. Deutsche Akademie der Landwirtschaftlicher Wissenschaft: Berlin (Tagungsberichte Nr 78).

Gruzdev, V. V. 1969a. The eastward extension of the range of the European hare. *Byulletin' Moskovskogo Obshchestva Ispÿtateleĭ Prirodÿ,* Otdel Biol. **74**(6): 19–29 [in Russian with English summary].

———. 1969b. The resettlement of the hare in the U.S.S.R. *Byulletin' Moskovskogo Obshchestva Ispÿtateleĭ Prirodÿ,* Otdel Biol. **74**(1): 25–36 [in Russian with English summary].

Gudmundson, F. 1952. Bird protection [from mink] in Iceland. *Bulletin of the International Council for Bird Preservation* **6**: 153–60.

Guiguet, C. J. 1975. Introduction of the gray squirrel to Vancouver Island, British Columbia. *Syesis* **8**: 399.

Gunderson, H. 1944. Notes on a heavy Norway rat population. *Journal of Mammalogy* **25**: 307–8.

H

Hadwen, S. & Palmer, L. J. 1922. Reindeer in Alaska. *US Department of Agriculture Bulletin* No. 1089: 1–74.

Hahn, D. E. 1966. The nine-banded armadillo in Colorado. *Southwest Naturalist* **11**: 303.

Haines, F. 1938. Where did the Plains Indians get their horses? *American Anthropologist* **40**: 112–17, 429–37.

Halder, U. 1974. Banteng and feral water buffalo in Java. In Kjerner, I. & Bjurholm, P. eds, *Proceedings of the 11th International Union of Game Biologists, 1973*, 159–64. Allmanna Förlaget: Stockholm (National Swedish Environmental Protection Board Publication).

Hall, E. R. 1963a. Introduction of exotic species of mammals. *Transactions of the Kansas Academy of Sciences* **66**: 516–18.

———. 1963b. Introduction of exotic species of mammals. *Proceedings of the 16th International Congress of Zoology* **1**: 267.

———. 1981. *The mammals of North America*, Vols 1 and 2. Wiley: New York.

Hall, L. S. 1977. Feral rabbits on San Juan Island, Washington. *Northwest Science* **51**: 293–7.

Hall, V. C. 1948. Wild stock. *Walkabout* **14**: 17–20.

Halloran, A. F. & Howard, J. A. 1956. Aransas refuge wildlife introductions. *Journal of Wildlife Management* **20**: 460–1.

Haltenorth, T. 1959. Bietrag zur Kenntnis des Mesopotamischen Damhirsches und Stammesund geschichte der Damhirsche allgemein [fallow deer]. *Säugetierkundliche Mitteilungen* **7**(Special No.): 1–89 (see also **9**: 15–39).

Haltenorth, T. & Diller, H. 1980. *A field guide to the mammals of Africa, including Madagascar.* Collins: London.

Hamann, O. 1975. Vegetational changes in the Galápagos Islands during the period 1966–1973. *Biological Conservation* **7**: 37–59.

———. 1979. Regeneration of vegetation on Santa Fé and Pinta Islands, Galápagos, after the eradication of goats. *Biological Conservation* **15**: 215–36.

Hamilton, W. R. 1952. The European hare in New York. *Bulletin to Schools, State University of New York* **38**: 189–90.

Hanson, H. C. 1952. Importance and development of the reindeer industry in Alaska. *Journal of Range Management* **5**: 243–51.

Hanson, R. P. & Karstad, L. 1959. Feral swine in the southeastern United States. *Journal of Wildlife Management* **23**: 64–74.

Hardcastle, E. 1906. *The deer of New Zealand.* New Zealand Tourist Department: Wellington.

Harmel, D. E. 1972. Food habits of exotics. Unpublished report, Texas Parks & Wildlife Department Job 17, Project W-76-R-15.

———. ed. 1975. Habitat preference of exotics. Unpublished report, Kerr Wildlife Management Area Project, Texas Parks & Wildlife Department Job 18, Project W-76-R-18.

*———. ed. 1980. Texas big game investigations; job No. 21. Statewide census of exotic big game animals; federal aid project No. W – 109–R–3. Unpublished report, Texas Parks & Wildlife Department, 33 pp.

Harney, W. E. 1951. The dingo. *Walkabout* (July).

Harris, D. R. 1962. The distribution and ancestry of the domestic goat. *Proceedings of the Linnean Society of London* **173**: 79–81.

———. 1965. *Plants, animals and man in the outer Leeward islands.* University of California Press: Berkeley (Publications in Geography No. 18).

Harris, L. H. 1974. *A hunting guide to introduced wild animals of New Zealand.* Government Printer: Wellington.

Harris, R. A. & Duff, K. R. 1970. *Wild deer in Britain.* David & Charles: Newton Abbot.

Harris, T. 1956. The nutria as a wild fur mammal in Louisiana. *Transactions of the 21st North American Wildlife Conference*, 474–86.

Harting, J. E. 1880. *British animals extinct within historic times.* Trübner: London.

———. 1898. *The rabbit.* Longmans Green: London.

Hausfater, G. & Walker, P. 1974. History of three little-known New World populations of macaques. *Laboratory of Primatology Newsletter* **13**(1): 16–18.

Hawksworth, D. L. ed. 1974. *The changing flora and fauna of Britain.* Academic Press: London (for Systematics Association).

Heidemann, G. 1973. [Fallow deer in West Germany.] *Zeitschrift für Säugetierkunde* **38**: 341–17.

Heidemann, G. & Witt, H. 1978. [Mouflon in West Germany.] *Zeitschrift für Jagdwissenschaft* **24**: 24–36.

Heller, E. 1903. Papers from the Hopkins Stanford Expedition 1898–99. XIV. Reptiles. [Dogs in Galápagos.] *Proceedings of the Washington Academy of Sciences* **5**: 39–98.

———. 1904. Mammals of the Galápagos Archipelago. *Proceedings of the California Academy of Sciences*, Ser. 3 **3**: 288–350.

Helminen, M. 1963. Riistatilanne 1962/1963 päättyessä. *Suomen Riista* **16**: 186–92.

———. 1967. Game animals in Finland. *Transactions of the 7th International Union of Game Biologists, Beograd, September 1965*, 135–8.

Helms, R. 1883. A Maori rat at Greymouth. *New Zealand Journal of Science* **1**: 466.

Hendricks, L. J. 1963. Observation of armadillo in east-central New Mexico. *Journal of Mammalogy* **44**: 581.

Henry, V. G. & Conley, R. H. 1978. Survival and mortality in European wild hogs [in the USA]. *Proceedings of the Annual Conference of the Southeastern Association of Fish & Wildlife Agencies* **32**: 93–9.

Heptner, V. G. 1967. The rehabilitation of certain mammal species in circumstances of environmental degradation. *Proceedings of the 10th IUCN Technical Meeting, Lucerne, 1966, IUCN Publications* n.s. No. 9: 194–6.

Heptner, V. G. & **Naumov, N. P.** 1961. *Mammals of the Soviet Union*. English transl. Israel Program for Scientific Translations: Jerusalem.

Herington, J. G. 1977. Wildlife introduced and imported into Papua-New Guinea. *Wildlife Papua New Guinea* No. 77(2): 1–9.

Hewson, R. 1971. Some aspects of the biology of feral mink in Banffshire [Scotland]. *Glasgow Naturalist* **18**: 539–46.

Hibben, F. C. 1964. The role of exotic animals in game management. *Proceedings of the 44th Annual Conference of the Western Association of State Game & Fish Commissioners*, 125–30.

Hicks, L. E. 1940. The role of exotics in the Ohio valley. *Transactions of the 5th North American Wildlife Conference*, 420–4.

Hill, J. E. 1959. Rats and mice from the islands of Tristan da Cunha and Gough, South Atlantic Ocean. *Results of the Norwegian Scientific Expedition to Tristan da Cunha, 1937–38*, No. 46.

Hingston, F. 1975. Deer in Spain. *Deer* **3**: 386–7.

***Hinton, H. E.** & **Dunn, A. M. S.** 1967. *Mongooses: their natural history and behaviour*. Oliver & Boyd: London & Edinburgh; University of California Press: Berkeley.

Hinton, M. A. C. 1931. *Rats and mice as enemies of mankind*. British Museum (Natural History): London (Economics Ser. No. 8).

Hobbs, D. F. 1954. Do newly introduced species present a separate problem? *Proceedings of the New Zealand Ecological Society* **2**: 12–14.

Hoffman, M. 1952a. *Die Bisamratte* [*muskrat*]. *Die Neu Brehm-Bücherei*. Akademische Verlagsgesellschaft, Geest und Portig: Leipzig.

———. 1952b. Vom Vorkommen des Sumpfbibers [coypus] in der freien Natur in Sachsen-Anhalt. *Nachrichtenblatt für den deutschen Pflanzenschutzdienst, Berlin* N. F. **6**: 216–19.

———. 1958. *Die Bisamratte* [*muskrat*]: *ihre Lebensgewohnheiten, Verbreitung Bekämptung und wirtshaftliche Bedeutung*. Akademische Verlagsgesellschaft, Geest und Portig: Leipzig.

———. 1970. Ein Beitrag über den Einflusz normaler und extrem kalter Winter auf die Bestände der Bisamratte [muskrat]. *Pelzgewerbe* **20**(4): 4–20.

———. 1977. Über die Verbrietung der Bisamratte [muskrat] in China, Mongolei, Korea und Japan. *Anzeiger für Schädlingskunde, Berlin* **50**: 86–8.

Holdgate, M. W. 1960. The fauna of mid-Atlantic islands. *Proceedings of the Royal Society* **B152**: 550–67.

———. 1965. The fauna of the Tristan da Cunha islands. *Philosophical Transactions of the Royal Society* B**249**: 361–402.

———. 1967. The influence of introduced species on the ecosystems of temperate oceanic islands. *Proceedings of the 10th IUCN Technical Meeting, Lucerne, June 1966, IUCN Publications* n.s. No. 9: 151–76.

———. 1970. Conservation in the Antarctic. In Holdgate, M. W. ed., *Antarctic ecology*, Vol. 2, 924–45. Academic Press: London and New York.

Holdgate, M. W. & **Wace, N. M.** 1961. The influence of man on the floras and faunas of southern islands. *Polar Record* **10**: 475–94.

Holdom, N. 1944. Additional notes on feral coypus. *Murrelet* **25**: 30.

Holland, C. W. 1923. Rabbits and their introduction into Australia. *Queensland Naturalist* (March).

Holloway, J. T. 1959. Noxious animal problems in the South Island watersheds. *New Zealand Scientific Review* **17**: 21–8.

Hood, G. A., Nass, R. D. & **Lindsey, G. D.** 1970. The rat in Hawaiian sugarcane. *Proceedings of the 4th Vertebrate Pest Conference, West Sacramento, California, March 1970*, 34–7. University of California: Davis (for Vertebrate Pest Conference).

Hosley, N. W. 1953. Fluctuations in the United States big game populations during the last decade. *Proceedings of the 7th Pacific Science Congress, New Zealand, February 1949* **4**: 620–6.

Howard, W. E. 1953. Nutria in California. *Journal of Mammalogy* **34**: 512–13.

———. 1959. The rabbit problem in New Zealand. *New Zealand DSIR Information Ser.* No. 16: 1–47.

———. 1964a. Introduced browsing animals and habitat stability in New Zealand. *Journal of Wildlife Management* **28**: 421–9.

———. 1964b. Modification of New Zealand's flora by introduced mammals. *Proceedings of the New Zealand Ecological Society* **11**: 59–62.

———. 1965. Control of introduced mammals in New Zealand. *New Zealand DSIR Information Ser.* No. 45: 1–95.

———. 1966. Interaction of behaviour, ecology, and genetics of introduced mammals. In Baker, H. G. & Stebbins, G. L. eds, *Genetics of colonizing species* (*1st IUBS Symposium on General Biology, Asilomar, 1964*). Academic Press: New York.

———. 1967. Ecological changes in New Zealand due to introduced mammals. *Proceedings of the 10th IUCN Technical Meeting, Lucerne, June 1966, IUCN Publications* n.s. No. 9: 219–40.

Howard, W. E. & **Marsh, R. E.** 1984. Ecological implications and management of mammals in California. In *Feral mammals – problems and potential* (*Proceedings of a SSC Caprinae Specialist Group Workshop, Helsinki, August 1982*), 31–41. IUCN: Morges.

Howe, T. D. & **Bratton, S. P.** 1976. Winter rooting activity of the European wild boar in the Great

Smoky Mountains National Park. *Castanea* **41**: 256–64.

Howell, P. G. 1977. Exotics in Oklahoma. *Outdoor Oklahoma* **33**(4): 11–15.

———. 1984. An evaluation of the biological control of the feral cat. *Proceedings of the 3rd International Theriological Congress, Helsinki, August 1982. Acta zoologica fennica* (in press).

Howitt, H. 1925. Another invasion of Canada. *Canadian Field Naturalist* **39**: 158–60.

Huey, W. S. 1968. New game, New Mexico's exotics. *New Mexico Wildlife*, 12–15, 22–3.

———. 1972. Exotic animals. *New Mexico Wildlife* (May/June), 30–31.

———. 1973. Old world perspective on New World introductions. *New Mexico Wildlife*, 18–21.

Humble, A. E. 1965. Banteng cattle in the Northern Territory. *Pastoral Review & Grazier's Record* **75**: 1097.

Humphrey, S. R. 1974. Zoogeography of the nine-banded armadillo in the United States. *BioScience* **24**: 457–62.

Hutton, F. W. 1876–7. Notes on the Maori rat. *Transactions of the New Zealand Institute* **9**: 348; **10**: 288.

Hutton, F. W. & **Drummond, J.** 1904. *The animals of New Zealand.* Whitcombe & Tombs: Christchurch.

I

Il'in, V. 1962. The European hare in Irkutsk Oblast. *Sel'skoe Khoz Sibiri* **12**: 71 (reprinted 1963 *Biologicheskii Zhurnal* **101**: 262).

Ingersoll, J. M. 1964. The Australian rabbit. *American Scientist* **52**: 265–73.

Isaković, I. 1970. Game management in Yugoslavia. *Journal of Wildlife Management* **34**: 800–12.

J

Jablonska, J. 1965. Remarks with regard to *Nyctereutes procyonoides. Przegląd Zoologiczny, Wrocław* **9**: 299–300.

Jackson, A. W. 1964. Texotics. *Texas Game & Fish* **22**(4): 7–11.

Jackson, O. F. (chairman). 1982. *Feral cats in the United Kingdom. Report of the Working Party on Feral Cats, 1977–81*, 38 pp. Royal Society for the Prevention of Cruelty to Animals: Horsham, Sussex.

Jarvis, P. J. 1979. The ecology of plant and animal introductions. *Progress in Physical Geography* **3**: 187–214.

———. 1980. The biogeography and ecology of introduced species. *Department of Geography, University of Birmingham, Working Paper Ser.* No. 1: 1–40.

Jenkin, J. F, Johnstone, G. W. & **Copson, G. R.** 1982. Introduced animal and plant species on Macquarie Island. *Colloque sur les Ecosystèmes Subantarctiques, 1981. Publications du Comité National Français des Recherches Antarctiques (CNFRA), Paris*, No. 51: 301–13.

Jenkins, C. F. H. 1977. *The Noah's ark syndrome: one hundred years of acclimatization and zoo development in Australia.* Zoological Gardens Board: Perth, W. Australia.

Jewell, P. A. ed. 1974. *Island survivors; the ecology the Soay sheep of St. Kilda.* Athlone Press: London.

Johnson, G. R. 1884. Remarks on the Maori rat. *Proceedings of the New Zealand Institute* **16**: 553–4.

Jones, E. 1977. Ecology of the feral cat on Macquarie Island. *Australian Journal of Wildlife Research* **4**: 248–62.

Jones, J. L. 1976. The spread of wild mink. *Country Life* **159**: 420–2.

Jones, P. 1959. The European wild boar in North Carolina. Report for the North Carolina Wildlife Resources Commission, Game Division, 30 pp.

Jones, R. D. & **Byrd, G. V.** 1979. Interrelations between seabirds and introduced animals. In *Conservation of marine birds of northern North America (Papers of the international symposium at Seattle, 1975).* US Fish & Wildlife Service (Wildlife Research Report No. 11).

Judd, H. P. 1930. The goat menace on Hawaii. *Friend* **100**: 193–4.

Jungius, H. 1978. Criteria for the reintroduction of threatened species into parts of their former range. In *Proceedings of the Workshop Meeting of the Survival Service Commission, Deer Specialist Group, 1977*, 342–52. IUCN: Morges.

K

Kaburaki, T. 1934 & 1940. Effect of some exotic plants and animals upon the flora and fauna of Japan. *Proceedings of the 5th Pacific Science Congress, Canada 1933* **1**: 801–5. University of Toronto Press. *Proceedings of the 6th Pacific Science Congress, California 1939* **4**: 229–30. University of California Press: Berkeley.

Kalela, O. 1942. Eräitä eläinten leviämistä koskevia nykytärkeitä kysymksiä. *Luonnon Ystävä* **46**: 53–64.

———. 1948. Hillerin levinneisyyden muutoksista Suomessa. *Suomen Riista* **2**: 77–92.

Kampmann, H. 1972. Der Waschbär [raccoon] in Deutschland. Dissertation, Faculty of Forestry, University of Göttingen.

Karl, B. J. & **Best, H. A.** 1982. Feral cats on Stewart Island: their foods and their effects on Kakapo. *New Zealand Journal of Zoology* **9**: 294–8.

Kastdalen, A. 1982. Changes in the biology of Santa Cruz Island between 1935 and 1965. *Noticias Galápagos* **35**: 7–12.

Kean, R. I. 1951. Estimation of the effects of deer and opossums in high country. *New Zealand Science Reviews* **9**: 146–9.

Kean, R. I. & **Pracy, L. T.** 1953. Effects of the Australian opossum on indigenous vegetation in New Zealand. *Proceedings of the 7th Pacific Science Congress, Auckland and Christchurch, 1949* **4**: 696–715.

Keiper, R. R. 1976. Social organization of feral ponies. *Proceedings of the Pennsylvania Academy of Sciences* **50**: 69–71.

Kepler, C. B. 1967. Polynesian rat predation on nesting Laysan albatrosses and other Pacific seabirds. *Auk* **84**: 426–30.

Kerschagl, W. 1964. Das Damwild [fallow deer]. In St Fürlinger, H. ed., *Jagd in Österreich*, 113–15. Fürlinger: Vienna.

***Kiddie, D. G.** 1962. The sika deer in New Zealand. *New Zealand Forest Service Information Ser.* No. 44: 1–37.

Kightley, S. P. J. & **Lewis Smith, R. I.** 1976. The influence of reindeer on the vegetation of South Georgia. *British Antarctic Survey Bulletin* No. 44: 57–76.

King, C.M. 1975. Stoats, weasels and ferrets. In Knox, R. ed., *New Zealand's nature heritage*, 345–53. Hamlyn: Hong Kong.

———. 1981. Studies on the control of stoats in the National Parks of New Zealand. In Chapman, A. & Pursley, D. eds, *Proceedings of the 1st World Furbearers Conference, Maryland, 1980*. University of Maryland Press: Adelphi.

King, C. M. & **Moody, J. E.** 1982. The biology of the stoat in the national parks of New Zealand. *New Zealand Journal of Zoology* **9**: 49–144 (collected offprints of a suite of seven papers).

King, C. M. & **Moors, P. J.** 1979. The life-history tactics of mustelids, and their significance for predator control and conservation in New Zealand. *New Zealand Journal of Zoology* **6**: 619–22. (See also *Oecologia* **39**: 129–50.)

King, W. 1968. As a consequence many will die. *Florida Naturalist* **41**: 99–103.

Kingsmill, W. 1920. Acclimatization. *Proceedings of the Royal Society of Western Australia* **5**: 33–8.

Kiris, I. D. ed. 1973. *Acclimatization of game and commercial animals and birds in the U.S.S.R.* Volgo-Vyatskoe Kn. Izdatelstvo: Kirov.

Kirk, G. 1973. Zur Geschichte der Einführung von Säugetieren und Vögeln in der Sowjetunion, in der Volksrepublik China und in der Mongolischen Volksrepublik. *Archiv für Naturschutz und Landschaftsforschung* **13**: 3–20.

Kirk, H. B. 1920. Opossums in New Zealand. *Appendices to the Journal of the New Zealand House of Representatives, Session 1*, H-28.

Kirk, T. 1895. The displacement of species in New Zealand. *Transactions of the New Zealand Institute* **28**: 1–27. (See also *Proceedings of the Wellington Philosophical Society.)*

Klein, D. R. 1959. St. Matthew island reindeer-range study. *US Fish & Wildlife Service Special Scientific Report, Wildlife* **43**: 1–48.

———. 1968. The introduction, increase and crash of reindeer on St. Matthew Island. *Journal of Wildlife Management* **32**: 350–67.

Klepperstück, J. 1954. *Der Sumpfbiber* [coypu]. *Neue Brehmbücherei.* A. Ziemsen Verlag: Wittenberg Lutherstadt.

Kluyver, H. M. 1937. De bisamrat. *Mededelingen Plantenziektenkundige Dienst, Wageningen* **85**: 8.

Knox, F. J. 1872. Observations on the kiore or indigenous rat of New Zealand. *Transactions of the New Zealand Institute* **4**: 362.

Knox, G. A. 1976. Zoological research in New Zealand. *New Zealand Journal of Zoology* **3** (1).

Koehler, J. W. 1960. The California undomesticated burro. *Bulletin of the Californian Department of Agriculture* **49** (1).

Koenders, J. W. 1964. De populatie beverratten [coypu] langr de Roer in Nederland voor en na de winter van 1962–63. *Lutra* **6**: 68–73.

Koersveld, E. van. 1953. De muskrat in Nederland en zijn bestrijding. *Jaarboek, 1951–2, Plantenziektenkundige Dienst, Wageningen*, No. 120: 229–49.

Koford, C. B. 1966. *Economic resources of the Galápagos Islands.* In Bowman, R. I. ed., *The Galápagos*, 286–90. (*Proceedings of the Symposium of the Galápagos International Scientific Project.*) University of California Press: Berkeley.

Kokes, O. 1976. [Seventy years since the introduction of muskrats in Bohemia.] *Časopis Národního Musea, Oddil Príroda* **145**: 107–14 [in Czech].

Korsakov, G. K. & **Sir'ajev, V. V.** 1978. Muskrats in the Ili river delta. *Proceedings of the 2nd International Theriological Congress, Brno, June 1978*, 303 [Abstract]. Czechoslovak Academy of Sciences.

Kotrlá, B. & Kotrlý, A. 1977. Helminths of wild ruminants introduced into Czechoslovakia. *Folia Parasitologica, Prague* **24**: 35–40.

Krafft, A. 1964. Management of moose in a Norwegian forest. *Meddelelser fra Statens Viltundersøkelser*, Ser. 2, No. 16: 1–61.

Kramer, P. 1984. Man and other introduced organisms [in the Galápagos Islands]. *Biological Journal of the Linnean Society* **21**: 253–8.

***Kramer, R. J.** 1971. *Hawaiian land mammals.* Charles E. Tuttle: Rutland, Vermont.

Krasnopevcev, N. V. 1953. Some results in the acclimatization of brown hares in the Novosibirsk region. *Izvestiya Tomskogo (Gosudarstvennogo) Universiteta* **123**: 119–28.

Kratochvil, J. 1964. Die ersten Nachrichten über das Vorkommen des Marderhundes [raccoon-dog] in der Tschechoslowakei. *Zoologické Listy, Brno* **13**: 174–5.

Kruuk, H. 1979a. Ecology and control of feral dogs in the Galápagos. Unpublished Report, Institute of Terrestrial Ecology, Banchory, Scotland, 59 pp.

———. 1979b. Ecology and control of feral dogs in Galápagos. *Annual Report of the Charles Darwin Research Station, Galápagos*, 100–8.

Kruuk, H. & **Snell, H.** 1981. Prey selection by feral dogs from a population of marine iguanas. *Journal of Applied Ecology* **18**: 197–204.

Kubiak, H. 1965. The appearance of raccoon dog in Cracow District (Poland). *Przegląd Zoologiczny, Wrocláw* **9**: 417–22.

Kuroda, N. 1939. Distribution of mammals in the Japanese empire. *Journal of Mammalogy* **20** (1).

———. 1955. The present status of the introduced mammals in Japan. *Journal of the Mammal Society of Japan* **1**(2): 13–18.

Kuschel, G. ed. 1975. *Biogeography and ecology in New Zealand*. Junk: The Hague.

Kuznetsov, B. A. 1944. In Boabrinskii, N. A., Kuznetsov, B. A. & Kuzyakin, A. P. [*Key to the mammals of the USSR.*] Moscow. [In Russian.]

L

Lahti, S. & **Helminen, M.** 1974. *Castor fiber* and *C. canadensis* in Finland. *Acta theriologica* **19**: 177–89.

———. 1976. [Beavers in Finland.] *Proceedings of the Voronezhsky State Reserve* **2** (21) [in Russian].

Lamond, H. G. 1950. Island venison [in Australia]. *Walkabout* **16**: 32–3.

———. 1951. Wild pigs [in Australia]. *Walkabout* **17**: 38–9.

———. 1960–1. Pests [in Australia]. *Pastoral Review & Grazier's Record* **70**: 469; **71**: 484, 735–6.

———. 1961. Camel teams. *Walkabout* **27**: 38–41.

Lane, J. N. 1969. Strange mammals in Florida. *Florida Naturalist* **42**(1): 50.

Langham, N. P. E., Wells, D. R. & **Charles, J. K.** 1975. Further notes on the vertebrates of Pulau Perak, July 1973. *Malayan Nature Journal* **28**: 13–17.

Länsimäki, A. 1951. Villi minkki tavattu Uudellamaalla. *Metsästys ja Kalastus* **40**: 341.

Lantis, M. 1950. The reindeer industry in Alaska. *Arctic* **3**: 27–44.

Lantz, D. E. 1909. The brown rat in the United States. *US Department of Agriculture Biological Survey Bulletin* No. 33.

Larrison, E. J. 1943. Feral coypus in the Pacific northwest. *Murrelet* **24**: 3–9.

Latham, R. M. 1955. The controversial San Juan rabbit. *Transactions of the 20th North American Wildlife Conference*, 406–14.

Laurie, A. 1983. Marine iguanas in Galápagos (predation by dogs, cats, rats and pigs). *Oryx* **17**: 18–25.

Laurie, E. M. O. 1946a. The reproduction of the house mouse (*Mus musculus*) living in different environments. *Proceedings of the Royal Society of London* B **133**: 248–81.

———. 1946b. The coypu in Great Britain. *Journal of Animal Ecology* **15**: 22–34.

Lavov, M. A. 1973. Acclimatization of beavers in the Krasnoyarsk Krays [USSR]. *Byulleten' Moskovskogo Obshchestva Ispȳtateleĭ Prirodȳ*, Otdel Biol. **78**(6): 40–44.

Lavrov, L. 1965. [Canadian beavers in the Karelian ASSR.] *Hunting & Hunting Economy Magazine, N.G.* [in Russian].

Lavrov, N. P. 1947. *New fur animals of the U.S.S.R.* Moscow Zoo Park.

———. 1955. [Dynamics of distribution and economic significance of muskrats in USSR.] *Zoologicheskiĭ Zhurnal* **34**: 441–53 [in Russian].

———. 1957. In Tsentrosoyuza, M. ed. *Acclimatization of muskrats in the U.S.S.R.* (Transl. by Department of Northern Affairs & Natural Resources, Canada.)

———. 1960. Acclimatization of muskrats in the U.S.S.R. *Translations of the Russian Game Reports* **7**: 1–150. Canadian Wildlife Service.

———. 1969. [Acclimatization of muskrats (in USSR).] *Príroda Moscow* **8**: 64–8 [in Russian].

———. 1978. Results of introduction of the muskrat in the U.S.S.R. *Proceedings of the 2nd International Theriological Congress*, 304 (abstract).

Lavrov, N. P. & **Pokrovsky, V. S.** 1967. Ecological relationships between local fauna and species commercially introduced in the USSR. *Proceedings of the 10th IUCN Technical Meeting, Lucerne, June 1966, IUCN Publications* n.s. No. 9: 181–93.

Lavsund, S. ed. 1977. *Proceedings from the Nordic Symposium on the Beaver, Ramsele, Sweden, 1975.* Royal College of Forestry: Stockholm (Institute of Forest Zoology Research Note No. 26).

Laycock, G. 1966. *The alien animals*. Ballantine: New York.

Layne, J. N. 1976. The armadillo. *Florida Naturalist* **49**(3): 8–12.

Le Souef, L. 1912. Acclimatization. In *Handbook of Western Australia*, 249–52.

Le Souef, W. H. D. 1980. Acclimatization in Victoria. *Report of the Second Meeting of the Australian Association for the Advancement of Science*, 476–82.

***Leader-Williams, N.** 1978a. The history of the introduced reindeer of South Georgia. *Deer* **4**: 256–61.

———. 1978b. Reindeer in the Antarctic. *Natural Environment Research Council News Journal* **2**(6): 4–6.

———. 1980a. Population ecology of reindeer on South Georgia. In Reimers, E., Gaare, E. & Skjenneberg, S. eds, *Proceedings of the 2nd International Reindeer/Caribou Symposium, Røros, 1979, Norway*, 664–76. Direktoratet for Vilt og Fersvannsfisk: Trondheim.

———. 1980b. Population dynamics and mortality of reindeer introduced into South Georgia. *Journal of Wildlife Management* **44**: 640–57.

Leader-Williams, N. & **Payne, M. R.** 1980. Status of *Rangifer* on South Georgia. In Reimers, E., Gaare, E. & Skjenneberg, S. eds, *Proceedings of the 2nd International Reindeer/Caribou Symposium, Røros, 1979, Norway*, 786–9. Trondheim.

Leader-Williams, N., Scott, T. A. & **Pratt, R. M.** 1981. Forage selection by introduced reindeer on South Georgia, and its consequences for the flora. *Journal of Applied Ecology* **18**: 88–106.

Lee, L. n.d. New Mexico's exotic game program. New Mexico Department of Game & Fisheries unpublished mimeo report, 7 pp.

Lehman, E. 1969. Über das Vorkommen des spanischen Rothirches [red deer] in Nordmorokko. *Säugetierkundliche Mitteilungen* **17**: 137–41.

Lehmann, V. W. 1948. Restocking on King ranch. *Transactions of the 13th North American Wildlife Conference*, 236–42.

Lennox, C. G. 1951. *Exotic animals in Hawaii: a history and analysis of the present day situation*. Hawaiian Academy of Sciences: Honolulu.

Leopold, A. S. & **Darling, F. F.** 1953. *Wildlife in Alaska: an ecological reconnaissance* [reindeer]. Ronald Press: New York.

Lesel, R. 1967. La renne aux Îles Kerguelen. In: Contribution à l'étude écologique de quelques mammifères importés aux Îles Kerguelen. *Terres Australes et Antarctiques Française* No. 38: 3–40.

Lesel, R. & **Derenne, P.** 1975. Introducing animals to Îles Kerguelen. *Polar Record* **17**: 485–94.

Letts, G. A. 1960. The Cobourg Peninsula native flora and fauna reserve. Report to the Division of Animal Industry, Northern Territory Administration, Australia.

———. 1961. Buffaloes – possible utilization in the Northern Territory. *Proceedings of the Northern Territory Scientific Liaison Conference, Melbourne*, 37–41.

———. 1962a. Buffaloes of the Northern Territory. *Australian Territorian* **2**: 11–16.

———. 1962b. [Buffaloes.] *Australian Veterinary Journal* **38**: 282.

———. 1962c. Early livestock introductions to the 'Top End' of the Northern Territory. *Australian Veterinary Journal* **38**: 282–7.

———. 1962d. Northern Territory buffalo industry. *Northern Australian Monthly* **9**: 6–7.

———. 1964. Feral animals in the Northern Territory. *Australian Veterinary Journal* **40**: 84–8.

***Letts, G. A., Bassingthwaighte, A.** & **de Vos, W. E. L.** 1979. *Feral animals in the Northern Territory: report of the board of enquiry*. Department of Primary Production, Northern Territory, 234 pp.

Leutscher, A. 1954. The edible dormouse. *Discovery* **15**: 70–1.

Lever, C. 1969. The invaders, and how they came here. *The Field* (28 August): 395.

———. 1977a. Britain's naturalized animals. *Illustrated London News* 45–7.

*———. 1977b. *The naturalized animals of the British Isles*. Hutchinson: London. (Paperback edn, Paladin Granada, 1979.)

———. 1978a. Are wild mink a threat? *New Scientist* **80**: 712.

———. 1978b. The not so innocuous mink? *New Scientist* **78**: 812–14.

———. 1979a. A question of survival. *Country Life* 400–1.

———. 1979b. The invaders. *The Listener* **101**: 716.

———. 1980a. Naturalized animals of the British Isles. *Animals* **3**: 4–7.

———. 1980b. No beavers for Britain. *New Scientist* **87**: 471–2.

———. 1980c. No to the beaver. *Ecos* **1**(2): 22–3.

———. 1980d. Unwanted visitors. *Wildlife Annual*. London Editions/World Wildlife Fund.

———. 1981a. The case against the beaver. *Illustrated London News* (April).

———. 1981b. Unwelcome guests. *The Living Countryside* **2**(21).

———. 1982. Introduced species: reasons, results and criteria – with special reference to mammals. *Paper presented at the British Herpetological Society, Fauna & Flora Preservation Society and Mammal Society Joint Symposium on Introduced Species*.

———. 1983. Golden hamsters in north-west London. *London Naturalist* **62**: 111.

Levi, H. J. 1952. Evaluation of wildlife importations. *Scientific Monthly* **74**: 315–22.

Levins, R. & **Heatwole, H.** 1973. [Palominitos Island.] *Ecology* **54**: 1056–64.

Lewin, R. 1978. Galápagos: the endangered islands. *New Scientist* **79**: 168–72.

Lewis, C. B. 1953. Rats and the mongoose in Jamaica. *Oryx* **2**: 170–2.

Liljestrom, C. 1954. Bisam [muskrat] i Norrbotten. *Svensk Jakt* **10**: 341.

Lindemann, W. 1956. Transplantations of game in Europe and Asia. *Journal of Wildlife Management* **20**: 68–70.

Lindsay, D. C. 1973. Effects of reindeer on plant communities in the Royal Bay area of South Georgia. *British Antarctic Survey Bulletin* No. 35: 101–9.

Linduska, J. P. 1956. Foreign game animals introduction. *Proceedings of the 46th Convention of the International Association of Game, Fish & Conservation Commissioners*, 111–16.

***Linn, I. L.** ed. 1979. *Wildlife introductions to Great Britain. Report by the Working Group on Introductions of the UK Committee for International Nature Conservation*, 32 pp. Nature Conservancy Council.

Linn, I. & **Chanin, P.** 1978a. Are mink really pests in Britain? *New Scientist* **77**: 560–2.

———. 1978b. More on the mink 'menace'. *New Scientist* **77**: 38–40.

Litjens, B. E. J. 1980a. Is de Amerikaanse nerts [mink] in Nederland al in het wild ingeburgerd? *De Pelsdierenfokker* No. 12: 359–61.

———. 1980b. [The coypu in the Netherlands. 1. Fluctuations in the population 1963–79.] *Lutra* **23**: 43–53 [in Dutch].

Lloyd, H. G. 1962. The distribution of grey squirrels in England and Wales. *Journal of Animal Ecology* **31**: 1157–66.

———. 1980. *The red fox.* Batsford: London.

———. 1983. Past and present distribution of red and grey squirrels [in the UK]. *Mammal Review* **13**(2–4): 69–80.

Lockhart, C. J. 1887. Rabbits in Australia. *Blackwood's Magazine* (December).

***Logan, P. C.** & **Harris, L. H.** 1967. Introduction and establishment of red deer in New Zealand. *New Zealand Forest Service Information Ser.* No. 55.

Lomen, C. J. 1919. The camel of the frozen desert [reindeer]. *National Geographic Magazine* **36**: 538–6.

Long, J. L. 1972. Introduced birds and mammals in Western Australia. *Agriculture Protection Board of Western Australia Technical Ser.* **1**: 1–30.

Low, J. B. 1946. Nutria introduced in Utah marshlands near Salt Lake, Utah. *Fish & Game Bulletin* **3**(4): 4–6.

Luit, M. van der. 1977. Enige ecologische aspeeten van de beverrat [coypu] in Nederland met betrekking tot een juist beheer. Doctoraal verslag, Rijksuniversiteit, Utrecht, 82 pp.

Lundgren, E. 1972. *Research report on the population of Javan rusa on the Bensbach plains, Western District, Papua.* Department of Agriculture, Stock & Fisheries, Wildlife Section Research Group: Port Moresby, Papua New Guinea.

M

MacArthur, R. H. & **Wilson, O. E.** 1967. *The theory of island biogeography.* Princeton University Press: Princeton, New Jersey.

McCarthy, F. D. 1954. Buffalo hunting in Arnhem Land. *Australian Museum Magazine* **11**: 252–6.

McDonald, P. J. 1959. The donkeys are doomed. *Journal of Agriculture of Western Australia* **3**: 180–2.

McGowan, N. 1964. Texotics. *Texas Game & Fish* **22**: 16–19.

McKelvey, P. J. 1959. Animal damage in North Island protection forests. *New Zealand Science Review:* **17**: 28–34.

McKenna, M. C. 1975. Towards a phylogenetic classification of the Mammalia. In Luckett, W. P. & Szalay, F. S. eds, *Phylogeny of the primates*, 21–46. Plenum Press: New York.

McKinnon, A. D. & **Coughlan, L.** 1960–4. *Data on the establishment of some introduced animals in New Zealand forests,* Vols I–VIII. New Zealand Forest Service: Wellington.

MacKinnon, J. 1981. The distribution and status of wild pigs in Indonesia. Unpublished report (mimeo) to the IUCN/SSC Pigs & Peccaries Specialist Group, 9 pp.

Mackintosh, G. R. 1941. The wild pig menace. *New Zealand Journal of Agriculture* **62**: 239–43.

***McKnight, T. L.** 1958. The feral burro in the United States: distribution and problems. *Journal of Wildlife Management* **22**: 163–79.

*———. 1959. The feral horses in Anglo-America. *Geographic Review* **49**: 506–25.

———. 1961. Feral livestock in California. *Yearbook of the Association of Pacific Coast Geographers* **23**: 28–42.

*———. 1964. Feral livestock in Anglo-America. *University of California Publications in Geography* No. 16.

*———. 1969. *The camel in Australia.* Melbourne University Press: Melbourne.

———. 1971. Australia's buffalo dilemma. *Annual of the Association of American Geographers* **61**: 759–73.

*———. 1976. Friendly vermin: a survey of feral livestock in Australia. *University of California Publication in Geography* No. 21.

MacRoberts, M. H. 1970. The social organization of Barbary apes on Gibraltar. *American Journal of Physical Anthropology* **33**: 83–100.

Magee, B. 1946. Rabbit history: who introduced them. *New Zealand Farmers Weekly* (Sept. 19).

Maingard, J. R. 1954. Sambar deer in Mauritius. *Journal of the Bombay Natural History Society* **50**: 648–9.

Mallach, N. 1967. Ein vereinfachtes Verfahren zur Bisammarkierung [muskrat]. *Nachrichtenblatt für den Deutschen Pflanzenschutzdienst* **19**: 90–1.

———. 1968. Beobachtungen und Erfahrungen bei der Bisambekämpfung [muskrat] 1967 in Bayern. *Bayerisches Landwirtschaftliches Jahrbuch* **45**: 460–5.

———. 1971. Markierungsversuche zur Analyse des Aktionsraum und der Ortsbewegungen des Bisams [muskrats]. *Anzeiger für Schädlungskunde u Pflavizensch* **44**: 129–36. (See also **45**: 40–4 (1972).)

Mallon, D. 1970. Britain's wild wallabies. *Animals* **13**: 256–7.

Mansueti, R. 1950. Extinct and vanishing mammals of Maryland. *Maryland Naturalist*: **20**: 3–48.

Manville, R. H. 1962. The nutria in the United States. *U.S. Bureau for Sport, Fisheries & Wildlife, Wildlife Leaflet* No. 445, 4 pp.

Maples, W. R., **Brown, A. B.** & **Hutchens, P. M.** 1976. Introduced monkey populations at Silver Springs, Florida. *Florida Anthropologist* **29**: 133–6. (See also *American Journal of Physical Anthropology* **41**: 492 (1974).)

Marchström, V. 1964. The muskrat in northern Sweden. *Viltrevy* **2**: 329–407.

Marlow, B. J. 1962. Dingoes. *Australian Natural History* **14**: 61–3.

———. 1983. *Marsupials of Australia*. Jacaranda Press: Brisbane.

Marples, R. R. 1955. *Rattus exulans* in western Samoa. *Pacific Science* **9**: 171–6.

Marques, A. 1905. Goats in Hawaii. *Hawaiian Annual for 1906*, 48–55.

Marshall, P. 1934. Effect of the introduction of exotic plant and animal forms into New Zealand. *Proceedings of the 5th Pacific Science Congress, June 1933* **1**: 811–16. University of Toronto Press: Toronto.

Marshall, W. H. 1961a. A note on the food habits of feral cats on Little Barrier Island, New Zealand. *New Zealand Journal of Science* **4**: 822–4.

———. 1961b. *Ecology of mustelids in New Zealand*. Department of Scientific and Industrial Research, Animal Ecology Division: Wellington (Research Information Series No. 1).

———. 1963. *The ecology of mustelids in New Zealand*. Department of Scientific and Industrial Research: Wellington (Research Information Series No. 38).

***Mason, I. L.** ed. 1979. *Inventory of special herds*. FAO/UNEP Project No. FP/1108-76-02 (833) Conservation of Animal Genetic Resources, 114 pp.

———. ed. 1984. *Evolution of domesticated animals*. Longman: London.

Matheson, C. 1931. *The brown and black rat in Wales*. National Museum of Wales: Cardiff.

———. 1932. *Changes in the fauna of Wales within historic times*. National Museum of Wales: Cardiff.

———. 1939. A survey of the status of *Rattus rattus* and its subspecies in the seaports of Great Britain and Ireland. *Journal of Animal Ecology* **8**: 76–93.

———. 1941. The rabbit and hare in Wales. *Antiquity* **15**: 371–81.

Matousek, B. 1966. *Nyctereutes procyonoides* ist bisher nicht ein Glied der slowakischen Fauna. *Sbornik Slovenského Národného Múzea, Bratislava* **12**: 167–8.

Matthams, J. 1921. *The rabbit pest in Australia*. Speciality Press: Melbourne.

Matthews, L. H. 1952. *British mammals*. Collins: London.

Mayne, R. 1935. Naturalization of the muskrat in Belgium. *Annales de la Société royale de Belgique* **66**: 13–19.

Mecke, M. B. 1979a. A discussion of exotic ungulates in Texas. Draft MS.

———. 1979b. Texotics – introduced ungulates of Texas rangelands. *Proceedings of the Annual Meeting of the Society for Range Management* **32**: 39.

Medeiros, J. S. 1954. Experimental introduction of the European mouflon. Unpublished report for Territory of Hawaii, Division of Fish & Game, Honolulu.

———. 1964. Lanai feral goat and mouflon survey (2 papers). Unpublished reports for Territory of Hawaii, Division of Fish & Game, Honolulu.

Meeson, J. 1885. The plague of rats in Nelson and Marlborough. *Transactions of the New Zealand Institute* **17**: 199–207.

Merton, D. V. 1978. Controlling introduced predators and competitors on islands. In Temple, S. A. ed., *Management techniques for preserving threatened species (proceedings of a symposium, 1977)*, 121–8. University of Wisconsin Press: Madison.

Middleton, A. D. 1930. The ecology of the American grey squirrel in the British Isles. *Proceedings of the Zoological Society of London*, 809–49.

*———. 1931. *The grey squirrel*. Sidgwick & Jackson: London.

———. 1932. The grey squirrel in the British Isles 1930–32. *Journal of Animal Ecology* **1**: 166–7.

———. 1935. The distribution of the grey squirrel in Great Britain in 1935. *Journal of Animal Ecology* **4**: 274–6.

***Millar, J. C. G.** 1980. Aspects of the ecology of the American grey squirrel in South Africa. Unpublished MSc thesis, University of Stellenbosch, 310 pp.

Miller, R. 1980. Social organization of feral horses in Wyoming's Red Desert. *Journal of the Colorado–Wyoming Academy of Sciences* **12**: 7–8.

Mills, S. 1982. Burros in Death Valley. *Oryx* **16**: 411–14.

Milner, C., Goodier, R. & **Crook, I. G.** 1968. Feral goats. *Nature in Wales* **11**: 3–11.

Mitchell, J. 1983. Strange beasts on the bonny banks. *Scottish Wildlife* **19**: 20–4.

Mittermeier, R. A. 1972. Jamaica's endangered species. *Oryx* **11**: 258–62.

Miura, S. 1976. Dispersal of nutria in Okayama prefecture, Japan. *Journal of the Mammal Society of Japan* **6**: 231–7.

Moehlman, P. D. 1974 & 1979. Behavior and ecology of feral asses. PhD dissertation, University of Wisconsin, 251 pp; *National Geographic Society Research Reports, 1970 Project*, 405–11.

Moens, R. 1971. Estimation de la densité de populations du rat musque. *Parasitica* **27**: 1–10.

———. 1978. Bio-ecological investigations on muskrats in Belgium. *Parasitica* **34**: 57–121.

***Mohr, E.** 1933. The muskrat in Europe. *Journal of Mammalogy* **14**: 58–63.

Moors, P. J. 1975. Introduced predators and the South Island [New Zealand] robin. *Wildlife – A Review* **6**: 26–31.

———. 1983. Predation by mustelids on the eggs and chicks of native and introduced birds in Kowhai Bush, New Zealand. *Ibis* **125**: 137–54.

———. ed. (in press). *Proceedings of the 18th world conference of the International Council for Bird Preservation at Cambridge, England, 1982: island management sympo-*

sium. ICBP: Cambridge (Technical Publication No. 3).

Morgan, C. L. 1898. *The rabbit question in Queensland.* Watson, Ferguson & Co.: Brisbane.

Mosanský, A. 1971. An introduction of strange faunistic elements in the Tatra national park. *Ochrana Fauny* **5**: 39–40 [English summary, 41].

Müller-Using, von D. 1965. Das Vorkommen der Nutria [coypu] in Deutschland. *Zeitschrift für Jagdwissenschaft* **4**: 161–4.

Mungall, E. C. 1978. *The Indian blackbuck antelope: a Texas view.* Caesar Kleberg Research Program in Wildlife Ecology, Texas A & M University: College Station, 184 pp.

Munro, D. 1923. The rabbit pest. *New Zealand Journal of Agriculture* **26**: 81–4.

Munro, D. & **Wright, R.** 1933. The rabbit pest and its control. *New Zealand Journal of Agriculture* **46**: 26–37.

Munro, G. C. 1970. Axis deer of Molokai and Lanai [Hawaii]. *Elepaio* **31**: 15–17.

Munro, H. 1917. The rabbit pest. *New Zealand Journal of Agriculture* **15**: 206–10.

Munton, P. N. (chairman). 1984. *Feral mammals – problems and potential (Proceedings of a workshop organized by the Species Survival Commission, Caprinae Specialist Group, Helsinki, August 1982)*, 152 pp. IUCN: Morges.

Murie, O. J. 1940. Wildlife in Alaska. *Transactions 5th North American Wildlife Conference*, 532–6.

———. 1951. Distribution and numbers of wapiti. *New Zealand DSIR Research Bulletin* No. 103: 17–21.

Murphy, R. C. 1951. The impact of man upon nature in New Zealand. *Transactions of the American Philosophical Society* **95**: 569–82.

Mursalogu, B. 1973. New records for Turkish rodents [coypu]. *Communications de la Faculté des Sciences de l'Université d'Ankara* Ser. C, *Natural Science* **17**: 213–19.

Myers, J. G. 1931. The present position of the mongoose in the West Indies. *Tropical Agriculture, Trinidad* **8**: 94–5.

Myers, K. 1971. The rabbit in Australia. In den Boer, P. J. & Gradwell, G. R. eds, *Dynamics of populations (Proceedings of the Advanced Study Institute on the dynamics of numbers in populations, Oosterbeck, 1970)*, 478–506. Centre for Agricultural Publishing and Documentation.

Myers, R. H. 1952. New game animal for Kentucky [*L. californicus melanotis*]. *Kentucky Happy Hunting Ground* **8**: 16–30.

Myrberget, S. 1971. Game management in Norway. *Meddelelser fra Statens Viltundersøkelser,* Ser. 2, No. 35: 1–22.

———. 1972. Utsetting av hare på svalbardi 1930 og 1931. *Fauna, Oslo* **25**: 126.

N

Nagahana, M., Hatsushika, R., Shimizu, M. & **Kawakami, S.** 1977. Does the nutria, *Myocastor coypus*, serve as the reservoir of liver fluke, *Clonorchis sinensis? Japanese Journal of Parasitology* **26**: 41–5 [in Japanese with English summary].

Naniragijn, D. 1977. Muskrat in the Mongolian Peoples' Republic. *Beiträge zur Jagd und Wildforschung* **10**: 342–52.

Napier, A. 1964. New game in New Mexico. *Sports Illustrated* (25 May): 57–60.

Nasimovich, A. A. 1961. Some general questions and results of the acclimatization of land animals. *Zoologichesikiĭ Zhurnal* **40**: 957–68.

Naumov, N. P. 1972. In Levine, N. D. ed., *The ecology of animals* (transl. by Plous, F. K.). University of Illinois Press: Champaign.

Naveda, B. H. 1950. *Galápagos a la vista.* Ultimas Noticias: Quito, Ecuador.

Nawa, A. 1965. On a common palm civet captured in Tokyo City. *Journal of the Mammal Society of Japan* **2**: 144–5.

Neill, W. T. 1952. The spread of the armadillo in Florida. *Ecology* **33**: 282–4.

Neitzel, R. 1963. Erster Nachweis des Marderhundes [raccoon dog] für Westdeutschland. *Säugetierkundliche Mitteilungen* **11**: 183–4.

Nellis, D. W., Eichholz, N. F., Regan, T. W. & **Feinstein, C.** 1978. Mongoose in Florida. *Wildlife Society Bulletin* **6**: 249–50.

Nelson, E. W. 1917. The rat pest. *National Geographic Magazine* **32**: 1–23.

Nevard, T. D. & **Penfold, J. B.** 1978. Wildlife conservation in Britain: the unsatisfied demand. *Biological Conservation* **14**: 25–44.

New Zealand acclimatization societies. *Annual reports.*

Newman, C. 1949. Florida's armored invasion [armadillos]. *Florida Wildlife* **3**(5): 3–5.

Nichols, L. 1960a. Ecology of the axis deer. Hawaiian Division of Fish & Game, Job Completion Report Project W-5-R-11, 32 pp.

———. 1960b. History of antelope introduction on Lanai Island, Hawaii. Unpublished (mimeo) report for Department of Agriculture and Conservation, Hawaii.

———. 1961–4. Ecology of the wild pig. Unpublished report for State of Hawaii Division of Fish & Game, Honolulu.

———. 1962. Management of axis deer in Hawaii. *Proceedings of the 42nd Annual Conference of the Western Association of State Fish & Game Commissioners*, 90–7.

***Niethammer, G.** 1963. *Die Einbürgerung von Säugetieren und Vögeln in Europa.* Paul Parey: Hamburg.

———. 1975. On the taxonomy and history of distribution of the black rat. *Zoologischer Anzeiger* **194**: 405–15.

Niethammer, J. 1964. Contribution à la connaissance des mammifères terrestres d'île Indefatigable, Galápagos. Resultats de l'expedition Allemande aux Galápagos 1962/63, No. 8. *Mammalia* **28**: 593–606.

Niethammer, J. & **Krapp, F.** eds. 1982. *Handbuch der Säugetiere Europas*. Akademische Verlagsgesellschaft: Wiesbaden.

Norman, F. I. 1970. The effects of sheep on the breeding success and habitat of the short-tailed shearwater, *Puffinus tenuirostris*. *Australian Journal of Zoology* **18**: 215–29.

———. 1975. The murine rodents *Rattus rattus, exulans* and *norvegicus* as avian predators. *Atoll Research Bulletin* No. 182: 1–13.

Norris, J. D. 1967. A campaign against feral coypus in Great Britain. *Journal of Applied Ecology* **4**: 191–9.

Northcott, T. H., Mercer, E. & **Menchenton, E.** 1974. The eastern chipmunk in insular Newfoundland. *Canadian Field Naturalist* **88**: 86.

Northcott, T. H., Payne, N. F. & **Mercer, E.** 1974. Dispersal of mink in insular Newfoundland. *Journal of Mammalogy* **55**: 243–8.

Noticias de Galápagos **35**: 1–28 (1982) (and other issues).

Novikov, G. A. 1962. *Carnivorous mammals of the fauna of the USSR*. Transl. from Russian by Israel Program for Scientific Translations: Jerusalem.

Novotny, I. 1963. Myvalovec Kuni [raccoon-dog] ve Slezsku. *Zprávy Slezského Ustavu CSAV, Opavě* **128-B:** 12.

———. 1964. Marderhund [raccoon-dog] – ein neus Wirbeltierfaunenelement des Gebietes der CSSR. *Zprávy Geografického Ustavo CSAV, Opavě* No. 5, **134-B:** 2–4.

Nowak, E. 1972. Ausbreitung des Marderhundes [raccoon-dog] in Europa. *Umschau, Frankfurt/Main* **72**: 766–7.

Nožička, J. 1966. Z historie chovu danka a obory Brezka [fallow deer]. *Výzkumný Ústav Lesniho Hospodárstvi a myslivosti, Zbraslav-Strnady, Prace* **32**: 45–65.

O

Oakley, C. 1976. Wyoming's free-roaming horses. *Wyoming Wildlife* **31**(7): 3–5.

O'Connor, J. 1953. The bearded Texan. *Outdoor Life* **112**(2): 44–5, 91–2.

O'Farrell, T. P. 1965. The rabbits of Middleton island, Alaska. *Journal of Mammalogy* **46**: 525–7.

O'Keefe, T. M. 1973. Spread of the aliens. *Florida Sportsman*, 44–51.

Ognev, S. I. 1928–50. [*Mammals of the USSR and adjacent countries (mammals of eastern Europe and northern Asia).*] Moscow. (English transl. 1962–66, Israel Program For Scientific Translations: Jerusalem.)

Ogren, H. A. 1962a. The Barbary sheep of the Canadian river gorge, New Mexico. Unpublished PhD dissertation, University of Southern California, Los Angeles, 234 pp.

———. 1962b. Barbary sheep in New Mexico. *New Mexico Department of Game & Fisheries Bulletin* No. 11, 1–32.

———. 1965. Barbary sheep. *New Mexico Department of Game & Fisheries Bulletin* No. 13: 1–117.

Oko, Z. & **Włodek, K.** 1978. Wystepowanie i dynamkia liczebności daniela w Polsce [fallow deer]. *Roczniki Akademii Rolniczej w Pocnaniu*, C. Zootech. **24**: 113–25.

Ólafsson, B. Ó. 1976. [Icelandic reindeer.] Unpublished BSc thesis, University of Iceland, 33 pp. [in Icelandic].

Oldeneel Tot Oldenzeel, J. van. 1969. The musk-rat position in the Netherlands. *EPPO Publications* Ser. A, No. 47: 39–41.

Oliver, W. L. R. 1984. Introduced and feral pigs. In *Feral mammals – problems and potential* (*Proceedings of a SSC Caprinae Specialist Group Workshop, Helsinki, August 1982*), 85–126. IUCN: Morges.

Olstad, O. 1930. Rats and reindeer in the Antarctic (South Georgia and Deception Islands). *Scientific Results of the Norwegian Antarctic Expedition 1927–1928, et seq.* No. 4: 1–20. Norwegian Academy of Science and Letters: Oslo.

Orwin, J. 1973. *Assessment and management of introduced animals in New Zealand forests*. New Zealand Forest Service (Forest Research Institute Symposia **14**).

Osborn, T. A. B. 1934. Effect of introduction of exotic plants and animals into Australia. *Proceedings of the 5th Pacific Science Congress, June 1933* **1**: 809–10. University of Toronto Press: Toronto.

Osgood, W. H. 1943. Mammals of Chile. *Field Museum of Natural History, Zoological Ser.* **30:** 1–268 (Publication No. 542).

Owadally, A. W. & **Butzler, W.** 1972. *The deer in Mauritius*. Alpha Printing: Port Louis, Mauritius.

Owen, I. L. 1977. Rusa deer as a host for cattle tick in Papua-New Guinea. *Journal of Wildlife Diseases* **13**: 208–17.

P

Packard, R. L. 1955. Release, dispersal, and reproduction of fallow deer in Nebraska. *Journal of Mammalogy* **36**: 471–3.

Page, F. J. T. & **Tittensor, A. M.** 1970–2. *The Sussex Mammal Report for 1969–71*. Sussex Naturalists Trust: Henfield.

Pages, M.-V. 1980. Essai de reconstitution de l'histoire du lapin de garenne en Europe. *Bulletin Mensuel de l'Office National de la Chasse*, No. Spec. Scien.

Tech., 13–21 [English summary].

Palmer, L. J. 1934. Raising reindeer in Alaska. *US Department of Agriculture Miscellaneous Publication* No. 207: 1–8.

Palmer, L. J. & **Rouse, C. H.** 1945. Study of the Alaska tundra with reference to its reactions to reindeer and other grazing. *Research Reports of the US Fish & Wildlife Service* **10**: 1–48.

***Palmer, T. S.** 1899. The danger of introducing noxious animals and birds. *US Department of Agriculture Yearbook, 1898,* 87–110.

Parker, J., Tustin, K. & **Stanley, L.** 1978. The history and control of red deer in the takahe area . . . [of New Zealand]. *New Zealand Journal of Ecology* **1**: 145–52.

Parsons, B. T. & **Middleton, A. D.** 1937. The distribution of the grey squirrel in Great Britain in 1937. *Journal of Animal Ecology* **6**: 286–90.

Pascal, M. 1980. Population structure and dynamics of feral cats on Kerguelen Island. *Mammalia* **44**: 161–82.

———. 1982. Les espèces mammaliennes introduites dans l'archipel des Kerguelen bilan des récherches entreprises sur ces espèces. *Colloque sur les Ecosystemes subantarctiques. Publications du Comité National Français des Recherches Antarctiques (CNFRA), Paris* No. 51: 269–88.

Pascal, M & **Meyer, J. A.** 1984. Modelling the cat population on Kerguelen archipelago. *Proceedings of the 3rd International Theriological Congress, Helsinki, August 1982. Acta zoologica fennica* (in press).

Patton, J. L., Yang, S. Y. & **Myers, P.** 1975. Genetic and morphological divergence among introduced rat populations of the Galápagos Archipelago, Ecuador. *Systematic Zooloogy* **24**: 296–310.

Pavlov, P. N. & **Hone, J.** 1982. The behaviour of feral pigs in flocks of lambing ewes. *Australian Wildlife Research* **9**: 101–9.

Payne, N. 1976. Red squirrel introduction to Newfoundland. *Canadian Field Naturalist* **90**: 60–4.

Pearson, J. 1938. The Tasmanian brush opossum. *Papers and Proceedings of the Royal Society of Tasmania, 1937,* 21–9.

Pedersen, J. A. 1964a. Is the raccoon-dog likely to colonize Norway? *Fauna, Blindern* **17**: 179–89.

———. 1964b. Villminken i Norge. *Naturen,* No. 9: 515–26.

———. 1970. Muskrat new in Norway. *Fauna, Oslo* **23**: 297.

Pegau, R. 1970. Effect of reindeer trampling and grazing on lichens. *Journal of Range Management* **23**: 95–7.

Pemberton, C. E. 1925. The field rat in Hawaii and its control [by mongooses]. *Bulletin of the Hawaiian Sugar Cane Planters Association,* Entomological Ser. No. 17.

———. 1933. Some food habits of the mongoose. *Hawaiian Planters Record* **37**: 12–13.

Perham, A. N. 1922. Deer in New Zealand. *Appendices to the Journal of the House of Representatives, New Zealand,* C–3A.

Perkins, R. C. L. 1913. Introduction. *Fauna Hawaiiensis,* Vol. 1. Cambridge University Press: London.

Pernetta, J. C. & **Watling, D.** 1978. The introduced and native terrestrial vertebrates of Fiji. *Pacific Science* **32**: 223–44.

Perry, R. 1969. Conservation problems in the Galápagos islands. *Micronesica* **5**: 275–81.

Petrides, G. A. 1950. The nutria comes to Texas. *Texas Game & Fish,* 4–5.

———. ed. 1968. Problems in species' introductions. *IUCN Bulletin* No. 2: 70–1.

———. 1975a. The importation of wild ungulates into Latin America. . . . *Publicaciones Biologicas, Instituto de Investigaciones Cientificas, UANL* **1**: 169–79.

———. 1975b. The importation of wild ungulates into Latin America, with remarks on their environmental effects. *Environmental Conservation* **2**: 47–51.

Petrides, G. A. & **Leedy, D. L.** 1948. The nutria in Ohio. *Journal of Mammalogy* **29**: 182–3.

Petrie, D. 1883. Some effects of the rabbit pest. *New Zealand Journal of Science* **1**: 412–14.

Phillipson, N. E. 1888–9. Camels in Australia. *Proceedings of the Royal Geographical Society of Australia (S. Australia Branch)* **3**: 83–92.

Pimental, D. 1955a. The control of the mongoose in Puerto Rico. *American Journal of Tropical Medicine & Hygiene* **4**: 147–51.

———. 1955b. Biology of the Indian mongoose in Puerto Rico. *Journal of Mammalogy* **36**: 62–8.

Pimlott, D. H. 1963. Newfoundland moose. *Transactions of the 28th North American Wildlife Conference,* 563–81.

Pinder, N. J. 1978. Faunal reintroductions: a policy, a proposal format and an example: beaver in Britain. Unpublished MSc thesis, University College London, 221 pp.

———. 1979. A policy for faunal reintroductions. *University College London Discussion Papers in Conservation* No. 23, 28 pp.

Pine, R. H., Angle, J. P. & **Bridge, D.** 1978. Mammals from the sea, mainland and islands at the southern tip of South America. *Mammalia* **42**: 105–14.

Pleiticha, P. J. R. 1973. Alpensteinböcke (*Capra ibex* L.) im Tiergarten Nürnberg. *Zeitschrift des Kölner Zoo* **14**: 115–21.

Plomley, N. J. B. 1972. Some notes on plagues of small mammals in Australia. *Journal of Natural History* **6**: 363–84.

Poglayen, I. 1962. Introduction of African antelope into New Mexico. *International Zoo Yearbook* **4**: 71–2.

Poirier, F. E. & **Smith E. O.** 1974. The crab-eating macaques of Angaur Island . . .. *Folia primatologia* **21**: 258–306.

Polard, F. 1976. The return of the native. *New Scientist* **69**: 616–17.

Pollard, J. 1945. Buffalo country. *Walkabout* **11**: 25–7.

Poole, A. L. 1959. The noxious animal problem. *New Zealand Science Review* **17**: 18–20.

———. 1968. *Distribution of game animals in New Zealand.* Government Printer: Wellington.

———. ed. 1973. *Wild animals in New Zealand.* A. H. & A. W. Reed: Wellington.

Poole, A. L. & **Holloway, J. T.** 1951. Deer and the vegetation. *New Zealand DSIR Research Bulletin* No. 103: 27–9.

Popov, B. H. & **Low, J. B.** Game, fur animal and fish introductions into Utah. *Utah Department of Fish & Game Miscellaneous Publications* **4**: 1–85.

Poppelwell, D. L. 1929. Some ecological effects of acclimatisation in New Zealand. *New Zealand Journal of Science & Technology* **11**: 125–7.

Poppelwell, D. L. & **Kean, R. I.** 1949. The opossum in New Zealand. *Wildlife Publication, New Zealand Department of Internal Affairs* No. 1: 1–19 (reprinted 1963 in *NZ Forest Service Publication* No. 40).

Powell, L. 1872. Remarks on the New Zealand rat. *Transactions of the New Zealand Institute* **4**: 401.

***Pracy, L. T.** 1962. Introduction and liberation of the opossum (*Trichosurus vulpecula*) into New Zealand. *New Zealand Forest Service Information Series* No. 45: 1–28.

———. 1964. *Opossum control in exotic forests.* New Zealand Forest Service, 14 pp.

———. 1974. Opossums (1). In Knox, R. ed., *New Zealand's Nature Heritage* **3**: 873–82. Hamlyn: Hong Kong.

———. 1977. Notes on opossums in New Zealand. *New Zealand Institute of Foresters Newsletter* **9**: 10–14.

Pracy, L. T. & **Kean, R. I.** 1949. The opossum in New Zealand. *Wildlife Publication New Zealand Department of Internal Affairs* No. 1 (reprinted 1963 in *NZ Forest Service Publication* No. 40).

Preble, E. A. 1923. Mammals of the Pribilof Islands. *North American Fauna* **46**: 102–20.

***Presnall, C. C.** 1958. The present status of exotic animals in the United States. *Journal of Wildlife Management* **22**: 45–50.

Pruthi, H. S. 1950. Consequences of uncontrolled introduction of aggressive exotic species to the native flora and fauna. *Proceedings and papers of International Technical Conference on the Protection of Nature, Lake Success, New York, August 1949.* Unesco: Paris.

Pullar, E. M. 1950. The wild (feral) pigs of Australia and their role in the spread of infectious disease. *Australian Vetarinary Journal* **26**: 99–110.

*———. 1953. The wild (feral) pigs of Australia; their origin, distribution and economic importance. *Memoirs of the National Museum, Melbourne* **18**: 7–23.

Puttonen, P. 1957. Minkki, kalastajien ja metsästajien yhteinen vihollinen. *Metsästys ja Kalastus* **46**: 75–7.

Pye, T. & **Bonner, W. N.** 1980. Feral brown rats in South Georgia. *Journal of Zoology, London* **192**: 237–55.

Q

Quednau, W. n.d. Sumpfbiber [coypus] in freier Wildbahn. *Die Pirsch* **5**: 806.

Quin, W. 1882. Rabbit question. *New Zealand Country Journal* (July).

Qvist, A. 1975. Minkkitarhan aitaaminen. *Turkistalous* **29**: 90–1.

R

Rackham, J. 1979. The introduction of the black rat into Britain. *Antiquity* **53**: 112–20.

Rainey, F. 1941. Native economy and survival in Arctic Alaska [reindeer]. *Applied Anthropology* **1**: 9–14.

Ramsey, C. W. 1967a. Texotics. *Texas Parks & Wildlife Department Bulletin* No. 49: 1–46.

———. 1967b. Texotic lineup. *Texas Parks & Wildlife* **25**(4): 2–7.

———. 1968. State view of governmental and private programs of introduction of exotic animals. In *Proceedings of a symposium on introduced exotic animals: ecological and socioeconomic considerations,* 9–10. Caesar Kleberg Research Program in Wildlife Ecology, Texas A & M University: College Station.

———. 1975. Exotic game on Texas rangelands. *Proceedings of the 30th Annual Conference of the Soil Conservation Society of America,* 123–7.

Ratcliffe, F. N. 1951. The rabbit problem. *CSIRO Publication* No. 1.

———. 1959. The rabbit in Australia. In *Biogeography and ecology in Australia,* Vol. 8, 545–64. Holland.

'Rawden'. 1917. Camels and donkeys in South Australia. *Pastoralists' Review* **27**: 447–8.

Redford, P. 1962. Raccoon in the U.S.S.R. *Journal of Mammalogy* **43**: 541–2.

Reichholf, J. 1977. [Herbivores.] *Zeitschrift für Säugetierkunde* **42**: 189–96.

Reischek, A. 1888. Notes on rats. *Transactions of the New Zealand Institute* **20**: 125–6.

Remfry, J. 1979. Feral cats in Britain. *Bulletin Feline Advisory Bureau* **17**(4): 11–12.

Repa, F. 1971. Introdukcia a aklimatizacia niektorých druhov vtakov a cicavcov na slovensku. *Ceskoslovanská Ochrana Prírody, Bratislava* **12**: 213–26.

Reynolds, J. K. 1952. The biology of the European hare in southwestern Ontario. Unpublished PhD thesis, University of Western Ontario, London.

———. 1955. Distribution and population of the European hare in southern Ontario. *Canadian Field Naturalist* **69**: 14–20.

Richmond, W. K. 1955. Wild goats. *Scottish Field* (May).

Riisalo, J. 1971. The coypu in Viljandi. *Eesti Loodus (Tartu)* **10**: 616.

Riney, T. 1956. Comparison of occurrence of introduced animals with critical conservation areas to determine priorities for control. *New Zealand Journal of Science & Technology* B**38**: 1–18.

———. 1957. Sambar in sandhill country. *Proceedings of the New Zealand Ecological Society* **5**: 26–7.

———. 1964. The impact of introductions of large herbivores on the tropical environment. *Proceedings of IUCN Technical Meeting, IUCN Publications* n.s. No. 4: 261–73.

———. 1967. Ungulate introductions as a special source of research opportunites. *Proceedings of the 10th IUCN Technical Meeting, Lucerne, 1966, IUCN Publications*, n.s. No. 9: 241–55. IUCN: Morges.

***Ritchie, J.** 1920. *The influence of man on animal life in Scotland*. Cambridge University Press: London.

Röben, P. 1975. Zur Ausbreitung des Waschbären [raccoon] und des Marderhundes [raccoon-dog] in der Bundesrepublik Deutschland. *Säugetierkundliche Mitteilungen* **40**: 93–101.

Robinson, D. T. & **Cowan, I. McT.** 1954. An introduced population of the gray squirrel in British Columbia. *Canadian Journal of Zoology* **32**: 261–82.

Rodier, W. 1902. *Rabbit pest in Australia*. W. Brooks & Co.: Sydney.

Roff, C. 1960. Deer in Queensland. *Queensland Journal of Agricultural Science* **17**: 43–58.

***Rolls, E. C.** 1969. *They all ran wild: the story of pests on the land in Australia*. Angus & Robertson: Sydney.

Roots, C. 1976. *Animal invaders*. David & Charles: Newton Abbot.

Rose, C. I. 1978. Introduced species and conservation. Unpublished MSc thesis, University College London.

———. 1979. Nature conservation and species introductions. *University College London Discussion Papers in Conservation*, No. 26: 1–47.

Rougeot, J. 1981. *Le lapin; aspects historiques, culturals et sociaux*. Societé d'Ethnozootechnité: Paris, 10 pp.

Rudge, M. R. 1970. Expected rate of increase of hunted populations of feral goats. *New Zealand Journal of Science* **13**: 256–9.

———. 1976. Feral goats in New Zealand. In Whitaker, A. H. & Rudge, M. R. eds, *The value of feral farm mammals in New Zealand*. New Zealand Department of Lands & Survey (Information Series No. 1).

———. 1983. A reserve for feral sheep on Pitt Island, New Zealand. *Ark* **10**: 211–12.

———. 1984. The occurrence and status of populations of feral goats and sheep throughout the world. In *Feral mammals – problems and potential (Proceedings of a SSC Caprinae Specialist Group Workshop, Helsinki, August 1982)*, 55–84. IUCN: Morges.

Rudge, M. R. & **Campbell, D. J.** 1977. The history and present status of goats on the Auckland Islands . . . . *New Zealand Journal of Botany* **15**: 221–53.

Rudge, M. R. & **Clark, J. M.** 1978. The feral goats of Raoul Island . . . . *New Zealand Journal of Zoology* **5**: 581–9.

Ruhl, H. D. 1940. Game introductions into Michigan. *Proceedings of the North American Wildlife Conference*, 424–7.

Russell, C. 1952. The apes of Gibraltar. *Oryx* **1**: 22–4.

Russell, J. P. 1964. Antelope of New Mexico. *Bulletin of the New Mexico Department of Game & Fish*, No. 12: 1–103.

Rutherford, A. J. 1894. Acclimatization. In *New Zealand Official Yearbook*. Government Printer: Wellington.

Rutland, J. 1890. On the habits of the New Zealand rat. *Transactions of the New Zealand Institute* **22**: 300–7.

S

Safonov, V. G. 1977. Prospects for relocation of beavers in the U.S.S.R. *Proceedings of the 13th International Congress of Game Biologists, Atlanta, March 1977*, 387–92. Wildlife Society: Washington, DC.

———. 1981. The status and reestablishment of fur resources in the U.S.S.R. In Chapman, A. & Pursley, D. eds, *Proceedings of the 1st World Furbearers Conference, Maryland, 1980*, 95–110. University of Maryland Press: Adelphi.

St Amant, J. A. 1977. Some problems with the exotic introductions. *Proceedings of the Annual Conference of the Western Association of State Game & Fish Commissioners* **57**: 247–51.

Salmi, A. M. 1949. The white-tailed deer in Finland. *Suomen Riista* **4**: 124.

Salvesen, S. 1928. The beaver in Norway. *Journal of Mammalogy* **9**: 99–104.

Salvin, O. 1876. On the avifauna of the Galápagos archipelago. *Transactions of the Zoological Society of London* **9**: 447–510.

Santini, L. 1980. La nutria [coypu] allo stato selvatico in Toscana. *Frustula Entomologica* n.s. **1**: 273–88 [in Italian with English summary].

Sartorius, O. 1970. *Stärkste Damschaufler der Welt* [fallow deer]. M. & H. Schaper: Hanover.

Saunders, R. M. 1935. The first introduction of European plants and animals into Canada. *Canadian Historical Review*, 388–406.

Scanlon, P. F., Teitt, T. R. & **Cross, G. H.** 1978. An overview of problems of introduced species . . . . *Proceedings of the Annual Conference of the South-East Association of Fish & Wildlife Agencies* **30**: 674–9.

Scheffer, V. B. 1951. The rise and fall of a reindeer herd. *Scientific Monthly* **73**: 356–62.

Schiffer, S. 1959. Nová zver v nasich revíroch. *Polovníctvo a rybárstvo* **11**: 7.

Schitoskey, F., Evans. J. & **La Voie, G. K.** 1972. Status and control of nutria in California. *Proceedings of the 5th Vertebrate Pest Conference, Fresno, California*, 15–17. University of California Press (for Vertebrate Pest Conference).

Schmidt, F. 1954. Peltierzucht in freier Wildbahn [furbearers]. *Kosmos, Stuttgart* **50**: 73–7.

Schneider, C. O. 1936. Noticia sobre la aclimatacion del *Cervus dama* en Chile. *Communicaciones del Museo de Concepción* **1**: 159–60.

Schoenichen, W. 1953. Südamerikanische National Parks [Argentina]. *Kosmos, Stuttgart* **49**: 433–8.

Schöps, P. 1963. Bisamratten für die Sowjetunion. *Pelzgewerbe* **2–3**: 4.

Schreier, O. 1956. Die gegenwärtige Verbreitung und wirtschaftliche Bedeutung der Bisamratte (*Fiber zibethicus* L.) in Österreich. *Pflanzenschutzberichte* **16**: 97–121.

Schwartz, E. & **Schwartz, H. K.** 1943. The wild and commensal stocks of the house mouse. *Journal of Mammalogy* **24**: 59–72.

Sclater, P. L. 1893. Mr. P. L. Sclater on the Gibraltar monkeys. *Transactions of the Zoological Society of London*, 325.

———. 1900. Mr. Sclater on *Macaca inuus*. *Transactions of the Zoological Society of London*, 773–4.

Scott, P. 1967. Cause and effect in the introduction of exotic species. *Proceedings of the 10th IUCN Technical Meeting, Lucerne, June 1966, IUCN Publications* n.s. No. 9: 120–3.

Seaman, G. A. 1952. The mongoose and Caribbean wildlife. *Transactions of the 17th North American Wildlife Conference*, 188–97.

Seaman, G. A. & **Randall, J. E.** 1962. The mongoose as a predator in the Virgin Islands. *Journal of Mammalogy* **43**: 544–6.

Sedgwick, L. E. 1968. The squirrels of South Perth. *Western Australian Naturalist* **11**(1): 1–4.

Segonzac, M. 1972. Données récentes sur la faune des Îles St. Paul et Nouvelle Amsterdam. *L'Oiseau et la Revue Française d'Ornithologie* **43**: 3–68.

Serventy, D. L. 1937. The menace of acclimatization. *Emu* **36**: 189–96.

***Shaw, A. C.** 1940. The European wild hog in America. *Transactions of the 5th North American Wildlife Conference*, 436–41.

Shaw, T. H. 1951. Occurrence of the muskrat in the Hu Ma district, North-eastern China. *Nature, London* **189**: 1215.

Sheail, J. 1971. *Rabbits and their history*. David & Charles: Newton Abbot.

Sheffield, W. J., Ables, E. D. & **Fall, B. A.** 1971. Geographic and ecological distribution of nilgai antelope in Texas. *Journal of Wildlife Management* **35**: 250–7.

Sherman, H. B. 1936. A list of the recent land mammals of Florida [armadillo]. *Proceedings of the Florida Academy of Sciences* **1**: 123.

———. 1943. The armadillo in Florida. *Florida Entomologist* **26**: 54–9.

———. 1954. Raccoons of the Bahama islands. *Journal of Mammalogy* **35**: 126.

Shorten, M. 1946. A survey of the distribution of the American grey squirrel and the British red squirrel in England and Wales in 1944–45. *Journal of Animal Ecology* **15**: 82–92.

———. 1953. Notes on the distribution of the grey squirrel and the red squirrel in England and Wales 1945–52. *Journal of Animal Ecology* **22**: 134–40.

*———. 1954. *Squirrels*. Collins: London.

———. 1975. Squirrels in England, Wales and Scotland in 1955. *Journal of Animal Ecology* **26**: 287–94.

Shult, M. J. 1973. An introduction to exotics in Texas. Unpublished report for Exotic Big Game Short Course, Texas Agricultural Extension Service, 15 pp.

Siefke, A. 1977. Verbreitung und Bestandgrössen des Damwildes [fallow deer]. *Säugetierkundliche Informationen* **1**: 43–65.

Siegfried, W. R. 1962. Introduced vertebrates in the Cape Province. *Report of the Cape Provincial Administration Department of Nature Conservation* No. 19: 80–7.

Siivonen, L. 1951. Suomen Riistanhoito – Säätaiön riistatiedustelujen pätevyydestä. *Suomen Riista* **6**: 149–54.

———. 1953. Raccoon-dog, the newest game species of Finland. *Suomen Riista* **8**: 117 [in Finnish with English summary p. 232].

———. 1958. [On the early history of the raccoon dog in Finland.] *Suomen Riista* **12**: 165–6 [in Finnish].

***Silver, J.** 1924. The European hare in North America. *Journal of Agricultural Research* **28**: 1133–7.

———. 1927. The introduction and spread of house rats in the United States. *Journal of Mammalogy* **8**: 58–60.

———. 1942. The house rat. *US Fish & Wildlife Service Wildlife Circular* No. 6.

Silverstein, A. & **Silverstein, V.** 1974. *Animal invaders: the story of imported wildlife*. Atheneum: New York.

Simonic, A. 1970. Acclimatization of mouflon in Slovenia. *Proceedings of the 8th International Congress of Game Biologists, Helsinki 1967. Rüstatieteellisiä Julkaisiya* No. 30, 282–8.

Simpson, C. D. ed. 1980. *Proceedings of the symposium on ecology and management of Barbary sheep (November 1979)*. Department of Range & Wildlife Management, Texas Tech University: Lubbock, 112 pp.

Simpson, C. D., Krysl, L. J., Hampy, D. B. & **Gray, G. G.** 1978. The Barbary sheep: a threat to desert bighorn survival. *Transactions of the Desert Bighorn Council* **22**: 26–31.

Simpson, F. L. 1944. Mongoose *Mele. Paradise of the Pacific* **56**(7): 10.

Simpson, G. G. 1945. The principles of classification and a classification of the mammals. *Bulletin of the American Museum of Natural History* **85**.

———. 1961. Historical zoogeography of Australian mammals. *Evolution* **15**: 431–6.

Singer, F. J. 1981. Wild pig populations in the national parks [of the USA]. *Environmental Management* **5**: 263–70.

Skalon, V. N. 1951. [*The beavers of northern Asia.*] Moscow Society of Naturalists.

Skinner, J. R., Condy, P. R., van Aarde, R. J., Bester, M. N. & **Robinson, T. J.** 1978. The mammals of Marion Island. *South African Journal of Antarctic Research* **8**: 35–8.

Skira, I. J. 1978. Reproduction of the rabbit on Macquarie Island. *Australian Wildlife Research* **5**: 317–26.

Skottesberg, C. J. F. 1920. The natural history of Juan Fernandez and Easter Island, 3 vols. Almqvist & Wiksells: Upsala.

Slevin, J. R. 1931. Log of the schooner *Academy. Occasional Papers of the California Academy of Sciences* **17**: 1–162.

———. 1959. The Galápagos Islands: a history of their exploration. *Occasional Papers of the California Academy of Sciences* **25**: 1–250.

Smal, C. M. & **Fairley, J. S.** 1978. The spread of the bank vole since 1970. *Irish Naturalists' Journal* **19**: 237–9.

Spencer, H. J. 1950. Mongoose control research project, Virgin Islands, U.S.A. Unpublished report, US Department of the Interior Fish & Wildlife Service.

Stead, D. G. 1928. *The rabbit menace in New South Wales.* (Privately printed.)

———. 1935. *The rabbit in Australia.* Winn & Co.: Sydney.

Stead, E. F. 1937. The Maori rat. *Transactions of the Royal Society of New Zealand* **66**: 178–81.

***Stebbins, H. G.** & **Baker, G. L.** eds. 1965. *The genetics of colonizing species.* Academic Press: New York.

Steen, M. O. 1949–55. Introduction of exotic animals. *Proceedings of the 39th, 42nd and 44th Convention of the International Association of Game, Fish & Conservation Commissioners*, 42–8; 78–81; 89–91.

Stefansson, V. 1923. The musk ox on Arctic islands. *Nature, London* **112**: 590.

Steffes, N. 1969. Musk-rat control in the Grand Duchy of Luxemburg. *EPPO Publications* Ser. A, No. 47: 36–7.

***Stegeman, L. C.** 1938. The European wild boar in the Cherokee National Forest, Tennessee. *Journal of Mammalogy* **19**: 279–90.

Steinbacher, G. 1978. Waschbär (*Procyon lotor*) bei Donauwörth erlegt. *Säugetierkundliche Mitteilungen* **26**: 240.

Steiniger, F. 1963. Über den Enok als neue Raubwildart in der westdeutschen Fauna. *Tier-und Naturphotographie, Baden-Baden* **6**: 8–10.

Stewart, A. E. 1933. Wild goats in Scotland. *The Field* (16 December).

Stillwell, H. 1955. Old-world hunt in Texas [blackbuck]. *Field & Stream* **60**: 68–9; 143–6.

Stocker, G. C. 1971. Water buffaloes and conservation in the Northern Territory *Wildlife in Australia* **8**: 10–12.

———. 1971. The effects of water buffaloes on paperbark forests in the Northern Territory. *Australian Forest Research* **5**: 29–34.

Stokes, J. F. G. 1917. Notes on the Hawaiian rat. *Occasional Papers of the Bernice P. Bishop Museum* **3**(4): 11–21.

Stone, W. 1917. The Hawaiian rat. *Occasional Papers of the Bernice P. Bishop Museum* **3**(4): 253–60.

Storer, T. I. 1931. Known and potential results of bird and animal introductions, with special reference to California. *California Department of Agriculture Monthly Bulletin* No. 20: 267–73.

———. 1934. Economic effects of introducing alien animals into California. *Proceedings of the 5th Pacific Science Congress,. Canada, June 1933* **1**: 779–84. University of Toronto Press: Toronto.

———. 1937. The muskrat as native and alien. *Journal of Mammalogy* **18**: 443–60.

———. ed. 1962. Pacific island rat ecology. *Bulletin of the Bernice P. Bishop Museum* No. 225: 1–274.

Strange, I. 1972. *The Falkland Islands.* David & Charles: Newton Abbot.

Street, J. M. 1962. Feral animals in Hispaniola. *Geographic Review* **52**: 400–6.

Street, P. 1955. The edible dormouse in England. *Country Life* **188**: 1484–5.

Studholme, E. C. 1954. *Te Waimate: early station life in New Zealand.* A. H. & A. W. Reed: Wellington.

Suminski, P. 1963. [Acclimatization of mammals in Poland.] *Chrońmy Przyrodę Ojczystą* **19**(3): 13–22.

Suomalainen, E. 1950. The raccoon dog found in Finland. *Archivum Societatis Zoologicae-botanicae fennicae 'Vanamo', Helsinki* **5**: 20–2.

Svihla, A. 1935. The Hawaiian rat. *Mid-Pacific Magazine* **48**: 344–6.

———. 1936. The Hawaiian rat. *Murrelet* **17**: 3–14.

Swank, W. G. & **Petrides, G. A.** 1954. Establishment and food habits of the nutria in Texas. *Ecology* **35**: 172–5.

Swanson, G. A. 1950. The need for research in the introduction of exotic animals. *Proceedings of an International Technical Conference on the Protection of Nature, Lake Success, New York, August 1949.* Unesco: Paris.

Sweatman, G. K. & **Williams, R. J.** 1962. Wild animals in New Zealand. *Transactions of the Royal Society of New Zealand (Zoology)* **26**: 222–50.

Swedberg, G. 1963, 1964 and 1967. Progress reports (unpublished) on introduction of the blacktail deer to Kauai, Hawaii for Department of Land & Natural Resources, Honolulu.

———. ed. 1968. The axis deer . . . in Hawaii. Unpublished report for Hawaiian State Division of Game & Fish, Honolulu.

Sykes, W. R. 1969. The effect of goats on the vegetation of the Kermadec Islands. *Proceedings of the New Zealand Ecological Society* **16**: 13–16.

Sympson, P. 1978. Red deer in Argentina. *Deer* **4**: 370.

Szunyoghy, J. 1958. The introduction and distribution of the rabbit in Hungary. *Annales Historico – Naturales Musei Nationales Hungarici* **50** (Ser. Nova 9): 349–58.

———. 1963. Ein neus Saugetier in Ungarn. *Állattani Közlemények, Budapest* **50**: 175–9.

T

Taber, F. W. 1939. Extension of the range of the armadillo. *Journal of Mammalogy* **20**: 489–93.

Taft, S. R. 1971. Aoudad sheep transplants in the Palo Duro canyon. Unpublished report for the Texas Chapter of the Wildlife Society.

Talmage, R. V. & **Buchanan, G. D.** 1954. The armadillo (*Dasypus novemcinctus*): a review of its natural history, ecology, anatomy and reproductive physiology. *Rice Institute Pamphlets, Monograph in Biology* **41**(2): 1–135.

Tate, G. H. H. 1935. Rodents of the genera *Rattus* and *Mus* from the Pacific islands. *Bulletin of the American Museum of Natural History* **68**: 145–78.

Taub, D. M. 1978. The Barbary macaque. *Oryx* **14**: 245–53.

Taylor, J. C. 1964. Implications from the grey squirrel experience. *Bulletin of the Mammal Society* **21**: 728.

Taylor, K. D. [1965]. The rat and mongoose problem in Grenada. Internal unpublished report for the Ministry of Agriculture, Fisheries & Food, London.

Taylor, P. 1969. Hog deer in Victoria [Australia]. *Victoria's Resources* **11**(2): 20–2.

Taylor, R. B. 1977. The burro or the bighorn? *National Parks Conservation Magazine* **51**(9): 10–14.

Taylor, R. H. 1968. Introduced mammals and islands: priorities for conservation and research. *Proceedings of the New Zealand Ecological Society* **15**: 61–7.

———. 1971. Influence of man on vegetation and wildlife of Enderby and Rose islands, Auckland islands. *New Zealand Journal of Botany* **9**: 225–68.

———. 1975a. The distribution and status of introduced mammals on the Auckland Islands, 1972–73. In Yaldwyn, J. C. ed., *Preliminary results of the Auckland Islands Expedition 1972–73*, 233–43. New Zealand Department of Lands & Survey: Wellington.

———. 1975b. What limits kiore distribution in New Zealand? *New Zealand Journal of Zoology* **2**: 473–8.

———. 1984. Distribution and interactions of introduced rodents and predators in New Zealand. *Proceedings of the 3rd International Theriological Congress, Helsinki, August 1982. Acta zoologica fennica* (in press).

Taylor, R. H., Bell, B. D. & **Wilson, P. R.** 1970. Royal albatrosses, feral sheep and cattle on Campbell island. *New Zealand Journal of Science* **13**: 78–88.

Taylor, W. L. 1939 and 1948. The distribution of wild deer in England and Wales. *Journal of Animal Ecology* **8**: 6–9; **17**: 151–4.

———. 1949. The distribution of wild deer in Scotland. *Journal of Animal Ecology* **18**: 187–92.

Teer, J. G. 1971. Introduction of exotic animals. In *Manual on wildlife conservation*, 90–2. Wildlife Society: Washington DC.

———. 1974. Game ranching in Texas. *Proceedings of the symposium on ungulates & related management, IUCN Publications* n.s. No. 24: 893–9.

———. 1975. Commercial uses of game animals on rangelands of Texas. *Journal of Animal Science* **40**: 1000–8.

Tegner, H. 1963. The wild goat of the British Isles. *Zeitschrift für Säugetierkunde* **28**: 15–19.

———. 1965. Wild goats of Britain. *Scottish Field.*

Tempel, W. & **Rau, E.** 1962. Bisam [muskrat] und Sumpfbiber [coypu], zwie neue 'Wildarten' in der Pfalz. *Pfälzer Heimat* **2**.

Tennant, W. 1949. Wildlife on South Georgia. *The Field* **193**: 405–6.

Tenovuo, S. 1963. Minkki: saaristoalueilla [mink in the archipelago areas of Finland]. *Suomen Riista* **16**: 92–9 [in Finnish, with English summary].

Teodoreanu, M. 1971. Muskrats in Romania. *Studia Universitatis Babeş-Bolyai*, Ser. Biol. **16**: 133–6.

Terry, M. 1965. The buffalo threat [in Australia]. *People* **16** (14 July): 24–6.

Thomas, A. D. & **Kolbe, F. F.** 1942. [Feral pigs.] *Journal of the South African Veterinary Medicine Association* **13**: 1–11.

Thomas, A. P. W. 1888. The rabbit nuisance. *Appendices to the Journal of the House of Representatives, New Zealand*, 8–18.

Thomas, O. 1910. [*Lepus europeaus.*] In: A collection of mammals from Eastern Buenos Ayres, with descriptions of related new mammals from other localities. *Annals and Magazine of Natural History*, Ser. 8, **5**: 246.

Thompson, H. V. 1953. The edible dormouse in England 1902–51. *Proceedings of the Zoological Society of London* **122**: 1017–24.

———. 1955. The wild European rabbit and possible

dangers of its introduction into the U.S.A. *Journal of Wildlife Management* **19**: 8–13.

———. 1956. The origin and spread of myxomatosis, with particular reference to Great Britain. *Terre et Vie* **3–4**: 137–52.

———. 1958. Rabbit control in Australia and New Zealand. *Agriculture, London* **65**: 388–92; 440–4.

———. 1962a. Mink at large in Britain. *Country Life* **131**: 645.

———. 1962b. Wild mink in Britain. *New Scientist* **13**: 130–2.

———. 1964. Wild mink. *Agriculture, London* **71**: 564–7.

———. 1965. Feral mink in Britain. *Animals* **6**: 240–4.

———. 1966. Animal immigrants. *Geographical Magazine* **38**: 762–72.

———. 1968a. British wild mink. *Annals of Applied Biology* **61**: 345–9.

———. 1968b. Mink, the new British beast ot prey. *The Field* **232**: 1094–5.

———. 1971. British wild mink – a challenge to naturalists. *Agriculture, London* **78**: 421–5.

Thompson, H. V. & **Peace, T. R.** 1962. The grey squirrel problem. *Quarterly Journal of Forestry* **56**: 33–41.

Thompson, H. V. & **Platt, F. B.** 1964. The present status of *Glis* in England. *Bulletin of the Mammal Society of the British Isles* **21**: 5–6.

***Thompson, H. V.** & **Worden, A. N.** 1956. *The rabbit*. Collins: London.

Thompson, K. 1952. New horizon for goats. *Montana Wildlife* **2**: 14–17.

Thompson, S. L. 1931. Some notes on the European hare. *Canadian Field Naturalist* **45**: 192–3.

Thomsen, J. J. & **Evans, C. A.** 1964. The feral San Juan rabbit ... *Laboratory Animal Care* **14:** 155–60.

Thomsen, P. J. 1978. On the occurrence of the raccoon dog in East Germany. *Säugetierkundliche Mitteilungen* **26**: 78–9.

Thomson, G. M. 1884. The rabbit pest. *New Zealand Journal of Science* **2**: 79–80.

———. 1891. On some aspects of acclimatization in New Zealand. *Report of the 3rd Meeting of the Australasian Association for the Advancement of Science*, 194.

*———. 1922. *The naturalization of animals and plants in New Zealand.* Cambridge University Press: London.

———. 1923. Naturalized animals and plants. *New Zealand Journal of Science and Technology* **6**: 223–31.

———. 1927. *Wildlife in New Zealand.* New Zealand Board of Science & Art (Booklets Nos 2 and 5).

Thórisson, S. 1980. Status of *Rangifer* in Iceland. In Reimers, E., Gaare, E. & Skjenneberg, S. eds, *Proceedings of the 2nd International Reindeer/Caribou Symposium, Røros, Norway, 1979*, 766–70. Trondheim.

———. 1983. *The history of reindeer in Iceland and reindeer study 1979–1981*. Orkustofnun: Reykjavík (OS-87073/VOD-06), pp. 210 [in Icelandic with English summary pp. 149–64].

Thorlacius, B. 1960. [Reindeer in Iceland.] *Freyr* **56**: 117–34.

Thornton, I. 1971. *Darwin's islands.* Natural History Press: New York.

Thorsteinsson, I., Gardarsson, A., Ólafsson, G. & **Gudbergsson, G. M.** 1970. [Reindeer in Iceland and their summer ranges.] *Náttúrufraedingurinn* **40**: 145–70 [in Icelandic with English summary].

Thorvaldsson, Ó. 1960. [*Reindeer in Iceland 1771–1960.*] Bókaútgafa Menningarsjóds: Reykjavík [in Icelandic].

Tierkel, E. S., Arbona, G., Rivera, A. & **de Juan, A.** 1952. Mongoose rabies in Puerto Rico. *Public Health Reports* **67**: 274–8.

Timofeeva, E. K. & **Fedotova, V. G.** 1973. [An experiment on the introduction of *Cervus nippon* on the Karelian isthmus, Leningrad district, U.S.S.R.] *Zoologicheskiĭ Zhurnal* **52**: 1046–54 [in Russian].

Tisdell, C. A. 1982. *Wild pigs: environmental pest or economic resource?* Pergamon Press: Sydney.

Titcomb, M. 1969. The axis deer [in Hawaii]. *Elepaio* **30**: 21–5, 52–7.

Tittensor, A. 1983. Myxomatosis in decline? *Country Life* **173**: 278–9.

Toldt, K. 1929. *Die Basamratte [muskrat] mit besonderer Berücksichtigung ihres Auftretens in Österreich.* Leipzig.

***Tomich, P. Q.** 1969. *Mammals in Hawaii: a synopsis and notational bibliography*. Bernice P. Bishop Museum Press: Honolulu (Special Publications No. 57).

Tomlinson, A. R. 1955. Wild dogs and dingoes in Western Australia. *Journal of Agriculture of Western Australia* **3**(4): 4–21.

———. 1960. Wild goats. *Journal of Agriculture of Western Australia* **4**: 85–6.

Towne, C. W. & **Wentworth, E. N.** *Pigs from cane to cornbelt.* University of Oklahoma Press: Norman.

Troughton, E. 1938. Australian mammals, their past and future. *Journal of Mammalogy* **19**: 401–11.

———. 1957. *Furred animals of Australia.* Angus & Robertson: Sydney.

Tsevgmid, D. & **Dashdorj, A.** 1974. Wild horses and other endangered wildlife in Mongolia [raccoon dog]. *Oryx* **12**: 361–70.

Tulloch, D. G. 1967. The distribution, density and social behaviour of the water buffalo in the Northern Territory. Unpublished MAg thesis, University of Queensland.

———. 1969a. Home range in feral water buffalo. *Australian Journal of Zoology* **17**: 143–52.

———. 1969b. A new look at the buffalo [in Australia]. *Turnoff* **2**: 117–21.

———. 1970. Seasonal movements and distribution of the sexes in the water buffalo in the Northern Territory. *Australian Journal of Zoology* **18**: 399–414.

———. 1972. *The principal habitat of the water buffalo in the Northern Territory*. Australian Society of Animal Production.

———. 1973. Water buffalo in the Northern Territory. *Proceedings of the 33rd Annual Conference of the Cold Storage Association of Australia*, 57–70.

———. 1974. The feral swamp buffaloes of Australia's Northern Territory. In Ross Cockrill, W. ed., *The husbandry and health of the domestic buffalo*, 493–505. FAO: Rome.

———. 1975. Buffalo in the northern swamplands [of Australia]. *Proceedings of the 3rd World Conference on Animal Production, May 1973, Theme 1*, 60–6.

*———. 1978. Water buffalo in Australia, *Australian Wildlife Research* **5**: 327–54.

———. 1979a. Redomestication of water buffaloes in the Northern Territory of Australia. *Animal Regulation Studies* **2**: 5–20.

*———. 1979b. The water buffalo in Australia. *Australian Wildlife Research* **6**: 265–87.

Turcek, F. J. 1951. Effect of introductions on two game populations in Czechoslovakia. *Journal of Wildlife Management* **15**: 113–14.

Twigg, G. I. 1978. The role of rodents in plague dissemination: a worldwide review. *Mammal Review* **8**: 77–110.

———. 1975. *The brown rat*. David & Charles: Newton Abbot.

Tyndale-Biscoe, C. H. 1955. Observations on the reproduction and ecology of the brush-tailed opossum in New Zealand. *Australian Journal of Zoology* **3**: 162–84.

———. 1960. The possum in New Zealand. *Australian Museum Magazine* **13**: 235–8.

U

Uchida, T. A. 1966. Appraisal of the efficiency of the Japanese weasel as a rat control agent in the Ryukyus. *WHO* 4, E.C. **67**: 46.

Ueckermann, E. & **Hansen, P.** 1968. *Das Damwild* [fallow deer] *Naturgeschichte Hege und Jagd*. Paul Parey: Hamburg.

Ulbrich, J. 1930. *Die Bisamratte* [muskrat]. Heinrich: Dresden.

Uloth, W. 1971. Vorkommen des Europaischen Damhirsches [fallow deer]. *Säugetierkundliche Mitteilungen* **19**: 202–6.

Upfield, A. W. 1935. An Australian camel station. *Walkabout* **1**: 27–9.

Urich, F. W. 1914. The mongoose in Trinidad and methods of destroying it. *Board of Agriculture, Trinidad, Circular* No. 12.

———. 1931. The mongoose in Trinidad. *Tropical Agriculture, Trinidad* **8**: 95–7.

V

Valtýsson, H. 1945. [*Reindeer in Iceland.*] Oddur Björnsson: Akureyri [in Icelandic].

Van Aarde, R. J. 1978. Reproduction and population ecology in the feral house cat on Marion island. *Carnivore Genetics Newsletter* **3**: 288–316.

———. 1979. The cats of Marion island: friend or foe? *African Wildlife* **32**(6): 30–2.

———. 1980. The diet and feeding behaviour of feral cats at Marion island. *South African Journal of Wildlife Research* **10** (3/4): 23–8.

———. 1981. Distribution and density of the feral house cat at Marion island. *South African Journal of Antarctic Research* No. 9: 14–19.

———. 1984. Population biology and the control of feral cats on Marion Island. *Proceedings of the 3rd International Theriological Congress, Helsinki, August 1982. Acta zoologica fennica* (in press).

Van Aarde, R. J. & **Skinner, J. D.** 1982. The feral cat population at Marion Island. *Colloques sur les Ecosystèmes Subantarctiques. Publications du Comité National Français des Recherches Antarctiques (CNFRA), Paris*, No. 51: 281–8.

Van Bemmel, A. C. V. 1949. Revision of rusine deer in the Indo-Australian archipelago. *Treubia* **20**: 191–262.

Van Gelder, R. G. 1979. Mongoooses on mainland North America. *Wildlife Society Bulletin* **7**: 197–8.

Van Vuren, D. & **Coblenz, B. E.** 1984. Impacts and adaptations of feral sheep on Santa Cruz Island, California. In *Feral mammals – problems and potential* (*Proceedings of a SSC Caprinae Specialist Group Workshop, Helsinki, August 1982*), 43–53. IUCN: Morges.

Van Wijngaarden, A. 1961. A wild raccoon trapped in the Netherlands. *Natuurhistorisch Maandblad, Maastricht* **50**: 54–5 [in Dutch, with English summary].

Van Wijngaarden, A. & **Bruijns, M. F. M.** 1961. De hermelijnen *Mustela erminea* van Terschelling. *Lutra* **3**: 35–42.

Varona, L. 1974. *Catálago de les mamiferos vivientes y extinguidos de las Antillas*. Academia de Ciencias de Cuba: Havana.

Veale, E. M. 1957. The rabbit in England. *Agricultural History Review* **5**: 85–90.

Velek, J. 1976. Chov danci zvere ve volnosti [fallow deer]. *Myslivost* **5**: 103–5 (see also **11**: 245–6).

Vereshchagin, N. K. 1941 and 1963. [Nutria in west Georgia.] *Trudÿ Zoologicheskogo Instituta, Academiya Nauk Gruzinskoĭ SSR, Tbilisi* **4**: 3–42; **29**: 1–66. [in Russian].

'Vigilans'. 1910. Raising buffalo – an industry for the Northern Territory [of Australia]. *Pastoralists' Review* **20**: 61–2.

Vincent, J. P. 1972. Quelques données sur la reproduction et sur la dynamique des populations du rat musqué dans le Nord de la France. *Annales de Zo-*

olgie – Ecologie Animale, Paris **4**: 395–415.

Vinson, J. 1950. Round island and Snake island. *Proceedings of the Royal Society of Arts & Science, Mauritius* **2** (3).

Voinstrenskii, M. A., Kistyakovskii, A., Parkhomenko, V. V., Salganskii, A. A. & **Sokur, I. T.** 1963. Results and prospects of acclimatization of game and fur animals in the Ukraine. In Jannsevic, A. I. ed., *Acclimatization of animals in the U.S.S.R.*, 47–51. Institute of Zoology, Kazakh SSR Academy of Sciences: Alma-Ata.

Voipio, P. 1948. Riistan laadun parantaminen ja sen biologiset edellytykset. *Suomen Riista* **2**: 7–72.

Voris, J. C., Yoakum, J. D. & **Yocum, C. F.** 1955. Notes on chinchilla liberations in California. *Journal of Mammalogy* **36**: 302.

W

Wace, N. M. & **Holdgate, M. W.** 1976. *Man and nature in the Tristan da Cunha Islands.* IUCN: Morges (IUCN Monograph No. 6, 114 pp.)

Waldo, E. 1958. The nutria story. *Louisiana Conservationist* **10**(3): 8–11.

Walker, L. W. 1945. The Hawaiian mongoose; friend or foe? *Natural History* **54**: 396–400.

Walker, R. L. 1959. Experimental introduction of the European mouflon. Unpublished report for Territory of Hawaii, Division of Fish & Game, Honolulu.

———. 1960. The hybridization of the mouflon with feral sheep [in Hawaii]. *Proceedings of the 40th Annual Conference of the Western Association of State Game & Fish Commissioners*, 148–56.

———. 1967. A brief history of exotic game bird and mammal introductions into Hawaii. *Proceedings of the 47th Annual Conference of the Western Association of State Game & Fish Commissioners*, 94–112.

Wallace, M. 1958. Camels for Coolgardie. *Walkabout* **24**: 32–5.

Wallace, N. F. 1960. Reaction and adaptability of introduced aoudad sheep in Palo Duro canyon. Unpublished report for Texas Parks & Wildlife Department Ref. Job No. 6, Project W-45-R-9.

———. 1961. Cliff-dwellers of the Palo Duro. *Texas Game & Fish* **19**(9): 16–17.

Wallis, F. P. & **James, I. L.** 1972. Introduced animal effects and erosion phenomena in the northern Urewera forests. *New Zealand Journal of Forestry* **17**: 21–36.

Walsh, P. 1893. The effect of deer on the New Zealand bush. *Transactions of the New Zealand Institute* **25**: 435–8.

Walters, B. 1983. The Australian dingo. *The Ark* **10**(1): 21–4.

Walther, F. 1961. Der Mesopotamische Damhirsch [fallow deer]. *Schweizer Naturschutz* **27**: 85–8.

Wammes, D. T. 1979. Zal de Amerikaanse nerts [mink] zich in Nederland kunnen vestigen? Script. in het kader van een doctoraal onderwerfp zoölog. oecol. *Rijks Instituut voor Natuurbeheer* **15** blz.

Warburton, C. 1944. *Buffaloes: adventure and discovery in Arnhem Land.* Angus & Robertson: Sydney.

Wardle, P. 1961. The effect of deer on sub-alpine forest and scrub. *Proceedings of the New Zealand Ecological Society* **8**: 52–4.

Wardle, P., Moor, N. T. & **Given, D. R.** 1978. Goats on Auckland Island. *New Zealand Journal of Botany* **16**: 291–2.

Warwick, T. 1934. The distribution of the muskrat in the British Isles. *Journal of Animal Ecology* **3**: 250–67.

———. 1935. Some escapes of coypus from nutria farms in Great Britain. *Journal of Animal Ecology* **4**: 146–7.

Watson J. S. 1951. Rat problem in Cyprus. *Colonial Research Publication* No. 9.

———. 1953. Aliens in the forest: the rat. *Forest & Bird* **109**: 10–12.

———. 1956. Present distribution of *Rattus exulans* in New Zealand. *New Zealand Journal of Science & Technology* B**37**: 560–70.

———. 1961a. Feral rabbit populations of Pacific islands. *Pacific Science* **15**: 591–3.

———. 1961b. Rats in New Zealand: a problem of interspecific competition. *Proceedings of the 9th Pacific Science Congress* **19**(Zool.): 15–16.

Watt, H. B. 1937. On the wild goat in Scotland. With supplement: Darling, F. F. Habits of wild goats in Scotland. *Journal of Animal Ecology* **6**: 15–22.

Wayre, P. 1969. Wildlife in Taiwan. *Oryx* **10**: 46–56.

Weber, D. 1971. Pinta, Galápagos: une île à sauver. *Biological Conservation* **4**: 8–12.

Werner, D. 1977. Introduced animal species in the Galápagos Islands and their effects on the endemic fauna. IUCN Species Survival Commission Agenda Paper CSS 77/1/22, 6 pp. (mimeo).

West, R. C. & **Ramsey, C. W.** 1968. Exotic artiodactyls in Texas. Unpublished Texas Parks & Wildlife Department Report Project W-76-R.

Westerskov, K. 1953. Acclimatization of new game species. *New Zealand Department of Internal Affairs Wildlife Publication* No. 17: 1–8. (Originally published in *New Zealand Fishing & Shooting Gazette* **19**(10): 4–8 (1952).)

Westermann, J. H. 1947. Fauna en natuurbescherming in Nederlandsch Nieuw Guinea. *Nederlandsche Commissie voor Internationale Natuurbescherming, Mededeelingen* **13**: 1–107.

———. 1953. Nature preservation in the Caribbean. *Uitgaven van de Natuurwetschappelijke Studiekring voor Suriname in de Nederlandse Antillen* **9**: 1–106.

Westmann, K. 1963. Occurrence of feral American minks in Finland. *Suomen Riista* **16**: 115–16.

———. 1966. Minkin Levinneisyydestä Suomessa.

Suomen Riista **18**: 101–16 [English summary].

Wexler, M. 1980. Monkeying around on the Rock [Gibraltar]. *Wildlife* 44–6.

Whitaker, A. H. & **Rudge, M. R.** 1976. *The value of feral farm mammals in New Zealand.* New Zealand Department of Lands & Survey (Information Series No. 1, 84 pp.)

White, T. 1890a. On the wild dogs of New Zealand. *Transactions of the New Zealand Institute* **22**: 327–30.

———. 1890b. [Rabbits, rats, weasels and mice.] *Transactions of the New Zealand Institute* **23**: 194–201.

———. 1894. Remarks on the rats of New Zealand. *Transactions of the New Zealand Institute* **27**: 240–61.

Whitehead, G. K. 1972a. *Deer of the world.* Constable: London.

———. 1972b. *The wild goats of Great Britain and Ireland.* David & Charles: Newton Abbot.

Wigg, H. C. 1889. The rabbit question [in Australia]. *Proceedings of the Royal Society of Victoria* n.s. **1**: 28–33.

Wildhagen, A. 1949. Minken i Norge. *Fauna, Oslo* **4**: 107–28.

———. 1956. Present distribution of North American mink in Norway. *Journal of Mammalogy* **37**: 116–18.

Wilkman, S. 1958. Pellingin villiintyneistä minkeistä. *Metsästys ja Kalastus* **47**: 61–2.

Willet, J. A. & **Mulloy, F.** 1970. Wild deer, their status and distribution. *Deer* **2**: 498–504.
2: 498–504.

Willey, K. 1962. Big game at the Top End [of Australia]. *Walkabout* **28**: 48–50.

⋆**Williams, G. R.** ed. 1973. *The natural history of New Zealand: an ecological survey.* A. H. & A. W. Reed: Wellington.

Williams, G. R. & **Rudge, M. R.** 1969. A population study of feral goats from Macauley Island, New Zealand. *Proceedings of the New Zealand Ecological Society* **16**: 17–28.

Wilson, M. L. & **Elicker, J. G.** 1976. Establishment, maintenance and behaviour of free-ranging chimpanzees on Ossabaw Island, Georgia, U.S.A. *Primates* **17**: 451–73.

Wilson, P. R. 1979. Inter-relationships of feral stock, sea birds and vegetation on Campbell Island. *New Zealand Journal of Ecology* **2**: 92–3.

Wilson, P. R. & **Orwin, D. F. G.** 1964. The sheep population of Campbell island. *New Zealand Journal of Science* **7**: 460–90.

Wirtz, W. O. 1972. Population ecology of the Polynesian rat on Kure Atoll, Hawaii. *Pacific Science* **26**: 433–64.

Withers, R. J. 1908. The rabbit pest in Australia. *Red Funnel* (March).

⋆**Wodzicki, K.** 1950. Introduced mammals of New Zealand: an ecological and economic survey. *New Zealand DSIR Bulletin* No. 98: 1–250.

———. 1953. Aliens in the forest. 4. Weasel-like animals (Mustelidae). *Forest & Bird* **109**: 4–5.

———. 1961. Ecology and management of introduced ungulates in New Zealand. *Terre et Vie* **1**: 130–57.

———. 1963. Introduced mammals in New Zealand forests. *Acta Biologica Cracoviensia,* Ser. Zool. **6**: 111–34.

———. 1965. The status of some exotic vertebrates in the ecology of New Zealand. In Baker, H. G. & Stebbins, G. L. eds., *The genetics of colonizing species* (*Proceedings of the 1st IUBS Symposium, General Biology*), 425–60. Academic Press: New York.

———. 1968. The use of Japanese weasels to control rats in Pacific islands. Unpublished Report for Tokelau Islands Administration.

———. 1973. Some problems arising from the invasion and deliberate introduction of exotic plant and animal species in the south-west Pacific. *Proceedings of the Regional Symposium on Conservation of Nature, Reefs & Lagoons* (*Noumea, New Caledonia*), 239–44.

⋆**Wodzicki, K.** & **Flux, J. E. C** 1967a. Guide to introduced wallabies in New Zealand. *Tuatara* **15**(2): 47–59.

———. 1967b. The rediscovery of the white-throated wallaby on Kawau island. *Australian Journal of Science* **29**: 429–30.

———. 1971. The parma wallaby and its future. *Oryx* **11**: 40–7.

Wodzicki, K. & **Taylor, R. H.** 1984. The distribution and status of the Polynesian rat (*Rattus exulans*). *Proceedings of the 3rd International Theriological Congress, Helsinki, August 1982. Acta zoologica fennica* (in press).

Wolcott, G. N. 1953. The food of the mongoose in St. Croix and Puerto Rico. *Journal of Agriculture, University of Puerto Rico* **37**: 241–7.

Wood, G. W. ed. 1977. *Research and management of wild hog populations.* Belle W. Baruch Forest Science Institute, Clemson University: Clemson, South Carolina.

Wood, G. W. & **Barrett, R. H.** 1979. Status of wild pigs in the United States. *Wildlife Society Bulletin* **7**: 237–46.

Wood, G. W. & **Lynn, T. E.** 1977. Wild hogs in southern [USA] forests. *Southern Journal of Applied Forestry* **1**(2): 12–17.

Wood, J. E., White, R. J. & **Durham, J. L.** 1970. Investigations preliminary to the release of exotic ungulates in New Mexico. *New Mexico Department of Game & Fish Bulletin* No. 13: 1–58.

Wood Jones, F. 1921. The status of the dingo. *Transactions of the Royal Society of South Australia* **45**: 254–63.

Woods, A. 1974. *Pest control: a survey.* McGraw Hill: Sydney.

Woods, W. 1975. Muskrat origin, distribution and range extension through the coastal areas of Del

Noste Co. Calif. & Curry Co. Oregon. *Murrelet* **55**(1): 1–4.

Woodward, S. L. 1979. The social system of feral asses. *Zeitschrift für Tierpsychologie* **49**: 304–16.

Wyman, W. D. 1945. *The wild horse of the west.* Caxton Printers: Caldwell, Idaho.

Y

Yalden, D. W. 1982. When did the mammal fauna of the British Isles arrive? *Mammal Review* **12**: 1–57.

⋆**Yalden, D. W.** & **Hosey, G. R.** 1971. Feral wallabies in the Peak District. *Journal of Zoology, London* **165**: 513–20.

⋆**Yanushevich, A. I.** ed. 1966. *Acclimatization of animals in the U.S.S.R.* (*Proceedings of a conference at Frunze, May 1963.*) Israel Program for Scientific Translations: Jerusalem.

Yocum, C. F. C. 1964. Ecological study of feral goats on U.S. National Park lands in Hawaii. Unpublished (mimeo) report for Humboldt State College, Division of Natural Resources, California, 70 pp.

———. 1967. Ecology of feral goats in Haleakala National Park, Maui, Hawaii. *American Midland Naturalist* **77**: 418–51.

Young, E. L. 1971–2. Habitat preference of exotics. Unpublished Texas Parks & Wildlife Report Job 18, Project W-76-R-14-15.

———. 1973. Exotics. *Texas Parks & Wildlife* **31**(8): 2–9.

Youngman, P. M. 1962. The spread of the European hare to the Ottawa region of Ontario. *Canadian Field Naturalist* **76**: 223.

Yurgenson, P. B. 1956. [*Studies on mammals in government preserves.*] Moscow. (English transl. 1961, Israel Program for Scientific Translations: Jerusalem.)

Z

Zarn, N., Heller, T. & **Collins, K.** 1977. Wild, free-roaming burros: an annotated bibliography. *US Bureau of Land Management Technical Note* No. 297.

Zeuner, F. E. 1952. Monkeys in Europe, past and present. *Oryx* **1**: 264–73.

———. 1963. *A history of domesticated animals.* Hutchinson: London.

Zhdanoff, A. P. 1957. Acclimatization of the beaver in the Ob river basin. *Translations of the Russian Game Reports* **1**: 46–59. Canadian Wildlife Service.

Zhitkov, B. 1930. Nutria raising in southern U.S.S.R. *Pushnoye Deolo, Narkom* **5.**

Zotov, V. D. 1947. Forest deterioration in the Tararuas due to deer and opossums. *Transactions and Proceedings of the Royal Society of New Zealand* **77**(5): 162–5.

Zotov, V. D., Elder, N. L., Beddie, A. D., Sainsbury, G. O. K. & **Hodgson, E. A.** 1939. An outline of the vegetation and flora of the Tararua Mountains. *Transactions of the Royal Society of New Zealand* **68**: 259–324.

Zygankow, D. S. 1955. Eine Methode zür Bestimmung des Alters und des Lebensdauer de *Ondatra* [muskrat]. *Zoologischeskiĭ Zhurnal* 640–51 [in Russian].

Index of vertebrate species

Main references to introduced and/or naturalized mammals are indicated in bold type

E

F

Q